MODERN
BUSINESS STATISTICS

MODERN
BUSINESS STATISTICS

SECOND EDITION

RONALD L. IMAN

SANDIA NATIONAL LABORATORIES

W. J. CONOVER

TEXAS TECH UNIVERSITY

WILEY

JOHN WILEY & SONS

New York / Chichester / Brisbane / Toronto / Singapore

To Rae, Deborah, and Susan

R. L. I.

To Patricia, Christopher, Robert,
Judith, Therese, and Bill

W. J. C.

Copyright © 1983, 1989 by John Wiley & Sons, Inc.

All rights reserved. Published simultaneously in Canada.

Reproduction or translation of any part of
this work beyond that permitted by Sections
107 and 108 of the 1976 United States Copyright
Act without the permission of the copyright
owner is unlawful. Requests for permission
or further information should be addressed to
the Permissions Department, John Wiley & Sons.

Library of Congress Cataloging in Publication Data:

Iman, Ronald L.
  Modern business statistics/Ronald L. Iman, W. J. Conover.
  p. cm.

  Includes bibliographies and indexes.
  ISBN 0-471-81116-5
  1. Commercial statistics.   2. Statistics.   I. Conover, W. J.
II. Title.
HF1017.I483 1989
519.5'024658—dc19   88-14897
                                                   CIP

Printed in the United States of America

10 9 8 7 6 5 4 3 2 1

ABOUT THE AUTHORS

RONALD L. IMAN is a statistical consultant at Sandia National Laboratories. He holds a Ph.D. in statistics from Kansas State University and has had several years of university teaching experience. His research papers are in nonparametric statistics, and uncertainty and sensitivity analysis techniques used with computer models in risk assessment applications. Dr. Iman is a Fellow of the American Statistical Association and has served on the ASA Board of Directors and the Executive Committee of the Board as District Representative, and currently serves on the Board as the Treasurer of the ASA. He has chaired the Budget Committee, the Committee on Long-Range Financial Planning for the ASA, and currently chairs the Finance and Audit Committees. He currently serves as chair of the ASA Section on Physical and Engineering Sciences, and has provided service on the editorial boards for several professional journals. He is the recipient of the 1988 Don Owen Award for excellence in research and service to the statistical community.

W. J. CONOVER is the Paul Whitfield Horn Professor of Statistics and former Area Coordinator for Information Systems and Quantitative Sciences, College of Business Administration, Texas Tech University. He holds a Ph.D. in mathematical statistics from the Catholic University of America. Professor Conover is also a consultant at Sandia National Laboratories and a visiting staff member at Los Alamos National Laboratory. An earlier book written by him is *Practical Nonparametric Statistics,* Second Edition, Wiley, 1980. His research papers include both theoretical and applied nonparametric statistics. He is a Fellow of the American Statistical Association, and the recipient of the Youden prize and the 1986 Don Owen Award for his contributions in the field of statistics.

PREFACE

This text reflects our extensive experience in teaching and consulting, especially with regard to the inadequacies and the incompleteness of most business statistics texts. Hence, in this text the emphasis is on training the student to proceed from a data-oriented situation to the proper statistical method. Most topics are introduced by means of a realistic problem setting. The method of solution is then unfolded in the subsequent discussion where the objective is to show students how to analyze a real situation. The problems, examples, and exercises further emphasize realistic business-related settings in which these methods are useful.

This text is an introduction to modern business statistical techniques. It is designed to be used in a one- or two-semester course, with the topics divided into lesson-sized sections of an easily digestible size. Each section concludes with a set of exercises; review exercises are also provided at the end of each chapter. The text contains 1000 exercises and numerous worked out examples. Answers to selected exercises are given at the end of the text. Throughout the text color-coded exercise numbers have been used to identify exercises with answers given in the text.

Each methodology in the text is presented in a self-contained format, complete with assumptions, an explanation of notation, statements of hypotheses, and decision rules. This format enables the reader to refer back easily to the various statistical procedures. We have found that most textbooks present methodology and assumptions in a manner that makes it difficult if not impossible for students to decide when to use a methodology. Often these texts fail to provide any guidance about what to do if the assumptions are not satisfied.

To avoid these shortcomings, the sections on testing procedures clearly state the assumptions for the validity of the testing procedure. Many of these testing procedures rely on the underlying assumption of normality. This assumption is easily checked by the graphs that have been specifically developed for this book with the Lilliefors test for normality. Thus, the cumbersome chi-square goodness-of-fit test for normality is avoided, enabling the assumption of normality to be checked easily throughout. For situations where the normality is not satisfied, guidance toward the appropriate nonparametric test is provided. This means that the parametric and nonparametric tests are presented side by side throughout the text. However, the transition to the nonparametric tests is made easy by presenting them as analogues to parametric tests. This is done by applying the parametric procedure to rank transformed data. The result is a statistic that is functionally equivalent to the usual presentation of the nonparametric test statistic. This way of presenting nonparametric tests makes them easier to learn than when the standard method is used, and it eases the burden on instructors who may not have a strong nonparametric background. We have described this approach in recent research papers.*

In writing the second edition of this text we have been guided by reviews provided to us by users of the first edition, input from others who have taught from the first edition, along with our own ideas of needed changes. The result is an extensive revision of several chapters with other chapters subjected to various levels of editorial work. Throughout the revision the emphasis has been on improving the clarity of the presentation and the usefulness of the text for the student. Some new sections have been added and some of the previous sections have been deleted, while some of the previous sections have been combined into a single section. These changes resulted in the second edition having four fewer sections than the first edition. A new feature in this edition is the labeling of many sections and some entire chapters as optional. We were pleased that the side-by-side presentation of parametric and non-parametric procedures was well received in the first edition. However, some reviewers thought that this format obligated them to teach these sections. Although we believe that the nonparametric sections are quite important and should be covered we have labeled each of them as optional to make it clear that the course can proceed in a more traditional manner by skipping these sections. The cartoons that appeared in each chapter of the first edition have been eliminated from the present edition.

Exercises. The exercises have been improved both by deleting many of the previous ones and adding many new ones, including many tutorials at the beginning of the exercises. The number of exercises has increased from 883 to 1000. Results of calculations, such as sums and sums of squares, are frequently given with the exercises to reduce the amount of computational effort required. An important change in the exercises is the reduction of the references to previous exercises. Data sets are repeated *unless* they already appeared in the same chapter. This greatly reduces the need for much page turning that was a problem in the first edition.

*See *The American Statistician*, August 1981. Relevant references appear in the bibliography for each chapter with additional references given in the Instructors Manual.

Quantile Notation. The notation for critical values used in the first edition has been changed throughout the text to correspond to the definition of quantile given in the text and to eliminate potential confusion. For example, critical values for upper-tailed tests previously referred to as $z_{.05}$ are now referred to in standard quantile notation as $z_{.95}$.

Tables. Three new tables have been added and three of the previous tables have been expanded. The new tables provide a new level of convenience for the reader. We believe that this expanded set of tables is the most complete and convenient that we have seen in any introductory statistics text.

Computer Supplements. Separate computer supplements for SAS® and Minitab® have been written to accompany the text and provide the student with instructions on how to use these popular software products. These supplements illustrate how to use the computer to implement the methods presented within each chapter.

 The support materials that accompany this text include an Instructor's Manual, a Student Study Guide, a Test Bank, a SAS® computer manual, and a Minitab® computer manual.

 We hope the you will find it a refreshing experience to teach from this text, and we welcome any correspondence regarding its strengths or weaknesses that could affect future editions.

RONALD L. IMAN
W. J. CONOVER

CONTENTS

8

TWO RELATED SAMPLES (MATCHED PAIRS)   321

9

ESTIMATION AND HYPOTHESES TESTING WITH TWO INDEPENDENT SAMPLES   344

10

CONTINGENCY TABLES (OPTIONAL)   375

11

CORRELATION   408

12

REGRESSION   437

13
TIME SERIES ANALYSIS (OPTIONAL)   487

14
FORMULATING GENERAL LINEAR MODELS FOR FITTING DATA   515

15
MULTIPLE LINEAR REGRESSION   558

16
ANALYSIS OF VARIANCE FOR ONE-FACTOR EXPERIMENTS   598

17
ANALYSIS OF VARIANCE FOR TWO-FACTOR EXPERIMENTS   629

ACKNOWLEDGMENTS

We express our appreciation to all of the individuals whose reviews of this text contributed greatly to smoothing and clarifying its presentation.

David F. Bauer, Virginia Commonwealth University

David Booth, Kent State University

John Bryant, University of Cincinnati

William Burrell, Wayne State University

Roger Champagne, Hudson Valley Community College

James E. Coleman, Coppin State College

John Dirkse, California State College, Bakersfield

John G. Foster, Jr., Montgomery College

Lydia Gans, California State Polytechnic University

Jeffrey J. Green, Ball State University

William S. Hamilton, Community College of Rhode Island

George Heitmann, Pennsylvania State University

Neil Henry, Virginia Commonwealth University

R. Carter Hill, University of Georgia

Stephen Hora, Texas Tech University

Tej Kaul, Western Illinois University

William Key, Loma Linda University

Martin Kotler, Pace University

Stephen Kubricki, Stockton State College

Barry W. Light, Fitchburg State College

Robin H. Lock, St. Lawrence University

William Mallios, California State University— Fresno

Jeffrey Mock, Diablo Valley College

W. J. Peacock, New Hampshire College

William Remus, University of Hawaii at Manoa

George F. Rhodes, Colorado State University

Don R. Robinson, Illinois State University

William Stein, Texas A & M University

Donna F. Stroup, University of Texas at Austin

Richard A. Sundheim, Kansas State University

Jack Suyderhoud, University of Hawaii at Manoa

Baldeo K. Taneja, Case Western Reserve University

Henry Tingey, University of Delaware

R. C. Tripathi, University of Texas at San Antonio

Terry A. Watkins, University of New Orleans

Mary M. Whiteside, University of Texas at Austin

Tom S. Witt, West Virginia University

Gary Yoshimoto, University of Oregon

R. L. I.
W. J. C.

THE RELATIONSHIP BETWEEN SAMPLING AND STATISTICS

1

A course in statistics is a degree requirement in many different fields of study, including all accredited programs in business in universities throughout this country. As a student entering your first statistics course, you may have wondered why that is true. One major reason is that the faculty, who have the responsibility of establishing the curriculum requirements, are well aware of the value of statistics in a wide range of areas, either through its use in their research, or through their experience with business and industry. For example, are you aware that statistics plays an important role in the following areas?

- Determining the list of top 40 hits played on the radio each week.
- Determining the royalties paid to the performers of those hits.
- Determining the Nielsen ratings for TV networks.
- Performing the quality-control techniques that have made Japanese products so popular with consumers.
- Determining the impact of late-breaking developments on the standings in the polls of the leading candidates running for president.
- Predicting the outcome of elections long before all of the votes are counted.
- Estimating monthly unemployment rates in the United States.
- Getting a new drug approved by the Food and Drug Administration.
- Conducting the census of the United States every ten years.

- Forecasting sales for the next quarter.
- Conducting audits of banks, government agencies, and other businesses.
- Determining marketing strategies for new products.

This list only scratches the surface of the wide-ranging applications to which the subject of statistics can be applied.

THE SUBJECT OF STATISTICS

Statistics is the science of data analysis. It is concerned with methods for collecting, organizing, summarizing, presenting, and analyzing data, as well as making valid conclusions about the characteristics of the sources from which the data were obtained.

Thus, **statistics** is intimately involved with data analysis. The types of analyses that can be performed are quite diverse and require statistical methods that are equally diverse. The statistical methods that are presented in this text have been selected on the basis of their proven usefulness in application after application, such as those previously mentioned. This text has been designed to serve as a handy future reference. Many of the statistical methods you are most likely to encounter in your work, in your own projects, and in reports written by colleagues in your field are treated here. The authors hope that you find the explanations easy to understand. Every effort has been made to present this technical subject in a way that is understandable to individuals who may not have a technical background, and to do so without sacrificing the accuracy and correctness of the material.

1.1

POPULATIONS AND SAMPLES

It has been stated, "You don't have to eat the whole ox to know that the meat is tough." This statement is one way of paraphrasing what the science of statistics is all about. On the basis of one serving, an inference is made regarding the quality of the meat of the entire animal, or indeed about the quality of ox meat in general. However, in statistics the serving and the ox are replaced, respectively, with the terminology **sample** and **population.**

POPULATION

A **population** consists of a collection of individual units, which may be persons, objects, or experimental outcomes, whose characteristics are to be studied.

SAMPLE

A **sample** is a portion of the population that is studied to learn about the population.

For example, populations and samples are involved in each of the following situations.

1. **Polls.** A candidate for political office hires a polling firm to assess his or her chances in the upcoming election. The population consists of all voters in the candidate's district. The sample consists of all persons contacted by the polling firm.

2. **Quality control.** The manufacturer of flashbulbs wants to be sure that the proportion of defective bulbs does not become too high. The population here is all flashbulbs made by the manufacturer. Since the test of the flash-bulbs is by its very nature destructive, only a sample of the bulbs is examined.

3. **General business.** Royalties for music played on the radio are determined by the frequency with which each piece of music is played. The population consists of all music played by all radio stations in the country. A sample of radio stations provides the information upon which the royalties are based.

WHAT WILL YOU LEARN FROM THIS TEXT?

Clearly if valid conclusions are to be drawn with respect to these populations, the sampling cannot be done in a haphazard manner. A classic example of poor sampling occurred in the 1936 presidential election when the publishers of *Literary Digest* conducted a survey and concluded that Republican Alf Landon would defeat Democrat Franklin Roosevelt. The results of the 1936 election showed Landon carrying only two states! The reason for this fiasco could have been that the individuals polled came from lists of telephone owners, magazine subscribers, car owners, most of whom were Republicans, or the reason could have been that the people who responded were not representative of the people polled.

In this text you will learn proper methods for the collection, display, summarization, and analysis of sample data. Specifically you will learn

1. proper methodology for obtaining samples.
2. how to summarize data and display data graphically as an aid to the correct interpretation of the data.
3. how to analyze data and make estimates of unknown population characteristics.
4. how to choose the correct statistical procedure for testing conjectures (hypotheses) about the population.

THE ROLE OF SAMPLING IN STATISTICS

The success or failure of almost any business, as a rule, depends on the wants and needs of its potential customers. Automobile manufacturers who increase production of large cars when the market is switching to more gasoline-efficient models find themselves overstocked with large cars and several months behind the demand in the small-car market. A television network that emphasizes situation comedies when the audience is interested in live sports coverage finds it difficult to obtain premium prices for its commercials. Many a political candidate has misread the pulse of the public and never been heard from again.

How is a large manufacturing company, headquartered in Lompoc, California, going to know what its potential customers in Fargo, North Dakota, need and want in terms of services or products? The answer lies in a powerful technique known as **sampling.** When it is impossible or impractical to survey all of the potential customers, known as the **population,** a properly obtained **sample** of customers can reveal a wealth of information about what the people want, need, or think they need, from that company.

The Gallup poll, the Harris poll, and the Nielsen ratings all rely on properly obtained samples to provide information about the wants, needs, and opinions of the population of people represented by the samples. This information is used by marketing departments, production managers, and even boards of directors to decide what products or services to offer, and when and where to offer them.

WHY USE A SAMPLE?

There are several standard situations where a sampling study is preferable to a **census,** or complete study of the population.

1. **A census may not be cost effective.** In a marketing survey in which free samples are given out in order to estimate consumer response to a new product in a large city, little or no additional information is obtained by giving free samples to everyone in the city rather than to only a portion of the population.

2. **There may not be enough time to obtain more than a sample.** In political polls information on voter preferences must be obtained on short notice to be useful in the campaign, or in projecting winners based on exit polls.

3. **A carefully obtained sample may be more accurate than a census.** In a large project such as a census, certain **biases** often appear systematically throughout the census because there are not enough resources available to give each person as much attention as he or she deserves. (A bias is a tendency to make errors in one direction or the other.) For example, in a large inventory census or in a complete audit, errors due to fatigue or carelessness on the part of the census taker may introduce a serious bias in the results. This may result in less accuracy than if a sample had been carefully examined and an accurate report had been made on the sample.

4. **In destructive testing of products a sample has to suffice.** If the quality-control test involves dismantling a car battery to see if it was made properly, exploding a firecracker to see if it works, or breaking a rope to check its strength, it is necessary to include only a few items in the sample. Under these circumstances a census would result in there being no products available for marketing.

5. **Sometimes a census is impossible.** If a new medicine is being tested with regard to its effectiveness, the population of interest is not only all people alive now but also all future generations as well. Clearly, a sample must

suffice in this case. Another example is the enumeration of the wildlife population, where a census is not a feasible alternative.

CENSUS   A **census** is a complete enumeration of the entire population, as opposed to a sample that consists of only a portion of the population.

BIAS   A **statistical bias** refers to the systematic tendency of a sample or method of analysis to give estimates of population characteristics that are either larger on the average (**positive bias**) or smaller on the average (**negative bias**) than the true quantity being estimated.

EXAMPLES OF SAMPLING

Here are some typical applications of sampling in the business world.

Management To find out how employees feel about certain management practices, such as Saturday overtime or procedures for determining merit pay increases, a sample of employees may be selected and interviewed.

Marketing The effectiveness of several different proposed marketing plans may be compared by testing them on small groups of consumers before the national plan is enacted. The final design on a product may not be decided until a marketing test indicates consumer preferences.

Accounting Many audits of a large firm's records involve merely sampling the accounts serviced by that firm. If the bookkeeping in those accounts is satisfactory, no further audit is necessary. Of course, complete audits may still be conducted at regular intervals.

Finance The stability and anticipated yield of a certain mix of stocks and bonds in a portfolio can be estimated by considering several model portfolios with similar types of stocks and bonds and by studying their behavior in recent periods of time. This represents not only a sample of portfolios with characteristics similar to the one being studied but also a sample in time, where past time periods are used to estimate behavior in future time periods.

Economics To obtain an improved estimate of the total number of unemployed people, a sample of the population is obtained and examined to see how many people are unemployed but not actively seeking work, and therefore not classified as unemployed under the federal government's definition of unemployment.

EXERCISES*

1.1 Explain briefly the difference between a sample and a population.

1.2 Give an example of a situation where a sample must be used because of the impossibility of examining the entire population.

*Solutions to exercise numbers indicated in color are given in the Solutions Appendix.

1.3 Why would you want to use a sample instead of a census in the following situations?

(a) Estimating the potential market of a new brand of soap by giving out free samples.

(b) Testing the effectiveness of a new flu vaccine.

(c) Estimating the nicotine content in a brand of cigarettes.

1.4 One popular radio show plays the week's top 40 hits based on sales throughout the country. How would you guess that the determination of the top sellers is made?

1.5 Which of the following procedures tend to give biased estimates? Would you expect the bias to be negative or positive? Why?

(a) A college placement office estimates the average starting salary of its graduates by computing the average of all of the offers made to its graduates.

(b) A homeowner estimates the price she can get for her home by finding out how much several of her neighbors are asking for their comparable homes that are for sale.

(c) The nation's unemployment figures are obtained by counting the number of people receiving unemployment benefits.

1.6 Which of the following estimates is likely to be biased? Is the bias positive or negative? Why?

(a) You estimate the average number of bank customers waiting for service at any time the bank is open by counting the number of customers waiting for service each time you go to that bank.

(b) You estimate the average number of months a person receives social security payments by examining all files that were closed out during the previous month and finding the average number of months social security payments have been received by those people.

(c) A highway patrolman parks next to a highway and records speeds on his radar in order to estimate the percentage of people who are exceeding the speed limit.

1.2

THE IMPORTANCE OF RANDOM SAMPLES IN STATISTICS

PROBLEM SETTING 1.1

An investment company handles about 8000 accounts for its clients. This involves transactions such as buying and selling stocks, bonds, options, futures, and other media for investment of money. To facilitate these transactions, the investment company keeps most of the investments in "street name," that is, officially in the name of the investment company instead of in the names of its clients. This places an additional responsibility on the investment company in terms of recordkeeping, timely distribution of dividend and interest checks to its client accounts, and many other tasks that tax laws require.

Every investment company hires an independent accounting firm to monitor its actions, to assure that all laws are being followed, and all of its clients' interests are being safeguarded. The accounting firm cannot look over the shoulder of every employee of the investment company, making sure each transaction is legal and proper. So the accounting firm, like all accounting firms, samples the accounts, looking in detail at all transactions in the accounts that are part of the sample, and not looking at the remainder of the 8000 or so accounts.

This sampling procedure satisfies all of the rules and regulations for proper accounting procedures, and since the investment company never knows which accounts are going to be examined, this sampling procedure also reduces the temptation to do anything illegal or unethical with client money. Of course, the effectiveness of this procedure depends on the inability of the investment company to predict which accounts will be selected for audit by the accounting firm. A simple procedure for removing all predictable patterns from the sampling procedure, called **random sampling,** assures the accounting firm that its sample will be a **random sample,** one in which every account is equally likely to be included in the sample, and one in which all predictable patterns of selection have been eliminated.

A SIMPLE RANDOM SAMPLE

A sample is called a **simple random sample,** or a **random sample** for short, if it was obtained in such a way that every possible sample, with the same number of observations, was just as likely to be selected. This "equal opportunity" method of sampling enables valid projections to be made to the entire population from which the sample was obtained.

SIMPLE RANDOM SAMPLE A sample is called a **simple random sample,** or a **random sample** for short, if it was obtained in such a way that every possible sample, with the same number of observations, was just as likely to be selected.

Notice that *random* refers to the method of selecting the sample. It is not possible to tell whether a sample is random by looking at the sample itself. One must examine the method by which the sample was obtained. Two possible methods for obtaining random samples are now described.

The names of each of the 8000 or so accounts in Problem Setting 1.1 can be put on a card, one account per card, and thoroughly mixed. Then the desired number of cards (accounts) are drawn from the pile of cards without first looking at the name on the card. If the mixing process is thorough, the sample will be random.

An easier and less expensive method of obtaining a random sample is with the aid of a table of random numbers. A table of random numbers (more precisely called random digits) is a table filled with digits 0 through 9 in no particular pattern whatsoever. These tables are usually obtained from well-tested computer programs designed to generate such numbers. See Figure 1.1 for a small table of random numbers.

One way to use a random number table to obtain a random sample of accounts is to number all of the accounts consecutively from 0001 through 8000, or however far is necessary. Any method for numbering the accounts is satisfactory. Then a starting place in a table of random numbers is selected "at random," such as by closing one's eyes and pointing "blind" to some number in the table (using a four-digit group as a number in this case) to be the starting number. Suppose the number 1484 formed from the first four digits in the fourth column and sixth row of Figure 1.1 is selected. Then the account with the same number is included in the sample. The four digits immediately below 1484 in the random number table are then selected to form the number 2801, and account number 2801 is included in the sample. By reading vertically down the column and continuing at the top of the next column, accounts are included until the sample is as large as desired. A random sample with five observations in it would consist of the following account numbers

1484   2801   (8672)   1106   4868   (9486)   7841

where the numbers in parentheses are not used because there are only 8000 accounts in the population.

Of course, once the starting point is obtained, additional random numbers may also be found by reading horizontally across the row to the next column and continuing on to the next row. If the same starting point is used again, a random sample obtained by reading horizontally across the row would contain the account numbers:

1484   6771   3762   6368   (8927)   3432

81080	67493	23666	22251	17616	60716	77125	18653
83272	18379	46498	60045	80649	35179	03185	57068
82844	85553	16852	57931	84063	57516	46529	47030
33097	46244	16769	48531	56618	90035	88363	04097
48477	33067	76572	84835	96208	68558	23560	89245
61186	63971	20547	14846	77137	62636	88927	34322
92545	83866	06895	28019	08547	04275	79277	28833
05172	25637	13665	86725	45970	42670	35291	22685
73850	99275	97475	11064	93492	05362	57562	99582
77978	42899	65518	48688	96755	83554	76916	15224
16463	00350	44697	94868	22697	33740	60701	04034
56564	40277	66044	78417	52968	52982	82340	92970
26355	51841	01235	15986	65898	74181	51391	11313
87582	80276	88583	30633	50721	65017	48735	04476
15659	86285	09579	07969	17850	88197	14309	25013

FIGURE 1.1
A BRIEF TABLE OF RANDOM NUMBERS.

It should be noted that any four-digit number already included in the sample, or any four-digit number greater than 8000, may be discarded because the corresponding accounts are not available for inclusion in the sample.

Many of the statistical methods presented in this text are designed for use on a random sample or several random samples. Probability statements can be made about the population being sampled only when randomized procedures are used in obtaining a sample. In actual practice other types of sampling are sometimes used, in which case the statements about the population may not be accurate if the methods in this text are used.

SAMPLING ERROR AND NONSAMPLING ERROR

It is obvious that all samples have one characteristic in common. None of the samples looks quite like the entire population looks. Therefore, a difference will most likely exist between an estimate of a population characteristic furnished by the sample and the corresponding population characteristic. If this difference can be attributed to the fact that there is a natural but inherent difference between a sample and the population, then that difference is said to be due to **sampling error,** or the error due to sampling. Statistical methods can supply an estimate of the amount of the sampling error. Sampling error refers to the natural variation from one sample to another. It does not imply a mistake or error on the part of anyone.

However, if the difference between the estimate of the population characteristic from the sample and the actual population characteristic is due to other causes, such as workers inventing data to fill in blank spots where real data are difficult to obtain, misleading answers supplied in a survey, or unintentional errors in transcribing or retranscribing the data, then the differences between the sample estimates and the true population values are said to be due to **nonsampling error.** A complete census of the population will not contain sampling errors, but even a census may contain nonsampling errors. This source for error can be avoided through careful planning and sound sampling procedures.

SAMPLING ERROR AND NONSAMPLING ERROR

Sampling error is the name given to natural variability inherent among samples from a population. It is always present when samples are obtained.

Nonsampling error is the name given to inaccuracies and actual errors (or mistakes) that can and should be avoided by using sound experimental techniques.

As an illustration of sampling error and nonsampling error, suppose it is desired to estimate the proportion of students who smoke marijuana. Each member of your statistics class obtains a random sample of 10 students and reports the proportion who smoke marijuana. One student may report 3 out of 10 or .3, another reports $\frac{2}{10} = .2$, and so on. The variability from one random sample to the next is sampling error.

However, there may be a tendency for students in each sample to answer the question, Do you smoke marijuana? with less than complete honesty. This introduces nonsampling error. Also some students in your class may have been too busy or too shy to complete the assignment, and may have fabricated some data, leading to another source of nonsampling error.

EXERCISES

1.7 A bank wishes to select a random sample of size 10 from 100 bank accounts for a sample audit. Assume these accounts can each be identified by two digits such as 00, 01, 02, 03, . . . , 99. Use a random starting point in the random number table of Figure 1.1 to select the accounts to be used in the audit.

1.8 A manufacturer has 745 employees, all of them belonging to the group medical insurance plan. A committee of employees wants to study the amount of claims made under the insurance plan and the types of claims made. Assume the employees are numbered from 1 to 745, and use Figure 1.1 to obtain a random sample of 20 employee accounts for examination by the committee.

1.9 Consider the first 10 single-digit columns of Figure 1.1 and record the frequency with which each of the 10 digits 0, 1, 2, . . . , 9 occurs. Each column contains 15 digits, so you will be summarizing 15 × 10 or 150 digits. It would be reasonable to expect each digit to appear the same number of times (i.e., 15 times). Your frequency tabulation represents the type of random variation you could expect to find in a sample due to sampling error.

1.10 In a survey the interviewer wrote the number 10 in response to a question, but the number later was misread as 16 for the analysis. Is this a sampling error or a nonsampling error?

1.11 Although 28% of a town's population is Catholic, a random sample of ten people happened to contain seven Catholics. Is this a sampling error or a nonsampling error?

1.12 Why would a sample obtained in Florida to estimate the percentage of people in the United States over age 65 likely lead to a biased result?

1.3
EXPERIMENTS, RANDOM VARIABLES, AND STATISTICS

When some people hear the word **experiment** they see a mental picture of a scientific laboratory with several technicians in white coats doing mysterious things with test tubes and chemicals. The word experiment is given a much broader interpretation in the sciences, however. In particular, a marketing survey is an experiment, trying a new employee incentive plan and recording the results is an experiment, and noting the length of time a clerk spends with each customer is an experiment.

An **experiment** is any planned process by which observations are made and/or data are collected.

SAMPLE SPACE

The observed result of an experiment is called the **outcome** of the experiment. An experiment produces one and only one observable outcome at a time. Several possible outcomes considered together are referred to as an **event.** The collection of all possible outcomes of an experiment is called the **sample space** of that experiment.

The usual notation for events is A, B, C, and other capital letters at the beginning of the alphabet, whereas S is usually reserved for the sample space. Notice that the sample space S is the set of all outcomes and hence is the largest possible event.

OUTCOME An **outcome** of an experiment is any possible result of the experiment.

SAMPLE SPACE The **sample space** of an experiment is the collection (or set) of all possible outcomes of the experiment.

EVENT An **event** is one or more outcomes considered as a group.

EXAMPLE

Job applicants in the personnel office of a large factory may be applying for any one of four types of positions: secretarial, assembly line, maintenance, or managerial. When an applicant arrives, the personnel office does not know what type of position will be sought until the application form is filled out and submitted. Since the result is uncertain, this falls under the definition of an *experiment.* The *sample space* consists of the four possible *outcomes,* the positions called secretarial, assembly line, maintenance, and managerial. Since S denotes the sample space,

$$S = \{\text{secretarial, assembly line, maintenance, managerial}\}$$

Any set of one or more outcomes constitutes an *event.* The sample space S is one event. Some other events are as follows.

$$A = \{\text{assembly line}\}$$
$$B = \{\text{maintenance, managerial}\}$$
$$C = \{\text{secretarial, maintenance, managerial}\}$$

When studying the theory of statistics, it is also convenient to consider the event that has *no* outcomes in it, called the *null event*.

How many different events can you find for this example? If a sample space has *n* outcomes, there are 2^n different events, including the null event. In this sample space with four outcomes there are $2^4 = 16$ different events, including the null event and the four events previously listed. Can you list them?

RANDOM VARIABLES

The outcomes of an experiment are much easier to work with when they can be associated with numbers. However, outcomes by themselves are not necessarily numerical. In a poll the response of an individual being polled may be entirely verbal, for example, consisting of an opinion regarding the quality of the public transportation system. Usually an attempt is made to associate the response with a number, such as

1—completely satisfied;

2—partially satisfied;

3—no opinion;

4—mildly dissatisfied;

5—strongly dissatisfied.

The numbers are easier to record, to tally, and to summarize in a report. Any rule for assigning numbers to possible outcomes of an experiment, such as the rule just described, is called a **random variable.**

RANDOM VARIABLE A **random variable** is a rule for assigning real numbers to the outcomes of an experiment.

EXAMPLE

When the personnel office receives an application for a job, it records the type of job being sought. But instead of writing down the complete name of the job, such as secretarial or assembly line, the personnel office uses a more convenient numerical coding system. Instead of "secretarial" the office writes the number "1." Instead of "assembly line" the office writes "2." The complete code is as follows.

1 = secretarial

2 = assembly line

3 = maintenance

4 = managerial

This rule for replacing the job names (outcomes of the experiment) with numerical values falls under the definition of a *random variable*. It is a *variable* because it takes on various numerical values, and it is *random* because the job name is unknown until the application is received.

THE MEANING OF STATISTICS

There is more than one meaning of the word **statistics.** The previous meaning presented in this chapter referred to **statistics** as a science. A second meaning for the word **statistics** is associated with numbers. For example, when you listen to a broadcast of a sports event, the announcer frequently reviews the game's *statistics,* mentioning numbers such as batting averages, numbers of free throws attempted, and total yardage by the star fullback. In this context the word **statistics** is used in its plural form, as opposed to the previous meaning where the word *statistics* was used in its singular form.

STATISTICS AS NUMBERS   **Statistics** (plural) are numbers, usually used as summaries of larger sets of data.

QUANTITATIVE AND QUALITATIVE DATA

Consider, for now, statistics as numbers. To understand some of the many forms statistics can take, consider a registration card that a customer might be asked to fill out following the purchase of an appliance (see Figure 1.2). This card has several types of data. First there are **qualitative data,** such as the name and address. Then there are data, such as the price of the item or the age of the main user that indicate how much or how many, that are called **quantitative data.**

Other data are not as easy to classify, such as the serial number or the product model number. If the serial number indicates how many items were manufactured up to that point, it would be quantitative. But if the serial number serves merely as an identification number for that particular appliance, it is a qualitative item of information.

All nonnumerical data can be converted to numerical data by using a random variable to assign numbers to the nonnumerical data. This conversion to numbers does not change the type of the data; if it was qualitative before the conversion, it remains qualitative after the conversion to numbers.

TWO TYPES OF DATA   **Quantitative data** tells how much or how many and is measured on a numerical scale.

Qualitative data identifies or names some quality, category, or characteristic, but is not quantitative.

CUSTOMER REGISTRATION

Mr.
Mrs. _____ Date Purchased _____
Ms.

Address _____ Apt. No. _____

City _____ State _____ ZIP Code _____

Product
Model No. _____ Serial No. _____ Price _____

Dealer's Name/City _____

SELECTED PRODUCT AS RESULT OF:	TYPES OF STORE WHERE PURCHASED
1. _____ Previous Ownership	1. _____ Discount Store
2. _____ Store Display	2. _____ Department Store
3. _____ Catalog	3. _____ Hardware Store
4. _____ Magazine	4. _____ Appliance Store
5. _____ Newspaper	5. _____ Other _____
6. _____ Other _____	The Age of the Main User Is _____

TO HELP US SERVE YOU BETTER, PLEASE MAIL THIS CARD IMMEDIATELY

FIGURE 1.2

A TYPICAL CUSTOMER REGISTRATION CARD.

On the customer registration card the only quantitative data are the *Price of the Appliance* and the *Age of the Main User*. The other information is qualitative. Note that some of the qualitative data are numerical, such as *ZIP Code* and *Serial Number,* whereas other qualitative data are nonnumerical, such as *Name.* Some qualitative data may be either numerical or nonnumerical, such as *Selected Appliance as Result of* or *Type of Store Where Purchased,* depending on whether the verbal description is used or the numerical designation for each category is used. The numbers 1 through 6 (in the case of *Selected Product as Result of*) merely serve as surrogate names for the six different categories. However, quantitative data must be numerical, because it tells either how much or how many.

Most of this text is devoted to handling quantitative data, because statistics is generally a quantitative science. But many important and useful statistical methods use qualitative data, as you will find throughout this text.

PREPARING A QUESTIONNAIRE

Consider once again the customer registration card. The idea of the card is for you, the customer, to fill it out and send it in. The company wants to know all about you, the paying customer. What inspired you to buy their product? How can they get you to buy more of their products? How may they get others to buy their products? Which of their marketing techniques are the most effective?

If the company requests too much information from you, you might simply throw the card (which might now resemble a five-page questionnaire) into the nearest wastebasket. So the company keeps the questionnaire brief and easy to answer. (Notice the multiple-choice questions.) By including a blank space for *Serial No.* the company takes advantage of the fact that many people will assume that they must fill out the card in its entirety and send it in to validate their warranty. Thus the company may get as many as 80% or 90% of its customers to fill out cards and send them in. This can be compared to responses to questionnaires that often are as low as 10% or 20%.

Sampling questionnaires, such as this one, require careful planning to be successful. A successful questionnaire is usually the result of a team of people with skills in fields such as marketing, psychology, and (you guessed it) statistics. Many of the types of questions a company hopes to get answered by such a questionnaire can be answered only by using statistical methods. The use of statistical methods requires the right kind of data, so questions need to be written with the statistical method in mind. Often a survey turns out to be a wasted effort when the person in charge of the survey brings collected data to a statistician and says, "Analyze these data," but the data are not capable of answering the questions the survey was intended to answer.

SUMMARIZING DATA FROM A QUESTIONNAIRE

The customer registration card in Figure 1.2 appears to be the result of good planning on the part of the company that issued it. Now suppose that you and many others like you have sent in their customer registration cards and the company now has stacks of cards available for study. The first thing the company might do is bring in its "information systems" specialist to put the data into its computer in such a way that all the interesting facts and figures can be extracted quickly and accurately.

After the cards are collected and compiled, many more quantitative measurements are available. Consider the following questions, whose answers involve quantitative measurements.

1. How many people sent in cards for Model No. TM-151?
2. Of the TM-151 cards, what percentage was purchased in discount stores?
3. What is the average price of the TM-151s sold in appliance stores?

Clearly there are many other measurements of a quantitative nature that are now available from this collection of information. It is always necessary to summarize the results of surveys such as this one, because there are simply too many data to comprehend in the present form. Recall that the plural usage of the word *statistics* referred to numbers, usually used as summaries of larger sets of data. A single *statistic* refers to a single rule, or random variable, that is used for summarizing data. This word *statistic* is used extensively in this text.

Region	Number of Appliances Sold
Northeast	1436
Southeast	1724
Central	1125
Midwest	838
Northwest	175
Southwest	1840

FIGURE 1.3
THE NUMBER OF APPLIANCES SOLD BY
GEOGRAPHIC REGION, AS REPORTED
BY THE CUSTOMER REGISTRATION
CARDS RECEIVED DURING 1980.

STATISTIC When a random variable is used to summarize data, it is called a **statistic.**

Suppose the company wants to know how many appliances (of all types) are sold in each of the six major geographic regions of the United States. There are six statistics involved, representing the number of appliances sold in each of the six regions. The results are given in Figure 1.3.

In the data summary given by Figure 1.3 the region represents a qualitative variable and the number of appliances sold is a quantitative variable. Often a pictorial representation of this information is easier to understand, and is especially helpful to people who do not have much time to spend studying numbers. Several graphical methods of displaying such information are given in the next chapter.

EXERCISES

1.13 What is the difference between quantitative data and qualitative data?

1.14 Identify below each item of numerical information as either quantitative or qualitative.

(a) Social security number

(b) Date of birth

(c) Age

(d) Weight

(e) Height

(f) Employee number

(g) Phone number

(h) Street address

(i) Number of dependents

(j) Years of professional experience

1.15 Given below is a reproduction of a product information card that accompanied an appliance. Which of the pieces of information requested represent quantitative data and which represent qualitative data?

Model Number ——————————————   Date Purchased—————

Name ————————————————   Address —————————————

City —————————————— State——————— ZIP ————

Name of Store ————————————————————————————

1. IS THIS APPLIANCE A REPLACEMENT? ☐ YES ☐ NO	4. AGE OF HEAD OF HOUSEHOLD ☐ 25-34 ☐ 45-54 ☐ OVER 54 ☐ 35-44 ☐ UNDER 25	6. WHICH FACTORS MOST INFLUENCED YOUR PURCHASE?
2. WHO WAS THE APPLIANCE PURCHASED BY? ☐ MAN ☐ WOMAN ☐ BOTH	5. IF GIFT, WHAT WAS OCCASION?	☐ MAGAZINE ☐ NEWS-AD ☐ FEATURES ☐ PRICE ☐ RECOM-MENDED ☐ WARRANTY ☐ RADIO/TV
3. IF YOU PURCHASED THIS APPLIANCE DID YOU SEE IT DEMONSTRATED? ☐ YES ☐ NO	☐ BIRTHDAY ☐ OTHER ☐ CHRISTMAS ☐ GRADUATION ☐ MOTHER'S DAY	☐ PREV. OWNER ☐ BRAND NAME ☐ SALES–PERSON

1.16 The appliance company had a specific purpose in mind for each of the questions on the product information card of Exercise 1.15 when it designed the card. A summary or tabulation will be made of the responses on this card as they are received from the customers. Suppose you are responsible for making recommendations aimed at improving future sales based on such a tabulation. How does each of the six questions provide specific help to you in making such recommendations?

1.17 A student takes a multiple-choice exam consisting of 20 multiple-choice questions. The exam score (outcome) consists of the total number of points, where five points are awarded for every correct answer and no points are awarded for incorrect answers. Describe the following:

(a) The experiment.

(b) The sample space for the experiment.

(c) Two possible outcomes of the experiment.

(d) The outcomes in the event "the grade is above 70."

1.18 A machine produces four parts during each hour of operation. Each manufactured part is then graded as either acceptable or not acceptable. Describe the sample space for one hour's production.

1.19 Employees in one factory have their choice of two unions to join. They can join union A, union B, or both, or neither. An experiment consists of selecting one employee at random and asking which union or unions the employee belongs to.

(a) Describe the sample space.

(b) Describe the outcomes in the event "belongs to only one union."

(c) Let the random variable be a rule for assigning the number of unions to which the employee belongs, to each point in the sample space. List the sample space in terms of the numbers assigned by this random variable.

1.20 A student takes a multiple-choice exam consisting of ten questions. The possible outcomes are the number of correct answers, which are the numbers 0, 1, 2, . . . , 10. Let the random variable be the percentage score, from 0 to 100. Give the values of the random variable assigned to each point in the sample space.

1.21 When you see the headline Business Statistics over a regular feature in your newspaper, does this refer to the singular or the plural definition of statistics?

1.22 Does the term Business Statistics in the title of this text refer to the singular or the plural definition of statistics?

1.4

SAMPLE SURVEYS (OPTIONAL)

PROBLEM SETTING 1.2

Several years ago the League of Women Voters in Manhattan, Kansas, wanted to assist the city in applying for federal funds for urban renewal. As a part of the application, the federal government required information concerning the location and extent of dilapidated housing in the city. Housing was considered dilapidated if and only if it met certain federal guidelines, which considered the appearance of the house, the number of bathrooms, the condition of the foundation, and numerous other factors. By carefully inspecting a residence and asking certain questions of its occupants, a league member could determine whether or not the house qualified for the term dilapidated according to federal guidelines.

Because of the time required to determine the condition of each residence, it is easy to see why the League of Women Voters, a volunteer organization, found it impossible to examine all of the residences in Manhattan, Kansas. They decided to take a sample.

SAMPLE SURVEYS

Unlike a census, a sample survey examines only a portion of the population of interest, which may be people, farms, businesses, accounts, or other units. A sample is selected and analyzed, and projections are often made with respect to the entire population. The validity of the projection depends on the validity of the sample. Samples consisting of volunteers, or a haphazard selection of individuals, do not provide valid projections with respect to the population. However, random samples and a few other sample types described in this section do allow valid statements to be made about the entire population even though the sample may be small relative to the size of the population. For example, the well-known national polls use a sample of only about 1500 persons to reflect the opinions and habits of a national population of over 200 million people, and the projections are accurate within about 2% of the true value!

TARGET POPULATIONS AND SAMPLED POPULATIONS

In Problem Setting 1.2 the **target population** of the league's survey was the collection of all residences within the city limits. In general the target population is the population about which information is desired. The **sampled population** was the list of all households given by the current city directory.

The city directory, if you are not familiar with this useful book, is available to businesses for a modest price, and usually may be found in the public library. It lists households geographically street by street, along with information such as the names of the inhabitants and their occupations if this information is known. The city directory was a valuable source of information in this project.

Sometimes the sampled population is called the **sampling frame.** In this case, the sampled population matched well with the target population, because the city directory was up to date and included most, if not all, of the households in the city. Households outside the city limits can be omitted from a study by checking addresses against a map containing the city limits.

TARGET POPULATION AND SAMPLED POPULATION

The **target population** is the population that is the ultimate object of study.

The **sampled population** is the population from which the sample was obtained.

The validity of the study depends on the target population and the sampled population being similar in the characteristics being studied.

One should be aware that the target population and the sampled population do not always agree. For example, university research projects often use students as subjects. Consider a research project that examines the effect of lack of sleep on the ability to concentrate. Subjects are obtained by selecting from students who have responded to an ad in the campus newspaper. The sampled population consists of young men and women who are above average in intelligence, who read the campus newspaper, and who happen to have more time than money. This may be quite unlike the target population, which may be people of all ages, from all walks of life.

SIX TYPES OF SAMPLES

There are several different types of sampling procedures the League of Women Voters could have used, although they used only one. Please keep in mind that there is a precise mathematical definition for each of these sampling procedures, but they are explained here only by example in the belief that this will lead to an easier understanding.

The first four sample types are called **probability samples,** because the probability of each element in the population being selected is known. The last two sample types are called **judgment samples** because personal judgment plays a major role in deciding which elements of the population are selected, and the probability of selection is not known.

1. **Simple random sample** This method of sampling enables valid projections to the entire sampled population to be made with confidence. A sample with n observations in it is considered to be a simple random sample of size n if it is chosen in such a way that every possible collection of n units from the sampled population is equally likely to be selected. Randomization is used on the population units to obtain a simple random sample. A more extensive discussion of random sampling was given in Section 1.2.

2. **Systematic sample** The League of Women Voters could have started at the beginning of the city directory, selected one household at random from the first 100 households listed, and then counted from that household, including every hundredth household in their sample. In this way a systematic sample of households would be drawn. If no bias results from this method of selecting a sample, the sample may be treated as if it were a random sample for purposes of making statistical analyses.

   This method of sampling is easy to use when the sampled population is listed, such as in a group of accounts in a file drawer or names on a list. Sampling from an assembly line to ensure quality control often uses a systematic sample. It forces the sample to span the entire population in a systematic manner, but prevents two items that are close together in the list from both being included in the sample.

   Systematic sampling can be good in one situation but undesirable in another. For example, a systematic housing survey may select only houses on the corner of each block, and corner houses may have characteristics different from those in the rest of the neighborhood. This would be undesirable. However, a systematic sample from an alphabetical list of college students to determine home states would ordinarily exclude brothers and sisters with the same surname, which might be desirable.

3. **Stratified sample** This is the sampling method used by the League of Women Voters in this section. Boundary lines were drawn on a map of the city outlining several sections of town that were known to be somewhat similar or homogeneous. One of the regions thus defined was the area around the business district, which was the area the league had in mind for urban renewal. Within each of these regions, called strata (singular, stratum), a random sample of households was selected using a random number table in the manner described in Section 1.2.

   The sections believed to contain a higher proportion of dilapidated residences were sampled much more heavily than the newer sections of town. In this way proportionately more time was spent in the neighborhoods that contained most of the dilapidated housing and less time was spent in the more affluent neighborhoods. The numbers obtained were weighted appropriately to reflect these proportions, and accurate estimates were obtained for the entire city.

   Stratified sampling improves accuracy when a heterogeneous (nonhomogeneous) population is divided into strata that are relatively homogeneous. The Gallup poll, the Harris poll, and most political polls attempt to use stratified sampling whenever possible, because of the greater accuracy from smaller sample sizes, as compared with simple random sampling.

Stratified samples require a different method of analysis than simple random samples. The methods in this text do not apply to stratified samples.

4. **Cluster sample** Once the League of Women Voters decided which households to include in its sample, it could have tripled its sample size by including the two households closest to the household selected as well. Thus the observations would be clusters of three each.

An advantage of cluster sampling is ease in obtaining additional observations, especially when distances between observations are large and time required to take the observations is short. In the league's survey, the distance from one observation to the next was only a few blocks, a minor inconvenience compared to the time and effort required to examine a household. Therefore cluster sampling was not used. In a survey of farm houses a different decision might have been reached.

A disadvantage of cluster sampling is that often units that are located close together are similar. Hence, these additional units may not contribute information much different from that of the first unit examined. As with stratified samples, the methods of this text should not be used with cluster samples.

5. **Convenience sample** By including each league member's household in the sample and throwing in the households of their friends they could easily obtain information on several hundred residences. Convenience samples are commonly used in "man on the street" broadcasts and supermarket surveys. The results of a convenience sample cannot be used to make valid conclusions about the target population because the sampled population is likely to be quite different from the target population.

6. **Representative sample** A search is conducted for several typical middle-income family dwellings, several typical low-income apartment households, and so on, until a sample is constructed that contains representatives of each of the household types likely to be encountered in the city. Without some estimate of the total number of each of these household types, such information does not answer the questions in the application of urban-renewal funds. Additionally, the method of determining which households to include or exclude from the sample invites personal bias to enter the selection process. The bias that is inherent in a sampling method such as this one should always be avoided. Probability sampling methods, unlike judgment sampling methods, use randomization to remove the unwanted bias, and therefore probability samples are the only samples that can be analyzed using statistical methods.

DESIGNING A SAMPLE SURVEY

The example involving the League of Women Voters illustrates the four steps in designing (i.e., planning) and conducting a sample survey.

1. **Planning** Considerable planning must go into any valid sample survey. All of this planning takes place before the actual survey begins. For example,

a listing of all units in the sampled population must be obtained. Questions such as: How well does the sampled population resemble the target population? How large a sample is needed? must be considered. A randomized procedure is needed to determine which units in the sampled population are to be included in the sample.

Questionnaires need to be designed and pretested on people similar to the people being surveyed. The people administering the questionnaires should be trained so that the interviews are conducted fairly, with no interviewer biases allowed to affect the answers to the questions. Alternative procedures need to be devised, in case no one is home, or the person being interviewed refuses to cooperate.

2. **Data collection** The survey is only as valid as the data. Difficulties that arise when collecting the data sometimes encourage shortcut methods that should be avoided. Therefore it is a good idea for the people in charge of the survey to participate in the data collection so that some of the problems involved can be observed firsthand. No matter how thorough the planning stage was, unforeseen problems may arise in the data-collection stage.

3. **Data analysis** An important part of the data analysis consists of simple charts, graphs, and data summaries. These and other, more sophisticated methods of data analysis are given in this text and can be used to make precise statements about the population from which the sample was obtained. A trained statistician is usually involved in all four stages of a sample survey, but especially in this stage, which involves analysis of the data.

4. **Conclusions** Often a survey will result in several recommendations for alternative courses of action, with an analysis of the likely consequences of each. The decision maker should consider these recommendations before making the final decision, but should not relegate the decision-making responsibility to the results of a survey.

The four steps necessary for a sample survey are:

DESIGN OF A
SURVEY AND
ANALYSIS OF
SURVEY DATA

1. Prior planning and design of the survey.
2. Collection of the data in the sample.
3. Analysis of the data collected.
4. Conclusions based on the data analysis.

EXERCISES

1.23 Suppose you are placed in charge of designing a sample survey to determine the number of hours the average student spends viewing television in one of the dormitories on your campus. Explain what would be involved in each of the four steps of conducting a sample survey.

1.24 Why do you think that stratified sampling adds to the greater accuracy of political polls such as the Gallup poll and Harris poll?

1.25 A congressional committee wishes to examine the effect of proposed legislation on the nation's high schools. It randomly selects five high schools from the Washington, D.C., area and conducts a study on those five schools.

(a) What is the target population?

(b) What is the sampled population?

1.26 A list of 100 bank accounts has been made available for sampling. These bank accounts are numbered from 00 to 99, and a sample of size five is drawn. From the following samples, identify the sample most likely to be a simple random sample, a convenience sample, a systematic sample, a cluster sample, or a stratified sample.

(a) Account numbers 18, 38, 58, 78, and 98.

(b) Account numbers 00, 01, 02, 03, and 04.

(c) Account numbers 07, 16, 43, 58, and 81.

(d) Account numbers 13, 36, 44, 77, and 90.

(e) Account numbers 12, 13, 75, 76, and 77.

1.27 Describe a method for estimating the following quantities, where the method is within reason with respect to cost, and yet can be expected to be fairly accurate.

(a) The percentage of people in the United States who are over age 65.

(b) The average amount of money a teenager in your community spends annually on records and tapes.

(c) Next year's gross sales in a small retail store that has been in existence five years.

1.28 To determine the prices to use in the Consumer Price Index, a random sample is obtained from each type of store, such as grocery stores, hardware stores, and clothing stores. What type of sample is this?

1.29 The manufacturing quality of a car battery is maintained by selecting every twenty-fifth battery to come off the assembly line and completely dismantling it to see if it has been constructed properly. What type of sample is this?

1.30 The students in your statistics class are being used to obtain an estimate of the average weight of students in your university. There are eight female and 24 male students in your class, but about 50% of the university students are female. How would you estimate the average university student's weight? What type of sample would be most appropriate here?

1.5

AN APPLICATION OF SAMPLING: THE CONSUMER PRICE INDEX (OPTIONAL)

PROBLEM SETTING 1.3

Statistical methods for obtaining random samples are used extensively in gathering business and economic data that are used to calculate quantitative measures of the general level of growth of prices, production inventory, and other quantities of economic interest. Such quantitative measures are commonly referred to as index numbers. Probably the best known of these index numbers is the Consumer Price Index, which is designed to reflect changes in the cost of living.

The Consumer Price Index is based on the prices of about 400 goods and services as compared with the prices of those goods and services during some previous period of time. Many wage contracts between management and labor are tied to the Consumer Price Index, as are veterans' benefits, social security payments, and government salary scales. Other types of price indexes are used in other applications, such as the Producer Price Index, which is used to measure the level of business prosperity. One major problem in computing any price index is that actual prices and quantities are not easily agreed upon. Therefore the prices and quantities used in computing the Consumer Price Index are the result of a statistical sampling procedure such as random sampling or stratified sampling.

SIMPLE INDEXES

Many indexes are much simpler than the Consumer Price Index and are derived from simple ratios involving only a small number of goods and services. This subsection considers some simple ratios because they serve as an introduction to the Consumer Price Index.

A quart of milk in 1980 cost 54 cents, whereas 10 years earlier a quart of milk cost only 38 cents. The 1980 price relative to the 1970 price, expressed as a percentage, is

$$I = \frac{.54}{.38} \cdot 100 = 142$$

The **price relative** is an example of a simple index number. It is a convenient way of expressing the price of milk in 1980 relative to 1970 prices. The index of 142 indicates that the price rose 42% during this 10-year time period.

The year 1970 serves in this case as a reference year or base year. In some widely used indexes the reference year is actually the average over several consecutive years, in which case it would be described more accurately as a reference period or base period.

The ratio of the two prices of a product, at two different time periods, expressed as a percentage is called a **price relative:**

PRICE
RELATIVE

$$I_t = \frac{P_t}{P_0} \cdot 100$$

where

$$P_t = \text{price at time } t$$
$$P_0 = \text{price at a base period}$$

Sometimes it is more meaningful to include quantity in the index, which may serve as a measure of each item's relative importance. Since quantity times price is the **value** of an item, indexes that incorporate the product of quantity and price are called **value relatives.** If per capita milk consumption is 128 qt per year in 1980, and 97 qt per year in 1970, the **value relative** of per capita milk consumption is

$$I_t = \frac{128(.54)}{97(.38)} \cdot 100 = 188$$

This 88% increase may reflect more accurately the change in the amount of money spent on milk on a per capita basis due to the change in the quantity consumed as well as the change in price.

The ratio of the value of a product (quantity × price) at time t to the value at a base time, expressed as a percentage, is called a **value relative:**

VALUE
RELATIVE

$$I_t = \frac{Q_t P_t}{Q_0 P_0} \cdot 100$$

where

Q_t = quantity at time t

Q_0 = quantity at a base time

WEIGHTED AGGREGATE INDEX

Even more quantitative data can be summarized into a single index number by combining information on several products. For example, a micro-market-basket price can be indexed by including several food items together. Consider the basic food items in Figure 1.4, with their prices and per capita annual consumption.

i	Food Item	1970 Q_{0i}	1970 P_{0i}	1970 Value	1980 Q_{ti}	1980 P_{ti}	1980 Value
1	Milk	97 qt	38¢/qt	$36.86	128 qt	54¢/qt	$ 69.12
2	Bread	45 lb	54¢/lb	24.30	22 lb	71¢/lb	15.62
3	Hamburger	21 lb	85¢/lb	17.85	42 lb	$1.88/lb	78.96
4	Corn	18 qt	78¢/qt	14.04	6 qt	89¢/qt	5.34
				$93.05			$169.04

FIGURE 1.4
PRICE AND QUANTITY CONSUMED FOR FOUR FOOD ITEMS IN 1970 AND 1980.

The total value of the micro-market basket in 1970 is found by summing the values for each item in the market basket. Thus the value in 1970 is

$$V_0 = \$36.86 + 24.30 + 17.85 + 14.04 = \$93.05$$

This may be compared with the value in 1980,

$$V_t = \$69.12 + 15.62 + 78.96 + 5.34 = \$169.04$$

The ratio of V_t to V_0 expressed as a percentage

$$
\begin{aligned}
I_t &= \frac{V_t}{V_0} \cdot 100 \\
&= \frac{169.04}{93.05} \cdot 100 \\
&= 181.7
\end{aligned}
$$

gives an index that reflects the current per capita consumption cost relative to that of 10 years earlier. In this case an 81.7% increase occurred.

This type of index is called **composite value index,** and is a special case of a **weighted aggregate index,** *weighted* because each price is given its own weight or multiplier (quantity in this case), and *aggregate* because several items are considered together. An unweighted aggregate index is one in which all of the weights are equal to 1.0.

The symbol Σ is the Greek letter "sigma" and is read "the sum of." Thus the sum of several terms $W_1P_1 + W_2P_2 + \cdots + W_nP_n$ may conveniently be written as ΣW_iP_i. More is said about this notation in Chapter 3.

A **weighted aggregate index** is a ratio of $\Sigma W_{ti}P_{ti}$ to $\Sigma W_{0i}P_{0i}$ expressed as a percentage:

$$I_t = \frac{\Sigma W_{ti}P_{ti}}{\Sigma W_{0i}P_{0i}} \cdot 100$$

WEIGHTED
AGGREGATE
INDEX where

W_{ti} = a weight assigned to the price P_{ti} of the ith item at time t

W_{0i} = a weight assigned to the price P_{0i} of the ith item at the base time 0

The weights are usually expressed as the quantities Q_{ti} and Q_{0i}, respectively.

THE CONSUMER PRICE INDEX

A general index based on the cost of many goods and services nationwide is the **Consumer Price Index** (CPI). The CPI is a statistical measure of change, over time, in the prices of goods and services in major expenditure groups such as food, housing, apparel, transportation, health, and recreation, typically purchased by urban consumers. Essentially it measures the purchasing power of consumers' dollars by comparing what a sample market basket of goods and services costs today with what a sample market basket cost at an earlier date.

Because the CPI is an attempt to reflect the changes in the cost of living, many wage contracts are tied directly to changes in it. Therefore, it is quite important that the CPI accurately reflect the change in the cost of living. This requires that many goods and services be included in the index; indeed the CPI uses about 400 such quantities. The decision about which goods and services to include in the computation of the CPI must be made carefully. The CPI should include enough quantities to prevent any one quantity from dominating the index.

The Consumer Price Index considers prices of 382 items in many categories. In 1978 the index was updated to reflect the following weights:

Items	Weights (%)
Food	19
Housing	44
Clothes	6
Transportation	18
Health and recreation	9
Miscellaneous	4

Social security and income taxes are not figured into the Consumer Price Index, so a change in those taxes would not change the Consumer Price Index, although it would affect buying habits.

Several weaknesses of the CPI should be noted to provide additional perspective. As prices change relative to each other, consumers typically react by substituting less expensive products for similar products that have gone up in relative price, such as by substituting chicken for beef. Many taxes are not computed in the CPI, although taxes are a major expenditure in most budgets. Also, the quantities used in computing the CPI may be unrealistic for figuring the cost of living of college students, elderly people, or other nonstandard family types.

Once the selection of the goods and services to be included in the CPI has been made, an accurate estimate of the cost of goods and services must be made. For example, a CPI based on the cost of goods and services in Miami Beach, Florida, could hardly be expected to reflect accurately the cost of goods and services in Butte, Montana. Hence, a sampling procedure must be used to get an estimate of the cost of goods and services that will be meaningful in all parts of the country. The CPI results from using a stratified sample on prices of food, clothing, shelter, fuels, transportation fares, doctors' and dentists' services, drugs, and the other goods and services that people buy for daily living. Prices are collected on a monthly basis in 85 urban areas across the country from over 18,000 tenants, 18,000 housing units, and about 24,000 establishments including grocery and department stores, hospitals, filling stations, and other types of stores and service establishments.

CHANGES IN THE CONSUMER PRICE INDEX

The Consumer Price Index (CPI) was 195.4 in 1978 and 217.4 in 1979, where 1967 = 100 is the **base year.** The difference between these two values is 217.4 − 195.4 = 22.0. Since the Consumer Price Index is expressed as a percentage of its 1967 value, it is correct to say that the Consumer Price Index went up 22 percentage points. It would be misleading, however, to say that the Consumer Price Index went up 22%, because that would suggest immediately that prices went up 22%, and they did not. The 1979 prices expressed as a percentage of the 1978 prices were

$$\frac{1979 \text{ CPI}}{1978 \text{ CPI}} \cdot 100\% = \frac{217.4}{195.4} \cdot 100\% = 111.3\%$$

so prices went up only 11.3%.

The Consumer Price Index has increased so much since its base year, where it is 100, that percentage changes in the Consumer Price Index can no longer be interpreted directly as percentage changes in consumer prices. Some conversion is required, such as the one just given. If the Consumer Price Index is close to its base value 100, such a conversion is not required. The change in the index from 100 in 1967 to 104.2 in 1968 represents 4.2 percentage points, and converts directly to a 4.2% increase in consumer prices. The following year's index rose from 104.2 to 109.8, a rise of 5.6 percentage points. The actual change in consumer prices is found from the ratio

$$\frac{1969 \text{ CPI}}{1968 \text{ CPI}} \cdot 100\% = \frac{109.8}{104.2} \cdot 100\% = 105.4\%$$

which shows a 5.4% increase in consumer prices. The further the index strays from 100, the greater the disparity between the percentage changes in the CPI and the actual percentage change in consumer prices. So the base year needs to be changed periodically to keep the index closer to the easily interpreted value 100.

CHANGING THE BASE YEAR

Changing the base year is a simple process. It involves dividing the old index by its value in the new base year. For example, the numbers in Figure 1.5 represent the Consumer Price Index under its old base period, which was an average of the values in 1957, 1958, and 1959. To change to the new base period, 1967, the old index numbers are divided by 116.3, which is the value of the old index in 1967. Then the new index will equal 100% in 1967, and all of the other values of the new index will be expressed relative to this new 1967 value. The base year of the CPI was changed in 1988 to a reference base of 1982–1984.

The choice of a base year is not easy. The base year should be a year in which unusual situations are absent, so that the values for that year can be considered rather "ordinary." No time period is completely free of unusual

	Consumer Price Index	
(1) Year	(2) 1957–59 = 100	(3) = (2) ÷ 116.3 1967 = 100
1955	93.3	80.2
1956	94.7	81.4
1957	98.0	84.3
1958	100.7	86.6
1959	101.5	87.3
1960	103.1	88.7
1961	104.2	89.6
1962	105.4	90.6
1963	106.7	91.7
1964	108.1	92.9
1965	109.9	94.5
1966	113.1	97.2
1967	116.3	100.0
1968	121.2	104.2
1969	127.7	109.8
1970	135.3	116.3
1971	141.1	121.3
1972	145.7	125.3
1973	154.8	133.1
1974	171.8	147.7
1975	187.5	161.2
1976	198.3	170.5
1977	211.1	181.5
1978	227.3	195.4
1979	252.8	217.4

FIGURE 1.5
CHANGING THE BASE YEAR ON THE CPI
TO 1967.

situations, so individual judgment is necessary in the determination of a base year. The base year should not be too far in the past, where the environment may have been considerably different from the current environment.

A table of index numbers usually indicates the base year by the code 1967 = 100 or a similar expression, often as a footnote. This means that the index in the base year (1967 in this case) is 100 and is the basis for comparisons made in other years.

THE PURCHASING POWER OF THE DOLLAR

Have you ever heard someone say that the dollar is worth only 42 cents (or some other figure) today as compared with 1967? Did you ever wonder how this figure was obtained? The answer is very simple. The **purchasing power** of the dollar is just the reciprocal of the Consumer Price Index, multiplied by

	Consumer Price Index (1967 = 100)	Purchasing Power of the Dollar (1/CPI) × 100
1967	100.0	1.00
1968	104.2	0.96
1969	109.8	0.91
1970	116.3	0.86
1971	121.3	0.82
1972	125.3	0.80
1973	133.1	0.75
1974	147.7	0.68
1975	161.2	0.62
1976	170.5	0.59
1977	181.5	0.55
1978	195.4	0.51
1979	217.4	0.46

FIGURE 1.6
COMPUTATION OF THE PURCHASING
POWER OF THE DOLLAR.

100. It expresses the purchasing power of the dollar each year in terms of the purchasing power of the 1967 dollar. Purchasing power is given in Figure 1.6, which makes it clear that the purchasing power of the dollar has decreased each year since the base period in 1967.

SPLICING INDEX NUMBERS

It has been mentioned that periodically the weights in the CPI are changed to reflect more current buying habits. In 1964 the index was changed to include new items that had become important, such as between-meal snacks, hotel and motel rooms, garbage-disposal units, moving expenses, college tuition, textbooks, and legal services. Further changes were made in 1978. When these items are added to the index and the weights of other items are adjusted, the index may take on a new value that may not resemble the former value at all. This new value must be adjusted to match the old value, so the new series of numbers will blend in with the old series of numbers. This is called **splicing the index numbers,** and consists of multiplying the new numbers by a constant so the new index equals the old index in the year in which the change occurred.

To illustrate the procedure, consider the hypothetical index series given in Figure 1.7 where the base year is 1972. The weights were changed in 1975 to update the index. The index with the new weights would ordinarily not be reported. Only the spliced values would be reported. The spliced values are obtained by multiplying the new index numbers by 1.1387, the ratio of the old to the new index numbers in 1975,

$$\frac{112.5}{98.8} = 1.1387$$

(1)	(2)	(3)	(4)
	Index	Index	Spliced Index
Years	(Old Weights)	(New Weights)	(4) = (3) × 112.5 ÷ 98.8
1970	96.9		
1971	98.3		
1972	100.0		
1973	104.8		
1974	107.1		
1975	112.5	98.8	112.5
1976		103.9	118.3
1977		108.0	123.0
1978		112.6	128.2
1979		117.7	134.0

FIGURE 1.7
EXAMPLE SHOWING SPLICED INDEX NUMBERS.

Now the series of index numbers has continuity. The old index values are given for years prior to 1975 and the spliced values are given from 1975 onward with no gaps occurring where the transition from the old to the new takes place.

EXAMPLE

The Dow Jones industrial average is used as an index of the performance of industrial stocks on the New York Stock Exchange, even though only 30 stocks are used in the Dow Jones average, while hundreds of industrial stocks are traded on the New York Exchange. In May 1984 the prices of one share from each of the following 30 companies were added together and divided by the constant 1.194 to obtain the Dow Jones average.

Allied Corporation	General Electric	Owens-Illinois
Aluminum Company	General Foods	Procter & Gamble
American Brands	General Motors	Sears Roebuck
American Can	Goodyear	Standard Oil of California
American Express	Inco	Texaco
AT&T	IBM	Union Carbide
Bethlehem Steel	International Harvester	United Technologies
Du Pont	International Paper	U.S. Steel
Eastman Kodak	Merck	Westinghouse Electric
Exxon	Minnesota M&M	Woolworth

On May 30 Allied Corporation gave each stockholder three new shares for every two shares they owned. This is called a 3 for 2 split, and resulted in an immediate change in the price of the stock from $50 per share to $33.25 per share. At the same time Westinghouse Electric declared a 2 for 1 split, thus

| 1984 | Dow Jones Industrial Average | | |
	Before Splits	After Splits	Spliced
May 21	1125.31		
May 22	1116.62		
May 23	1113.80		
May 24	1103.43		
May 25	1107.10		
May 29	1101.24		
May 30	$1102.59 = \dfrac{\$1316.49}{1.194}$	$1071.49 = \dfrac{\$1279.36}{1.194}$	$1102.59 = \dfrac{\$1279.36}{1.16032}$
May 31			1104.85
June 1			1124.35
June 4			1131.57
June 5			1124.89
June 6			1133.84
June 7			1131.25

FIGURE 1.8
EXAMPLE SHOWING SPLICING OF THE DOW JONES INDUSTRIAL AVERAGE.

reducing the price per share from $40.75 to 20.37 on the New York Exchange. The total price of the 30 industrial shares was $1316.49 before the splits, but only $1279.36 after the splits, reflecting the reduction in price of those two shares. Clearly a new divisor is needed to splice the new series smoothly onto the old series.

The new divisor, 1.160, was obtained by multiplying the old divisor, 1.194, by the ratio of the new total price to the old total price.

$$1.16032 = 1.194 \left(\frac{1279.36}{1316.49} \right)$$

The sequence of averages before and after the stock splits is shown in Figure 1.8. By continual splicing like this, the Dow Jones industrial average maintains its continuity through stock splits and through changes in the companies that comprise the index.

ADJUSTING A TIME SERIES USING INDEX NUMBERS

Consider the numbers in Figure 1.9, which gives the disposable personal income for the years 1961 to 1978. Disposable personal income is the total personal income less tax and nontax payments to governments.

The numbers in Figure 1.9 are called a time series because they represent a series of observations taken at successive time points. The purpose of the numbers in Figure 1.9 is to convey information, but this time series does a poor job. Not many people get meaning out of the statement, "The disposable personal income in the United States was 362.9 billion dollars in 1961."

Year	Amount	Year	Amount	Year	Amount
1961	$362.9	1967	$544.5	1973	$ 901.7
1962	383.9	1968	588.1	1974	984.6
1963	402.8	1969	630.4	1975	1,086.7
1964	437.0	1970	685.9	1976	1,184.4
1965	472.2	1971	742.8	1977	1,303.0
1966	510.4	1972	801.3	1978	1,451.2

FIGURE 1.9
DISPOSABLE PERSONAL INCOME (BILLIONS).

The first step toward converting these numbers into more understandable figures would be to divide them by the population size (another time series) each year to get per capita personal disposable income. These new numbers, although more meaningful than before, can be adjusted further to reflect the current *purchasing power* of the income. This is easily accomplished by dividing by the Consumer Price Index. The result, multiplied by 100 because the CPI is expressed as a percentage, is the *purchasing power of the per capita disposable income,* expressed in terms of 1967 dollars, since 1967 is the base period of the Consumer Price Index. The computation of purchasing power for the years 1961–1978 is illustrated in Figure 1.10.

(1) Year	(2) Disposable Personal Income (Billions)	(3) Population of the U.S. (Thousands)	(4) = (2) ÷ (3) Per Capita Disposable Personal Income	(5) Consumer Price Index (1967 = 100)	(6) = [(4) ÷ (5)] × 100 Per Capita Purchasing Power (1967 Dollars)
1961	$ 362.9	183,691	$1,975.6	89.6	$2,204.9
1962	383.9	186,538	2,058.0	90.6	2,271.5
1963	402.8	189,242	2,128.5	91.7	2,321.2
1964	437.0	191,889	2,277.4	92.9	2,451.5
1965	472.2	194,303	2,430.2	94.5	2,571.6
1966	510.4	196,560	2,596.7	97.2	2,671.5
1967	544.5	198,712	2,740.1	100.0	2,740.1
1968	588.1	200,706	2,930.2	104.2	2,812.1
1969	630.4	202,677	3,110.4	109.8	2,832.8
1970	685.9	204,878	3,347.8	116.3	2,878.6
1971	742.8	207,053	3,587.5	121.3	2,957.5
1972	801.3	208,846	3,836.8	125.3	3,062.1
1973	901.7	210,410	4,285.4	133.1	3,219.7
1974	984.6	211,901	4,646.5	147.7	3,145.9
1975	1,086.7	213,559	5,088.5	161.2	3,156.6
1976	1,184.4	215,142	5,505.2	170.5	3,228.9
1977	1,303.0	216,817	6,009.7	181.5	3,311.1
1978	1,451.2	218,502	6,641.6	195.4	3,399.0

FIGURE 1.10
COMPUTATION OF PURCHASING POWER.

The numbers representing purchasing power in Figure 1.10 are much easier to interpret than those representing disposable income. In comparing the five years from 1973 to 1978, for instance, the total disposable income rose 60.9%

$$\frac{1451.2}{901.7} \cdot 100\% = 160.9\%$$

but the per capita purchasing power rose only 5.6%

$$\frac{3399.0}{3219.7} \cdot 100\% = 105.6\%$$

The latter figure reflects more accurately the increase in the general level of prosperity during those five years.

EXERCISES

1.31 Assuming the price of a new Toyota pickup in 1973 was $2400 and in 1980 was $6900, find the 1980 price relative to the 1973 price.

1.32 Supermarket shoplifters in 1966 took items averaging $3.05 in value, according to a Commercial Service System survey. By 1979 the average value of stolen supermarket goods was $5.99. Find the 1979 price of stolen goods relative to the 1966 price. By what percentage did the average value increase? Now adjust both average values by the CPI. What is the percentage increase after adjusting for changes in the CPI?

1.33 From 1973 to 1978 the average length of a hospital stay decreased from 7.8 days to 7.6 days, but the cost per day increased from $114 to $222. Find the value relative for the cost of hospital care.

1.34 The prices and quantities of four products sold by a lumber company during 1979 and 1980 are given below. Compute a weighted aggregate index for 1980 relative to 1979 using quantities as weights.

	Price (In Dollars)		Quantity	
Product	1979	1980	1979	1980
Cement	$2.75	$3.25	17 tons	14 tons
Nails	0.45	0.55	120 boxes	128 boxes
Wire	0.25	0.33	5100 ft	5500 ft
Paint	6.37	7.29	870 gal	920 gal

1.35 Consider the following prices for a market basket of food, typical items of clothing, and one unit of housing.

	Food	Clothing	Housing
1980	$62.30	$23.60	$ 82.40
1981	64.80	25.20	86.60
1982	67.10	24.40	96.00
1983	73.20	28.30	104.00
1984	75.60	32.70	112.70

(a) Find the simple price index of clothing for 1982 relative to 1980.

(b) If a typical family requires two market baskets of food, one unit of clothing, and two housing units per week, find the index number of living expenses in 1982 relative to 1980 prices.

(c) What is the percentage of increase in the cost of living from 1982 to 1983 for the typical family of part (b)?

1.36 An irrigation pipe manufacturer produces three grades of pipe. The price of each (dollars per pound) and the average daily quantity sold are shown below.

Grade	1985 Price	1985 Quantity	1986 Price	1986 Quantity	1987 Price	1987 Quantity
1	$3.41	280 lb	$3.60	210 lb	$3.93	174 lb
2	1.20	964 lb	1.41	1026 lb	1.68	1490 lb
3	0.81	340 lb	0.90	350 lb	1.10	345 lb

(a) For each product grade, construct a simple price index for 1987 with 1985 = 100.

(b) For each product grade, construct a simple quantity index for 1986 with 1985 = 100.

(c) Using quantities as weights, construct a weighted aggregate price index for 1987 with 1985 = 100.

1.37 Joan was making $11,600 in 1975 and her salary increased to $15,200 in 1979. By what percentage did her purchasing power change?

1.38 What would a salary of $10,000 in 1969 be worth in 1979 in terms of purchasing power?

1.39 If the Consumer Price Index changed from 121.3 in 1971 to 125.3 in 1972, how many percentage points did it increase? By what percentage did prices increase between these two years?

1.40 Refer to the CPI values given in Figure 1.6 where the base year is 1967. Change the CPI value for 1979 to reflect a change to a base year of 1975. What is the purchasing power of the dollar for 1979 using this new base year of 1975?

1.41 For the 12-month period from April 1979 to April 1980 the CPI rose 14.7%. However, where you live and the kind of life you lead are major factors in determining whether the costs you face are rising faster or slower than the national price index. For example, a Citibank study showed the following percentage changes for the 12 months ending in April 1980.

Gasoline	+59.5%	Used cars	−1.7%
Fuel oil, coal, and bottled gas	+59.1%	Utilities other than electricity and gas	+2.2%
Home financing, taxes, and insurance	+33.7%	Fruits and vegetables	+6.4%
Public transportation	+22.5%	Tobacco products	+6.8%

Gas (piped) and electricity	+17.4%	Entertainment admission and fees	+7.1%
Sugar and sweets	+16.5%	Fats and oils	+7.1%
Home purchase	+14.4%	Apparel and upkeep	+7.2%
Meats, poultry, fish, and eggs	−2.0%		

How would your personal inflation rate compare to the CPI for each of the following situations if you

(a) Commuted by car and spent more of your budget on gasoline than the average consumer?

(b) Bought a new house and took out a big, high-rate mortgage?

(c) Lived in an old, uninsulated house in the northern part of the country and used oil heat?

(d) Lived in a moderate climate?

(e) Walked to work?

(f) Allocated more of your budget to food and clothing than the average family?

1.42 The yearly high values of the Associated Press weighted wholesale price index of 35 commodities are given below for the years 1977 to 1980. The base year for this index is 1926 = 100. Reexpress this index with the base year changed to 1977. What is the advantage of using this new base year?

	1980	1979	1978	1977
High	546.95	489.62	428.21	394.90

1.6

REVIEW EXERCISES

1.43 The following is a list of deceased U.S. presidents, the age at which they were inaugurated for the first time, and the age at which they died.

Name	Inaugurated	Died	Name	Inaugurated	Died
1. Washington	57	67	12. Taylor	64	65
2. J. Adams	61	90	13. Fillmore	50	74
3. Jefferson	57	83	14. Pierce	48	64
4. Madison	57	85	15. Buchanan	65	77
5. Monroe	58	73	16. Lincoln	52	56
6. J.Q. Adams	57	80	17. A. Johnson	56	66
7. Jackson	61	78	18. Grant	46	63
8. Van Buren	54	79	19. Hayes	54	70
9. Harrison	68	68	20. Garfield	49	49
10. Tyler	51	71	21. Arthur	50	56
11. Polk	49	53	22. Cleveland	47	71

Name	Inaugurated	Died	Name	Inaugurated	Died
23. Harrison	55	67	30. Hoover	54	90
24. McKinley	54	58	31. F. Roosevelt	51	63
25. T. Roosevelt	42	60	32. Truman	60	88
26. Taft	51	72	33. Eisenhower	62	78
27. Wilson	56	67	34. Kennedy	43	46
28. Harding	55	57	35. L. Johnson	55	64
29. Coolidge	51	60			

Use a table of random numbers and this list to obtain a random sample of size 5 of the ages in which the U.S. presidents were first inaugurated.

1.44 Use a table of random numbers to obtain a systematic sample of size 7 from the names of the presidents given in Exercise 1.43.

1.45 Use a table of random numbers and the list in Exercise 1.43 to obtain a stratified sample of seven ages at which U.S. presidents died. Divide the list of presidents into seven strata of equal size, according to the order in which they appear.

1.46 The Consumer Price Index uses 1967 as the base year. Find the CPI figures for 1975–1979 when 1975 is used as the base year (1975 = 100).

1.47 The average hourly earnings for workers in three broad employment categories are given below, for 1960, 1965, 1970, and 1975. (Source: Bureau of Labor Statistics)

Year	Construction	Retail Trade	Manufacturing
1960	$3.08	$1.52	$2.26
1965	3.70	1.82	2.61
1970	5.24	2.44	3.36
1975	7.24	3.33	4.81

(a) What is the price relative of construction wages in 1975 compared with 1960?

(b) What is the price relative of manufacturing wages in 1975 compared with 1960?

(c) In comparing your answers to parts (a) and (b), what conclusions can you reach?

1.48 The number of workers (in millions) in the same three employment categories for the same years given in Exercise 1.47 is given below. (Source: Bureau of Labor Statistics)

Year	Construction	Retail Trade	Manufacturing
1960	3	11	17
1965	3	13	18
1970	4	15	19
1975	3	17	18

    (a) Find the value relative of wages paid in 1975 in construction compared with 1960.

    (b) Find the value relative of wages paid in 1975 in manufacturing compared with 1960.

    (c) Compare your answers to parts (a) and (b). What conclusions can you draw? How does this comparison of value relatives relate to a similar comparison of price relatives?

1.49 Adjust the hourly earnings in Exercise 1.47 for inflation by means of the CPI using 1967 = 100. Then find the purchasing power relative index for 1975 relative to 1960. How does this adjustment affect the relative picture of hourly earnings in the three employment categories?

1.50 Describe how you might set up an approximate random sampling method for drawing a sample of

    (a) Ten employees out of 147 employees in a company.

    (b) Forty parts, out of a shipment of metal parts, where the shipment consists of 83 boxes and each box has 144 parts in it.

    (c) Twenty students currently taking a statistics course, where the course has 15 sections of students and the number of students per section ranges from 23 to 48.

    (d) Twenty-five businesses in a city that has about 500 businesses.

    (e) Thirty accounts in a department store that has 1500 accounts.

1.51 Restaurants often have comment cards available for their customers to use. Comment on the advantages and disadvantages of this method of obtaining a sample.

1.52 A newsletter by a conservative women's organization requested that the readers send their opinions regarding a proposed bill to their local congressman. Comment on the advantages and disadvantages of this method of obtaining a sample.

1.53 A government safety inspector is in charge of seeing that all offshore oil rigs adhere to certain regulations. She does not have the staff or the resources to inspect every rig every month, so she sets up a stratified sampling scheme where the strata are geographic areas, and the oil rigs within each stratum are selected at random. All site visits are unannounced. Discuss the advantages and disadvantages of this method over a systematic census where each oil rig is visited on a regular basis according to a published schedule.

1.54 The federal tax receipts (millions) for the years 1974 through 1978 are given by the U.S. Bureau of the Census as follows:

1974	251,790
1975	267,321
1976	283,794
1977	338,754
1978	381,776

Adjust these figures for inflation and for the changes in U.S. population to see how the real per capita federal tax receipts in terms of 1967 dollars have changed over the years. Which years show a real decline in the purchasing power of the per capita federal tax receipts?

1.55 A travel brochure for San Diego advertises that San Diego has 360 days of sunshine a year. You decided to rent a beachfront motel room for your two-week vacation. While you were there the fog obscured the sun for six of the 14 days. What are some reasons that your experience did not agree with the brochure information, assuming that the statement in the brochure is correct?

1.56 A newly formed committee has its initial meeting on Tuesday evening to decide on a regular meeting time. What kind of a bias is likely to be evidenced?

1.57 **(a)** In order to see if one university practiced discrimination in its faculty hiring, records were kept of 100 recent applicants. Thirty of these applicants were women, and only 4 were hired, while 70 applicants were men, and 33 were hired. Assuming the men and women were equally qualified for the jobs they were seeking, does this indicate the possibility of sex discrimination?

(b) Further investigation revealed that in the Colleges of Engineering and Business Administration only one applicant was female and she was hired, while only 30 of the 40 male applicants were hired. In the College of Arts and Sciences there were 29 female applicants and 30 male applicants. Three women and 3 men were hired. This accounted for all of the 100 applicants. Which of these colleges, if any, appears to have discriminatory hiring practices?

(c) What lesson can you learn by comparing your answers to parts (a) and (b)?

1.58 Adjust the CPI figures given in Figure 1.6 to the new base year, 1977 = 100. This reflects the change made in February 1982 for the CPI.

1.59 In a household survey if no one is home at one house, is it better to (a) return to that home a second time, or third time if necessary until someone is home, or (b) go to the next door, or next house, if necessary until finding someone at home?

1.60 In a household survey, if the respondent does not understand the question, is it better to (a) repeat the question, using essentially the same wording that is in the questionnaire, or (b) use your own words in repeating the question, using a different vocabulary than is in the questionnaire?

1.61 When recording a response to a household survey, is it better for the interviewer to (a) rephrase the response in the interviewer's words to correct bad grammar and maintain respondent anonymity, or (b) record the response in the respondent's exact words, as nearly as possible?

1.62 Which of the following situations represent sampling errors and which represent nonsampling errors?

(a) A pollster interviews 100 voters and finds 57% support for a political candidate. The next day the pollster interviews 100 more voters and finds 46% support for the candidate.

(b) In a household survey an interviewer fills in some of the answers left blank by the respondent.

(c) In a household survey the interviewer rewords the questions.

BIBLIOGRAPHY

Additional material on the topics presented in this chapter can be found in the following publications.

American Statistical Association (1980). "What Is a Survey?" Publication of the subcommittee of the section on survey research methods.

CPI Detailed Report April 1980. U.S. Department of Labor, Bureau of Labor Statistics.

2

DISPLAYING SAMPLE DATA

The display of various types of data or information has become so widely used by the various news media that it is almost impossible to find a daily newspaper, news magazine, or television newscast that does not routinely employ various **graphical** techniques. Their motivation is quite simple. They want to **summarize** and **display** a body of information in a readily digestible form. One of the most common ways of graphing data is in the form of a simple x-y plot with one variable on the x-axis and another variable on the y-axis, with straight lines or curves. The x-y plot is similar to the same type of graph used extensively in mathematics and economics to plot functions or show economic trends over time. An example of an x-y plot is illustrated in Figure 2.1 to show the worldwide market shares of computer chips. In this figure both the Japanese and the U.S. shares are plotted together to allow an instant, and dramatic, comparison of the trends for those two countries.

Another popular method used by the media to present data, especially economic time series data where economic information is shown changing over time, is in the form of a **bar graph.** While lines or curves are used to connect data points in an x-y plot, each data point in a bar graph is represented as either a horizontal bar or a vertical bar. An x-y plot is relatively uncluttered so that more than one trend can be pictured on the same graph, as in Figure 2.1, but a bar graph uses more of the free space in the graph, so it is not convenient to show more than one or two data sets simultaneously. As an example, a vertical bar graph is used in Figure 2.2 to represent the same data

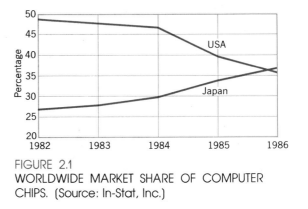

FIGURE 2.1
WORLDWIDE MARKET SHARE OF COMPUTER
CHIPS. (Source: In-Stat, Inc.)

shown in Figure 2.1. Note that there is room in Figure 2.1 to show one or two more countries, but such an addition to Figure 2.2 would clutter the graph to the point where it would be difficult to read.

Although the widest use for a bar graph is to represent time series, as in Figure 2.2, a bar graph is also a convenient method for representing *relative frequencies of observations*. For example, suppose the owner–manager of a men's clothing store does an analysis of the repeat business done by the store. That is, he determines that during the past year 43% of the store's sales were to customers who made only one purchase during the year, 27% of the customers made two purchases, 21% made three purchases, and 9% made four or more purchases. These data are shown as a horizontal bar graph in Figure 2.3. The length of the bar represents the percentage of the store's customers making each of the indicated number of purchases.

As you might guess, the science of statistics provides many opportunities for innovative ways of summarizing and displaying *quantitative* data. In this chapter some basic and easily used graphical techniques are presented. Many of these graphical procedures are used throughout this text to display sample data prior to an analysis of the data. Such persistent usage of these methods should aid the reader in attaining some perspective regarding the appropriate application of these various types of graphs. If the general form of the group performance of the data is of interest, the methods of the first two sections of this chapter are more appropriate. If the primary interest concerns the relative

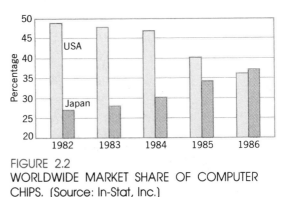

FIGURE 2.2
WORLDWIDE MARKET SHARE OF COMPUTER
CHIPS. (Source: In-Stat, Inc.)

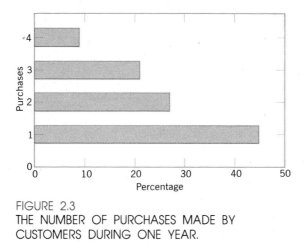

FIGURE 2.3
THE NUMBER OF PURCHASES MADE BY
CUSTOMERS DURING ONE YEAR.

location or size of each particular observation as compared with the group, then the methods of Section 2.3 are more appropriate. Showing the relationship between two variables calls for the use of the plots shown in Section 2.4. Computer-generated methods in Section 2.5 demonstrate how multivariate data can be displayed so that they are more easily interpreted.

2.1

STEM AND LEAF PLOTS

PROBLEM SETTING 2.1

As director of personnel at a large company you are in charge of hiring secretarial help for the company. As part of the interview process the secretarial candidates are given typing tests. The director receives scores indicating each candidate's speed (i.e., net words per minute) and accuracy (i.e., total errors). Twenty recent applicants made the following scores for typing speeds.

68, 72, 91, 47, 52, 75, 63, 55, 65, 35
84, 45, 58, 61, 69, 22, 46, 55, 66, 71

How can these data be displayed so that they are easily interpreted?

THE STEM AND LEAF PLOT

A casual examination of these data shows the highest score to be 91 and the lowest to be 22; however, it is difficult for even a trained eye to determine much else from these data. Displaying quantitative data in a **stem and leaf plot** will help. This procedure, in addition to being an easy and quick way of displaying the data, is also extremely useful for arranging the observations from smallest to largest, and there will be many opportunities to use it through-

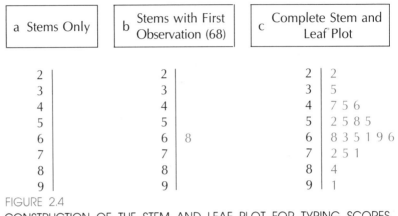

a Stems Only	b Stems with First Observation (68)	c Complete Stem and Leaf Plot
2	2	2 | 2
3	3	3 | 5
4	4	4 | 7 5 6
5	5	5 | 2 5 8 5
6	6 | 8	6 | 8 3 5 1 9 6
7	7	7 | 2 5 1
8	8	8 | 4
9	9	9 | 1

FIGURE 2.4
CONSTRUCTION OF THE STEM AND LEAF PLOT FOR TYPING SCORES.

out this text. The *stem and leaf plot* is formed by starting with the **stem** and then putting on the **leaves.** In this case, since the numbers have only two digits, the first digit of each number, the tens digit, is the *stem* and the second digit, the ones digit, is the *leaf*. A vertical line is drawn with the *stem* on the left of the line and the *leaves* on the right of the line. Figure 2.4a shows the placing of the stems for the previous typing scores; Figure 2.4b shows how the first score (68) is placed. Each new typing score added to the stem and leaf plot results in one new leaf. The completed stem and leaf plot for the typing scores is given by Figure 2.4c.

To read the typing scores from Figure 2.4c, start at the first row and read the score 22. The second row contains 35; the third row contains three scores; 47, 45, and 46, and so on for the other rows. Note that the number of leaves must be equal to the number of observations. From Figure 2.4c, the largest (91) and smallest (22) scores can easily be seen. In addition, an entire picture of how the scores are distributed (or scattered) emerges. For example, it is readily apparent that there are more scores in the sixties than in any other group; only five scores are less than 50, and only two scores are above 80. Additionally, some of the numbers on the stem may have no corresponding leaves. That is, in Figure 2.4c the stem position "3" would have no corresponding leaf if the observation "35" were removed from the data set.

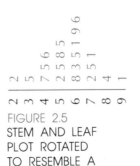

FIGURE 2.5
STEM AND LEAF
PLOT ROTATED
TO RESEMBLE A
HISTOGRAM.

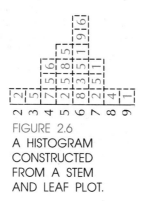

FIGURE 2.6
A HISTOGRAM
CONSTRUCTED
FROM A STEM
AND LEAF PLOT.

CONSTRUCTING A HISTOGRAM FROM
A STEM AND LEAF PLOT

An interesting result is obtained by rotating the stem and leaf plot 90° counterclockwise so that it appears as in Figure 2.5. If each of the leaves is thought of as a building block, the stem and leaf plot resembles a graph known as a **histogram.** Histograms are considered in detail in the next section; however, in general, a histogram is a graphical method for presenting data, where the observations are located on the horizontal axis (usually grouped into intervals) and the frequency of those observations is depicted in some way along the vertical axis. In fact, a stem and leaf plot is about the easiest way to generate a histogram. The stem and leaf "building block" histogram would appear as in Figure 2.6.

MORE ON STEM AND LEAF PLOTS

There are data sets for which the generation of the stem and leaf plot is not as straightforward as for the data presented in Figure 2.4. For example, if the leading digit assumes only two or three different values, some modifications are usually helpful. Consider the following set of 20 sample values.

14, 16, 22, 31, 17, 37, 26, 24, 18, 20
29, 33, 36, 13, 15, 22, 30, 28, 25, 24

Since the leading digits of these numbers are either 1, 2, or 3, the stem and leaf plot could appear as follows.

```
1 | 4 6 7 8 3 5
2 | 2 6 4 0 9 2 5 4
3 | 1 7 3 6 0
```

This stem and leaf plot is not as revealing as it could be. Thus, a modification

```
1*  |  4 3
1·  |  6 7 8 5
2*  |  2 4 0 2 4
2·  |  6 9 8 5
3*  |  1 3 0
3·  |  7 6
```

FIGURE 2.7
A STEM AND
LEAF PLOT
WITH SINGLE
SPLITTING OF
STEMS.

is made by splitting each stem. For example, the stem 1 is split into two stems 1* and 1· as follows.

```
1*  |  with only the digits 0, 1, 2, 3, and 4 as leaves
1·  |  with only the digits 5, 6, 7, 8, and 9 as leaves
```

The revised stem and leaf plot using split stems appears in Figure 2.7.

Other data sets may require splitting the leaves even more. For example, consider the following 20 sample values.

22, 30, 26, 31, 29, 32, 34, 25, 28, 27

30, 28, 31, 29, 35, 33, 21, 24, 29, 27

Since the leading digits in this set are all either 2 or 3, some splitting will be helpful. However, a single splitting of each stem may not be enough. That is,

```
2*  |  1
2t  |  2
2f  |  5 4
2s  |  6 7 7
2·  |  9 8 8 9 9
3*  |  0 1 0 1
3t  |  2 3
3f  |  4 5
3s  |
3·  |  8
```

FIGURE 2.8
A STEM AND
LEAF PLOT
USING FIVE-
WAY SPLITTING
OF STEMS.

a more meaningful stem and leaf plot may be made by splitting each stem five times as follows for the leading digit 2.

2*	followed by the digits 0 or 1
2t	followed by the digits 2 or 3
2f	followed by the digits 4 or 5
2s	followed by the digits 6 or 7
2˙	followed by the digits 8 or 9

Note that in this splitting the symbol *t* is used for the digits *two* and *three*; the symbol *f* is used for the digits *four* and *five*; and finally the symbol *s* is used for the digits *six* and *seven*. The completed stem and leaf plot would appear as in Figure 2.8.

HOW MANY LINES TO USE IN A STEM AND LEAF PLOT

The previous examples appear to require some judgment on the part of the person making the stem and leaf plot as to whether or not the stems should be split, and if so, how much splitting is necessary. This question can be resolved by using a rule suggested in a book by Hoaglin, Mosteller, and Tukey for the number of lines to use with a stem and leaf plot.

For a sample of size n, a reasonable number of lines to use in a stem and leaf plot is given approximately by the integer portion of $2\sqrt{n}$ for $n \leq 100$ and by the integer portion of $10 \log_{10} n$ for $n > 100$. For example:

SUGGESTED RULE FOR THE NUMBER OF LINES TO USE WITH A STEM AND LEAF PLOT

n	Number of Lines	n	Number of Lines
10	6	70	16
15	7	80	17
20	8	90	18
30	10	100	20
40	12	150	21
50	14	200	23
60	15	300	24

Thus, in the previous examples with $n = 20$, a reasonable number of lines to use is eight. In Figure 2.4 exactly eight lines were used. In Figure 2.7 six lines were used after single splitting, while ten lines were used in Figure 2.8 after multiple splitting. In this latter case single splitting would have resulted in four lines, while multiple splitting provides a number closer to eight. In cases like this the reader is reminded that these rules are offered only as guidelines and should not be considered as hard and fast rules.

MORE SIGNIFICANT DIGITS IN A STEM AND LEAF PLOT

If data occur with more than two significant digits, the stem and leaf plot can be constructed in slightly different ways. For example, consider the measured capacitances of twenty .6-microfarad capacitors.

.6116, .6241, .6266, .6296, .6301, .6309, .6320, .6329, .6359

.6362, .6397, .6399, .6428, .6430, .6445, .6458, .6483, .6499

.6685, .6711

For these sample values the decimal point and leading digit 6 are common to all values, so they can be ignored in determining the stem of the plot. Thus, the criterion for determining the stem is the second digit, so the stems consist of the two-digit numbers .61, .62, .63, .64, .65, and .66. The leaves can be added in one of two ways. First, the third digit can be used as the leaf after dropping the fourth digit (do not round the number first). This representation is given in Figure 2.9. In the second method the leaves appear as two-digit numbers as in Figure 2.10.

UNUSUAL OBSERVATIONS IN THE DATA

It is quite common for data sets to contain one or more values that appear unusual when compared to the bulk of the values in the sample. Values that are substantially different from the others in the sample are commonly referred to as **outliers.**

OUTLIERS   Observations that are much larger or much smaller than the rest of the observations in the sample are called **outliers.**

Outliers can greatly influence the scale of a stem and leaf plot. To illustrate this point consider the following set of twenty sample observations.

121, 125, 126, 128, 130, 131, 132, 133, 134, 135
137, 139, 141, 141, 144, 147, 148, 153, 205, 213

```
.61 | 1
.62 | 4 6 9
.63 | 0 0 2 2 5 6 9 9
.64 | 2 3 4 5 8 9
.65 | 8
.66 | 1
```

FIGURE 2.9
STEM AND LEAF
PLOT WITH THE LAST
SIGNIFICANT DIGIT
DROPPED.

```
.61 | 16
.62 | 41 66 96
.63 | 01 09 20 29 59 62 97 99
.64 | 28 30 45 58 83 99
.65 | 85
.66 | 11
```

FIGURE 2.10
STEM AND LEAF PLOT WITH
THE LEAVES FORMED BY THE
LAST TWO SIGNIFICANT DIGITS.

The observations 205 and 213 are outliers in the sense that they are much larger than the rest of the data. Figure 2.11 gives a stem and leaf plot of these data utilizing 10 lines. This figure clearly shows the impact of the outliers on the scale of the stem and leaf plot with the bulk of the observations appearing as leaves on three stems. An alternative approach is to ignore the two outliers and construct a stem and leaf plot utilizing seven lines after single splitting of the stems 12, 13, 14, and 15. The outlying observations are added to the stem and leaf plot with a stem labeled as either *high* or *low* followed by the leaves consisting of the outlying observations separated by commas inside a single set of parentheses. Figure 2.12 shows the revised stem and leaf plot. Note that the vertical line separating the leaves from the stems in Figure 2.12 has been broken before the high stem is added. If this data set also contained low observations that were identified as outliers, then they would appear at the top of the stem and leaf plot labeled *low*.

WHEN TO USE A STEM AND LEAF PLOT

A stem and leaf plot is a very useful technique for ordering a set of observations from smallest to largest and is used frequently for that purpose throughout this text. The stem and leaf plot is also quite useful for grouping the data into intervals while retaining the identity of each sample observation. The techniques of the next section group the data into intervals without retaining the

```
12 | 1 5 6 8
13 | 0 1 2 3 4 5 7 9
14 | 1 1 4 7 8
15 | 3
16 |
17 |
18 |
19 |
20 | 5
21 | 3
```

FIGURE 2.11
A STEM AND LEAF
PLOT WITH TWO
OUTLIERS.

```
12* | 1
12' | 5 6 8
13* | 0 1 2 3 4
13' | 5 7 9
14* | 1 1 4
14' | 7 8
15* | 3
```

high | (205, 213)

FIGURE 2.12
A REVISED STEM
AND LEAF PLOT
WITH TWO
OUTLIERS.

individual value of each data point. If only a grouped summary of the data is available, then a stem and leaf plot cannot be used; rather the techniques of the next section would be more appropriate.

EXERCISES

2.1 An arithmetic achievement test was given to 100 pupils. The scores ranged from 23 to 65.

(a) How many lines would be used to represent the test scores in a stem and leaf plot if no stems are split?

(b) If each stem is split once, how many lines would be used?

(c) If each stem is split five times, how many lines would be used?

(d) What is a reasonable number of lines to use in a stem and leaf plot if the rule given in this section is applied?

2.2 Daily high temperatures recorded on April 24, 1984, for 38 cities selected from across the United States are given below. Summarize these data in a stem and leaf plot.

67 78 48 70 52 47 47 51 47 91 51 69 51 85 61 65

70 83 86 70 87 72 77 83 68 93 83 64 88 66 50 94

53 77 66 63 50 61

2.3 The owner of a small business has recorded the number of customers for each of the last 40 business days. Make a stem and leaf plot for these data.

186	121	143	159	180
125	178	215	166	158
187	148	151	162	153
128	133	188	170	168
153	123	134	184	208
178	186	162	202	160
201	200	175	126	150
174	218	165	166	185

2.4 The earnings per share for 25 stocks randomly selected from a recent issue of *Standard and Poor's Stock Guide* are given below. Summarize these data in a stem and leaf "building block" histogram.

$5.44	$2.25	$0.68	$4.70	$2.30
0.95	6.85	0.41	1.34	0.45
3.01	1.60	2.11	6.79	2.57
3.55	0.20	1.80	7.30	2.07
1.60	1.43	2.34	4.76	1.31

2.5 A survey of 617 companies in 1983 by Michigan State University produced the average expected starting salaries listed below by academic field. Summarize these values in a stem and leaf plot.

Electrical Engineering	$26,643	Agriculture	$17,586
Chemical Engineering	26,164	Marketing	17,550
Mechanical Engineering	25,888	Social Science	16,763
Computer Science	25,849	Business Administration	16,650
Metallurgy	24,445	Personnel Administration	15,908
Physics	22,852	Communications	15,636
Civil Engineering	21,266	Hotel Management	15,447
Mathematics	19,539	Education	14,779
Accounting	18,684	Liberal Arts	14,179
Financial Administration	18,122	Human Ecology	13,917

2.6 Summarize the measured capacitances of twenty .6-microfarad capacitors in a stem and leaf plot.

.6365, .6235, .6452, .6438, .4584, .6315, .6277
.6129, .6327, .6388, .6475, .6395, .6291, .4477
.6305, .6327, .6498, .6418, .6355, .6449

2.2

HISTOGRAMS

PROBLEM SETTING 2.2

The heights of 45 female high school seniors have been recorded in centimeters as follows.

170	151	154	160	158	154	171	156	160
157	160	157	148	165	158	159	155	151
152	161	156	164	156	163	174	153	170
149	166	154	166	160	160	161	154	163
164	160	148	162	167	165	158	158	176

How could these data be summarized in a graph to show the frequencies associated with different intervals of height so that the heights are easily compared with other schools?

CONSTRUCTING A HISTOGRAM WITH EQUAL CLASS WIDTHS

The data in Problem Setting 2.2 could be summarized easily in a stem and leaf plot; however, the stem and leaf plot is a fairly recent development and is not as commonly used as the more traditional methods discussed in this section. Additionally, there are settings where stem and leaf plots cannot be used at all. For example, if only a summary of the data is available, a stem and leaf plot could not be used. A more general technique for displaying sample data is called a **histogram,** and its construction is considered in this section. First, however, several terms associated with histograms need to be defined.

TERMS ASSOCIATED WITH HISTOGRAMS

A **class** is an interval containing sample observations. Each sample observation is classified into one and only one class.

The **class boundaries** are the endpoints or limits for each class.

The **class width** is the distance between the class boundaries of a class.

The **class mark** is the midpoint of a class and is found midway between the class boundaries.

The **frequency** of a class is the number of sample observations associated with a class.

The **relative frequency** of a class is its frequency divided by the total number of observations in the sample.

The data in Problem Setting 2.2 have been summarized in a histogram in Figure 2.13. This figure can be used to help illustrate each of the terms associated with histograms. There are six classes in that histogram. The first class has class boundaries 147.5 and 152.5, the second class has boundaries of

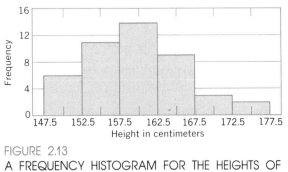

FIGURE 2.13
A FREQUENCY HISTOGRAM FOR THE HEIGHTS OF 45 FEMALE HIGH SCHOOL SENIORS.

152.5 and 157.5, and so on. The class widths are each 5.0. The class marks are not noted in the figure but are 150, 155, and so forth, and lie exactly midway between the class boundaries. The frequencies are 6, 11, 14, 9, 3, and 2, respectively, whereas the relative frequencies are obtained by dividing these numbers by the total number of observations, 45, to get .13, .24, .31, .20, .07, and .04. The relative frequencies always sum to 1.00, except for roundoff error.

The information contained in the histogram in Figure 2.13 can be summarized as follows.

Class	Boundaries	Class Mark	Frequency	Relative Frequency
1	147.5 to 152.5	150	6	.13
2	152.5 to 157.5	155	11	.24
3	157.5 to 162.5	160	14	.31
4	162.5 to 167.5	165	9	.20
5	167.5 to 172.5	170	3	.07
6	172.5 to 177.5	175	2	.04
			45	1.00

The numbers on the horizontal axis may represent class boundaries, as in Figure 2.13, or the class marks as in Figure 2.15. Additionally, the range of each interval is sometimes indicated as shown in Figure 2.14. Any of these labeling schemes can be used. The choice is determined by whichever may be easiest for the intended reader of the graph to interpret.

HISTOGRAM WITH EQUAL CLASS WIDTHS A **histogram** for classes with **equal class widths** is a graphical presentation of the sample data using classes on the horizontal axis and either frequency or relative frequency as the vertical axis.

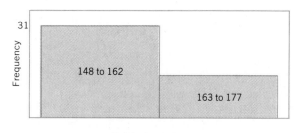

Height in centimeters

FIGURE 2.14
A FREQUENCY HISTOGRAM USING ONLY TWO CLASSES.

Formal rules for the construction of a histogram are stated for the case where all class widths are the same. The case of unequal class widths is considered later in this section. Many computer programs are available for constructing histograms and for constructing stem and leaf plots.

RULES FOR
GRAPHICAL
DISPLAY OF
QUANTITATIVE
DATA IN A
HISTOGRAM

1. If the data set is of a reasonable size (say 200 observations or less), a stem and leaf plot can be quickly constructed by hand and may provide a good starting point for constructing a histogram. For larger data sets, a computer could be used to make a stem and leaf plot.

2. Decide upon the number of classes to be used in grouping the data, usually from 5 to 20, depending on the number of observations. (See the suggested rule in this section.)

3. Decide upon class boundaries so that each class has the same width and every observation can be classified uniquely in exactly one class. The approximate class width can be found as follows:

$$\text{Class width} = \frac{X_{max} - X_{min}}{\text{number of classes}}$$

where X_{max} is the largest observation in the sample and X_{min} is the smallest observation in the sample. This approximate class width is usually rounded to a number that is convenient to work with.

4. In a frequency histogram, the height of each vertical bar in the histogram represents the number of observations in each class. In a relative frequency histogram, the height of each vertical bar in the histogram represents the proportion of observations in each class. For example, a class with ten observations should be twice as high as a class with five observations, but only one-tenth as high as a class with 100 observations.

NUMBER OF CLASSES

The rules for displaying sample data in a histogram are not very specific with respect to the number of classes to use in the construction of a histogram. The reason for this vagueness is that a certain amount of artistry is involved and there are no hard and fast rules that are universally accepted by statisticians. Rather, statisticians are usually aiming at an accurate representation of the

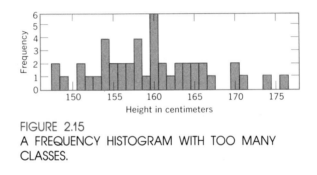

FIGURE 2.15
A FREQUENCY HISTOGRAM WITH TOO MANY
CLASSES.

data. That is, a histogram should not misrepresent the data. Using too few classes gives an inaccurate picture by smoothing out too many details. (An example of this situation is given in Figure 2.14.) Too many classes present too many details and obscure the overall view. (See Figure 2.15.) With this thought in mind, a rule is offered, which although somewhat arbitrary in nature does allow some uniformity to be applied in the construction of histograms. For some types of data, such as integer-valued data, the integers themselves may be logical classes to use and rules for the number of classes to use may be inappropriate.

The number of classes k to be used in the construction of a histogram for sample data is the smallest integer value of k such that $2^k \geq n$, where n is the number of observations in the sample. For example:

SUGGESTED RULE FOR FINDING THE NUMBER OF CLASSES TO USE IN THE CONSTRUCTION OF A HISTOGRAM

Number of Observations n	Number of Classes k
8 or less	3
9 to 16	4
17 to 32	5
33 to 64	6
65 to 128	7
129 to 256	8
257 to 512	9
513 to 1024	10
More than 1024	The smallest value of k such that $2^k \geq n$

EXAMPLE

The data given in Problem Setting 2.2 was displayed in a histogram in Figure 2.13. The construction of that histogram is explained in this example. From the suggested rule, 6 classes are used with 45 observations. The smallest observation is 148 and the largest is 176. By the suggested rule, the approximate number of classes is found as $(176 - 148)/6 = 4.67$. This number is rounded up to 5, which is more convenient to use. To make each class a convenient multiple of 5, the lower bound for the first class is defined as 147.5 and the upper bound for the last class is defined as 177.5. That is, the total width of the histogram is $177.5 - 147.5 = 30$, and since division of the width by 5 gives $30/5 = 6$, there are 6 classes, each having a width of 5. A frequency summary of the data is helpful in completing the construction of the histogram.

Class	Frequency
At least 147.5 but less than 152.5	6
At least 152.5 but less than 157.5	11

At least 157.5 but less than 162.5	14
At least 162.5 but less than 167.5	9
At least 167.5 but less than 172.5	3
At least 172.5 but less than 177.5	2
	$\overline{45}$

FREQUENCY POLYGON

A useful variation of the histogram is known as a **frequency polygon.** A histogram is easily converted to a frequency polygon by connecting the midpoints (class marks) of the tops of each rectangle by straight-line segments. The frequency polygon is completed by adding two line segments, one at each extreme class in the histogram, which connect the class marks of these classes with the value of zero in the classes that would be adjacent to but outside of these extreme classes. This is done to reflect the fact that the frequency returns to zero outside of the intervals covered by the histogram. The selection of these endpoints is arbitrary in the case of histograms with unequal class interval widths, so caution must be observed when interpreting frequency polygons in those situations. The frequency polygon is demonstrated with the 20 typing scores given in Problem Setting 2.1. These scores can be summarized in five classes.

Typing Score	Frequency
22 but less than 36	2
36 but less than 50	3
50 but less than 64	6
64 but less than 78	7
78 but less than 92	2

FREQUENCY POLYGON   A **frequency polygon** is a plot of the class frequency versus the class mark for each class in a histogram. The successive points in this plot are then connected with straight lines with frequencies of zero used for the classes immediately on either side of the histogram.

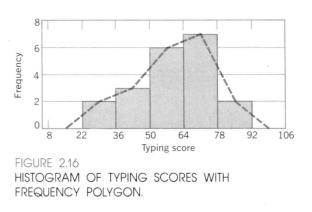

FIGURE 2.16
HISTOGRAM OF TYPING SCORES WITH
FREQUENCY POLYGON.

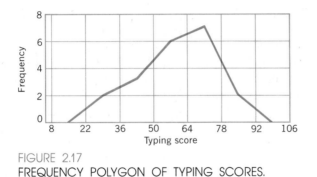

FIGURE 2.17
FREQUENCY POLYGON OF TYPING SCORES.

The histogram with two extra classes added to accommodate the frequency polygon is given in Figure 2.16. Of course, the histogram does not need to be part of the graph. The frequency polygon appears by itself in Figure 2.17.

HISTOGRAMS WITH PROPORTIONAL AREAS

There are settings where equal class widths are not appropriate, as well as settings where only a summary of the sample data is available. Such is the case with the projected population of the United States for the year 2000, which appears in Figure 2.18. The projections are broken down by age groups where the groupings follow natural areas of interest, such as preschool, grade school, high school, college, adults in four age groups covering 10 years each, and retirement age. Note that the last interval has no upper limit indicated, that is, it is open-ended. To plot these data as a histogram it will be necessary to broaden the definition of a histogram to include a series of rectangles where the area of the rectangle is proportional to the number of observations in the interval covered by the rectangle.

Age Group	Projected Number (Millions)
Under 5 years	17.9
5 to 13 years	35.1
14 to 17 years	16.0
18 to 24 years	24.7
25 to 34 years	34.4
35 to 44 years	41.3
45 to 54 years	35.9
55 to 64 years	23.3
65 and over	31.8
	260.4

FIGURE 2.18
POPULATION PROJECTION FOR THE UNITED STATES IN THE YEAR 2000. (Source: Department of Commerce, Bureau of the Census.)

HISTOGRAM WITH PROPORTIONAL AREAS
A **histogram** for classes with **unequal class widths** is a graphical presentation of the sample data that consists of rectangular boxes plotted on each class interval where the *areas* of the rectangles are proportional to the number of observations in each class. The height of the rectangle for each class is found by dividing that class frequency by its width.

EXAMPLE

Make a histogram for the population projections given in Figure 2.18. First the summary of Figure 2.18 is reorganized to reflect the class widths. This means that an arbitrary upper bound must be assumed for the last class, such as 100. Using the ratio of frequency to class width, the height of the histogram in each interval is easily determined. The height represents the density in each interval.

Class	Frequency (Millions)	Class Width	Height = Frequency/Class Width
Less than 5	17.9	5	3.58
5 but less than 14	35.1	9	3.90
14 but less than 18	16.0	4	4.00
18 but less than 25	24.7	7	3.53
25 but less than 35	34.4	10	3.44
35 but less than 45	41.3	10	4.13
45 but less than 55	35.9	10	3.59
55 but less than 65	23.3	10	2.33
65 to 100	31.8	35	0.91
	260.4		

The histogram is now constructed from this summary.

It is very important to note that the vertical axis used in this histogram has no direct interpretation, and that the heights of the rectangular boxes have been constructed to reflect the proportion of individuals in each class. This feature makes it easy for the eye to make accurate comparisons. For example, Figure 2.18 shows that there are about twice as many individuals in the 5-to-13 age group as there are in groups on either side, but the class width is also about twice the size of the classes on either side. The histogram in Figure 2.19 is adjusted for this difference and shows that the population density in the 5-to-13 interval is about the same as in the adjacent classes. The scale used to construct the height can be inches, centimeters, or whatever is desired as long as it is consistent throughout. The histogram will always look the same.

A WARNING

Histograms with unequal class-interval widths are often misconstructed and misinterpreted. Sometimes the unequal interval widths are ignored so that higher bars are used to represent larger frequencies, giving the false impression

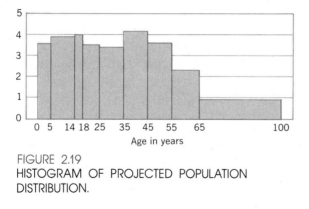

FIGURE 2.19
HISTOGRAM OF PROJECTED POPULATION
DISTRIBUTION.

of a greater density of observations in those intervals. Also the open ends usually have arbitrary endpoints selected, and the selection of endpoints influences the appearance of the graph.

WHEN TO USE A HISTOGRAM

A histogram can be used on the same type of data sets as a stem and leaf plot, as well as for displaying data that have been summarized into intervals. When a data set is too large for a stem and leaf plot, a histogram can still be used. Unlike the stem and leaf plot, the histogram does not retain the identity of individual sample observations. Histograms are in much wider use than the more recently developed stem and leaf plots, and for that reason are more likely to appear in reports summarizing quantitative information.

One possible drawback to histograms is the fact that different individuals may construct different looking histograms for the same set of data as shown in Figures 2.13 to 2.15. The reason for this variation is the choice of the number of classes to use. Another problem with histograms is the adverse effect caused by outliers in the data. In the presence of outliers some artistic modification is required on the histogram. In the next section an alternative method for displaying sample data is presented that avoids many of the problems associated with histograms.

EXERCISES

2.7   The number of phone calls made in one week by each of 30 randomly selected salespersons shows a total of 928 calls.

   (a) Approximately how many classes would be used to represent the weekly phone call frequencies in a histogram?

   (b) If the fewest number of calls made by any salesperson was 21 and the most was 60, what would be the width of each of the classes in part (a)?

   (c) What class boundaries would be used for each of the classes based on the information in parts (a) and (b) if equal width classes are used?

   (d) What are the class marks of the intervals determined in part (c)?

2.8 Applicants for secretarial positions at a large company are required to take a grammar test prior to receiving a personal interview. The frequencies of errors made by 230 recent applicants are summarized in the accompanying histogram. Answer the following questions by referring to the histogram.

(a) How many classes were used in the summary?

(b) What are the class boundaries for each class?

(c) What is the class width?

(d) What is the class mark for each class?

(e) What is the frequency of each class?

(f) What is the relative frequency of each class?

(g) What percentage of the applicants made 10 or fewer errors?

(h) What percentage of the applicants made more than 20 errors?

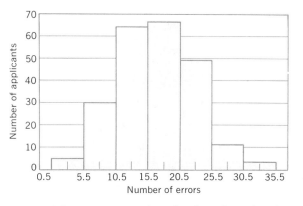

2.9 A discount store wishes to put in a fast checkout lane for shoppers who purchase only a small number of items. The store manager observes 102 customers selected at random and counts the number of items that each has. The sample provides the following frequency summary.

Number of Items	Frequency	Number of Items	Frequency
1	4	14	7
2	1	15	4
3	9	16	2
4	7	17	1
5	6	18	2
6	10	21	1
7	6	22	1
8	5	23	1
9	12	25	1
10	7	26	1
11	3	29	1
12	4	36	1
13	4	42	1

Construct a histogram for these data.

2.10 An office supply store examined its charge accounts for the month of March and noted how many purchases were recorded for each of its charge customers.

Number of Purchases	Number of Customers
0	3
1	7
2	14
3	8
4	2
7	1

That is, one customer charged seven purchases, two customers charged four purchases, and so on. Represent these data by a histogram.

2.11 A medium-sized factory is trying to make its workers more aware of the danger of accidents. A survey of the last 50 weeks shows the number of accidents per week to be as follows.

```
0   5    9    8    2   10   4   12   14    8
3   8    4    4    0    9   3   10   10    9
1   1    2   14    4    9   1    8    2    7
2   6    2    9    8    5   0   18    8   11
4   6   12    1   12    8   6    6    6   11
```

(a) Display these data in a frequency histogram with five classes of equal width.

(b) What are the class boundaries for the middle class in your histogram?

(c) What is the class mark for the middle class in your histogram?

2.12 Summarize the data in Exercise 2.3 in a histogram.

2.13 Magazine advertisements frequently use histogram-type representations to compare a client's sales with the competition. However, the construc-

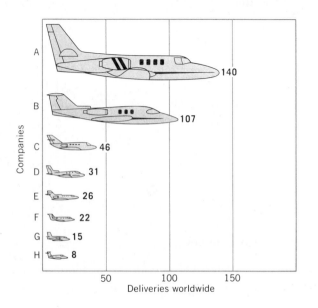

tion of some of these graphs can be misleading. What is misleading about the way the graph in this exercise was constructed?

2.14 The histograms given below compare salaries for executives with four different functions. The rectangles have been constructed so that their areas are proportional to the percentages they represent. However, there is one basic fault involved in the construction of the histogram; what is it?

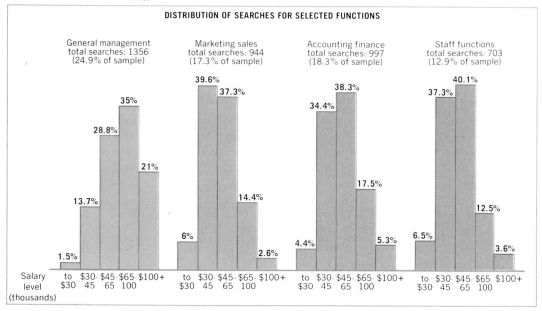

2.15 A frequency distribution summary of adjusted gross income compiled from 1976 income tax returns is given below. Summarize these data in a histogram.

Adjusted Gross Income (thousands of dollars)	Number of Taxpayers (in thousands)
0 to 3	15,015
3 to 5	8,837
5 to 10	19,891
10 to 15	14,182
15 to 20	11,182
20 to 25	6,662
25 to 30	3,611
30 to 50	3,632
50 to 100	945
100 to 500	221
500 to 1,000	4
1,000 and over	1

2.16 Construct a frequency polygon for the data in Exercise 2.15.

2.17 Given below are the population projections made in 1980 for the year 1985. Summarize these data in a histogram similar to the one in Figure 2.19 and comment about any possible differences in the relative frequency distributions for the years 1985 and 2000.

Age Group	Projected Number (Millions)
Under 5 years	18.8
5 to 13 years	29.1
14 to 17 years	14.4
18 to 24 years	27.9
25 to 34 years	39.9
35 to 44 years	31.4
45 to 54 years	22.8
55 to 64 years	21.7
65 and over	27.3
	233.3

2.18 Construct a frequency polygon for the data in Exercise 2.17.

2.19 Prior to putting a new tire on the market a tire company conducts tread-life tests on a random sample of 100 of these new tires. The test results are summarized below. Construct a frequency polygon for these data.

Number of Miles	Frequency
10,000 but less than 15,000	6
15,000 but less than 20,000	21
20,000 but less than 25,000	38
25,000 but less than 30,000	19
30,000 but less than 35,000	16

2.3

CUMULATIVE FREQUENCY DISTRIBUTIONS

CONSTRUCTING AN OGIVE

The graphs of the previous section were based on either the premise of frequency or relative frequency indicating either the frequency of observations or the proportion of observations in each class. In this section methods are presented for displaying sample data in a **cumulative frequency** graph.

CUMULATIVE (RELATIVE) FREQUENCY	The **cumulative frequency** is the total number of observations less than or equal to a given number or class.
	The **cumulative relative frequency** is the cumulative frequency divided by the total number of observations.

Typing Score	Frequency	Relative Frequency
22 but less than 36	2	.10
36 but less than 50	3	.15
50 but less than 64	6	.30
64 but less than 78	7	.35
78 but less than 92	2	.10

Typing Score	Cumulative Frequency	Cumulative Relative Frequency
21 or less	0	.00
35 or less	2	.10
49 or less	5	.25
63 or less	11	.55
77 or less	18	.90
91 or less	20	1.00

FIGURE 2.20
TYPING SCORES ORGANIZED FOR A CUMULATIVE
FREQUENCY GRAPH.

Cumulative frequency and cumulative relative frequency are illustrated by the typing scores of Problem Setting 2.1, which are grouped into intervals and displayed in Figure 2.20. In that figure the frequencies of observations in each class are 2, 3, 6, 7, and 2 for a total of 20 observations. The relative frequencies are found by dividing each of these frequencies by the sample size 20. For a plot based on cumulative frequency rather than relative frequency, the typing scores need to be summarized as in Figure 2.20.

The first method for displaying a cumulative relative frequency is called an **ogive** (pronounced o-jive). The ogive is useful for estimating the proportion of observations in the sample that are less than or equal to any selected value.

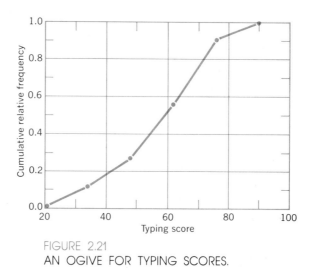

FIGURE 2.21
AN OGIVE FOR TYPING SCORES.

It may also be used to visualize where each observation lies relative to the rest of the sample.

OGIVE An **ogive** is a plot of cumulative relative frequency versus the upper class boundary for each class. The successive points in this plot are then connected with straight lines.

A plot of the cumulative relative frequencies from Figure 2.20 is given in the ogive in Figure 2.21.

REAL NUMBER LINE PLOT

The typing scores of Problem Setting 2.1 are used to illustrate another simple but informative way to display the data graphically. For this approach a simple concept is borrowed from mathematics, and the data are plotted on a real number line as illustrated in Figure 2.22. This plot is easily made and makes clear how the spacing differs between pairs of scores. Clearly the scores of 22 and 35 are inferior to the others, since they fall well below the otherwise tightly clustered group in the middle. At the other extreme the scores of 84 and 91 are well above their nearest competitors. The addition of a second dimension to a real-number-line plot to make information more readily discernible is now illustrated.

THE EMPIRICAL DISTRIBUTION FUNCTION

Notice that the ogive depends on the class boundaries selected. Another type of graphing technique somewhat akin to the cumulative frequency plot but that does not depend on class boundaries is based on the **empirical distribution function** (e.d.f.). Here the word *empirical* means the same as the word *sample;* hence, an e.d.f. is a plot made from sample data. The e.d.f. is graphed by adding a second dimension to a real-number-line plot. The second dimension is formed by building a step function or "staircase" from left to right on the real number line, adding another step to the staircase every time a sample data point is encountered. The height of the step at each data point is the reciprocal of the sample size. For a sample of size 20, each step is of height $\frac{1}{20} = .05$. In the event there are two identical values in the sample data the step height at that value is twice as high as an ordinary step. Likewise, steps with heights three times the ordinary step height would indicate three identical sample data values and so on. This means that the vertical axis of an e.d.f. plot will always be labeled from 0 to 1. The staircase will now be added to the real number line plot of Figure 2.22 to generate the e.d.f. This e.d.f. appears in Figure 2.23.

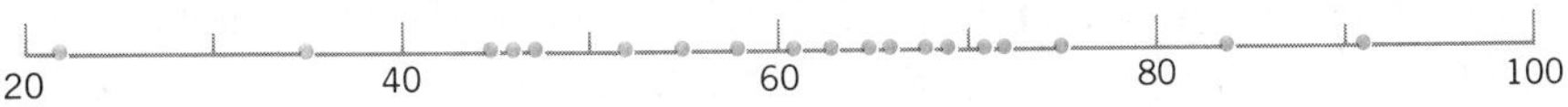

FIGURE 2.22
REAL-NUMBER-LINE PLOT OF TYPING SCORES.

66 CHAPTER TWO / DISPLAYING SAMPLE DATA

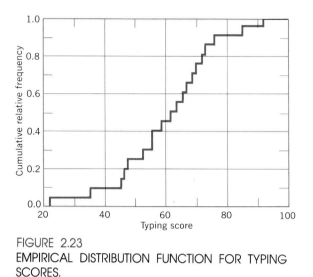

FIGURE 2.23
EMPIRICAL DISTRIBUTION FUNCTION FOR TYPING
SCORES.

Although the vertical axis for the e.d.f. has the same label as the ogive, the e.d.f. differs from the ogive not only because it is a step function but also because every data point is represented in the e.d.f. rather than being grouped together in classes as in the ogive. Extensive use of the e.d.f. is made throughout this text, and it is treated more formally in the next chapter. However, some of its usefulness may readily be seen. For example, the top (or bottom) 5%, 10%, and so forth, scores are easily obtained from reading the e.d.f. plot, as are other percentages of interest. Formal rules for plotting an e.d.f. are now given.

EMPIRICAL DISTRIBUTION FUNCTION An **empirical distribution function** is a function of x, which equals the proportion of the sample less than or equal to x, for x from $-\infty$ to $+\infty$. In graphical form it is a graph of the cumulative relative frequency of a sample.

RULES FOR GRAPHICAL DISPLAY OF DATA IN AN EMPIRICAL DISTRIBUTION FUNCTION

1. The sample data points must first be ordered from smallest to largest. A stem and leaf plot can be quite helpful in ordering the sample data. A real-number-line plot can also be used to order the data.

2. The empirical distribution function looks like a stairstep function as it proceeds from left to right, having a value of 0 to the left of the smallest data point and a value of 1 to the right of the largest data point. In between, the cumulative step heights will start at $1/n$ at the smallest data point and increase to $2/n$, $3/n$, and so on to $n/n = 1$ as additional sample data points are encountered from left to right along the horizontal axis. In the event of identical sample values, the step heights should be increased to indicate exactly the multiplicity of the data point.

WHEN TO USE OGIVES AND EMPIRICAL DISTRIBUTION FUNCTIONS

Since an ogive is derived from a histogram, it can be used with any set of data for which a histogram is suitable, or it can be used when only a data summary is available. The shape of an ogive is influenced by the choice of class boundaries for the histogram. As with a histogram, the identity of individual observations is lost when using an ogive.

On the other hand, an empirical distribution function retains the individual identity of all sample observations. It can be used with all types of data and is not adversely affected by the presence of outliers. Empirical distribution functions are not designed to be used with data appearing in summary form such as in a histogram. However, if only a summary or histogram is available, an approximate empirical distribution function could be constructed using class marks and class frequencies at each of the steps of the empirical distribution function.

EXERCISES

2.20 Which of the following points apply to empirical distribution functions? Which apply to histograms?

(a) Uses all of the data values and retains their identity.

(b) Does not require an arbitrary grouping of the data.

(c) Is easy to construct.

(d) Percentages are easy to read from its graph.

(e) Is not adversely influenced by outliers.

2.21 Construct an ogive for the tread-life data given in Exercise 2.19.

2.22 Construct an ogive for the population prediction data for the year 1985 that were presented in Exercise 2.17.

2.23 Construct an e.d.f. for the charge-account-customer data given in Exercise 2.10.

2.24 Summarize the data given in Exercise 2.3 in an e.d.f. Use the e.d.f. to determine the percentage of the time that the number of customers exceeded 200.

2.25 Refer to the data given in Exercise 2.9 and summarize these data in an e.d.f. If the discount store desires to have 10% or less of its customers eligible for the fast checkout lane, what is the maximum number of items a person can have and still use the fast lane?

2.26 Use an e.d.f. to summarize the data given in Exercise 2.4. What percentage of the companies would you estimate to have earnings of less than $1 per share?

2.27 Construct an ogive for the accident survey data given in Exercise 2.11.

2.28 Construct an ogive to represent the data on the following page, which were obtained from a random sample of 50 farm incomes.

Income	Number of Farms
$0 to 1,999	2
2,000 to 3,999	5
4,000 to 5,999	11
6,000 to 7,999	9
8,000 to 11,999	15
12,000 to 27,999	6
28,000 to 43,999	1
44,000 to 75,999	1

2.4

SCATTERPLOTS AND TRANSFORMATIONS

UNIVARIATE AND BIVARIATE DATA

The type of data considered thus far is called **univariate** because it involves only one variable. However, much quantitative data are **bivariate** in nature, that is, a *pair* of measurements is recorded for each sample unit.

UNIVARIATE DATA AND BIVARIATE DATA

Univariate sample data consist of values of a single variable measured on each unit in the sample.

Bivariate sample data consist of values of a pair of variables measured on each unit in the sample.

The following examples illustrate bivariate quantitative data.

1. Bank assets and deposits for last year in a city with 12 banks.
2. Spring and fall enrollments at 50 major universities.
3. Expenditures in each of two fiscal years by 22 federal agencies in a particular region of the country.
4. The number of tourists visiting each of several National Park Service areas for two consecutive years.
5. Height and weight measurements for a group of individuals.
6. Age at inauguration and age at death of U.S. presidents.

In Problem Setting 2.1 the statement was made that you, as director of personnel, had available, in addition to the typing scores, the number of errors that each secretarial applicant made. This is another example of bivariate data. The complete set of bivariate data is given in Figure 2.24.

Applicant Number	Typing Score	Number of Errors
1	68	8
2	72	2
3	35	9
4	91	14
5	47	9
6	52	13
7	75	12
8	63	3
9	55	0
10	65	14
11	84	0
12	45	14
13	58	14
14	61	12
15	69	2
16	22	2
17	46	5
18	55	5
19	66	13
20	71	2

FIGURE 2.24
TYPING SCORES AND NUMBER OF ERRORS FOR
EACH OF 20 APPLICANTS.

SCATTERPLOT

The methods of the previous two sections could also be used to display only
the errors as was done for typing speeds. However, consider a joint display
of the bivariate observations in a *scatterplot*. This means that a plot needs to
be constructed where the horizontal axis represents the typing speeds and the
vertical axis represents the number of errors. Each applicant is represented as
a single point in the two-dimensional display given in Figure 2.25. The x
coordinate is the typing speed and the y coordinate is the number of errors.

The usefulness of the scatterplot is apparent when comparing the various
applicants. Previously (see Figures 2.4, 2.5, and 2.6) there was no way to
choose between the two applicants with typing speeds of 55. Now it is clear
that one applicant had five errors, whereas the other had none. Likewise, the
applicant with a typing speed of 91 previously appeared to be far superior to
all others. However, this applicant had 14 errors, whereas the applicant with
a speed of 84 had no errors. In this particular application auxiliary lines could
be added to the scatterplot to make identification of desirable candidates easy.
For example, if it is desired to consider only applicants with two or fewer
errors and a speed of 65 or higher, a line drawn parallel to the horizontal axis
and intersecting the vertical axis between 2 and 3 shows six candidates with

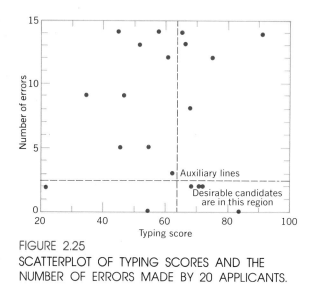

FIGURE 2.25
SCATTERPLOT OF TYPING SCORES AND THE
NUMBER OF ERRORS MADE BY 20 APPLICANTS.

two or fewer errors. Likewise, a vertical line at 65 shows nine candidates with speeds of at least 65. However, in the lower right-hand corner of the plot there are only four viable candidates satisfying both conditions.

The real-number-line plot for the typing scores, given in Figure 2.22, may be obtained from Figure 2.25 by moving each point vertically to the x-axis. Clearly, Figure 2.25 exhibits the relationship that exists between the two variables, and is more informative than the real-number-line plot of Figure 2.22.

AIDS IN MAKING SCATTERPLOTS

One major area of concern when making a graphical display of quantitative data is that all points be identifiable in the graph. For example, draw a real number line and try to plot the following numbers on it: 1, 10, 100, 1000, 10,000, and 100,000. After 100,000 is set as the upper limit on the scale on the graph you will find that all of the numbers less than 1000 are so close together that they cannot be distinguished. A solution that is frequently offered in this situation is to plot the *logarithms* of the numbers, which are in this case 0, 1, 2, 3, 4, and 5. Recall that a logarithm, $\log_{10} X$, is the exponent to which 10 must be raised to equal X. If the logarithms of these numbers are plotted, each of these points is easily identifiable. This is known as *transforming* the data. For convenience in reading the graph, however, the original numbers frequently appear as labels on the axis even though a logarithm scale is used to plot the data.

THE LOG TRANSFORM   The **log transformation** of the sample data $X_1, \ldots, X_n$ replaces the sample data with their corresponding base 10 logarithms, that is, $\log_{10} X_1, \ldots, \log_{10} X_n$. The log transform can be used only when all sample values are greater than zero.

	Deposits (in Thousands)	Yearly Growth of Deposits (%)
Albuquerque National Bank	$675,709	5.1
First National Bank	457,085	12.0
Bank of New Mexico	242,682	7.1
American Bank of Commerce	99,615	−2.9
Rio Grande Valley Bank	78,947	22.3
Citizens Bank	39,603	−7.5
Fidelity National Bank	38,213	6.9
Southwest National Bank	42,813	13.5
Republic Bank	33,445	−2.3
Western Bank	30,271	5.1
Plaza del Sol National Bank	10,370	13.9
El Valle State Bank	5,635	−6.0

FIGURE 2.26
DEPOSIT AND DEPOSIT GROWTH FOR 12 ALBUQUERQUE, NEW
MEXICO, BANKS.

As an example of the usefulness of logs in graphical display, consider data from a study of 12 Albuquerque, New Mexico, banks given in Figure 2.26. Information is available on the year-end deposits (in thousands of dollars) of the bank as well as the percentage growth or decline of deposits over the previous year.

Plotting the deposit data presents some problems because of the large differences in the amount of deposits held by the different banks. To illustrate this point consider a real-number-line plot of deposits (Figure 2.27). It is clear that some points are so close together as to be indistinguishable. This is caused by the few extremely large values or outliers that create the scaling problem in the first place. One way to moderate the influence of outliers in the real-number-line plot is to replace the data with their logs. That is, use a log transform on the data. A graph of a real-number-line plot based on logarithms of bank deposits is given in Figure 2.28. A comparison of Figures 2.27 and 2.28 shows clearly that using a logarithmic scale makes the data points become more distinguishable without completely losing a sense of distance or relative ordering among the points.

THE RANK TRANSFORMATION

The log transformation cannot be used directly when the data contain negative numbers such as the percentages associated with the bank deposit data in Figure 2.26. A transformation that works equally well on all types of numbers

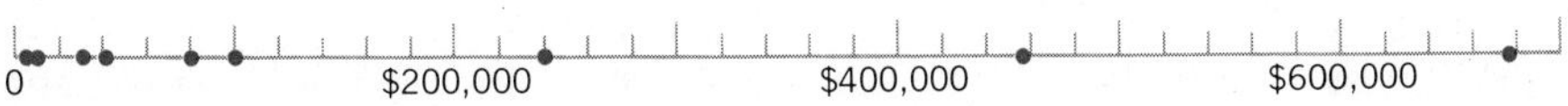

0          $200,000 $400,000 $600,000

FIGURE 2.27
REAL-NUMBER-LINE PLOT OF BANK DEPOSITS.

FIGURE 2.28
REAL-NUMBER-LINE PLOT OF BANK DEPOSITS ON A LOGARITHMIC SCALE.

is the **rank transformation.** The rank transformation will be used extensively in this text and has many uses beyond simple plotting. It merely replaces each of the data points by its respective **rank.** The rank of an observation is its relative position when the observations are arranged in order from smallest to largest.

THE RANK
TRANSFORM

The **rank transformation** for a sample of n observations replaces the smallest observation by the integer 1 (called the rank), the next smallest by *rank* 2, and so on until the largest observation is replaced by *rank n.*

Consider the following data set and the corresponding set of ranks.

Data:	5.6	7.2	3.4	11.8	6.9	8.1
Ranks:	2	4	1	6	3	5

The rank transformation is easily performed and can be used on positive or negative numbers. There is only one additional item to remember and that is how to handle ties in the sampling data; ties are sample observations that are equal to one another. Ranks within each group of tied observations are determined by adding together the ranks that would have been assigned within that group had there been no ties, and then dividing this total by the number of tied observations and assigning the result to each tied observation in that group. This is called *the method of average ranks.* Consider the following revision of the previous data set.

Data:	6.9	7.2	3.4	11.8	6.9	8.1
Ranks:	2.5	4	1	6	2.5	5

Here the previous data set has been changed to include two values of 6.9. Had these differed only slightly, they would have been assigned ranks of 2 and 3, so each is assigned a rank of $(2 + 3)/2 = 2.5$ to settle the tie. If the original data set is further modified so that the 7.2 is also a 6.9, then the tie is settled by assigning to each tied sample value the rank $(2 + 3 + 4)/3 = 3$.

USE OF THE STEM AND LEAF PLOT TO ASSIGN RANKS

A stem and leaf plot may be used to order the sample data from smallest to largest. This ordering makes the assignment of ranks easy. For example, refer back to the original stem and leaf plot of typing scores in Figure 2.4 and see how quickly the ranks can be assigned from this plot.

RANK TRANSFORM SCATTERPLOT

As an example of the usefulness of ranks in a graphical display of data, the banking data given in Figure 2.26 are used. Since the yearly percentage growth contains some negative values, a log transformation cannot be used; however, this is not a problem for a rank transformation. First the original data are replaced with their corresponding ranks 1 to 12 and the axis labeled accordingly.

Deposits	Percentage Growth	Ranks of Deposits	Ranks of Percentages
675,709	5.1	12	5.5
457,085	12.0	11	9
242,682	7.1	10	8
99,615	−2.9	9	3
78,947	22.3	8	12
39,603	−7.5	6	1
38,213	6.9	5	7
42,813	13.5	7	10
33,445	−2.3	4	4
30,271	5.1	3	5.5
10,370	13.9	2	11
5,635	−6.0	1	2

Note that the deposits are almost in descending order and that the percentages have two values of 5.1. These values are tied for positions 5 and 6, so each

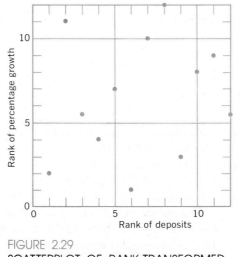

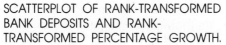

FIGURE 2.29
SCATTERPLOT OF RANK-TRANSFORMED BANK DEPOSITS AND RANK-TRANSFORMED PERCENTAGE GROWTH.

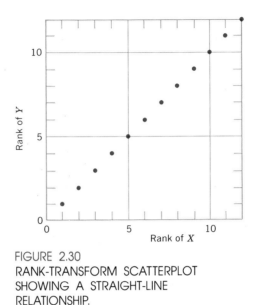

FIGURE 2.30
RANK-TRANSFORM SCATTERPLOT
SHOWING A STRAIGHT-LINE
RELATIONSHIP.

is assigned the rank of $(5 + 6)/2 = 5.5$. The bivariate scatterplot appears in Figure 2.29.

Figure 2.29 presents an easy-to-read graph with the points not clustered, and there is no longer a scaling problem. It makes apparent that there is no discernible pattern present; that is, the points appear to be scattered at random in the plot. This random scattering is in contrast to other possible plots that might show some relationship, such as one connecting the pairs (1,1), (2,2), (3,3), . . . , (12,12) and forming a strong straight-line relationship from the lower left-hand corner of the plot to the upper right-hand corner. (See Figure 2.30.) Such relationships are discussed in a later chapter.

WHEN TO USE SCATTERPLOTS

Scatterplots are clearly appropriate for use with any set of data that occurs as bivariate observations, that is, points that simultaneously represent two variables. Their use produces a visual display of such bivariate points that makes relationships between the two variables easy to see. The pattern in the scatterplot often aids in interpreting the results of a statistical analysis of such data, and scatterplots will be used throughout this text with reference to paired data, correlation, regression, and time series analysis.

Scaling can be a problem in the construction of scatterplots if either or both of the variables represented by the bivariate observations contain outliers or values that differ greatly from the majority of the data, such as the deposit data in Figure 2.26. A log transformation of the deposit data may prove to be quite helpful in scaling data such as these, as illustrated in Figures 2.27 and 2.28. A rank transformation of the deposit data works equally well, as illustrated in Figure 2.29. Moreover, the rank transformation can be used on all types of data, whereas the log transformation cannot. For example, the negative numbers contained in the growth data in Figure 2.26 prevent a log transformation from being used, but cause no problems for a rank transformation. However, changing the scale on graphs with any type of transformation has the potential of distorting the data. Thus, if transformations are used, one must be careful to interpret the scatterplot in terms of the type of transformation used, such as logarithms or ranks.

EXERCISES

2.29 The Bureau of Labor Statistics reported expenditures on clothing and personal care for low-, intermediate-, and high-budget four-person families in the northcentral portion of the United States for the autumn of 1979 as listed below. Plot these data in a scatterplot taking care to use a different symbol (*, ○, ●) with each budget so that the points are easily identified within the plot. What facts does the scatterplot make readily apparent? Do the values for Cincinnati and Milwaukee appear to fit the pattern within each budget type?

	TYPE OF BUDGET					
	Low		Intermediate		High	
	Clothing	Personal Care	Clothing	Personal Care	Clothing	Personal Care
Chicago	797	326	1153	426	1684	586
Cincinnati	994	305	1430	401	2083	550
Cleveland	920	409	1323	544	1930	756
Detroit	830	339	1195	452	1750	619
Kansas City	927	391	1329	516	1940	724
Milwaukee	973	357	1390	465	2034	645
Minneapolis	872	357	1251	470	1820	651
St. Louis	820	334	1184	426	1747	573

2.30 A summary of financial statistics of institutions of higher education compiled by the National Center for Education Statistics gives a breakdown of revenues. Given below is a summary by different regions of the country of the percentages of revenue obtained from tuition and fees and

Region	Percentage of Total Revenue From	
	Tuition and Fees	Government Appropriations
New England	33.0	17.1
Mideast	29.1	27.2
Great Lakes	22.8	33.3
Plains	19.7	35.4
Southeast	18.0	37.9
Southwest	13.6	45.8
Rocky Mountains	17.0	36.3
Far West	11.9	41.9

government appropriations (mostly state government). Plot these percentages in a scatterplot. Does your plot indicate any relationship between tuition and fees and the size of government appropriations?

2.31 The National Center for Education Statistics has compiled data on the type of degree granted to males and females for 11 consecutive academic years. Data regarding the awarding of first professional degrees is given below. Use these data to make a scatterplot of the number of first professional degrees granted to males versus the number of first professional degrees granted to females. What fact does the graph make apparent regarding the awarding of first professional degrees during the last few years of the study?

Academic Year	Number of First Professional Degrees to	
	Men	Women
1966–1967	31,064	1,429
1967–1968	33,083	1,645
1968–1969	34,069	1,612
1969–1970	33,344	1,908
1970–1971	35,797	2,479
1971–1972	41,021	2,753
1972–1973	46,827	3,608
1973–1974	48,904	5,374
1974–1975	49,230	7,029
1975–1976	53,210	9,851
1976–1977	52,668	12,112

2.32 Market share trends of the liquor industry are given below for the years 1959 and 1974. During this time, demand shifted away from domestic bourbons and blends. In this setting it would be desirable to make a

	Percentage of Market	
	1959	1974
Whiskey types:		
Straights	25.9	16.3
Spirit blends	32.1	13.8
Scotch	7.7	13.7
Canadian	5.0	11.4
Bonds	4.6	1.0
Other	0.3	0.3

Vodka	7.3	16.9
Gin	9.2	10.0
Cordials	3.5	6.2
Brandy	2.4	4.0
Rum	1.5	3.7
Other	0.5	2.7

graph to show how this shift has affected preference for various liquors. That is, how does the ranking of liquors in 1959 compare with the ranking of liquors in 1974 in a scatterplot? Make this rank transform scatterplot and determine which liquors have made the greatest change in position in the market.

2.33 One way to evaluate a stock is to see how many shares are held by financial institutions, that is, banks and investment and insurance companies. Given below is such a listing for 10 discount variety chain stores along with their recent earnings per share as obtained from an issue of *Standard and Poor's Stock Guide*. A plot is needed that shows if the amount of institutional holdings is associated with the size of the stock's earnings, but representing the data listed below in a scatterplot would be difficult because of the large variation in the size of the numbers of shares of stock held. However, the desired information can be obtained from a rank-transform scatterplot. Make such a plot for these data.

Company	Number of Institutional Shares Held (Thousands)	Earnings per Share
Danners Inc.	197	$1.28
Duckwall-Alco	24	2.20
Fed-Mart	99	4.80
Grand Central	2	1.22
Hartfield-Zodys	377	1.97
K-Mart	53,170	2.96
Thrifty Corp.	544	1.46
Wal-Mart	3,452	2.34
Woolworth	6,587	4.84
Zayre	632	2.64

2.34 Do a log transformation of the number of institutional shares held using data from Exercise 2.33 and then make a scatterplot of these logs versus earnings per share. Compare your results with those from Exercise 2.33.

2.35 The monetarist school of economic thought states that all monetary expansion in excess of real economic growth is eventually transmitted to inflation. In other words, money supply (cash and bank deposits) should only be permitted to grow as fast as the rate of increase of output of goods and services. Any more rapid growth of money supply will merely create excess dollars to chase after the same supply of goods and ser-vices—and send prices soaring. Given below are 20-year-summary data for 14 countries. Plot the anticipated inflation versus actual inflation in a scatterplot to see if the monetarist hypothesis is supported by these data.

| Country | Annual Rates of Change 1960–1979 | | | |
	(1) Money Supply Growth (%)	(2) Real Economic Growth (%)	(1)–(2) Anticipated Inflation (%)	Actual Inflation (%)
Brazil	44.7	9.1	35.6	32.1
Japan	16.8	8.4	8.4	6.1
Italy	16.5	4.3	12.2	9.1
Spain	16.2	6.3	9.9	10.5
Mexico	15.8	6.1	9.7	8.4
France	10.4	4.1	6.3	6.4
Netherlands	9.5	4.2	5.3	6.4
Sweden	9.1	3.1	6.0	6.6
Germany	8.7	3.8	4.9	4.4
U.K.	8.1	2.5	5.6	8.3
Canada	8.1	4.8	3.3	5.3
Switzerland	7.8	2.8	5.0	5.1
Belgium	7.1	4.2	2.9	5.2
U.S.A.	5.3	3.6	1.7	4.7

2.36 Convert the scatterplot of Exercise 2.35 to a rank-transform scatterplot.

2.37 A survey of the 14 counties in southwest Florida revealed the percent of their residents 65 years of age or older, and the percent of their residents who were nonwhite. These data are given below. Present these data in a scatterplot. What interesting information does this graph convey to you? How would you account for the phenomenon?

County	Percent Over 65	Percent Nonwhite	County	Percent Over 65	Percent Nonwhite
Citrus	29.4	3.6	Hillsborough	11.5	14.2
Hernando	24.4	6.7	Pinellas	27.8	8.2
Pasco	30.7	2.5	Manatee	27.1	9.4
Polk	14.3	15.6	Sarasota	30.0	5.7
Hardee	11.4	8.7	Charlotte	33.9	2.6
DeSoto	16.2	19.2	Lee	22.3	8.4
Highlands	26.3	14.5	Collier	19.0	5.8

2.38 The effective buying income is estimated each year for each county in

Mississippi. The percent change in estimated effective income is given below for a sample of eight counties, for both 1981 and 1982. Make a scatterplot of these data. Is there any apparent relationship between the 1981 and 1982 percent increase?

County	1981	1982	County	1981	1982
Benton	0.96	13.62	Madison	9.75	41.56
Coplah	2.52	22.91	Pearl River	10.19	29.31
Hinds	8.70	4.07	Smith	6.67	15.20
Kemper	4.72	25.09	Warren	6.34	30.21

2.5

USING CHERNOFF FACES TO DISPLAY MULTIVARIATE DATA (OPTIONAL)

PROBLEM SETTING 2.3

Data on 13 variables relating to the federal offshore leasing and producing activities of 12 company groups have been obtained. The 13 variables are described in Figure 2.31 and the values of each of these variables are listed in Figure 2.32 for each of the 12 company groups. How can these data be displayed so that the various company groups can easily be compared?

Note that the problem here is the same as before—too much data for the average reader to digest. In previous sections there were many observations on just one or two variables, and stem and leaf plots, histograms, empirical distribution functions, and scatterplots were used to simplify the presentation of the data. In this section many variables are involved—13 variables in this problem setting. Therefore the problem becomes more difficult, and some innovative techniques are required for displaying the data.

1. Net number of leases won
2. Number of leases won/number bid on
3. Average percent of ownership of leases
4. Total net dollars paid
5. Net acreage leased
6. Net number of owned leases that produced
7. Average number of years between lease year and first production
8. Revenue earned
9. Revenue/bonus ratio
10. Net gas production
11. Net liquid production
12. Net revenue per year of production
13. Percentage of owned leases terminated

FIGURE 2.31

VARIABLES USED TO DESCRIBE THE FEDERAL OFFSHORE LEASING AND PRODUCING ACTIVITIES OF 12 COMPANY GROUPS.

COMPANY GROUP NAME	VARIABLE												
	1	2	3	4	5	6	7	8	9	10	11	12	13
ARCO	233.80	0.31	0.52	922.94	1.17	25.00	4.40	965.48	1.05	0.70	113.30	167.07	0.21
UNION	111.60	0.29	0.49	666.09	0.56	18.50	4.20	992.36	1.49	1.70	133.60	91.55	0.43
GETTY	78.10	0.29	0.30	667.15	0.36	19.00	4.00	582.97	0.87	1.00	67.70	69.33	0.28
MOBIL	121.50	0.38	0.48	1400.40	0.58	29.70	3.80	1055.96	0.75	1.60	104.10	210.16	0.28
TEXACO	123.70	0.33	0.65	1363.64	0.60	31.50	7.10	774.13	0.57	0.60	86.80	164.45	0.39
SUNOCO	105.80	0.21	0.73	612.13	0.47	6.60	4.00	301.44	0.49	0.30	24.80	79.13	0.43
CHEVRON	288.40	0.35	0.68	970.23	1.28	67.30	5.30	1339.86	1.38	1.10	233.40	168.78	0.36
TENNECO	87.60	0.29	0.70	670.01	0.40	33.00	5.00	1292.52	1.93	2.10	95.30	263.08	0.22
GULF	174.50	0.36	0.60	1583.88	0.83	50.70	3.70	2300.18	1.45	2.30	422.20	281.78	0.36
AMOCO	140.90	0.26	0.53	690.33	0.69	26.10	6.30	591.64	0.86	0.80	55.60	106.66	0.39
SHELL	359.30	0.38	0.91	1365.32	1.79	67.90	3.60	3539.47	2.59	2.60	566.30	507.79	0.51
EXXON	284.60	0.37	0.88	1930.72	1.43	36.10	5.00	2017.68	1.05	1.60	339.30	259.74	0.31

FIGURE 2.32

VALUES OF THE VARIABLES GIVEN IN FIGURE 2.31 FOR EACH OF THE 12 COMPANY GROUPS.

MULTIVARIATE DATA

In the previous sections of this chapter either a single (univariate) variable or a pair (bivariate) of variables were associated with each sample unit. There are many sets of data, such as in Problem Setting 2.3, where two or more variables are associated with each sample unit. Such data are referred to as **multivariate.** Thus, bivariate data are the simplest type of multivariate data.

MULTIVARIATE DATA — **Multivariate** sample data consist of two or more variables measured on each unit in the sample.

Some attempts at graphical representation of multivariate data are extensions of the bivariate scatterplot. For example, the dot in the scatterplot can be replaced by a circle whose diameter is varied to be proportional to the

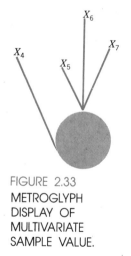

FIGURE 2.33
METROGLYPH
DISPLAY OF
MULTIVARIATE
SAMPLE VALUE.

magnitude of a third variable. Additional variables may be represented by splines extending from the circle like that shown in Figure 2.33. The length of the splines represents the magnitude of four additional variables X_4 to X_7. Displays like the one in Figure 2.33 are called **metroglyphs.**

Some more innovative attempts at representing multivariate data use pictures with several characteristics to represent each multivariate data point. For example, trees can be used where characteristics such as trunk width, tree height, and density of foliage, represent the magnitude of the various variables. Other examples include castles or large trucks. One of the more popular methods uses cartoon faces. This method is now considered in detail.

CHERNOFF FACES

One of the most successful methods for the display of multivariate data has been the representation of each multivariate sample unit (or data point) as a cartoon face. Individual variables associated with the sample unit are represented by various facial features, such as size of the eyes, position of the pupils, size of the ears, length and width of the nose, and curvature of the mouth. This technique was published in 1973 by Herman Chernoff and has since become known as **Chernoff faces.** The reason for selecting a face for displaying multivariate data is that it makes it easy for the human mind to grasp many of the essential regularities and irregularities present in the data.

Chernoff faces can represent as many as 20 different variables for each sample unit. Figure 2.34 shows a neutral Chernoff face.

Chernoff faces are most easily explained by use of a simple example. Suppose that it is desired to represent the following three observations of a pair of variables using a Chernoff face.

	Variable 1	Variable 2
Pair 1	.20	1.30
Pair 2	.50	1.60
Pair 3	.80	1.90

Each variable must be assigned to a facial feature in Figure 2.34. For purposes of illustration, Variable 1 is assigned to the position of the pupils of the eyes

FIGURE 2.34
A NEUTRAL CHERNOFF FACE.

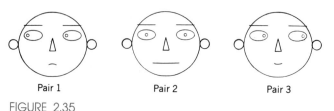

Pair 1   Pair 2   Pair 3

FIGURE 2.35
CHERNOFF FACES USING POSITION OF PUPILS AND
CURVATURE OF THE MOUTH TO REPRESENT THREE
OBSERVATIONS ON A PAIR OF VARIABLES.

and Variable 2 is assigned to the curvature of the mouth. The results are shown
in three Chernoff faces (one face for each pair of observations) in Figure 2.35.
Note the change in the position of the pupils in these three faces. One face
shows the eyes looking to the right, another shows the eyes looking straight
ahead, and one shows the eyes looking toward the left. These three pupil
positions represent, respectively, the values of .20, .50, and .80 for Variable
1. At the same time, the three values of Variable 2 are represented by the
curvature of the mouth, first as a frown, then as a horizontal mouth, and finally
as a smile.

The exact positions of the pupils and shape of the mouth are based on
scaling the values of the variables against some predetermined standard position, such as in Figure 2.34. Such scaling, as well as the actual construction
of the Chernoff face, is best done through use of a computer. For that reason,
such a computer program is now briefly discussed.

COMPUTER-GENERATED CHERNOFF FACES

An examination of Figures 2.34 and 2.35 provides some indication of the
amount of work necessary to construct a single Chernoff face. Even more work
is required to represent the multivariate data in Figure 2.32, as it becomes
necessary to construct 12 Chernoff faces, one for each of the 12 company
groups, with each face utilizing 13 facial features, one for each variable.
Because of the amount of work involved, the construction of a Chernoff face

Variable	Facial Feature	Variable	Facial Feature
1	Eye separation	8	Mouth width
2	Eye length	9	Mouth curvature
3	Pupil position	10	Ear radius
4	Face width	11	Ear level
5	Face height	12	Eye slant
6	Nose length	13	Eyebrow height
7	Nose width		

FIGURE 2.36
PAIRING OF THE MULTIVARIATE DATA VARIABLES IN
FIGURE 2.31 AND THE CHERNOFF FACIAL FEATURES
USED TO CONSTRUCT THE FACES IN FIGURE 2.37.

needs to be done with the aid of a computer, where the user of the computer program specifies which multivariate variables are to be associated with which facial features.

Programs for the construction of Chernoff faces exist at many different computer facilities. The program used to generate the faces in this section was written by Lawrence A. Bruckner and others at Los Alamos National Laboratories. This program automatically scales each variable and allows for the easy assignment of facial features to data variables. The Los Alamos program was used to construct Chernoff faces for the multivariate data in Figure 2.32 based on the variable-facial-feature pairings indicated in Figure 2.36. The resulting faces appear in Figure 2.37.

INTERPRETING THE CHERNOFF FACES
FOR COMPANY GROUPS

From Figures 2.31 and 2.36 it can be seen that the net number of owned leases that produced is represented by the length of the nose; the greater the number, the longer is the nose. Based on this facial feature it can be seen in Figure 2.37 that Chevron and Shell have fared better than all other companies, while Sunoco has the fewest number of producing leases. This interpretation can easily be verified by the numbers given in Figure 2.32. Likewise, the width and curvature of the mouth are joint indicators of revenue. A wide smile

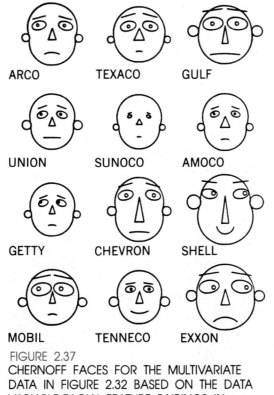

FIGURE 2.37
CHERNOFF FACES FOR THE MULTIVARIATE
DATA IN FIGURE 2.32 BASED ON THE DATA
VARIABLE-FACIAL-FEATURE PAIRINGS IN
FIGURE 2.36.

represents high revenue and a large revenue/bonus ratio, while a narrow frown represents low revenue, and a small revenue/bonus ratio. Shell and Sunoco represent the extremes in this feature.

In other comparisons in Figure 2.37 it can be seen that eye separation, representing variable 1 (net number of leases won), seems to increase as face height, representing variable 5 (net acreage leased), increases. The same is true of the pair mouth width (revenue earned) and ear level (net liquid production), as well as for the pair eye length (percentage of leases won) and face width (total net dollars paid).

WHEN TO USE A CHERNOFF FACE

If the data are multivariate in form, the Chernoff face may provide a convenient way to display up to 20 variables simultaneously. For two variables (bivariate data), which is the simplest case of multivariate data, a scatterplot will provide a more useful and easily interpreted display of relationships that may exist between the two variables. Chernoff faces are not recommended for use with bivariate data. Likewise, a three-dimensional display for multivariate data having three variables may be preferable to a Chernoff face. Chernoff faces are probably most useful with four or more variables, but the faces become increasingly difficult to interpret as the number of variables increases.

It is important to keep in mind that it is necessary to construct a Chernoff face for each multivariate data point (sample unit). Since the construction of a Chernoff face is cumbersome, it is almost essential to use a computer-generated graphics program in the construction.

Chernoff faces can differ greatly depending on how the facial features and variables are paired. This can be a disadvantage if the reader makes a subjective interpretation of the faces (such as "happy" or "sad") instead of relating each facial characteristic back to the variable that it represents.

EXERCISES

2.39 If a multivariate data set has 19 observations on each of 15 variables, how many Chernoff faces would be required to represent the data set? How many facial features would be required?

2.40 What is the maximum number of variables that can be represented by a Chernoff face? What is the maximum number of observations that can be represented?

2.41 With what type of multivariate data is a Chernoff face most useful?

2.42 Other than ease of construction, why is it preferable to use a scatterplot to represent bivariate data rather than using Chernoff faces?

2.43 Net gas production is represented in Figure 2.37 based on the radius of the ear. If the radius increases as gas production increases, which company would appear to have the largest gas production? Which company appears to have the lowest gas production? Compare your selections with the numeric values given in Figure 2.32 for gas production.

2.44 Variable number 7 in Figure 2.31 is the average number of years between the year in which the lease was granted and the year in which production first started. In Figure 2.37 this variable is represented by the width of the nose, with the width increasing as the average time between granting of the lease and start of actual production increases. Which company in Figure 2.37 would appear to have the largest average time lag? Check your answer with the numeric values for variable number 7 in Figure 2.32.

2.45 In Figure 2.37 Sunoco has very small eyes while several companies all have large eyes. What information is being conveyed by this result?

2.46 Make a scatterplot of variable 1 (net number of leases won) versus variable 5 (net acreage leased) for the data given in Figure 2.32. Label each of the points in the scatterplot with the corresponding name of the oil company group that it represents. How would you describe the results shown in the scatterplot? In Figure 2.37 variable 1 is represented by eye separation and variable 5 by face height. How can the relationship in the scatterplot be seen in Figure 2.37?

2.6

REVIEW EXERCISES

2.47 Given below is a frequency summary of the weights of 114 students. Summarize these data in a histogram. Why is a stem and leaf plot not appropriate for these data?

Weight	Frequency
104.5 to 114.5	1
114.5 to 124.5	5
124.5 to 134.5	16
134.5 to 144.5	29
144.5 to 154.5	24
154.5 to 164.5	19
164.5 to 174.5	12
174.5 to 184.5	4
184.5 to 194.5	3
194.5 to 204.5	1

2.48 Use a stem and leaf plot to represent the number of hours worked by 50 employees in one week in a manufacturing company.

47	50	52	55	46	50	53	49	43	48
51	54	52	47	49	45	51	53	49	47
52	61	50	51	53	49	50	45	55	51
52	46	47	50	48	57	50	48	55	49
53	49	54	44	47	51	45	51	50	52

2.49 Plot the data in Exercise 2.48 as an empirical distribution function.

2.50 Plot the data given below in a scatterplot for heights (X) and weights (Y) of 10 college students.

Height (X):   64   73   71   69   66   69   75   71   63   72

Weight (Y):   121   181   156   162   142   157   208   169   127   165

2.51 Make a rank transform scatterplot of the data in Exercise 2.50.

2.52 Construct an ogive for the sample data given in Exercise 2.47.

2.53 Given below are the weight losses (in grams) of laboratory rats 24 hours after they were injected with an experimental drug. Summarize these data with a stem and leaf plot.

4.2	4.4	4.8	4.9	4.4
3.9	4.3	4.5	4.8	3.9
3.6	4.1	4.3	3.9	4.2
4.1	4.0	4.0	3.8	4.6
3.8	4.7	3.9	4.0	4.2
4.4	4.6	4.4	4.9	4.4
4.1	4.3	4.2	4.5	4.4
4.2	4.1	4.0	4.7	4.1
4.7	4.2	4.1	4.4	4.8
4.3	4.6	4.5	4.6	4.0

2.54 Summarize the sample data given in Exercise 2.53 in a histogram.

2.55 Construct an empirical distribution function for the sample data in Exercise 2.53.

2.56 The GPAs of sophomore students are summarized in the following table.

Overall GPA	Number of Students
0.00 to 1.99	382
2.00 to 2.49	381
2.50 to 2.99	297
3.00 to 4.00	321

(a) Present the data in a histogram.

(b) Present the data in an ogive.

(c) Approximately what percentage of students has a GPA lower than 3.3?

2.57 The 12 finalists for the steer roping competition are selected on the basis of their total time for the first three rounds of competition. Then their times for a fourth and final round are obtained.

Competitor	Three-Round Total (sec)	Fourth Round (sec)
Walter	40.40	13.69
Melvin	47.26	15.22
Rusty	47.45	17.55
Jim	48.43	14.64
Rocky	51.28	22.93
Sonny	51.46	24.09
Gip	51.83	14.89
Pokey	56.27	16.96
Bill	57.70	20.24
Charley	63.55	14.92
Rod	66.71	14.73
Roy	69.73	11.83

Make a scatterplot of these data. Does the scatterplot reveal any interesting information?

BIBLIOGRAPHY

Additional material on the topics presented in this chapter can be found in the following publications.

Anderson, E. E. (1957). "A Semigraphical Method for the Analysis of Complex Problems." *Proceedings of the National Academy of Science,* **13,** 923–27. Reprinted, with an appended note, in 1960 in *Technometrics,* **2,** 387–91.

Andrews, D. F. (1972). "Plots of High-Dimensional Data." *Biometrics,* **28,** 125–36.

Bruckner, L. A. and Mills, C. F. (1979). "The Interactive Use of Computer Drawn Faces to Study Multidimensional Data." Informal Report LA-7752-MS, Los Alamos National Laboratory, Los Alamos, N. Mex.

Chernoff, H. (1973). "The Use of Faces to Represent Points in k-Dimensional Space Graphically." *Journal of the American Statistical Association,* **68,** 361–68.

Freni-Titulaer, L. W. J. and Louv, W. C. (1984). "Comparisons of Some Graphical Methods for Exploratory Data Analysis." *The American Statistician,* **38,** 184–88.

Hoaglin, D. C., Mosteller, F. and Tukey, J. W., Eds. (1983). *Understanding Robust and Exploratory Data Analysis.* Wiley, New York.

Kleiner, B. and Hartigan, J. A. (1981). "Representing Points in Many Dimensions by Trees and Castles." *Journal of the American Statistical Association,* **76,** 272–75.

Tukey, J. W. (1977). *Exploratory Data Analysis.* Addison-Wesley, Reading, Mass.

3

DESCRIPTIVE SAMPLE STATISTICS

PRELIMINARY REMARKS

There is a popular riddle that goes something like this. One person is talking about a photograph, and says,

> *"Brothers and sisters,*
> *I have none;*
> *But that man's father*
> *is my father's son."*

What is the relationship between the speaker and the person in the photograph?

Confusing, isn't it? Well, a report with too many numbers in it may be equally confusing. Just as the speaker could have made things clearer by stating that the picture was of his son, a report can be made clearer by summarizing the data. The previous chapter described methods of summarizing the sample data with graphs. This chapter shows how to summarize the data with statistics.

The word **statistic** originally referred to numbers published by the state, where the numbers were the result of a summarization of data collected by the government. Thus some people think of a statistic as a number that is based on several numbers in a sample, the proportion of a population that is in a particular category, and so on. In this sense a statistic is just a number. However, if one stops to consider that the numbers being summarized may vary from one sample to the next or that the population may change from one year to the next, one can justify extending the idea of a statistic from being only a *number* to being a function, or rule, for associating a number with the sample.

Suppose several numerical observations in a sample are being summarized by stating their average value. Then "the average of the numbers in the sample" is the statistic, and the actual average obtained in one sample is a value of the statistic. A statistic is a rule for obtaining a number, and therefore it meets the requirements of being a random variable, a function that assigns numbers to the points in the sample space (for an appropriately defined sample space). A statistic also conveys the idea of a summarization of data, so usually a statistic is considered to be a random variable that is a function of several other random variables. Then a value taken by the statistic is implicitly assumed to be the result of some arithmetic operations performed on other numbers (the data) that, in turn, are the values assumed by several random variables. Note that this is the third use of the word *statistic*, as defined in Chapter 1.

STATISTIC   A **statistic** is a random variable that is a function of other random variables that are associated with the observations in a sample.

The sample statistics presented in this chapter are all standard, well-known, and widely used methods of summarizing data. They include the sample mean, sample standard deviation, sample mode, sample percentiles, sample proportion, and other sample statistics closely related to these.

3.1

THE MEAN FOR THE SAMPLE

PROBLEM SETTING 3.1

A nationally known company has an active program of in-service training courses for its employees. Each year thousands of the company's employees take these courses. At the completion of each course the employees are required to take a final examination. Forty final-exam scores are selected at random from all employees completing a class in advanced marketing techniques during the past 12 months. Let $X_1, X_2, \ldots, X_{40}$ be random variables representing the test scores of the 40 employees. These scores are as follows.

77	68	86	84	95	98	87	71
84	92	96	83	62	83	81	85
91	74	61	52	83	73	85	78
50	81	37	60	85	100	79	81
75	92	80	75	78	71	64	65

Take a minute or two to study these numbers. Were there many perfect papers (100 points)? Were there many very poor papers? How can these data be presented so they can be more easily understood?

A MEASURE OF LOCATION

In Problem Setting 3.1 a sample of size 40 has been randomly selected from a much larger population of unknown or unspecified size. The question of interest is how to make a concise and meaningful summary of the numerical values presented.

In the case of sample data, a logical response to the question of how to summarize the numerical values might be to start with a graphical display of the sample values, such as a stem and leaf plot, a histogram, or an empirical distribution function. Another way to summarize the data would be through a single numerical value that represents the center of the data.

Numerical values that represent the center of the sample (or the population) are referred to as **measures of location.** Methods are presented in this section for quantifying the measure of location for the sample. The method of quantification is preceded by two graphical displays of the sample data that provide an intuitive feeling for a measure of location for the sample data.

A STEM AND LEAF PLOT OF THE SAMPLE DATA

A stem and leaf plot for the sample data in Problem Setting 3.1 is given in Figure 3.1. It is clear from this plot that more scores fall in the 80s than in any other class, with the 70s a clear second. Since these classes are also toward the center of the data, a good guess at a measure of location for the center of these data might be in the upper 70s or lower 80s.

THE EMPIRICAL DISTRIBUTION FUNCTION

The empirical distribution function for the sample data in Problem Setting 3.1 is given in Figure 3.2. From the empirical distribution function it again appears that the sample observations are clustered around 80 as a measure of location.

Both the empirical distribution function and the stem and leaf plot provide valuable insight, but at this point they both have to be supplemented with subjective judgment to determine a numerical value that represents the center

```
 3 | 7
 4 |
 5 | 0  2
 6 | 8  1  0  2  4  5
 7 | 7  5  4  5  8  3  1  9  1  8
 8 | 4  1  6  0  4  3  3  5  3  7  1  5  5  1
 9 | 1  2  2  6  5  8
10 | 0
```

FIGURE 3.1
A STEM AND LEAF PLOT OF THE 40 EXAM SCORES.

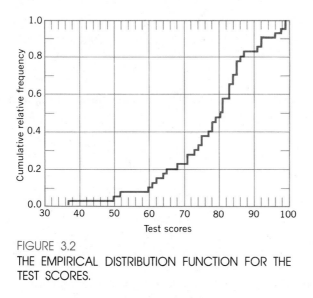

FIGURE 3.2
THE EMPIRICAL DISTRIBUTION FUNCTION FOR THE
TEST SCORES.

of the sample. Methods are now presented that remove such subjectivity from
the quantification process.

THE SAMPLE MEAN

Even a person accustomed to handling numbers will find the 40 exam scores
in Problem Setting 3.1 difficult to interpret all at once. Statistics to summarize
these data would be helpful. In this case some measure of the "average" value
is needed. One technical term for average value is the **mean,** or in this case
the **sample mean,** since these 40 employees may be thought of as a sample
from the population of all employees taking this course. The sample mean is
obtained by adding all of the observations together and dividing by the sample
size. The usual notation for the sample mean is $\overline{X}$, pronounced, "ex bar."

$$\overline{X} = (77 + 68 + 86 + \cdots + 65)/40$$
$$= 3102/40$$
$$= 77.55$$

Note how this sample mean 77.55 fits right into the middle of both the stem
and leaf plot in Figure 3.1 and the e.d.f. in Figure 3.2. This illustrates that the
sample mean is often a good statistic for locating the middle of a sample.

The equation just presented is awkward because there are so many numbers. The triple-dot notation "$\cdots$" is used to represent the numbers that are
omitted for brevity. A much more convenient notation involves the summation
notation Σ.

Let the individual test scores be random variables denoted by X_1, X_2, X_3,
and so on up to X_{40}. Thus the test scores are represented by X_i, $i = 1, \ldots ,$
40. The formula for the sample mean is conveniently expressed as

$$\overline{X} = \frac{1}{n}(X_1 + X_2 + \cdots + X_n) = \frac{1}{n}\sum_{i=1}^{n} X_i$$

where n is the sample size. In this case $n = 40$. If the sample had been denoted by $Y_1, Y_2, \ldots, Y_{40}$, then $\overline{Y}$ would be used for the sample mean.

The **sample mean** $\overline{X}$ is a statistic that measures the center of a sample, and is given by the equation

SAMPLE MEAN

$$\overline{X} = \frac{1}{n} \sum_{i=1}^{n} X_i \qquad (3.1)$$

where n is the sample size and $X_1, X_2, \ldots, X_n$ are random variables that represent the sample observations.

THE SAMPLE MEAN FOR REPEATED OBSERVATIONS

It is not uncommon for several observations to each occur many times in a sample. For example, suppose a sample of size 20 contains the observation 2.1 six times, the observation 1.8 four times, the observation 3.2 five times, the observation 3.5 four times, and the observation 4.2 once. Use of Eq. 3.1 to find the sample mean produces the following result.

$\overline{X} = (2.1 + 2.1 + 2.1 + 2.1 + 2.1 + 2.1 + 1.8 + 1.8 + 1.8 + 1.8$
$\qquad + 3.2 + 3.2 + 3.2 + 3.2 + 3.2 + 3.5 + 3.5 + 3.5 + 3.5$
$\qquad + 4.2)/20$
$\quad = 54.0/20 = 2.7$

The equation for finding the **sample mean for data with repeated observations** is

$$\overline{X} = \frac{1}{n} \sum_{i=1}^{k} f_i x_i \qquad (3.2)$$

SHORTCUT
FORMULA FOR
$\overline{X}$ WHEN
REPEATED
OBSERVATIONS
OCCUR

where the sample size is

$$n = \sum_{i=1}^{k} f_i$$

and

$\qquad k$ = the number of unique observations
$\qquad x_i$ = the sample observations
$\qquad f_i$ = the frequency with which x_i occurs in the sample

Clearly, many terms must be added together to find the sum of the observations. When the individual observations are repeated many times, the amount of arithmetic can be shortened by noting the frequency with which each observation occurs and writing the following expression involving multiplication and addition.

$$\overline{X} = [2.1(6) + 1.8(4) + 3.2(5) + 3.5(4) + 4.2(1)]/20$$
$$= 54.0/20 = 2.7$$

Thus, each different value x_i in the sample is simply multiplied by its frequency of occurrence f_i to shorten the amount of arithmetic necessary to find the sum of the observations. This example generalizes to the expression in Eq. 3.2. In the case where each $f_i = 1$, Eq. 3.2 is the same as Eq. 3.1.

EXAMPLE

To estimate the volume of repeat business, a fast photo-developing firm asks its customers how many times they have used its photo-developing services in the past month. The responses from the first 25 customers were 1, 2, 1, 0, 0, 1, 0, 2, 1, 0, 0, 1, 2, 1, 3, 0, 1, 1, 1, 0, 2, 0, 1, 1, 2. These numbers may be summarized in a frequency distribution.

x_i	f_i
0	8
1	11
2	5
3	1
	$n = 25$

The sample mean is easily computed by using Eq. 3.2.

$$\overline{X} = \frac{1}{n} \sum_{i=1}^{k} f_i x_i$$
$$= (1/25)[8(0) + 11(1) + 5(2) + 1(3)]$$
$$= (24/25) = .96$$

THE WEIGHTED SAMPLE MEAN

When sample data contain repeated observations, it is common for the data to be summarized using relative frequencies. The sample mean can also be calculated using relative frequencies. First note that Eq. 3.2 can be rearranged to look like

$$\overline{X} = \sum_{i=1}^{k} x_i \left(\frac{f_i}{n} \right) \qquad (3.3)$$

In this form, the relative frequencies, which sum to 1.0, act as weights on each sample observation and $\overline{X}$ is referred to as **weighted sample mean.** The weighted version of the sample mean is used in some special situations when some observations are considered more important than others, and this relative importance needs to be reflected appropriately in the mean.

WEIGHTED SAMPLE MEAN
A **weighted sample mean** for the sample observations x_i is $\Sigma\, w_i x_i$, where the weights w_i are the relative frequencies associated with the sample observations and as such the w_i sum to 1.0.

EXAMPLE

The previous example is reworked using Eq. 3.3. The data in the previous example are summarized using relative frequencies as follows.

x_i	$f_i/n = w_i$
0	8/25 = .32
1	11/25 = .44
2	5/25 = .20
3	1/25 = .04
	1.00

Using Eq. 3.3, the sample mean is found as

$$\overline{X} = 0(.32) + 1(.44) + 2(.20) + 3(.04) = .96$$

which is exactly the same value as found before. Note that the weights assigned to the four observations, 0, 1, 2, 3, are proportional to the frequencies with which those observations occurred.

CALCULATING THE MEAN FROM GROUPED DATA

Often the data are grouped when first seen, because the complete data set may have been too lengthy to report and somebody has summarized the data for easier understanding by the reader. The grouped data might have the form shown in Figure 3.3 for the 40 test scores in Problem Setting 3.1.

How can the sample mean be found *when only the grouped results in Figure 3.3 are known?* The answer is that the *exact value* for the sample mean cannot be found. However, *an approximate value* can be found by assuming that all of the observations in an interval happened to fall right at the midpoint of that interval, called the **class mark.** The midpoint is selected because it is somewhat representative of all possible values in the interval. For instance, the seven As could all be taken at the midpoint of that interval, 95.5, which is found by averaging the largest and the smallest possible numbers in the interval. Thus

Grade	Interval	Number of Employees
A	91–100	7
B	81–90	13
C	71–80	11
D	61–70	5
F	≤60	4

FIGURE 3.3
THE TEST SCORES PRESENTED AS
GROUPED DATA.

the As convert to seven 95.5s, the Bs become 13 scores of 85.5 each, the Cs are 11 values of 75.5, and the Ds convert to five 65.5s. What should be done about the Fs? If the standard procedure is used, the largest, 60, and the smallest, presumably zero, scores would be averaged to get a *class mark* (midpoint) of 30. However, intuition suggests that the four Fs are more likely to be closer to 60 than to zero, and in fact it might be more accurate to treat that class as being from 51 to 60, with a midpoint at 55.5. Remember, it is assumed that the only information available is the grouped data as it appears in Figure 3.3, and the actual test scores are not available for examination.

The equation for finding the **mean from grouped data** is as follows.

$$\bar{X} = \frac{1}{n} \sum_{i=1}^{k} f_i m_i \tag{3.4}$$

FINDING THE MEAN FROM GROUPED DATA

where the sample size is

$$n = \sum_{i=1}^{k} f_i$$

and

k = the number of intervals
f_i = the frequency in the ith interval
m_i = the midpoint of the ith interval

The end classes in grouped data often pose difficult problems for estimating the sample mean. There is no substitute for good judgment, and the judgment of the person doing the analysis is needed in situations like this. The only guideline offered here is: Always tell the reader the basis for your calculations, including your reasoning, if possible.

The midpoint in each class interval will be selected as a basis for the calculations, except for the Fs, which will be taken to equal 55.5, the midpoint of the interval from 51 to 60. The sample mean is found approximately by Eq. 3.4. Figure 3.4 illustrates the calculations.

(1) Grade	(2) Interval	(3) Frequency (f_i)	(4) Midpoint (m_i)	(5) (f_i)(m_i)
A	91–100	7	95.5	668.5
B	81–90	13	85.5	1111.5
C	71–80	11	75.5	830.5
D	61–70	5	65.5	327.5
F	60 or less	4	55.5	222.0
		40		3160.0

$$\sum_{i=1}^{40} X_i \approx \sum_{i=1}^{5} f_i m_i = 3160 \qquad \overline{X} \approx \frac{1}{40}(3160) = 79.00$$

FIGURE 3.4
THE WORKSHEET FOR APPROXIMATING $\overline{X}$ FROM GROUPED DATA.

The term $\sum X_i$ is approximated by $\sum f_i m_i$ at the bottom of Column (5). The resulting sample mean is 79.00, slightly larger than the more accurate value 77.55, which was obtained from the original observations. This method of approximating the sample mean is not completely satisfactory because of the use of interval midpoints or other numbers instead of the original observations. But when the original observations are not available, this may be the best substitute.

EXERCISES

3.1 Pharmaceutical companies use large numbers of mice in experiments designed to screen new chemical compounds for effectiveness in treating some types of disease in developing a new drug to market. These experiments are carefully run and monitored and the mice are routinely weighed as part of the quality control process. Find the sample mean of the following data that represent the weight in grams of five randomly selected three-week-old laboratory mice.

15.7   14.8   13.7   16.1   15.2

3.2 Production figures for a company for 12 consecutive quarters are given below. Find the average production for these 12 quarters.

Year	Quarter	Total Production (Number of Units)
1978	1	1048
	2	964
	3	833
	4	1265
1979	1	1117
	2	848
	3	769
	4	1306

Year	Quarter	Total Production (Number of Units)
1980	1	1082
	2	968
	3	812
	4	1240

3.3 A random sample of homes in a neighborhood revealed the length of time that each resident had lived in that particular neighborhood. These data were: 4 years, 8 months, 2 years, 27 years, and 14 months. Find the sample mean for the number of years lived in the neighborhood.

3.4 A sample of six employees selected at random reported the following monthly rents (in dollars): 245, 195, 250, 280, 215, 345. Find the sample mean for these data.

3.5 A class was given a test with 10 questions on it. Two students got all 10 correct, eight got 9 correct, and so on as in the following table.

Number Correct	Number of Students
10	2
9	8
8	16
7	12
6	5
5 or less	0

Find the mean number correct first using Eq. 3.2 and then Eq. 3.3. Compare your answers.

3.6 Find the average mileage based on tread-life tests of a random sample of 100 tires.

Number of Miles	Frequency
10,000–15,000	6
15,000–20,000	21
20,000–25,000	38
25,000–30,000	19
30,000–35,000	16

3.7 Data on the number of items purchased by randomly selected shoppers in a discount store are given below. Find the mean number of items purchased by shoppers.

Number of Items	Frequency	Number of Items	Frequency
1	4	14	7
2	1	15	4
3	9	16	2
4	7	17	1
5	6	18	2

(Continued on next page)

Number of Items	Frequency	Number of Items	Frequency
6	10	21	1
7	6	22	1
8	5	23	1
9	12	25	1
10	7	26	1
11	3	29	1
12	4	36	1
13	4	42	1

3.8 Six sales slips selected randomly from the sales slips for the morning business in a neighborhood hardware store are as follows: $1.36, $18.40, $183.79, $2.65, $1.95, and $7.16. Find the sample mean for these sales data. Do you think the sample mean gives a meaningful measure of the center of this sample?

3.9 A frequency table of salary levels for a random sample of all full-time employees at a local utility company furnished the following grouped data.

Salary Group	Number of Employees
$ 0– 5,000	0
5,000–10,000	27
10,000–15,000	73
15,000–20,000	48
20,000–25,000	13
25,000–30,000	8
30,000–35,000	1

Find the sample mean salary for these data.

3.10 In a manufacturing plant 50-foot rolls of steel are used in the manufacturing process, but the portion at the end of the roll cannot be used and is thrown away. Seven rolls are selected at random, and the unused amounts are 1.4, 2.1, 0.7, 0.9, 1.6, 1.8, and 1.3 feet, respectively. What is the average unused amount per roll? If 1000 rolls are used each week, what is your estimate of the percentage waste in one month?

3.2
THE STANDARD DEVIATION FOR THE SAMPLE

A MEASURE OF SPREAD

The sample mean is frequently the first statistic computed on a set of sample data, as it provides a measure of location for the sample. Another statistic usually computed along with the sample mean is the **sample standard deviation,** which provides a measure of "spread" or variability in the sample. For example, if all of the sample values are nearly equal to each other, the spread or variability is quite small and the standard deviation will be close to zero as an indication of this situation. On the other hand, if the sample values are spread out in a quite diverse manner, the sample standard deviation will tend to be large to reflect this different situation. Methods are presented in this section for quantifying the measure of spread for the sample data.

CALCULATION OF THE SAMPLE STANDARD DEVIATION

The computation of the sample standard deviation is given first, followed by its interpretation. Since the sample standard deviation is a measure of the variability in the sample data, it seems reasonable to start by noting how far each observation lies from the mean of the sample. The expression $X_i - \overline{X}$ represents how far "out" each observation lies. These quantities are squared to make all of them positive, and the squared values are averaged to find the **sample variance.** It is customary to use $n - 1$ as the divisor in finding the sample variance instead of n, because n tends to underestimate the population variance (i.e., has a negative bias), and dividing by $n - 1$ eliminates this bias.

The square root of the sample variance is the **sample standard deviation** and is denoted by s. Use of the square root allows the sample standard deviation to be expressed in the same units as the original observations. The sample standard deviation is a measure of distance; in particular it is a measure of the distance from the observations in a sample to the middle of that sample. Consider the sample values 3, 4, 5, 6, and 7, which have $\overline{X} = 5$ and $s = 1.58$, whereas a more spread-out group of numbers, 1, 2, 5, 8, 9 has the same sample mean $\overline{X} = 5$, but has a larger sample standard deviation, $s = 3.54$.

One formula for the sample standard deviation is

$$s = \left[\frac{1}{n-1} \sum_{i=1}^{n} (X_i - \overline{X})^2 \right]^{1/2} \tag{3.5}$$

If there are several samples and therefore several standard deviations involved in a discussion, one may be denoted by s_x, another by s_y, and a third by s_z, or s_1, s_2, and s_3 may be used as notation.

SAMPLE STANDARD DEVIATION
  The **sample standard deviation** is a statistic that measures the amount of variability within a sample. It may be computed using either Eq. 3.5 or Eq. 3.6.

SAMPLE VARIANCE
  The **sample variance** s^2 is the square of the sample standard deviation s.

Another formula for the sample standard deviation is generally used because it is easier to compute. It is exactly equivalent to the one just given.

$$s = \left\{ \frac{1}{n-1} \left[\sum_{i=1}^{n} X_i^2 - \left(\sum_{i=1}^{n} X_i \right)^2 \Big/ n \right] \right\}^{1/2} \tag{3.6}$$

To use Eq. 3.6 each observation X_i is squared and then summed to get ΣX_i^2. A common error occurs if the X's are first summed and then squared, the reverse of the correct order. Equation 3.6, though usually easier to use, is more susceptible to roundoff error than Eq. 3.5, so the user of Eq. 3.6 is cautioned to carry many significant digits in the calculations, especially if $\overline{X}$ is very large compared with the estimated value of s.

EXAMPLE

Suppose your telephone bill for the last five months is given in column (2) of the following table.

(1)	(2)	(3)	(4)	(5)
(i) Month	Bill (X_i)	X_i^2	$X_i - \overline{X}$	$(X_i - \overline{X})^2$
(1) January	$12.37	153.02	-0.85	0.7225
(2) February	10.43	108.78	-2.79	7.7841
(3) March	13.66	186.60	-0.44	0.1936
(4) April	18.14	329.06	4.92	24.2064
(5) May	11.51	132.48	-1.71	2.9241
Totals	66.11	909.94	0.01 (roundoff error)	35.8307

$$\overline{X} = \frac{1}{5}(66.11) = \$13.22$$

$$s = \left\{ \frac{1}{4}\left[909.94 - \frac{(66.11)^2}{5} \right]\right\}^{1/2} = [8.9584]^{1/2} = \$2.99$$

or

$$s = \left[\frac{1}{4}(35.8307) \right]^{1/2} = [8.9577]^{1/2} = \$2.99$$

The sample mean is $13.22. Actually the sample mean is $13.222, but it has been rounded off to the nearest cent.

The total for Column (3) is used in Eq. 3.6 along with the total for Column (2), to get a standard deviation of $2.99, which is also rounded off to the nearest cent. This standard deviation represents, in a sense, the average amount of variability of these five observations.

Columns (4) and (5) are not needed, except to illustrate how Eq. 3.5 is used. The total for Column (4) will always be zero, except for roundoff error caused by using $13.22 for $\overline{X}$ instead of the more accurate value $13.222. The total for Column (5) is used in Eq. 3.5 to obtain a standard deviation of $2.99.

BACK TO THE TEST SCORES

The computations for the standard deviation of the 40 test scores given in Problem Setting 3.1 yield the following intermediate calculations.

$$\sum_{i=1}^{40} X_i^2 = [(77)^2 + (68)^2 + (86)^2 + \cdots + (65)^2]$$

$$= 247{,}714$$

Equation 3.6 for the standard deviation gives

$$s = \left\{ \frac{1}{39} \left[247{,}714 - \frac{(3102)^2}{40} \right] \right\}^{1/2}$$
$$= [183.4333]^{1/2}$$
$$= 13.54$$

A USEFUL RULE OF THUMB

The sample standard deviation is a measure of how widely scattered the observations are from the sample mean. As a rule of thumb, often about two-thirds of the sample observations are within one standard deviation of the mean. That is, for these 40 test scores about 26.67 observations (two-thirds of 40) can be anticipated as being between 64.01 and 91.09. Those values are obtained from the calculations

$$\overline{X} - s = 77.55 - 13.54 = 64.01$$

and

$$\overline{X} + s = 77.55 + 13.54 = 91.09$$

Every test score that is within one standard deviation of the mean appears with an asterisk in the following diagram.

77*	68*	86*	84*	95	98	87*	71*
84*	92	96	83*	62	83*	81*	85*
91*	74*	61	52	83*	73*	85*	78*
50	81*	37	60	85*	100	79*	81*
75*	92	80*	75*	78*	71*	64	65*

There are 27 observations between 64.01 and 91.09, in close agreement with the rule of thumb prediction.

Another rule of thumb says that about 95% of the observations are within two standard deviations of the mean. Since $\overline{X} - 2s = 50.47$ and $\overline{X} + 2s = 104.63$, it can be seen that all but two observations are between 50.47 and 104.63, or within two standard deviations of $\overline{X}$, namely the two test scores 50 and 37. The 38 out of 40 test scores that are between 50.47 and 104.63 happen to be precisely 95% of the total observations.

An approximate rule of thumb, applicable to many samples, is that

RULE OF
THUMB ON
SPREAD

1. about two-thirds of the sample observations are within one sample standard deviation of the sample mean (i.e., between $\overline{X} - s$ and $\overline{X} + s$), and
2. about 95% of the observations are within two sample standard deviations of the sample mean (i.e., between $\overline{X} - 2s$ and $\overline{X} + 2s$).

Although the results agreed extremely well with the two-thirds rule and the 95% rule, these rules should be considered merely rules of thumb. These rules apply better to some types of data sets than to others. If the histogram has a bell-shaped appearance, these rules are quite accurate, but they are still surprisingly close for many other types of data sets.

A QUICK CHECK OF THE COMPUTATIONS

Since it is easy to make an error somewhere in the process, the computations should always be checked against the sample, to see if $\overline{X}$ is near the middle of the observations and to see if about two-thirds of the observations are between $(\overline{X} - s)$ and $(\overline{X} + s)$. This simple test should reveal whether $\overline{X}$ or s contains a gross error in calculation.

THE SAMPLE COEFFICIENT OF VARIATION

The **coefficient of variation** is a single statistic that combines both $\overline{X}$ and s. It provides a measure of the standard deviation in units of the sample mean. Thus it has the advantage of being free of dimensions (such as dollars or feet). Also a sample with larger numbers tends to have more variability than a sample with smaller numbers. The weights of a sample of men, for example, tend to have a larger standard deviation than the weights of a sample of women, simply because the weights tend to be larger for men. Division by $\overline{X}$ tends to equalize the measures of spread in the two samples, so that more meaningful comparisons of their *relative* spread can be made.

The ratio of the standard deviation to the mean is called the **sample coefficient of variation,**

THE SAMPLE
COEFFICIENT
OF VARIATION

$$CV = \frac{s}{\overline{X}}$$ (3.7)

and is often expressed as a percent, CV × 100%.

This descriptive statistic is used only when all of the observations are positive (so $\overline{X}$ cannot be zero) as a single statistic that combines both s and $\overline{X}$. The coefficient of variation for the test score data is CV = 13.54/77.55 = .175.

STANDARDIZED OBSERVATIONS, OR z SCORES

An observation by itself does not indicate as much information as when the mean and the standard deviation of the entire sample are also given. Then the observation can be placed relative to the other observations in the sample. There is a simple method for conveying all of this information in a single number. It involves *standardizing* the observations. Standardizing involves first subtracting the sample mean from each observation, and then dividing by the sample standard deviation. The resulting numbers are called **z scores.**

$$Z = \frac{X - \overline{X}}{s} \qquad (3.8)$$

If X is greater than $\overline{X}$, then Z is positive. If X is less than $\overline{X}$, then Z is negative. The size of the z score indicates how many standard deviations separate X and $\overline{X}$. If the z score is greater than 2.0, for instance, this indicates that the X value is more than two standard deviations above the mean. The rule of thumb would suggest that the X value is in the outer 5% of the sample, or in the upper 2.5% of the sample.

AN APPLICATION OF z SCORES

Suppose one of your instructors said that every person in the class could drop his or her lowest hour-exam score from inclusion in the course grade. Also, one of the exams was unusually difficult and the class mean on that exam was low, only 50, with a standard deviation of 10. However, you had studied extra for the exam and got a score of 80, the highest grade in the class. Even though that 80 may be your lowest exam grade, it may represent your best effort. An examination of z scores will show this

$$Z = \frac{80 - 50}{10} = 3.0$$

which is your highest z score. You may want to try to convince your instructor that it would be more fair to figure grades by averaging z scores and dropping the lowest z score when computing averages. In this way, the grade being dropped would depend more on the individual's performance relative to the rest of the class, and less on the level of difficulty of the exam. This represents the philosophy behind using z scores instead of "raw scores," and should help to explain the popularity of z scores in many applied sciences.

MORE ABOUT z SCORES

Some interesting properties of z scores are listed below.

1. The sample mean of the z scores is always zero. This is because the sample mean $\overline{X}$ was first subtracted from the raw scores X_i.

2. The sample standard deviation of the z scores is always one. This is because the raw scores are divided by the sample standard deviation s.

3. The empirical distribution function of the z scores looks exactly like the empirical distribution function of the original sample, except for new "standardized" numbers along the horizontal axis.

THE SAMPLE STANDARD DEVIATION
FOR REPEATED OBSERVATIONS

As with the sample mean, the amount of arithmetic can be shortened in calculating the sample standard deviation when individual observations are repeated many times by noting the frequency f_i with which each observation occurs. Equations 3.9 and 3.10 provide the necessary formulas for making the calculation.

Equation 3.9 requires carrying only 3 or 4 significant digits in the calculations, and is more convenient to use because of this when calculations are done by hand. If a calculator is being used, Eq. 3.10 is more convenient to use, but at least 6 or 7 significant digits should be used in the calculations in order to maintain three- or four-digit accuracy in the result.

The **sample standard deviation for repeated observations** may be found from either

$$s = \left[\frac{1}{n-1} \sum_{i=1}^{k} f_i(x_i - \overline{X})^2 \right]^{1/2} \tag{3.9}$$

or

SAMPLE
STANDARD
DEVIATION
FOR REPEATED
OBSERVATIONS

$$s = \left\{ \frac{1}{n-1} \left[\sum_{i=1}^{k} f_i(x_i)^2 - \left(\sum_{i=1}^{k} f_i x_i \right)^2 \Big/ n \right] \right\}^{1/2} \tag{3.10}$$

where

$$n = \sum_{i=1}^{k} f_i = \text{the total number of observations}$$

k = the number of unique observations

f_i = the frequency of the ith unique observation

x_i = the ith unique observation

For example, suppose a sample of size 20 contains the observation 2.1 six times, the observation 1.8 four times, the observation 3.2 five times, the observation 3.5 four times, and the observation 4.2 once. Use of Eq. 3.10 to find the sample standard deviation produces the following result.

$$\sum f_i x_i^2 = 6(2.1)^2 + 4(1.8)^2 + 5(3.2)^2 + 4(3.5)^2 + 1(4.2)^2 = 157.26$$

$$\sum f_i x_i = 6(2.1) + 4(1.8) + 5(3.2) + 4(3.5) + 1(4.2) = 54.00$$

$$\sum f_i = 6 + 4 + 5 + 4 + 1 = 20$$

$$s = \{[157.26 - (54)^2/20]/19\}^{1/2} = .7766$$

Thus, each different value x_i in the sample is simply multiplied by its frequency of occurrence f_i to shorten the amount of arithmetic necessary to find the sum of the observations and the sum of the squares of the observations. In the case where each $f_i = 1$, Eqs. 3.9 and 3.10 are the same as Eqs. 3.5 and 3.6, respectively.

CALCULATING THE SAMPLE STANDARD DEVIATION FROM GROUPED DATA

If the sample data are already grouped into intervals and the individual observations are not available, the exact sample standard deviation cannot be obtained. However, an approximate value can be obtained by treating all of the observations in an interval as if they were equal to the midpoint of that interval, as was done for obtaining an approximate value of the mean in the previous section. This method will now be illustrated using the same example that was given in Figure 3.3 with the aid of Eq. 3.12.

The grouped data are given again in Figure 3.5 with the calculations needed to find $\overline{X}$ as they were presented in Figure 3.4, plus an additional column that is needed to compute the sample standard deviation from grouped data. The term $\sum X_i^2$, needed to find s, is approximated by $\sum f_i(m_i)^2$, the total for the sixth column. This approximation results in a **standard deviation for grouped data** equal to 12.10, which is slightly smaller than the exact standard deviation for the individual test scores, 13.54, computed earlier in this section. In general, the standard deviation of the grouped data may be larger or smaller than the standard deviation of the original observations. As was stated earlier for $\overline{X}$

Grade	Interval	Frequency (f_i)	Midpoint (m_i)	(f_i)(m_i)	$f_i(m_i)^2$
A	91–100	7	95.5	668.5	63,841.75
B	81–90	13	85.5	1111.5	95,033.25
C	71–80	11	75.5	830.5	62,702.75
D	61–70	5	65.5	327.5	21,451.25
F	≤60	4	55.5	222.0	12,321.00
		40		3160.0	255,350.00

$$\sum X_i \approx \sum f_i m_i = 3160 \qquad \overline{X} \approx \frac{1}{40}(3160) = 79.00$$

$$\sum X_i^2 \approx \sum f_i(m_i)^2 = 255,350$$

$$s = \left\{\frac{1}{39}\left[255,350 - \frac{(3160)^2}{40}\right]\right\}^{1/2}$$

$$= 12.10$$

FIGURE 3.5
THE WORKSHEET FOR APPROXIMATING $\overline{X}$ AND s FROM GROUPED DATA.

computed from the grouped data, if only the grouped data are available, this may be the best method available for estimating the sample standard deviation.

The **sample standard deviation for grouped data** may be found from either

$$s = \left[\frac{1}{n-1} \sum_{i=1}^{k} f_i (m_i - \overline{X})^2 \right]^{1/2}$$ (3.11)

or

SAMPLE STANDARD DEVIATION FOR GROUPED DATA

$$s = \left\{ \frac{1}{n-1} \left[\sum_{i=1}^{k} f_i (m_i)^2 - \left(\sum_{i=1}^{k} f_i m_i \right)^2 \middle/ n \right] \right\}^{1/2}$$ (3.12)

where

$n = \sum_{i=1}^{k} f_i =$ the total number of observations

$k =$ the number of intervals

$f_i =$ the frequency in the ith interval

$m_i =$ the midpoint of the ith interval

EXERCISES

3.11 Given the following set of 10 sample observations:

         1   4   4   10   10   10   13   13   16   19

(a) Find ΣX_i and $\overline{X}$.

(b) Find $(\Sigma X_i)^2$.

(c) Find ΣX_i^2.

(d) Explain why your answer to part (b) is not the same as your answer to part (c).

(e) Find $\Sigma (X_i - \overline{X})^2$.

(f) Find $\Sigma X_i^2 - (\Sigma X_i)^2/n$.

(g) Explain why your answers to parts (e) and (f) are identical.

(h) Find s using Eq. 3.5.

(i) Find s using Eq. 3.6.

3.12 Convert the sample data given in Exercise 3.11 to z scores. Compute the mean and standard deviation for the z scores. Do your results agree with the properties of z scores given in this section?

3.13 Compute the standard deviation for the data originally given in Exercise 3.1 using both formulas for ungrouped data (Eq. 3.5 and Eq. 3.6). Compare your answers for the two formulas. The data are repeated here as follows:

         15.7   14.8   13.7   16.1   15.2

3.14 Calculate the coefficient of variation for the sample data given in the previous exercise.

3.15 Data are presented below on the number of customers for 40 business days by a small business. Calculate the mean and standard deviation for these data. How many observations do you find within one standard deviation of the mean? How many observations do you find within two standard deviations of the mean? How do these results compare with the rules of thumb given in this section? ($\Sigma X_i = 6669$, $\Sigma X_i^2 = 1,138,039$).

186	121	143	159	180
125	178	215	166	158
187	148	151	162	153
128	133	188	170	168
153	123	134	184	208
178	186	162	202	160
201	200	175	126	150
174	218	165	166	185

3.16 The data given in Exercise 3.2 are repeated below. Calculate z scores for these data. What fact do the z scores make clear about the production in the second and third quarters of each year?

Year	Quarter	Total Production (Number of Units)
1978	1	1048
	2	964
	3	833
	4	1265
1979	1	1117
	2	848
	3	769
	4	1306
1980	1	1082
	2	968
	3	812
	4	1240

3.17 Grouped data on tread-life tests for 100 randomly selected tires are presented below. Use the formula for grouped data to calculate the sample standard deviation for these data.

Number of Miles	Frequency
10,000–15,000	6
15,000–20,000	21
20,000–25,000	38
25,000–30,000	19
30,000–35,000	16

3.18 The data given in Exercise 3.5 are repeated below. Compute the standard deviation for these data using the formula for repeated observations.

Number Correct	Number of Students
10	2
9	8
8	16
7	12
6	5
5 or less	0

3.19 The data given in Exercise 3.9 are repeated below. Find the standard deviation for these data.

Salary Group	Number of Employees
$ 0– 5,000	0
5,000–10,000	27
10,000–15,000	73
15,000–20,000	48
20,000–25,000	13
25,000–30,000	8
30,000–35,000	1

3.3
MODES, QUANTILES, PROPORTIONS, AND BOXPLOTS

PROBLEM SETTING 3.2

The owner–manager of a shoe store is considering ordering a new style of ladies' fashion boot, but she is concerned about overstocking since she is unsure about how popular that style of boot will be. To get some idea of what sizes to order, she examines sales records that indicate foot sizes for the 20 most recent sales of women's shoes. Those foot sizes are given in Figure 3.6. As she sorts through the records containing these foot sizes, she automatically keeps a tally by first writing down the foot sizes she knows to be reasonable sizes and then placing a check mark beside each size as it occurs. When she has gone through the records, her final tally looks like Figure 3.7.

THE SAMPLE MODE

The store owner has obtained the frequency distribution of a sample of size 20. The most frequently occurring size, called the **sample mode,** is size $6\frac{1}{2}$. If she were to order only one pair of this new style boot, the owner–manager would probably want to order the **modal** size, which is $6\frac{1}{2}$ in this sample, because past records indicate that it is the size most likely to fit a customer.

$$
\begin{array}{ccccc}
9\tfrac{1}{2} & 7 & 5\tfrac{1}{2} & 7\tfrac{1}{2} & 9 \\
5\tfrac{1}{2} & 5 & 7 & 5\tfrac{1}{2} & 8 \\
7 & 6\tfrac{1}{2} & 6\tfrac{1}{2} & 6 & 6\tfrac{1}{2} \\
6\tfrac{1}{2} & 9 & 7\tfrac{1}{2} & 6\tfrac{1}{2} & 6
\end{array}
$$

FIGURE 3.6
TWENTY
HYPOTHETICAL
WOMEN'S SHOE SIZES.

SAMPLE MODE   The **sample mode** is the observation that occurs the most frequently.
The **modal class** in grouped data is the class with the greatest frequency.

A sample mode is usually considered to be the most frequently occurring value. However, sometimes a value that occurs more frequently than the values on either side of it is also called a mode. Thus size $5\tfrac{1}{2}$, which occurs more frequently than the sizes 5 or 6 on either side of it in the sample, could be considered to be a mode. It would be called a **secondary mode** because its frequency is less than the frequency at size $6\tfrac{1}{2}$, which is the **primary mode.** Do you see another secondary mode in the sample?

The term **mode** is also used with histograms and other graphs. The mode of the histogram is the highest point in the graph. Graphs can also have secondary modes. A graph with one mode is called **unimodal,** whereas a graph with two modes is called **bimodal.**

THE SAMPLE MEDIAN

The owner–manager is not interested in ordering just one pair, however. She is more interested in ordering one dozen pairs, of various sizes. That way she will have a better selection for her regular customers; she can always sell the

Size	Tally	Frequency
5	√	1
$5\tfrac{1}{2}$	√√√	3
6	√√	2
$6\tfrac{1}{2}$	√√√√√	5
7	√√√	3
$7\tfrac{1}{2}$	√√	2
8	√	1
$8\tfrac{1}{2}$		0
9	√√	2
$9\tfrac{1}{2}$	√	1
10		0

FIGURE 3.7
A FREQUENCY
DISTRIBUTION OF
SHOE SIZES.

leftover sizes in a clearance sale; and besides that she gets a substantial price break when she orders in lots of a dozen.

Which sizes should she order? She feels safest in ordering boots that are somewhere near the middle of the distribution of sizes. The exact middle value, when the observations are arranged from smallest to largest, is called the **sample median.** In this case there is no exact middle value, since there is an even number of observations in the sample. When the sample size is an even number, the two middle observations are averaged to get the sample median. Here, this is simple to do because both middle observations are $6\frac{1}{2}$. Thus the median size is $X_{.5} = 6\frac{1}{2}$.

SAMPLE MEDIAN
The **sample median** $X_{.5}$ is the middle observation when the sample is ordered from smallest to largest. If the number of observations is even, the median is the average of the two middle observations.

The sample median is a useful statistic for locating the middle of a sample. It will always be one of the observations in the sample or halfway between two observations, unlike the sample mean, which will rarely equal one of the observations.

For example, suppose you have five uncles, and the number of children each has is as follows: 1, 0, 0, 4, 3. To find the sample median these numbers are arranged from smallest to largest: 0, 0, 1, 3, 4. The median is 1 child. The sample mean is 1.6 children, larger than the median. If your father is counted along with the five uncles, and your father has six children, the sample median becomes $(1 + 3)/2 = 2$ children. One half of the families have more than two children and one half have less. The sample mean is 2.3, which coincidentally has the same property in this sample.

INSENSITIVITY OF THE SAMPLE MEDIAN TO OUTLIERS

The sample median is not as sensitive as the sample mean to observations that are much larger or much smaller than the rest of the sample. That is, one very large observation, when averaged in with the other observations, may increase the sample mean until it no longer is a reasonable measure of the middle of the sample, but that one large observation will not change the sample median as much. If your family, instead of having six children as in the previous example, had 16 children, the sample median would remain unchanged. One half of the families would still have more than two children and one half would have fewer than two children. But the mean is now increased to 4.0. Only one family is larger than the mean and that family is your family, which produced the "outlying" observation 16.

Sometimes, however, the sample mean is preferred because it does reflect the total size of all of the observations. The choice of whether to use the mean or the median, or perhaps both, depends largely on personal preference and what happens to be customary in the particular field of application.

THE MEDIAN VERSUS THE MEAN

For some sets of data or populations there can be a problem in computing the mean. For example, suppose it is desired to compute the mean anticipated price–earnings ratio (P/E) for the more than 5300 common and preferred stocks listed in *Standard and Poor's Stock Guide*. Some of the listed stocks are not expected to earn anything in the coming year, or they may be expected to lose money. This means that P/E involves dividing by zero, which is not possible, or dividing by a negative number, which gives a meaningless price–earnings ratio. In those cases, no price–earnings ratio is listed in the *Standard and Poor's Stock Guide*. This causes problems in calculating the mean. This is a typical problem one encounters when collecting data. In this situation it would be advisable to use the median to locate the "center" instead of the mean, and to consider these troublesome values of P/E to be simply "very large," in keeping with the idea that the price of these stocks is very large relative to the earnings per share expected. As long as only a small percentage of the population consists of these numbers, the median can still be found without any difficulty.

EXAMPLE

Several years ago the U.S. government started a policy of reporting median incomes rather than mean incomes. A median income of $20,000 implies that one half of the incomes are less than $20,000 and one half are greater than $20,000. If your income is $20,000 you know exactly how you rank with the rest of the population.

A mean income of $20,000 is more difficult to interpret, however, since a few extremely large salaries may be responsible for most of the $20,000 average, whereas the great majority of salaries may be considerably less than $20,000.

SKEWNESS

The bar graph for the shoe sizes is given in Figure 3.8. It is said to be **skewed to the right** because the bar graph extends farther to the right of the mode than it does to the left. The modal size is $6\frac{1}{2}$, and there are six sizes to the right of the modal size, but only three to the left.

SKEWNESS A histogram or bar graph is said to be **skewed to the right** if it extends farther to the right of the mode than it does to the left. It is said to be **skewed to the left** if the reverse is true.

When histograms and bar graphs are skewed to the right, the sample mean is typically larger than the sample median, and both are typically larger than the mode. This rule holds true in most cases but not always, as is illustrated

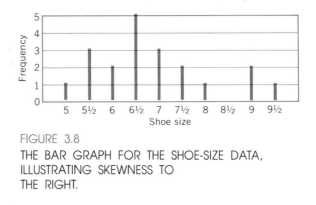

FIGURE 3.8
THE BAR GRAPH FOR THE SHOE-SIZE DATA,
ILLUSTRATING SKEWNESS TO
THE RIGHT.

in this data set where the mode and the median are both equal to $6\frac{1}{2}$. The sample mean, incidentally, is 6.88, which does follow this rule, since it is larger than the median. For graphs that are skewed to the left the opposite relationship usually holds.

THE SAMPLE PROPORTION

The **sample proportion** is the fraction of the sample that meets some stated criterion. If the store manager orders only boot sizes between size $5\frac{1}{2}$ and size 9, but not including those sizes, she would be ordering sizes belonging to only 65% of the sample values. This is because only 13 of the 20 sample values were greater than $5\frac{1}{2}$ and less than 9, and $13/20 = 65\%$. The *sample proportion* between sizes $5\frac{1}{2}$ and 9 is 65%. The notation for sample proportion is $\hat{p}$. If the sizes $5\frac{1}{2}$ and 9 are included in the calculations, the sample proportion becomes $\hat{p} = 18/20 = 90\%$.

SAMPLE
PROPORTION
The **sample proportion** $\hat{p}$ is the fraction of the sample that meets some stated criterion.

Sample proportions often may be read from the e.d.f. To find the sample proportion less than or equal to any number, first find that number on the horizontal scale, then read up to the e.d.f. to the top of the step if a step is involved. The ordinate of that point, read from the vertical scale, is the desired proportion.

EXAMPLE

The proportion of sample shoe sizes less than or equal to size $5\frac{1}{2}$ is the height at the top of the step above $5\frac{1}{2}$ in Figure 3.9, which is $\hat{p} = .2$. The proportion less than *but not including* $5\frac{1}{2}$ is read from the bottom of the step above $5\frac{1}{2}$, which is $\hat{p} = .05$. The height of the step itself, .15, is the proportion of the sample that equals size $5\frac{1}{2}$. Proportions of the sample *greater than* given values may be found by subtracting from 1.0 the sample proportion less than or equal to that number. The sample proportion greater than size $5\frac{1}{2}$ is $1.0 - .2$, which is 80%.

SAMPLE PERCENTILES

Can you remember taking a standard achievement exam such as the ACT or SAT exams? In addition to your exam score you were probably given a **percentile.** If you scored in the eighty-seventh percentile, your counselor may have explained that it meant 87% of the people taking the exam got scores lower than or equal to yours. **Percentiles** are also statistics that locate different portions of the sample. Percentiles generally range from the *first* percentile to the *ninety-ninth* percentile, although fractional percentiles such as the 87.3 percentile are sometimes used. The fiftieth percentile is the median. If the shoe-store manager wanted to order boots from the middle part of the distribution of sizes, she might have found the tenth percentile and the ninetieth percentile from her sample, and ordered boot sizes only between these two values. In this way she would not have been ordering any very small or very large boots that could be hard to sell.

SAMPLE PERCENTILE   If r is any whole number from 1 to 99, then the **rth percentile** $X_{r/100}$ for a sample is any value that has r% or less of the observations less than that value, and $(100 - r)$% or less of the observations greater than that value.

OTHER SAMPLE QUANTILES

Some percentiles are used more than others and are given their own names. The twenty-fifth percentile is called the **lower quartile** and the seventy-fifth is called the **upper quartile.** The tenth, twentieth, thirtieth, etc., percentiles are called **deciles.**

SAMPLE QUARTILES   The twenty-fifth percentile $X_{.25}$ is called the lower quartile or **first quartile,** and the seventy-fifth percentile $X_{.75}$ is called the upper quartile or **third quartile.**

These percentiles, deciles, and quartiles are all special cases of **quantiles,** the more general term. Quantiles are associated with a number between 0 and 1. For example the .50 quantile is the fiftieth percentile, the fifth decile, or the median. The one-third quantile is between the thirty-third and thirty-fourth percentiles.

SAMPLE QUANTILE   A **qth sample quantile** X_q, where q is a quantity between 0 and 1, is a value that has a proportion q or less of the sample observations less than X_q and a proportion $(1 - q)$ or less of the sample observations greater than X_q.

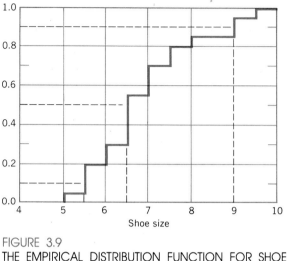

FIGURE 3.9
THE EMPIRICAL DISTRIBUTION FUNCTION FOR SHOE
SIZES.

One way to find the qth sample quantile is to multiply q by n, the sample size. Then round $q \cdot n$ upwards to the next higher integer, and find the observation with that rank in the sample. That observation is the qth sample quantile. If $q \cdot n$ is an integer, then the sample quantile is the average of the value with that rank and the next value with the next higher rank.

An easier way to find sample quantiles is from the empirical distribution function. The e.d.f. for the 20 sample observations on shoe sizes is given in Figure 3.9. Any sample quantile may be found easily by locating the desired value of q on the vertical axis, moving horizontally to the e.d.f., and reading the corresponding quantile from the horizontal axis. The dotted lines in Figure 3.9 show the tenth percentile to be size $5\frac{1}{2}$, the median to be $6\frac{1}{2}$, and the ninetieth percentile to be size 9. Any number between $5\frac{1}{2}$ and 6 may be used to represent the twentieth percentile; however, it is recommended that you use the midpoint of these two numbers or 5.75 to represent the twentieth percentile. For data that are grouped, the ogive should be used instead of the e.d.f. to obtain approximate sample quantiles.

THE INTERQUARTILE RANGE AND THE SAMPLE RANGE

Just as the median may be used as a measure of location instead of the mean, quantiles may be used to measure the spread of the observations instead of the standard deviation. One way to do this is to use the **interquartile range,** which is the distance from the lower quartile to the upper quartile. From the e.d.f. in Figure 3.9 the upper sample quartile is easily found to be $X_{.75} = 7\frac{1}{2}$ and the lower sample quartile is $X_{.25} = 6$, so the sample interquartile range is $1\frac{1}{2}$ sizes. The interquartile range will usually tend to be a little larger than the standard deviation. However, the sample interquartile range will always be less than or equal to the **sample range,** the largest observation in the sample minus the smallest observation, which is $9\frac{1}{2} - 5 = 4\frac{1}{2}$, in the previous examples.

SAMPLE The **sample interquartile range** is the distance from the lower quartile to the
INTERQUARTILE   upper quartile, $X_{.75} - X_{.25}$.
RANGE AND The **sample range** is the largest observation in the sample minus the smallest
SAMPLE   observation in the sample.
RANGE

EXAMPLE

The sample median and the sample interquartile range can be compared with the sample mean and the sample standard deviation for the shoe-size data. Preliminary calculations give

$$n = 20 \qquad \sum_{i=1}^{20} X_i = 137.5 \quad \text{and} \quad \sum_{i=1}^{20} X_i^2 = 974.75$$

The sample mean is

$$\bar{X} = \frac{1}{n} \sum_{i=1}^{20} X_i = \frac{1}{20}(137.5) = 6.875$$

which is fairly close to the sample median

$$X_{.5} = 6.5$$

as would be expected because there are no *outliers* in the sample. Outliers affect the sample mean and sample standard deviation much more than they affect the sample median and the sample interquartile range. The sample standard deviation is

$$s = \left\{ \frac{1}{n-1} \left[\sum_{i=1}^{20} X_i^2 - \left(\sum_{i=1}^{20} X_i \right)^2 \middle/ n \right] \right\}^{1/2}$$

$$= \left\{ \frac{1}{19} \left[974.75 - \frac{(137.5)^2}{20} \right] \right\}^{1/2}$$

$$= \sqrt{1.5493}$$

$$= 1.2447$$

which, as expected, is slightly smaller than the sample interquartile range 1.5. Again, the presence of outliers can greatly affect the sample standard deviation, making it much larger than the sample interquartile range.

THE BOXPLOT FOR A SET OF SAMPLE DATA

Several methods were presented in Chapter 2 for the graphical representation of a set of sample data. An additional graphical method that *summarizes* a set

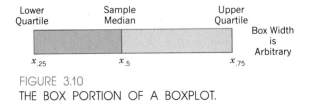

FIGURE 3.10

THE BOX PORTION OF A BOXPLOT.

of sample data is called the **boxplot.** The boxplot is based on quantiles. The construction of a boxplot is shown in general terms in Figures 3.10 and 3.11. In Figure 3.10 the "box" portion of the boxplot is formed by constructing a rectangle with one end at the lower quartile $(X_{.25})$ and the other end at the upper quartile $(X_{.75})$. Thus, the length of the box is the interquartile range (IQR $= X_{.75} - X_{.25}$). The width of the rectangle is sometimes arbitrary and as such has no interpretation, or the width sometimes represents the sample size. A line crosses the interior of the rectangle at the median $(X_{.5})$.

In Figure 3.11 lines are added that extend from either end of the box. The line on the left extends to the smallest element in the sample (X_{min}) if X_{min} is greater than $X_{.25} - 1.5$ IQR. Otherwise the line extends to $X_{.25} - 1.5$ IQR. The line on the right extends to the largest element in the sample (X_{max}) if X_{max} is less than $X_{.75} + 1.5$ IQR. Otherwise the line extends to $X_{.75} + 1.5$ IQR. Any sample values less than $X_{.25} - 1.5$ IQR or greater than $X_{.75} + 1.5$ IQR are called *outliers* and are marked with an "X" at their actual location. The next example illustrates the construction of a boxplot.

EXAMPLE

A boxplot is constructed for the data given in Figure 3.6 for Problem Setting 3.2. The following values have previously been found in this section for these data:

$$X_{.25} = 6, X_{.5} = 6.5, X_{.75} = 7.5, \text{ and IQR} = 7.5 - 6 = 1.5$$

Since $X_{min} = 5$ is greater than $X_{.25} - 1.5$ IQR $= 6 - 1.5(1.5) = 3.75$, the line on the left side of the box extends to $X_{min} = 5$. Also, since $X_{max} = 9.5$ is less than $X_{.75} + 1.5$ IQR $= 7.5 + 1.5(1.5) = 9.75$, the line on the right side of the box extends to $X_{max} = 9.5$. The completed boxplot as given in Figure 3.12 makes for an easy identification of $X_{min}, X_{.25}, X_{.5}, X_{.75}$, and X_{max}. Since there are no points beyond the endmarks, there are no outliers identified in the sample.

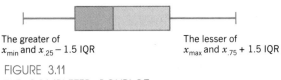

The greater of x_{min} and $x_{.25} - 1.5$ IQR   The lesser of x_{max} and $x_{.75} + 1.5$ IQR

FIGURE 3.11

THE COMPLETED BOXPLOT.

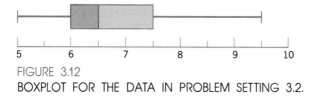

FIGURE 3.12
BOXPLOT FOR THE DATA IN PROBLEM SETTING 3.2.

In the next example a boxplot is constructed for a set of data containing two outliers.

EXAMPLE

A boxplot is constructed for the data summarized in a stem and leaf plot in Figure 2.11. For ease of reference these data are repeated as follows.

121, 125, 126, 128, 130, 131, 132, 133, 134, 135
137, 139, 141, 141, 144, 147, 148, 153, 205, 213

For these data X_{min} = 121, X_{max} = 213, $X_{.25}$ = (130 + 131)/2 = 130.5, $X_{.50}$ = (135 + 137)/2 = 136, $X_{.75}$ = (144 + 147)/2 = 145.5, and IQR = 145.5 − 130.5 = 15. Since X_{min} = 121 is larger than $X_{.25}$ − 1.5 IQR = 130.5 − 1.5(15) = 108, the line on the left of the box extends to 121. However, since X_{max} = 213 is larger than $X_{.75}$ + 1.5 IQR = 145.5 + 1.5(15) = 168, the line on the right side of the box extends only to 168. Thus, two sample values are identified as outliers; namely, X = 205 and 213. The completed boxplot is given in Figure 3.13. As in the previous example, various sample characteristics are easily identified from the boxplot, but it is clear that the sample contains two outliers as denoted by the x's at 205 and 213.

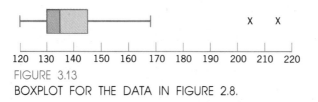

FIGURE 3.13
BOXPLOT FOR THE DATA IN FIGURE 2.8.

EXERCISES

3.20 The test scores given in Problem Setting 3.1 are repeated below. Find the mode for these data. Is the mode unique for these data or does more than one number satisfy the definition of the mode?

77	68	86	84	95	98	87	71
84	92	96	83	62	83	81	85
91	74	61	52	83	73	85	78
50	81	37	60	85	100	79	81
75	92	80	75	78	71	64	65

3.21 The data of Problem Setting 3.1 are summarized in a frequency distribution in Figure 3.3. What is the modal class for this summary?

3.22 The data of Problem Setting 3.1 are summarized in an e.d.f. in Figure 3.2. Refer to Figure 3.2 to find the following sample statistics.

(a) The median.

(b) The lower quartile.

(c) The upper quartile.

(d) The interquartile range.

(e) The ninth decile.

(f) The eightieth percentile.

3.23 Refer to your answers to Exercise 3.22 and construct a boxplot for the data given in Problem Setting 3.1. Are there any outliers in these data?

3.24 Find the median for each of the following data sets.

$$\text{Data set 1:} \quad 1 \quad 2 \quad 3 \quad 4 \quad 3 \quad 2 \quad 1$$
$$\text{Data set 2:} \quad 1 \quad 2 \quad 3 \quad 3 \quad 2 \quad 1$$

3.25 Find the mean and median for the following data sets. Why does the median change very little from set to set while the mean changes a great deal?

$$\text{Data set 1:} \quad 7.2 \quad 5.1 \quad 8.4 \quad 6.3 \quad 9.1$$
$$\text{Data set 2:} \quad 7.2 \quad 5.1 \quad 8.4 \quad 6.3 \quad 9.1 \quad 70.8$$

3.26 Data for the number of institutional shares held (in thousands) for 10 discount variety stores are given below. For these data find the median and the mean as well as the interquartile range and the standard deviation. Why do each of these respective measures differ so greatly from one another?

Company	Number of Institutional Shares Held (Thousands)
Danners Inc.	197
Duckwall-Alco	24
Fed-Mart	99
Grand Central	2
Hartfield-Zodys	377
K-Mart	53,170
Thrifty Corp.	544
Wal-Mart	3,452
Woolworth	6,587
Zayre	632

3.27 Refer to your answers to Exercise 3.26 and construct a boxplot for the data in Exercise 3.26. Do these data contain any outliers?

3.28 Find the modal class for the data in Exercise 3.9.

3.4

REVIEW EXERCISES

3.29 Data for the number of hours worked by 50 employees in one week are given below. Calculate the sample mean and sample standard deviation for these data. ($\Sigma X = 2506$, $\Sigma X^2 = 126,194$).

47	50	52	55	46
50	53	49	43	48
51	54	52	47	49
45	51	53	49	47
52	61	50	51	53
49	50	45	55	51
52	46	47	50	48
57	50	48	55	49
53	49	54	44	47
51	45	51	50	52

3.30 If the sample data given in Exercise 3.29 were converted to z scores, how many of the z scores would be negative? How many would be zero? How many would be positive?

3.31 Find the sample median and interquartile range for the data given in Exercise 3.29. (Recall that the empirical distribution function was plotted for these data in Exercise 2.49.)

3.32 Find the sample mode for the data given in Exercise 3.29.

3.33 Construct a boxplot for the data in Exercise 3.29. Are there any outliers in these data?

3.34 On your way to work you pass by four electronic signs that display the current temperature. On one winter day in particular you noticed that the first sign said $43°$ and the second sign said $31°$. The third sign said $41°$ and the fourth sign said $42°$. Would you estimate the true temperature by using the mean or the median of these four readings? Why?

3.35 The doctoral graduates of 1975 were asked ten years later to report their total income for 1985. This is what they reported: $43,140, $40,260, $56,500, $36,225, $38,743, $44,840, $61,284, $44,182, $41,870, $38,910. The masters degree graduating class that year had 112 graduates, so ten were selected at random and asked for their total income in 1985 with the following results: $28,142, $32,112, $17,821, $142,621, $34,477, $29,840, $98,700, $36,888, $34,519 and $110,429. (Doctoral: $\Sigma X = 445,954$, $\Sigma X^2 = 2.04729 \times 10^{10}$, Masters: $\Sigma X = 565,549$, $\Sigma X^2 = 4.9049 \times 10^{10}$.)

(a) Find the mean, median, and standard deviation for each group of ten graduates. What are the main differences between the two groups?

(b) Graph the empirical distribution function of each group of graduates and compare the two graphs.

3.36 The data given represent the weight losses (in grams) of laboratory rats 24 hours after they were injected with an experimental drug. Calculate the sample mean and sample standard deviation for these data. ($\Sigma X = 214.8$, $\Sigma X^2 = 927.66$.)

4.2	4.4	4.8	4.9	4.4
3.9	4.3	4.5	4.8	3.9
3.6	4.1	4.3	3.9	4.2
4.1	4.0	4.0	3.8	4.6
3.8	4.7	3.9	4.0	4.2
4.4	4.6	4.4	4.9	4.4
4.1	4.3	4.2	4.5	4.4
4.2	4.1	4.0	4.7	4.1
4.7	4.2	4.1	4.4	4.8
4.3	4.6	4.5	4.6	4.0

3.37 If the sample data given in Exercise 3.36 were converted to z scores, what would be the values of the mean and standard deviation of the z scores?

3.38 Find the sample median and interquartile range for the data given in Exercise 3.36. (Recall that the empirical distribution function was plotted for these data in Exercise 2.55.)

3.39 Find the sample mode for the data given in Exercise 3.36.

3.40 Construct a boxplot for the data in Exercise 3.36. Are there any outliers in these data?

3.41 A department store manager is interested in the number of refunds granted customers in the men's wear department. The daily records for 30 days are summarized below. Find the mean number of refunds per day, and the daily median number of refunds. Is the distribution skewed to the right or to the left?

Number of Refunds	Number of Days That Number of Refunds was Made
0	6
1	12
2	7
3	3
4	1
7	1

3.42 A high school football coach recorded a sample of yards gained by the offensive team during a scrimmage as follows:

$$2, \ -3, \ 0, \ 1, \ 6, \ 27, \ 4, \ 1, \ 3, \ -12, \ 8, \ 12, \ 3, \ 0, \ 1$$

Find the sample mean and the sample median. Is the distribution skewed to the right or to the left?

3.43 A frequency summary of the weights of 114 students is given below. Find the sample mean and sample standard deviation for the data in this summary.

Weight	Frequency
104.5 to 114.5	1
114.5 to 124.5	5
124.5 to 134.5	16
134.5 to 144.5	29
144.5 to 154.5	24
154.5 to 164.5	19
164.5 to 174.5	12
174.5 to 184.5	4
184.5 to 194.5	3
194.5 to 204.5	1

3.44 Use the formula for repeated observations to calculate the sample standard deviation for the sample data in Exercise 3.7.

4
PROBABILITY, POPULATIONS, AND RANDOM VARIABLES

PRELIMINARY REMARKS

The weather bureau has made probability a household word by including probabilities in the weather forecasts. Yet probability is often misunderstood in this context, as evidenced by people who maintain that the weather bureau was wrong because it rained when the probability of rain was only 20%. Probability is only a measure of the uncertainty associated with events. It can indicate which events are more likely to occur, but it does not predict the future with certainty.

There is a solid and extensive mathematical development of the theory of probability. The mathematics of probability theory originated with problems related to games of chance, where the objective was to determine whether certain gambling games were favorable or unfavorable to the player. Therefore the discussion of probability often involves dealing cards, tossing coins, and rolling dice.

Not all probability lends itself to a rigorous mathematical development, however. In the business world probability often takes the form of an expert opinion, as when the financial officer estimates the probability of a bond issue selling out within the first week, or the marketing specialist assesses the probability of a new product selling more than a given number of units.

Probability theory provides the foundation on which much of the science of statistics is built. It is also used extensively in analyzing decision-making problems involving risk or incomplete information. In this chapter a brief introduction to probability theory is given. It is discussed only in relationship to the populations from which samples are obtained. The closely related concept of random variables is also discussed.

4.1

PROBABILITY

PROBLEM SETTING 4.1

All 706 employees of a company are classified according to two variables. One variable is the type of job they hold and the second variable is their sex. The total number of employees and the number that fall into each category are given in the *frequency table* in Figure 4.1. This frequency table represents the *population* of employees.

Consider an employee selected *at random,* which means that each of the 706 employees is *equally likely* to be selected.

(a) What is the probability that the employee is a clerical employee?
(b) What is the probability that the employee is male?
(c) What is the probability that the employee is both clerical and male?
(d) If the employee is clerical, what is the probability that the employee is male?
(e) Does the employee's job classification depend on whether the employee is male or female?

These and other questions will be answered in this section.

THE RELATIVE FREQUENCY DEFINITION OF PROBABILITY

If an experiment could be repeated many times, then the relative frequency associated with each outcome will tend toward some number between 0 and 1 inclusive, as the number of repetitions gets larger and larger. This long-term relative frequency approaches the **probability of the outcome** if the nature of

	Male	Female	Total
Clerical	27	51	78
Administrative	24	12	36
Production	314	261	575
Other	14	3	17
Total	379	327	706

FIGURE 4.1
FREQUENCY TABLE OF EMPLOYEES.

the experiment remains unchanged. The probabilities associated with the outcomes in a sample space must satisfy two requirements:

1. The probability of each outcome must be a number between 0 and 1, inclusive.
2. The sum of the probabilities of all of the outcomes in the sample space must equal 1.0.

PROBABILITY
OF AN
OUTCOME

The **probability of an outcome** is the long-term relative frequency with which that outcome could be expected to occur, if the experiment could be repeated many times. This relative frequency will always be a number between 0 and 1, inclusive.

OTHER DEFINITIONS OF PROBABILITY

The **relative frequency** definition of probability given previously is not the only definition of probability. The most rigorous and logically consistent definition involves using a set of *axioms* to define probability. This **axiomatic** definition of probability is found in most mathematical texts on statistics and is used in some applied texts also. Although it is more rigorous, it is less intuitive and more difficult to understand than the relative frequency definition.

If probabilities are obtained from the opinions of one or more people, then the probabilities are called **subjective** probabilities. Odds-makers in Las Vegas attach subjective probabilities to the various outcomes of future sports events, and these are used to form betting odds that customers may use for making wagers. Participants in corporate planning sessions are often asked to make subjective estimates of the chances of particular events occurring in the future, like, "What are the chances Jones will quit if we promote Smith?"

There is a danger in attaching too much confidence in the use of past events to determine probabilities of the occurrence of future events. For example, if taxes were raised in five out of the last six years, it would be precarious to say that the probability of taxes being raised next year is 5/6. Conditions change over time, and other factors may need to be considered in estimating probabilities. The relative frequency approach to probability should be used with common sense when assigning probabilities to events. True probabilities are never known; they are either assumed or estimated.

PROBABILITY MODELS

Probabilities may be obtained on the basis of subjective estimates, or estimated from relative frequencies. They may even result from a set of assumptions. In any event, the result is called a **probability model,** which provides numbers to use as probabilities. Later in this text procedures will be provided for checking the set of assumptions on which the probability model is based.

PROBABILITY
MODEL A **probability model** is a set of assumptions that leads to the assignment of precise probabilities to the outcomes in a sample space.

An **event** was defined in Chapter 1 as a set of one or more outcomes. The **probability of an event** equals the sum of the probabilities associated with the outcomes that constitute the event.

PROBABILITY
OF AN EVENT The **probability of an event** is the sum of the probabilities of the outcomes that comprise the event.

Much of the usefulness of probability theory lies in considering the relationship between two different events. If one event has occurred, does the probability of the other event remain unchanged? If the probability of each event is known, what can be said about the probability that both events occur? These are often useful questions to consider.

JOINT PROBABILITY TABLE

There are 706 different outcomes of the experiment in Problem Setting 4.1, representing the 706 employees that could be selected. Since each employee is equally likely to be selected, it is reasonable to assume that if this experiment (i.e., selecting one employee at random from the 706 employees) were repeated infinitely many times, each employee would be selected about $\frac{1}{706}$th of the time. Therefore the *relative frequency* for each outcome of this experiment would approach $\frac{1}{706}$, the *probability* for each outcome.

One possible *event* is that the randomly selected employee is *male*. Since 379 employees are male, the probability of this event is $\frac{379}{706}$. Another possible event is that the randomly selected employee is *clerical*. The probability of this happening is $\frac{78}{706}$, because there are 78 clerical employees. The two events could occur simultaneously, by selecting one of the 27 male clerical workers. The probability of selecting a *male clerical* worker is $\frac{27}{706}$, which equals .038.

The simultaneous occurrence of two events is called a **joint event,** and its probability is called a **joint probability.** A frequency table can be converted to a **relative frequency table** by dividing each entry in the frequency table by the total population size. If an element in the population is selected at random, then the relative frequencies are also probabilities, and the relative frequency table is also a **joint probability table.** The joint probability table represents the probabilities of the various joint events occurring simultaneously in one trial of the experiment.

JOINT
PROBABILITY The probability associated with the simultaneous occurrence of two events in an experiment is called a **joint probability.** The joint probability of two events A and B is denoted by $P(A \text{ and } B)$ or by $P(AB)$.

EXAMPLE

The frequency table in Problem Setting 4.1 is converted to a relative frequency table by dividing each entry in the table by 706, the total population size. This relative frequency table provides the probabilities of the various joint events that can occur when one employee is selected at random from the population. The joint probability of the randomly selected employee being both *male* and *clerical* is seen from the table in Figure 4.2 to be .038, as before.

MARGINAL PROBABILITIES

The probabilities associated with one outcome of one classification variable (job type) by itself, irrespective of the values of the other classification variable (sex), may be obtained by summing all of the joint probabilities associated with the value of that variable (job type), where the sum is taken over all outcomes of the other variable (sex). This is easy to see in Figure 4.2. There the probability of selecting a clerical employee is .110, which is obtained by summing the two joint probabilities for male clerical employees and female clerical employees. Because these sums of probabilities typically appear in the margins of joint probability tables, as in Figure 4.2, they are called **marginal probabilities.**

Another marginal probability from Figure 4.2 is the probability that a randomly selected employee will be male, .537. This marginal probability is found by summing the joint probabilities for male clerical employees, .038, male administrative employees, .034, male production employees, .445, and other male employees, .020. The collection of male employees in these four categories constitutes the entire group of male employees, so the sum of the probabilities from those mutually exclusive categories results in the probability of a randomly selected employee being male.

MARGINAL PROBABILITY   The sum of several joint probabilities over all values of one classification variable is called a **marginal probability.**

Employee	Male	Female	Total
Clerical	.038	.072	.110
Administrative	.034	.017	.051
Production	.445	.370	.814
Other	.020	.004	.024
Total	.537	.463	1.000

FIGURE 4.2
JOINT PROBABILITY TABLE OF EMPLOYEES.

CONDITIONAL PROBABILITY

Suppose that an employee is selected at random. It has just been shown that the probability of a randomly selected employee being male is .537. However, if some additional information is known about the selected employee this probability may change. For example, what is the probability that the employee is male, given the information (or condition) that the employee is a clerical employee? Since there are 78 clerical employees and only 27 of them are male, the probability is $\frac{27}{78} = .346$ of a randomly selected employee being male, given that the employee is clerical. This is called a **conditional probability.** In this case, it is different than the marginal probability .537 given earlier as the probability of an employee being male when no conditioning information is considered.

There is another way of computing conditional probabilities, shown in Eq. 4.1, that is equivalent to the one just demonstrated. If both the numerator and the denominator of the preceding fraction, $\frac{27}{78}$, are divided by the total population size, 706, then the new numerator, $\frac{27}{706}$, is the *joint probability* of being *male* and *clerical*, and the new denominator, $\frac{78}{706}$, is the *marginal probability* of being *clerical*. These probabilities may be obtained directly from the joint probability table in Figure 4.2 as .038 and .110, respectively, and their ratio .038/.110 gives the desired conditional probability, except for roundoff error. This demonstrates the reasoning behind Eq. 4.1.

CONDITIONAL PROBABILITY

The **conditional probability** of one event, given that another event or condition is true, is given by the joint probability of the two events divided by the marginal probability of the given event. Notation for this equation is

$$P(A|B) = \frac{P(A \text{ and } B)}{P(B)} \tag{4.1}$$

where $P(A|B)$ is read "the probability of A given B."

INDEPENDENCE

When the conditional probability is different than the unconditional probability, the two events are said to be **dependent.** Conversely, two events are **independent** if the conditional probability of one event given the other is the same as the unconditional probability of the one. The events "selecting a male employee" and "selecting a clerical employee" are dependent events because the conditional probability of the first event, selecting a male employee, given the second, selecting a clerical employee, is .346, whereas the unconditional probability of the first event is .537.

To see how Eq. 4.2 works, consider the joint probability of "selecting a male employee" and "selecting a clerical employee" that is given in Figure 4.2 as .038.

$$P(male \text{ and } clerical) = .038$$

Two events A and B are **independent** if their joint probability equals the product of their marginal probabilities. Notation for this is

**INDEPENDENT
EVENTS**

$$P(A \text{ and } B) = P(A) \cdot P(B) \qquad (4.2)$$

Equivalently, two events are independent if the conditional probability of A given B is equal to the unconditional probability of A. Notation for this is

$$P(A|B) = P(A) \qquad (4.3)$$

The product of the two marginal probabilities

$$P(male) = .537$$
$$P(clerical) = .110$$

is $(.537)(.110) = .059$, which does not equal the joint probability of the same two events. Therefore the events are dependent.

The same conclusion is reached using Eq. 4.3. The probability of a randomly selected employee being male (event A) given that the employee is clerical (event B) was found previously to be .346.

$$P(A|B) = .346$$

This is not equal to the probability of an employee being male, which is .537.

$$P(A|B) \neq P(A) = .537$$

Therefore the events A and B are not independent.

COMBINED EVENTS AND THE ADDITION RULE

The **combined event A or B** consists of all outcomes that are in at least one of the two events. At first thought it may seem that the probability of A or B equals $P(A) + P(B)$. However, closer examination reveals that some outcomes are counted twice by simply adding $P(A)$ and $P(B)$; those are the outcomes that are in both A and B. Subtracting the effect of those outcomes, so they are counted only once, gives the **addition rule** for combined events.

**THE ADDITION
RULE**

$$P(A \text{ or } B) = P(A) + P(B) - P(AB) \qquad (4.4)$$

If two events A and B have no outcomes in common, they are **mutually exclusive.** This means their joint probability $P(AB)$ is zero, and the addition rule simplifies to

$$P(A \text{ or } B) = P(A) + P(B) \qquad \text{(mutually exclusive events)} \qquad (4.5)$$

MUTUALLY EXCLUSIVE	Two events are **mutually exclusive** if the occurrence of one excludes the occurrence of the other.

Find the probability that a randomly selected employee in Problem Setting 4.1 is either *male* or *clerical*. From Figure 4.2 $P(male) = .537$, $P(clerical) = .110$, and $P(male$ and $clerical) = .038$, so from Eq. 4.4

$$P(male \text{ or } clerical) = P(male) + P(clerical) - P(male \text{ and } clerical)$$
$$= .537 + .110 - .038$$
$$= .609$$

As a check on these calculations consider Figure 4.1, which shows that 379 employees are male. In addition there are 51 female clerical workers. Therefore, in total, 430 employees are either *male* or *clerical*. The probability of selecting one of these employees at random is $\frac{430}{706} = .609$, in agreement with the addition rule.

PROBABILITY TREES

A **probability tree** provides an alternative to the joint probability table for illustrating joint events. The probability tree can be used to illustrate three or more events occurring jointly, where the joint probability table illustrates only two events jointly at a time. Also the probability tree is useful to illustrate a sequence of events. A disadvantage of most probability trees is that they quickly become large and complex, making them cumbersome to present and difficult to read. It is sometimes useful to note that each time a branch in a probability tree divides into several subbranches, the conditional probabilities on the subbranches sum to 1.00.

PROBABILITY TREE	A **probability tree** is a diagrammatic method of illustrating sequences of events. Each branch of the tree carries the conditional probability of the event given the events on the tree leading to that branch.

In Problem Setting 4.1 an employee is selected at random and classified as *male* or *female*, with respective probabilities .537 and .463. See the first step in the probability tree in Figure 4.3 and note how the probabilities on the two branches on the left sum to 1.

Then the job classification is determined, with conditional probabilities as given on the secondary branches in Figure 4.3. Each path from left to right determines one joint event, and multiplication of the branch probabilities

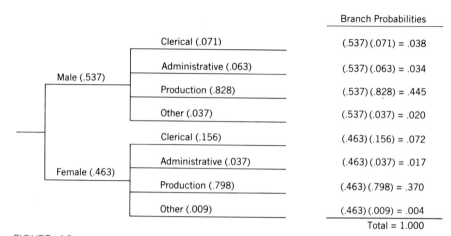

FIGURE 4.3

PROBABILITY TREE FOR EMPLOYEE CLASSIFICATION.

along that path gives the joint probability of the joint event represented by that path. For example, the event "male and clerical" has the probability $(.537)(.071) = .038$, and is represented by the path along the top of the tree in Figure 4.3.

Note that these eight branches correspond to the eight classification categories in Figure 4.2. However, unlike Figure 4.2, it is convenient to add

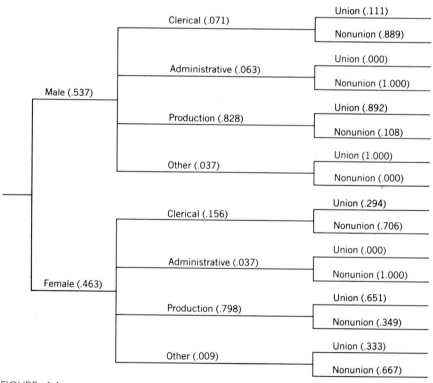

FIGURE 4.4

EXTENDED PROBABILITY TREE FOR EMPLOYEE CLASSIFICATION.

another series of branches to the tree in Figure 4.3 to represent other classification categories, such as union versus nonunion membership. This additional set of branches is shown in Figure 4.4. The branch probabilities on the
third set of branches are the conditional probabilities of union or nonunion
membership, given the sex and job classification of the path leading to each
branch.

From Figure 4.4 it is easy to see that the joint probability of selecting a male
clerical union member is given by the product of the three probabilities along
that branch.

$$
\begin{aligned}
P(\text{male, clerical, union}) &= P(\text{male})\, P(\text{clerical}\,|\,\text{male}) \\
&\quad \cdot P(\text{union}\,|\,\text{male, clerical}) \\
&= (.537)(.071)(.111) \\
&= .004
\end{aligned}
$$

There are sixteen branches in the probability tree in Figure 4.4, representing
the sixteen different possible events (or classification categories). The probability for each event is found by taking the product of the three probabilities
along that branch representing the event. The sum of the branch probabilities
is 1.0, illustrating that the sum of the probabilities of all outcomes in the sample
space is 1.0.

EXERCISES

4.1 A survey of all employees regarding marital status and the choice of
three medical plans produces the following frequencies.

	Married	Single	Total
Plan 1	20	50	70
Plan 2	140	10	150
Plan 3	45	15	60
Total	205	75	280

Assume that one employee is selected at random from the 280 employees.

(a) Convert the frequency table to a joint probability table.

(b) Find the probability of Plan 1 being preferred by the employee selected.

(c) Find the probability that the selected employee is married.

(d) Is the preference for Plan 1 independent of being married?

(e) Find the conditional probability of Plan 1 being preferred given that
the employee is married.

(f) Find the conditional probability of Plan 1 being preferred given that
the employee is single.

(g) Find the conditional probability of Plan 2 being preferred given that the employee is married.

(h) Find the conditional probability of Plan 2 being preferred given that the employee is single.

(i) Based on your answers for (e) through (h), do you think plan preference is independent of marital status?

4.2 Stocks in a large investment portfolio are classified according to how much gain (or loss) they have experienced in the last 12 months, and on whose recommendation they were bought, resulting in the following table of frequencies.

		Broker			Total
		A	B	C	
Change in Stock Value	>20%	1	2		6
	10% to 20%	13	12		45
	0% to 10%	20	20		60
	−10% to 0%	2	2		20
	< −10%				12
	Total	48	36	59	143

Assume that one stock is selected at random from the 143 stocks in the portfolio.

(a) Fill in the blanks in the frequency table and convert it to a joint probability table.

(b) Find the probability that the selected stock had a gain of over 20%.

(c) Find the conditional probability that Broker C had recommended that stock, given that the stock increased over 20%.

(d) Find the probability that the stock decreased in value.

(e) Find the conditional probability that Broker A had recommended the stock given that the stock decreased in value by more than 10%.

4.3 The probability of a student getting an A on the first exam is .25, and the probability of his getting an A on the second exam is .25. The probability of his getting As on both exams is .15. Are the two events independent?

4.4 The probability that a customer buys a service contract on her new automobile is .4, and the probability of the automobile requiring major repair is .2. Assume the two events are independent. What is the probability that the customer buys a service contract on her new automobile, and the automobile requires major repair?

4.5 A men's clothing store has a policy of extending credit to its customers. A study of all of its accounts revealed that 12% of the accounts were delinquent and the other 88% were not. This study also revealed that 66% of the accounts belonged to married men. Only 4% of the accounts were both delinquent and belonging to married men. There were 500 accounts in all.

(a) Construct a joint probability table for a randomly selected account.

(b) Find the joint probability of an account being nondelinquent and belonging to an unmarried man.

(c) Find the conditional probability of an account being delinquent, given that the account belongs to an unmarried man.

(d) Is account status independent of marital status?

4.6 Employees voluntarily terminating with a large company have been classified according to sex and reason for termination. The reason for termination can be resignation (RES), early retirement (ER), retirement (RET), disability retirement (DR), or leave of absence (LOA). The frequencies are as follows:

	RES	ER	RET	DR	LOA	Total
Male	82	78	18	10	6	194
Female	48	17	7	0	15	87
Total	130	95	25	10	21	281

Assume that one of the 281 employees is selected at random.

(a) Convert the frequency table to a joint probability table.

(b) The employee is known to have terminated by LOA. Find the conditional probability that this employee is a female.

(c) What is the probability of termination by some type of retirement?

4.7 A magazine vendor classifies 100 customers by types of magazine purchased and age of the customer. The following table of frequencies is a summary of these classifications.

	<30 yr	30–40 yr	41–50 yr	>50 yr	Total
News	12	10	11	14	47
Sports	10	7	8	6	31
Hobby	1	3	5	13	22
	23	20	24	33	100

Assume that one of the 100 customers is selected at random.

(a) Convert the frequency table to a joint probability table.

(b) Find the probability that the customer is over 40.

(c) Find the conditional probability that the customer is over 50 if a hobby magazine is sold.

(d) Find the conditional probability that the customer is over 40 if a hobby magazine is sold.

(e) Find the probability that a customer is under 41 and purchases a sports magazine.

4.8 If the joint probability of a resident taking both the morning paper and the evening paper is .2 and the marginal probability of taking the evening paper is .6, find the conditional probability of a resident taking the morning paper given that the resident takes the evening paper.

4.9 A survey shows that 80% of all households have a color television set and 30% have a microwave oven. If 20% have both a color television set and a microwave oven, what percent have neither?

4.10 A survey shows that 40% of all convenience store shoppers buy milk and 30% buy bread. If 25% buy both bread and milk, what percent buy neither?

4.11 A well is being drilled for oil. The probability of drilling through shale is .4, in which case the conditional probability of striking oil is .3. If the well does not go through shale, the probability of striking oil is only .1. Draw a probability tree where the first branches represent hitting shale or not hitting shale, and the second set of branches represents striking oil or not striking oil.

4.12 Draw a probability tree for the situation described in Exercise 4.1. Let the first set of branches represent whether the employee is married or single, and let the second set of branches represent the medical plan the employee selects.

4.13 Draw a probability tree for the situation described in Exercise 4.4. Let the first set of branches represent whether the customer buys a service contract or not, and let the second set of branches represent the probability of the automobile requiring major repair.

4.14 Draw a probability tree for Exercise 4.5, where the first set of branches represents the customer's marital status, and the second set of branches represents the status of the account.

4.2

DISCRETE RANDOM VARIABLES

In Chapter 1 a random variable was defined as a rule for assigning real numbers to the outcomes of an experiment. The purpose of a random variable is to simplify the results of an experiment, so that instead of referring to the outcomes of the experiment in lengthy verbal terms, simple numbers are used. Random variables are often given names like X, Y, or Z, with or without subscripts, for ease in identification.

All of the random variables considered so far in this text have had one characteristic in common; the numbers they used were all separate and distinct, like 0, 1, 2. Random variables with this characteristic are called **discrete random variables.** Some experimental situations that naturally relate to discrete random variables include:

1. The number of customers per hour (0, 1, 2, 3, etc.).

2. The number of days of sick leave taken by an employee (0, 1, 2, 3, etc.).

3. The proportion of customers, out of 43 surveyed, who prefer Brand X (0/43, 1/43, 2/43, . . . , 43/43).

4. The price–earnings ratio of stock randomly selected from a portfolio of 50 stocks.

5. The number of defective items in a lot.

Notice that in each of these examples the random variable assumes values that are separate and distinct from one another. For example, in the first illustration previously given, the number of customers could be 1 or 2, but not any number between 1 and 2.

DISCRETE A random variable is called **discrete** if all of its possible values are separate
RANDOM and distinct.
VARIABLE

PROBABILITY FUNCTION OF A DISCRETE RANDOM VARIABLE

A random variable associates numbers with the outcomes of an experiment. Sometimes those numbers are equally likely, but sometimes they are not. The function that gives the probability associated with each possible value of the random variable is called a **probability function.**

PROBABILITY The **probability function** of a discrete random variable is the function that
FUNCTION associates probabilities with the various numerical values the random variable
can assume.

A bar graph of the probability function is often a useful device for displaying the relative likelihood of obtaining the various possible values of the random variable. The probabilities, taken as a whole, sum to 1.0 in a probability function, because all possible outcomes of an experiment must be considered. The probability function describes the *population probabilities,* as opposed to the relative frequency histogram of Chapter 2, which describes the estimates of those probabilities as obtained from a sample. The next example provides a graph of a probability function.

EXAMPLE

Company records indicate the number of college degrees earned by each of its employees. Out of a population consisting of 150 employees, the number of degrees earned by each is given as follows:

Number of Degrees	Number of Employees	Relative Frequency
0	75	.50
1	36	.24
2	18	.12
3	15	.10
4	6	.04
Total	150	1.00

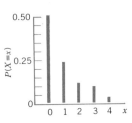

FIGURE 4.5

A GRAPH OF THE
PROBABILITY FUNC-
TION FOR THE RANDOM
VARIABLE X.

Let the random variable X equal the number of college degrees earned by an employee who is randomly selected from the 150 employees. Then X is a discrete random variable that can take on the values 0, 1, 2, 3, or 4. The event "the selected employee has no college degrees" is represented as $X = 0$, while "the selected employee has one college degree" is represented as $X = 1$, and so on. The probabilities associated with each value of the random variable X in the population are given by the population relative frequencies. These relative frequencies are used to form the probability function for the random variable X:

$$P(X = 0) = \ .50$$
$$P(X = 1) = \ .24$$
$$P(X = 2) = \ .12$$
$$P(X = 3) = \ .10$$
$$P(X = 4) = \ \underline{.04}$$
$$1.00$$

A graph of the probability function for the random variable X is given in Figure 4.5.

THE CUMULATIVE DISTRIBUTION FUNCTION

Another way of expressing the probabilities associated with a random variable is by means of a **cumulative distribution function** (c.d.f.), which merely accumulates all of the probabilities of values less than or equal to the value of X of interest. The notation for a c.d.f. is $P(X \leq x)$, or sometimes simply $F(x)$. Since every c.d.f. is a cumulative probability function, it increases from 0 to 1 as x increases.

CUMULATIVE
DISTRIBUTION
FUNCTION

The **cumulative distribution function** (c.d.f.) of a random variable X is the function that gives the probability that X is less than or equal to x, written as $P(X \leq x)$, or simply $F(x)$.

The c.d.f. of a discrete random variable has a graph that identifies it immediately as being associated with a discrete random variable; the graph appears to be a series of stairsteps, where the heights of the stairs are not necessarily equal (see Figure 4.6). Any time the c.d.f. has a steplike appearance, the random variable is discrete. The vertical lines in the graph are not part of the c.d.f., but are usually included to make the graph easier to read.

Incidentally, the height of each step of the c.d.f. equals the probability associated with the value x at the step. Since the sum of all of the probabilities in a probability function equals 1.0, the c.d.f. goes to 1.0 as x increases, because the c.d.f. is the cumulative probability.

The steplike appearance of the c.d.f. of a discrete random variable should not cause it to be confused with an e.d.f., which also has a steplike appearance. The step-heights in an e.d.f. represent the relative frequencies with which each observation in a sample occurs, while the step-heights in the c.d.f. of a discrete random variable represent the exact probabilities associated with each possible value of the random variable.

EXAMPLE

The cumulative distribution function for the probabilities given in Figure 4.5 is obtained as follows. For each value of x, the probabilities associated with all numbers assigned by the random variable that are less than or equal to that value of x are collected and added together to get $P(X \leq x)$. For example, in order to obtain $P(X \leq 2.1)$, all probabilities associated with numbers assigned by the random variable that are less than or equal to 2.1 are collected and summed. These include

$$
\begin{aligned}
P(X = 0) &= .50 \\
P(X = 1) &= .24 \\
P(X = 2) &= \underline{.12} \\
.86 &= P(X \leq 2.1)
\end{aligned}
$$

This value .86 holds not only for $P(X \leq 2.1)$ but also for $P(X \leq 2.0)$, $P(X \leq 2.2)$, and in fact for all values of x from 2.0 to 3.0, but not including 3.0. At $x = 3.0$ a new probability enters into the calculation, so

$$P(X \leq 3.0) = .86 + .10 = .96$$

because .10 is the probability that X equals 3.0. A summary of the c.d.f. is given below.

$$
\begin{aligned}
F(x) &= P(X \leq x) = 0 \text{ for all } x < 0 \\
F(0) &= P(X \leq 0) = .50, \text{ also for } 0 \leq x < 1 \\
F(1) &= P(X \leq 1) = .74, \text{ also for } 1 \leq x < 2 \\
F(2) &= P(X \leq 2) = .86, \text{ also for } 2 \leq x < 3 \\
F(3) &= P(X \leq 3) = .96, \text{ also for } 3 \leq x < 4 \\
F(4) &= P(X \leq 4) = 1.00, \text{ also for } 4 \leq x
\end{aligned}
$$

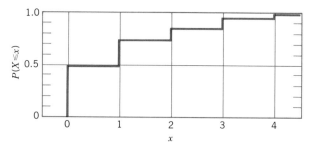

FIGURE 4.6
A GRAPH OF THE CUMULATIVE DISTRIBUTION
FUNCTION FOR THE RANDOM VARIABLE X.

The final probability is 1.00 for all x greater than 4 because when x is greater than 4, the probabilities for $X \leq x$ include all of the individual probabilities associated with X. A graph of the cumulative distribution function of X is given in Figure 4.6.

THE POPULATION c.d.f. VERSUS THE e.d.f.

Because every cumulative distribution function is associated with a population of values, it is also called the population c.d.f., or sometimes the **population distribution function.**

POPULATION DISTRIBUTION FUNCTION

The **population distribution function,** or population c.d.f., is the cumulative distribution function of a population.

The empirical distribution function, called the e.d.f., represents the cumulative relative frequencies for a sample. While the c.d.f. gives the true probabilities in a population, the e.d.f. merely presents estimates of those probabilities based on a random sample from that population.

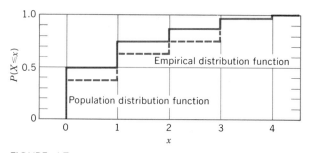

FIGURE 4.7
THE EMPIRICAL DISTRIBUTION FROM A SAMPLE OF
SIZE 20, AND THE DISCRETE CUMULATIVE
DISTRIBUTION FUNCTION OF THE POPULATION FROM
WHICH THE SAMPLE WAS OBTAINED.

Consider again the example in which the random variable is the number of degrees earned by each of 150 employees of a company. The random variable X equals 0, 1, 2, 3, or 4. A random sample of 20 employees is observed with the following results.

Number of Degrees	Number of Employees in Sample	Sample Relative Frequency	e.d.f.	Population Relative Frequency	c.d.f.
0	8	.40	.40	.50	.50
1	5	.25	.65	.24	.74
2	2	.10	.75	.12	.86
3	4	.20	.95	.10	.96
4	1	.05	1.00	.04	1.00
	20	1.00		1.00	

The empirical distribution function of these 20 sample data points appears in Figure 4.7, along with the population c.d.f. from Figure 4.6. Note that the e.d.f., with its multiplicity of points at $x = 0$, 1, 2, 3, and 4, has five steps, just like the population c.d.f. Since there are only five values possible in the population the e.d.f. is forced to bear a strong resemblance to the population c.d.f., other than in the step-heights.

As the sample size increases, the heights of the steps in the e.d.f., representing relative frequencies, will be closer to the heights of the steps in the c.d.f., representing probabilities, so the two curves will tend to be virtually indistinguishable with large sample sizes.

DISCRETE BIVARIATE RANDOM VARIABLES

Experiments frequently have more than one random variable involved. Once an item from the population is selected, or an outcome from the experiment is observed, several characteristics are usually noted, and each characteristic uses its own random variable. In the previous example each of the 150 employees might have characteristics recorded by variables such as sex (male or female), number of years with the company (0, 1, 2, . . .), or job classification (clerical, administrative, . . .), in addition to the number of degrees. Each of the characteristics could be denoted by a random variable;

X = number of degrees

$Y = 0$ if male, $Y = 1$ if female

Z = number of years with the company

and so on. Several random variables considered jointly, like (X, Y, Z), are called *multivariate random variables*. The simplest multivariate random variable involves only two random variables, such as (X, Y), and is called a *bivariate random variable*. Joint probabilities for a bivariate random variable take the form $P(X = x, Y = y)$, and represent the probability of the joint event $X = x$ and $Y = y$.

EXAMPLE

For each of the 150 employees in a company, the number of degrees (denoted by X) and the sex (Y = 0 for male, Y = 1 for female) are noted in Figure 4.8. The frequency table given in Figure 4.8 is easily converted to a probability table, representing the joint probabilities of a randomly selected employee falling into each of the eight categories, as shown in Figure 4.9.

MARGINAL PROBABILITIES

Marginal probabilities are defined for discrete random variables just as they were defined for events. That is, the marginal probabilities for one random variable are found by summing the joint probabilities over all of the other random variables.

MARGINAL PROBABILITY OF X (DISCRETE)

The **marginal probability** of a discrete random variable X is the sum of the joint probabilities over all values of one or more variables (not including X)

$$P(X = x) = \sum_{\text{all } y} P(X = x, Y = y) \tag{4.6}$$

For a bivariate random variable (X, Y) the marginal probabilities for X are found for each value of X by summing the joint probabilities for the various values of Y. The resulting distribution of probabilities is called the **marginal distribution** of X. This is the same probability distribution for X given in Figure 4.5 before Y was taken into consideration.

		Sex (Y) 0	Sex (Y) 1	Total
	0	35	40	75
	1	30	6	36
Number of	2	16	2	18
Degrees (X)	3	14	1	15
	4	5	1	6
Total		100	50	150

FIGURE 4.8
FREQUENCY TABLE OF EMPLOYEES.

	Sex (Y) 0	Sex (Y) 1	Total
0	.233	.267	.500
1	.200	.040	.240
Number of 2	.107	.013	.120
Degrees (X) 3	.093	.007	.100
4	.033	.007	.040
Total	.667	.333	1.000

FIGURE 4.9
JOINT PROBABILITY TABLE OF EMPLOYEES.

EXAMPLE

In Figure 4.9 the marginal probabilities of X are obtained by summing the joint probabilities within each row, that is, over all values of Y. The result is

$$P(X = 0) = P(X = 0, Y = 0) + P(X = 0, Y = 1) = .233 + .267 = .500.$$

Likewise,

$$P(X = 1) = .240$$
$$P(X = 2) = .120$$
$$P(X = 3) = .100$$
$$P(X = 4) = .040$$

as in Figure 4.5.

Similarly the marginal probabilities for Y are found by summing the joint probabilities down each column in Figure 4.9, that is over all values of X.

$$P(Y = 0) = .233 + .200 + .107 + .093 + .033 = .667$$

The marginal distribution of Y is then

$$P(Y = 0) = .667$$
$$P(Y = 1) = .333$$

Note that these marginal distributions for X and Y appear in the right margin and the lower margin, respectively, of Figure 4.9. This suggests why they are called marginal probability distributions.

CONDITIONAL PROBABILITY

The **conditional probabilities** for one random variable, given the value of another random variable, are found in the same way conditional probabilities for events are found.

The **conditional probability** of $X = x$ given $Y = y$ is given by the formula

CONDITIONAL
PROBABILITY

$$P(X = x|Y = y) = \frac{P(X = x, Y = y)}{P(Y = y)}$$ (4.7)

when $P(Y = y) > 0$.

The collection of conditional probabilities of X, for the various values x and one given value $Y = y$, is called the **conditional distribution** of X given $Y = y$.

EXAMPLE

The probability of $X = 1$, representing one college degree, given $Y = 0$, male, is given by Eq. 4.7 as

$$P(X = 1|Y = 0) = \frac{P(X = 1, Y = 0)}{P(Y = 0)} = \frac{.200}{.667} = .30$$

This result is verified by referring to Figure 4.8, which shows that 30 of the 100 male employees have one degree. Since the selection is random, the conditional probability is $30/100 = .30$.

The entire conditional probability distribution of X given $Y = 0$ is obtained in a similar manner, with the result

$$P(X = 0|Y = 0) = .35$$
$$P(X = 1|Y = 0) = .30$$
$$P(X = 2|Y = 0) = .16$$
$$P(X = 3|Y = 0) = .14$$
$$P(X = 4|Y = 0) = \underline{.05}$$
$$\text{Total } 1.00$$

INDEPENDENCE

Two random variables are **independent** if all of their joint probabilities are equal to the product of their marginal probabilities. More commonly, in applied statistics, two random variables are *assumed* to be independent, and then their joint probabilities are found conveniently from their marginal probabilities using Eq. 4.8.

EXAMPLE

To show the dependence between X and Y in Figure 4.9 it suffices to look at only one joint probability, $P(X = 1, Y = 0) = .200$. Note that this is not equal to the product of the two marginal probabilities

$$P(X = 1) = .240 \quad \text{and} \quad P(Y = 0) = .667,$$
$$P(X = 1, Y = 0) \neq P(X = 1)P(Y = 0)$$
$$.200 \neq (.240)(.667) = .160$$

and therefore Eq. 4.8 does not hold. In this case, X and Y are not independent. In order to prove independence between X and Y, the relationship in Eq. 4.8 would have to be true for all ten joint probabilities in Figure 4.9.

Two discrete random variables X and Y are **independent** if the following equation holds

INDEPENDENT
DISCRETE
RANDOM
VARIABLES

$$P(X = x, Y = y) = P(X = x)P(Y = y) \qquad (4.8)$$

for all values of x and y. An equivalent definition of independence is

$$P(X = x | Y = y) = P(X = x) \qquad (4.9)$$

for all values of x and y.

EXERCISES

4.15 An employment agency gets calls for unskilled laborers in groups of 1, 2, 3, 4, 5, and more than 5. The probability model used for the number of laborers requested is .40, .23, .11, .09, .08, and .09 for each respective group size. Let the random variable X equal the number of laborers requested in a group and find the following probabilities for this discrete random variable.

(a) $P(X = 1)$

(b) $P(X \leq 3)$

(c) $P(1 < X < 3)$

(d) $P(1 \leq X \leq 3)$

(e) $P(X > 3)$

(f) $P(X = 0)$

4.16 Use a vertical bar graph to construct a graph of the probability function of the random variable X in Exercise 4.15. Add the individual probabilities to get the total probability.

4.17 Graph the cumulative distribution function for the random variable X in Exercise 4.15.

4.18 A sample of 10 calls received by the employment agency in Exercise 4.15 showed the following requested group sizes: 1, 3, 1, 1, 2, 5, 10, 1, 4, 1. Plot the empirical distribution function of these data on the same graph with the c.d.f. graphed in Exercise 4.17. How do the two graphs compare?

4.19 A bank classifies its borrowers as high risk ($X = 3$), medium risk ($X = 2$), or low risk ($X = 1$). Half of its loans are made to low-risk borrowers and 30% to medium-risk borrowers. One-third of the medium-risk borrowers default, while only 4% of the low-risk borrowers default. Overall, 22% of the loans end up in the default category. First construct the joint probability table and then answer the following questions.

(a) What is the probability of a borrower defaulting on a loan given the borrower is a high-risk borrower?

(b) Let $Y = 1$ if the borrower defaults and let $Y = 0$ if the borrower doesn't default. Find $P(Y = 0|X = 3)$.

(c) Plot the marginal probability function of Y.

(d) Plot the conditional probability function of Y, given $X = 3$.

(e) Plot the marginal c.d.f. of X. On the same axes, plot (using dotted lines) the c.d.f. of X given that $Y = 1$.

4.20 Official government statistics recently showed that 47% of all bachelor's degrees granted one year were earned by women, but only 27% of all bachelor's degrees in business administration were earned by women. Also, 17.5% of all bachelor's degrees were in business. Let $X = 1$ for women, $X = 0$ for men. Also let $Y = 1$ for business administration majors, and $Y = 0$ for others. Let one graduate be selected at random.

(a) Find $P(Y = 1|X = 1)$.

(b) Find $P(Y = 0|X = 1)$.

(c) Graph the c.d.f. of Y, given $X = 1$.

(d) Graph the c.d.f. of Y, given $X = 0$.

(e) What does a comparison of the answers to parts (c) and (d) suggest?

4.21 Television viewers are classified as to income category (high, middle, and low) and asked whether they regularly, or occasionally, or never watch the Johnny Carson show. The population breakdown is as follows.

They watch	High ($X = 3$)	Middle ($X = 2$)	Low ($X = 1$)
Regularly ($Y = 3$)	.03	.20	.22
Occasionally ($Y = 2$)	.06	.25	.08
Never ($Y = 1$)	.10	.05	.01

(a) What percent of the high-income viewers regularly watch the show?

(b) Find the marginal probability distribution of Y and plot the c.d.f.

(c) On the same axes as in part (b) plot the conditional distribution of Y given $X = 1$. (Use dotted lines.)

(d) What does a comparison of the answers to parts (b) and (c) suggest?

4.3

CONTINUOUS RANDOM VARIABLES

Recall that the values possible for a discrete random variable are all separate and distinct, such as 1 or 2 but no numbers between 1 and 2. Not all measurements are of this type, however. For instance, measurements of time, such as the time required to serve a customer, may take the value 1 or 2 minutes, *or any number between 1 and 2 minutes.* There are *no* separate and distinct values in this case.

When actually recording the amount of time required to serve a customer, the measurement may be recorded as 1 minute or 2 minutes if the amount of time is rounded to the nearest integer. Or the time may be recorded at 1.3 minutes if the measurement is rounded to the nearest tenth. Recorded measurements may look like measurements on a discrete random variable because the rounding represents the interval within which the observation lies. That is, due to rounding, the recorded measurement, 1.3 minutes, represents all of the actual times between 1.25 and 1.35 minutes. The recorded measurement appears to be discrete, but the actual time is not a discrete random variable.

CONTINUOUS RANDOM VARIABLE A random variable is **continuous** if none of its possible values, considered as a single point, has a positive (greater than zero) probability; only *intervals* of possible values can have probabilities greater than zero.

Random variables that have no separate and distinct values are called **continuous random variables;** their possible values form one or more continuous intervals of numbers and probabilities greater than zero are associated only with intervals of observations, not individual points as with discrete random variables. Where the typical discrete random variable is counting some quantity, the typical continuous random variable is measuring some quantity, such as time, weight, height, and volume. Examples of continuous random variables include:

1. The time it takes to serve a customer.

2. The time an employee arrives at work.

3. The actual weight of a loaf of bread.

4. The length of a steel reinforcement rod.

5. The actual volume of medicine in a bottle marked 3 cc.

HISTOGRAMS VERSUS DENSITY FUNCTIONS

In Chapter 2 histograms were recommended as a way of displaying sample data graphically. Because histograms, like continuous random variables, cover all values of x in an interval, histograms are more appropriate for observations on continuous random variables than they are for observations on discrete random variables, although they can be used on either. Vertical bar graphs are sometimes more natural for displaying data from discrete random variables.

Suppose a histogram is drawn for a large number of sample data points from a continuous population. As the sample gets large, the histogram tends to resemble more closely a population function called a **density function.**

DENSITY FUNCTION   The **density function** of a continuous random variable is the function that a histogram of a random sample will tend to resemble as the sample size increases and the class widths decrease. Density functions are often denoted by $f(x)$.

EXAMPLE

A random sample of 17 days records is obtained and the daily amounts of machine downtime are recorded. A histogram for these 17 observations is given in Figure 4.10. Recall from Chapter 2 that a histogram of 17 observations suggests five classes, so if the observations range from 0 to 125, the classes are 0 to 25, 25 to 50, etc., for this illustration.

The population density function is also given in Figure 4.10 for comparison. Note that the differences between the histogram and the density in Figure 4.10 are due to sampling variability. Also note that no scale is given on the vertical axis, since the vertical scales tend to be different for histograms and density functions.

As the sample size increases, the number of classes in a histogram increases also, and the histogram tends to become smoother and more similar in appearance to a density function. However, even for a sample of size 2000 the recommended number of classes is still only 11, and a graph consisting of 11 rectangles may be quite unlike a smooth density function.

In general, histograms tend to show more variability from sample to sample than empirical distribution functions. Also they do not tend to converge as fast

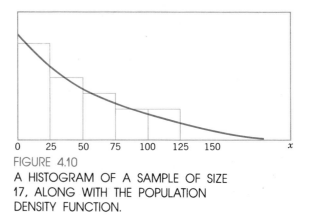

FIGURE 4.10
A HISTOGRAM OF A SAMPLE OF SIZE 17, ALONG WITH THE POPULATION DENSITY FUNCTION.

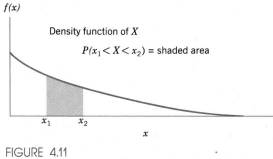

FIGURE 4.11
FINDING PROBABILITIES FROM A DENSITY
FUNCTION.

to the density function as the empirical distribution function tends to converge to the population distribution function. For this reason empirical distribution functions are often preferred over histograms.

OBTAINING PROBABILITIES FROM DENSITY FUNCTIONS

Density functions are always adjusted in height so that the total area under the density function, above the x-axis, is 1.0. That way, probabilities can be represented by areas under the density function. That is, to find the probability of a random variable X being between x_1 and x_2, calculate the area under the density function between x_1 and x_2. This may not be easy to do. (See Figure 4.11.) It is always easier to find probabilities from the cumulative distribution function, as described later in this section.

TWO PRIMARY USES FOR CONTINUOUS RANDOM VARIABLES

All measurements in the real world are recorded as discrete valued simply because measuring instruments must round the true value to some finite number of decimal places. While your true weight is a continuous random variable, the scales you weigh yourself on always round your weight, maybe to the nearest pound, maybe even to the nearest tenth of a pound. One of the primary uses of continuous random variables is to serve as a model for the real underlying value. That is, scientific problems can often be solved by assuming some continuous distribution of values for your weight, or your height, or the change in the Dow-Jones average, or the market price of a house, and so on. Working with continuous distributions requires calculus, but is actually much easier than working with models involving discrete distributions.

A second primary use for continuous random variables is to serve as an approximation for discrete random variables, again because calculus enables continuous random variables to be handled more easily than discrete random variables. In this case, the model uses a discrete random variable, but the solution to related scientific problems is approximate, based on a continuous distribution approximating a discrete distribution. All of the discrete distributions used in the following chapters will use a continuous distribua as an approximation for those cases where the exact probabilities are too difficult to obtain.

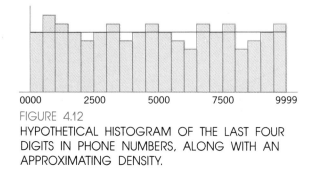

FIGURE 4.12
HYPOTHETICAL HISTOGRAM OF THE LAST FOUR
DIGITS IN PHONE NUMBERS, ALONG WITH AN
APPROXIMATING DENSITY.

EXAMPLE

What is a good model to use for the probability distribution of the last four
digits in a telephone number? If you selected a telephone number at random
from a large directory, what is the probability that the last four digits would
form a number larger than 7500?

To select a good model, try to imagine what a histogram of the entire
population of the numbers in that directory would look like. It might look like
a series of rectangles, all about the same height, ranging from 0000 to 9999,
as shown in Figure 4.12.

Also shown in Figure 4.12 is an approximating density function, uniform
in height from 0000 to 9999. The equation for that density function can be
written as

$$f(x) = .0001 \quad \text{for} \quad 0 < x < 10{,}000$$
$$f(x) = 0 \quad \text{for all other values of } x$$

To approximate the probability of a randomly drawn number being larger than
7500, it is convenient to consider the proportion of the total area under the
density curve that lies to the right of 7500, which of course is very close to
.25. Therefore the desired probability is estimated at .25.

CONTINUOUS CUMULATIVE DISTRIBUTION FUNCTIONS

The c.d.f. of a continuous random variable is distinctive because its graph has
no steps in it, in contrast with the c.d.f. of a discrete random variable that
graphs as a series of stairsteps. That is, the graph of the c.d.f. of a continuous
random variable appears as a smooth continuous function, such as the graphs
in Figures 4.13 and 4.14.

In Figure 4.13 probabilities may be read directly from the graph. The probability of X being less than or equal to 2 is obtained by reading the height of
the c.d.f. directly above the point $x = 2$, which is shown to be 0.82 by the
dashed lines in Figure 4.13. This is written as

$$P(X \leq 2) = 0.82$$

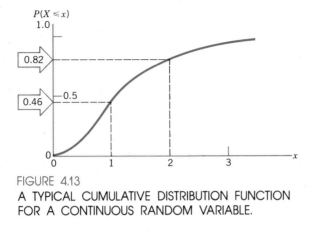

FIGURE 4.13
A TYPICAL CUMULATIVE DISTRIBUTION FUNCTION
FOR A CONTINUOUS RANDOM VARIABLE.

Similarly the probability that X is less than or equal to 1.0 is read from Figure 4.13 as 0.46.

$$P(X \leq 1.0) = 0.46$$

Probabilities over intervals of the x-axis, such as the probability that X is between 1 and 2, are found by measuring the corresponding interval on the vertical axis. The interval from $x = 1$ to $x = 2$ on the x-axis corresponds to the interval from .46 to .82 on the vertical (probability) axis, so the probability of X being between 1 and 2 is the distance between .46 and .82, or .36. Mathematically this is written as

$$P(1 < X < 2) = P(X < 2) - P(X < 1)$$
$$= .82 - .46$$
$$= .36$$

Notice that no distinction is made between $X < 2$ and $X \leq 2$ with continuous random variables, because the event $X = 2$ has zero probability attached to it. Notice also that the probability of exceeding a certain number, called an *exceedance probability*, is easily found using the method just described. For example, the probability of X exceeding 2 is given by the portion of the vertical axis that corresponds to $x > 2$, which is

$$P(X > 2) = 1 - P(X \leq 2)$$
$$= 1 - .82$$
$$= .18$$

All c.d.f.s are read in the way just described for continuous random variables, except that if the random variable is not continuous, distinction needs to be made between the probabilities $P(X \leq x)$ and $P(X < x)$ in those cases where $P(X = x)$ is not equal to zero. In general, the latter two probabilities sum to the first

$$P(X < x) + P(X = x) = P(X \leq x)$$

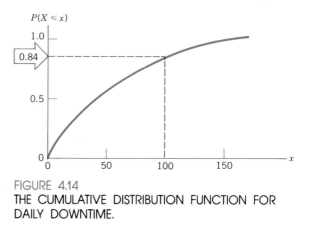

FIGURE 4.14
THE CUMULATIVE DISTRIBUTION FUNCTION FOR
DAILY DOWNTIME.

and this should be considered when finding probabilities from noncontinuous c.d.f.s.

EXAMPLE

Let the random variable X represent the daily downtime due to machine breakdowns. Suppose that the c.d.f. of X is given by Figure 4.14. From the c.d.f. the probability of the daily downtime being less than 100 minutes for any particular day is seen to be .84. The probability of the downtime exceeding 100 minutes is $1 - .84 = .16$. This means that in the long run about 16% of the days will involve downtime greater than 100 minutes.

EMPIRICAL DISTRIBUTION FUNCTIONS VERSUS CUMULATIVE DISTRIBUTION FUNCTIONS

In Chapter 2 the empirical distribution function was introduced for sample data. It is a function that increases in height from 0 to 1 as the value of x increases. In that way it resembles the c.d.f.s introduced in this section.

However, there is a difference between the two, analogous to the difference between population and sample. The c.d.f. is a population curve that represents true probabilities of getting observations in any given range, whereas the e.d.f. is a sample curve that represents actual relative frequencies of sample data in any given range. Since relative frequencies tend to get closer to probabilities as the sample sizes get larger, the empirical distribution functions also tend to get closer to their respective cumulative distribution functions as the sample sizes increase. This fact is the basis for using e.d.f.s from sample data to estimate unknown c.d.f.s. in later chapters of this text.

EXAMPLE

A random sample of 20 days' records is obtained and the daily amounts of machine downtime are recorded. The empirical distribution function of these

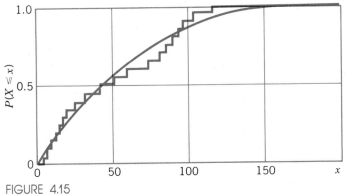

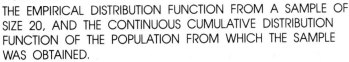

FIGURE 4.15
THE EMPIRICAL DISTRIBUTION FUNCTION FROM A SAMPLE OF
SIZE 20, AND THE CONTINUOUS CUMULATIVE DISTRIBUTION
FUNCTION OF THE POPULATION FROM WHICH THE SAMPLE
WAS OBTAINED.

20 sample data points is plotted in Figure 4.15, along with the c.d.f. Notice
that the e.d.f. shows some sampling error, that is, it does not follow the c.d.f.
as closely as is possible for a sample of size 20. Yet there is a tendency for
the e.d.f. to follow along the general direction of the c.d.f. The e.d.f. of a
larger sample will tend to follow the c.d.f. closer than that of a smaller sample.
A sample of size 200, for example, will almost certainly produce an e.d.f.
that will be almost indistinguishable from the c.d.f.

BIVARIATE CONTINUOUS RANDOM VARIABLES

Two random variables (X, Y) considered jointly form a bivariate random variable. If both random variables are continuous, then (X, Y) has a joint density
function. A graphical representation of a joint density function is difficult because it requires three dimensions, one for X, one for Y, and a third dimension
for the height of the density function. Figure 4.16 represents a joint density

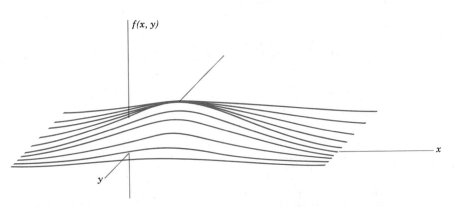

FIGURE 4.16
A JOINT DENSITY FUNCTION $f(x,y)$.

| | Weight | | | | | Marginal |
	125–144	145–164	165–184	185–204	205–224	Totals
Height 77–79	.000	.000	.000	.007	.005	.012
75–77	.001	.016	.027	.022	.010	.076
73–75	.016	.031	.036	.037	.008	.128
71–73	.018	.057	.068	.052	.015	.210
69–71	.036	.112	.110	.049	.015	.322
67–69	.017	.088	.061	.023	.006	.195
65–67	.009	.036	.010	.002	.000	.057
Totals	.097	.340	.312	.192	.059	1.000

FIGURE 4.17
A JOINT RELATIVE FREQUENCY TABLE FOR ARMY TRAINEES.

function. The volume under the entire function, above the plane $z = 0$, equals 1.0, so that the probability associated with any set of values of X and Y is given by the volume bounded by those values of X and Y, and under the joint density function.

Just as a single continuous random variable is often used as an approximation to a discrete random variable, a bivariate continuous random variable is sometimes used as an approximation to a discrete bivariate random variable.

EXAMPLE

An organization that supplies an army training camp with uniforms is interested in the joint distribution of height and weight of the 1000 men most recently assigned to that camp. The relative frequency table in Figure 4.17 represents the observed data.

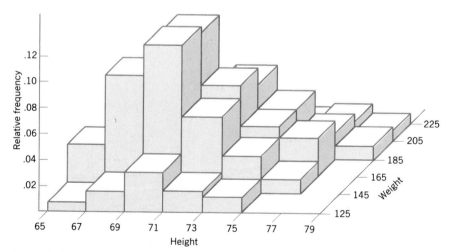

FIGURE 4.18
A BIVARIATE HISTOGRAM REPRESENTING RELATIVE FREQUENCIES OF 1000 SOLDIERS ACCORDING TO HEIGHT AND WEIGHT.

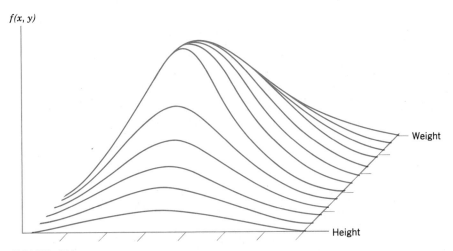

$f(x, y)$

Weight

Height

FIGURE 4.19
A BIVARIATE DENSITY FUNCTION REPRESENTING THE HEIGHT AND WEIGHT
OF ARMY TRAINEES.

The joint relative frequency table in Figure 4.17 can be depicted graphically
in three dimensions where one dimension is height, one dimension is weight,
and the third dimension is the relative frequency for each height and weight.
(See Figure 4.18.)

Imagine, if you will, what happens to Figure 4.18 as the interval widths get
narrower, and the number of observations gets larger, so that the bivariate
histogram becomes smoother. It will approach a bivariate density function,
which may look like Figure 4.19.

INDEPENDENT RANDOM VARIABLES

In most of this text, random variables are treated as if they are independent of
one another. Independence is defined for continuous random variables in
almost the same way as it was for discrete random variables in the previous
section, or for independent events in the first section of this chapter. That is,
random variables or events are **independent** if the probabilities associated with
one do not depend in any way on the outcomes associated with the other.
This means that the probability for each joint event can be obtained merely
by multiplying the probabilities for the individual events. This statement is
expressed in terms of c.d.f.s in Eq. 4.10.

INDEPENDENT RANDOM VARIABLES

Two random variables X and Y are **independent** if the values they assume
satisfy the definition of independent events, that is, if

$$P(X \leq x, Y \leq y) = P(X \leq x)P(Y \leq y) \qquad (4.10)$$

for all possible values of x and y.

EXAMPLE

If your semester grade-point average is independent of the performance of the basketball team, then the probability that you get below a 3.0 can be multiplied by the probability that the basketball team loses less than half its games, to get the probability of the joint event that you get below a 3.0 and the team loses less than half its games.

THE NEED FOR CALCULUS TO CONTINUE

This text does not require calculus as a prerequisite. Most of the applications of statistical methods do not require calculus, but occasionally the lack of calculus severely limits the discussion of certain topics. This is one of those times. In order to discuss marginal distributions and conditional distributions for continuous random variables some knowledge of calculus is required. The presentation is not difficult. It consists primarily of the same presentation given in the previous section for discrete random variables, but using an integral instead of a summation, and a density function instead of a probability function.

The science of statistics is based on probability theory. Anyone wishing to pursue the study of statistics in depth needs to have a working knowledge of calculus, because much of the probability theory uses calculus. The interested reader should consult one of the many excellent books on the subject of probability.

EXERCISES

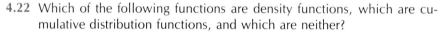

4.22 Which of the following functions are density functions, which are cumulative distribution functions, and which are neither?

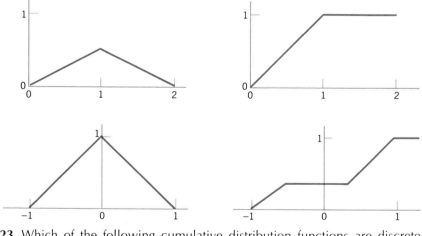

4.23 Which of the following cumulative distribution functions are discrete and which are continuous?

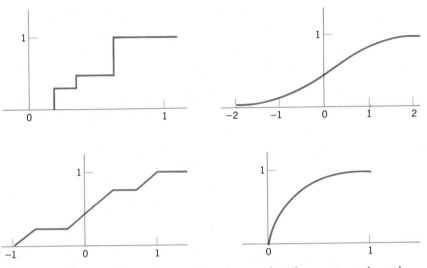

4.24 A business that specializes in used tires knows that the amount of tread on the tires in its inventory ranges from no tread at all to about one-half inch of tread. Assume that the amount of tread X on a randomly selected tire has the following density function.

$$f(x) = 0 \quad \text{for } x < 0$$
$$f(x) = 2 \quad \text{for } 0 < x < .5$$
$$f(x) = 0 \quad \text{for } .5 < x$$

Draw a graph of this density function. (This density function is called a *uniform density function*. Can you guess why?)

4.25 Use the density function in Exercise 4.24 to find the probability that a randomly selected tire has more than $\frac{3}{8}$-inch tread on it.

4.26 The c.d.f. for the density function in Exercise 4.24 is given by

$$F(x) = 0 \quad \text{for } x < 0$$
$$F(x) = 2x \quad \text{for } 0 < x < .5$$
$$F(x) = 1 \quad \text{for } .5 < x$$

Draw a graph of this c.d.f.

4.27 Use the c.d.f. in Exercise 4.26 to find the probability that a randomly selected tire has more than $\frac{3}{8}$-inch tread on it.

4.28 The probability of rain, as reported by a meteorologist in Pennsylvania, varies from day to day like a random variable with the following density function.

$$f(x) = 0 \quad \text{for } x < 0$$
$$f(x) = 2 - 2x \quad \text{for } 0 < x < 1$$
$$f(x) = 0 \quad \text{for } x > 1$$

Draw a graph of this density function.

4.29 Use the density function in Exercise 4.28 to find the probability that, on a randomly selected day, the probability of rain will be less than .25.

4.30 The c.d.f. for the density function in Exercise 4.28 is

$$F(x) = 0 \qquad \text{for } x < 0$$
$$F(x) = 2x - x^2 \quad \text{for } 0 < x < 1$$
$$F(x) = 1 \qquad \text{for } x > 1$$

Draw a graph of this c.d.f.

4.31 Use the c.d.f. in Exercise 4.30 to find the probability that, on a randomly selected day, the probability of rain will be less than .25.

4.32 The length of time between telephone calls (X = time in minutes) in an office has the following cumulative distribution function.

$$P(X \leq x) = 0 \qquad \text{for } x < 0$$
$$= \sqrt{x/10} \quad \text{for } 0 \leq x \leq 10$$
$$= 1 \qquad \text{for } x > 10$$

Use the cumulative distribution function to find the following probabilities for this continuous random variable.

(a) $P(X \leq 10)$

(b) $P(X \leq 5)$

(c) $P(X < 5)$

(d) $P(X > 5)$

(e) $P(1 < X < 5)$

(f) $P(X = 4)$

4.33 The density function for the random variable X in Exercise 4.32 is given as

$$f(x) = \frac{.158}{\sqrt{x}} \quad \text{for } 0 < x \leq 10$$
$$= 0 \quad \text{for } x \leq 0 \text{ or } x > 10$$

Construct a graph of this density function.

4.34 Graph the cumulative distribution function given in Exercise 4.32.

4.35 A sample of 10 calls arriving in the office mentioned in Exercise 4.32 showed the following lengths of time between calls in minutes: 6.3, 4.0, 0.2, 1.1, 7.1, 1.5, 9.0, 0.8, 0.6, 2.8. Plot the empirical distribution function for these data on the same graph with the c.d.f. graphed in Exercise 4.34. How do the two graphs compare?

4.4

POPULATION PARAMETERS

DEFINITION OF A PARAMETER

In Chapter 3 sample statistics were introduced. These included the sample mean $\overline{X}$, the sample standard deviation s, and sample quantiles X_q such as the sample median, sample quartiles, and sample percentiles. In addition to being used to summarize data, these statistics are also used to make inferences about the fixed characteristics of the populations from which the samples were obtained.

Some of the population characteristics of interest include the **population mean** μ (mu), which locates the middle of the population, the **population standard deviation** σ (sigma), which measures the spread of the population, and **population quantiles** λ_q (lambda-sub-q), such as the **population median** $\lambda_{.5}$, **population quartiles** $\lambda_{.25}$, and $\lambda_{.75}$, and **population percentiles,** which divide the population into two parts, one of relative size q and the other of relative size $1 - q$.

A fixed population characteristic is called a **parameter.** Unlike a statistic, which is a random variable and changes from one sample to another, a parameter is a constant and does not change. When a population parameter's value is unknown, sample statistics are used to estimate the parameter.

PARAMETER   A **parameter** is a fixed numerical quantity that describes a population characteristic.

THE POPULATION MEAN

As with the sample, a single numeric value needs to be identified that represents a measure of location for the population. The sample mean provides a measure of location for the sample with a weighted mean, where each observed value of X is weighted (multiplied) by its relative frequency of occurrence in the sample. In a similar manner the **population mean** is a weighted mean, where each possible value of X is weighted by its probability in the population.

Population means are easily demonstrated using discrete random variables. For continuous random variables the concept is the same, but the actual calculation of the population mean involves an integral instead of a summation. The Greek letter μ is used to denote the population mean. Sometimes the notation $E(X)$, "the expected value of X," is also used to denote the population mean. Both notations appear in Eq. 4.11.

POPULATION
MEAN

In populations where X equals the values $x_1,\ x_2,\ x_3,\ \ldots$, with respective probabilities $p_1,\ p_2,\ p_3,\ \ldots$, the **population mean** may be computed from the formula

$$E(X) = \mu = \sum_i x_i p_i \tag{4.11}$$

EXAMPLE

In Section 4.2 an example of a discrete random variable X was presented where X equals the number of college degrees earned by a randomly selected employee in a company with 150 employees. In that example the probabilities for X are given by the population relative frequencies as follows.

x	$P(X = x)$
0	.50
1	.24
2	.12
3	.10
4	.04

The population mean is found using Eq. 4.11 as

$$\mu = 0(.50) + 1(.24) + 2(.12) + 3(.10) + 4(.04) = .94$$

which provides .94 as a measure of the middle of the probability distribution. Of course, no employee has .94 degrees, but the *average* number of degrees in the population of employees is .94.

The *sample* mean may or may not agree exactly with the mean of the population from which the sample was obtained. For example, suppose a random sample of size 20 is obtained from the preceding population as it was in Section 4.2. The sample values are as follows.

Number of Degrees	Observed Frequency
0	8
1	5
2	2
3	4
4	1

The sample mean is

$$\bar{X} = \sum x_i(f_i/n)$$
$$= 0(8/20) + 1(5/20) + 2(2/20) + 3(4/20) + 4(1/20)$$
$$= 1.25$$

which is slightly larger than the population mean .94. A different sample will yield a different sample mean, which may be larger or smaller than the population mean. However, in a large sample the sample mean $\bar{X}$ will tend to be fairly close to the population mean μ, since in a large sample the relative frequency of occurrence of sample values will tend to be close to the relative frequency of the corresponding values in the population.

THE POPULATION STANDARD DEVIATION

The **population standard deviation** is denoted by the Greek letter σ and is given by Eq. 4.12. The population standard deviation is a measure of the variability within the population. The sample standard deviation s is the statistic customarily used for estimating σ when σ is not known. The **population variance** σ^2 is simply the square of the population standard deviation σ.

In populations where X assumes the values $x_1, x_2, x_3, \ldots$, with respective probabilities $p_1, p_2, p_3, \ldots$, the **population standard deviation** σ may be computed using either of the following two equations:

POPULATION
STANDARD
DEVIATION

$$\sigma = \sqrt{\sum_i (x_i - \mu)^2 p_i} \qquad (4.12)$$

or

$$\sigma = \sqrt{\sum_i x_i^2 p_i - \mu^2} \qquad (4.13)$$

EXAMPLE

In the previous example the population mean μ was found to be .94. The population standard deviation will be found using both equations. First, Eq. 4.12 gives

$$\sigma = [(0 - .94)^2(.50) + (1 - .94)^2(.24) + (2 - .94)^2(.12)$$
$$+ (3 - .94)^2(.10) + (4 - .94)^2(.04)]^{1/2}$$
$$= \sqrt{1.3764} = 1.173$$

Equation 4.13 is usually a little easier to use, and gives

$$\sigma = [0^2(.50) + 1^2(.24) + 2^2(.12) + 3^2(.10)$$
$$+ 4^2(.04) - (.94)^2]^{1/2}$$
$$= \sqrt{1.3764} = 1.173$$

as before. This value, 1.173 college degrees, gives a measure of the spread of the values in the population, and represents in one sense the mean distance between the population values and the population mean. Incidentally, the population variance is 1.3764, the square of the standard deviation.

For the sample of size 20 given in the previous example the sample standard deviation is

$$s = \left\{ \left[\sum x_i^2 f_i - \left(\sum x_i f_i \right)^2 \bigg/ n \right] \bigg/ (n-1) \right\}^{1/2}$$
$$= \{[0^2(8) + 1^2(5) + 2^2(2) + 3^2(4) + 4^2(1) - (25)^2/20]/19\}^{1/2}$$
$$= \sqrt{1.7763} = 1.333$$

which is slightly larger than the population standard deviation 1.173. Like all sample statistics, the sample standard deviation will vary from sample to sample, and is unlikely to be exactly equal to the population parameter σ that it estimates.

THE POPULATION COEFFICIENT OF VARIATION

The **population coefficient of variation,** σ/μ, provides a standardized measure of the population standard deviation in terms of the population mean. As with the sample coefficient of variation, it is free of dimension and is usually expressed as a percent $(\sigma/\mu)\cdot100\%$. The population coefficient of variation in the preceding example is

$$\frac{\sigma}{\mu} = \frac{1.173}{.94} = 1.248 = 124.8\%$$

and represents a standardized, dimensionless measure of spread in the population.

THE POPULATION MEDIAN

The population counterpart to the sample median is called, naturally, the **population median.** Just as the sample median is the middle observation in a sample, the population median is the middle value of the random variable, in a probability sense. That is, if $\lambda_{.5}$ is the median of a random variable X, then the probability of $X < \lambda_{.5}$ is 0.5 or less, and the probability of $X > \lambda_{.5}$ is also 0.5 or less. In practice, the population median is easily obtained from the population distribution function in the same way that the sample median is easily obtained from the empirical distribution function.

POPULATION MEDIAN

The **median of a population** is the number $\lambda_{.5}$ (read "lambda-sub-.5"), which has the property that a randomly obtained observation will be less than $\lambda_{.5}$ with probability 0.5 or less, and will exceed $\lambda_{.5}$ with probability 0.5 or less.

EXAMPLE

The population distribution function for the previous example in this section appeared in Figure 4.6, and is repeated in Figure 4.20 for the reader's convenience. Figure 4.20 also demonstrates how to find the population median. From 0.5 on the vertical scale a horizontal line is drawn until it intersects the c.d.f. The first point of intersection is at $x = 0$, but it continues to intersect all along the step from $x = 0$ to $x = 1$, so any value of x from 0 to 1 satisfies the definition of the population median. To avoid ambiguities, the middle value of the range from 0 to 1, $x = \frac{1}{2}$ in this case, is customarily used as $\lambda_{.5}$.

    This population median $x = \frac{1}{2}$ represents the middle of the population in the sense that the probability of $X < \frac{1}{2}$ is 0.5 or less, and the probability of

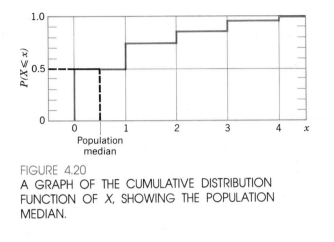

FIGURE 4.20
A GRAPH OF THE CUMULATIVE DISTRIBUTION
FUNCTION OF X, SHOWING THE POPULATION
MEDIAN.

$X > \frac{1}{2}$ is 0.5 or less. To be specific in this case the former probability equals one-half exactly, as does the latter probability.

POPULATION QUANTILES

The qth **population quantile,** $0 < q < 1$, called λ_q, is most easily found by locating q on the vertical scale of the c.d.f. and reading horizontally to find the x coordinate with that ordinate. The probability associated with values in the population less than λ_q is q or less, and the probability in the other direction is $(1 - q)$ or less. The terminology for population quantiles parallels that of sample quantiles, with $\lambda_{.25}$ and $\lambda_{.75}$ being the lower and upper **population quartiles,** respectively, $\lambda_{.43}$ being the forty-third **population percentile,** and so on. Figure 4.21 illustrates the method for finding the population quartiles, which are $\lambda_{.25} = 0$ and $\lambda_{.75} = 2$ in the previous example.

POPULATION QUANTILE   The qth **population quantile** λ_q is the number that has probability q or less associated with population values less than λ_q and probability $(1 - q)$ or less associated with population values greater than λ_q.

THE INTERQUARTILE RANGE

Just as the median may be used as a measure of location instead of the mean, quantiles may be used to measure the spread of the observations instead of the standard deviation. One way to do this is to use the **interquartile range,** which is the distance from the lower quartile to the upper quartile. The population interquartile range, obtained for the example in this section, is seen from Figure 4.21 to equal $2 - 0 = 2$. The interquartile range will usually tend to be a little larger than the standard deviation. However, the population interquartile range will always be less than or equal to the **population range,** the largest possible observation in the population minus the smallest possible observation in the population. The population range for the previous example is $4 - 0 = 4$.

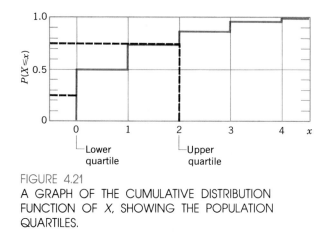

FIGURE 4.21
A GRAPH OF THE CUMULATIVE DISTRIBUTION
FUNCTION OF X, SHOWING THE POPULATION
QUARTILES.

POPULATION INTERQUARTILE RANGE	The **interquartile range** is the distance from the lower quartile to the upper quartile, $\lambda_{.75} - \lambda_{.25}$ for populations.

POPULATION PROPORTION

Population proportions are defined and found in a way similar to that used to find sample proportions, only the population is used instead of the sample.

POPULATION PROPORTION	The **population proportion** p is the fraction of the population that meets some stated criterion.

EXERCISES

4.36 *Consumer Reports* conducted a survey of insurance claims for 31 insurance companies based on 31,640 claims reported between January 1976 and June 1979. Even with a large sample such as this, the entire population is not accounted for; however, the estimates from this sample may be quite close to the true population values. As part of their summary, *Consumer Reports* gave the table below summarizing dollar amounts of claim settlements. Treat the relative frequencies (percentages) reported in this table as true population probabilities and find the population mean dollar amount for these claims. *Hint:* Use the class mark to represent each class. The arbitrary choice of 75,000 was made for the class mark in the last class.

Dollar Amount of Claim Settlement	Class Mark (in dollars)	Percentage of Claims (%)[a]
1–99	50	13.8
100–199	150	15.8
200–299	250	10.9

Dollar Amount of Claim Settlement	Class Mark (in dollars)	Percentage of Claims (%)[a]
300–399	350	7.9
400–499	450	5.0
500–999	750	16.8
1,000–1,499	1,250	6.9
1,500–1,999	1,750	4.0
2,000–2,499	2,250	3.0
2,500–4,999	3,750	5.9
5,000–9,999	7,500	4.0
10,000–49,999	30,000	5.0
50,000 or more	75,000	1.0

[a]These percentages were normalized to total 100% because of the rounding off in the *Consumer Reports* article.

4.37 Find the population standard deviation for the frequency distribution reported in Exercise 4.36.

4.38 Planning for television broadcasts of events such as the World Series presents a number of scheduling and advertising problems for television networks because the series winner is determined by the first team to win four games played between the two participating teams. This could take as few as four games or as many as seven games. Summary data are given below for all World Series from 1923 to 1986. Since there are data for 62 years, treat the relative frequencies as probabilities and find the mean for this population.

World Series 1923 to 1986

Number of Games Played	Frequency
4	11
5	12
6	13
7	26
	Total 62

4.39 Graph the population distribution function for the frequency distribution given in Exercise 4.38. Your graph should appear as a step function, as given in Figure 4.6.

4.40 Find the population standard deviation for the frequency distribution given in Exercise 4.38.

4.41 Find the population median for the frequency distribution reported in Exercise 4.38.

4.42 A restaurant has established the following probability model for groups of eight or less requiring seating.

Group size:	1	2	3	4	5	6	7	8
Probability:	.10	.30	.10	.20	.08	.11	.03	.08

Find the population mean for the group size.

4.43 Graph the population distribution function for the probability model given in Exercise 4.42.

4.44 Find the population standard deviation for the probability model given in Exercise 4.42.

4.45 Determine the population median from the graph of the population distribution function made in Exercise 4.43.

4.46 Find the population interquartile range from the graph of the population distribution function made in Exercise 4.43. How does this value compare with the population standard deviation found in Exercise 4.44?

4.47 The number of customers waiting for service in a barber shop is described by the probability model given below. Find the population mean for the number of customers waiting to be served.

Number waiting:	0	1	2	3	4	5	6	7
Probability:	.10	.20	.25	.15	.10	.08	.06	.06

4.48 Graph the population distribution function for the probability model given in Exercise 4.47.

4.49 Find the population standard deviation for the probability model given in Exercise 4.47.

4.50 A restaurant manager has been offering a buffet lunch, all you can eat, for $4.75. Half of her customers eat the equivalent of a $4.00 lunch, one-fourth eat a $5.00 lunch, and the other one-fourth eat a $6.00 lunch. Should the manager raise her prices to come out even on her buffet lunch?

4.51 An insurance company needs to charge $100 per year just for maintenance and fees associated with each policy. In addition they need to charge enough to cover the expected loss associated with the policy. Suppose that seven out of every 1000 men who are 50 years old will die within one year, on the average. How much should the insurance company charge for a one-year $10,000 policy on a 50-year-old man? How much should they charge for a $20,000 policy?

4.52 A frequency table of salary levels for all full-time employees at a local utility company furnished the following grouped data. Find the population mean and the population standard deviation for these data. (Use the midpoint of each interval as the representative point for each interval.)

Salary Group	Number of Employees
$ 0–5,000	0
5,000–10,000	27
10,000–15,000	73
15,000–20,000	48
20,000–25,000	13
25,000–30,000	8
30,000–35,000	1

4.53 A used car dealer finds that 40% of the days no cars are sold, 20% of the time one car is sold, 15% of the time two are sold, 10% of the time three are sold, 8% of the time four are sold, 6% of the time five cars are sold, and 1% of the time six cars are sold.

(a) Plot the distribution function for the number of cars sold.

(b) Find the mean and standard deviation of the number of cars sold.

(c) What percent of the time can the dealer expect the number of cars sold to be within one standard deviation of the mean?

(d) The dealer's salesman, Mike, gets a straight commission of $50 on the first car sold and $100 for each additional car sold, no matter who sells it. Mike gets $30 on days when no cars are sold. What is Mike's expected daily earnings?

4.5

REVIEW EXERCISES

4.54 A university has conducted a survey of 250 employees to obtain their opinion on collective bargaining by a teacher's union. The frequency summary is given in the following table.

Employee Type	Opinion Opposed	Opinion Undecided	Opinion In Favor	Total
Staff	18	19		75
Faculty	63	12		125
Administrator				50
Total	100	35	115	250

Assume that one of the 250 employees is selected at random.

(a) Fill in the empty cells in the frequency table and then convert it to a joint probability table.

(b) What is the marginal probability of a favorable opinion?

(c) What is the probability of an opinion being favorable and from a faculty member?

(d) Find the conditional probability of an opinion being favorable given that the opinion was expressed by a faculty member.

4.55 Assuming the joint probability of a motorist being both uninsured and having an accident is .3 and the marginal probability of being uninsured is .5, find the conditional probability of a motorist having an accident given that the motorist is uninsured.

4.56 The number of magazines sold each hour at a street corner newsstand is described by the probability model given below.

Number of magazines sold:	0	1	2	3	4	5	6
Probability:	.14	.27	.27	.18	.09	.04	.01

Let the random variable X equal the number of magazines sold each hour and find the following probabilities.

(a) $P(X \leq 1)$

(b) $P(2 \leq X \leq 4)$

(c) $P(2 < X < 4)$

(d) $P(X = 3)$

(e) $P(X > 4)$

(f) $P(X \geq 4)$

4.57 Construct a vertical bar graph of the probability function of the random variable X in Exercise 4.56.

4.58 Graph the cumulative distribution function for the random variable in Exercise 4.56.

4.59 From past experience an insurance company knows that 1 check in 500 received for payment of premiums will be drawn on insufficient funds and will be postdated. Additionally, 1 check in 50 will be postdated. A postdated check is received by the company. What is the probability that the check will be drawn on insufficient funds?

4.60 A joint probability table is given below for color blindness by sex. Use this table to find the conditional probability that an individual will be male given that the individual is color-blind. Also find the conditional probability that an individual will be female given that the individual is color-blind. Is color blindness independent of sex?

	Normal Vision	Color-blind
Male	.475	.025
Female	.495	.005

4.61 Two factories manufacture the same machine part. Each part is classified as having either 0, 1, 2, or 3 manufacturing defects. The joint probability distribution for this setting is as follows.

	Number of Defects 0	1	2	3
Manufacturer A	.1250	.0625	.1875	.1250
Manufacturer B	.0625	.0625	.1250	.2500

(a) A part is observed to have no defects. What is the conditional probability that it was produced by manufacturer A?

(b) A part is known to have been produced by manufacturer A. What is the conditional probability that the part has no defects?

(c) A part is known to have two or more defects. What is the conditional probability that it was manufactured by A?

4.62 At rifle practice a soldier fires 30 rounds from each of three positions. In the standing position he hits the target 18 times, in the sitting position 27 times, and in the prone position 24 times.

(a) Select a round at random. What is the probability that it hit the target?

(b) Select a round at random. What is the probability that it was fired from the prone position?

(c) Given that a round hit the target, what is the probability that it was fired from a prone position?

(d) Are the events "hitting the target" and "firing from a prone position" independent events? Why?

(e) Draw a probability tree to represent this situation, where the first set of branches represents the position, and the second set of branches represents hitting or not hitting the target.

(f) Use the probability tree in part (e) to answer the questions in parts (a) and (b). Is this method of finding the answers any simpler than the method you used before?

4.63 The manager of a quick-stop store knows that customers with out-of-state license plates on their cars will purchase only gasoline 25% of the time, only in-store products 30% of the time, and both gasoline and in-store products the remaining 45% of the time. In-state customers, however, purchase only gasoline 40% of the time, only in-store products 50% of the time, and both 10% of the time. One-third of her customers are from out-of-state.

(a) What is the overall probability of a customer purchasing only gasoline?

(b) What percent of the manager's customers purchase gasoline, without regard to whether other products were purchased also?

(c) Are the events "a customer purchases gasoline" and "a customer purchases in-store products" mutually exclusive? Why?

(d) Are the events in part (c) independent? Why?

(e) Draw a probability tree to represent this situation, where the first set of branches represents in-state or out-of-state customers, and the second set of branches represents the type of purchase.

(f) Use the probability tree in part (e) to answer the questions in parts (a) and (b). Is this method of finding the answers any simpler than the one you used before?

4.64 A summary of motor vehicle accident data is given below. The accidents are classified according to the vehicle size and whether or not there was serious injury to the driver.

	Driver Injury	
Vehicle Size	Serious	Not Serious
Small	200	807
Medium	214	1068
Large	250	1638

(a) Which size of vehicle had the most accidents involving a serious injury to the driver?

(b) If an accident occurs, which size of vehicle is most likely to have a serious driver injury?

(c) What additional information is needed in order to say which size vehicle is the safest with regard to serious driver injury?

4.65 An airline boards 36 passengers. These passengers are classified according to sex and according to whether they are smokers or not. Twenty-four passengers are smokers. Eight of the 12 female passengers are smokers.

A pen is left at the check-in counter and the flight attendant feels that it is equally likely to belong to any of the passengers on the airplane.

(a) What is the probability the pen belongs to a male passenger?

(b) What is the probability the pen belongs to a smoker?

(c) Given the pen is a lady's pen, what is the probability the pen belongs to a smoker?

(d) Are the events "the pen belongs to a lady" and "the pen belongs to a smoker" independent? Why?

4.66 A family has three children.

(a) List all the points in the sample space where each outcome consists of the sex of the first child, the second child, and the third child, in that order, such as (F,M,M).

(b) What is a reasonable probability for each outcome? Why?

(c) What is the probability that there are two boys and a girl in the family?

(d) Given that the first child is a boy, what is the probability that there are two boys and a girl in the family?

(e) Given that there is at least one boy in the family, what is the probability that there are two boys and a girl in the family?

(f) What are the differences in the questions in parts (c), (d), and (e)?

(g) Draw a probability tree for this situation, where the first set of branches represents the sex of the first child, the second set of branches represents the sex of the second child, and the third set of branches represents the sex of the third child.

(h) Use the probability tree in part (g) to answer the questions in parts (c), (d), and (e). Is it any clearer now why the answers to those three questions are not the same?

4.67 A family has four children.

(a) List the points in the sample space, where each outcome is the sex of the first child, of the second child, of the third child, and of the fourth child, in that order, such as (F,F,M,M), that means the first two children were girls and the last two were boys.

(b) What is a reasonable probability for each point in the sample space?

(c) Let the random variable X be the number of girls in the family. What are the possible values of X?

(d) Find $P(X = 0)$ and $P(X = 1)$.

(e) Graph the probability distribution of X.

(f) Graph the c.d.f. of X.

(g) Is X discrete or continuous?

4.68 An automobile dealer receives a shipment of 10 cars. He randomly selects two of those cars and inspects them for major defects. If both cars pass the inspection, he accepts the entire shipment; otherwise, he refuses the shipment. If the cars are "Monday" cars (i.e., made under less than the best conditions) there will be five cars with major defects out of the 10. What is the probability that the dealer accepts the shipment when they are "Monday" cars?

4.69 Over a 12-year period the inflation-adjusted yield on a utility's bonds was as follows:

-2.37%, -3.19%, -8.83%, -4.18%, -6.66%, -14.35%
-10.41%, 4.81%, 5.74%, 10.48%, 6.46%, 8.07%

Find the population mean yield and the population median yield for a randomly selected year.

4.70 A popular sports magazine allocates its advertising space such that each page either contains no ads or has $\frac{1}{3}$, $\frac{2}{3}$, or the entire page devoted to advertising. The probability model below describes the allocation used by the magazine on their ads. Find the mean proportion of each page devoted to advertisements.

Fraction of the page devoted to ads:	0	1/3	2/3	1
Probability:	.410	.015	.025	.550

4.71 Find the population standard deviation for the probability model given in Exercise 4.70.

4.72 Graph the cumulative distribution function for the probability model given in Exercise 4.70 and use it to find the population median.

BIBLIOGRAPHY

Additional material on the topics presented in this chapter can be found in the following publication.

Mosteller, F., Rourke, R., and Thomas, G. (1970). *Probability with Statistical Applications*, 2nd ed. Addison-Wesley, Reading, Mass.

5
SOME USEFUL DISCRETE AND CONTINUOUS DISTRIBUTIONS

PRELIMINARY REMARKS

Thus far in this text much effort has been devoted to the discussion of *sample* data, display of *sample* data, and summarization of *sample* data. In addition, references have been made to *populations* with *population* means and *population* standard deviations defined. One of the primary reasons for studying *sample statistics* is to obtain some information about the corresponding *population parameters*.

For example, the empirical distribution function for a random sample is useful because it approximates the distribution function of the population. That is, for reasonably large sample sizes, the empirical distribution function resembles the population distribution function, and as the sample size gets larger the approximation tends to get better. Since the population distribution function (or mean, or standard deviation) is the item of interest, whenever it is unknown the empirical distribution function (or sample mean, or sample standard deviation) can be used to make inferences about the unknown population.

In this chapter some distributions are introduced that are used in many situations—as population distribution functions, as distribution functions of sample statistics, or as models of common random occurrences. There are countless numbers of different distribution functions. These are just a few of the most useful ones.

Much of this material relates to Chapter 4. Random variables are discussed freely throughout this chapter. Recall that a random variable is a rule for assigning numbers to the possible outcomes of an experiment. The terms *the mean of X* or *the standard deviation of X* are used interchangeably with the mean and standard deviation of the population distribution function associated with the random variable X. Also the term *distribution function* is often used instead of the more cumbersome expressions *population distribution function* or *cumulative distribution function*.

EXAMPLE

Consider the output of a production process. Let X be a random variable denoting *the number of defective parts* manufactured in one day. Then the expression X = 0 means the number 0 is assigned to the outcome for a day on which *no* defective parts were manufactured. Likewise, X = 9 is used for a day on which exactly 9 defective parts were manufactured. It may be helpful to think of the following expressions as saying the same thing.

The number of defective parts is 6

$$X = 6$$

Thus, the use of the letter X is a shortcut notation (for convenience in formulas) for the expression "the number of defective parts." This random variable is a discrete random variable. Two discrete distributions that are useful as probability models for this type of random variable are the **binomial** and the **Poisson** distributions. Both are introduced in this chapter.

EXAMPLE

Let X be a random variable representing the amount of time (in weeks) elapsed from the time a light bulb is first turned on until it fails in a continuous test, that is, X is the failure time. Thus, X = 11.89 means a bulb burned for 11.89 weeks before it failed. This is an example of a continuous random variable. Two popular continuous distributions introduced in this chapter are the **normal** and the **exponential** distributions.

5.1

THE BINOMIAL DISTRIBUTION

PROBLEM SETTING 5.1

In an attempt to increase the volume of business at your restaurant, you invest in a two-week advertising campaign on TV. After the TV ads start, you inquire of your customers when they pay their bills if they came to your restaurant because of the TV ads. Each customer's response can be classified as either a simple yes or no. At the end of the ad campaign you want to know how many customers were influenced by the ads so you can determine the effectiveness of the ads and, in particular, if the increased volume of business will offset the cost of the ads.

A BINOMIAL RANDOM VARIABLE

For this study the **experiment** is the process of asking all of the customers if they came to the restaurant because of the TV ads. The random variable is defined as the number of customers who indicated they were influenced by the TV ads. Clearly the random variable can assume only the values 0, 1, 2, 3, . . . , n, where n is the total number of customers, and therefore X is a **discrete** random variable.

Consider the characteristics that define this experiment. First, the experiment consists of n independent trials. Trials are repetitions under identical conditions. Each trial is one customer's response, and the responses are assumed to be independent, that is, each customer's response is not influenced by the responses of other customers. Second, each customer gives a dichotomous response, that is, either yes or no. Third, it can be assumed that the probability of a yes answer is the same from customer to customer. This is an example of an experiment in which the **binomial distribution** can be used as the probability distribution of X. A random variable whose distribution is the binomial distribution is called a **binomial random variable.**

A random variable has the **binomial distribution** if the following conditions exist.

A BINOMIAL
RANDOM
VARIABLE

1. There are one or more trials. (The number of trials is denoted by n, and is a known number.)

2. Each trial results in one of two outcomes. (The two outcomes are usually called **success** and **failure** for convenience.)

3. The outcomes from trial to trial are independent. That is, the probability of an outcome for any particular trial is not influenced by the outcomes of the other trials.

4. The probability of success, denoted by p, is the same from trial to trial. (So is the probability of failure q, where $q = 1 - p$.)

5. The random variable equals the number of successes in the n trials. (Thus the random variable may equal any integer value from 0 to n.)

EXAMPLE

For purposes of illustration, suppose the sample data obtained in Problem Setting 5.1 are analyzed after the first four customers, that is, $n = 4$. If X is a random variable representing the total number of affirmative responses among these four customers, then the possible values of X are 0, 1, 2, 3, and 4.

Further, if the TV ads influenced 25% of all customers to come to the restaurant, the probability of each person responding yes is $p = .25$. Note that p is usually unknown. The probability that the first customer says no is

$$P(\text{no}) = 1 - p = q = .75$$

The joint probability of the first customer saying no and the second customer saying yes is

$$P(\text{no, yes}) = P(\text{no}){\cdot}P(\text{yes}) = q{\cdot}p = (.75)(.25)$$

because of the assumed independence between responses as indicated in Eq. 4.2. The probability of a no from the first customer, then yes from the second customer, and then no from the third customer is

$$P(\text{no, yes, no}) = q{\cdot}p{\cdot}q = (.75)(.25)(.75)$$

The probability of only the second customer saying yes, with the first, third, and fourth customers responding no is

$$P(\text{no, yes, no, no}) = q{\cdot}p{\cdot}q{\cdot}q = p{\cdot}q^3 = (.25)(.75)^3$$

In this latter outcome the random variable $X = 1$ since only one customer said yes. How many other ways are there to get $X = 1$, or one yes and three no responses? The first customer could be the one to respond yes, with probability

$$P(\text{yes, no, no, no}) = p{\cdot}q{\cdot}q{\cdot}q = p{\cdot}q^3 = (.25)(.75)^3$$

The yes could come from the third customer

$$P(\text{no, no, yes, no}) = q{\cdot}q{\cdot}p{\cdot}q = p{\cdot}q^3 = (.25)(.75)^3$$

Finally the yes could be from the fourth customer

$$P(\text{no, no, no, yes}) = q{\cdot}q{\cdot}q{\cdot}p = p{\cdot}q^3 = (.25)(.75)^3$$

Thus there are four different ways in which there could be exactly one yes

$$
\begin{aligned}
f(0) &= 1(.25)^0(.75)^4 = .316 = P(X = 0)\\
f(1) &= 4(.25)^1(.75)^3 = .422 = P(X = 1)\\
f(2) &= 6(.25)^2(.75)^2 = .211 = P(X = 2)\\
f(3) &= 4(.25)^3(.75)^1 = .047 = P(X = 3)\\
f(4) &= 1(.25)^4(.75)^0 = \underline{.004} = P(X = 4)\\
&\ \text{Total} = 1.000
\end{aligned}
$$

FIGURE 5.1
BINOMIAL PROBABILITIES FOR $n = 4$, $p = .25$.

response, so the probability of $X = 1$ is the sum of the four probabilities

$$P(X = 1) = 4pq^3 = 4(.25)(.75)^3 = .422$$

Probabilities for the other possible values of X are given in Figure 5.1. An illustration is given later in this section that shows how to get these probabilities from Table A1 in the Appendix, so these calculations will not be necessary in most cases.

THE BINOMIAL PROBABILITY FUNCTION

When a random variable satisfies the requirements to be a binomial random variable, it takes one of the possible values 0, 1, 2, up to n, the number of trials. The probability associated with each possible value is denoted by $f(x)$, and is given by

$$f(x) = \binom{n}{x} p^x q^{n-x} \quad \text{for } x = 0, 1, 2, \ldots, n \tag{5.1}$$

The term $\binom{n}{x}$, called the **binomial coefficient,** is commonly referred to as a counting formula or as a **combination** of n trials selected x at a time.

The number of ways that exactly x successes can occur in n trials is referred to as a **combination** of n trials selected x at a time and is calculated as follows:

$$\binom{n}{x} = \frac{n!}{x!(n - x)!} \tag{5.2}$$

for $x = 0, 1, 2, \ldots, n$.

COMBINATION   The factorial notation $a!$ ("a factorial") represents the product

$$a! = a(a - 1)(a - 2) \cdots (2)(1)$$

for any integer a.

Note By convention 0! is taken to be equal to 1.

The combination formula is useful because it can be used with the binomial probability function to count (compute) the number of ways that exactly x successes can occur in n trials of a binomial random variable.

The term p^x in Eq. 5.1 represents the probability of x successes in x trials, the term q^{n-x} represents the probability of $n - x$ failures in $n - x$ trials, and they are multiplied together because the trials are independent. The binomial coefficient represents the number of possible ways in which n trials can result in exactly x successes and $n - x$ failures. The function $f(x)$ is called the **probability function.** This formula for $f(x)$ is just another way of assigning probabilities to the various outcomes of X, as was done in Chapter 4.

This example continues the previous example. The probabilities in Figure 5.1 agree with the formula for binomial probabilities given in Eq. 5.1. Since $0! = 1$, the first binomial coefficient is

$$\binom{n}{x} = \binom{4}{0} = \frac{4!}{0!4!} = \frac{24}{1(24)} = 1$$

resulting in

$$f(0) = \binom{4}{0}(.25)^0(.75)^4 = .316$$

The other binomial coefficients are

$$\binom{4}{1} = \frac{4!}{1!3!} = 4 \qquad \binom{4}{2} = \frac{4!}{2!2!} = 6$$

$$\binom{4}{3} = \frac{4!}{3!1!} = 4 \qquad \binom{4}{4} = \frac{4!}{4!0!} = 1$$

which shows how the other probabilities in Figure 5.1 were computed using the formula for binomial probabilities given in Eq. 5.1.

A graph of the binomial probabilities of Figure 5.1 is given in Figure 5.2. It shows that the most probable response is exactly one yes out of the four customers. Note that the binomial probability function with $p = .25$ is skewed to the right.

The probabilities given by the probability function in Figure 5.1 are accumulated in Figure 5.3 to get $P(X \leq x)$, the cumulative distribution function. The probability of getting two or fewer yes responses is $P(X \leq 2) = .949$, meaning that almost all of the time the number of yes responses will be 0, 1,

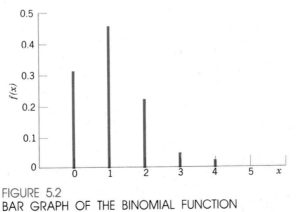

FIGURE 5.2
BAR GRAPH OF THE BINOMIAL FUNCTION
FOR $n = 4$, $p = .25$.

x	$P(X \leq x)$	
0	.316	= .316
1	.738	= .316 + .422
2	.949	= .316 + .422 + .211
3	.996	= .316 + .422 + .211 + .047
4	1.000	= .316 + .422 + .211 + .047 + .004

FIGURE 5.3

VALUE OF $P(X \leq x)$ IN THE BINOMIAL DISTRIBUTION FOR
$n = 4$, $p = .25$.

In the **binomial distribution** the probabilities are given by the probability function

$$P(X = x) = f(x) = \binom{n}{x} p^x q^{n-x}, \qquad x = 0, 1, 2, \ldots, n$$

where

X is the number of successes in n trials
n is the number of trials
p is the probability of success on any one trial
$q = 1 - p$ and is the probability of failure on any one trial

THE BINOMIAL
DISTRIBUTION

The distribution function

$$F(x) = P(X \leq x) = \sum_{i \leq x} f(i)$$

is given in Table A1 for $n \leq 30$ and for selected values of p. The parameters of the binomial distribution are n and p.

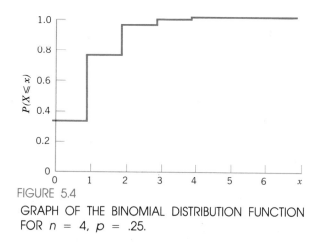

FIGURE 5.4

GRAPH OF THE BINOMIAL DISTRIBUTION FUNCTION
FOR $n = 4$, $p = .25$.

or 2. The probability of three or fewer yes responses is .996, even closer to 1, and finally the probability of four or fewer yes responses is exactly 1.0.

A graph of the probabilities from Figure 5.3 appears in Figure 5.4, where several points are illustrated. First note that the graph is that of a step function, illustrating that the binomial distribution is discrete. Next, note that the height of each step equals the probability associated with that value, as given in Figure 5.1. Finally, note that this distribution function, like all distribution functions, increases from 0 on the left to 1 on the right.

TABLES FOR THE BINOMIAL DISTRIBUTION FUNCTION

The formula for computing binomial probabilities is seldom used in practice. Tables are usually used to look up the desired probabilities. One such table is given in the Appendix of this book as Table A1 for values of $n \leq 30$ and selected values of p. Figure 5.5 reproduces a small portion of Table A1. When using Table A1 first find the section for the number of trials n (sample size). Next, for that value of n, find the column corresponding to p. The numbers in that column are the cumulative probabilities, $P(X \leq x)$, for each value of x from 0 to n. If the value for p falls between two of the columns given in Table A1, the probability can be calculated *exactly* using the probability function, or linear interpolation between values in the table can be used to *approximate* the desired probabilities. (See the appendix to this chapter for an example of linear interpolation.)

EXAMPLE

Figure 5.5 shows the probabilities for $n = 4$. The column for $p = .25$ gives more exact probabilities than were given in Figure 5.3. Individual probabilities, such as $P(X = 1)$, may be found by subtracting successive entries in the table

$$P(X = 1) = P(X \leq 1) - P(X \leq 0)$$
$$= .7383 - .3164$$
$$= .4219$$

which is in agreement with the number in Figure 5.1.

				p			
n	x	.05	.10	.15	.20	.25	.30
4	0	.8145	.6561	.5220	.4096	.3164	.2401
	1	.9860	.9477	.8905	.8192	.7383	.6517
	2	.9995	.9963	.9880	.9728	.9492	.9163
	3	1.0000	.9999	.9995	.9984	.9961	.9919
	4	1.0000	1.0000	1.0000	1.0000	1.0000	1.0000

FIGURE 5.5
A PORTION OF TABLE A1 OF THE BINOMIAL DISTRIBUTION FUNCTION.

POPULATION MEAN

The *population mean* for the binomial distribution may be obtained by direct application of the definition given by Eq. 4.11. It can be shown mathematically that for binomial distributions the definition of the population mean

$$\mu = \sum x_i f(x_i)$$

reduces to the much simpler form

$$\mu = np \tag{5.3}$$

For example, if $n = 4$ and $p = .25$, the mean of X is $4(.25) = 1.0$. This agrees with intuition. If four customers are asked a certain question and the probability is .25 of getting a yes from any one of them, then one would expect, on the average, about 1.0 yes responses from the four customers.

POPULATION STANDARD DEVIATION

The *population standard deviation* is almost as easy to find as the population mean. The standard deviation for the binomial distribution is given by the equation

$$\sigma = \sqrt{npq} \tag{5.4}$$

which may be obtained by direct application of Eq. 4.12. Thus the *variance* is npq. In the example with $n = 4$ and $p = .25$, the standard deviation is

$$\sigma = \sqrt{4(.25)(.75)}$$
$$= \sqrt{.75}$$
$$= .866$$

The actual numerical value for the population standard deviation is not as easy to verify intuitively as the population mean.

ESTIMATING p IN THE BINOMIAL DISTRIBUTION

The value of n is usually known in the binomial distribution because one can usually count the number of trials. But the value of p is seldom known. The sample is obtained primarily to estimate p.

In Problem Setting 5.1, as owner of the restaurant you want to estimate the proportion p of customers you get because of the TV ads, so you ask several customers whether they came because of the ads. If 28% of the customers you ask indicate that they came because of the TV ads, you would estimate the unknown p to be about .28. The value .28 is called the **sample proportion** $\hat{p}$ and is found by dividing the number of favorable responses X by the number of customers n:

$$\hat{p} = \frac{X}{n} = \text{sample proportion} \tag{5.5}$$

Just as $\hat{p}$ may be used to estimate p, it seems reasonable to use $n\hat{p}$ to estimate the population mean np, and $\sqrt{n\hat{p}(1-\hat{p})}$ to estimate the population standard deviation $\sqrt{npq}$ of the number of successes X. More will be said about estimating parameters in the following chapter.

EXAMPLES OF BINOMIAL RANDOM VARIABLES

1. **Accounting** A sample audit involves examining 100 checking accounts from a bank that has over 5000 checking accounts. Among other things, the auditor notes whether the accounts have been overdrawn at any time in the last 30 days. In this case n equals 100, X equals the number of overdrawn accounts, and p (unknown) represents the proportion of all accounts that have been overdrawn in the last 30 days. The unknown value of p is estimated using $\hat{p}$. If the total number of overdrawn accounts is to be estimated, the value for $\hat{p}$ is simply multiplied by the total number of accounts. That is, suppose 7 of the 100 accounts in the sample have been overdrawn in the last 30 days. Then $\hat{p} = \frac{7}{100} = .07$ is an estimate of the overall proportion p of overdrawn accounts. To estimate the total number of overdrawn accounts, .07 is multiplied by 5000 to get 350.

    Note that the X has the binomial distribution, because there are n trials (each account is a trial) where each trial is "overdrawn" or "not overdrawn," presumably independently of each other. If the sampling is done at random, then each account selected has the same probability p of being overdrawn.

2. **Marketing** A sample tube of toothpaste "A" is mailed to hundreds of households. One month later 25 of those households are contacted to see if they bought any tubes of toothpaste A because of the mailing. The number of favorable responses is called X. The number of trials n is 25, and the probability of a favorable response p is unknown. The sample proportion $\hat{p}$ may be used to estimate p so that the marketing team can judge whether or not this marketing technique is worthwhile. If 12 out of the 25 households bought toothpaste A because of the mailing, then the estimate of the proportion of families buying toothpaste A out of the entire population of families in the mailing is $\hat{p} = X/n = \frac{12}{25} = .48$. Moreover, this estimate could apply to the population of families in anticipated mailings as well. As before, the binomial distribution applies to the probability distribution of X, the number of favorable replies.

3. **Management** A random sample of 10 firms from the Fortune 500 list of major firms is selected to see how many CEOs (chief executive officers) were formerly CEOs of other firms, not necessarily in the Fortune 500 list. Here $n = 10$, and $X = $ the number of firms in those 10 that match the preceding criterion. The unknown population proportion p may be estimated using $\hat{p}$.

    Suppose 40% of all Fortune 500 CEOs were formerly CEOs of other firms. The probability of a randomly selected firm having this property is then $p = .40$. There is a slight dependence in the successive selection of additional firms, but that dependence can be ignored in cases like this where the population size (500) is much larger than the sample size (10). That is, if the first firm selected is one of the 200 that have the stated property (40% of 500 is 200), then the probability of the second firm selected also having the stated property is only .399, since only 199 of the remaining 499 firms have the stated property, $\frac{199}{499} = .399$. But .399 is close enough to .40 that .40 can safely be used for all trials. Therefore, the binomial distribution is a reasonable probability distribution to use for the probabilities of X.

The probability of getting *two or fewer* firms with the stated property, out of a sample of size 10, is given in Table A1 ($n = 10$, $p = .40$) as

$$P(X \leq 2) = .1673$$

The probability of getting *two or more* firms with the stated property is found by subtracting probabilities in Table A1

$$P(X \geq 2) = 1 - P(X \leq 1) = 1 - .0464 = .9536$$

The probability of getting *exactly two* firms with the stated property, in a sample of size 10, is also found by subtraction

$$P(X = 2) = P(X \leq 2) - P(X \leq 1) = .1673 - .0464 = .1209$$

EXERCISES

5.1   Use Table A1 to find the following binomial probabilities.

(a) $P(X \leq 7)$ if $n = 10$ and $p = .65$

(b) $P(X \leq 7)$ if $n = 20$ and $p = .65$

(c) $P(X \leq 7)$ if $n = 30$ and $p = .65$

(d) $P(X < 6)$ if $n = 8$ and $p = .40$

(e) $P(X \geq 6)$ if $n = 8$ and $p = .40$

(f) $P(X > 6)$ if $n = 8$ and $p = .40$

(g) $P(6 \leq X \leq 14)$ if $n = 28$ and $p = .35$

(h) $P(6 \leq X < 14)$ if $n = 28$ and $p = .35$

(i) $P(6 < X \leq 14)$ if $n = 28$ and $p = .35$

(j) $P(6 < X < 14)$ if $n = 28$ and $p = .35$.

5.2   A judge is scheduled to hear 20 appeals for traffic tickets. Each appeal has a probability of .4 of being approved, independent of the other appeals. Find the probability that less than half (9 or fewer) of the appeals are approved.

5.3   What is the expected number of people with high blood pressure in a random sample of 80 people if approximately 10% of all people have high blood pressure? What is the population standard deviation?

5.4   A student in marketing who has not had time to prepare for a 20-question multiple choice test decides to use a random guess on each question. If each question on the exam has five choices, what is the student's chance of getting the correct answer on any single question? What mean score should the student expect from this type of an approach to taking an exam? What is the probability that the student will get more than half (11 or more) of the questions correct? What is the probability that the student will get more than five correct? What is the probability that the student will get between two and six (inclusive) correct?

5.5 If a machine has a probability .1 of manufacturing a defective part, what is the expected number of defective parts in a random sample of 30 parts manufactured by this machine? What is the probability of finding four or more defective parts in the sample?

5.6 In the 1972 presidential election approximately one-third of the voters chose McGovern and two-thirds chose Nixon. Suppose five people are selected at random from those who voted. Define the binomial random variable to be the number of people in this sample who voted for McGovern. Since the value of one-third is not covered in Table A1, you will have to use the binomial probability function and make a table similar to the one found in Figure 5.1 to answer the following questions.

 (a) What is the probability that no one in the sample voted for McGovern?

 (b) What is the probability that everyone in the sample voted for McGovern?

 (c) What is the probability that a majority of those questioned in the sample voted for McGovern?

 (d) What is the most probable number of people who voted for McGovern? That is, what is the mode of this probability function?

5.7 A tollbooth operator is wondering how much change (in $1 bills) she will need to begin her shift. She knows that 10% of all customers want change for a $5 bill or larger. Suppose the first 20 customers are like a random sample from the population.

 (a) What is the probability that three or fewer customers want change for a $5 bill or larger?

 (b) What is the probability that three or more customers want change for a $5 bill or larger?

 (c) What is the probability that exactly three customers want change for a $5 bill or larger?

5.8 If 5% of the checks received in an all-night convenience store are returned by the bank, what is the probability that the first eight checks received by the store are all cleared by the bank?

5.9 A marketing survey in Sao Paulo shows that approximately 70% of the upper-class shoppers purchased their TV sets in department stores. If a random sample of 15 upper-class shoppers is questioned, what is the probability that eight or fewer of those questioned purchased their sets from stores other than department stores?

5.10 Use Table A1 to aid in making a graph of the binomial probability function for $n = 10$ and $p = .5$ as was done in Figure 5.2. Also make a similar graph for $n = 20$ and $p = .5$. Make a graph of the binomial distribution function for both of these cases as was done in Figure 5.4. Are these distributions skewed, and if so, in what direction?

5.11 A wildcat oil driller plans to drill eight holes. He must hit oil in at least two of them to make money on the venture. The holes will be far enough apart so that he can assume independence from hole to hole. If there is a probability of .3 of hitting oil for each hole, what is the probability that he will make money on the venture?

5.12 A new cough medicine has probability .85 of curing a person's cough. Twenty people are given the new cough medicine. What is the probability that all twenty are cured of their coughs?

5.13 A construction company places sealed bids on seven different jobs. It knows that it can handle three jobs at a time, but there is only a .3 probability of getting each job, and getting or not getting each job is independent of the other jobs. What is the probability that the company will get more jobs than it can handle?

5.14 An insurance salesman knows that in the long run 40% of his calls will result in sales. He has set a goal for the next two weeks of making 28 calls. What is the probability that he will make at least 14 sales?

5.15 A university is filling six secretarial vacancies. It will hire applicants as they show up for interviews provided they are qualified. Forty percent of the qualified applicants will be over 35 years of age, based on previous experience. What is the probability that an equal number of applicants over 35 and 35 or under will be hired?

5.2

THE NORMAL DISTRIBUTION

PROBLEM SETTING 5.2

The Environmental Protection Agency (EPA) has the responsibility of establishing estimates of miles per gallon (mpg) for all new cars sold in the United States. These estimates are printed on the window stickers. Television ads for new cars flash these figures on the screen usually with an asterisk referring to smaller type that cannot readily be seen or else with a somewhat soft warning that your own mileage may vary. The reason for this is that the EPA tests are performed in the laboratory and not on the road. Therefore, without wind resistance the resulting mpg estimates are high. These estimates cause problems for both the car dealer and customers because the fuel-cost-conscious public would like an accurate estimate of the mpg so that it can make the best informed judgment possible with respect to mpg when purchasing an automobile.

A new car dealership decides to perform a large-scale test of the automobiles they sell as well as those of the competition. In this way they hope to be able to quote more realistic averages for use in comparison.

HISTOGRAMS OF THE DATA

Clearly the random variable of interest in this problem setting is the mpg and the random variable will have properties that differ from one car model to another and from city to highway driving. Therefore, to reduce the magnitude

FIGURE 5.6
TWO HYPOTHETICAL HISTOGRAMS FOR CITY MILEAGE OF ONE MODEL OF CAR.

of the problem, only one model of car driven under city conditions will be considered. The properties of this random variable may be considered by examining a histogram of hypothetical values of this random variable. In Figure 5.6 two histograms are presented. One shows what might be obtained from 20 mileage tests, and the other shows possible results from 2000 mileage tests. As the number of mileage tests increases, the class widths get smaller and the histogram becomes smoother. In fact, it seems reasonable to assume that as the number of observations becomes very large, the histogram might approach a graph like the one in Figure 5.7. Figure 5.7 is the graph of a **normal density function.**

NORMAL RANDOM VARIABLES

When a histogram of the data appears to follow a symmetric, unimodal bell-shaped pattern, the normal density function may be a reasonable curve to use as the population curve. In such a case the random variable is said to have a **normal distribution,** and the random variable is called a **normal random variable.**

The word *normal* does not imply that other distributions are abnormal. It is simply the name customarily given to this distribution. Sometimes the word **Gaussian** is used instead of **normal** to avoid confusion. The term *normal* is used in this text because it is the most widely used term.

The mpg is a good example of a *continuous* random variable. Theoretically it can assume any number within a certain range. The normal distribution function is a continuous distribution function.

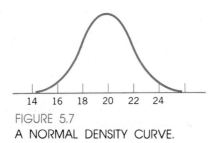

FIGURE 5.7
A NORMAL DENSITY CURVE.

THE NORMAL DISTRIBUTION

You may have noticed that a histogram is a way of presenting the density of observations in the various intervals. More observations contained in an interval mean a higher density in that interval, and the histogram is taller in that interval. Since the histogram approaches the shape of some population curve, that curve is usually called a density curve.

The **normal density function** is given in Eq. 5.6 and the cumulative distribution function is given in Eq. 5.7. Both functions are very difficult to work with. Therefore, extensive tables, such as Tables A2 and A2* in the Appendix, have been developed to enable any desired probabilities for normal distributions to be found quickly and easily. To understand the use of Tables A2 and A2*, you must first learn more about the different types of normal distributions.

The **normal density function** is given by

$$f(x) = \frac{1}{\sigma\sqrt{2\pi}} \, e^{-(x-\mu)^2/2\sigma^2} \qquad -\infty < x < \infty \tag{5.6}$$

where μ and σ are parameters equal to the population mean and standard deviation, respectively, and where $\pi = 3.14159\ldots$ and $e = 2.71828\ldots$ are well-known constants.

THE NORMAL DISTRIBUTION

The **normal distribution function** is given by

$$F(x) = P(X \leq x) = \int_{\infty}^{x} f(t) \, dt \tag{5.7}$$

where the integral is used to represent the area under the density curve, to the left of x. Calculus is not needed here, because Tables A2 and A2* may be used to find $F(x)$.

POPULATION PARAMETERS AND THEIR ESTIMATES

As can be seen from Eq. 5.6 the normal distribution has two parameters, μ and σ. Each different value of μ and σ represents a different normal distribution, so the normal distribution is actually a family of distributions indexed by μ and σ. There is a reason for choosing these particular Greek letters as the two parameters of the normal distribution. It can be shown that the mean of the normal distribution equals μ, the one parameter, and the standard deviation of the normal distribution equals σ, the other parameter. The mean and standard deviation of a continuous distribution such as the normal distribution are analogous to their counterparts for discrete distributions as defined in Section 4.4. So, if a particular normal distribution has parameters $\mu = 7$ and $\sigma = 2$, the mean is 7 and the standard deviation is 2 for that distribution.

Suppose a population has a normal distribution with parameters μ and σ, and a random sample is drawn from that population. In practice the values of

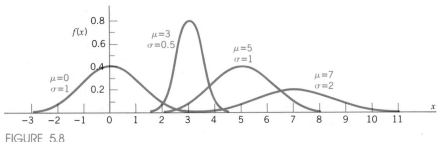

FIGURE 5.8
GRAPHS OF SEVERAL NORMAL DENSITY FUNCTIONS FOR VARIOUS VALUES
OF μ AND σ.

μ and σ are usually not known. The population mean μ can be estimated by
the sample mean $\overline{X}$. The population standard deviation σ can be estimated by
the sample standard deviation s. More is said about this method of estimating
parameters in the next chapter.

A normal density curve can be graphed for any particular combination of
μ and σ by using the normal density function given in Eq. 5.6. Figure 5.8
shows four such curves corresponding to four pairs of values for μ and σ. The
mean μ locates the center of the normal distribution. Note that the mean
equals the median in normal distributions because the distributions are symmetric. The standard deviation σ describes how widely the distribution is
spread around the mean. The larger the value of σ, the more spread out the
distribution.

Corresponding to each normal density curve is a distribution function whose
graph can be developed from Eq. 5.7. Figure 5.9 shows the distribution functions corresponding to each of the normal density functions shown in Figure
5.8.

THE STANDARD NORMAL DISTRIBUTION

The normal distribution shown in Figures 5.8 and 5.9, corresponding to μ =
0 and σ = 1, is called the **standard normal distribution.** Quite often the term
standard or **standardized** is used in statistics to denote distributions or random

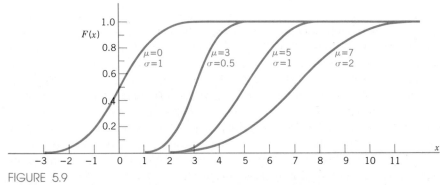

FIGURE 5.9
GRAPHS OF SEVERAL NORMAL DISTRIBUTION FUNCTIONS FOR VARIOUS
VALUES OF μ AND σ.

variables that have zero mean and unit standard deviation (i.e., $\sigma = 1$). Although the standard normal distribution is defined for all values of x from $-\infty$ to ∞, Figure 5.8 makes it clear that the distribution is concentrated between -3 and 3. The standard normal distribution function has been tabulated extensively in Tables A2 and A2*. Although these tables are only for the standard normal distribution, they can be used to find probabilities or quantiles for any normal distribution. The letter Z is customarily used to denote a standard normal random variable, to emphasize the fact that it has mean 0 and standard deviation 1 (like z scores).

TABLES A2 and A2*

Two common questions occur repeatedly when using standard normal distributions.

1. Given a probability p, what is the value of the standard normal deviate z_p such that $P(Z \le z_p) = p$? That is, find the pth quantile of the standard normal distribution. For example, if $p = .95$, find $z_{.95}$.
2. Given a value for the standard normal deviate z, what is $P(Z \le z)$? For example, if $z = 1.80$, find $P(Z \le 1.80)$.

It is common practice in statistics to provide only one standard normal table to address both types of problems. However, it is easier for the inexperienced reader if one style of table (Table A2) is provided to answer the first type of problem and a second style of table (Table A2*) is used to address the second type of problem. The experienced reader will be able to use either Table A2 or A2* with equal facility to answer either type of problem, with the choice of tables depending on personal preference. Discussions are now presented to aid the reader in the use of the standard normal tables.

HOW TO FIND QUANTILES FROM TABLE A2

The pth quantile z_p of a continuous distribution is that number that satisfies the equation

$$P(Z \le z_p) = p$$

The pth quantile of a standard normal distribution, $0 < p < 1$, is illustrated in Figure 5.10. Such quantiles are given in Table A2 for the standard normal distribution. A portion of this table is reproduced in Figure 5.11. To find the pth quantile in Table A2, the value of p must be expressed to three decimal places, for example, $p = .958$. The first two decimal places, .95, are used to find the correct row in Table A2. This row is highlighted in Figure 5.11. The third decimal place in p, in this case .008, indicates which column to use. The .008 appears at the top of the column. This column is highlighted in Figure 5.11. Finally, the table entry in that row and column is the pth quantile, z_p, of the standard normal distribution. For $p = .958$, this value is shown in Figure 5.11 as $z_{.958} = 1.7279$, thus

$$P(Z \le 1.7279) = .958.$$

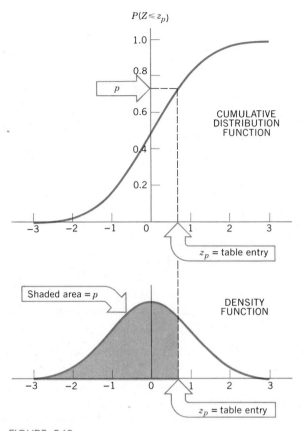

FIGURE 5.10
THE pTH QUANTILE OF THE STANDARD NORMAL
RANDOM VARIABLE, Z, AS SHOWN WITH THE
DENSITY FUNCTION FOR Z AND THE DISTRIBUTION
FUNCTION FOR Z.

p	.000	.001	.002	.003	.004	.005	.006	.007	.008	.009
.90	1.2816	1.2873	1.2930	1.2988	1.3047	1.3106	1.3165	1.3225	1.3285	1.3346
.91	1.3408	1.3469	1.3532	1.3595	1.3658	1.3722	1.3787	1.3852	1.3917	1.3984
.92	1.4051	1.4118	1.4187	1.4255	1.4325	1.4395	1.4466	1.4538	1.4611	1.4684
.93	1.4758	1.4833	1.4909	1.4985	1.5063	1.5141	1.5220	1.5301	1.5382	1.5464
.94	1.5548	1.5632	1.5718	1.5805	1.5893	1.5982	1.6072	1.6164	1.6258	1.6352
.95	1.6449	1.6546	1.6646	1.6747	1.6849	1.6954	1.7060	1.7169	1.7279	1.7392
.96	1.7507	1.7624	1.7744	1.7866	1.7991	1.8119	1.8250	1.8384	1.8522	1.8663
.97	1.8808	1.8957	1.9110	1.9268	1.9431	1.9600	1.9774	1.9954	2.0141	2.0335
.98	2.0537	2.0749	2.0969	2.1201	2.1444	2.1701	2.1973	2.2262	2.2571	2.2904
.99	2.3263	2.3656	2.4089	2.4573	2.5121	2.5758	2.6521	2.7478	2.8782	3.0902

FIGURE 5.11
A PORTION OF TABLE A2.

EXAMPLE

Use Table A2 to find the following quantiles from a standard normal distribution:

(a) $z_{.975}$ (b) $z_{.90}$ (c) $z_{.20}$ (d) $z_{.005}$ (e) $z_{.9254}$

(a) Enter Table A2 on the row beginning with .97 and the column headed by .005. The table entry in that row and column is 1.9600 corresponding to the .975 quantile of the standard normal distribution, that is

$$P(Z \le 1.9600) = .975$$

(b) Enter Table A2 on the row beginning with .90 and the column headed by .000 (since the third decimal place is a zero). The table entry in that row and column is 1.2816, corresponding to the .900 quantile of the standard normal distribution, that is

$$P(Z \le 1.2816) = .900$$

(c) Enter Table A2 on the row beginning with .20 and the column headed by .000. The table entry in that row and column is $-.8416$, corresponding to the .200 quantile of the standard normal distribution, that is

$$P(Z \le -.8416) = .200$$

Note that this quantile is negative as are all quantiles of the standard normal distribution for choices of $p < .5$.

(d) Enter Table A2 on the row beginning with .00 and the column headed by .005. The table entry in that row and column is -2.5758, corresponding to the .005 quantile of the standard normal distribution, that is

$$P(Z \le -2.5758) = .005$$

Note that this quantile is the negative of the .995 quantile, since the standard normal distribution is symmetric about 0.

(e) In this case the user must either round the value of p to .925, in which case the answer is found directly in Table A2 as $z_{.925} = 1.4395$ or use linear interpolation as explained in the Appendix to this chapter to interpolate between the quantiles $z_{.925} = 1.4395$ and $z_{.926} = 1.4466$. Interpolation yields the approximate value of $z_{.9254} = 1.4423$. Since values of p are almost never known accurately to four decimal places, rounding p to three decimal places should suffice for almost all applications.

HOW TO FIND PROBABILITIES FROM TABLE A2*

The second type of problem mentioned previously concerns finding $P(Z \le z)$ for a given value of z. Table A2* has been constructed to make this probability easy to find. In Table A2* the roles of the row and column headings and table entries as used in Table A2 have been reversed. That is, in Table A2 the value of p was used to find the proper row and column to enter the table, and the table entry was the quantile z corresponding to p. In Table A2* the row and column headings are based on the value of z, while the table entry is the value of p corresponding to z. A portion of Table A2* is shown in Figure 5.12.

z	.00	.01	.02	.03	.04	.05	.06	.07	.08	.09
1.0	.8413	.8438	.8461	.8485	.8508	.8531	.8554	.8577	.8599	.8621
1.1	.8643	.8665	.8686	.8708	.8729	.8749	.8770	.8790	.8810	.8830
1.2	.8849	.8869	.8888	.8907	.8925	.8944	.8962	.8980	.8997	.9015
1.3	.9032	.9049	.9066	.9082	.9099	.9115	.9131	.9147	.9162	.9177
1.4	.9192	.9207	.9222	.9236	.9251	.9265	.9279	.9292	.9306	.9319
1.5	.9332	.9345	.9357	.9370	.9382	.9394	.9406	.9418	.9429	.9441
1.6	.9452	.9463	.9474	.9484	.9495	.9505	.9515	.9525	.9535	.9545
1.7	.9554	.9564	.9573	.9582	.9591	.9599	.9608	.9616	.9625	.9633
1.8	.9641	.9649	.9656	.9664	.9671	.9678	.9686	.9693	.9699	.9706
1.9	.9713	.9719	.9726	.9732	.9738	.9744	.9750	.9756	.9761	.9767

FIGURE 5.12
A PORTION OF TABLE A2*.

To find $P(Z \leq z)$ for a given value of z in Table A2*, the value of z must be expressed to two decimal places, for example 1.67. The first two digits, 1.6, are used to find the correct row in Table A2*. This value is highlighted in Figure 5.12. The third digit in z, in this case .07, indicates which column to use, and has been highlighted in Figure 5.12. Finally, the table entry in that row and column is the value of p such that $P(Z \leq z) = p$. For z = 1.67, this value is shown in Figure 5.12 as .9525, thus $P(Z \leq 1.67) = .9525$. If more than two decimal places are used with the value of z, the user can either round z to two decimal places or use linear interpolation in the table.

EXAMPLE

Use Table A2* to find $P(Z \leq z)$ for a standard normal random variable, when z takes on the following values:

(a) 1.96   (b) 2.5   (c) 0.19   (d) −1.82   (e) 1.874

(a) Enter Table A2* on the row beginning with 1.9 and the column headed by .06. The table entry in that row and column is .9750. Thus, $P(Z \leq 1.96) = .9750$.

(b) Enter Table A2* on the row beginning with 2.5 and the column headed by .00 (since the second decimal place is zero). The table entry in that row and column is .9938. Thus, $P(Z \leq 2.50) = .9938$.

(c) Enter Table A2* on the row beginning with 0.1 and the column headed by .09. The table entry in that row and column is .5753. Thus, $P(Z \leq 0.19) = .5753$.

(d) Enter Table A2* on the row beginning with −1.8 and the column headed by .02. The table entry in that row and column is .0344. Thus, $P(Z \leq -1.82) = .0344$.

(e) In this case the user must either round the value of z to two decimal places, in which case the answer is found directly in Table A2* as

$P(Z \leq 1.87) = .9693$, or use linear interpolation to interpolate between the z values of 1.87 and 1.88. Interpolation yields the approximate value of $P(Z \leq 1.874) \approx .9695$. As in the last example, values of p accurate to four decimal places are rarely, if ever, justified in practice, so rounding z to two decimal places should suffice for almost all applications.

As mentioned previously, Table A2 can also be used to find $P(Z \leq z)$ for a given value of z by finding the value z in the heart of Table A2 and then reading out the value of p in the row and column in which the value of z is found. For example, in Figure 5.11, $z = 1.4255$ is found in the .92 row and the column headed by .003; thus

$$P(Z \leq 1.4255) = .92 + .003 = .923$$

If the exact value of z cannot be found in Table A2, then selection of the value in the table closest to the given value of z will provide an answer accurate to three decimal places that is adequate for most applications. If more accuracy is desired, linear interpolation can be used to find a more precise answer.

Likewise, Table A2* can be used to find z_p such that $P(Z \leq z_p) = p$ for a given value of p by finding p in the heart of Table A2* and then reading out the value of z in the row and column in which the value of p is found. For example, in Figure 5.12, $p = .9484$ is found in the 1.6 row and the column headed by .03; thus

$$P(Z \leq 1.63) = .9484$$

If the exact value of p cannot be found in Table A2*, then selection of the value in the table closest to the given value of p will provide an answer accurate to two decimal places that is adequate for most applications. If more accuracy is desired, linear interpolation can be used to find a more precise answer.

The real usefulness of Tables A2 and A2* lies in the fact that they can be used to find quantiles and probabilities for *any* normal distribution through knowledge of μ and σ. To use Table A2 or A2* with any normal distribution it is first necessary to convert a normal random variable X to a standard normal random variable Z.

CONVERTING TO A STANDARD NORMAL DISTRIBUTION

Consider the random variable X with a normal distribution and parameters μ and σ. Subtraction of μ converts X to a new random variable, Y, which is also normal with standard deviation σ, but with mean 0:

$$Y = X - \mu$$

That is, if the mean of X is 7, each observation on X is converted to an observation on Y simply by subtracting 7 from the observation on X. The mean of Y is then 0.

Subsequent division by σ converts Y to a standard normal random variable Z:

$$Z = \frac{Y}{\sigma} = \frac{X - \mu}{\sigma} \qquad (5.8)$$

If $\sigma = 2$, then after subtracting 7 from each observation on X, the result is divided by 2 to get an observation on Z. The mean of Z is zero, and the standard deviation is 1.

This is analogous to the conversion of a sample of observations to z scores, by subtracting the sample mean $\overline{X}$ and dividing by the sample standard deviation s. The sample of z scores always has a sample mean of zero and a sample standard deviation of 1. By subtracting the *population* mean μ and dividing by the *population* standard deviation σ, the population mean and standard deviation of Z are 0 and 1, respectively. This does not imply that a sample of observations on Z will have a sample mean equal to zero, or a sample standard deviation equal to 1. In fact, the sample mean and standard deviation will probably not equal 0 or 1, respectively, because of the sampling variability always present in a sample.

HOW TO FIND QUANTILES FOR ANY NORMAL DISTRIBUTION

To find the pth quantile x_p for a normal distribution with mean μ and standard deviation σ, first find the standard normal quantile z_p and then use the relationship (obtained by solving Eq. 5.8)

$$x_p = \mu + \sigma z_p \qquad (5.9)$$

to convert the standard normal quantiles to quantiles for the distribution with μ and σ. Quantiles from a standard normal distribution are used as intermediate values in converting to quantiles from other normal distributions.

The **pth quantile** x_p, $0 < p < 1$, **of a normal distribution** with known mean μ and standard deviation σ is found from Table A2 as follows.

TO FIND
NORMAL
QUANTILES

1. Round p to three decimal places.
2. Enter the row and column corresponding to p, and find z_p, the pth quantile of a standard normal distribution.
3. Convert to x_p using the relationship

$$x_p = \mu + \sigma z_p$$

EXAMPLE

To find the $\frac{2}{7}$ quantile from a normal distribution with $\mu = 6$ and $\sigma = 2$, first find the $\frac{2}{7} = .286$ quantile for a standard normal distribution from Table A2 as $z_{.286} = -.5651$. This standard normal quantile is substituted into Eq. 5.9

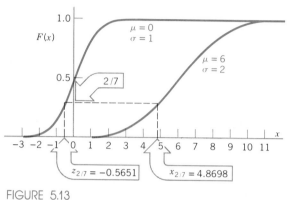

FIGURE 5.13
THE $\frac{2}{7}$TH QUANTILE OF A STANDARD NORMAL
DISTRIBUTION, AND THE $\frac{2}{7}$TH QUANTILE OF A
NORMAL DISTRIBUTION WITH $\mu = 6$, $\sigma = 2$.

to get

$$X_{.286} = 6 + 2(-.5651)$$
$$= 6 - 1.1302$$
$$= 4.8698$$

as the .286 quantile of a normal distribution with $\mu = 6$ and $\sigma = 2$. That is, there is a probability of .286 of getting a value less than 4.8698 when observing a value from a normal distribution with $\mu = 6$ and $\sigma = 2$. Figure 5.13 shows the relationship between $z_{2/7}$ and $x_{2/7}$.

HOW TO FIND PROBABILITIES FROM TABLE A2*

To find $P(X \leq x)$ from Table A2*, where X has a normal distribution with mean μ and standard deviation σ, two steps are involved. The first step is to **standardize** x by subtracting μ and dividing by σ

$$z = \frac{x - \mu}{\sigma} \tag{5.10}$$

In the second step z is rounded to two decimal places and that value is used to enter Table A2* in the usual manner.

EXAMPLE

Suppose it is desired to find the probability that a randomly drawn value from a normal population with $\mu = 10$ and $\sigma = 3$ is less than or equal to 14, that is, the $P(X \leq 14)$ is needed. First 14 is standardized,

$$z = \frac{14 - \mu}{\sigma}$$

$$= \frac{14 - 10}{3}$$

$$= 1.3333$$

Rounded to two decimal places the z value used to enter Table A2* is 1.33. Therefore the desired probability is .9082. The probability of a normal random variable with $\mu = 10$ and $\sigma = 3$ being less than or equal to 14 is .9082.

To find $P(X > x)$ use the relationship

$$P(X > x) = 1 - P(X \leq x)$$

and find $P(X \leq x)$ in the usual way. To find $P(x_1 < X \leq x_2)$ use the relationship

$$P(x_1 < X \leq x_2) = P(X \leq x_2) - P(X \leq x_1)$$

and find $P(X \leq x_2)$ and $P(X \leq x_1)$ in the usual way.

TO FIND
NORMAL
PROBABILITIES

To find the probability $P(X \leq x)$, where X is a normal random variable with known mean μ and standard deviation σ, and x is given:

1. Standardize x using the relationship

$$z = \frac{x - \mu}{\sigma}$$

2. Round the z value to two decimal places and use this value to enter Table A2*. The table entry in the row and the column determined by z is $P(X \leq x)$.

In addition, $P(X > x)$ and $P(x_1 < X \leq x_2)$ may be found using the preceding procedures and the relationships

$$P(X > x) = 1 - P(X \leq x)$$

and

$$P(x_1 < X \leq x_2) = P(X \leq x_2) - P(X \leq x_1)$$

EXAMPLE

Refer to Problem Setting 5.2 and consider a particular four-cylinder model, with a manual transmission tested in city driving, that gets an average of 24.32 miles per gallon, with a standard deviation of .58 for a tankful of gas.

Suppose you buy a car of the type being tested. What is the probability that your mileage will be less than 23 mpg? If X is a random variable representing miles per gallon, and if X is normally distributed with population mean $\mu = 24.32$ and population standard deviation $\sigma = .58$, then $P(X \leq 23)$ is found as follows from Table A2*:

$$P(X \leq 23) = P\left(Z \leq \frac{23 - 24.32}{.58}\right)$$

$$= P(Z \leq -2.28) = .0113$$

Therefore, your chances are only about 1 in 100 of getting less than 23 mpg.

What mpg would be required to be considered in the upper 5%? This question is worded such that the upper 5% z value must first be found from Table A2, which is 1.6449, the .95 quantile. Next, this value is substituted into Eq. 5.9, which is then solved for $x_{.95}$:

$$x_{.95} = 24.32 + (.58)(1.6449)$$

$$= 25.27$$

Therefore, 25.27 is the upper 5% point in the distribution of gas mileage. Only 5% of all cars of this particular model can be expected to get better than 25.27 mpg.

EXERCISES

5.16 For a normal random variable with $\mu = 0$ and $\sigma = 1$ use Table A2 to find the following quantiles: $z_{.01}$, $z_{.05}$, $z_{.50}$, $z_{.95}$, and $z_{.99}$.

5.17 For a normal random variable with $\mu = 10$ and $\sigma = 2$ use Table A2 to find the quantiles $x_{.025}$ and $x_{.975}$.

5.18 Assuming Z is a normally distributed random variable with $\mu = 0$ and $\sigma = 1$, use Table A2* to find the following probabilities:

(a) $P(Z \leq -2.33)$

(b) $P(Z \leq -1.65)$

(c) $P(Z \leq 1.65)$

(d) $P(Z \leq 2.33)$

5.19 Assume that X is a normally distributed random variable with $\mu = 10$ and $\sigma = 2$ and use Table A2* to find the following probabilities:

(a) $P(X \leq 6.08)$

(b) $P(X \leq 13.92)$

(c) $P(6.08 \leq X \leq 13.92)$

5.20 The daily output of a production line varies according to a normal distribution with a mean of 163 units and a standard deviation of 3.5 units.

(a) Find the probability that the daily output will be 160 units or less.

(b) Find the probability that the daily output will exceed 170 units.

(c) The production manager wants to tell his boss, "Eighty percent of the time our production is at least x units." What number should he use for x?

5.21 Suppose that X is a random variable representing the reaction time in seconds in a simulated driving experiment for persons who have been given a specified amount of alcohol. Assuming X has a normal distribution with a mean of 0.8 sec and a standard deviation of 0.2 sec, find the following probabilities.

(a) The probability that the reaction time is less than one-half second.

(b) $P(X \geq 1.2)$

(c) $P(.75 \leq X \leq 1.25)$

5.22 Assume the errors of measurement observed on individual wristwatches are normally distributed with a mean of 0 and a standard deviation of 60 sec. What is the probability of a randomly selected watch being within 30 sec of the correct time of day?

5.23 A student with an IQ of 140 claims that his IQ score is in the top 5% of students at his university. Is his claim true if IQ scores are normally distributed at his university with a mean of 125 and a standard deviation of 10?

5.24 Suppose that the present speeds on Michigan highways are normally distributed with a mean of 63 mph and a standard deviation of 5 mph. If the state police decide to ticket the fastest 20% of the motorists, how fast could you drive without risking a ticket?

5.25 The manager of a gas station compares the total gasoline sales in gallons with the amount of gasoline delivered by a tanker every time a delivery is made. She notes that this difference is normally distributed with a mean 0 and standard deviation 45 gallons. On one delivery the tanker delivered 67 gallons less than was sold since the previous delivery, and the manager wonders if the tanks are full. Find the probability of a difference at least this large in this direction.

5.26 Two holes are made in a sheet of metal for the insertion of a two-pronged plug in an assembly process. The holes can be anywhere from 0.67 to 0.71 inches apart and the plug will still fit. If the distance between holes is normally distributed with a mean of 0.68 inches and a standard deviation of 0.01, find the proportion of metal sheets that the plugs will not fit.

5.3

THE IMPORTANCE OF THE NORMAL DISTRIBUTION

NORMAL APPROXIMATION TO THE BINOMIAL

A question that may have occurred to the reader is how to find binomial probabilities for those values of n and p not covered in Table A1. As was shown in Exercise 5.10, the plot of the probability function of a binomial random variable with $p = .5$ is symmetric, and as n gets larger the graph starts to appear bell-shaped. This suggests that the normal distribution might serve as an approximation to the binomial distribution. Indeed, it is true that for large sample sizes and a range of values for p, the normal distribution provides a good approximation to binomial probabilities. However, it turns out that this approximation is also fairly good for small sample sizes if p is not near 0 or 1. A formal statement of this approximation is now presented.

If X is a binomial random variable with parameters n and p, an approximation to the probability $P(a \leq X \leq b)$, where a and b are integers, can be found using the standard normal distribution, as follows:

1. Compute

$$z_1 = \frac{a - np - .5}{\sqrt{npq}} \quad \text{and} \quad z_2 = \frac{b - np + .5}{\sqrt{npq}}.$$

NORMAL
APPROXIMATION
TO THE
BINOMIAL

2. Use Table A2* in the usual way to find

$$P(a \leq X \leq b) \approx P(Z \leq z_2) - P(Z \leq z_1) \tag{5.11}$$

where Z is a standard normal random variable. This is the normal approximation to the binomial distribution

3. If $a = 0$, use the approximation

$$P(0 \leq X \leq b) \approx P(Z \leq z_2) \tag{5.12}$$

4. If $b = n$, use the approximation

$$P(a \leq X \leq n) \approx 1 - P(Z \leq z_1) \tag{5.13}$$

Note This approximation should be used only when np and nq are greater than 5.

Notice that to use the normal distribution to approximate the distribution of X, the values a and b are standardized by subtracting the mean np and dividing by the standard deviation $\sqrt{npq}$. The .5 is a **continuity correction,** to adjust for the fact that a *continuous* distribution is being used to approximate a *discrete* distribution that has probabilities at both a and b.

EXAMPLE

If X is a binomial random variable with $n = 20$ and $p = .3$, find $P(3 \leq X \leq 8)$. First note that $np = 20(.3) = 6$ and $nq = 20(.7) = 14$ are both greater than 5, so the normal approximation to the binomial can be used.

$$z_1 = (3 - 6 - .5)/\sqrt{4.2} = -1.71$$
$$z_2 = (8 - 6 + .5)/\sqrt{4.2} = 1.22$$
$$P(3 \leq X \leq 8) \approx P(Z \leq 1.22) - P(Z \leq -1.71)$$
$$= .8888 - .0436$$
$$= .8452$$

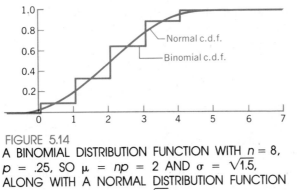

FIGURE 5.14
A BINOMIAL DISTRIBUTION FUNCTION WITH $n = 8$,
$p = .25$, SO $\mu = np = 2$ AND $\sigma = \sqrt{1.5}$,
ALONG WITH A NORMAL DISTRIBUTION FUNCTION
WITH $\mu = 2$ AND $\sigma = \sqrt{1.5}$.

This approximate answer is reasonably close to the exact answer of .8512 found by using Table A1.

When $p = .5$ the normal approximation to the binomial distribution will provide good answers even for small n. When $p \neq .5$, the approximation is considered adequate if both $np > 5$ and $nq > 5$. See Figure 5.14 for a graphical comparison of the cumulative distribution functions for binomial and normal random variables with the same mean and standard deviation.

EXAMPLE

What is the probability of obtaining 50 heads on the toss of 100 fair coins? Let X be a random variable representing the number of heads. Then X has a binomial distribution with $n = 100$ and $p = .5$, and $P(X = 50) = \binom{100}{50}(.5)^{50}(.5)^{50}$ $= .0796$. This is the exact answer (to four decimal places), but do not try to check the calculations by hand. The normal approximation for the binomial is obtained as follows:

$$z_1 = (50 - 50 - .5)/\sqrt{25} = -.10$$
$$z_2 = (50 - 50 + .5)/\sqrt{25} = .10$$
$$P(X = 50) \approx P(Z \leq .10) - P(Z \leq -.10)$$
$$= .5398 - .4602$$
$$= .0796$$

The exact value, obtained from special tables, when rounded off (to four decimal places) agrees exactly with the number obtained using the normal approximation.

THE CONTINUITY CORRECTION

The previous example illustrates the principle behind the use of .5 as a continuity correction. The discrete random variable X has no probability between 49 and 50, or between 50 and 51. All of the probability is concentrated at the integers 49, 50, and 51. However, the continuous random variable has zero probability at the points 49, 50, and 51. It has nonzero probability only over intervals. So the probability for X at 50 corresponds to the interval on either side of 50 for Z. It seems reasonable to include all of the interval that is closer to 50 than to 49 or 51, so the interval extends from $50 - .5$ to $50 + .5$. This is where the .5 comes from, and this is the idea behind this continuity correction.

As another way of looking at the normal approximation and the continuity correction, refer again to Figure 5.14. The exact binomial probability equals the height of the jump at each integer, such as at $x = 4$. But the normal curve rises gradually. The rise in the normal distribution function from $x = 3.5$ to 4.5 is about equal to the rise in the binomial distribution function at $x = 4$. Therefore, it could be said that the probability of the normal random variable being between 3.5 and 4.5 is about equal to the probability of the binomial random variable exactly equaling 4.

EXAMPLE

During one week, 20% of all companies showed a decrease in the value of their stock prices. The portfolio in one trust has 40 different companies' stocks represented. Find the probability that five or fewer stocks in the portfolio went down in price.

Let X be the number of stocks that went down in price. If the companies in the portfolio resemble a random sample of all companies, then X has a binomial distribution with $n = 40$ and $p = .2$. The desired probability may be approximated as follows:

$$z_2 = (5 - 8 + .5)/\sqrt{6.4} = -.99$$

$$P(X \le 5) \approx P(Z \le z_2)$$

$$= P(Z \le -.99)$$

$$= .1611$$

from Table A2*. This result compares with an exact answer of .1613.

LILLIEFORS TEST FOR NORMALITY

The normal distribution is the most important and useful distribution in the study of statistics. In later chapters of this text the normal distribution appears over and over again because many of the procedures to be presented depend on the population being at least approximately normal. Therefore, it is essential to have a procedure for checking that the sample data could have come from a normal distribution. Many procedures exist in statistics for doing this, but the one presented here is quite easy to use. It simply compares the empirical distribution function of the standardized sample values against the standard normal distribution function. This procedure is called the **Lilliefors test for normality.** In Figure 5.15 the basic diagram is presented on which the e.d.f. of the standardized sample values is to be plotted.

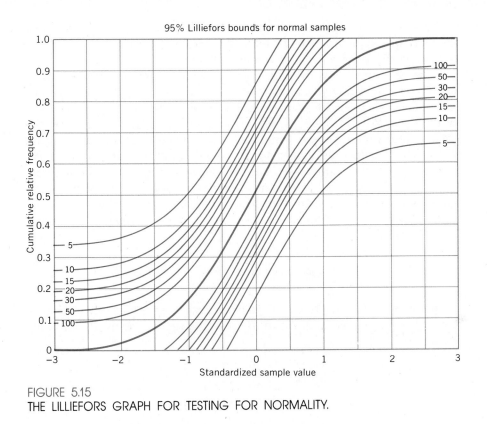

FIGURE 5.15
THE LILLIEFORS GRAPH FOR TESTING FOR NORMALITY.

RULES FOR
USING THE
LILLIEFORS
GRAPH AS A
CHECK FOR
NORMAL
DATA

1. Find the sample mean $\overline{X}$ and sample standard deviation s.

2. Standardize each sample observation X_i as
$$Z_i = (X_i - \overline{X})/s.$$

3. Plot an e.d.f. of these standardized values Z_i in the Lilliefors diagram.

4. The heavy curve in the middle of the Lilliefors diagram represents a standard normal distribution. The lighter curves immediately on either side of the heavy curve represent Lilliefors bounds for a sample of size 100 from a normal distribution. That is, if an e.d.f. plot for $n = 100$ strays outside either of these bounds, then the population should be considered nonnormal. Likewise for other sample sizes use the corresponding bounds provided for $n = 5, 10, 15, 20, 30,$ and 50.

5. For large sample sizes not covered by Figure 5.15, the normality assumption may be tested as follows. Measure the largest distance (measured in a vertical direction) between the e.d.f. and the heavy curve in Figure 5.15. If this distance is larger than $.886/\sqrt{n}$, where n is the sample size, then the population should be considered nonnormal.

Note Lilliefors graph paper is available in the study guide that accompanies this text.

The e.d.f. is based on the observations within a sample, and therefore every e.d.f. shows sampling variability. For some samples the graph of an e.d.f. may tend to be above the population distribution function, whereas for other samples the e.d.f. may show the opposite tendency. The probability of the graph of an e.d.f. wandering very far from the c.d.f. should be small. As the sample size gets larger, the e.d.f. will tend to be closer to the c.d.f. This is the basis of the diagram in Figure 5.15. For a sample of size 100, the e.d.f. of the standardized values will stay completely within the indicated bounds around the standard normal distribution for about 95% of all samples, if in fact the samples are from a normal distribution. However, if the sample is taken from a nonnormal population, there is no reason for the e.d.f. to stay within those bounds. This is a convenient method for comparing an e.d.f. with a normal distribution function. Other curves in Figure 5.15 serve as boundaries for other sample sizes.

AN APPLICATION OF THE LILLIEFORS TEST

A random sample of 20 observations of the mpg for the four-cylinder model considered in the last example of the previous section yields the following data:

$$
\begin{array}{cccc}
24.21 & 24.35 & 23.82 & 24.21 \\
24.14 & 24.60 & 23.75 & 25.01 \\
24.66 & 24.72 & 24.47 & 24.38 \\
23.08 & 23.88 & 23.09 & 24.57 \\
25.16 & 24.62 & 24.62 & 25.14
\end{array}
$$

The sample mean for these 20 observations of mpg is 24.32 and the sample standard deviation is .58. The standardized sample values $(X_i - 24.32)/.58$ are as follows:

$$
\begin{array}{cccc}
-0.20 & 0.04 & -0.87 & -0.20 \\
-0.32 & 0.48 & -0.99 & 1.19 \\
0.58 & 0.69 & 0.25 & 0.10 \\
-2.15 & -0.77 & -2.13 & 0.43 \\
1.45 & 0.51 & 0.51 & 1.41
\end{array}
$$

The e.d.f. plot for these data is given in Figure 5.16. The check for normality is concerned with the fourth curve from both the top and bottom of the graph, that is, the curves with a label of 20. Since the e.d.f. does not cross either of these curves at any point, it may be assumed that the sample could have come from a normal population. This does not mean that the population is actually normal. Rather, the evidence is not sufficient to conclude that the population is nonnormal. If the population is not exactly normal it is close enough to resembling a normal population so that the difference is not evidenced in this sample. In the absence of convincing evidence that the population is nonnormal, the population may be treated as if it were normal. The Lilliefors test is used throughout this book to check for normality.

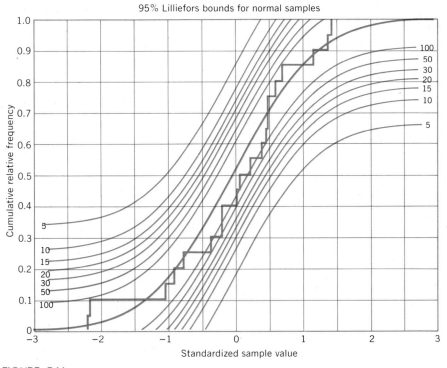

95% Lilliefors bounds for normal samples

FIGURE 5.16
USING THE LILLIEFORS TEST TO CHECK THE NORMALITY ASSUMPTION FOR
THE MPG DATA.

EXERCISES

5.27 Suppose X is a binomial random variable with $p = .25$ and $n = 4$. Find $P(1 \leq X \leq 3)$ by exact and approximate methods. The difference between these two answers represents the error encountered by using the normal approximation to the binomial.

5.28 You are testing your roommate's ability in ESP. You draw a card from a deck of cards and concentrate on the suit (clubs, diamonds, hearts, or spades). If your roommate is guessing, he or she has a probability .25 of stating the correct suit each time. Use the normal approximation to find the probability that he or she will be correct 30 or more times in 100 independent trials.

5.29 A candidate for public office is thought to have the support of 60% of the voters. For a random sample of 50 voters, what would be the answers to the following questions?

   (a) How many supporters should the candidate expect to find in this sample?

   (b) What is the probability that the number of supporters in the sample will be between 26 and 35 inclusive?

   (c) What is the probability that the candidate will fail to obtain support from at least a majority of those in the sample?

5.30 Work Exercise 5.2 using the normal approximation to the binomial. How does your answer compare to the exact answer?

5.31 Data for the number of hours worked by 50 employees in one week are converted to z scores and are given below. Test these data for normality using the Lilliefors graph.

-0.90	-0.03	0.54	1.40	-1.18	-0.03	0.83	-0.32	-2.05
-0.61	0.25	1.12	0.54	-0.90	-0.32	-1.47	0.25	0.83
-0.32	-0.90	0.54	3.13	-0.03	0.25	0.83	-0.32	-0.03
-1.47	1.40	0.25	0.54	-1.18	-0.90	-0.03	-0.61	1.98
-0.03	-0.61	1.40	-0.32	0.83	-0.32	1.12	-1.76	-0.90
0.25	-1.47	0.25	-0.03	0.54				

5.32 The following data represent the 1987 net earnings (in dollars) of common stocks for 20 representative corporations. Use the Lilliefors graph to test these data for normality. (Use $\bar{X} = 4.85$ and $s = 2.20$.)

1.68	1.72	2.50	2.90	3.11
3.35	3.80	3.85	3.89	4.36
4.64	4.76	5.35	5.81	6.11
6.35	6.69	8.41	8.83	8.97

5.33 On one day 87 new applications were made for driver's licenses, and 53 of these were male. What is the probability of getting a proportion of male applicants at least this high if each applicant is equally likely to be male or female?

5.34 An airline that is known to overbook, has booked 122 passengers for a flight that only has 90 seats. If the regular no-show rate for that flight is 30%, what is the probability that the airline will have enough seats for the passengers who show up for the flight?

5.35 A telephone company has 1500 subscribers being served by one telephone exchange. At the peak usage time each subscriber has probability .03 of being on the telephone. How many calls must the exchange be able to handle at one time in order to be at least 95 percent sure that all callers can be accommodated during the peak usage time? Assume independence of calls among subscribers.

5.4
SAMPLING DISTRIBUTIONS AND THE CENTRAL LIMIT THEOREM

THE SAMPLING DISTRIBUTION OF A STATISTIC

The idea of sampling variability has been mentioned frequently in this text. That is, random samples from a population are by their very nature unpredictable in each individual instance, and two random samples cannot be expected to be exactly alike even if they are drawn from the same population. Any statistic computed on the basis of the observations in a sample can be expected to vary in its value from sample to sample for the same reason, sampling variability. Although the individual values of a statistic are usually unpredictable, the **probabilities** associated with the possible values of a statistic can be determined if the population probabilities are known.

It was mentioned in Chapter 3 that a statistic is a type of random variable, and therefore it has a cumulative distribution function, a mean, a standard deviation, and all of the other characteristics of a random variable. In the case of statistics, however, a slightly different terminology is used. The probability distribution of a statistic is often called its **sampling distribution.** The **standard error** of a statistic is another name for the **standard deviation** of the statistic. The idea of a sampling distribution is discussed in the following example.

SAMPLING The probability distribution of a statistic is often called the **sampling distri-**
DISTRIBUTION **bution** of that statistic.

STANDARD The standard deviation of a statistic is often called the **standard error** of that
ERROR statistic.

EXAMPLE

As a class exercise, each member of a statistics class was asked to determine by interview the weight of each of 10 randomly selected male students at their university. Each class member then reported back to the class the average weight of the 10 students interviewed, that is, the value of $\overline{X} = \Sigma X_i/10$ was reported. If the statistics class has 25 members, then the data reported would appear as

$$\overline{X}_1, \overline{X}_2, \ldots, \overline{X}_{25}$$

where $\overline{X}_i$ is the sample mean reported by the ith member of the class.

Some relevant questions about the class exercise include the following.

1. Will each member of the class report the same sample mean, that is, does $\overline{X}_1 = \overline{X}_2 = \cdots = \overline{X}_{25}$, or will there be some variation in the sample means?
2. If all of the sample means are indeed not the same, is there a distribution that describes the sample means?
3. If there is a distribution describing the sample means, does it depend on the distribution of the parent population from which the sample data were obtained?

Of course, the answers to these questions follow directly from the previous discussion on sampling distributions.

1. No, every member of the class will not report the same sample mean. Yes, there will be a variation in the sample means, so that it would be surprising

in this case to have two students report exactly the same sample mean, unless they happened to choose the same samples.

2. Yes. The distribution that describes the sample means is called its *sampling distribution.*

3. Yes, the sampling distribution depends on the population from which the sample was obtained. In fact, if the population distribution function is known, it is possible to find the exact distribution function of the sample statistic, although it is not always very easy to do.

THE SAMPLING DISTRIBUTION OF THE SAMPLE MEAN

The previous example uses the sample mean to illustrate the sampling distribution of a statistic. The sample mean is one of the most important statistics, and so its sampling distribution deserves further discussion. The mean and standard deviation of the sampling distribution are given by Eq. 5.14 and Eq. 5.15, respectively. If the population is at least 10 times as large as the sample size, Eq. 5.16 is used instead of Eq. 5.15 for the standard deviation of $\overline{X}$. This is usually the real situation, so Eq. 5.16 is encountered far more often than Eq. 5.15, such as whenever the population has a continuous distribution function.

Let X_1, X_2, . . . , X_n be a random sample of size n from a population of size N. If μ and σ denote the population mean and population standard deviation, respectively, then the **mean $\mu_{\overline{x}}$ and standard error** (standard deviation) $\sigma_{\overline{x}}$ **of the sample mean** are given by

$$\mu_{\overline{x}} = \mu \quad \text{(The mean of } \overline{X} \text{ equals the population mean)} \quad (5.14)$$

and

THE MEAN
AND
STANDARD
DEVIATION
OF $\overline{X}$

$$\sigma_{\overline{x}} = \frac{\sigma}{\sqrt{n}} \left[\frac{N - n}{N - 1} \right]^{1/2} \quad \begin{matrix} \text{(Sampling from a} \\ \text{finite population} \\ \text{without replacement)} \end{matrix} \quad (5.15)$$

Equation 5.16 can be used in place of Eq. 5.15 if any of the following conditions apply.

1. N is large compared to n (since $(N - n)/(N - 1) \approx 1$).
2. The population distribution function is continuous.
3. The sampling is done with replacement.

$$\sigma_{\overline{x}} = \frac{\sigma}{\sqrt{n}} \quad \begin{matrix} \text{(Sampling with replacement,} \\ \text{or from a large population)} \end{matrix} \quad (5.16)$$

Equation 5.16 is more commonly used in practice for the standard error of $\overline{X}$ than is Eq. 5.15.

EXAMPLE

A soft drink bottling company employs 584 people. A complete search of the personnel records would show that the length of employment for those 584 people at that company has a population mean value μ = 6.37 years with a standard deviation σ = 2.58 years.

A random sample consisting of 15 employees is drawn, and $\overline{X}$ is about to be computed. Even before $\overline{X}$ is computed it is possible to say something about the distribution of $\overline{X}$. The mean of $\overline{X}$ is μ = 6.37 years, the same as the population mean. This gives an indication of where the middle of the distribution of $\overline{X}$ lies. The spread of the distribution of $\overline{X}$ about its mean, 6.37, is measured by the standard error of $\overline{X}$, which is obtained using Eq. 5.16

$$\sigma_{\overline{X}} = \frac{\sigma}{\sqrt{n}} = \frac{2.58}{\sqrt{15}} = .666$$

rather than Eq. 5.15 because the population size is more than 10 times as large as the sample size. If Eq. 5.15 had been used instead, no harm would result; it is simply more cumbersome to use, and the result

$$\sigma_{\overline{X}} = \frac{\sigma}{\sqrt{n}} \left[\frac{N - n}{N - 1}\right]^{1/2} = \frac{2.58}{\sqrt{15}} \left[\frac{584 - 15}{584 - 1}\right]^{1/2} = .658$$

is not much different than before. Notice that although the mean of $\overline{X}$ is always equal to the mean of the population from which the sample was taken, the standard error of $\overline{X}$ is always smaller than the population standard deviation, and continues to get smaller as the sample size gets larger.

THE DISTRIBUTION OF $\overline{X}$ WHEN THE POPULATION IS NORMAL

The sampling distribution of $\overline{X}$ is a normal distribution when the population from which the sample was drawn has a normal distribution. This result is useful because it enables exact probability statements to be made concerning $\overline{X}$ when sampling from a normal population. Thus, when the population is normally distributed, with mean μ and standard deviation σ, then the mean of a sample of size n is also normally distributed, with mean μ and standard error $\sigma/\sqrt{n}$.

EXAMPLE

The exact weight of a particular type of gold coin has been found to be approximately normally distributed, with mean weight 1.000 g and standard deviation .0038 g. A collector buys 10 gold coins and determines their total weight to be 9.984 g. What is the probability of getting a total weight this small or smaller if the 10 coins are truly a random sample of all coins of that type?

To answer this question note that questions about the total weight are closely related to questions about average weight, since the average is merely the total divided by the sample size. So the question translates to, What is the probability of $\overline{X}$ being .9984 or less (9.984/10 = .9984) when the sample came from a normal population with $\mu = 1.000$ and $\sigma = .0038$? Since $\overline{X}$ is normal, the desired probability can be found from Table A2*. The mean of $\overline{X}$ is

$$\mu_{\overline{X}} = \mu = 1.000$$

and the standard error of $\overline{X}$ is given by Eq. 5.16 because of the continuous distribution

$$\sigma_{\overline{X}} = \frac{\sigma}{\sqrt{n}} = \frac{.0038}{\sqrt{10}} = .0012$$

The value .9984 is standardized to

$$Z = \frac{\overline{X} - \mu_{\overline{X}}}{\sigma_{\overline{X}}} = \frac{.9984 - 1.000}{.0012} = -1.33$$

which corresponds to the probability .0918 in Table A2*. Thus there is probability .0918, or about 1 chance in 11, of a random sample of 10 gold coins having an average weight as small as the observed average weight, or equivalently having a total weight of 9.984 g or less. Such a small probability might be the cause for suspicion that the source of gold coins for this dealer may not have a full 1.000 as a population mean weight. More discussion on this type of conclusion appears in subsequent chapters.

THE DISTRIBUTION OF $\overline{X}$ WHEN THE POPULATION IS NOT NORMAL

When the population distribution is not normal, then the distribution of $\overline{X}$ is not normal. However, one of the central results in statistics is that in most situations the distribution of $\overline{X}$ becomes **approximately normal** as n increases. This enables Tables A2 or A2* to be used to obtain *approximate* probabilities for $\overline{X}$ when the sample size is large. Just *how* large the sample size must be depends on the population distribution function and on how close an approxi-

If $X_1, \ldots, X_n$ is a random sample from any infinite population having a finite variance, then the distribution of the random variable $\overline{X}$ becomes normal in form as n increases. More precisely, the distribution of

CENTRAL LIMIT
THEOREM

$$\frac{\overline{X} - \mu}{\sigma/\sqrt{n}} \tag{5.17}$$

becomes the standard normal distribution as n goes to infinity, where μ and σ denote the population mean and standard deviation.

mation the user requires. Therefore, it is not possible to say how large the sample size must be before the normal approximation may be used. Often, however, sample sizes of 20 or 30 meet many users' requirements for accuracy.

AN ILLUSTRATION OF THE CENTRAL LIMIT THEOREM IN ACTION

An order for a magazine ad is given to five advertising specialists, who work independently to draw up plans for the ad. One of the advertising specialists is Sue, who was just hired last week. The five ads are shown to a consultant, who is asked to rank the ads from 1 (best) to 5 (worst). If all five advertising specialists are equally capable, then Sue's ad should have an equal chance of being ranked first, second, third, fourth, or fifth. Let X_1 equal the rank Sue gets from the consultant. Then the probability distribution of X_1 is given by

$$f(x) = \tfrac{1}{5} \qquad x = 1, 2, 3, 4, 5$$
$$f(x) = 0 \qquad \text{otherwise} \tag{5.18}$$

A probability distribution like this, where the probabilities are equal for the various possible values of the random variable, is called a **uniform probability distribution.** A bar graph of these probabilities is given in Figure 5.17. The mean and standard deviation of this distribution are 3 and $\sqrt{2}$, respectively. A normal population c.d.f. with a mean of 3 and a standard deviation of $\sqrt{2}$ is plotted on top of the uniform c.d.f. in Figure 5.18. It should be immediately apparent that the distribution of X_1 is not even approximately a normal distribution.

Now suppose the same ads are shown to a second consultant who ranks them independently of the first consultant. Let X_2 be the rank that the second consultant gives to Sue's ad. Then X_1 and X_2 have independent outcomes, and each has the marginal probability distribution given by Eq. 5.18. Let $\overline{X}$ be the sample mean

$$\overline{X} = \frac{X_1 + X_2}{2}$$

It may be helpful to think of the population as consisting of five ping-pong balls that have the numbers 1, 2, 3, 4, 5 painted on them and then are placed

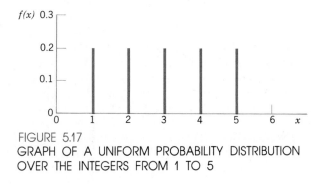

FIGURE 5.17
GRAPH OF A UNIFORM PROBABILITY DISTRIBUTION
OVER THE INTEGERS FROM 1 TO 5

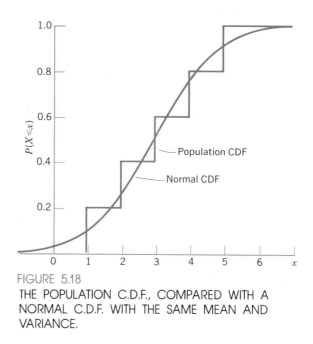

FIGURE 5.18
THE POPULATION C.D.F., COMPARED WITH A
NORMAL C.D.F. WITH THE SAME MEAN AND
VARIANCE.

in a box. The balls are mixed, a ball is randomly drawn, and the number recorded. The ball is replaced, and after mixing again a second ball is selected with the number again recorded, thus completing one sample. A listing of every possible pair of outcomes obtained in this manner, along with the corresponding sample mean, is given in Figure 5.19.

A list of every possible sample mean along with the probability of occurrence is given in Figure 5.20. The mean for this newly defined distribution is given by

$$\mu_{\bar{x}} = \sum \bar{x} P(\bar{X} = \bar{x})$$
$$= 1.0(0.4) + 1.5(0.8) + 2.0(.12) + \cdots + (5.0)(.04) = 3.0$$

using the method described in Section 4.4. This result is in agreement with Eq. 5.14 because the mean of $\bar{X}$ always equals the population mean. The

		X_1 = rank according to first consultant				
		1	2	3	4	5
X_2 = rank	1	1.0	1.5	2.0	2.5	3.0
according	2	1.5	2.0	2.5	3.0	3.5
to second	3	2.0	2.5	3.0	3.5	4.0
consultant	4	2.5	3.0	3.5	4.0	4.5
	5	3.0	3.5	4.0	4.5	5.0

FIGURE 5.19
A TABLE OF SAMPLE MEANS.

$\bar{x}$	$P(\bar{X} = \bar{x})$
1.0	$(1/5)(1/5) = .04$
1.5	$2(1/5)(1/5) = .08$
2.0	$3(1/5)(1/5) = .12$
2.5	$4(1/5)(1/5) = .16$
3.0	$5(1/5)(1/5) = .20$
3.5	$4(1/5)(1/5) = .16$
4.0	$3(1/5)(1/5) = .12$
4.5	$2(1/5)(1/5) = .08$
5.0	$(1/5)(1/5) = .04$

FIGURE 5.20
THE SAMPLING
DISTRIBUTION OF $\bar{X}$ AS
GIVEN IN FIGURE 5.19.

variance of this distribution of $\bar{X}$ is

$$\sigma_{\bar{X}}^2 = \sum \bar{x}^2 P(\bar{X} = \bar{x}) - \mu_{\bar{X}}^2$$
$$= (1.0)^2(.04) + (1.5)^2(.08) + (2.0)^2(.12) + \cdots + (5.0)^2(.04) - 3^2$$
$$= 10 - 9 = 1$$

using the methods described in Section 4.4. This result is in agreement with Eq. 5.16 because the standard deviation of the distribution of the sample mean is given by $\sigma/\sqrt{n}$ or $\sqrt{2}/\sqrt{2} = 1$. The graph in Figure 5.21 represents the sampling distribution of the random variable $\bar{X}$ when $n = 2$. It is roughly triangular in shape, which is closer to a normal density function in appearance than was the original population probability distribution of Figure 5.17. The c.d.f. of $\bar{X}$ in Figure 5.22 also appears to be closer to the normal c.d.f. than is the population c.d.f. in Figure 5.18.

To proceed one more step, suppose four consultants are each asked to rank the ads independently of each other. Let X_1, X_2, X_3, and X_4 represent the rank each of the four gives to Sue's ad, and consider the distribution of the sample mean $\bar{X}$ for this random sample of size 4. The sampling distribution of $\bar{X}$ for

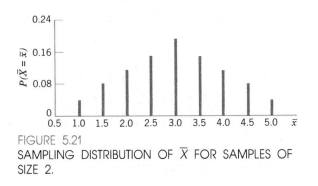

FIGURE 5.21
SAMPLING DISTRIBUTION OF $\bar{X}$ FOR SAMPLES OF
SIZE 2.

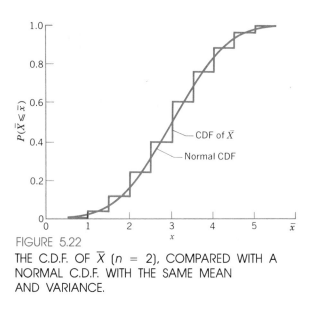

FIGURE 5.22

THE C.D.F. OF $\overline{X}$ ($n = 2$), COMPARED WITH A
NORMAL C.D.F. WITH THE SAME MEAN
AND VARIANCE.

a sample of size 4 is obtained in the same way as for a sample of size 2, but
the computations are somewhat lengthy, so they are omitted. The result is the
probability distribution of $\overline{X}$ given in Figure 5.23 for a sample of size 4 from
the population given by Figure 5.17. Notice that this distribution of probabilities in Figure 5.23 and the corresponding c.d.f. given in Figure 5.24, are
closer to the appearance of a normal distribution than the corresponding Figures 5.21 and 5.22 for samples of size 2. As the sample size increases, the
appearance will be more and more like a normal distribution, according to
the **central limit theorem.** The following should now be noted:

1. The rating X_1 given by the first consultant is a discrete random variable
   defined for five values of x. So is the rating given by each subsequent
   consultant, X_2, X_3, X_4. They each have a uniform distribution as shown in
   Figure 5.17.
2. $\overline{X}$ is also a discrete random variable, but it is defined for nine unique values
   when $n = 2$, and 17 unique values when $n = 4$.
3. The probability distribution of $\overline{X}$ is not the same as that of X_1, X_2, X_3, or
   X_4, the individual rating of each consultant.

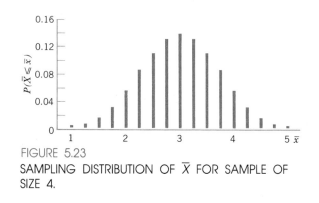

FIGURE 5.23

SAMPLING DISTRIBUTION OF $\overline{X}$ FOR SAMPLE OF
SIZE 4.

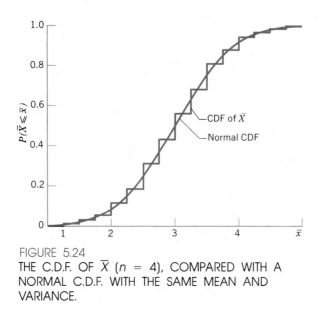

FIGURE 5.24
THE C.D.F. OF $\overline{X}$ (n = 4), COMPARED WITH A
NORMAL C.D.F. WITH THE SAME MEAN AND
VARIANCE.

4. $\overline{X}$ is not normally distributed; however, the graph of the probabilities of $\overline{X}$ looks more like a bell-shaped normal graph than does the graph of the original population probabilities.

5. The central limit theorem does not say that the distribution of $\overline{X}$ should be normal, since samples only of size n = 2 and n = 4 were used. However, it does say that as n increases, the normal approximation to the probabilities of $\overline{X}$ becomes more accurate.

In summary, it may be said that as the sample sizes get larger, the normal distribution becomes a better approximation to the distribution of $\overline{X}$. For some population distributions, the normal distribution provides a good approximation for the distribution of $\overline{X}$ even for small sample sizes. In fact, if the population has a normal distribution, then the distribution of $\overline{X}$ is *exactly* normal no matter how small n might be, as was pointed out earlier in this section. However, since exact normal distributions do not exist with real-world data, this point is more of academic interest than it is of practical value.

THE APPROXIMATE NORMALITY OF THE SAMPLE PROPORTION

The central limit theorem may also be used to justify the normal approximation to the distribution of the sample proportion $\hat{p}$. Recall, from Eq. 5.5, that the sample proportion is X/n, where X is the number of "successes" in n trials. If the result of each trial is recorded separately, by the random variable Y_i for the ith trial, then Y_i = 1 if the trial results in a success, which has probability p; Y_i = 0 if the trial does not result in a success, and X is merely the sum of the Y_i. Since all of the Y_i are independent and they all have the same distribution, the central limit theorem applies to their mean

$$\overline{Y} = \sum_{i=1}^{n} \frac{Y_i}{n}$$

which is just another way of writing the sample proportion

$$\hat{p} = \frac{X}{n} = \sum_{i=1}^{n} \frac{Y_i}{n}$$

since X is the sum of the Y_i.

The mean of each Y_i may be shown to be p, and the standard deviation may be shown to equal

$$\sqrt{pq}$$

since Y_i is merely a binomial random variable with $n = 1$. Therefore, the mean of $\hat{p}$ is p, its standard error is $\sqrt{pq}/\sqrt{n}$, and the statistic

$$\frac{\hat{p} - p}{\sqrt{pq}/\sqrt{n}}$$

may be approximated by the standard normal distribution if n is large. However, the preceding statistic may be written in a more familiar form,

$$\frac{\hat{p} - p}{\sqrt{pq}/\sqrt{n}} = \frac{(X/n) - p}{\sqrt{pq}/\sqrt{n}} = \frac{X - np}{\sqrt{npq}}$$

which is the standardized form of a binomial random variable. This is the justification behind the normal approximation to the binomial distribution, which was introduced in the previous section.

EXERCISES

5.36 Verify that the population mean and standard deviation of the uniform probability distribution given in Eq. 5.18 are 3 and $\sqrt{2}$, respectively.

5.37 The mean and standard deviation of the sampling distribution of $\overline{X}$ for samples of size 2 represented in Figure 5.21 are 3 and 1, respectively. What are the values of the mean and standard deviation for the sampling distribution of $\overline{X}$ for samples of size 4 as represented in Figure 5.23?

5.38 At one time, the number of shares sold per day on the New York Stock Exchange followed a normal distribution with $\mu = 86$ million and $\sigma = 5$ million. Find the probability that the average daily shares sold in a one-week (5-day) period exceeds 88 million shares.

5.39 The daily output of a production line varies according to a normal distribution with a mean of 163 units and a standard deviation of 3.5 units. Find the probability that the average production for 10 days will be 160 units or less.

5.40 Consider a probability model that assigns the probabilities .3, .15, .1, .15, and .3 to the values of the random variable $X = 1, 2, 3, 4,$ and 5, respectively. Graph this probability function and compute the population mean and the standard deviation.

5.41 Refer to the probability model given in Exercise 5.40. Means of samples of size 2 for this model would appear as in Figure 5.19. However, the sampling distribution of $\overline{X}$ would not appear as in Figure 5.20. Find the correct sampling distribution of $\overline{X}$ for samples of size 2 and make a summary similar to Figure 5.20.

5.42 Make a graph of the sampling distribution of $\overline{X}$ found in Exercise 5.41 and compare it with the one found in Figure 5.21.

5.43 Compute the mean and the standard deviation of the sampling distribution found in Exercise 5.41 using the definitions given in Section 4.4, and compare them with population mean and variance found in Exercise 5.40. Does this comparison show agreement with Eqs. 5.14 and 5.16, respectively?

5.5

THE EXPONENTIAL AND POISSON DISTRIBUTIONS (OPTIONAL)

PROBLEM SETTING 5.3

The manager of a quality assurance division of an electronics firm has the responsibility of determining the length of time the firm should offer on the warranty of an amplifier they manufacture. His decision will be based on laboratory tests made by the firm on the **time to failure** of the amplifier. This setting is typical of decisions or policies made on the basis of experimental results. Such experiments will usually produce some numerical results (i.e., sample data), in which case the decision-making process reduces to one of applying scientific judgment to the numerical results.

Assume that the manager knows that the competition offers a warranty of 50 months on their amplifier and wonders if this is a reasonable warranty to use. Suppose the manager is forced to make a "seat of the pants" decision (i.e., no scientific judgment allowed) for each of the following sets of laboratory tests.

Test set 1 120, 115, 170, 65, 143, 186, 174, 150, 43, 60

Manager's decision It may be reasonable to offer a warranty of 50 months.

Test set 2 50, 30, 95, 58, 42, 19, 27, 65, 89, 75

Manager's decision A warranty of 50 months may be too high.

Do you agree with the manager's decision? At this point his decisions may appear to be correct because test set 1 certainly has times well above 50 months in nine of 10 cases, whereas test set 2 may be too close to 50 for comfort. What the manager really needs to know is what percentage of the amplifiers his firm will have to replace because they fail in the first 50 months of operation.

THE EXPONENTIAL RANDOM VARIABLE

An attempt is now made to associate some scientific judgment with this problem setting and to evaluate the manager's decisions. The random variable of interest is the length of life of the amplifier or, equivalently, its time to failure. Many different probability distributions have been used to describe the length of life of a product, one of which, used most often, is the **exponential distribution.** It is used for products that have the same probability of failing throughout their lifetime. That is, all through the life of the product, their probability of failing during the next minute (or next hour or next day) remains unchanged from when the product was new. If the probability of the amplifier failing during the next 24 hours is the same now as it was at any time in the past including when it was new, and will be the same at any time in the future given that the amplifier is still operating properly at that future time, then the amplifier is said to have a **constant failure rate,** and the exponential distribution is appropriate to use as a model. Other examples where the exponential probability distribution is used to model random variables include the time between phone calls arriving in an office, the time between accidents at a street intersection, and the time between calls at a fire department. Because the random variable frequently involves waiting times between events, it is also called the **waiting-time distribution.**

The probability distribution function and population mean and standard deviation for this continuous random variable are given as follows.

PROBABILITY DISTRIBUTION FOR AN EXPONENTIAL RANDOM VARIABLE

If a random variable X has an **exponential distribution** with parameter λ, then the density function is given by

$$f(x) = \lambda e^{-\lambda x} \qquad x > 0, \quad \lambda > 0$$

and the distribution function is given by

$$\begin{aligned} P(X \leq x) &= 0 && \text{for } x \leq 0 \\ &= 1 - e^{-\lambda x} && \text{for } x > 0 \end{aligned} \qquad (5.19)$$

Population mean $= 1/\lambda$.

Population standard deviation $= 1/\lambda$.

The parameter λ is sometimes called the *failure rate.*

Note that this is another example of a continuous random variable, but unlike the normal random variable, the exponential random variable is always greater than zero. Also, the population mean equals the population standard deviation in exponential populations, so the coefficient of variation always equals 1. Some typical graphs of the probability density function and distribution function for various values of λ are given in Figures 5.25 and 5.26. Note that the exponential distribution is not symmetrical.

For answering probabilistic questions about exponential random variables, use Eqs. 5.19, 5.20, and 5.21.

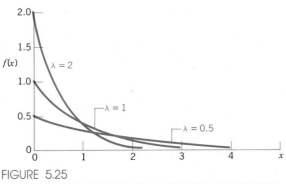

FIGURE 5.25
EXAMPLES OF EXPONENTIAL DENSITY FUNCTIONS.

If X is an exponential random variable with parameter λ, then

FINDING
PROBABILITIES
FOR AN
EXPONENTIAL
RANDOM
VARIABLE

$$P(X \geq x) = e^{-\lambda x} \qquad \text{for } x \geq 0 \tag{5.20}$$

and

$$P(x_1 \leq X \leq x_2) = e^{-\lambda x_1} - e^{-\lambda x_2} \tag{5.21}$$

for $x_1 > 0$ and $x_2 > 0$.

ESTIMATING THE PARAMETER OF AN EXPONENTIAL POPULATION

One way to estimate the parameter λ from sample data is first to find the sample mean $\overline{X}$, which is an estimate of the population mean $1/\lambda$. Then the estimator $\hat{\lambda}$ of λ is given by $1/\hat{\lambda} = \overline{X}$ or $\hat{\lambda} = 1/\overline{X}$. The manager's decisions, associated with Problem Setting 5.3, are now evaluated based on the assumption of an exponential distribution.

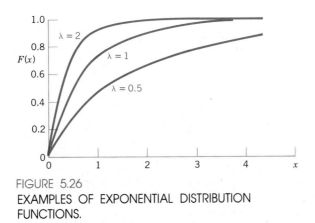

FIGURE 5.26
EXAMPLES OF EXPONENTIAL DISTRIBUTION
FUNCTIONS.

Test set 1 120, 115, 170, 65, 143, 186, 174, 150, 43, 60

$$\overline{X} = 122.6 \qquad \hat{\lambda} = \frac{1}{122.6} = .00816$$

What proportion of the amplifiers will fail within 50 months, the figure used on the warranty, under the assumption $\lambda = .00816$?

$$P(X \le 50) = 1 - e^{-.00816(50)}$$
$$= 1 - e^{-.408}$$
$$= .335$$

The number $e^{-.408}$ may be found on most pocket calculators, so tables are not needed to find probabilities for the exponential distribution. The result shows 33.5% of the amplifiers could be expected to fail in the first 50 months. Therefore, 50 appears to be too high a number to put on the warranty. The analysis for the second test set is as follows.

Test set 2 50, 30, 95, 58, 42, 19, 27, 65, 89, 75

$$\overline{X} = 55 \qquad \hat{\lambda} = \frac{1}{55} = .0182$$
$$P(X \le 50) = 1 - e^{-.0182(50)}$$
$$= 1 - e^{-.9091}$$
$$= .597$$

This test set indicates that there is an estimated probability of 59.7% that each amplifier will fail in the first 50 months. Based on the second test set a warranty of 50 months is again too high.

THE POISSON DISTRIBUTION

There is a second way of looking at items failing over time, or events like telephone calls occurring over time. That is by counting *how many* items fail, or *how many* telephone calls occur, in a fixed interval of time, such as in one

Time	X_i	Time	X_i	Time	X_i
9:03	3	9:29	5	10:06	7
9:05	2	9:43	14	10:13	7
9:16	11	9:54	11	10:15	2
9:23	7	9:56	2	10:27	12
9:24	1	9:59	3		

FIGURE 5.27
THE CUSTOMER ARRIVAL TIMES, AND X_i (THE TIME INTERVAL BETWEEN ARRIVALS).

hour, or one day, or two days. When the *time between failures* has an exponential distribution, the *number of failures* in a fixed interval of time has a distribution known as the **Poisson distribution.** Notice that although the time between failures (or telephone calls) is a continuous random variable, the number of failures (or telephone calls) in a fixed time interval takes on only the integer values 0, 1, 2, 3, etc., and is therefore a *discrete* random variable.

If a random variable Y has the **Poisson distribution,** then it takes 0, 1, 2, . . . as possible values, with probabilities given by the probability function

POISSON
DISTRIBUTION

$$P(Y = y) = \frac{e^{-\lambda}\lambda^y}{y!} \qquad y = 0, 1, 2, \ldots \qquad (5.22)$$

where λ is the parameter of the Poisson distribution

Population mean = λ

Population standard deviation = $\sqrt{\lambda}$

EXAMPLE

The receptionist in a bank notes the time of arrival of customers seeking special services, such as opening a new account and obtaining a loan. This is part of a study designed to determine how many employees are needed to provide sufficient service so that customers do not have to wait very long.

The arrival times are given in Figure 5.27, starting from 9:00 A.M. when the records were begun. Also given are the time between successive arrivals, or the time until the first arrival X_i. The first arrival is at 9:03, so $X_1 = 3$ min. The second arrival is at 9:05, 2 min later, so $X_2 = 2$. If the probability of a customer arriving during a particular minute is independent of how long it has been since the previous customer arrived, then X_i has an exponential distribution. (Even though X_i is measured in minutes, this measurement is merely rounded off to the nearest minute for convenience. The time between arrivals is actually a continuous random variable.)

The real-number-line plot in Figure 5.28 displays the arrival times, translated so 9:00 A.M. corresponds to time zero and time is measured in minutes. The random variable Y_i represents the number of arrivals in the ith 10-min interval after 9:00. Two customers arrived in the first 10 min, so $Y_1 = 2$; only one customer arrived in the next 10 min interval, so $Y_2 = 1$, and so on. It is

Y_i: 2   1   3   0   1   3   1   2   1

FIGURE 5.28
A REAL-NUMBER-LINE PLOT OF ARRIVAL TIMES (9:00 = TIME 0), AND Y_i
(THE NUMBER OF ARRIVALS IN 10-MIN INTERVALS).

clear that Y_i is a discrete random variable, taking only the values 0, 1, 2, etc. Because the distribution of X is the exponential distribution, the probability distribution of Y is the Poisson distribution.

THE RELATIONSHIP BETWEEN POISSON AND EXPONENTIAL DISTRIBUTIONS

The probability function for the Poisson distribution is indexed by a parameter λ, which also equals the mean and the variance of the distribution. It is no coincidence that the same parameter is used for both the exponential distribution and the Poisson distribution, because of the close connection between the two distributions. If the time between arrivals X has an exponential distribution with parameter λ, then the number of arrivals in *one unit* of time Y has a Poisson distribution, with the same numerical value for λ as in the exponential distribution of X.

EXAMPLE

Suppose that the time between arrivals of airplanes at Chicago's O'Hare Airport follows an exponential distribution with $\lambda = 1.7$, where time is measured in minutes. Then the number of airplanes arriving each minute follows a Poisson distribution with parameter $\lambda = 1.7$. The mean time between arrivals is found from the mean of the exponential distribution,

$$\mu = \frac{1}{\lambda} = \frac{1}{1.7} = .588 \text{ min} \qquad \text{(Mean time between arrivals)}$$

with a standard deviation of .588 also. This translates to about 35 seconds between arrivals, on the average. The number of planes arriving per minute is found from the mean of the Poisson distribution with the same λ, which has a mean of $\lambda = 1.7$ planes, with a standard deviation of $\sqrt{\lambda} = 1.30$. This agrees with the previous calculation, in that the mean number of planes arriving is somewhere between 1 and 2 per minute, corresponding to the average time between arrivals of about 35 seconds.

The next example is presented as another application of the Poisson distribution to a process involving a demand for a service.

EXAMPLE

Two enterprising college students decide to form a small business by renting out their two cars on a daily basis. If the number (Y) of requests for a car is distributed as a Poisson random variable with a mean of 1.5 requests per day, find the following:

(a) The proportion of days that neither car is required.

(b) The proportion of days that more than two cars are requested.

The answer to part (a) is found from Eq. 5.22. If the mean is 1.5, then $\lambda = 1.5$, because λ equals the mean in a Poisson distribution. The probability that neither car is required is

$$P(Y = 0) = \frac{e^{-1.5}(1.5)^0}{0!} = e^{-1.5} = .223$$

The answer to part (b) is found as follows:

$$
\begin{aligned}
P(Y > 2) &= 1 - P(Y \leq 2) \\
&= 1 - P(Y = 0) - P(Y = 1) - P(Y = 2) \\
&= 1 - \frac{e^{-1.5}(1.5)^0}{0!} - \frac{e^{-1.5}(1.5)^1}{1!} - \frac{e^{-1.5}(1.5)^2}{2!} \\
&= 1 - .223 - .335 - .251 = .191
\end{aligned}
$$

Based on the Poisson distribution, the students can expect no requests on 22.3% of the days, one request on 33.5% of the days, two requests on 25.1% of the days, and only 19.1% of the time will the number of requests be expected to exceed the number of cars available.

EXERCISES

5.44 If the interarrival times in minutes at an emergency receiving room in a large city hospital can be regarded as following an exponential distribution, use the following interarrival times to estimate the parameter λ of an exponential distribution.

9,   13,   10,   7,   10,   13,   9,   5,   3,   14

5.45 Construct a real-number-line plot to represent time from 0 to 100 minutes with tick marks at every 10 minutes. Plot the arrival times from the interarrival times of Exercise 5.44 on the real-number line. The first data point will be plotted at 9, the second at 22 ($=9 + 13$), the third at 32 ($=9 + 13 + 10$), etc. How many of the 10 intervals had no arrivals? 1 arrival? 2 arrivals? 3 arrivals? Use the values you have calculated to estimate the parameter λ of a Poisson distribution.

5.46 The time in minutes between incoming phone calls in a business office has been recorded for 10 calls: 1.8, 0.3, 4.5, 9.8, 3.2, 15.7, 4.8, 1.0, 2.7, 6.2.

(a) Estimate the mean waiting time between incoming calls.

(b) Estimate the parameter λ of an exponential distribution using these data.

(c) Use the value of $\hat{\lambda}$ from part (b) to estimate the probability that the waiting time between incoming calls will be less than 1 min. Also estimate the probability that the waiting time will exceed 10 min.

5.47 A fast-food chain finds that the average time their customers must wait before they are served is 45 sec. If the waiting time can be treated as an exponential random variable, find the exponential parameter λ and use this value to find the probability that a customer will have to wait more than 3 min to be served.

5.48 At one of the busiest airports in the United States there is an average of 1 min between the arrival of planes. If these interarrival times can be treated as an exponential random variable, find the probability that there is more than 5 min between plane arrivals.

5.49 A small business uses a computer for its accounting. The computer experiences a failure of some type on the average of once every 2 hours of operation. If the time to failure follows an exponential distribution, what is the probability of the computer operating properly for an 8-hour period?

5.50 New automobiles are routinely covered by an unconditional warranty for 90 days. If the average time before a repair is required is 300 days, what proportion of cars can a new-car dealer expect to have back within the 90-day warranty period? Assume time to repair follows an exponential distribution.

5.51 If $\lambda = 2$ for a Poisson random variable X, which represents the number of pieces of junk mail a manager receives each day, find the following probabilities:

(a) $P(X = 0)$

(b) $P(X = 1)$

(c) $P(X = 2)$

(d) $P(X = 3)$

(e) $P(X > 3)$

(f) Find the mean of X.

(g) Find the standard deviation of X.

5.52 The number of patients arriving at an emergency receiving room of a large city hospital has been recorded in 15-min intervals during a 10-hr period. The results are as follows, where Y is the number of patients arriving in 15-min intervals.

Y:	0	1	2	3	4	5	6	7
Frequency:	5	11	11	7	4	1	0	1

Assuming that a Poisson distribution is appropriate for describing these data, estimate the parameter λ for a Poisson distribution.

5.53 Use the estimate of λ in Exercise 5.52 to estimate the probability that the number of patients arriving in a 15-min interval will be more than three.

5.54 A submarine must carry enough spare parts so that there is less than .01 chance of not having enough to last the length of a mission. Suppose a radio circuit, XL-18, needs replacement at random times, but an average of 0.3 times per month. How many circuits should the submarine stock for a six-month cruise?

5.55 During the 96-year period from 1837 to 1932 there were 59 years in which no vacancies occurred in the U.S. Supreme Court, 27 years with one vacancy, 9 years with two vacancies, and 1 year with three vacancies. On the assumption that a Poisson distribution is appropriate for the number of vacancies per year, and conditions do not change, find the probability that an appointment will not be made during the first four years of a presidential term beginning in 1933. Assume the number of appointments is independent from year to year.

5.6

REVIEW EXERCISES

5.56 Of all TV sets sold by an appliance firm, 10% require repair before the warranty expires. Use the normal approximation to the binomial to find the probability that the firm will have to repair at least 20 of their last 100 sales. Find the probability that the number of sets requiring repair is between 6 and 14 inclusive.

5.57 Faculty salaries at a certain university are normally distributed with a mean of $27,500 and a standard deviation of $3000. Find the probability that one faculty member chosen at random will have a salary less than $23,500; a salary greater than $30,000.

5.58 The distribution of test scores for an examination is normal with a mean of 75 and a standard deviation of 6. What is the probability that the mean score obtained by 25 students taking this exam will be between 73 and 77?

5.59 The monthly earnings of all employees of a large company are normally distributed with a mean of $1500 and with a standard deviation of $900. A random sample of 25 employees is obtained and their average salary $\overline{X}$ is computed. Find the following probabilities.

(a) $P(\overline{X} \leq \$1230)$

(b) $P(\overline{X} \leq \$1600)$

(c) $P(\$1230 \leq \overline{X} \leq \$1600)$

5.60 Of the professional employees of a large company, 20% have a doctorate degree. A random sample of 50 employees is obtained. Find the probability that the number of employees in the sample who have a doctorate is between 6 and 15 inclusive.

5.61 A machine that fills 16-oz boxes of cereal is set to fill the boxes with a mean of 16.3 oz and a standard deviation of .15. Assume that the weights of the cereal contents are normally distributed and find the probability of a box having less than 16 oz.

5.62 Of the incoming freshmen at a midwestern university, 25% drop out by their sophomore year. Find the probability of more than 140 students dropping out from a class of 500.

5.63 The shelf life for a brand of film has an exponential distribution with a mean of one year, and a 14-month expiration date on it. What is the probability that the film will be unsatisfactory prior to its expiration date?

5.64 Twenty 10-lb bags of flour were randomly selected from the flour stock and weighed. Check these weights for normality: 9.8, 9.9, 9.9, 10.1, 9.9, 9.8, 10.1, 10.0, 10.0, 9.7, 10.5, 10.1, 9.5, 10.1, 9.9, 10.1, 10.2, 9.9, 10.0, 9.6. (Use $\overline{X} = 9.96$ and $s = .22$.)

5.65 If the time to failure in hours of electronic components follows an exponential distribution, use the following data to estimate the parameter λ of the exponential distribution.

42.7, 63.1, 91.7, 55.8, 120.5, 48.6, 51.7, 36.9, 61.9, 57.8, 75.4, 81.2, 44.5, 47.2, 65.9

5.66 Graph the binomial probability function with $n = 4$ and $p = .75$. Is it skewed, and if so, in what direction?

5.67 An airline deliberately overbooks on some flights because it knows some passengers who make reservations will not show up for a flight. Suppose each passenger has .85 probability of showing up for a particular flight that holds 122 passengers. How many reservations can the airline make and still be at least 90% certain that every person who shows up for the flight with a reservation can be accommodated with a seat on the plane?

5.68 In order to decide whether to accept a shipment of transistors a company selects a random sample of 20 transistors for testing from the lot. If more than one transistor is defective among the 20 tested, then the 'entire shipment is rejected. Draw a graph of the probability of accepting the shipment (y-axis) versus the probability of a transistor being defective (x-axis).

5.69 There is exactly a 5% probability that a supermarket's customer will buy a newspaper. How many newspapers should the supermarket stock if it wants to be 99% sure that newspapers will be available to the first 1000 customers?

5.70 An insurance company sells a group theft insurance policy on ten cars. For each car that is stolen the company will pay $10,000. The probability of each car being stolen is .05. What should the premium rate be on each car so that the company will be at least 80% sure that it will not lose any money on the policy? Assume independence of events.

5.71 Customers arrive randomly at a busy store at the average rate of two per minute. What is the probability that a 2-min time period lapses with no customers arriving?

5.72 A builder's supply store sells hot water heaters. The manager of the store can sell the heaters for $400 each, but he wants to add enough to the price to cover a 10-year money-back guarantee in case of failure of the heater. If there is a 20% chance of failure of each heater, independently of the others, how much does he need to add to the price of the heater to cover the expected cost of the guarantee?

INTERPOLATION IN THE STANDARD NORMAL TABLES

Since Tables A2 and A2* cannot contain every possible value of z and p, it may be necessary to use interpolation to get a more exact answer in some situations. Interpolation is illustrated by using Table A2 to find $P(Z \leq 1.0000)$. Note this value can be read directly from Table A2* as .8413, so a check is available for the adequacy of the linear interpolation. The process of interpolation begins by noting that the two values of z closest to 1.0000 in Table A2 are .9986 and 1.0027 for which

$$P(X \leq .9986) = .84 + .001 = .841$$

and

$$P(Z \leq 1.0027) = .84 + .002 = .842$$

Clearly, the correct answer is between .841 and .842. Since three-decimal-place accuracy is sufficient for *most applications,* the tabled value closest to the z value of 1.0000 would be used to obtain the probability .841. However, linear interpolation can be used here if four-decimal-place accuracy is desired. The rectangle in Figure A5.1 is useful in understanding how linear interpolation works. If it is desired to find $P(Z \leq z)$, two values z_1 and z_2 for which probabilities are given in Table A2 are found such that $z_1 < z < z_2$. Of course, z_1 and z_2 should be as close as possible to z. These three values of z are plotted along the bottom of the rectangle in Figure A5.1, with z_1 and z_2 plotted in the corners as shown. The left side of the rectangle is labeled in the corners with the corresponding probabilities, that is, $P(Z \leq z_1) = p_1$ and $P(Z \leq z_2) = p_2$. Linear interpolation between p_1 and p_2 for a given value of z such that $z_1 < z < z_2$ is accomplished by projecting a line upward from z on the z-axis to the line connecting the corners of the rectangle and then projecting this point horizontally to the vertical p-axis. The resultant intersection with the p-axis is the desired answer. Algebraically this projection is expressed as follows:

$$P(Z \leq z) \approx p_1 + \left[\frac{z - z_1}{z_2 - z_1} \right] (p_2 - p_1) \qquad (A5.1)$$

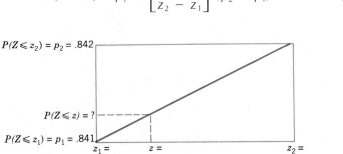

FIGURE A5.1
A DIAGRAM FOR AIDING IN LINEAR INTERPOLATION.

For the present example use of Eq. A5.1 gives

$$P(Z \leq 1.000) \approx .841 + \left[\frac{1.0000 - .9986}{1.0027 - .9986} \right] (.842 - .841)$$

$$= .841 + .0003 = .8413$$

Note that this value agrees exactly with the value obtained from Table A2* earlier. Although the results of linear interpolation will not always give perfect agreement, they should be close.

BIBLIOGRAPHY

Additional material on the topics presented in this chapter can be found in the following publications.

Conover, W. J. (1980). *Practical Nonparametric Statistics,* 2nd ed. Wiley, New York.

Iman, R. L. (1982). "Graphs for Use with the Lilliefors Test for Normal and Exponential Distributions." *The American Statistician,* **36**(2), 108–112.

Lilliefors, H. W. (1967). "On the Kolmogorov-Smirnov Test for Normality with Mean and Variance Unknown." *Journal of the American Statistical Association,* **62,** 399–402.

6

ESTIMATION (ONE SAMPLE)

In the previous chapters statistics have been presented for the summarization of sample data. These statistics include the sample mean, sample variance and standard deviation, the sample median, and sample quantiles, as well as other sample statistics. Besides their usefulness in summarizing sample data, these same summary statistics often provide *estimates* for population parameters. For example, the sample mean $\bar{X}$ provides an estimate of the population mean μ, whereas the sample standard deviation s provides an estimate of the population standard deviation σ. The population proportion p is estimated by the sample proportion $\hat{p} = X/n$, where X is the number of successes observed in a random sample of size n.

TYPES OF ESTIMATES

Each of the preceding sample statistics is expressed as a single number that in turn provides a **point estimate** of a population parameter. Point estimates are quite useful, but they do not always provide the type of information that is desired in a particular setting. For example, consider the board of directors of a company needing an estimate of a competitor's current volume of sales. An example of a point estimate might be that the sales are $2.7 million. Another response might be that with 95% confidence the sales are between $2.5 million and $2.9 million. This latter estimate is an example of an **interval estimate** in that an upper bound and a lower bound are provided along with a stated degree of confidence for the interval. In this chapter methods for finding interval estimates for the population mean, the population proportion, the population median, and the population standard deviation are explained.

POINT ESTIMATE | A **point estimate** is a single number that is used as an estimate of a population parameter or population characteristic. Usually a point estimate is derived from a random sample from the population of interest.

INTERVAL ESTIMATE, OR CONFIDENCE INTERVAL | An **interval estimate** is an interval that provides an upper and lower bound for a specific population parameter whose value is unknown. This interval estimate has an associated **degree of confidence** of containing the population parameter. Such interval estimates are also called **confidence intervals** and are calculated from random samples.

A FORMAT FOR PRESENTING THE PROCEDURES

The procedures for making statistical inferences in the rest of this book will be stated in a concise format, to make it as easy as possible for the reader to follow the procedure without getting confused by long verbal descriptions. The format includes a very important statement of the assumptions necessary for a correct analysis. Techniques are provided for checking these assumptions, and appropriate alternative procedures are provided when assumptions are not satisfied (which occurs frequently).

6.1

GENERAL REMARKS ABOUT ESTIMATION

POINT ESTIMATES AND INTERVAL ESTIMATES

This section presents some definitions associated with estimation. Samples may be used to estimate population parameters, such as μ and σ, which represent the population mean and standard deviation, or other population characteristics such as the median or other quantiles. Estimates may take the form of a single number, called a **point estimate,** or an interval of values, called an **interval estimate.** Some examples of point estimates follow.

EXAMPLES OF POINT ESTIMATES

Marketing A marketing survey determines the number of persons in a random sample who will probably buy a product offered by the company. A sample proportion can be calculated to estimate the proportion of all people in the population who will probably buy the product. A sample proportion of .28, for example, is a point estimate of the population proportion.

Management A random sample consisting of 35 employees is studied in detail to determine the actual level of productivity of each employee in the sample. The average level of productivity in the sample can be used as a point estimate of the population (all employees) mean productivity.

Accounting Rather than take a complete inventory of all parts in a warehouse in an audit, a stratified sample is obtained. (See Section 1.4 for a definition of stratified sample.) The stratification is based on the cost of the item. All items costing more than $50,000 are inventoried, 50% of the items between $10,000 and $50,000 are inventoried, and so on, until only 1% of items costing less than $1 are inventoried. In this case, a weighted sample mean is used to estimate the total inventory value.

Finance Banks obtain assets in the form of customer savings accounts, customer checking accounts, and money borrowed from other financial institutions. Some of this money is loaned out at a higher interest rate. This is how the bank makes money. But a percentage of the assets must be kept on hand to cover customer withdrawals or checks cashed, and a percentage must be kept in non-interest-bearing accounts at the Federal Reserve. The percentages vary from day to day within a bank, and from bank to bank. A random sample of banks is examined to get an idea of the probability distribution of the percentage of assets on hand. The empirical distribution function obtained from the random sample of banks is examined to get an idea of the probability distribution of percentage of assets on hand. The empirical distribution function obtained from the sample percentages can be used to estimate the population distribution function. Sample percentiles are used as point estimates of the population percentiles. The sample median provides a point estimate of the population median percentage—the value exceeded by 50% of all banks in the population.

Economics Finding the exact unemployment percentage is not possible because the actual value of the population percentage changes over time. Therefore, the population unemployment percentage is estimated on the basis of random samples from the population.

EXAMPLES OF INTERVAL ESTIMATES

In many situations where point estimates are obtained, interval estimates can also be obtained. The interval estimate is an interval that includes the point estimate. That is, in a marketing survey where the point estimate is .28 for the population proportion, an interval estimate might be of the form: "You can be 95% confident that the true value for the population proportion lies between .25 and .31." Clearly this interval contains the point estimate .28. If the point estimate for the median percentage of assets held in reserve by the banks is 36%, an interval estimate might read: "You can be 90% confident that the median percentage of assets held in reserve by all banks is between 33% and 39%."

ESTIMATOR An **estimator** is a random variable calculated from a random sample that provides either a point estimate or an interval estimate for some population parameter.

EXAMPLES OF ESTIMATORS

The sample proportion $\hat{p} = X/n$, where X is the number of successes in a random sample of size n, is a random variable calculated from sample data. Hence, $\hat{p}$ is an **estimator** that is used to estimate p, the population proportion of successes or the probability of getting a success on any one trial. The value of $\hat{p}$ resulting from the calculation is the *point estimate*.

The two *estimators* $\overline{X} - 1.9600s/\sqrt{n}$ and $\overline{X} + 1.9600s/\sqrt{n}$, where $\overline{X}$ and s are the sample mean and standard deviation in a random sample of size n, are random variables calculated from sample data. As is shown later in this chapter, these two random variables form an approximate 95% *interval estimator* for the population mean when the sample size n is large (say, greater than 30). The actual numbers resulting from the calculations for a specific set of observations form the *interval estimate*.

PROPERTIES OF ESTIMATORS

Statisticians have studied properties of estimators to aid in evaluating and comparing the usefulness of various estimators. The two most commonly stated properties are those of **unbiasedness** and **minimum variance.** Some estimators, such as $\overline{X}$, have both properties when used to estimate μ for some distributions, such as the normal distribution. Other estimators, such as s, used to estimate σ, seldom satisfy either one of these properties and still are considered very useful estimators. The following discussion is not intended to make you an expert on these properties, but is intended to broaden your perspective regarding the selection of a good estimator.

UNBIASEDNESS

An estimator is **unbiased** if its mean or expected value is the population parameter being estimated. That is, every estimator has a probability distribution, called its *sampling distribution,* and if the mean of the sampling distribution is the same as the parameter being estimated, then the estimator is unbiased. An unbiased estimator does not have a tendency to underestimate the population parameter, nor does it have a tendency to overestimate the population parameter in the long run. The sampling distribution of $\overline{X}$ has as its mean the same μ that is the mean of the population. Therefore $\overline{X}$ is an unbiased estimator for μ. Also the mean of $\hat{p}$ is p, so $\hat{p}$ is an unbiased estimator for the population proportion p.

Bias is the difference (by subtraction) between the mean of the estimator and the parameter being estimated. Thus unbiased estimators have a bias of zero, which is another way of saying the same thing.

UNBIASED An estimator is **unbiased** if the mean of its sampling distribution is the popu-
ESTIMATOR lation parameter being estimated.

MINIMUM VARIANCE

The best estimator is the estimator that comes closest, in some sense, to the
population parameter being estimated. Since "closeness" is often measured
in terms of standard deviation, or variance, the best estimator is often considered to be the estimator that has the smallest variance associated with its
sampling distribution, provided that the estimator is unbiased. Such estimators
are called **minimum variance unbiased estimators.** For many distributions,
including the normal and the Poisson, the sample mean is a minimum variance
unbiased estimator of the population mean. Such a property is usually difficult
to prove mathematically. Although several estimators may be unbiased for
some population parameter, only one will also have the minimum variance.

MINIMUM An estimator is a **minimum variance unbiased estimator** if the variance of its
VARIANCE sampling distribution is the smallest of the variances of the sampling distri-
UNBIASED butions of all other unbiased estimators.
ESTIMATOR

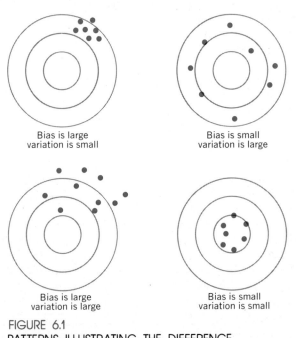

Bias is large
variation is small

Bias is small
variation is large

Bias is large
variation is large

Bias is small
variation is small

FIGURE 6.1
PATTERNS ILLUSTRATING THE DIFFERENCE
BETWEEN SMALL BIAS AND SMALL VARIATION.

Roughly speaking, the properties of unbiasedness and minimum variance can be associated with targets, as in Figure 6.1. Think of the bull's eye as the parameter being estimated. If the shot pattern on the target is centered on the bull's eye, then the bias is small. If the shot pattern is tightly clustered, then the variation is small. A desirable situation is to have small bias with small variation, as the following discussion illustrates.

EXAMPLE TO ILLUSTRATE PROPERTIES

Assume that the number of hours worked per week by managers of individual stores belonging to a fast-food chain follows a normal distribution with a mean of 60 and a standard deviation of 4. Also assume that the distribution of hours worked is the same from city to city. Sample data for 10 stores selected at random in each of five large cities are given in Figure 6.2. These observations are arranged in order, from smallest to largest within each city, for ease in reading the values.

EXAMPLE OF UNBIASEDNESS

The sample mean $\bar{X}$ is the estimator that is best for estimating the mean of a normal population, because it has both properties of unbiasedness and minimum variance. The sample means have been computed and listed with each of the samples in Figure 6.2. Clearly none of these sample means is exactly equal to the population mean of 60. However, each is quite close, as is the mean of all 50 observations taken together, which is 59.73. The property of unbiasedness does not require that individual sample means be equal to 60, rather it requires that the value to be expected for any one sample mean prior to taking the sample is 60 and not 60.1, nor 59.99, nor any other number.

EXAMPLE OF MINIMUM VARIANCE

To understand the property of minimum variance, it is useful to note that in symmetric populations, such as the normal, several estimators for the mean are unbiased. These estimators include the sample mean and the sample median as well as others. If the variability associated with some of these competing unbiased estimators were known, then the one with the minimum variance could be used. This point is now illustrated.

City	Number of Hours Worked										Mean
L.A.	55.9	56.3	56.8	57.2	61.2	61.9	62.5	63.8	64.4	68.2	**60.82**
N.Y.	55.7	55.8	57.0	57.4	59.0	59.5	59.9	60.4	64.2	67.7	**59.66**
K.C.	53.0	54.6	54.7	54.8	57.6	58.6	62.4	63.5	65.5	66.6	**59.13**
D.C.	57.3	58.1	58.6	58.7	59.0	61.9	62.6	64.4	64.9	66.7	**61.22**
S.F.	50.5	51.4	54.8	56.3	58.3	59.0	61.2	61.6	62.2	63.1	**57.84**

FIGURE 6.2

NUMBER OF HOURS WORKED PER WEEK BY THE STORE MANAGER AT 50 DIFFERENT STORES.

Another unbiased estimator for the mean of a normal population is now introduced. This statistic, L, is the minimum observation in the sample plus the maximum observation in the sample divided by 2, or simply

$$L = (X_{min} + X_{max})/2.$$

This is the average of the smallest and the largest observations in the sample. This estimator is denoted by L (for lazy), as it reduces the amount of work by ignoring all sample observations except for two values. The values of L are given as follows.

City:	L.A.	N.Y.	K.C.	D.C.	S.F.
$L = (X_{min} + X_{max})/2$:	62.05	61.70	59.80	62.00	56.80

Since these calculations are so easy to do you should check them to make sure you understand how they were obtained. One way to get a feeling about the variability of the estimators $\overline{X}$ and L is to plot the five values of $\overline{X}$ and five values of L on the same real-number line as is given in Figure 6.3.

Figure 6.3 shows that both sets of five values are clustered about 60, but the observed variation associated with the five values of $\overline{X}$ is less than it is with the five values of L. Although different samples could give results that contradict the preceding plot, the actual variance associated with $\overline{X}$ is σ^2/n. Texts on mathematical statistics show that the variance of all other estimators for μ in a normal population must be greater than σ^2/n. The advantage of this property is the greater precision obtained in the estimate of μ by $\overline{X}$ rather than other estimators.

INTERVAL ESTIMATES

Point estimates are useful, but interval estimates convey more information. A point estimate does not indicate how much uncertainty might be associated with the estimate. For example, "the estimate for the population mean is 60.82" does not indicate how close 60.82 might be to the population mean. An interval estimate indicates an interval around 60.82, and a degree of con-

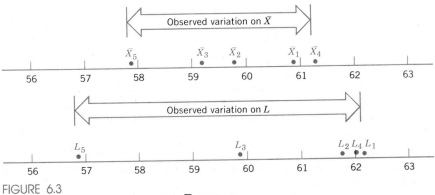

FIGURE 6.3
VARIATION OF THE ESTIMATORS $\overline{X}$ AND L.

fidence that the interval contains the population mean. For example, "the population mean is between 58.74 and 62.90, with 90% confidence" indicates a strong likelihood of the population mean being between the two numbers 58.74 and 62.90. This is why interval estimates are often preferred over point estimates.

CONFIDENCE AND PROBABILITY

An interval estimator provides an upper and lower bound for some population parameter based on sample data. The resulting interval is referred to as a *confidence interval* because of the existence of an associated level of confidence that accompanies the interval. The idea of confidence associated with an interval estimate requires a bit of explanation to clarify the difference between confidence and probability.

Confidence intervals are derived from probabilistic statements involving random variables. For example, when sampling from a normal population, the random variable $\sqrt{n}(\overline{X} - \mu)/\sigma$ has the standard normal distribution, so the probability statement

$$P(-1.6449 \leq \sqrt{n}(\overline{X} - \mu)/\sigma \leq 1.6449) = .90$$

is true. Algebraic rearrangement inside the parentheses gives

$$P(\overline{X} - 1.6449\ \sigma/\sqrt{n} \leq \mu \leq \overline{X} + 1.6449\ \sigma/\sqrt{n}) = .90$$

which is a true probability statement, since it involves the random variable $\overline{X}$.

However, once a value is substituted for $\overline{X}$ the statement is no longer a probability statement, because the random variable has been replaced by a number. The interval that was formerly part of a probability statement now becomes a confidence interval, and the number that formerly represented the probability now becomes the level of confidence. The calculations and diagram in Figure 6.4 show confidence intervals for the five samples of Figure 6.2.

There are now five different intervals for which a basic question can be asked. Does each of these intervals have a probability of .90 of containing the population mean μ? The answer, quite simply, is no, for any one interval either contains the population mean or it does not. In this example the true population mean is 60, and four of the five intervals provide bounds around 60 but one does not. Hence, this is where the idea of confidence comes into play.

It is correct to state about any one of these intervals that the confidence is 90% that the population mean is contained in the interval. What is meant is that if the experiment were repeated many times, then 90% of the intervals would be expected to contain μ, whereas the remaining 10% would not be expected to contain μ. When a 90% confidence interval is computed from sample data, one can never be certain that the interval actually contains μ. Therefore, the statement is made, "You can be 90% confident that the interval contains μ."

City	$\overline{X}$	$\overline{X} - 1.6449(4)/\sqrt{10}$	$\overline{X} + 1.6449(4)/\sqrt{10}$
L.A.	60.82	58.74	62.90
N.Y.	59.66	57.58	61.74
K.C.	59.13	57.05	61.21
D.C.	61.22	59.14	63.30
S.F.	57.84	55.76	59.92

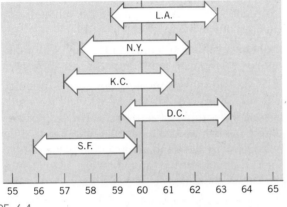

FIGURE 6.4
CONFIDENCE INTERVALS FOR THE POPULATION MEAN
DERIVED FROM FIVE DIFFERENT SAMPLES. THE TRUE
POPULATION MEAN IS SHOWN BY A VERTICAL LINE AT 60.

Although this was an example of confidence intervals for means of normal populations, the discussion holds for the interpretation of all confidence intervals. In the rest of this chapter confidence intervals are developed for the population proportion, for the mean of a normal population where σ is known and where σ is unknown, for the population median, and for σ in a normal population.

EXERCISES

6.1 A random sample of pumpkin weights yields $\overline{X} = 16.3$ lb. Which, of $\overline{X}$ or 16.3 lb, is the estimate and which is the estimator for the population mean?

6.2 The sample standard deviation of apple weights is 2.3 oz. Which, of s or 2.3 oz, is the estimate and which is the estimator of the population standard deviation?

6.3 You are trying to estimate the mean number of typographical errors per page in the rough draft of a 300-page book. You can carefully examine two pages and compute the average number of errors per page, or you can carefully examine ten pages and compute the average per page. Which plan involves more work? Which plan probably has the estimator with the smaller variance?

6.4 Your English instructor has asked you to write a 5000-word essay. Your partially completed hand-written manuscript has 35 pages at 27 lines per page, and you want to estimate the number of words you have written so far. Which of the following plans is most likely to be unbiased? Why?

(a) You count all of the words on the first page and multiply by 35.

(b) You count the number of words on the bottom line of each page, and multiply the total by 27.

(c) You randomly select one line on the second page and on every fifth page thereafter (eyes closed, finger stab method) and multiply the total number of words by $5 \times 27 = 135$.

(d) You count the number of words on the line 5 in. from the bottom of the page for each page, and multiply the total by 27.

6.5 In Figure 6.2 the five sample means from normal populations are given as 60.82, 59.66, 59.13, 61.22, and 57.84. The standard deviation calculated on these five sample means is 1.36. For each of these five samples find the sample median and then calculate the standard deviation for the five medians. Compare the standard deviation of the sample medians with the value of 1.36 found for the sample means. Comment on the result of this comparison for these normal populations.

6.6 Refer to the table of random numbers given in Figure 1.1. Treat the last five columns as 15 samples of size 5 so that the first group of five digits, 18653, becomes the first sample of 1, 8, 6, 5, and 3. The second group of five digits, 57068, becomes the second sample of 5, 7, 0, 6, and 8. Continue on down these columns until the last group of five digits, 25013, is used to form the fifteenth sample of size 5.

(a) For each of these 15 samples calculate the sample mean.

(b) For each of these 15 samples calculate the sample median.

(c) Plot the 15 sample means and medians on a real-number-line plot similar to Figure 6.3 and indicate the observed variation for each statistic.

(d) Calculate the standard deviation of the 15 sample means.

(e) Calculate the standard deviation of the 15 sample medians.

(f) Which statistic, the sample mean or the sample median, do you believe has the smaller variance associated with it for this nonnormal distribution?

6.2
ESTIMATING THE TRUE PROPORTION IN A POPULATION

PROBLEM SETTING 6.1

Companies often use surveys to obtain information about employee job satisfaction. The answers to many of these questions are tabulated and expressed as percentages or proportions. For example, consider the following questions from such a survey.

1. Do you know of any safety hazards that could cause an accident in your working area?

   Yes _____ No _____

2. Do you feel your present salary accurately reflects your contribution to the company?

   Yes _____ No _____

3. Do you believe you would be better off if the company were unionized?

   Yes _____ No _____

4. In the future, would you like to see the company place more emphasis on salary or increased job security?

   _____ More job security.

   _____ Bigger raises are needed.

All of the responses to these questions are dichotomous in nature, and in each case an estimate is needed of the true proportion in the population giving a particular response to a certain question. In this section a point estimate is obtained for the true population proportion, and then two methods (one exact and one approximate) are presented for finding confidence intervals for the true population proportion.

EXACT CONFIDENCE INTERVALS

If the answers to a survey question are dichotomous, for instance, yes or no, and if the sample is a random sample, then the random variable X representing the number of favorable responses has a binomial distribution. That is, if the sample is random, then each response has the same probability of being favorable and the responses are independent of one another. Therefore the random variable X has a binomial distribution with n, the sample size, known, and p, the probability of a favorable response, unknown. A point estimate for p is given by $\hat{p} = X/n$. This point estimate is unbiased, which as you may recall means that the expected value of $\hat{p}$ is p, the population parameter being estimated.

POINT ESTIMATE OF THE POPULATION PROPORTION

An **unbiased point estimate of the true population proportion** p is given by the sample proportion

$$\hat{p} = \frac{X}{n}$$

where X is a random variable representing the number of successes observed in a random sample of size n from the population.

Although large sample sizes should be obtained whenever possible, there are times when small sample sizes are unavoidable. When recording the number of "hits" scored by torpedoes fired from a submarine, the probability of obtaining a hit is of interest, but small samples are an economic necessity. A franchising company may wish to estimate the probability of a store in its chain going out of business within the first two years of operation, but the number of stores available for examination is quite limited. The success rate of the solid rocket boosters on the space shuttle is of interest, but the number of flights is quite small. In cases such as these, there is a need for an accurate method of obtaining a confidence interval that can be used with small samples.

INTERVAL ESTIMATE OF THE POPULATION PROPORTION p (EXACT SOLUTION FOR SMALL SAMPLES, i.e., $n \leq 30$)

Let X be a random variable representing the number of successes in a random sample of size n. To find an exact 90%, 95%, or 99% confidence interval for the population proportion p when $n \leq 30$ consult Table A6. First find the portion of the table corresponding to the sample size n. Then look at the row that matches the observed value x of the random variable X. Read across the row to get the exact lower and upper bounds on the interval estimate for p corresponding to the desired level of confidence found in the table heading.

The exact confidence intervals given in Table A6 were found by solving two very difficult equations obtained from the distribution function for a binomial random variable. The lower bound of the interval estimate for p is found by solving the following equation for p_L:

$$\sum_{i=x}^{n} \binom{n}{i} p_L^i (1 - p_L)^{n-i} = \frac{\alpha}{2} \tag{6.1}$$

Note that n is the observed sample size, x is the observed value of the random variable X, and α is related to the desired size of the confidence interval, $100(1 - \alpha)\%$. Therefore p_L is the only unknown in Eq. 6.1. Equation 6.1 states that the upper tail (from x to n) of the binomial distribution with parameters p_L and n has probability $\alpha/2$.

The upper bound for the interval estimate for p is found by solving the equation

$$\sum_{i=0}^{x} \binom{n}{i} p_U^i (1 - p_U)^{n-i} = \frac{\alpha}{2} \tag{6.2}$$

for p_U, where Eq. 6.2 represents the probability in the lower tail (from 0 to x) of the binomial distribution with parameters p_U and n. In most cases, the only way to solve these equations for p_L and p_U is with the aid of a computer. A computer was used to obtain 90%, 95%, and 99% confidence limits for p, for all $n \leq 30$, and all possible values x of the random variable X from 0 to n. The results are given in Table A6. Figure 6.5 shows a portion of Table A6 with $n = 20$. A method for obtaining approximate upper and lower bounds for p, which can be used when n is larger than 30, is given following the next example.

		90%		95%		99%	
n	x	Lower	Upper	Lower	Upper	Lower	Upper
20	0	.000	.139	.000	.168	.000	.233
	1	.003	.216	.001	.249	.000	.317
	2	.018	.283	.012	.317	.005	.387
	3	.042	.344	.032	.379	.018	.449
	4	.071	.401	.057	.437	.036	.507
	5	.104	.456	.087	.491	.058	.560
	6	.140	.508	.119	.543	.085	.610
	7	.177	.558	.154	.592	.114	.657
	8	.217	.606	.191	.639	.146	.701
	9	.259	.653	.231	.685	.181	.743
	10	.302	.698	.272	.728	.218	.782
	11	.347	.741	.315	.769	.257	.819
	12	.394	.783	.361	.809	.299	.854
	13	.442	.823	.408	.846	.343	.886
	14	.492	.860	.457	.881	.390	.915
	15	.544	.896	.509	.913	.440	.942
	16	.599	.929	.563	.943	.493	.964
	17	.656	.958	.621	.968	.551	.982
	18	.717	.982	.683	.988	.613	.995
	19	.784	.997	.751	.999	.683	1.000
	20	.861	1.000	.832	1.000	.767	1.000

FIGURE 6.5
A PORTION OF TABLE A6 SHOWING EXACT 90%, 95%, AND 99% CONFIDENCE INTERVALS FOR THE POPULATION PROPORTION p WHEN n = 20.

EXAMPLE

A random survey of 20 employees shows 12 preferring more job security instead of raises. Find a 95% confidence interval for the proportion of all employees in favor of more job security at the expense of raises. Based on Eq. 6.1, the lower confidence bound p_L is found as the solution to the following equation:

$$\sum_{i=12}^{20} \binom{20}{i} p_L^i (1 - p_L)^{20-i} = .025$$

The solution to this equation is found in Figure 6.5 (or in Table A6) in the row x = 12 for n = 20 and in the column labeled "Lower" under the heading of 95%. This value is p_L = .361. Figure 6.6 illustrates the binomial distribution with n = 20 and p = .361, showing the upper tail (from 12 to 20) with a probability of only .025.

Based on Eq. 6.2, the upper confidence bound p_U is found as the solution

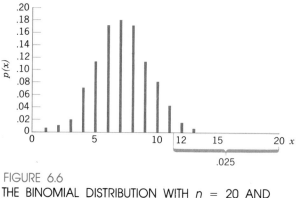

FIGURE 6.6
THE BINOMIAL DISTRIBUTION WITH $n = 20$ AND
$p = .361$, SHOWING THAT $P(X \geq 12) = .025$.

to the following equation:

$$\sum_{i=0}^{12} \binom{20}{i} p_U^i (1 - p_U)^{20-i} = .025$$

The solution to this equation is found in Figure 6.5 (or in Table A6) in the row $x = 12$ for $n = 20$ and in the column labeled "Upper" under the heading of 95%. This value is $p_U = .809$. Figure 6.7 illustrates the binomial distribution with $n = 20$ and $p = .809$, showing the lower tail (from 0 to 12) with a probability of .025.

Thus, based on 12 positive responses in the random sample of 20 employees the point estimate of the proportion in the population favoring increased job security over raises is $\hat{p} = 12/20 = .60$. A 95% confidence interval for the population proportion is from .361 to .809. Note this confidence interval is not symmetric about the point estimate. Also note that this confidence interval is rather wide, and therefore not very informative. Since the width of a confidence interval gets smaller as the sample size gets larger, this example illustrates the need for large sample sizes when estimating population proportions.

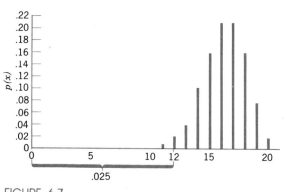

FIGURE 6.7
THE BINOMIAL DISTRIBUTION WITH $n = 20$ AND
$p = .809$, SHOWING THAT $P(X \leq 12) = .025$.

CONFIDENCE INTERVALS BASED ON
THE NORMAL APPROXIMATION

For values of $n > 30$, which are not given in Table A6, an approximate method based on the normal approximation to the binomial distribution can be used. The approximation is based on the fact that the central limit theorem applies to the sampling distribution of $\hat{p}$ when n is large. The sampling distribution of $\hat{p}$ is approximately normal with a mean p and standard deviation $\sqrt{pq/n}$. The unknown standard deviation $\sqrt{pq/n}$ is approximated by its estimate $\sqrt{\hat{p}\hat{q}/n}$. Then Table A2 is used to approximate the probabilities for the statistic $(\hat{p} - p)/\sqrt{\hat{p}\hat{q}/n}$. This leads to the approximate probability

$$P\left(\hat{p} - z_{1-\alpha/2}\sqrt{\frac{\hat{p}\hat{q}}{n}} < p < \hat{p} + z_{1-\alpha/2}\sqrt{\frac{\hat{p}\hat{q}}{n}}\right) \approx 1 - \alpha$$

which is the basis for the large sample confidence interval approximation.

The confidence interval for the population proportion p with a level of confidence of $100(1 - \alpha)\%$ is given approximately as

INTERVAL
ESTIMATE
OF THE
$$\hat{p} \pm z_{1-\alpha/2}\sqrt{\frac{\hat{p}\hat{q}}{n}}$$

POPULATION   where n is the size of the random sample, $\hat{p}$ is the sample proportion in the
PROPORTION p random sample, $\hat{q} = 1 - \hat{p}$, and $z_{1-\alpha/2}$ is the $1 - \alpha/2$ quantile of the normal
(APPROXIMATE   distribution, found in Table A2. Some very commonly used confidence inter-
SOLUTION   vals are 90%, 95%, and 99%, summarized as follows:
APPROPRIATE
FOR LARGE   90% confidence interval for p: $\hat{p} \pm 1.6449\sqrt{\hat{p}\hat{q}/n}$
SAMPLES   95% confidence interval for p: $\hat{p} \pm 1.9600\sqrt{\hat{p}\hat{q}/n}$
$n > 30$)   99% confidence interval for p: $\hat{p} \pm 2.5758\sqrt{\hat{p}\hat{q}/n}$

Note This approximate solution should be used with caution if $n\hat{p} < 5$ or $n\hat{q} < 5$.

EXAMPLE

The federal government has threatened to withdraw federal highway funds from states not complying with the 55-mph speed limit on interstate highways. Noncompliance is judged on a sliding scale, and at one time was set at 60%. That is, if more than 60% of the drivers are exceeding the 55-mph speed limit, then the state is in violation of the federal requirement.

A random survey was taken at different times and different locations and it was found that 6250 drivers out of 10,000 were driving over 55 mph. Find a 99% interval for the true proportion of drivers exceeding 55 mph. From Table

A2, $z_{.995} = 2.5758$. Since $\hat{p} = .625$, the 99% confidence interval is

$$.625 \pm 2.5758\sqrt{.625(.375)/10000} = .625 \pm .012$$

or

$$61.3\% \text{ to } 63.7\%$$

Do you think there is a need for concern about compliance with the federal requirement based on these results?

Confidence intervals for the population proportion p can be used to monitor the quality of a product being produced, as shown in the next example.

EXAMPLE

Television sets coming off an assembly line are automatically checked to make sure that they are not defective. The company manufacturing these TV sets wants an interval estimate of the percentage of sets that fail the testing procedure. A 95% confidence interval is computed based on a random sample of size 50 in which 7 sets failed the testing procedure.

For $n = 50$ the normal approximation is used to find the confidence interval. The estimate of the proportion of sets with defects is given as $\hat{p} = 7/50 = .14$, and the confidence interval is given as

$$.14 \pm 1.9600\sqrt{.14(.86)/50} = .14 \pm .096$$

or

$$.044 \text{ to } .236$$

The percentage of defective sets coming off the assembly line is between 4.4% and 23.6% with 95% confidence.

DETERMINING THE SAMPLE SIZE REQUIRED

Larger sample sizes generally result in shorter confidence intervals. Note that the approximate confidence interval for p goes from $\hat{p} - z_{1-\alpha/2}\sqrt{\hat{p}\hat{q}/n}$ to $\hat{p} + z_{1-\alpha/2}\sqrt{\hat{p}\hat{q}/n}$

$$\hat{p} - z_{1-\alpha/2}\sqrt{\hat{p}\hat{q}/n} \qquad \hat{p} \qquad \hat{p} + z_{1-\alpha/2}\sqrt{\hat{p}\hat{q}/n}$$

If a confidence interval of a given width is desired for the unknown population proportion p, the sample size required to achieve that width can be obtained by noting that the total width w of the confidence interval is approximately

$$w = 2z_{1-\alpha/2}\sqrt{\frac{\hat{p}\hat{q}}{n}}$$

By squaring both sides and solving for n this becomes

$$n = 4\hat{p}\hat{q}\left(\frac{z_{1-\alpha/2}}{w}\right)^2$$

This expression still depends on the values of $\hat{p}$ and $\hat{q} = 1 - \hat{p}$, which are not known until after the sample is obtained. Therefore the most conservative values may be used: $\hat{p} = \frac{1}{2}$ gives $\hat{p}\hat{q} = \frac{1}{4}$, its largest possible value. So n must be at least equal to $(z_{1-\alpha/2}/w)^2$ to be sure that the interval will have width w or less.

SAMPLE SIZE
REQUIRED
FOR A
CONFIDENCE
INTERVAL FOR
p OF
SPECIFIED
WIDTH

A sample size n where

$$n \geq (z_{1-\alpha/2}/w)^2 \tag{6.3}$$

will ensure that a $100(1 - \alpha)\%$ confidence interval for the binomial parameter p will have a total width of w or less, where $z_{1-\alpha/2}$ is the $1 - \alpha/2$ quantile from Table A2.

If p is not close to $\frac{1}{2}$, the preceding formula is too conservative, and it may be worthwhile to obtain an estimate $\hat{p}$ and use the more accurate formula

$$n \geq 4\hat{p}\hat{q}(z_{1-\alpha/2}/w)^2 \tag{6.4}$$

EXAMPLE

Suppose that in the previous example the manager wants to know, within $\pm.04$, the probability of a television set being defective. What sample size is required for a 95% confidence interval?

In this case, the total width is .08, which allows for a .04 error on either side of $\hat{p}$. Therefore the sample size should be at least

$$\left(\frac{z_{1-\alpha/2}}{w}\right)^2 = \left(\frac{1.9600}{.08}\right)^2 = 600$$

to satisfy the manager.

If a sample of size 600 results in 83 defective sets, the 95% confidence interval is

$$\hat{p} \pm z_{1-\alpha/2}\sqrt{\frac{\hat{p}\hat{q}}{n}} = \frac{83}{600} \pm 1.9600\sqrt{\frac{(83/600)(517/600)}{600}}$$

$$= .138 \pm .028$$

The point estimate, .138, is within $\pm.028$ of the true probability p, with 95% confidence. Note that the actual width .055 is less than the targeted value .08, because the actual value of $\hat{p}\hat{q}$ is .119, less than the conservative value .25 used in developing Eq. 6.3. If past records indicated that the proportion should be somewhere near the value $p = .12$, then Eq. 6.4 produces the estimate

$$4\hat{p}\hat{q}\left(\frac{z_{1-\alpha/2}}{w}\right)^2 = 4(.12)(.88)\left(\frac{1.9600}{.08}\right)^2$$

$$= 253.5$$

which is considerably less than the $n = 600$ obtained earlier, but which may prove to be too small if the true proportion defective turns out to be larger than the estimated value .12.

EXERCISES

6.7   A random survey of 18 employees indicates that six are in favor of being represented by a union. Find a 95% confidence interval for the true proportion of employees favoring unionization of the company.

6.8   Use the approximate solution appropriate for large samples to find the desired confidence interval in Exercise 6.7. Compare your answer with the exact solution obtained in that exercise.

6.9   A random survey shows that 15 out of 20 employees indicate that their present salary does not compensate them adequately for their contribution to the company. Find a 90% confidence interval for the true proportion of employees expressing dissatisfaction with their salary treatment.

6.10 Use the approximate solution appropriate for large samples to find the desired confidence interval in Exercise 6.9. Compare your answer with the exact solution obtained in that exercise.

6.11 A random sample of 50 checking accounts at the First National Bank is examined to see how many were overdrawn. Six of the 50 accounts were found to be overdrawn. Find a 95% confidence interval for the population proportion of overdrawn accounts.

6.12 Fourteen people said they preferred the news on Channel 4, of the 88 people randomly selected in a survey. Find a 99% confidence interval for the proportion of the population that prefers the news on Channel 4.

6.13 A national poll based on interviews with 1200 voters in the 1972 presidential campaign gave Nixon 57% of the vote. Calculate a 99% confidence interval for the true proportion of voters supporting Nixon.

6.14 How large a sample size is required to estimate the percentage of the population watching the World Series on television, where the width of the 90% confidence interval should be .04 or less?

6.15 In an election where preliminary estimates give the incumbent 70% of the vote, how many voters are needed in the survey to estimate the true percent of the votes garnered by the incumbent? A 95% confidence interval of width less than 10% is desired.

6.16 A TV survey states that an estimate of CBS's share of the viewing audience one night was 36%, with a maximum error of ±2%, at 90% confidence. What sample size was used?

6.17 The germination rate for some flower seeds was stated as a 90% confidence interval from 61% to 69%. How large a sample was used to produce this interval?

6.3

ESTIMATING THE MEAN

PROBLEM SETTING 6.2

The recruitment and retention of employees is a problem continually faced by employers. Recruitment is initially expensive and by the time relocation expenses are paid, a company has made a substantial cash investment in the recruit, so new employees must be retained long enough to recover the money spent in hiring them. Also, as employees gain experience they become more valuable to the company and the company loses money when they lose good employees.

In general, companies do not want to absorb the cost of training employees just so they can advance to better paying positions with competing companies. Therefore, companies that are conscientious about retaining their employees need to maintain a realistic and competitive salary structure. One way a company can assess its salary position is to examine the distribution of salaries throughout the industry. Unless some summary is available a sample will probably be used to estimate the mean of such a distribution. In addition, a confidence interval for the mean will be helpful.

CONFIDENCE INTERVAL FOR THE MEAN

The central limit theorem states that for all populations with a finite variance the sampling distribution of the random variable $\bar{X}$ based on a random sample of size n becomes approximately normal in form as n increases. Also, for normal populations the sampling distribution of $\bar{X}$ is exactly normal. This is the basis for forming a confidence interval for the mean of a population, based on $\bar{X}$.

For the standard normal distribution the relationship

$$P(-z_{1-\alpha/2} < Z < z_{1-\alpha/2}) = 1 - \alpha$$

holds, where $z_{1-\alpha/2}$ is the $1 - \alpha/2$ quantile from Table A2. Since the mean and standard deviation of $\bar{X}$ are μ and $\sigma/\sqrt{n}$, respectively (from Section 5.4), the sampling distribution of $(\bar{X} - \mu)/(\sigma/\sqrt{n})$ is approximately standard normal. Therefore

$$P\left(-z_{1-\alpha/2} < \frac{\bar{X} - \mu}{\sigma/\sqrt{n}} < z_{1-\alpha/2}\right) \approx 1 - \alpha$$

holds true. By working only with the inequality within the parentheses, some elementary algebra shows that the inequality can be written as

$$\overline{X} - z_{1-\alpha/2}\frac{\sigma}{\sqrt{n}} < \mu < \overline{X} + z_{1-\alpha/2}\frac{\sigma}{\sqrt{n}}$$

so the entire probability statement becomes

$$P\left(\overline{X} - z_{1-\alpha/2}\frac{\sigma}{\sqrt{n}} < \mu < \overline{X} + z_{1-\alpha/2}\frac{\sigma}{\sqrt{n}}\right) \approx 1 - \alpha$$

This statement forms the basis for the confidence interval for the mean of a population. Such a confidence interval is a good approximation for nonnormal populations if n is large, and is exact for normal populations for all sample sizes.

The **confidence interval for a population mean,** with a level of confidence of $100(1 - \alpha)\%$, is given for large samples ($n > 30$) as

$$\overline{X} \pm z_{1-\alpha/2}\frac{\sigma}{\sqrt{n}}$$

A CONFIDENCE INTERVAL FOR THE POPULATION MEAN ($n > 30$) where $\overline{X}$ is the sample mean in a simple random sample, σ is the population standard deviation (if σ is unknown, s may be used instead of σ as an approximation when n is large), and $z_{1-\alpha/2}$ is found in Table A2 as the $(1 - \alpha/2)$ quantile.

Some commonly used confidence intervals are 90%, 95%, and 99%. They are summarized as follows:

90% confidence interval for μ: $\overline{X} \pm 1.6449\, \sigma/\sqrt{n}$
95% confidence interval for μ: $\overline{X} \pm 1.9600\, \sigma/\sqrt{n}$
99% confidence interval for μ: $\overline{X} \pm 2.5758\, \sigma/\sqrt{n}$

If the population is normal and σ is known, this procedure provides an exact confidence interval for all sample sizes.

EXAMPLE

A company would like to compare its annual salaries (including fringe benefits and overhead costs) for its professional employees against an industrywide distribution of salaries. A random sample of 25 industry salaries is selected and the average is found to be $62,000. From past experience the company knows this distribution is normal, and the standard deviation of these salaries is $10,000. Since the population is normal and σ is known, the previous procedure may be used even though n is small. Hence, a 95% confidence interval for the mean salary is

$$\$62{,}000 \pm 1.9600 \frac{10{,}000}{\sqrt{25}}$$

or

$$\$62{,}000 \pm 3920$$

or

$$\$58{,}080 \text{ to } \$65{,}920$$

This interval is a 95% confidence interval for the mean salaries throughout the industry and the interpretation is that the company can be 95% confident that the interval from \$58,080 to \$65,920 contains the true mean salary for the industry.

FINDING THE SAMPLE SIZE FOR A CONFIDENCE INTERVAL

A common question that occurs to individuals when they are first exposed to confidence intervals is: Why not always use a 99% or a 99.9% confidence interval and have the higher degree of confidence? Examination of the preceding confidence intervals shows that the price paid for this greater confidence is wider intervals, and confidence intervals that are wide are often useless. For instance, it may be stated with 99.9% confidence that a new drug is between 8 and 92% effective; however, this interval is so wide that it loses any value it might otherwise have had. The width of a confidence interval can be decreased either by decreasing the level of confidence or by increasing the sample size.

If a confidence interval is to be calculated for the population mean on the basis of a random sample of size n from a normal population for which σ is known, the width of the confidence interval can be determined as a function of the sample size. Conversely the sample size may be determined as a function of the desired width. The following example illustrates how the sample size is determined.

DETERMINATION OF THE SAMPLE SIZE FOR A GIVEN CONFIDENCE INTERVAL WIDTH
The sample size required for a confidence interval of total width w, for the mean of a normal distribution with standard deviation σ, is given by

$$n = \left[2z_{1-\alpha/2} \frac{\sigma}{w} \right]^2 \qquad (6.5)$$

where $z_{1-\alpha/2}$ is the $1 - \alpha/2$ quantile from Table A2.

Note If $n > 30$, this answer is approximate for nonnormal populations also.

EXAMPLE

How large a sample would be needed to form a 90% confidence interval for the mean nicotine content of a brand of cigarettes if the nicotine content has a normal distribution with $\sigma = 8.5$ mg and the width of the interval must be 6 mg to satisfy testing requirements? The upper and lower limits of the confidence interval are given, respectively, as

$$\bar{X} + 1.6449 \frac{8.5}{\sqrt{n}} \quad \text{and} \quad \bar{X} - 1.6449 \frac{8.5}{\sqrt{n}}$$

Since the difference of the limits is desired to be 6 mg, the following equation can be solved for n:

$$(\text{Upper limit}) - (\text{lower limit}) = \text{width}$$

$$\left(\bar{X} + 1.6449 \frac{8.5}{\sqrt{n}}\right) - \left(\bar{X} - 1.6449 \frac{8.5}{\sqrt{n}}\right) = 6$$

or

$$\sqrt{n} = 2(1.6449)\frac{8.5}{6} = 4.66055$$

or, after squaring both sides of the equation,

$$n = 21.72$$

Since a sample size of 21.72 cannot be obtained, the required size is $n = 22$, which is the same as would be obtained using Eq. 6.5.

STUDENT'S t DISTRIBUTION

If the population is normal but the standard deviation is unknown and has to be estimated, confidence intervals for μ that use standard normal quantiles are only approximate. The quantiles that provide exact confidence intervals are obtained from Table A3 for a distribution called **Student's t distribution.**

Student's t distribution is the name given to a family of distributions indexed by a parameter called degrees of freedom (d.f.). As the degrees of freedom change, the distribution changes, which accounts for the different quantiles in Table A3 for the various degrees of freedom. As the degrees of freedom increases, the Student's t distribution approaches the standard normal distribution in form. This is the reason standard normal values are appropriate for large sample sizes.

Figure 6.8 compares the density function of a Student's t distribution having five degrees of freedom with a standard normal density function. Note that all t distributions are symmetric around zero, so the lower quantiles t_p are just the negative of the upper quantiles, $t_p = -t_{1-p}$ as with the normal distribution.

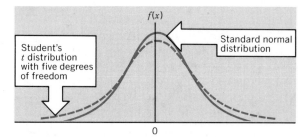

FIGURE 6.8
A COMPARISON OF STUDENT'S t DISTRIBUTION
WITH THE STANDARD NORMAL DISTRIBUTION.

CONFIDENCE INTERVAL FOR THE MEAN OF NORMAL POPULATIONS

When the population is normal, the probabilities for $(\overline{X} - \mu)/(\sigma/\sqrt{n})$ may be obtained from Table A2 and the probabilities for $(\overline{X} - \mu)/(s/\sqrt{n})$ may be obtained from Table A3, with d.f. $= n - 1$. Note that the only change is that s is used instead of σ. The use of s to estimate σ introduces additional variability in the statistic $(\overline{X} - \mu)/(s/\sqrt{n})$, and this results in a larger variance in the distribution of the statistic, as indicated by Figure 6.8. The use of quantiles from Student's t distribution in Table A3 instead of normal quantiles from Table A2 leads to the confidence interval for μ in normal populations when σ is unknown.

CONFIDENCE INTERVAL FOR THE MEAN OF A NORMAL POPULATION WHEN σ IS UNKNOWN

The **exact confidence interval for the mean of a normal population,** with a level of confidence of $100(1 - \alpha)\%$, based on a random sample, is given as

$$\overline{X} \pm t_{1-\alpha/2,n-1}\frac{s}{\sqrt{n}}$$

where $\overline{X}$ is the sample mean that estimates the unknown value of μ, s is the sample standard deviation that estimates the unknown standard deviation σ, n is the sample size, and $t_{1-\alpha/2,n-1}$ is the $1 - \alpha/2$ quantile from a Student's t distribution with $n - 1$ degrees of freedom.

ERROR INVOLVED WHEN USING APPROXIMATE CONFIDENCE INTERVALS WITH NORMAL SAMPLES

Use of the Student's t distribution will give wider confidence intervals than are obtained using the inappropriate standard normal values, because of the larger variance of the Student's t distribution. Conversely, this means that if quantiles of Z are used when values of t should be used, then the actual degree of confidence associated with the confidence interval will be less than is intended.

	Desired Level of Confidence for Intervals from Normal Populations (%)		
n	90%	95%	99%
5	82.4	87.8	93.8
10	86.5	91.8	97.0
15	87.8	92.9	97.8
20	88.3	93.5	98.1
25	88.7	93.8	98.3
30	88.9	94.0	98.4
50	89.3	94.4	98.6

FIGURE 6.9
ACTUAL LEVEL OF CONFIDENCE WHEN STUDENT'S t VALUES ARE ERRONEOUSLY REPLACED WITH STANDARD NORMAL VALUES.

Suppose a 90% confidence interval is desired for μ when σ has been estimated from a sample of size 5. From Table A3 $t_{.95,4} = 2.1318$, and the proper form of the confidence interval is $\overline{X} \pm 2.1318s/\sqrt{5}$. If the standard normal distribution were used in this problem, the confidence interval would appear as $\overline{X} \pm 1.6449s/\sqrt{5}$, which is obviously not as wide as the one using Student's t distribution. Therefore, the confidence associated with this last interval is less than 90%. In fact, the confidence is only 82.4%, which can be found by interpolation in the Student's t table with $5 - 1 = 4$ degrees of freedom.

Figure 6.9 shows the relationship between the actual level of confidence and the sample size when the standard normal values are erroneously substituted for Student's t values. Figure 6.9 illustrates that there is not much discrepancy for samples of size 50 or larger.

Two examples are now presented to demonstrate how the confidence interval for the mean of a normal population is found when the standard deviation is estimated from sample data.

EXAMPLE

In the first example of this section the mean salary throughout the industry was estimated by taking a random sample of size 25. The sample mean was $62,000 and the population standard deviation was assumed to be known.

Assume that the population standard deviation was not known and was estimated from the sample standard deviation to be $10,000. The 95% confidence interval is constructed by using the .975 quantile $t_{.975,25-1} = 2.0639$ from the Student's t distribution with 24 degrees of freedom.

$$\$62{,}000 \pm 2.0639 \frac{10{,}000}{\sqrt{25}}$$

or

$$\$62{,}000 \pm 4128$$

or

$$\$57{,}872 \text{ to } \$66{,}128$$

The total width is \$416 more than the width of the confidence interval obtained by using the standard normal distribution. This additional width can be attributed to the uncertainty associated with having to estimate the population standard deviation.

EXAMPLE

The manager of the shipping department of a manufacturing company has been receiving customer complaints about the length of time it takes to ship one of their products across country. Before the manager confronts the individuals responsible for the delivery, he selects a random sample of 20 items over a period of a few weeks and follows up on them to see how many days it takes for each item in the sample to be delivered. The number of days to delivery is as follows: 10, 9, 10, 11, 16, 15, 8, 6, 18, 17, 4, 12, 15, 14, 15, 9, 7, 8, 16, 14. Application of the Lilliefors test shows that the shipment time could be approximated by a normal distribution. A 95% confidence interval for the average delivery time is found as follows:

$$\sum X_i = 234 \quad \sum X_i^2 = 3048$$

$$\bar{X} = \frac{234}{20} = 11.7 \quad s = \sqrt{\frac{3048 - (234)^2/20}{19}} = 4.041$$

From Table A3 $t_{.975,19} = 2.0930$, and the confidence interval is

$$11.7 \pm 2.0930 \frac{4.041}{\sqrt{20}} = 11.7 \pm 1.9$$

or

$$9.8 \text{ to } 13.6 \text{ days}$$

Therefore, the manager can be 95% confident that the mean number of days required for delivery is between 9.8 and 13.6.

SMALL-SAMPLE CONFIDENCE INTERVALS IN NONNORMAL POPULATIONS

The previous procedures cover all cases for finding a confidence interval for the mean, except when the sample size is small and the population is non-normal. In this case, it is easier to work with the sample median. Procedures are given in Section 6.4 for finding a confidence interval for the population median.

EXERCISES

6.18 Consider the Student's t distribution with 14 degrees of freedom.

(a) Find the .95 quantile.

(b) Find the .05 quantile.

(c) Which quantile is .6924?

(d) Which quantile is 1.52?

(e) What is the probability of getting an observation as large as 1.52?

6.19 Consider the Student's t distribution with 26 degrees of freedom.

(a) Find the .90 quantile.

(b) Find the .10 quantile.

(c) Which quantile is 2.0555?.

(d) Which quantile is 1.96? Which quantile is 1.96 in the standard normal distribution? How do the two answers differ?

(e) What is the probability of getting an observation as large as 1.96?

6.20 A random sample of 50 checking accounts out of the 5525 accounts at the First National Bank is examined to see what the "average daily balance" was for the previous month. This figure is routinely calculated once a month on each account. The mean for the 50 observations was $2100 and the standard deviation was $400. Find a 95% confidence interval for the population mean "average daily balance" for all accounts at the First National Bank and explain the meaning of this interval.

6.21 A random sample of nine light bulbs shows a mean burning time of 175 hours. The population standard deviation for the burning time of the bulb under consideration is $\sigma = 16$. Find 90%, 95%, and 99% confidence intervals for the true mean burning time of the type of bulb being tested, assuming a normal distribution. What is the meaning of the different widths associated with the confidence intervals you have found?

6.22 How many households would it be necessary to survey to form a $400-wide 99% confidence interval for the mean income in a suburb of a large city if the standard deviation is $500?

6.23 If you are sampling from a population with a standard deviation of 15, how large a sample do you need to have a 95% chance that the sample mean will fall within two units of the population mean?

6.24 A past analytical study has shown that the standard deviation of an order for a small home radio is $25. How large a sample must be taken so that a 95% confidence interval for the average order size will have a width of $3?

6.25 Assume that the weekly dollar volume at a Pancake House is normally distributed. Find a 95% confidence interval for the population mean weekly dollar volume and interpret it assuming the observed volumes for five weeks were as follows: $22,130, $18,465, $25,616, $22,440, $19,869.

6.26 Daily receipts at Beverly's Gallery of Fashion were $684, $972, $740, $868, and $777 for five successive days. Find a 90% confidence interval for the mean of the population of daily receipts, assuming that the population is normal and that these five observations resemble a random sample. Explain the meaning of the interval you have found.

6.27 A manufacturing company claims that its new floodlight will last at least 1000 hours. A random sample of 10 lights gave an average life of 980 hours with a standard deviation of 15.8. Find a 90% confidence interval for the mean of the population assuming a normal distribution. Based on your confidence interval, do you think that the company's claim is justified?

6.28 A random sample of scores on an exam yielded the scores 85, 78, 52, 66, and 74. Assume normality and find a 95% confidence interval for the mean of all scores on the exam. Explain the meaning of the interval you have found.

6.29 A sample of size 25 from a normal population with unknown mean and variance yields $\overline{X} = 20$ and $s = 5$. An individual who claims to have some statistical training computes a 95% confidence interval for the population mean as $\overline{X} \pm 1.9600s/\sqrt{25}$ and obtains 18.04 to 21.96 as his confidence interval. He then claims that he can be 95% confident that this interval will contain the true value of the population mean. Approximately what is the confidence that should be associated with this interval? What is the correct 95% confidence interval for the population mean?

6.4

ESTIMATING THE POPULATION MEDIAN (OPTIONAL)

ESTIMATING THE MEDIAN

In the previous section of this chapter a method was given for finding a confidence interval for the population mean. In some applications the population median is more appropriate than the population mean as the primary statistic for describing the population. This is especially true with highly skewed populations, such as individual incomes, where a few large incomes in the population influence the mean so that it is usually too large to reflect accurately the middle of the distribution of incomes.

In other cases, the sample size is small and the population is nonnormal, so the usual methods for finding a confidence interval for the mean cannot be used. Then a confidence interval for the population median is appropriate. The population median is estimated from the sample median.

POINT
ESTIMATE The **sample median** is used as a **point estimate of the population median.** The
OF THE sample median is defined in Section 3.3, and the population median in Section
POPULATION 4.4.
MEDIAN

Examples of distributions where a few large observations may distort the mean so that it no longer represents the middle of the distribution as well as the median include the following.

1. The time to repair a boat may be very long for a few boats that require the ordering of special parts.

2. The time for a letter to reach its destination may include some very long times for a few letters that go astray.

| | | | | Target Confidence Level | | | | | | | |
| | 90% | | | | 95% | | | | 99% | | |
n	S_1	S_2	Actual Percentage	S_1	S_2	Actual Percentage	S_1	S_2	Actual Percentage
4	1	4	87.50						
5	1	5	93.75						
6	2	5	78.13	1	6	96.87			
7	2	6	87.50	1	7	98.44			
8	3	6	71.09	2	7	92.97	1	8	99.22
9	3	7	82.03	2	8	96.09	1	9	99.61
10	3	8	89.06	2	9	97.85	1	10	99.80
11	3	9	93.46	2	10	98.83	1	11	99.90
12	4	9	85.40	3	10	96.14	2	11	99.37
13	4	10	90.77	3	11	97.75	2	12	99.66
14	5	10	82.04	4	11	94.26	3	12	98.71
15	5	11	88.15	4	12	96.48	3	13	99.26
16	5	12	92.32	4	13	97.87	3	14	99.58
17	6	12	85.65	5	13	95.10	4	14	98.73
18	6	13	90.37	5	14	96.91	4	15	99.25
19	6	14	93.64	5	15	98.08	4	16	99.56
20	7	14	88.47	6	15	95.86	5	16	98.82

FIGURE 6.10
A SHORT TABLE FOR QUICKLY OBTAINING AN EXACT CONFIDENCE INTERVAL FOR A POPULATION MEDIAN FOR $n \leq 20$. TABLE A7 CONTAINS CORRESPONDING VALUES FOR $n \leq 50$.

3. The amount of daily receipts in a store may be highly skewed because of a few sale days and the Christmas rush.

4. The amount contributed by each family to a particular church may include amounts from a few families that contribute large sums of money.

In these situations the mean does not represent the middle of the distribution as accurately as the median. Also the first two may necessarily involve small sample sizes, so methods for handling small sample sizes are useful.

EXACT CONFIDENCE INTERVALS

Two methods for finding a confidence interval for the median are presented. The first method is exact and uses special tables. It is, therefore, necessarily restricted to those limited cases for which tables are available. Table A7 is given in the Appendix for $n \leq 50$, and a shorter version ($n \leq 20$) is given in this section. The second method is approximate and is appropriate for larger sample sizes. It is important to realize that these procedures may be used for all types of populations, unlike the confidence interval for the population mean, which requires the assumption of a normal distribution when the sample size is small.

INTERVAL ESTIMATE OF THE POPULATION MEDIAN (EXACT SOLUTION FOR SMALL SAMPLES, i.e., $n \leq 50$)

To find a confidence interval for the population median from a random sample, proceed as follows.

1. Order the n sample observations from smallest to largest. Use the following notation $X^{(1)}$, $X^{(2)}$, $X^{(3)}$, . . . , $X^{(n)}$, where $X^{(1)}$ denotes the smallest observation, $X^{(2)}$ denotes the next smallest, and so on until $X^{(n)}$ denotes the largest observation in the sample.

2. Enter Table A7 with the appropriate sample size, for $n \leq 50$. Figure 6.10 may be used for $n \leq 20$.

3. For the column representing the target confidence level, find S_1, S_2, and the actual confidence level.

4. The lower bound for the confidence interval is the ordered sample observation found in position S_1, that is, $X^{(S_1)}$.

5. The upper bound for the confidence interval is the ordered sample observation found in position S_2, that is, $X^{(S_2)}$.

6. The exact level of confidence is found in Table A7, next to the values of S_1 and S_2.

Note that in the exact method the actual sample values are used to form the upper and lower bounds for the confidence interval. Also, the level of confidence is seldom exactly $100(1 - \alpha)\%$ because of the discrete nature of the exact distribution of S_1 and S_2. The following example demonstrates the procedure for finding the exact confidence interval for a population median.

EXAMPLE

To control their quality of production, an electronics company randomly selects 16 electronic components to determine their time to failure in hours. Time-to-failure measurements frequently produce some observations that are much larger or much smaller than the bulk of the rest of the observations (i.e., outliers).

Since the mean is very sensitive to outliers, the median may provide a better point estimate of the middle of the distribution of the time to failure. For this reason it is desired to find a point estimate and a 90% confidence interval for the median time to failure. Since n is an even number, the point estimate for the median is found by averaging the numbers in the $\frac{16}{2}$ = eighth and ($\frac{16}{2}$ + 1) = ninth positions in the ordered array of sample values listed below, or

$$X_{.5} = \text{sample median} = \frac{X^{(8)} + X^{(9)}}{2}$$

$$= \frac{63.2 + 63.3}{2} = 63.25$$

The step-by-step procedure for finding the confidence interval is as follows.

1. The 16 sample observations are first ordered from smallest to largest. The ordered sample observation found in the ith position is $X^{(i)}$.

$X^{(1)} = 46.7$	$X^{(5)} = 56.8$	$X^{(9)} = 63.3$	$X^{(13)} = 67.1$
$X^{(2)} = 47.2$	$X^{(6)} = 59.2$	$X^{(10)} = 63.4$	$X^{(14)} = 67.7$
$X^{(3)} = 49.1$	$X^{(7)} = 59.9$	$X^{(11)} = 63.7$	$X^{(15)} = 73.3$
$X^{(4)} = 56.5$	$X^{(8)} = 63.2$	$X^{(12)} = 64.1$	$X^{(16)} = 78.5$

2. Either Figure 6.10 or Table A7 is entered, with $n = 16$.
3. For the 90% column, $S_1 = 5$, $S_2 = 12$, and the exact confidence level is 92.32%.
4. The lower bound for the confidence interval is found in the fifth position or $L = X^{(5)} = 56.8$.
5. The upper bound for the confidence intervals is found in the twelfth position or $U = X^{(12)} = 64.1$. The confidence interval is from 56.8 hr to 64.1 hr. However, because of the discrete distribution the actual level of confidence is 92.32%.

CONFIDENCE INTERVALS BASED ON THE NORMAL APPROXIMATION

For sample sizes larger than 50 the values of S_1 and S_2 can be found using the normal approximation. Since the exact procedure is based on the binomial distribution with $p = .5$, the normal approximation works quite well for finding S_1 and S_2. The exact level of confidence can also be approximated using the normal distribution. This point is illustrated in the next example.

INTERVAL ESTIMATE OF THE POPULATION MEDIAN (APPROXIMATE SOLUTION FOR LARGE SAMPLES, i.e., $n > 50$)

1. Order the n observations in the random sample from smallest to largest. Use the following notation with the ordered sample; $X^{(1)}, X^{(2)}, \ldots, X^{(n)}$, where $X^{(1)}$ denotes the smallest observation, $X^{(2)}$ denotes the next smallest, and so on until $X^{(n)}$ denotes the largest observation in the random sample.

2. For a target confidence level of $100(1 - \alpha)\%$, use Table A2 to find the critical value $z_{1-\alpha/2}$. Some widely used values of $z_{1-\alpha/2}$, and their associated level of confidence, are $z_{.95} = 1.6449$ (90%), $z_{.975} = 1.9600$ (95%), and $z_{.995} = 2.5758$ (99%).

3. Compute $S_1^* = (n - z_{1-\alpha/2}\sqrt{n})/2$.

4. Let S_1 be the integer obtained by rounding S_1^* upward to the next higher integer. (This is necessary, since S_1^* will seldom be an integer.)

5. Let $S_2 = n - S_1 + 1$.

6. The lower bound for the confidence interval is the ordered sample observation found in position S_1, that is $X^{(S_1)}$.

7. The upper bound for the confidence interval is the ordered sample observation found in position S_2, that is, $X^{(S_2)}$.

8. To find the appropriate level of confidence, compute

$$z = \frac{2S_1 - n - 1}{\sqrt{n}}$$

and use Table A2*. The value of p corresponding to this entry is the probability $\alpha_1 = P(Z \leq z)$. The approximate level of confidence is $100(1 - 2\alpha_1)\%$.

EXAMPLE

The procedure for large samples is demonstrated by reworking the previous example.

1. The ordered sample observations appear exactly as they did in the previous example.

2. From Table A2, $z_{.95} = 1.6449$.

3. $S_1^* = (16 - 1.6449\sqrt{16})/2 = (16 - 6.58)/2 = 4.71$.

4. $S_1 = 5$ (the next higher integer above $S_1^* = 4.71$).

5. $S_2 = 16 - 5 + 1 = 12$.

6. The lower bound for the confidence interval is found in the fifth position, or $L = X^{(5)} = 56.8$.

7. The upper bound for the confidence interval is found in the twelfth position, or $U = X^{(12)} = 64.1$.

8. The confidence interval is again from 56.8 hr to 64.1 hr. The approximate level of confidence is obtained by using Table A2* with the entry

$$z = \frac{2(5) - 16 - 1}{\sqrt{16}} = -1.75$$

to find $p = .0401 = \alpha_1$. The approximate level of confidence is
$$100(1 - 2\alpha_1)\% = 91.98\%$$
which agrees well with the exact value 92.32% found earlier.

EXAMPLE

In Section 6.3 an example was given where a 95% confidence interval was found for the mean number of days required for a company to ship their product across the country. The same sample data are used to find a 95% confidence interval for the median number of days required to ship the product across country.

1. The ordered sample data appear as follows:

$$
\begin{array}{llll}
X^{(1)} = 4 & X^{(6)} = 9 & X^{(11)} = 12 & X^{(16)} = 15 \\
X^{(2)} = 6 & X^{(7)} = 9 & X^{(12)} = 14 & X^{(17)} = 16 \\
X^{(3)} = 7 & X^{(8)} = 10 & X^{(13)} = 14 & X^{(18)} = 16 \\
X^{(4)} = 8 & X^{(9)} = 10 & X^{(14)} = 15 & X^{(19)} = 17 \\
X^{(5)} = 8 & X^{(10)} = 11 & X^{(15)} = 15 & X^{(20)} = 18
\end{array}
$$

Note The sample median of these data is $(X^{(10)} + X^{(11)})/2 = (11 + 12)/2 = 11.5$, which compares with the sample mean, which is 11.7.

2. Since $n = 20$, either Figure 6.10 or Table A7 can be used to find the confidence interval.

3. In both Figure 6.10 and Table A7 under column 95%, $S_1 = 6$, $S_2 = 15$, and the exact confidence level is 95.86%.

4. The lower bound for the confidence interval is the ordered sample observation found in position 6, or $L = X^{(6)} = 9$.

5. The upper bound for the confidence interval is the ordered sample observation found in position 15, or $U = X^{(15)} = 15$.

6. Therefore, the 95.86% confidence interval is from 9 to 15 days. This compares with 9.8 to 13.6 days, which was the 95% confidence interval for the mean.

In this example the confidence intervals for the mean and median are in good agreement with one another. However, it is easy to change the example so that this agreement disappears. For example, if the largest four sample observations are changed to $X^{(17)} = 31$, $X^{(18)} = 33$, $X^{(19)} = 38$, and $X^{(20)} = 40$, the new sample mean and standard deviation become 15.45 and 10.87, respectively, with a corresponding 95% confidence interval for the mean from 10.4 to 20.5 days. This interval reflects the new sample mean but hardly reflects the center of the distribution of sample values, because the lower end of the confidence interval (10.4 days) is larger than 9 of the 20 sample observations. However, since $X^{(6)}$ and $X^{(15)}$ remain unchanged, the confidence interval for the median is the same as before. This indicates the insensitivity of the sample median to the presence of outliers and shows that the confidence interval for the median still reflects the center of the distribution of sample values.

WHICH PROCEDURE SHOULD BE USED?

If the population is normal, the mean equals the median because the distribution is symmetric, and interval estimates for the mean and the median are estimating the same quantity, that is, the middle of the population. In normal populations the confidence interval for the mean tends to be shorter, and therefore better, than the confidence interval for the median, because the latter confidence interval is valid for all populations, not just normal populations. However, for distinctly nonnormal populations the confidence interval for the mean may be wider, and therefore worse, than the confidence interval for the median. The principle to observe is to use the procedure that is appropriate for the situation. If the data pass the normality test, use the confidence interval for the mean. In nonnormal situations the confidence interval for the median is preferred.

In summary, three different methods for forming confidence intervals have been introduced. In the previous section two methods were introduced.

Method 1 (normal) Uses $\overline{X}$ and a standard normal quantile

Method 2 (Student's t) Uses $\overline{X}$ and a Student's t quantile

In this section a third method was introduced.

Method 3 (median) Uses two selected order statistics from the sample

The normal method is preferred only in the unlikely situation where the population is normal and σ is known. However, it is valid when the sample size is large and may be used merely because it is a convenient procedure to use, even though it may not be preferred. That is, if the population is normal, or approximately normal as shown by the Lilliefors test for normality, then the Student's t method is preferred, no matter what the sample size may be. If the

	The Population Is	
	Normal	Nonnormal
($n \leq 30$)	Student's t (Normal, if σ is known)	Median
($n > 30$)	Student's t (Normal, if σ is known) (Second choice: Normal)	Median (Second choice: Normal)

FIGURE 6.11
THE PREFERRED METHOD FOR FINDING A CONFIDENCE INTERVAL FOR A LOCATION PARAMETER.

data fail the test for normality, then the median procedure is often a better procedure to use even if the sample size is large. Of course, if the sample size is small, the median procedure is the only option available for nonnormal populations. A summary of which procedure to use is given in Figure 6.11.

EXERCISES

6.30 For each city listed in Figure 6.2 find the sample median and use either Figure 6.10 or Table A7 to aid in constructing a 90% confidence interval for the population median number of hours worked by managers of individual stores of a fast-food chain.

Note: The data displayed in Figure 6.10 have already been ordered from smallest to largest for each city.

Compare these 90% confidence intervals for the median with the five 90% confidence intervals for the mean given in Section 6.1.

6.31 Refer to the sample data displayed in Figure 6.2 and find the median number of hours worked for all five cities ($n = 50$). Also find a 90% confidence interval for the population median based on all 50 observations.

6.32 Consider the following 20 observations on mpg for a four-cylinder model car with a manual transmission, tested in city driving.

24.21	24.35	23.82	24.21
24.14	24.60	23.75	25.01
24.66	24.72	24.47	24.38
23.08	23.88	23.09	24.57
25.16	24.62	24.62	25.14

Find the median for these data and a 95% confidence interval for the population median mpg for this model.

6.33 Rework Exercise 6.32 using the large sample approximation and find the approximate level of confidence for this interval. Compare your answer with the exact level of significance given in Figure 6.10 or Table A7.

6.34 Find a 99% confidence interval for the median 1987 net earnings of common stocks for corporations based on the data given below.

1.68	1.72	2.50	2.90	3.11
3.35	3.80	3.85	3.89	4.36
4.64	4.76	5.35	5.81	6.11
6.35	6.69	8.41	8.83	8.97

6.35 Find a 95% confidence interval for the median time between incoming phone calls based on the following data: (in minutes) 1.8, 0.3, 4.5, 9.8, 3.2, 15.7, 4.8, 1.0, 2.7, 6.2.

6.36 The following stem and leaf plot represents the exam scores for a random sample of all employees taking an exam. Find a 95% confidence interval for the population median score for that exam.

3	7
4	
5	0 2
6	8 1 0 2 4 5
7	7 5 4 5 8 3 1 9 1 8
8	4 1 6 0 4 3 3 5 3 7 1 5 5 1
9	1 2 2 6 5 8
10	0

6.37 Use the data in the following random sample to find a 90% confidence interval for the median shoe size sold to women.

$9\frac{1}{2}$ 7 $5\frac{1}{2}$ $7\frac{1}{2}$ 9 $5\frac{1}{2}$ 5

7 $5\frac{1}{2}$ 8 7 $6\frac{1}{2}$ $6\frac{1}{2}$ 6

$6\frac{1}{2}$ $6\frac{1}{2}$ 9 $7\frac{1}{2}$ $6\frac{1}{2}$ 6

6.5

ESTIMATING THE POPULATION STANDARD DEVIATION (NORMAL POPULATIONS) (OPTIONAL)

PROBLEM SETTING 6.3

A canning factory produces cans containing peach halves. The label on the cans says that the can contains 32 ounces. Because it is impossible to put exactly 32 ounces into each can, some variability is always present. Twelve cans are selected at random and their contents weighed. How can these 12 measurements be used to find a confidence interval for σ, the population standard deviation for the actual contents of the cans? What assumptions underlie this procedure?

COMMENT

Methods for finding a confidence interval for μ were given earlier in this chapter. A confidence interval for σ is described in this section. This procedure applies only to normal distributions, and it is not very accurate when it is used with nonnormal data. Therefore it is important to check the data for normality before using this procedure, no matter how large the sample size may be. Corresponding confidence intervals for nonnormal populations are available but they are beyond the scope of this book.

THE CHI-SQUARE DISTRIBUTION

In order to find a confidence interval for σ it is first necessary to be familiar with the chi-square distribution, whose quantiles are given in Table A4. Like the Student's t distribution, the chi-square distribution is indexed by a param-

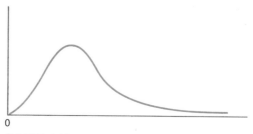

FIGURE 6.12
A TYPICAL CHI-SQUARE DENSITY FUNCTION.

eter, known as degrees of freedom (d.f.). Therefore the chi-square distribution is actually a family of distributions, one distribution for each value of its parameter. Like the t distribution, the degrees-of-freedom parameter assumes only positive integer values 1, 2, 3, 4, etc., and like the t distribution, only a few selected quantiles are given for each chi-square distribution in Table A4.

Unlike the t distribution, the chi-square distribution covers positive values only, so it cannot be used to describe a random variable that can have negative values. This is easy to remember by noting that the name refers to the square of the Greek letter chi, and squares of numbers are never negative. Also unlike the t distribution, the chi-square distribution is not symmetric, so the lower quantiles must be given in the table in addition to the upper quantiles. See Figure 6.12 for a typical chi-square density function, illustrating the chi-square distribution is not symmetric, and takes only positive values.

One way of considering the chi-square distribution is as follows. If the random variable Z has a standard normal distribution, its square

$$Y = Z^2$$

has a chi-square distribution with one degree of freedom. If several independent random variables $Z_1, Z_2, \ldots, Z_n$ have standard normal distributions, the sum of their squares

$$Y = Z_1^2 + Z_2^2 + \cdots + Z_n^2$$

has a chi-square distribution with n degrees of freedom. The number of degrees of freedom always equals the number of variables in the sum. The variables Z_i must have the standard normal distribution, $\mu - 0$ and $\sigma = 1$, or this characterization of the chi-square distribution doesn't hold.

HOW TO READ TABLE A4

Table A4 for the chi-square distribution is read just like Table A3 for the Student's t distribution. That is, first you need to know which chi-square distribution (degrees of freedom) you are working with. That information is usually obtained from the data and the description of the procedure you want to use. Find the row in Table A4 that corresponds to the correct degrees of freedom. Read across the row to see the various quantiles for that distribution. The quantiles given in the table range from the .0001 quantile to the .9999 quantile, at irregular intervals. These are all of the quantiles you usually need. The pth quantile from a chi-square distribution with k degrees of freedom is customarily denoted by $\chi^2_{p,k}$, where χ^2 is the square of the lowercase Greek letter chi.

EXAMPLE

What quantiles are given in Table A4 for the chi-square distribution with 10 degrees of freedom? The first page in Table A4 gives quantiles from $\chi^2_{.0001,10}$ to $\chi^2_{.40,10}$. The second page continues, giving quantiles from $\chi^2_{.60,10}$ to $\chi^2_{.9999,10}$. A portion of that table is reproduced in Figure 6.13. The .05 quantile is $\chi^2_{.05,10} = 3.940$, which means

$$P(Y \le 3.940) = .05$$

for a random variable Y that has a chi-square distribution with 10 degrees of freedom.

The density function and the distribution function of a chi-square random variable with 10 degrees of freedom are given in Figure 6.14. The shaded portion in the density function represents the fact that 5% of the total area under the density function is to the left of 3.940, the .05 quantile. It is easier to find the .05 quantile from the distribution function, simply by finding the value of χ^2, 3.940 in this case, that has an ordinate of .05 on the curve.

Suppose the situation is reversed. The value $Y = 19.7$ is observed from a chi-square distribution with 10 degrees of freedom. Which quantile is this? In Figure 6.13 the observed value $Y = 19.7$ falls between the .95 and .975 quantiles, which are 18.31 and 20.48, respectively. Therefore 19.7 is somewhere between being the .95 quantile and being the .975 quantile. A more precise determination can be found by using linear interpolation as an approximate procedure:

$$P(Y \le 19.7) \approx .95 + (.975 - .95)(19.7 - 18.31)/(20.48 - 18.31)$$
$$= .966$$

Thus 19.7 is approximately equal to the .966 quantile.

			Quantiles				
d.f.	.01	.025	.05	$\cdots$	.95	.975	.99
9	2.088	2.700	3.325	$\cdots$	16.92	19.02	21.67
10	2.558	3.247	3.940	$\cdots$	18.31	20.48	23.21
11	3.053	3.816	4.575	$\cdots$	19.68	21.92	24.72

FIGURE 6.13
A PORTION OF TABLE A4, HIGHLIGHTING THE CHI-SQUARE DISTRIBUTION WITH 10 DEGREES OF FREEDOM.

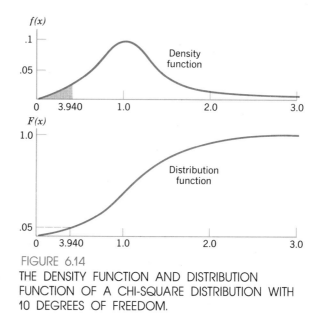

FIGURE 6.14
THE DENSITY FUNCTION AND DISTRIBUTION
FUNCTION OF A CHI-SQUARE DISTRIBUTION WITH
10 DEGREES OF FREEDOM.

A CONFIDENCE INTERVAL FOR σ

A **confidence interval for the standard deviation σ in a normal population** is
based on the following fact. Let $X_1, X_2, \ldots , X_n$ be a random sample from a
normal distribution with unknown parameters μ and σ. Then the theory of
statistics states that the sampling distribution of the random variable

$$\frac{(n - 1)s^2}{\sigma^2}$$

is a chi-square distribution with $n - 1$ degrees of freedom, where s^2 is the
sample variance.

Thus, for appropriately chosen quantiles $\chi^2_{\alpha/2,n-1}$ and $\chi^2_{1-\alpha/2,n-1}$ from a chi-
square distribution with $n - 1$ degrees of freedom, the following probability
statement is true:

$$P(\chi^2_{\alpha/2,n-1} < \frac{(n - 1)s^2}{\sigma^2} < \chi^2_{1-\alpha/2,n-1}) = 1 - \alpha \qquad (6.6)$$

Some algebraic manipulation on the expression inside the parentheses in Eq.
6.6 gives the following equivalent representation.

$$P\left(\frac{(n - 1)s^2}{\chi^2_{1-\alpha/2,n-1}} < \sigma^2 < \frac{(n - 1)s^2}{\chi^2_{\alpha/2,n-1}}\right) = 1 - \alpha \qquad (6.7)$$

Thus the sample variance s^2 and the two quantiles $\chi^2_{\alpha/2,n-1}$ and $\chi^2_{1-\alpha/2,n-1}$ are sufficient to find a lower limit

$$L = \frac{(n-1)s^2}{\chi^2_{1-\alpha/2,n-1}} \tag{6.8}$$

and an upper limit

$$U = \frac{(n-1)s^2}{\chi^2_{\alpha/2,n-1}} \tag{6.9}$$

for the $100(1-\alpha)\%$ confidence interval for the population variance σ^2.

Most of the time an interval estimate for the standard deviation is needed rather than an interval estimate for the variance. This is obtained by taking square roots of the quantities inside the parentheses of Eq. 6.7 to get

$$P\left(\left\{\frac{(n-1)s^2}{\chi^2_{1-\alpha/2,n-1}}\right\}^{1/2} < \sigma < \left\{\frac{(n-1)s^2}{\chi^2_{\alpha/2,n-1}}\right\}^{1/2}\right) = 1 - \alpha \tag{6.10}$$

so the lower and upper limits are merely the positive square roots of L and U in Eqs. 6.8 and 6.9.

To find a $100(1-\alpha)\%$ **confidence interval for σ, the standard deviation from a normal population,** proceed as follows.

INTERVAL ESTIMATE FOR σ (NORMAL POPULATIONS ONLY)

1. From a random sample $X_1, X_2, \ldots, X_n$ find

$$(n-1)s^2 = \sum_{i=1}^{n}(X_i - \bar{X})^2 = \sum_{i=1}^{n} X_i^2 - \left(\sum_{i=1}^{n} X_i\right)^2 / n$$

2. Find the quantiles $\chi^2_{\alpha/2,n-1}$ and $\chi^2_{1-\alpha/2,n-1}$ from Table A4.
3. The lower limit is $\{(n-1)s^2/\chi^2_{1-\alpha/2,n-1}\}^{1/2}$.
4. The upper limit is $\{(n-1)s^2/\chi^2_{\alpha/2,n-1}\}^{1/2}$.

EXAMPLE

Twelve cans are selected at random from the cans containing peach halves, mentioned in Problem Setting 6.3. Their contents are weighed with the following results.

Weights (oz): 32.6 33.1 32.1 32.7 31.9 31.8
32.4 32.3 32.8 33.3 32.7 32.6

First these data are checked for normality using the Lilliefors test, and they easily pass the test. To find a 95% confidence interval for σ the sum ΣX and the sum of squares ΣX^2 are calculated:

$$\Sigma X = 390.3 \qquad \Sigma X^2 = 12{,}696.75 \qquad n = 12$$
$$\text{d.f.} = 11 \qquad \chi^2_{.025,11} = 3.816 \qquad \chi^2_{.975,11} = 21.92$$
$$[(n-1)s^2/\chi^2_{.975,11}]^{1/2} = [(12696.75 - (390.3)^2/12)/21.92]^{1/2} = .3198$$
$$[(n-1)s^2/\chi^2_{.025,11}]^{1/2} = [(12696.75 - (390.3)^2/12)/3.816]^{1/2} = .7666$$

A 95% confidence interval for the standard deviation of the weight in a can of peach halves marked 32 oz is from .3198 oz to .7666 oz.

Note that the point estimate for σ is s, whether the population is normal or not normal. However, only when the population is normal may the chi-square distribution be used to find a confidence interval for σ as described in this section.

EXERCISES

6.38 Is the chi-square distribution discrete or continuous?

6.39 Would it be appropriate to use the chi-square distribution for the random variable X, where X is the profit (or loss) of a business? Why?

6.40 Find the .75 quantile of a chi-square distribution with 17 degrees of freedom.

6.41 Find the .90 quantile of a chi-square distribution with 29 degrees of freedom.

6.42 What is the approximate lower-tail probability associated with 7.93 in a chi-square distribution with 15 degrees of freedom?

6.43 What is the approximate lower-tail probability associated with 16.93 in a chi-square distribution with 21 degrees of freedom?

6.44 A chi-square random variable with 18 degrees of freedom has taken the value 27.3. What is the probability of observing a value this large or larger?

6.45 What is the exceedance probability associated with the value 50.8 in a chi-square distribution with 39 degrees of freedom?

6.46 In a large factory the internal mail is delivered by a minibus that makes its rounds every hour. The bus is scheduled to arrive at your office at 13 minutes past the hour, but sometimes it is early and sometimes it is late. Part of your job is to build a model that represents the communication system in the entire factory, and the mail bus is an important part of that system. It is reasonable to model the arrival time of the bus as a normal random variable with parameters μ and σ. So you collect some data on the actual arrival time of the mail bus. From these data you would like to estimate μ and σ.

The arrival times for the mail minibus are recorded for 16 trips. These are given on the next page, expressed as minutes past the hour. What is a 95% confidence interval for the population standard deviation?

Times: 13   15   14   18   17   14   11   15
         13   14   16   14   13   15   13   17

6.47 The diameter of a manufactured part must be consistently close to the stated value or the part will not fit properly. An interval estimate of the standard deviation provides a good measure of the consistency in the manufacturing process. Eighteen parts are selected at random, and their diameters are measured with the following results.

Diameters (cm):   0.897   0.899   0.898   0.901   0.902   0.900
                  0.900   0.898   0.897   0.900   0.896   0.902
                  0.902   0.903   0.899   0.899   0.903   0.900

Find a 95% confidence interval for the population standard deviation. Assume the data are normally distributed. Note that

$$\sum_{i=1}^{18} X_i = 16.196 \quad \text{and} \quad \sum_{i=1}^{18} X_i^2 = 14.572876$$

6.48 Consistency is important in manufacturing baseballs, so the bounce of the balls is close to the official standard, without individual balls being too lively or too dead. Baseballs are tested by dropping them from a standard height onto a hard surface and the height of their bounce is measured. Find a 90% confidence interval for the population standard deviation, if 20 observations yielded

$$\sum_{i=1}^{20} X_i = 52.18 \quad \text{and} \quad \sum_{i=1}^{20} X_i^2 = 137.21884$$

Assume the data are normally distributed.

6.49 When mixing ingredients for manufacturing medicated capsules it is important to get a thorough mixing so that each capsule contains nearly the same amount of active ingredient. Thirty capsules were selected at random to see how consistent they were with regard to an anticongestant ingredient. The percentage X of the ingredient was determined for each of the 30 capsules, with the following results.

$$n = 30 \quad \sum_{i=1}^{30} X_i = 2137 \quad \sum_{i=1}^{30} X_i^2 = 154,416$$

Find a 95% confidence interval for the standard deviation of X. Assume the data are normally distributed. Translate the bounds into a percentage variation from the mean dosage. Would you be satisfied with this much variation?

6.6

REVIEW EXERCISES

6.50 Police records on 30 randomly selected individuals booked on assault showed the ages recorded on the next page. Estimate the population median age for these individuals and find a 95% confidence interval for the population median age.

24	20	18	32	16	25
21	18	38	22	18	16
15	14	21	17	17	17
17	23	22	16	24	20
27	21	18	26	20	15

6.51 A random sample of 40 cars from a fleet of cars driven by employees of a large company showed an average of 2870 miles driven in a month with a standard deviation of 425 miles. Find a 95% confidence interval for the average number of miles per car driven by the fleet each month.

6.52 A reading comprehension test given to 10 randomly selected fourth grade students showed an average score of 84.2 with a standard deviation of 12.2. Assume that the population is normal and find a 90% confidence interval for the mean reading comprehension score for fourth graders.

6.53 Assume normality for the ages given in Exercise 6.50 and find a 95% confidence interval for the population mean age. Compare these results with the confidence interval found for the median in that exercise. (Use $\Sigma X = 618$ and $\Sigma X^2 = 13,536$.)

6.54 Of 16 randomly selected cars inspected during a safety campaign, 6 were found to be unsafe. Find a 95% confidence interval for the proportion of unsafe cars in the population.

6.55 A civic group reported to the town council that a random sample of 100 residents showed 48 in favor of an upcoming school bond issue. Find a 90% confidence interval for the true proportion of residents supporting the bond issue.

6.56 A random sample of tenth grade boys resulted in the weights listed here. Estimate the median weight for the population of tenth grade boys and find a 95% confidence interval for the population median.

142	134	98	119	131
103	154	122	93	137
86	119	161	144	158
165	81	117	128	103

6.57 The standard deviation of the viscosity of a brand of car oil is .02. How large a sample would be needed for a 95% confidence interval for μ to have a width of .01?

6.58 Two television commercials are shown to a group of 85 people. Fifty-seven people prefer commercial B. Assume that this group is like a random sample and find a 90% confidence interval for the population proportion of people preferring commercial B.

6.59 A social worker wants to estimate the mean number of years of education for the adult residents of one district. A random sample of 80 residents resulted in a mean of 11.3 years with a standard deviation of 2.2 years. Find a 95% confidence interval for the population mean. Do these results depend on the population (years of education) being normal?

6.60 Find a 95% confidence interval for the standard deviation of the population in Exercise 6.59. Do these results depend on the population (years of education) being normal?

6.61 A movie theater wants to estimate the median age of its customers. It randomly selects ten customers and gets the following ages: 19, 17, 18, 16, 47, 33, 23, 20, 17, 19. Find a 90% confidence interval for the median age of its customers.

6.62 A cereal manufacturer wants to know how many raisins go into each box of cereal, by weight. A random sample of 70 boxes yields a mean weight of raisins per box of 1.43 oz and a standard deviation of .121. Find a 90% confidence interval for the mean weight per box.

6.63 Find a 90% confidence interval for the population standard deviation of the weight of raisins per box in Exercise 6.62. Assume the population of raisins weights per box is normally distributed.

BIBLIOGRAPHY

Additional material on confidence intervals for nonnormal populations can be found in the following publication.

Conover, W. J. (1980). *Practical Nonparametric Statistics,* 2nd ed. Wiley, New York.

7

HYPOTHESIS TESTING

PRELIMINARY REMARKS

One of the primary purposes for studying statistics is to learn how to make *statistical inferences* on the basis of a random sample obtained from a population. Examination of the random sample allows inferences to be made regarding parameters in the population such as the population mean, the population median, the population standard deviation, or the population proportion, since parameters such as these are almost always unknown. The *sample* mean, median, and standard deviation may be used to make inferences about the *population* mean, median, and standard deviation, respectively. The sample proportion may be used to make inferences about the population proportion.

Statistical inference takes two forms. One is **estimation**, which was introduced in the previous chapter. The other is **hypothesis testing**, which is the subject of this chapter.

1. **Estimation** A random sample is obtained from some population and used to make an estimate of some parameter in the population. This estimate may be a point estimate or an interval estimate. A typical result might be expressed as follows: "On the basis of a random sample consisting of 200 consumers, a point estimate of the population percentage of consumers who prefer the flavor of the new brand of coffee is 69% with a 95% confidence interval estimate from 63% to 75%. There is a 5% probability that intervals established in this manner will not contain the true population percentage."

2. **Hypothesis testing** A random sample is obtained from some population, and a judgment is made as to whether the sample is consistent with a preconceived statement, or hypothesis, about some parameter of the population. If the sample is not consistent with the hypothesis, the hypothesis is rejected (that is, it is deemed to be unacceptable). Otherwise, the hypothesis is not rejected.

A typical conclusion from hypothesis testing might be as follows: "On the basis of a random sample consisting of 200 consumers the hypothesis, 'Thirty percent of all consumers prefer the new flavor of coffee,' is rejected. However, there is approximately one chance in 20 that hypothesis tests conducted in this manner will mistakenly reject the hypothesis when it is actually true."

Thus both forms of statistical inference use a random sample to make some conclusion about an unknown population parameter, and both provide some probability that the conclusion is incorrect. As long as only samples are examined, not entire populations, there will always be some possibility that the resulting inference about the population parameter is incorrect.

The probability of making an incorrect conclusion with confidence intervals was explained in the previous chapter, while the probability of making an incorrect conclusion in hypothesis testing is examined in detail in this chapter. Statistical theory is concerned with evaluating this probability of error.

The results of many theoretical studies in statistics are presented in this chapter and subsequent chapters, in the form of useful hypothesis tests for various situations that frequently occur in surveys and other research projects. Methods of forming confidence intervals are also presented for population parameters not already covered in the previous chapter.

7.1
HYPOTHESES, TEST STATISTICS, AND p-VALUES

PROBLEM SETTING 7.1

A company that manufactures snack foods has developed a new variety of snack, currently known only as Mix 37. The company's management would like to know how the consumer rates Mix 37 when compared with other snack foods currently on the market.

Management decides on the following plan. Mix 37 will be offered to the public in a party atmosphere, along with another popular snack food known as Brand X. A commercial cocktail-hour caterer will place bowls of both products on the tables at the next 100 catering jobs it has, and at the end of the cocktail hour the caterer will determine how much of each snack was consumed. After 100 trials the management expects to be able to make a decision regarding whether or not to market Mix 37.

Clearly, if Mix 37 is preferred in almost all 100 trials, then Mix 37 has been proven to be a preferred product and it should be marketed. Just as clearly, if almost none of the 100 trials show Mix 37 to be preferred, then it would be a mistake to try to put it on the market in its present form. But how does one decide which results of the experiment indicate that the product is marketable? A formal procedure for hypothesis testing provides one solution to this problem, as shown in this chapter.

THE ALTERNATIVE HYPOTHESIS

The **alternative hypothesis**, sometimes called the *research hypothesis*, is the statement that describes the desired goal of the experiment of study. It states what the researcher suspects (or merely hopes) to be true. The goal of an experiment is to try to establish that the alternative hypothesis is a true statement. In this text H_1 is used to identify the alternative hypothesis.

ALTERNATIVE HYPOTHESIS   The **alternative hypothesis H_1** is the statement that the researcher would like to show to be true.

EXAMPLE

The management in Problem Setting 7.1 would like to demonstrate that Mix 37 is preferred over Brand X. Therefore, H_1 is, "The population of consumers prefers Mix 37 over Brand X." If p represents the probability that Mix 37 is the preferred snack out of the two choices, then the alternative hypothesis could be written simply as $H_1: p > \frac{1}{2}$, because being preferred more than one-half of the time is the same as being preferred over Brand X.

THE NULL HYPOTHESIS

The **null hypothesis**, sometimes called the *test hypothesis* because it is the hypothesis being tested, is a statement that is generally believed (or hoped) by the experimenter to be false. It is usually the opposite of the alternative hypothesis, and is denoted by H_0.

If the sample data are not consistent with the null hypothesis, the null hypothesis is rejected. On the other hand, if the sample data are not inconsistent with the null hypothesis, the null hypothesis is not rejected.

The purpose of an experiment may be to see if a change has occurred in some population or if there is a difference between two populations. Perhaps the purpose is to see if there is a relationship between two variables. The null hypothesis is often a version of the statement, "There is no change or difference in the population(s)" or "There is no relationship between the variables." That is why it is called a *null* hypothesis, or hypothesis of no change or difference.

NULL HYPOTHESIS   The **null hypothesis H_0** is the statement being tested. The experiment usually attempts to show that the null hypothesis is false.

Examples of null and alternative hypotheses include the following. A research project supported by a health organization might have these hypotheses.

H_0: Smoking is not harmful to your health.
H_1: Smoking is harmful to your health.

Note that this alternative hypothesis suggests that smoking is harmful to health, and the experiments that accompany this hypothesis test probably would not have been conducted without the presumption that some health problems are caused by smoking. Thus, this presumption becomes the alternative hypothesis. The null hypothesis, on the other hand, states that smoking is not harmful to your health, and is suspected of being false. The experiment is designed to show this null hypothesis to be false.

For most statistical hypothesis tests the hypotheses are stated in terms of some population parameter such as the mean, median, standard deviation, or proportion. Consider a test by a light bulb manufacturer to examine the life of a new long-life type of light bulb it hopes to market. The hypotheses are

H_0: The mean burning time of the new type of light bulb is not greater than the mean of the well-established type of light bulb.

H_1: The mean burning time of the new type of light bulb is greater than the mean of the well-established type of light bulb.

Note again how H_1 represents the situation believed to be true, or the test would not have occurred. The null hypothesis represents the hypothesis of "no improvement" over the current bulb.

Sometimes the null hypothesis represents the statement of no dependence between variables, when it is suspected that the variables are related to each other, such as in these hypotheses that are concerned with the relationship between the degree of job satisfaction of an employee and the level of education attained by the employee.

H_0: The degree of job satisfaction is independent of the level of education.

H_1: The degree of job satisfaction is dependent upon the level of education.

As a final example, the Food and Drug Administration requires that the null hypothesis

H_0: The new drug is not effective

be rejected in a statistical hypothesis test before a new drug can be marketed. The pharmaceutical company is, of course, interested in showing that the alternative hypothesis,

H_1: The new drug is effective

is true. Rejection of the null hypothesis is possible only if there is conclusive evidence pointing to effectiveness of the drug. In this way the public is protected against worthless products being marketed as effective drugs.

THE TEST STATISTIC

Every statistical hypothesis test uses a statistic to measure the agreement between the sample data and the null hypothesis. When used in this manner, the statistic is called a **test statistic**.

A test statistic is some function of the sample values, and it must be especially sensitive to the difference between H_0 and H_1. That is, a test statistic should tend to take on certain values when H_0 is true, and different values when H_1 is true. Thus, the experimenter may decide whether or not to reject H_0 merely by noting the value of the test statistic.

TEST STATISTIC
A **test statistic** is a statistic computed from the sample data, and it is used in the testing of the null hypothesis. In this text the test statistic is usually denoted by T, with or without subscripts.

If the hypotheses are concerned with the population mean, the test statistic is usually a function of the sample mean. If the hypotheses are statements about the population standard deviation, the test statistic is usually closely related to the sample standard deviation. When each test is presented, the appropriate test statistic will be given as part of the protocol for the test.

EXAMPLE

In Problem Setting 7.1 the hypotheses are concerned with whether or not the population of consumers prefers Mix 37 over Brand X. The experiment is set up to see how many times out of 100 trials Mix 37 is the more popular of the snacks.

The null hypothesis could be stated in terms of the population proportion p of consumers that prefer Mix 37 over the other brand. If there really is no difference in preference for the two brands, then p equals one-half, because each person in the population is equally likely to prefer either of the two products. Therefore the null hypothesis becomes

$$H_0: p = \tfrac{1}{2}$$

The alternative hypothesis, that the population prefers Mix 37, is equivalent to saying that more than one-half of the population prefers Mix 37:

$$H_1: p > \tfrac{1}{2}$$

Since the hypotheses are statements about the population proportion p, the test statistic should be closely related to the sample proportion $\hat{p}$ of the 100 trials that resulted in a preference for Mix 37. In fact, the *test statistic* T in this situation is merely *the number of times that Mix 37 is the preferred mix*. Note that $\hat{p}$ is simply $T/100$, so the two statistics $\hat{p}$ and T are very closely related.

THE DISTRIBUTION OF THE TEST STATISTIC WHEN H_0 IS TRUE

The test statistic is used to summarize the sample data as a single number, and that number is the basis on which the decision whether or not to reject H_0 is made. If the data are inconsistent with the null hypothesis, this must be reflected by the test statistic. Which values of the test statistic are not likely to occur when H_0 is true? If those same values indicate that H_1 is likely to be true, then they should result in rejection of the null hypothesis. In order to determine which values of the test statistic should result in rejection of the null hypothesis, it is necessary to know the sampling distribution of the test statistic when H_0 is true.

The sampling distribution of the test statistic under the assumption H_0 is true is called the **null distribution** of the test statistic. No statistical test is possible without knowing the null distribution of the test statistic. The test statistics used for the tests in this text will use the null distributions given in the tables in the back of the text.

NULL **The** sampling distribution of the test statistic when the null hypothesis is true
DISTRIBUTION is called the **null distribution**.

EXAMPLE

In the previous example the test statistic

T = the number of trials in which Mix 37 is preferred

is used to test H_0: $p = \frac{1}{2}$ against H_1: $p > \frac{1}{2}$, where p is the proportion of the population that prefers Mix 37. There are 100 trials, and it seems reasonable to assume that the results are independent from trial to trial. Under these assumptions:

1. $n = 100$ trials, all mutually independent, and
2. $p = \frac{1}{2}$ on each trial (the null hypothesis),

the null distribution of T is the binomial distribution with parameters $n = 100$ and $p = \frac{1}{2}$. The mean of this binomial distribution is

$$\mu = np = 100(\tfrac{1}{2}) = 50$$

and the standard deviation is

$$\sigma = \sqrt{npq} = \sqrt{100\ (\tfrac{1}{2})(\tfrac{1}{2})} = 5.$$

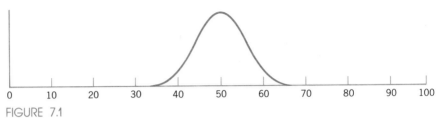

FIGURE 7.1
A NORMAL DENSITY FUNCTION, WITH $\mu = 50$ AND $\sigma = 5$, WHICH
CLOSELY APPROXIMATES THE NULL DISTRIBUTION OF T.

Because n is large, this sampling distribution may be closely approximated by the normal distribution with $\mu = 50$ and $\sigma = 5$. Therefore the null distribution of T is given approximately by the density in Figure 7.1.

ONE-TAILED AND TWO-TAILED TESTS

If large values of the test statistic indicate that H_1 rather than H_0 is true, then the test is called an **upper-tailed test**, because large values of the test statistic are located in the upper tail (right-hand tail) of the null distribution (see Figure 7.2). Similarly, if small values of the test statistic indicate agreement of the data with H_1 rather than H_0, the test is called a **lower-tailed test**. Sometimes H_1 is worded so that values of the test statistic in either tail of its null distribution indicate that H_1 is true. In that case the test is called a **two-tailed test**, as opposed to the upper-tailed and lower-tailed tests, which are **one-tailed tests**.

EXAMPLE

In the example used in this section the test statistic T is the number of trials in which Mix 37 is preferred. Large values of T indicate a clear preference for Mix 37, and hence indicate H_1 is true. Therefore the test is upper-tailed.

THE p-VALUE

Once a value for the test statistic is calculated from the data, it is possible to calculate the probability of getting a value *at least that extreme* (in the direction of H_1) under the null distribution. This is called the **p-value** and is a measure

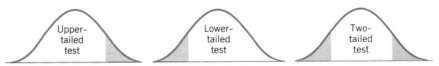

FIGURE 7.2
AN UPPER-TAILED TEST (LEFT), A LOWER-TAILED TEST (CENTER), AND A TWO-TAILED TEST (RIGHT).

of how well the observed sample agrees with the null hypothesis. A small p-value (close to zero) indicates that the sample is not consistent with the null hypothesis and the null hypothesis should be rejected. On the other hand a p-value larger than .05 or .10 generally indicates a reasonable level of agreement between the sample and the null hypothesis.

Although the notation is similar, there is no direct connection between the p-value and the parameter p in the binomial distribution.

---

p-VALUE   The **p-value** associated with an observed value of a test statistic is the probability of getting the observed value or a value more extreme (in the direction of H_1) assuming that H_0 is true. The p-value is calculated using the null distribution of the test statistic.

EXAMPLE

In continuation of the previous example, suppose the catering service found that 61 of the 100 trials resulted in Mix 37 being preferred over Brand X. Therefore T equals 61. What is the p-value?

The p-value is the probability of getting $T = 61$ or a more extreme value in the direction of H_1. Large values of T indicate Mix 37 is preferred (H_1), so the p-value is the probability of getting $T = 61$ or larger, when p equals $\frac{1}{2}$ as given by the null hypothesis. The null distribution of T is the binomial distribution with $p = \frac{1}{2}$, and $n = 100$, but because n is large, the normal approximation to the binomial distribution is used:

$$p\text{-value} = P(T \geq 61) \approx 1 - P\left(Z \leq \frac{61 - 100(\frac{1}{2}) - .5}{\sqrt{100(\frac{1}{2})(\frac{1}{2})}}\right)$$

$$= 1 - P(Z \leq 2.1)$$

$$= 1 - .9821$$

$$= .0179$$

Thus the probability of getting a value of T as extreme as 61, in the direction of H_1, is .0179 if H_0 is true. This small probability (equivalent to less than one chance in 50), is a reasonable indication that the sample is not consistent with

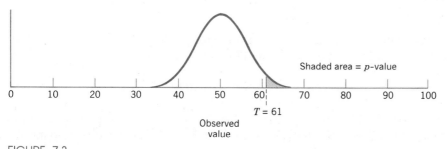

FIGURE 7.3
THE SHADED AREA REPRESENTS THE p-VALUE ASSOCIATED WITH $T = 61$.

the null hypothesis. The decision in this case is to reject the null hypothesis in favor of the alternative, and conclude that Mix 37 is preferred over Brand X by the population of consumers.

The p-value is shown in Figure 7.3 as the shaded area under the null density function, from $T = 61$ to the right, toward the values more likely under H_1.

HOW TO FIND THE p-VALUE

Three items of information are needed to find a p-value:

1. The observed value of the test statistic.
2. The null distribution of the test statistic.
3. Whether the test is upper-tailed, lower-tailed, or two-tailed.

If the test is an upper-tailed test, the p-value is the probability of getting a value equal to or greater than the observed value, found from tables for the null distribution. Similarly, in a lower-tailed test the p-value is the probability of getting a value less than or equal to the observed value. If the test is a two-tailed test, the one-tailed p-value is found first (the smaller of the upper-tailed or lower-tailed p-values is used) and doubled to get the two-tailed p-value. However, the p-value is never allowed to exceed 1.0, so if this procedure results in a p-value greater than 1, simply set the p-value equal to 1, as demonstrated in the next example.

To find the p-value for an observed value of T, say t_{obs}, use the null distribution of T, and

TO FIND
p-VALUES

1. for an upper-tailed test, p-value $= P(T \geq t_{obs})$.
2. for a lower-tailed test, p-value $= P(T \leq t_{obs})$.
3. for a two-tailed test, p-value $= 2P(T \geq t_{obs})$ or p-value $= 2P(T \leq t_{obs})$, whichever is less than 1; or, if both are greater than 1, as sometimes occurs with a discrete null distribution, the p-value equals 1.

EXAMPLE

Suppose the null distribution of T is binomial, with $n = 11$ and $p = .6$. The p-value in a two-tailed test, where the observed value is $T = 3$, is given by

$$p\text{-value} = 2P(T \leq 3) = 2(.0293) = .0586$$

where $P(T \leq 3)$ is obtained from Table A1 with $n = 11$ and $p = .6$. Note that $2P(T \geq 3) = 1.9882$. Since $2P(T \geq 3) > 1$, the p-value is found as $2P(T \leq 3)$.

If the observed value of T is 7 instead of 3, then Table A1 yields

$$P(T \leq 7) = .7073$$

and

$$
\begin{aligned}
P(T \geq 7) &= 1 - P(T \leq 6) \\
&= 1 - .4672 \\
&= .5328
\end{aligned}
$$

In both cases doubling the one-tailed probability results in a two-tailed p-value greater than 1, so the p-value is simply equal to 1 in a two-tailed test for an observed value of 7.

EXERCISES

7.1 Formulate an alternative hypothesis for each of the following null hypotheses.

  (a) H_0: Employee typing speeds are not increased as a result of taking a specialized training course.

  (b) H_0: Daily sales are unchanged after the start of the use of TV commercials.

  (c) H_0: The average number of sickness absentees for Monday and Friday is not higher than the average number of sickness absentees for Tuesday, Wednesday, and Thursday.

  (d) H_0: The proportion of viewers watching the local news on Channel 4 is less than or equal to 40%.

  (e) H_0: The median grade-point average of students at this university is 2.613.

7.2 An enterprising high school teacher has started an evening business giving courses that prepare students for the Standardized Achievement Test (SAT). She would like to advertise, truthfully, that the mean score for students who complete her course is above the national average. The national average is 800. She intends to examine the scores of several randomly selected students who have completed her course and use a statistical hypothesis test. What should H_0 and H_1 be in her test?

7.3 A quality-control engineer in a factory that manufactures sheet metal checks the thickness of the sheet metal to make sure the machines are in proper adjustment. There is always some variability in measurements of thickness, so statistical procedures are used. If the engineer is checking sheet metal that is supposed to be one-quarter inch thick, what is H_0 and what is H_1?

7.4 A quality-control manager in a cannery will order the readjustment of all equipment involved in canning a particular food if the mean weight per can is less than specifications require. For cling peaches, the contents must be at least 16 oz (the null hypothesis). A sample of cans is weighed, and the sample mean is used as the test statistic.

(a) Formulate the null and the alternative hypotheses in terms of specific values for the population mean.

(b) Is the test upper-tailed or lower-tailed?

7.5 The null distribution of the test statistic is the standard normal. The observed value is $T = 1.33$. Find the p-value in an upper-tailed test.

7.6 Suppose that when H_0 is true the test statistic has a Student's t distribution with 14 degrees of freedom. The observed value of T is 1.7613. Find the p-value in a two-tailed test.

7.7 Suppose the distribution of T is the normal distribution with mean μ and standard deviation 1. Also, if H_0 is true, then $\mu = 0$. The observed value of T in a two-tailed test is -1.88. Find the p-value.

7.8 A test consists of 12 independent trials. Further, when H_0 is true, the probability of success on any one trial is .25. The test statistic T is the number of successes observed in the 12 trials. What is the null distribution of the test statistic? Find the p-value in an upper-tailed test if the observed value of T is 7.

7.2

THE DECISION RULE AND POWER

LEVEL OF SIGNIFICANCE AND TYPE I ERROR

A **Type I (type one) error** is the error of rejecting H_0 when H_0 is true. The experimenter selects a small, but acceptable, probability of making a Type I error, such as .05, calls this probability the **level of significance** and denotes it by α. The choice of the level of significance usually depends on the consequences of making a Type I error. If a Type I error is very serious, a small value of α, such as .01 or smaller, is customary. For a moderately serious Type I error, $\alpha = .05$ is selected, and if a Type I error is not at all serious, a larger value such as $\alpha = .10$ may be used.

THE DECISION RULE

The **decision rule** identifies those values of the test statistic T that result in rejection of the null hypothesis, and calls that set of values the **rejection region**. Then if T is in the rejection region, the null hypothesis is rejected; otherwise, it is not rejected.

The rejection region is formed so the probability of the test statistic being in the rejection region is α when H_0 is true. Easy-to-follow instructions for finding the rejection region are given with each hypothesis test.

TYPE ONE ERROR A **Type I (type one) error** refers to rejecting H_0 when H_0 is true.

LEVEL OF SIGNIFICANCE The **level of significance**, denoted by α, is the probability of making a Type I error.

CRITICAL VALUES

If H_1 indicates that an upper-tailed test is needed, the decision rule is as follows:

Reject H_0 if $T > T_{\text{upper}}$; otherwise, do not reject H_0

where T_{upper}, or T_U for short, is a **critical value** for T obtained from the null distribution of T as indicated in Figure 7.4a. The critical value T_U is the $1 - \alpha$ quantile in the null distribution of T, so the probability of making a Type I error is equal to the preselected value α, that is, $P(T > T_U) = \alpha$.

Similarly, the decision rule for a lower-tailed test takes the form:

Reject H_0 if $T < T_{\text{lower}}$; otherwise, do not reject H_0

where T_{lower}, or T_L for short, is a critical value selected from the null distribution of T such that α is the level of significance (see Figure 7.4b). In this case, T_L is the α quantile in the null distribution of T, so $P(T < T_L) = \alpha$.

In a two-tailed test the decision rule takes the form:

Reject H_0 if $T > T_U$ or $T < T_L$; otherwise, do not reject H_0

where T_U and T_L are critical values obtained from the distribution function of

---

In a one-tailed test the **decision rule** is stated as:

Reject H_0 if and only if $T > T_U$	(For an upper-tailed test)
Reject H_0 if and only if $T < T_L$	(For a lower-tailed test)

DECISION RULES AND CRITICAL VALUES A two-tailed test has the decision rule:

Reject H_0 if $T > T_U$ or if $T < T_L$.

The values T_U and T_L are **critical values** that are selected so the test will have the desired level of significance.

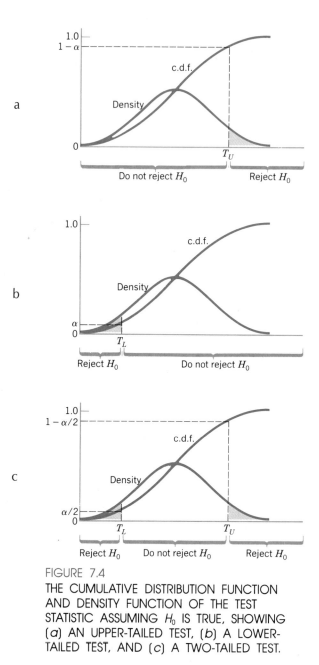

FIGURE 7.4

THE CUMULATIVE DISTRIBUTION FUNCTION
AND DENSITY FUNCTION OF THE TEST
STATISTIC ASSUMING H_0 IS TRUE, SHOWING
(a) AN UPPER-TAILED TEST, (b) A LOWER-
TAILED TEST, AND (c) A TWO-TAILED TEST.

T when H_0 is true, as indicated in Figure 7.4c. The values T_U and T_L are usually
selected so that there is equal probability (or nearly equal) associated with
each tail. In a two-tailed test $P(T > T_U) + P(T < T_L) = \alpha$.

EXAMPLE

In the previous section the test statistic T was the number of trials in which
Mix 37 was preferred over Brand X. When the null hypothesis of "no prefer-

ence" is true, T has a binomial distribution with $n = 100$ and $p = \frac{1}{2}$, which can be approximated by a normal distribution with $\mu = 50$ and $\sigma = 5$. Suppose α is set at .05. What is the rejection region? What is the decision rule?

Note that large values of T indicate Mix 37 is preferred over Brand X, in agreement with H_1. Therefore the test is an upper-tailed test. The decision rule is to reject H_0 if $T > T_U$, where T_U is the .95 quantile of the binomial distribution with $n = 100$ and $p = \frac{1}{2}$. An approximation for T_U uses the standard normal distribution, and is given at the end of Table A1 as

$$T_U = np + z_{.95}\sqrt{npq} = 50 + 1.6449(5) = 58.2$$

The rejection region of size $\alpha = .05$ includes all values of T greater than $T_U = 58.2$. For an observed value $T = 61$, the decision rule leads to rejection of H_0. The p-value associated with $T = 61$ was found in the previous section to be .0179. Note that whenever the decision is to reject H_0, the p-value is always less than or equal to the level of significance, α.

WHAT DOES THE DECISION MEAN?

If the sample is clearly not consistent with the null hypothesis, then the null hypothesis is rejected and the alternative hypothesis is considered to be the true statement. On the other hand, failure to reject H_0 does *not* imply H_0 is true. It merely indicates that the sample is *not inconsistent* with the null hypothesis. Perhaps the sample size is too small to detect any inconsistency with H_0. Perhaps the true state of nature is not H_0 but is so close to H_0 that it is very difficult to detect any difference between the two. And there is always the possibility that H_0 is actually true. So, while rejection of H_0 is fairly conclusive evidence that H_0 is not true, failure to reject H_0 does not provide conclusive evidence that H_0 is true.

A SUMMARY OF HYPOTHESIS TESTING

To summarize, there are eight steps in any hypothesis test. The first three steps are primarily the responsibility of the experimenter and the next five steps are primarily the responsibility of the statistician.

Responsibility of the experimenter:

1. State the alternative hypothesis, H_1.
2. State the null hypothesis, H_0.
3. Decide on a level of significance α.

Responsibility of the statistician:

4. Choose an appropriate testing procedure.
5. State the decision rule.
6. Compute the test statistic from the sample data.

7. Make the decision either to reject the null hypothesis, or to fail to reject it.

8. Compute the p-value from the null distribution of the test statistic and interpret it.

The examples given in this and the previous sections are combined in the following example, to show how a typical hypothesis test proceeds.

EXAMPLE

A snack food manufacturer is interested in knowing whether its new product, called Mix 37, has customer appeal. The company employs a commercial caterer to test Mix 37 against Brand X in 100 independent trials, and to record how many of the trials result in a preference for Mix 37 over the other brand.

Step 1 State the alternative hypothesis, H_1 Let p represent the population proportion that prefers Mix 37 over Brand X. The statement that the manufacturer would like to prove is that the consumers have a preference for Mix 37. This translates to

$$H_1: p > \tfrac{1}{2}$$

Step 2 State the null hypothesis, H_0 The statement "the consumers have no preference for either brand" translates to

$$H_0: p = \tfrac{1}{2}$$

Step 3 Decide on a Type I error rate, α Because one chance in 20 of falsely rejecting the null hypothesis seems like a reasonably small chance to take, and also because hypothesis tests frequently use this error rate, α is set at $\tfrac{1}{20}$, or .05.

Step 4 Choose an appropriate testing procedure The conditions of this experiment, and the hypotheses, agree with those of the binomial test, which is formally presented in Section 7.3. Therefore the binomial test is selected as the appropriate procedure to use in this case.

Step 5 State the decision rule The binomial test states that the decision rule is to reject H_0 if T is greater than T_U, where T is the number of trials in which Mix 37 is preferred, and where T_U is given by

$$T_U = np + z_{.95} \sqrt{npq} = 100(\tfrac{1}{2}) + 1.6449 \sqrt{100(\tfrac{1}{2})(\tfrac{1}{2})} = 58.2$$

The value for T_U is the .95 quantile from the null distribution of T. In this case, the null distribution of T is the binomial distribution with $n = 100$ and $p = \tfrac{1}{2}$, and the normal approximation is used because n is large.

Step 6 Compute the test statistic from the sample data The test statistic T equals 61, because 61 of the 100 trials resulted in Mix 37 being the preferred mix.

Step 7 Make the decision The decision rule in step 5 is to reject H_0 if T is greater than 58.2. Since $T = 61$ in step 6, the decision is to reject H_0. The conclusion is that consumers prefer Mix 37.

Step 8 Compute the *p*-value and interpret it The probability of getting an experimental result as extreme as $T = 61$, when H_0 is true, can be found from the null distribution of T, which is the binomial distribution with $n = 100$ and $p = \frac{1}{2}$. The normal approximation to the binomial distribution gives

$$P(T \geq 61) \approx 1 - P\left(Z \leq \frac{61 - 100(\frac{1}{2}) - .5}{\sqrt{100(\frac{1}{2})(\frac{1}{2})}} \right)$$

$$= 1 - P(Z \leq 2.1)$$

$$= 1 - .9821$$

$$= .0179$$

The probability of getting a result as extreme as the observed result is only .0179 when H_0 is true. This small probability supports the conclusion that H_0 is likely to be false.

THE LILLIEFORS TEST

At this point you may have realized that the Lilliefors test of Chapter 5 falls into this framework of hypothesis testing. That is, the null hypothesis is that the sample comes from a normal population, whereas the alternative hypothesis is that the population is nonnormal. The only level of significance permitted in the graph (see Figure 5.15) is $\alpha = .05$. Graphs for additional levels and more information about the Lilliefors test appear in the references at the end of this chapter.

The Lilliefors test as presented here is a two-tailed test. The null hypothesis is rejected if the empirical distribution function crosses the bounds corresponding to the sample size (see Figure 5.16). It is not possible to compute *p*-values from the limited information presented here.

POWER

So far all of the emphasis in hypothesis testing has been on the null distribution—the distribution of the test statistic when H_0 is true. The rejection region is found using the null distribution, and the *p*-value is found using the null distribution. However, some probabilities cannot be found from the null distribution. Consider the probability of rejecting H_0 when H_0 is false. This is called the **power** of a test. A good test has good (high) power.

The opposite event, accepting H_0 when H_0 is false, is a **Type II (type two) error**. The probability of making a Type II error is denoted by the Greek letter β (beta), so the power equals $1 - \beta$. Figure 7.5 summarizes the terminology and symbolism associated with Type I and Type II errors in hypothesis testing.

TYPE TWO ERROR A **Type II (type two) error** refers to the error of accepting H_0 when H_0 is false.

True Situation

	H_0 is true	H_0 is false
Do Not Reject H_0	Correct decision P(Correct decision) $= 1 - \alpha$	Type II error P(Type II error) $= \beta$
Reject H_0	Type I error P(Type I error) $= \alpha$ (Level of significance)	Correct decision P(Correct decision) $= 1 - \beta$ (Power of the test)

Decision

FIGURE 7.5
A SUMMARY OF POSSIBLE DECISIONS IN HYPOTHESIS TESTING AND THEIR PROBABILITIES.

POWER　The **power** of a test is the probability of rejecting a false null hypothesis, and is denoted by $1 - \beta$, where β is the probability of making a Type II error.

To find the power of a test, two things must be known. First, the rejection region must be known. This is the same rejection region used in the hypothesis test, and the method for finding it is described earlier in this section. Second, the sampling distribution of the test statistic *when H$_1$ is true* must be known. This sampling distribution is often vague and ambiguous, so the power is often not easy to determine, as the next example illustrates.

EXAMPLE

In the example used earlier in this section, the rejection region is $T > 58.2$, where T is the number of times in 100 trials that Mix 37 is preferred. What is the power of the test when $p = .7$?

The sampling distribution of T when H_0 is true is the binomial distribution with $n = 100$ and $p = \frac{1}{2}$. This is the null distribution of T. However, when H_1 is true, the sampling distribution of T changes. It is still the binomial distribution with $n = 100$, but with $p > \frac{1}{2}$. The power depends on the distribution of T when H_1 is true, which in this example means the power depends specifically on the exact value of p. Every value of p greater than $\frac{1}{2}$ provides a different distribution for T and therefore a different power. One of the values possible under H_1 is $p = .7$. If $p = .7$, the power is the probability of $T > 58.2$, the rejection region, when T has the binomial distribution with $n = 100$ and $p = .7$. This is found using the normal approximation to the binomial distribution as follows.

$$\text{Power} = P(T > 58.2 \mid n = 100, p = .7)$$
$$= P(T \geq 59 \mid n = 100, p = .7)$$
$$\approx 1 - P\left(Z \leq \frac{59 - 100(.7) - .5}{\sqrt{100(.7)(.3)}}\right)$$
$$= 1 - P(Z \leq -2.51)$$

$$= 1 - .0060$$
$$= .9940$$

This means there is probability .994 (very high) of rejecting H_0 with this test if the actual proportion of the population preferring Mix 37 is 70%. With a high power (.994) and a low α (.05) the test is a good one. The reader can readily substitute other values of p into the preceding calculations to show that the actual power will vary depending on the true proportion (actual value of p) of the population preferring Mix 37.

QUESTIONS FREQUENTLY ASKED ABOUT HYPOTHESIS TESTING

Some questions that are frequently asked about hypothesis testing and the corresponding answers are as follows.

Question What is the difference between the null hypothesis and the alternative hypothesis?

Answer The hypothesis that the experimenter wants to prove is the alternative hypothesis and the negation of that hypothesis is the null hypothesis.

Question How can I tell if the test is supposed to be upper-tailed, lower-tailed, or two-tailed?

Answer Ask yourself which values of the test statistic indicate H_1 is most likely to be true. Those are the values that determine the rejection region, and whether the test is upper-tailed, lower-tailed, or two-tailed. For your convenience, and to help eliminate errors, the instructions for the hypothesis tests in this text will always provide this information for you.

Question How do I know which test to use?

Answer Check the assumptions associated with each test. These are listed with each test description. If the assumptions of a test are satisfied, then that test is valid to use. If more than one test is valid, choose the one with the most power.

Also be sure the hypotheses in the test are the hypotheses of interest to you. If they are not, you should find another test, one that matches with the hypotheses you want to investigate.

Question Of what use is the p-value, once the decision has been made?

Answer Some data analysts stop after rejecting or failing to reject H_0, without reporting the p-value. Others skip the decision process and report only the p-value. The recommendation in this text is to make a decision *and* report a p-value, in order to provide the most complete analysis possible.

The p-value informs the reader whether the decision was clearly indicated by the data, or whether the decision was borderline and could easily have gone either way. Thus the p-value adds information about how well the data agree with the null hypothesis.

Question Will all of the null distributions be given in tables in the back of this text, or will I have to compute some probabilities myself?

Answer Although you have enough background in probability at this point to compute some of the null distribution probabilities yourself, most of the null distributions are very difficult to find. Therefore, to reduce the chance for error, all of the null distributions used in this text are found in tables in the Appendix.

CONCLUDING REMARKS

As stated previously, an objective in hypothesis testing is to have α small (close to zero) and the power large (close to 1). A very small value of α (close to zero) will result in a test with less power than the same test with a larger value of α. Therefore, there is always a trade-off between the level of significance and the power. An increase in the sample size usually is accompanied by an increase in power if α remains unchanged.

EXERCISES

7.9 Suppose the test statistic T is normally distributed with mean μ and standard deviation 1. Further suppose that when H_0 is true, μ equals 0. What must the decision rule be in order to have an upper-tailed test with $\alpha = .05$?

7.10 Suppose the test statistic T has a Student's t distribution with 17 degrees of freedom when H_0 is true. What is the decision rule in a two-tailed test with $\alpha = .05$?

7.11 If the test statistic T has a Student's t distribution with 28 degrees of freedom as the null distribution, find the critical value for a lower-tailed test with $\alpha = .10$.

7.12 The null distribution of a test statistic T is the standard normal distribution. Find the decision rule for a two-tailed test with $\alpha = .05$.

7.13 Suppose the test statistic T has a binomial distribution with $n = 12$. Also, when H_0 is true, $p = .5$. Find the decision rule for a lower-tailed test with $\alpha = .0193$.

7.14 Suppose the test statistic T has a binomial distribution with $n = 100$. Also, when H_0 is true, $p = .5$. Find the decision rule for a lower-tailed test with $\alpha = .05$.

7.15 Suppose that when H_0 is true, the test statistic T has a binomial distribution with $n - 24$ and $p = .8$. Find the two-tailed rejection region of approximate size $\alpha = .05$. If the observed value of the test statistic is $T = 11$, what is the decision? Find the p-value.

7.16 The courtroom is the stage for a test of the null hypothesis

H_0: The defendant is not guilty

against the alternative hypothesis

H_1: The defendant is guilty.

In U.S. courtrooms the defendant is presumed innocent until proven guilty. Yet in statistical hypothesis testing the alternative hypothesis is always the statement presumed to be true. Should the hypotheses be reversed in this case? Explain.

7.17 A bank is considering a policy of paying passbook interest on checking accounts that maintain a minimum monthly balance of at least $500. Such accounts are said to qualify for the "500 Club." The bank manager feels that more than 40% of the checking accounts will qualify for earned interest under this plan. The manager of a branch office of the bank wants to test the null hypothesis

H_0: The proportion of accounts qualifying for the "500 Club" is at most 40%.

The manager randomly selects 20 of the accounts at his branch and lets the test statistic T equal the number of accounts that qualify for the "500 Club."

(a) Formulate the appropriate alternative hypothesis.

(b) The test statistic has a binomial distribution with $n = 20$ and $p \leq .4$ when H_0 is true. Consult Table A1 with $p = .4$ and $n = 20$ to construct a decision rule having a level of significance of approximately .05 and state the exact level of significance associated with your decision rule.

(c) If the branch manager finds $T = 11$ of the 20 accounts qualify for the "500 Club," what decision is made regarding the null hypothesis?

(d) What is the p-value associated with the observed value of the test statistic $T = 11$? Interpret this p-value.

7.18 A political candidate believes that he is ahead of his opponent, but needs to have some proof of this to show potential contributors to his campaign. He takes a random sample of the registered voters, asking 200 people whom they intend to vote for in the coming election. After discarding 31 people with no opinion, there are 169 people left in the sample, and 98 of them intend to vote for him.

(a) What are H_0 and H_1 in terms of p, the proportion of all registered voters with an opinion, who intend to vote for this candidate?

(b) Let the test statistic T be the number of people in the random sample who intend to vote for this candidate. Under the null distribution, T has the binomial distribution with $n = 169$ and $p = .5$. Find the rejection region of size $\alpha = .05$, and state the decision rule.

(c) For the observed value of T, state the decision.

(d) Find the p-value.

7.19 A machine that fills cans with ground coffee is working properly if the standard deviation of the net weight in the cans is small. If the standard deviation gets large, the machine needs to be repaired. The production manager wants to repair the machine only when there is convincing evidence to indicate that the standard deviation is above 2.6, the normal operating value.

(a) State H_0 and H_1.

(b) Suppose the test statistic T has a chi-square distribution with 28 degrees of freedom when H_0 is true. Also suppose large values of T tend to indicate that H_1 is true. Find the decision rule for $\alpha = .10$.

(c) Suppose the observed value of T is 41.1. What is the decision?

(d) Find the approximate p-value.

7.20 Suppose the test statistic T is normally distributed with mean μ and standard deviation 1. Further suppose that when H_1 is true μ equals 3. Find the power of a test that has $T > 1.6449$ as its rejection region.

7.21 If the sampling distribution of T is the binomial distribution with $n = 30$ and $p = .75$ when H_1 is true, and if the decision rule is to reject H_0 if $T > 22$, what is the power?

7.22 If the test statistic T has a binomial distribution with parameters $n = 15$ and p, where $p = \frac{1}{2}$ when H_0 is true and $p = .8$ when H_1 is true, find the power of the test with the two-tailed rejection region $T < 4$ and $T > 11$.

7.23 The sampling distribution of T is the normal distribution with mean -2 and standard deviation 1 when H_1 is true. Find the power for the two-tailed test that has as its decision rule, "reject H_0 if $T < -1.96$ or if $T > 1.96$."

7.3
THE BINOMIAL TEST, ACCEPTANCE SAMPLING, AND QUALITY-CONTROL CHARTS

PROBLEM SETTING 7.2

The American Institute of Certified Public Accountants (AICPA) was concerned when it learned that only 40% of the small accounting member firms were using sampling techniques in their audits. To remedy this situation it issued a pamphlet emphasizing the importance of using sampling methods in audits, and offered short courses to its members on how to use sampling methods in audits.

Two years later the AICPA took a small survey to see if the percentage of the small accounting member firms using sampling techniques in their audits had increased from the 40% figure of two years earlier. Twelve small firms were obtained in a random sample of all of its small-firm members. Eight of those 12 firms indicated that they were using sampling techniques in their audits. Does this represent a change from the historical 40% figure, or is this increase in percentage attributable to sampling error?

TESTING PROCEDURE

The problem setting is concerned with testing a hypothesis about the true population proportion p. In this case, p represents the proportion of all small accounting firms that use sampling techniques in their audits. Some other situations where the population proportion is the central issue include the following.

1. A hardware store has counted on only 25% of its customers being women, but it would like to appeal more to this market. The store undergoes a remodeling project, followed by a change in advertising strategy. Then it observes that in a random sample of its customers, 32 out of 78, or 41%, are women. Is this conclusive evidence to indicate that the percentage of female customers has increased?

2. A graduate school has had 40% of its students coming from outside of the state. This year the tuition rate is increased for all students, with the increase being greater for out-of-state students. Of the 86 new graduate students this year, only 26, or 30.2%, are from out of state. Does this offer conclusive evidence of a decrease in the percentage of graduate students coming from out of state?

In each of these situations the sample proportion can be used to see if the population proportion has changed. The only requirement in this procedure is that the sample be a random sample. That is, the items being examined such as classification of customers as male or female, or incoming graduate students being in-state or out-of-state, need to be independent of one another. Additionally, these items need to have the same probability of being in the category of interest such as female customer, or out-of-state student, as the entire population.

THE TEST STATISTIC

If the sample is random, then the random variable T, representing the number of observations in the category of interest, has a binomial distribution with parameters p and n, the sample size, and serves as a natural test statistic. It is sensitive to changes in p and furnishes a test with good power.

The null hypothesis is

$$H_0: p = p_0$$

where p_0 is some specified number. It is tested using

$$T = \text{Number of observations in the category of interest}$$

as the test statistic. The alternative hypothesis

$$H_1: p > p_0$$

leads to an upper-tailed test that rejects H_0 if T is greater than the $1 - \alpha$ quantile T_U obtained from Table A1 for the binomial distribution with parameters n and p_0. The alternative hypothesis

$$H_1: p < p_0$$

leads to a lower-tailed test that rejects H_0 if T is less than the α quantile T_L obtained from the same table. The alternative hypothesis

$$H_1: p \neq p_0$$

leads to a two-tailed test that uses both tails of the binomial distribution as the rejection region. The lower and upper critical values are chosen as close as possible to the $\alpha/2$ and $1 - \alpha/2$ quantiles, respectively. Because the binomial distribution is discrete, the actual level of significance, α, may differ slightly from the targeted level of significance, such as .05, or .10.

The normal approximation to the binomial distribution is used for n greater than 30 because the exact tables are given only for $n \leq 30$ in this text. In the

---

PROCEDURE FOR TESTING HYPOTHESES ABOUT THE POPULATION PROPORTION p (Binomial Test)

Data

The sample consists of n trials where each trial results in a success or a failure.

Assumptions

1. The trials are independent of one another.

2. The probability of success p remains the same from trial to trial.

Null Hypothesis

H_0: $p = p_0$, where p_0 is some specified number such that $0 \leq p_0 \leq 1$.

Test Statistic

T is the number of successes observed in n trials. The null distribution of T is the binomial distribution with parameters n and p_0.

Decision Rules

The decision rule depends on the alternative hypothesis. Let α be the level of significance.

(a) H_1: $p > p_0$. Reject H_0 if $T > T_U$. The value of T_U is found by using Table A1 with n and p_0. The critical value T_U is the value such that $P(T > T_U) = \alpha$, or equivalently $P(T \leq T_U) = 1 - \alpha$. For large sample sizes ($n > 30$) the value of T_U is found as $T_U = np_0 + z_{1-\alpha}\sqrt{np_0q_0}$, where $z_{1-\alpha}$ is the $1 - \alpha$ quantile from Table A2.

(b) H_1: $p < p_0$. Reject H_0 if $T < T_L$. The value of T_L is found by using Table A1 with n and p_0. The critical value T_L is the value such that $P(T < T_L) = \alpha$, or equivalently $P(T \leq T_L - 1) = \alpha$. For large sample sizes ($n > 30$) the value of T_L is found as $T_L = np_0 + z_\alpha\sqrt{np_0q_0}$, where z_α is the α quantile from Table A2.

(c) H_1: $p \neq p_0$. Reject H_0 if $T > T_U$ or if $T < T_L$. The values of T_U and T_L are found by using Table A1 with n and p_0. The critical value T_U is the closest value such that $P(T > T_U) = \alpha/2$, or equivalently $P(T \leq T_U) = 1 - \alpha/2$, and T_L is the closest value such that $P(T < T_L) = \alpha/2$, or equivalently $P(T \leq T_L - 1) = \alpha/2$. The actual level of significance is the sum of the two tail probabilities:

$$\alpha = P(T < T_L) + P(T > T_U)$$

For large sample sizes ($n > 30$) the values of T_U and T_L are given as $T_U = np_0 + z_{1-\alpha/2}\sqrt{np_0q_0}$ and $T_L = np_0 + z_{\alpha/2}\sqrt{np_0q_0}$, where $z_{\alpha/2}$ and $z_{1-\alpha/2}$ are the respective $\alpha/2$ and $1 - \alpha/2$ quantiles from Table A2.

Warning

If $np_0 < 5$ or $nq_0 < 5$, the large-sample approximation is not very accurate.

normal approximation, normal quantiles z_α from Table A2 are substituted into the equation,

$$x_\alpha = np_0 + z_\alpha \sqrt{np_0(1 - p_0)}$$

to obtain approximate quantiles x_α for the binomial distribution.

THE p-VALUE

Once the sample is observed and T is determined for that sample, the p-value can be found from the null distribution of T. The binomial distribution in Table A1, for the sample size n and the value of p_0 given in the null hypothesis, is the null distribution. Let t_{obs} be the observed value of T. For an upper-tailed test the p-value is $P(T \geq t_{obs})$, which is found using the relationship

$$
\begin{aligned}
p\text{-value} &= P(T \geq t_{obs}) \\
&= 1 - P(T < t_{obs}) \\
&= 1 - P(T \leq t_{obs} - 1)
\end{aligned}
$$

since the tables are set up to yield cumulative probabilities. Therefore $x = t_{obs} - 1$ is used in Table A1 and substituted into the preceding relationship to get the p-value for an upper-tailed test.

A lower-tailed test's p-value is simply $P(T \leq t_{obs})$, which may be obtained directly from Table A1 by letting $x = t_{obs}$. For a two-tailed test the p-value is found by doubling the smaller of the two one-tailed p-values as long as the result is less than 1. If the result is still larger than 1, the p-value simply equals 1.

For $n > 30$ the normal approximation, as described in Section 5.3, is used to find p-values. That is, the p-value in an upper-tailed test is found using Table A2* and the approximation

$$P(T \geq t_{obs}) \approx 1 - P(Z \leq z_1)$$

where

$$z_1 = \frac{t_{obs} - np_0 - .5}{\sqrt{np_0(1 - p_0)}}$$

The p-value in a lower-tailed test is approximated using

$$P(T \leq t_{obs}) \approx P(Z \leq z_2)$$

where

$$z_2 = \frac{t_{obs} - np_0 + .5}{\sqrt{np_0(1 - p_0)}}$$

and the p-value in a two-tailed test is approximately twice the smaller of the one-sided p-values.

EXAMPLE

In Problem Setting 7.2, $n = 12$ small accounting firms are used to test H_0: $p = .4$ versus H_1: $p > .4$. The approximate level of significance desired is $\alpha = .05$. Therefore, the decision rule is to reject H_0 if $T > T_U$, where T_U is obtained from Table A1 for $n = 12$ and $p = .40$.

To find T_U use Table A1 with $n = 12$ and $p = .40$. Read down the column of cumulative probabilities to the value closest to .95. This is the cumulative probability .9427, which is in the row $x = 7$, so $T_U = 7$. The exact probability associated with the upper tail is

$$
\begin{aligned}
P(T > T_U) &= P(T > 7) \\
&= 1 - P(T \leq 7) \\
&= 1 - .9427 \\
&= .0573
\end{aligned}
$$

Thus the exact level of significance is .0573 rather than the target level .05. Figure 7.6 illustrates that target levels of significance are usually not attainable when the distribution of the test statistic is discrete as in the binomial test.

The random sample results in $T = 8$, or 8 out of 12 firms using sampling techniques in their audits, which is greater than $T_U = 7$, so the null hypothesis is rejected.

Since $t_{obs} = 8$, the p-value is $P(T \geq 8)$, where the probability is found using the null distribution of T, the binomial distribution with $n = 12$ and $p = .4$.

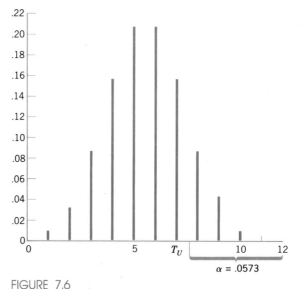

FIGURE 7.6
THE BINOMIAL DISTRIBUTION WITH $n = 12$ AND $p = .4$, SHOWING THE UPPER-TAILED REJECTION REGION $T > 7$ AND $\alpha = P(T > 7) = .0573$.

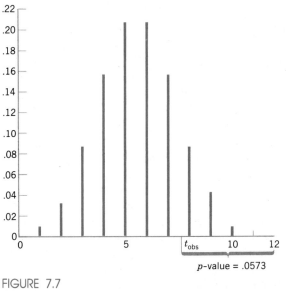

FIGURE 7.7
THE BINOMIAL DISTRIBUTION WITH $n = 12$ AND
$p = .4$, SHOWING THE OBSERVED VALUE $T = 8$
AND THE UPPER-TAILED p-VALUE $P(T \geq 8) =$
.0573.

From Table A1 this probability is given as

$$P(T \geq 8) = 1 - P(T \leq 7)$$
$$= 1 - .9427$$
$$= .0573$$

See Figure 7.7 for an illustration of the p-value.

Note that the p-value equals α in this case because the observed value of T barely fell into the rejection region. If the observed value of T had been 9 instead of 8, the p-value would have been

$$P(T \geq 9) = 1 - P(T \leq 8) = 1 - .9847 = .0153$$

instead of .0573.

THE NORMAL APPROXIMATION FOR T

The procedure for using the normal approximation for the distribution of T is straightforward. If used in the previous example, the normal approximation results in the decision rule

$$\text{Reject } H_0 \text{ if } T > T_U$$

where

$$T_U = np_0 + z_{.95}\sqrt{np_0q_0}$$
$$= 12(.4) + 1.6449\sqrt{12(.4)(.6)}$$
$$= 4.80 + 2.79$$
$$= 7.59$$

Thus the decision rule would be

Reject H_0 if $T > 7.59$

which has an approximate level of significance of $\alpha = .05$. The decision for $T = 8$ is to reject H_0. Of course, the exact level of significance associated with these numbers was found earlier, $\alpha = .0573$, from the binomial distribution tables. This illustrates the fact that exact tables should always be used when they are available.

The p-value associated with $X = 8$ may be approximated by using the normal distribution and Table A2*.

$$P(T \geq 8) \approx 1 - P\left(Z \leq \frac{8 - 12(.4) - .5}{\sqrt{12(.4)(.6)}}\right)$$
$$= 1 - P(Z \leq 1.59)$$
$$= 1 - .9441$$
$$= .0559$$

The approximate tail probability .0559 is reasonably close to the exact p-value of .0573 given in the example.

EXAMPLE

A hardware store has recently remodeled, and changed its advertising strategy in order to increase its percentage of female customers from the traditional 25% that has been true in the past. The hypotheses are

$$H_0: p = .25$$
$$H_1: p > .25$$

where p is the proportion of customers that are female.

A random sample of 78 customers is observed. The decision rule is to reject H_0 at $\alpha = .05$ if T, the number of female customers in the sample, exceeds $T_U = 25.8$. This decision rule was obtained from the normal approximation for the .95 quantile in the null distribution of T.

$$T_U = np_0 + z_{.95}\sqrt{np_0(1 - p_0)}$$
$$= 78(.25) + 1.6449\sqrt{78(.25)(.75)}$$
$$= 25.8$$

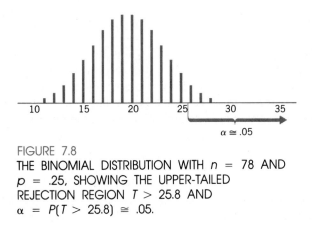

FIGURE 7.8
THE BINOMIAL DISTRIBUTION WITH $n = 78$ AND
$p = .25$, SHOWING THE UPPER-TAILED
REJECTION REGION $T > 25.8$ AND
$\alpha = P(T > 25.8) \cong .05$.

The rejection region and α are illustrated in Figure 7.8.

There are 32 female customers in the sample, so the null hypothesis is rejected and the conclusion is that the proportion of customers that are female is larger than 25%, the former proportion.

The p-value is found using the normal approximation

$$P(T \geq 32) \approx 1 - P\left(Z \leq \frac{32 - 78(.25) - .5}{\sqrt{78(.25)(.75)}}\right)$$

$$= 1 - P(Z \leq 3.14)$$

$$= 1 - .999$$

$$= .001$$

Actually, since 3.14 is larger than the .999 quantile, 3.09, in Table A2*, the p-value is less than .001, providing strong evidence that the proportion of female customers is now larger than 25%. See Figure 7.9 for an illustration of the p-value.

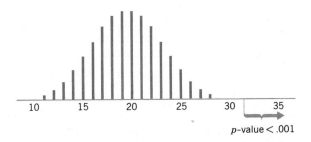

FIGURE 7.9
THE BINOMIAL DISTRIBUTION WITH $n = 78$ and
$p = .25$, SHOWING THE OBSERVED VALUE $T = 32$ AND THE UPPER-TAILED p-VALUE
$P(T \geq 32) < .001$.

POWER

The power for the binomial test is found as follows. First the rejection region must be known. Next the alternative value of p under H_1 must be known. Since many values of p are usually possible under H_1, one of these values must be selected. Then the power is the probability of getting a value of T in the rejection region, using the selected value of p in the binomial distribution.

FINDING THE POWER OF THE BINOMIAL TEST

The power of the binomial test with level of significance α, is found as follows. Let p_0 be the population proportion as stated in H_0 and let p_1 be the population proportion when H_1 is true.

(a) Upper-tailed test Let T_U denote the critical value. For $n \leq 30$,

$$\text{Power} = P(X > T_U) = 1 - P(X \leq T_U)$$

from Table A1 using n and p_1.

For $n > 30$, replace the decimal portion of T_U with .5. (For example, $T_U = 25.8$ becomes $T_U = 25.5$.) Then use Table A2* to find

$$\text{Power} = P\left(Z > \frac{T_U - np_1}{\sqrt{np_1(1 - p_1)}}\right)$$

(b) Lower-tailed test Let T_L denote the critical value. For $n \leq 30$,

$$\text{Power} = P(X < T_L) = P(X \leq T_L - 1)$$

from Table A1 using n and p_1.

For $n > 30$, replace the decimal portion of T_L with .5. (For example, $T_L = 25.8$ becomes $T_L = 25.5$.) Then use Table A2* to find

$$\text{Power} = P\left(Z \leq \frac{T_L - np_1}{\sqrt{np_1(1 - p_1)}}\right)$$

(c) Two-tailed test Let T_U and T_L denote the critical values. For $n \leq 30$,

$$\text{Power} = P(X > T_U) + P(X < T_L)$$
$$= 1 - P(X \leq T_U) + P(X \leq T_L - 1)$$

from Table A1 using n and p_1.

For $n > 30$, replace the decimal portion of both T_U and T_L with .5. (For example, $T_U = 25.8$ becomes $T_U = 25.5$.) Then use Table A2* to find

$$\text{Power} = P\left(Z > \frac{T_U - np_1}{\sqrt{np_1(1 - p_1)}}\right) + P\left(Z \leq \frac{T_L - np_1}{\sqrt{np_1(1 - p_1)}}\right)$$

EXAMPLE

In the previous example suppose the true percentage of female customers is now 50%. What is the probability that the binomial test will detect this amount of change? (That is, what is the power?)

The rejection region is $T > 25.8$. The power is found using the normal approximation to the binomial distribution with parameters $n = 78$ and $p = .5$.

$$\text{Power} = P(T > 25.8 | n = 78, p = .5)$$
$$= P(T \geq 26) \quad \text{(due to the discrete nature of } T)$$
$$= 1 - P\left(Z \leq \frac{26 - 78(.5) - .5}{\sqrt{78(.5)(.5)}}\right)$$
$$= 1 - P(Z \leq -3.06)$$
$$= 1 - .0011$$
$$= .9989$$

See Figure 7.10 for an illustration of the power. This high power suggests the test is a good one if the true percentage of female customers is as high as 50%.

QUALITY MONITORING

Attention is now focused on two techniques that are useful for monitoring the quality of a product when that quality is measured by a proportion p, such as the proportion of defective items produced. The rationale behind quality monitoring is quite simple: If the manufacturing process gets out of control and starts producing defective items, then it will be to the company's advantage to catch it as soon as possible. Otherwise, they face the prospect of replacing defective items or of losing their customers.

For manufacturing processes that are geared to mass production the quantity of the product manufactured makes inspection of every detail in every item impossible. Therefore a sampling procedure is needed that will have the capability of quickly detecting manufacturing irregularities. It is also used by a buyer to be able to determine whether or not to accept a shipment without testing every item in the shipment.

ACCEPTANCE SAMPLING

The first method of monitoring quality is known as acceptance sampling. Suppose that a manufacturing process is set up so that at different time periods a random sample of size n can be obtained. Each of the sample items is inspected for defects. The manufacturing process is declared out of control if more than k defectives are observed. Otherwise the process is accepted as

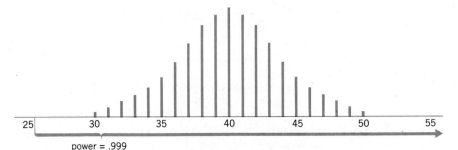

power = .999

FIGURE 7.10
THE BINOMIAL DISTRIBUTION WITH $n = 78$ AND $p = .50$, SHOWING THE POWER ASSOCIATED WITH THE UPPER-TAILED REJECTION REGION $T > 25.8$, POWER $= P(T > 25.8) = .999$.

satisfactory. In this setting the probability of detecting a faulty manufacturing process, as the population proportion p of defective products increases, is of interest.

The random variable is T, the number of defective items in a random sample of size n. The manufacturing process is accepted as being in control if $T \le k$, and is regarded as being out of control if $T > k$. From Table A1 the probabilities $P(T \le k)$ for different values of p, n, and k are easily found. Figure 7.11 presents values of $P(T \le k)$ for $k = 0$ and $n = 5$, as p varies from 0 to .90.

The values in Figure 7.11 can be used as x and y coordinates in a graph that is referred to as an **operating characteristic (OC) curve.** The OC curve is represented in Figure 7.12 and can be used to find the probability of accepting the process as being satisfactory, for any value of p. Note this curve is plotted for $n = 5$ and an acceptance rule based on $T = 0$, but the curve is easily constructed for any n and any acceptance rule based on $T \le k$.

Acceptance sampling is closely related to hypothesis testing. The null hypothesis is, "The process is in control" or "The shipment of items is satis-

Population Proportion of Defectives p	$P(T = 0)$
0.00	1.0000
0.05	.7738
0.10	.5905
0.15	.4437
0.20	.3277
0.25	.2373
0.30	.1681
0.35	.1160
0.40	.0778
0.45	.0503
0.50	.0312
0.55	.0185
0.60	.0102
0.65	.0053
0.70	.0024
0.75	.0010
0.80	.0003
0.85	.0001
0.90	.0000

FIGURE 7.11
THE PROBABILITY THAT A SAMPLE OF SIZE 5 CONTAINS NO DEFECTIVE ITEMS, AS A FUNCTION OF THE POPULATION PROPORTION OF DEFECTIVE ITEMS.

(e) State the decision.

(f) Find the p-value.

7.25 A random sample of eight automobile drivers showed that only two of the drivers were wearing their seat belts. Is this sufficient evidence to reject the null hypothesis that at least 60% of all drivers are wearing their seat belts (i.e., H_0: $p \geq .6$ is to be tested versus H_1: $p < .6$) if a level of significance of $\alpha = .05$ is used? State the p-value and interpret it.

7.26 A random sample of 20 businesspeople showed that 15 favored a Kodak copy machine over a Xerox. If there is really no difference between the machines, then each has a chance of $\frac{1}{2}$ of being selected as the best when one must be selected. Let the hypotheses to be tested be expressed as follows:

$$H_0: p = .5 \quad \text{versus} \quad H_1: p \neq .5$$

The random variable T represents the number of business persons selecting the Kodak machine. Answer the questions below based on the following decision rule: Reject H_0 if $T < 6$ or if $T > 14$.

(a) What is the exact level of significance associated with this decision rule?

(b) What decision is made based on the sample data?

(c) Find the p-value and interpret its meaning.

(d) Assuming the true proportion of business persons selecting a Kodak copier is .8, find the power associated with this test.

7.27 A national magazine says that 70% of the owners of new Plymouths are satisfied with their purchase. To see if this is true for Davis, California, a new-car dealer obtained a random sample of 10 recent customers to see if the population proportion is different from .7. That is, H_0: $p = .7$ is tested versus H_1: $p \neq .7$. The following decision rule is used: Reject H_0 if $T < 5$ or $T > 9$, where the random variable T is the number of successes in 10 trials.

(a) What is the exact level of significance for this test?

(b) What is the power of this test if the true value of p is .5?

7.28 Fourteen out of 88 people interviewed in a random survey said they preferred the news on Channel 13. Use a level of significance of .10 to test the hypothesis that exactly one-third of the population prefers the news on Channel 13 against the two-sided alternative. State the p-value and interpret it.

7.29 An authority on drug abuse, appearing on a local TV talk show, claims that at least 78% of college students are either using or have tried drugs of some kind. After the show has aired, a newspaper takes a random survey at a local university and finds that 64 out of the 92 students questioned said they had either tried or were currently using drugs. Based on this evidence, can the newspaper refute the authority's claim if they are willing to take a 5% chance of wrongly refuting his claim? State the p-value and interpret it.

7.30 Suppose that normally the percentage of persons with high blood pressure is at most 10%, but that for people over 40 years old it is suspected to be higher. A random sample of 50 individuals over 40 indicates that 10 have high blood pressure. Test the appropriate hypothesis using a level of significance of $\alpha = .05$. State the p-value and interpret it.

7.31 A manufacturing process is monitored by periodically taking random samples of size 10 and observing the number of defective items. The manufacturing process is said to be in control if $T \leq 1$, where T is the number of defective items observed in the sample. Construct an operating characteristic curve for this process similar to the one given in Figure 7.12.

7.32 A machine producing typing elements for electric typewriters has been observed over a period of time to produce defective parts 3% of the time. Random samples of size 50 are obtained periodically and inspected for defective items. Construct a 99% quality-control chart for this process. If a sample of size 50 shows five defective items, would the manufacturing process be considered to be out of control?

7.33 Suppose it is desired to test H_0: $p = .7$ versus H_1: $p = .8$ for the population proportion. A random sample of size 30 is taken from the population and the null hypothesis is to be rejected if the number of successes is greater than 23.

(a) What is the significance level for this test?

(b) What is the power of this test?

7.34 The operating characteristic curve in Figure 7.12 was constructed based on random samples of size 5 and acceptance only if $T = 0$. What effect will increasing the sample size to 10 have on the curve if the acceptance remains at $T = 0$? What effect will changing the acceptance to $T \leq 1$ have on the curve if the sample size remains at 5?

7.35 A manufacturing process is known to produce 4% defective pieces. The quality-control chart for the process has the upper control line located at $p + 2.3263\sqrt{pq/n}$ based on samples of size 100. Make a graph of the control chart for this process. Random samples of size 100 are taken at 10 successive time periods and show the following number of defective pieces in each sample: 3, 2, 6, 4, 3, 5, 7, 5, 1, and 4. Add these points to the quality-control chart. Does the process appear to be in control?

7.4

TESTING HYPOTHESES ABOUT THE POPULATION MEAN AND STANDARD DEVIATION

PROBLEM SETTING 7.3

Wholesalers of French fried potatoes to fast-food service chains operate in a very competitive market. The demand for the product is high, and the competition is very keen. To get and maintain a contract with a fast-food service chain requires that the product be of high quality. As potatoes age the sugar content increases and this, in turn, decreases the eating quality of the potato. This means that the wholesaler must buy quality brands of potatoes with low sugar content.

Shipments of potatoes are received daily by these wholesalers, who must decide whether a particular shipment is acceptable by sampling the potatoes and measuring the mean percent content of the potatoes. A random sample of 13 potatoes from one shipment provides the following percentages of sugar content: 15.22, 12.36, 15.44, 13.53, 15.94, 17.06, 16.90, 14.31, 19.06, 18.23, 15.54, 16.28, 20.00. The mean of this sample is 16.14. On the basis of this sample the wholesaler needs to decide whether the entire shipment is acceptable or not.

A TEST FOR THE MEAN

The problem setting suggests a hypothesis test regarding the mean percent sugar content of shipments of potatoes. If the sample mean percent sugar content is determined by averaging many readings, the sampling distribution of the sample mean is approximately normal by the central limit theorem, as discussed in Section 5.4. Because the distribution of the sample mean is approximately normal for large samples, the standardized sample mean can be used as a test statistic for *all* populations when the sample sizes are large. This test is called the Z-test because the test statistic is compared with quantiles z_p from the standard normal distribution.

The sample mean can also be used as a test statistic when sample sizes are small if the samples come from normal populations where σ is known. In this case, the sampling distribution of the sample mean is exactly normal for all sample sizes. If σ is not known, a different procedure is used. It is described later in this section.

EXAMPLE

The potato wholesaler referred to in Problem Setting 7.3 needs a one-tailed test that rejects a shipment of potatoes when the mean percent sugar content is judged to be unacceptably high. Previous experience shows that a sugar content above 15% is unacceptable, and since the wholesaler needs to detect this situation when it exists in the potatoes, this becomes the alternative hypothesis. Previous experience also shows that the population standard deviation of sugar content is $\sigma = 2.6\%$. The hypotheses are H_0: $\mu \leq 15$ versus H_1: $\mu > 15$.

Because the sample size is small, $n = 13$, the sample data must be checked for normality using the Lilliefors test. The reader should verify that these data easily pass the Lilliefors test.

As stated in Problem Setting 7.3, the mean percent sugar content for this sample is 16.14. The value of the test statistic is

$$T_1 = \frac{16.14 - 15}{2.6/\sqrt{13}} = 1.5809$$

PROCEDURE FOR TESTING HYPOTHESES ABOUT THE POPULATION MEAN (One-sample Z-Test)

Data

The data are represented by $X_1, X_2, \ldots, X_n$ with mean $\overline{X}$ and standard deviation s.

Assumptions

1. The observations $X_1, X_2, \ldots, X_n$ represent a random sample of size n from some population.
2. Either the sample size is large ($n > 30$) or the population is approximately normal with a known value for σ, the population standard deviation.

Null Hypothesis

$H_0: \mu = \mu_0$, where μ_0 is some specified number. Note that this null hypothesis is equivalent to using $\mu \leq \mu_0$ or $\mu \geq \mu_0$, respectively, for the one-sided alternatives (a) or (b) below.

Test Statistic

$$T_1 = \frac{\overline{X} - \mu_0}{\sigma/\sqrt{n}} = \frac{\sqrt{n}\,(\overline{X} - \mu_0)}{\sigma} \qquad (7.1)$$

For large samples, $n > 30$, the sample standard deviation s may be used in place of σ when σ is unknown. The null distribution of T_1 is approximated by the standard normal distribution.

Decision Rules

The decision rule depends on the alternative hypothesis. Let z_p represent the pth quantile of a standard normal distribution obtained from Table A2. Let α be the level of significance.

(a) $H_1: \mu > \mu_0$. Reject H_0 if $T_1 > z_{1-\alpha}$.

(b) $H_1: \mu < \mu_0$. Reject H_0 if $T_1 < z_\alpha$.

(c) $H_1: \mu \neq \mu_0$. Reject H_0 if $T_1 < z_{\alpha/2}$ or if $T_1 > z_{1-\alpha/2}$.

Some commonly used critical values are: $z_{.90} = 1.2816$, $z_{.95} = 1.6449$, $z_{.975} = 1.9600$, $z_{.99} = 2.3263$, $z_{.995} = 2.5758$. By symmetry, $z_\alpha = -z_{1-\alpha}$.

If the wholesaler uses $\alpha = .05$, the decision rule is

$$\text{Reject } H_0 \text{ if } T_1 > 1.6449$$

since $z_{.95} = 1.6449$.

The null hypothesis is accepted and the shipment is declared to be satisfactory. However, it should be kept in mind that the sample mean is 16.14%, above the 15% level of acceptability. This suggests that perhaps H_0 is accepted only because the sample size is too small to detect a difference, if there is a difference.

To find the p-value, consider that the null distribution of T_1 is the standard normal distribution, the test is upper-tailed, and the observed value of T_1 is

1.5809. Therefore the p-value is found from Table A2* as follows:

$$P(T_1 > 1.5809) = 1 - P(T_1 \leq 1.5809)$$
$$= 1 - .9429$$
$$= .0571$$

Thus the p-value is .0571, which indicates that the null hypothesis was almost rejected.

SAMPLES FROM NORMAL POPULATIONS WITH σ UNKNOWN

In almost all cases the population standard deviation σ is not known. If a sample is large, then the previous procedure can be used, because s can be used to approximate σ in large samples. But a better method is available for samples from normal populations. This procedure uses the Student's t distribution and is exact. It applies to samples of all sizes from normal populations.

PROCEDURE FOR TESTING HYPOTHESES ABOUT THE MEAN OF A NORMAL DISTRIBUTION WHEN σ IS UNKNOWN
(One-sample t-Test)

Data

The data are represented by $X_1, X_2, \ldots, X_n$ with mean $\overline{X}$ and standard deviation s.

Assumptions

1. The observations $X_1, X_2, \ldots, X_n$ represent a random sample of size n.
2. The sample comes from a normal population. (This assumption should always be verified using the Lilliefors test of Section 5.3. If this assumption is not satisfied but n is greater than 30, the method presented earlier in this section can be used, or the method given in Section 7.5 for testing the median in nonnormal populations can be used.)

Null Hypothesis

H_0: $\mu = \mu_0$, where μ_0 is some specified number.

Test Statistic

$$T_2 = \frac{\overline{X} - \mu_0}{s/\sqrt{n}} = \frac{\sqrt{n}(\overline{X} - \mu_0)}{s} \qquad (7.2)$$

The null distribution of T_2 is Student's t distribution with $n - 1$ degrees of freedom.

Decision Rules

Let $t_{1-\alpha,n-1}$ represent the $1 - \alpha$ quantile of a Student's t distribution with $n - 1$ degrees of freedom obtained from Table A3 for a given level of significance α. The decision rule depends on the alternative hypothesis.

(a) H_1: $\mu > \mu_0$. Reject H_0 if $T_2 > t_{1-\alpha,n-1}$.
(b) H_1: $\mu < \mu_0$. Reject H_0 if $T_2 < -t_{1-\alpha,n-1}$.
(c) H_1: $\mu \neq \mu_0$. Reject H_0 if $T_2 < -t_{1-\alpha/2,n-1}$ or $T_2 > t_{1-\alpha/2,n-1}$.

This set of rules is very similar to those given before for normal populations with σ known, the difference being the use of s in the calculation of the test statistic and the use of Student's t distribution to find critical values for a prespecified value of α.

In summary, when the population is normal the statistic T_1 has normal distribution, whereas the statistic T_2 has a Student's t distribution. However, as the sample size gets large, the difference between the two distributions becomes negligible. This fact can be observed by referring to the last line in Table A3 where it can be seen that the t distribution with large degrees of freedom is virtually the same as a normal distribution.

The previous example will now be reworked using the sample standard deviation s instead of σ.

EXAMPLE

In the previous example the population standard deviation was given as 2.6. Had this value not been known the sample data would be used to provide an estimate for σ. The sample standard deviation is $s = 2.15$. For the previous choice of $\alpha = .05$ and hypotheses H_0: $\mu \leq 15$ versus H_1: $\mu > 15$ the test statistic is

$$T_2 = \frac{16.14 - 15}{2.15/\sqrt{13}} = 1.9118$$

and the decision rule is reject H_0 if $T_2 \geq t_{.95,12} = 1.7823$. Clearly the decision is to reject H_0 and the shipment is declared to be unsatisfactory. The p-value found by interpolating in Table A3 with 12 degrees of freedom is .042. This indicates that the null hypothesis was barely rejected at $\alpha = .05$.

THE EFFECT OF ESTIMATING σ

It would be very reasonable to ask why there is a difference in the conclusions in the preceding examples when the same data are used in both cases. The answer is quite simply that the statistics T_1 and T_2 have different distributions and, therefore, each answer is correct for its respective setting. To explore this question one step further, suppose in the second example that the sample standard deviation turned out to be 2.6, then the value of T_2 would be changed to 1.5809, which is the same value that was calculated for T_1. However, this new value of T_2 would still be compared against the Student's t distribution value of 1.7823, whereas T_1 would be compared against 1.6449. Therefore, since the denominators $2.6/\sqrt{13}$ are now the same for both T_1 and T_2, the difference $\overline{X} - \mu_0$ must be larger for T_2 to detect a significant difference than it is for T_1.

That is, if the standard deviation is estimated, then the sample mean $\overline{X}$ must deviate more from the hypothesized value μ_0 than is the case when σ is known, in order to be significant. This phenomenon might be thought of as a penalty (or a safety measure) built into the test when the standard deviation has to be estimated, because of the additional variability introduced by esti-mating σ.

FINDING THE POWER

Recall that the power is the probability of rejecting H_0 when H_0 is false. Computing the power depends on knowing the rejection region and the sampling distribution of the test statistic when H_0 is false. In the first example of this section the rejection region was $T_1 > 1.6449$. The sampling distribution of T_1 when H_0 is false must be found in order to compute the power.

First, when H_0 is false, many alternative values of μ are possible, as long as they exceed 15 as stated in H_1. For the sake of illustration, assume that the true value of μ is 16.5. What is the sampling distribution of T_1 under this assumption?

Recall that the sampling distribution of

$$T_1 = \frac{\overline{X} - 15}{2.6/\sqrt{13}}$$

is the standard normal distribution when $\mu = 15$. When $\mu = 16.5$ the sampling distribution of T_1 is still normal, with standard error $= 1$, but the mean of T_1 has shifted to 2.0801. This is because T_1 is no longer a standard normal random variable, as it was when H_0 was true, but is a standard normal random variable plus a constant, which is shown as follows:

$$T_1 = \frac{\overline{X} - 15}{2.6/\sqrt{13}} = \frac{\overline{X} - 16.5 + 16.5 - 15}{2.6/\sqrt{13}} = \frac{\overline{X} - 16.5}{2.6/\sqrt{13}} + \frac{16.5 - 15}{2.6/\sqrt{13}}$$

$$= Z + 2.0801$$

The power, which is the probability of $T_1 > 1.6449$ when $\mu = 16.5$, is now merely the probability of a normal random variable, with mean 2.0801 and standard deviation 1.0, exceeding 1.6449. This probability is found as follows, with the aid of Table A2*:

$$
\begin{aligned}
\text{Power} &= P(T_1 > 1.6449 | \mu = 16.5) \\
&= P(Z + 2.0801 > 1.6449) \\
&= P(Z > 1.6449 - 2.0801) \\
&= P(Z > -.4352) \\
&= 1 - P(Z \leq -.44) \\
&= 1 - .3300 \\
&= .6700
\end{aligned}
$$

Therefore, if the new shipment has a population mean percent sugar content of 16.5, there is a 67% chance that the shipment will be correctly refused when a sample of size 13 and a level of significance of $\alpha = .05$ are used. Of course, these calculations are valid only for $n = 13$ and $\mu = 16.5$, but clearly similar calculations could be done for any combination of n and μ. To show how the power increases as the sample size increases, consider what happens as the sample size increases from 13 to 50, for example. The test statistic is now

$$T_1 = \frac{\overline{X} - 15}{2.6/\sqrt{50}}$$

which has the standard normal distribution if $\mu = 15$. However, if $\mu = 16.5$, the sampling distribution of T_1 changes to a normal distribution with mean 4.0795.

$$T_1 = \frac{\overline{X} - 15}{2.6/\sqrt{50}} = \frac{\overline{X} - 16.5}{2.6/\sqrt{50}} + \frac{16.5 - 15}{2.6/\sqrt{50}} = Z + 4.0795$$

Now the power is calculated with the aid of Table A2* as follows:

$$
\begin{aligned}
\text{Power} &= P(T_1 > 1.6449 | \mu = 16.5) \\
&= P(Z + 4.0795 > 1.6449) \\
&= P(Z > 1.6449 - 4.0795) \\
&= P(Z > -2.4346) \\
&= 1 - P(Z \le -2.43) \\
&= 1 - .0075 \\
&= .9925
\end{aligned}
$$

Thus the power goes from only 67% for a sample of size 13 to over 99% for a sample of size 50. Looking at it another way, there is about one chance in three of making a Type II error when $n = 13$, but less than one chance in 100 of making a Type II error when $n = 50$, when the real mean percent sugar is 16.5.

FINDING THE POWER OF THE ONE-SAMPLE Z-TEST

The power of the one-sample Z-test, with level of significance α, is found from Table A2* as follows. Let μ_0 be the mean as stated in H_0 and let μ_1 be the mean when H_1 is true.

(a) Upper-tailed test

$$\text{Power} = P\left(Z > \frac{\mu_0 - \mu_1}{\sigma/\sqrt{n}} + z_{1-\alpha}\right)$$

(b) Lower-tailed test

$$\text{Power} = P\left(Z < \frac{\mu_0 - \mu_1}{\sigma/\sqrt{n}} + z_{\alpha}\right)$$

(c) Two-tailed test

$$\text{Power} = P\left(Z > \frac{\mu_0 - \mu_1}{\sigma/\sqrt{n}} + z_{1-\alpha/2}\right) + P\left(Z < \frac{\mu_0 - \mu_1}{\sigma/\sqrt{n}} + z_{\alpha/2}\right)$$

A TEST FOR THE STANDARD DEVIATION

Although the primary focus in data analysis is often on the mean, or a location parameter, sometimes the standard deviation is the focus of attention. A test of the hypothesis that the standard deviation in a normal population equals some specified value, σ_0, is based on the chi-square distribution, just as the confidence interval for σ in Chapter 6 is based on the chi-square distribution. Tests concerning the standard deviation and the variance are equivalent, since one is the square root of the other. *This test is valid only if the population is normal, no matter what the sample size, so the sample should always be checked for normality using the Lilliefors test.*

PROCEDURE FOR TESTING HYPOTHESES ABOUT THE STANDARD DEVIATION OF A NORMAL POPULATION (Chi-square Test)

Data

The data are represented by $X_1, X_2, \ldots, X_n$ with sample variance s^2.

Assumptions

1. The observations $X_1, X_2, \ldots, X_n$ represent a random sample of size n.

2. The sample comes from a normal population.

Null Hypothesis

H_0: $\sigma = \sigma_0$, where σ_0 is some specified number.

Test Statistic

$$T_3 = \frac{(n-1)s^2}{\sigma_0^2} \tag{7.3}$$

where s^2 is the sample variance. The null distribution of T_3 is the chi-square distribution with $n - 1$ degrees of freedom.

Decision Rules

Let $\chi^2_{p,n-1}$ represent the p quantile of a chi-square distribution with $n - 1$ degrees of freedom obtained from Table A4. The decision rule depends on the alternative hypothesis and the level of significance α.

(a) H_1: $\sigma > \sigma_0$. Reject H_0 if $T_3 > \chi^2_{1-\alpha,n-1}$.

(b) H_1: $\sigma < \sigma_0$. Reject H_0 if $T_3 < \chi^2_{\alpha,n-1}$.

(c) H_1: $\sigma \neq \sigma_0$. Reject H_0 if $T_3 < \chi^2_{\alpha/2,n-1}$ or if $T_3 > \chi^2_{1-\alpha/2,n-1}$.

EXAMPLE

In the example presented in this section a sample consisting of 13 observations on percentage sugar content yielded a sample standard deviation of $s = 2.15$. Past history indicates that the population standard deviation has been 2.6. Does this sample standard deviation indicate a change in the historical value?

The null hypothesis H_0: $\sigma = 2.6$ is tested against the two-sided alternative H_1: $\sigma \neq 2.6$ using the test statistic T_3.

$$T_3 = \frac{(n-1)s^2}{\sigma_0^2} = \frac{12(2.15)^2}{(2.6)^2} = 8.206$$

The decision rule is to reject H_0 at $\alpha = .05$ if $T_3 < \chi^2_{.025,12} = 4.404$ or if $T_3 > \chi^2_{.975,12} = 23.34$. Therefore the null hypothesis is not rejected. The observed value of T_3 is 8.206, which is between the .10 quantile and the .25 quantile in the chi-square distribution with 12 degrees of freedom, so the p-value is between .20 and .50, since this is a two-tailed test.

EXERCISES

7.36 The quality control manager in a cannery will order the readjustment of all equipment involved in canning a particular food if the mean weight per can μ is less than specifications require. For cling peaches, the contents must be at least 16 oz. Previous testing has established a standard deviation of 0.5 oz per can. A random sample of 100 cans is to be selected for testing.

　(a) Formulate a decision rule for testing the null hypothesis

$$H_0: \mu \geq 16 \quad \text{versus} \quad H_1: \mu < 16$$

　　with a level of significance of $\alpha = .05$.

　(b) Based on your answer to (a), indicate whether the null hypothesis should be rejected or not for each of the following results. (Also, indicate for each whether or not the equipment will be readjusted.)

$$\overline{X} = 15.5 \text{ oz} \qquad \overline{X} = 15.95 \text{ oz}$$
$$\overline{X} = 16.1 \text{ oz} \qquad \overline{X} = 15.90 \text{ oz}$$

7.37 A random sample of laboratory mice is taken to see if the population mean weight is 25.0 g. The sample yields the following data: 24.2, 25.1, 23.0, 22.8, 24.5, 23.8. Assume normality.

　(a) Test the null hypothesis $H_0: \mu = 25$ versus $H_1: \mu \neq 25$ using a level of significance of $\alpha = .10$.

　(b) What is the p-value associated with the value of the test statistic? Interpret this p-value.

7.38 Replacement of paint on highways and streets represents a large investment of funds by state and local governments each year. A new, cheaper brand of paint is tested for durability after one month's time by means of reflectometer readings. For the new brand to be acceptable it must have a mean reflectometer reading of at least 19.6. The sample data based on 25 randomly selected readings show $\overline{X} = 19.8$ and $s = 1.5$, and are accepted as being normally distributed.

　(a) State the appropriate null and alternative hypotheses.

　(b) Test the null hypothesis in (a) with $\alpha = .01$.

　(c) State the p-value associated with the value of the test statistic and interpret it.

7.39 Ten high school seniors taking the ACT test received the following scores: 28, 26, 30, 24, 25, 29, 31, 26, 23, 27. Past ACT scores at their high school have shown the scores to be normally distributed with $\sigma = 4$ and $\mu = 25$. Test the null hypothesis $H_0: \mu = 25$ versus $H_1: \mu > 25$

with $\alpha = .05$. State the p-value associated with the value of the test statistic and interpret it.

7.40 A random sample of six recent graduates of a local business school, all accounting majors, revealed salaries $25,000, $21,000, $22,000, $21,000, $23,000, and $26,000 per year. Assume normality and test the null hypothesis H_0: $\mu \geq \$25,000$ versus H_1: $\mu < \$25,000$ with $\alpha = .01$. State and interpret the p-value.

7.41 A random sample of 200 employees at a large corporation showed their average age to be 42.8 years with $s = 6.89$ years. Test H_0: $\mu = 40$ versus H_1: $\mu > 40$ with $\alpha = 01$. State the p-value associated with the value of the test statistic and interpret it.

7.42 A survey of 40 senior citizens selected at random showed that they watched TV an average of 24 hr per week with a standard deviation of 10 hr. Test H_0: $\mu = 30$ versus H_1: $\mu < 30$ with $\alpha = .05$. State and interpret the p-value.

7.43 Truckloads of fill dirt arriving at a construction site are contracted to carry 6.3 cu yd of dirt per load. A random sample of 25 loads showed $\overline{X} = 6.0$ cu yd and $s = 1.5$ cu yd. Assuming these data were accepted as normal, test the null hypothesis H_0: $\mu = 6.3$ versus H_1: $\mu < 6.3$ with $\alpha = .05$. State the p-value associated with the value of the test statistic and interpret it.

7.44 The manager of a dry cleaning shop believes the average charge for cleaning is more than $1.80. A random sample of 100 orders shows $\overline{X} = \$1.85$ and $s = \$0.25$. Test H_0: $\mu = \$1.80$ versus H_1: $\mu > \$1.80$ with $\alpha = .05$. What is the probability that the null hypothesis will be rejected if the true mean is really $1.85; that is, what is the power associated with this test?

7.45 In Exercise 7.39 test H_0: $\sigma = 4$ versus H_1: $\sigma \neq 4$ with $\alpha = .05$.

7.46 In Exercise 7.40 test H_0: $\sigma = 1000$ versus H_1: $\sigma \neq 1000$ with $\alpha = .05$.

7.47 In Exercise 7.41 assume the population is normal and test H_0: $\sigma = 8$ versus H_1: $\sigma < 8$ with $\alpha = .01$.

7.48 In Exercise 7.42 assume the population is normal and test H_0: $\sigma = 8$ versus H_1: $\sigma \neq 8$ with $\alpha = .10$.

7.49 In Exercise 7.43 test H_0: $\sigma = 1$ versus H_1: $\sigma > 1$ with $\alpha = .10$.

7.50 In Exercise 7.44 assume normality and test H_0: $\sigma = \$0.30$ versus H_1: $\sigma < \$0.30$ with $\alpha = .01$.

7.5

TESTING HYPOTHESES ABOUT THE POPULATION MEDIAN (OPTIONAL)

THE SAMPLE MEDIAN

The sample mean is not always a meaningful way to summarize a set of sample data. For example, a few years ago a small town in Kansas reported an average family income of $50,000. A rich oilman lived in the town and his income was in the millions, whereas the other families in the town had an average income of about $10,000. This situation is typical of cases where the median rather than the mean should be used as a summary statistic.

In this section a procedure is given for testing hypotheses about the median of a population. This procedure can be used with all populations and all sample sizes, unlike the procedures for the population mean presented in the previous section, which are valid only for large samples or for samples from normal distributions.

AN INTUITIVE APPROACH

Prior to the presentation of the test procedure consider the real-number-line plots of four different samples in Figure 7.14. In each case the null hypothesis is that the median of the population is 15.

The hypothesized median value of 15 is marked on each of the real-number-line plots. Do you think 15 is a good population median for each one of these samples? Each of these samples will be examined one at a time in order to help answer this question.

Sample 1 All five sample observations are above the hypothesized median; therefore it would seem likely that the true median is greater than 15.

Sample 2 All five sample observations are below the hypothesized median; therefore it would seem likely that the true median is less than 15.

Sample 3 This sample is similar to sample 2 except that all five observations are closer to the hypothesized median. Should this make a difference? Not really, as all sample observations are still below 15. Being closer to the median could merely reflect a scale change, such as pennies to dollars, which should not affect the decision.

Sample 4 The decision on this sample becomes more difficult to make because some sample observations are above the hypothesized median and some are below. Samples like this one require a more formal statistical analysis to aid in the decision-making process.

These four examples suggest that a test statistic could be based on counting the number of sample observations on either side of the hypothesized median. All that is missing for the hypothesis test is a probability distribution for the test statistic that will allow the computation of the probability of Type I and Type II errors. The binomial distribution adapts easily to this situation.

Suppose the median grade point average (GPA) of students at your school

FIGURE 7.14
REAL-NUMBER-LINE PLOTS FOR HYPOTHETICAL SAMPLES.

is 2.613. If you were to obtain a random sample of 20 students, how would you expect these points to appear on a real-number-line plot with respect to 2.613? Would you be surprised if all 20 were either above or below 2.613? Of course, all 20 on one side or the other of 2.613 is a possibility if 2.613 is the true median, but this is not a very likely outcome (in fact, it is like tossing 20 coins in the air and having them show all heads when they land, or all tails).

What you should expect is 10 on either side of 2.613 or some kind of split not too far away from this. This model with its independent observations or trials naturally leads to a binomial random variable with parameter $p = .5$. That is to say, if the random variable X is the number of observations greater than or equal to the median, then the binomial distribution in Table A1 for $n = 20$ and $p = .5$ can be used to find critical values for the test statistic.

THE MEDIAN TEST

The test for the population median uses the number of observations greater than the hypothesized median as the test statistic. The exact distribution of this statistic when H_0 is true is the binomial distribution that can be obtained from Table A1 for $n \leq 30$. Critical values for the median test for $n \leq 20$ from Table A1 have been summarized in Figure 7.15 for ready reference. Further, critical values of the median test for $n \leq 50$ are given in Table A8. The normal approximation is recommended for larger sample sizes. Actually, the normal distribution provides a reasonable approximation for $n > 30$, but Table A8 was expanded to include values of $n \leq 50$ because of the frequent use of these sample sizes. The use of Table A8 is illustrated in the following example.

PROCEDURE FOR TESTING HYPOTHESES ABOUT THE MEDIAN OF A POPULATION (One-sample Median Test)

Data

The data are represented by $X_1, X_2, \ldots, X_n$.

Assumptions

The observations $X_1, X_2, \ldots, X_n$ represent a random sample of size n from some population.

Null Hypothesis

H_0: The population median is M, where M is a number specified in the context of the experiment.

Test Statistic

Let T_1 equal the number of observations in the sample that are *greater than or equal to* M. Let T_2 equal the number of observations in the sample that are *greater than* M. *Note:* If no sample observations are equal to M, then $T_1 = T_2$.

The null distribution of T_1 and T_2 is the binomial distribution with parameters $p = .5$ and n.

Decision Rules

The decision rule depends on the alternative hypothesis. Let α be the approximate level of significance desired.

(a) H_1: The population median is less than M. Reject H_0 if $T_1 < T_L$. The value of T_L is found by using Table A8 (see Figure 7.15). The critical value T_L is the closest value such that $P(T_1 < T_L) \approx \alpha$. For large sample sizes ($n > 50$) the value of T_L is found as $T_L = (n + z_\alpha\sqrt{n})/2$, where z_α is the α quantile from Table A2.

(b) H_1: The population median is greater than M. Reject H_0 if $T_2 > T_U$. The value of T_U is found by using Table A8 (see Figure 7.15). The critical value T_U is the closest value such that $P(T_2 > T_U) \approx \alpha$. For large sample sizes ($n > 50$) the value of T_U is found as $T_U = (n + z_{1-\alpha}\sqrt{n})/2$, where $z_{1-\alpha}$ is the $1 - \alpha$ quantile from Table A2.

(c) H_1: The population median is not equal to M. Reject H_0 if $T_1 < T_L$ or if $T_2 > T_U$. The values of T_U and T_L are found by using Table A8 (see Figure 7.15). The critical value T_L is the closest value such that $P(T_1 < T_L) \approx \alpha/2$, and T_U is the closest value such that $P(T_2 > T_U) \approx \alpha/2$. For large sample sizes ($n > 50$) the values of T_L and T_U are given as $T_U = (n + z_{1-\alpha/2}\sqrt{n})/2$ and $T_L = n - T_U$, where $z_{1-\alpha/2}$ is the $1 - \alpha/2$ quantile from Table A2.

EXAMPLE

This example has four parts, to illustrate the median test on a sample of size 10 with α as close as possible to .05. All parts are concerned with testing the following null hypothesis:

$$H_0: \text{The population median is 15.}$$

Real-number-line plots of the sample values are used in calculating the test statistic without listing the actual sample values, since all that is needed for the calculation of T_1 and T_2 is the position of the points relative to the hypothesized median.

Part 1 H_1: The population median is greater than 15.

```
───────────────── x  x ──────────┼── x  xx  x    xx      x  x ────────────
Sample 1                        15
```

From Table A8 (or Figure 7.15) the sample size $n = 10$ leads to $T_U = 7$ for a test with $\alpha = .0547$.

Decision rule: Reject H_0 if $T_2 > 7$.
Test statistic: $T_2 = 8$.

H_0 is rejected in favor of the alternative. Notice that almost all of the sample points are greater than 15. The p-value is $P(T_2 \geq 8) = .0547$ from Table A1 for $n = 10$, $p = .5$.

Part 2 H_1: The population median is less than 15.

```
———————×—××——×—××———————×——×—————×————————+———×——————————
Sample 2                                        15
```

From Table A8 (or Figure 7.15) $T_L = 3$ for $\alpha = .0547$.

Decision rule: Reject H_0 if $T_1 < 3$.
Test statistic: $T_1 = 1$.

H_0 is rejected in favor of the alternative. Almost all of the sample points are less than 15; thus, it would appear that the population median is also less than 15. The p-value is $P(T_1 \le 1) = .0107$ from Table A1 for $n = 10$ and $p = .5$.

Part 3 H_1: The population median is less than 15.

```
                        Two sample values at 15
———————×——×———×————×—×————×———×————————×———×——————————
Sample 3                                    15
```

Decision rule: Reject H_0 if $T_1 < 3$. (Same as in part 2)
Test statistic: $T_1 = 3$.

H_0 is not rejected as it was in part 2. Note that T_1 and T_2 are not equal this time and that a test statistic with a value of 1 would have led to rejection of H_0, whereas one with a value of 3 does not. That is why it is important to account for the sample values equal to the hypothesized median. The p-value is $P(T_1 \le 3) = .1719$, which is found from Table A1 with $n = 10$ and $p = .5$.

Part 4 H_1: The population median does not equal 15.

```
———————×————×××——×—×————×—×——×—————+———×——————————
Sample 4                              15
```

From Table A8 (or Figure 7.15) $T_L = 2$ and $T_U = 8$, with $\alpha = .0215$. (This is the two-tailed test having a value of α closest to .05.)

Decision rule: Reject H_0 if $T_1 < 2$ or $T_2 > 8$.
Test statistic: $T_1 = 1$, $T_2 = 1$.

H_0 is rejected. The p-value is $2 \cdot P(T_1 \le 1) = .0215$.
 The normal approximation is now demonstrated on part 4 of the example. It gives $T_U = (10 + 1.9600\sqrt{10})/2 = 8.1$ and $T_L = 10 - 8.1 = 1.9$. Therefore the decision rule is to reject H_0 if $T_1 < 1.9$ or $T_2 > 8.1$, which is equivalent to the previous decision rule. The p-value, using the normal approximation to the binomial distribution, is found as follows, with the assistance of Table A2*.

$$p\text{-value} = 2 \cdot P(T_1 \le 1)$$

$$\approx 2 \cdot P\left(Z \le \frac{1 - 10(.5) + .5}{\sqrt{10(\frac{1}{2})(\frac{1}{2})}}\right)$$

$$= 2 \cdot P(Z \le -2.21)$$

$$= 2(.0136)$$

$$= .0272$$

This is close to the exact value .0215 found earlier.

n	T_L	T_U	α Level One-Tailed Test	α Level Two-Tailed Test	n	T_L	T_U	α Level One-Tailed Test	α Level Two-Tailed Test
4	1	3	.0625	.1250	14	2	12	.0009	.0018
						3	11	.0065	.0129
5	1	4	.0312	.0625		4	10	.0287	.0574
						5	9	.0898	.1796
6	1	5	.0156	.0313					
	2	4	.1094	.2187	15	2	13	.0005	.0010
						3	12	.0037	.0074
7	1	6	.0078	.0156		4	11	.0176	.0352
	2	5	.0625	.1250		5	10	.0592	.1185
8	1	7	.0039	.0078	16	3	13	.0021	.0042
	2	6	.0352	.0703		4	12	.0106	.0213
						5	11	.0384	.0768
9	1	8	.0020	.0039		6	10	.1051	.2101
	2	7	.0195	.0391					
	3	6	.0898	.1797	17	3	14	.0012	.0023
						4	13	.0064	.0127
10	1	9	.0010	.0020		5	12	.0245	.0490
	2	8	.0107	.0215		6	11	.0717	.1435
	3	7	.0547	.1094					
					18	4	14	.0038	.0075
11	1	10	.0005	.0010		5	13	.0154	.0309
	2	9	.0059	.0117		6	12	.0481	.0963
	3	8	.0327	.0654		7	11	.1189	.2379
12	1	11	.0002	.0005	19	4	15	.0022	.0044
	2	10	.0032	.0063		5	14	.0096	.0192
	3	9	.0193	.0386		6	13	.0318	.0636
	4	8	.0730	.1460		7	12	.0835	.1671
13	1	12	.0001	.0002	20	4	16	.0013	.0026
	2	11	.0017	.0034		5	15	.0059	.0118
	3	10	.0112	.0225		6	14	.0207	.0414
	4	9	.0461	.0923		7	13	.0577	.1153

FIGURE 7.15
CRITICAL VALUES FOR THE MEDIAN TEST FOR $n \le 20$. TABLE A8 CONTAINS CRITICAL VALUES FOR $n \le 50$.

	The population is	
	Normal	Nonnormal
$(n \leq 30)$	Student's t-test (Z-test, if σ is known)	Median test
$(n > 30)$	Student's t-test (Z-test, if σ is known) (Second choice: Z-test)	Median test (Second choice: Z-test)

FIGURE 7.16
THE PREFERRED METHOD FOR TESTING HYPOTHESES ABOUT A
LOCATION PARAMETER.

WHEN SHOULD THE MEDIAN TEST BE USED?

If the population distribution is normal, then the tests of Section 7.4 are more powerful and should be preferred. If the population is not approximately normal (which can be examined by making a Lilliefors test on the sample data), then the median test in this section may have more power than the tests of Section 7.4, especially if the sample tends to have outliers. The Z-test is easier to use than the median test, especially when the sample size is large, and therefore may be used in the large-sample case merely for convenience. See Figure 7.16 for a summary of which test is preferred.

For symmetric probability distributions the mean equals the median, so a test for the median also tests for the mean, and vice versa. Often the population probability distribution is not symmetric; consequently, a test for the median and a test for the mean are actually testing slightly different parameters. However, this seldom causes any difficulty, since both the mean and the median are measures of central tendency in a population.

EXERCISES

7.51 Shoe sizes for a random sample of 20 sales made to women are given below. Use a level of significance of $\alpha = .05$ to test the following hypothesis.

H_0: The population median shoe size sold to women is $7\frac{1}{2}$

versus

H_1: The population median shoe size sold to women is less than $7\frac{1}{2}$.

$$9\frac{1}{2} \quad 7 \quad 5\frac{1}{2} \quad 7\frac{1}{2} \quad 9$$
$$5\frac{1}{2} \quad 5 \quad 7 \quad 5\frac{1}{2} \quad 8$$
$$7 \quad 6\frac{1}{2} \quad 6\frac{1}{2} \quad 6 \quad 6\frac{1}{2}$$
$$6\frac{1}{2} \quad 9 \quad 7\frac{1}{2} \quad 6\frac{1}{2} \quad 6$$

Also find the p-value and interpret it.

7.52 The exam scores of 40 randomly selected employees in an in-service course are summarized below. Use a level of significance of $\alpha = .01$ to test the following hypothesis.

H_0: The population median score for this exam is 70

versus

H_1: The population median score for this exam is greater than 70.

77	68	86	84	95	98	87	71
84	92	96	83	62	83	81	85
91	74	61	52	83	73	85	78
50	81	37	60	85	100	79	81
75	92	80	75	78	71	64	65

7.53 The mpg recorded on 20 randomly selected models of a four-cylinder automobile in city driving are given below. Use a level of significance of $\alpha = .05$ to test the following hypothesis. State the p-value and interpret it.

H_0: The population median mpg for this model is 24

versus

H_1: The population median mpg for this model is greater than 24.

24.21	24.14	24.66	23.08	25.16
24.35	24.60	24.72	23.88	24.62
23.82	23.75	24.47	23.09	24.62
24.21	25.01	24.38	24.57	25.14

7.54 The 1988 net earnings of common stocks for 10 randomly selected corporations are reported below. Use a significance level of $\alpha = .01$ to test the following hypothesis. State the p-value and interpret it.

H_0: The population median net earnings on common stocks for corporations is $2.00

versus

H_1: The population median net earnings on common stocks for corporations is greater than $2.00.

| $1.71 | 2.17 | 2.25 | 2.43 | 2.32 |
| 3.15 | 3.30 | 5.52 | 3.32 | 3.76 |

7.55 The time in minutes between incoming phone calls in a business office has been recorded for 10 calls:

1.8, 0.3, 4.5, 9.8, 3.2, 15.7, 4.8, 1.0, 2.7, 6.2

Assume this sample behaves like a random sample. Use these data to test the following hypothesis with $\alpha = .10$.

H_0: The population median time between calls is 7 min

versus

H_1: The population median time between calls is not 7 min.

Find the p-value and interpret it.

7.56 The number of overtime hours worked in one week by 13 randomly selected employees in a large department store during the Christmas season is as follows: 19.5, 16.6, 16.7, 17.8, 20.2, 23.3, 21.2, 18.6,

23.3, 22.5, 19.8, 20.5, 29.3. Use these data to test the following hypothesis with $\alpha = .05$.

H_0: The population median number of overtime hours for workers in this store is 20

versus

H_1: The population median number of overtime hours for workers in this store is greater than 20.

Find the p-value and interpret it.

7.57 Data are given that represent the survival time in days for 13 randomly selected patients with cancer of the bronchus: 39, 427, 17, 460, 90, 187, 58, 52, 100, 200, 42, 167, 33. These patients were diagnosed as terminally ill and were then given a treatment of 10 g of vitamin C daily. Use a level of significance of $\alpha = .01$ to test the following hypothesis. State the p-value and interpret it.

H_0: The population median survival time is 30 days

versus

H_1: The population median survival time is greater than 30 days.

7.6

REVIEW EXERCISES

7.58 Police records on 30 randomly selected individuals booked on assault showed the ages recorded, and are given below.

24	20	18	32	16	25
21	18	38	22	18	16
15	14	21	17	17	17
17	23	22	16	24	20
27	21	18	26	20	15

Test the hypothesis that the median age for these individuals is 22 versus the alternative that the median age is less than 22. Let $\alpha = .05$ and state and interpret the p-value.

7.59 Some random samples of pure iron are obtained and their melting point is determined in a laboratory with the results given below. Test these data for normality. Note that $\Sigma X = 23,960$ and $\Sigma X^2 = 35,881,820$.

1486	1502	1478	1497	1499	1504	1489	1503
1493	1484	1509	1494	1507	1513	1514	1488

7.60 For the sample data in Exercise 7.59 test H_0: $\mu = 1492$ versus H_1: $\mu > 1492$. Let $\alpha = .05$. State and interpret the p-value.

7.61 For the sample data in Exercise 7.59 test H_0: $\sigma = 10$ versus H_1: $\sigma > 10$ at $\alpha = .05$. State the p-value and interpret it.

7.62 A standard insulin (known potency) is injected into 30 randomly selected laboratory test animals, and the sample mean percentage decrease in blood sugar four hours after injection is found to be 35.3%. The standard deviation of the dosage is known to be $\sigma = 15.7\%$. Use these sample results to test H_0: $\mu = 30\%$ versus H_1: $\mu > 30\%$ with $\alpha = .01$. Assume the population is normal.

7.63 Assuming the true mean percentage decrease in blood sugar in Exercise 7.62 is 35%, find the power of the test used in that exercise.

7.64 State officials have expressed the opinion that at least 15% of the cars on the road would not pass a safety inspection. A random sample of 20 cars shows 6 to be unsafe. Use these results to test H_0: $p = .15$ versus H_1: $p > .15$. Let $\alpha = .05$ and state and interpret the p-value.

7.65 The results of a voter preference poll involving 200 randomly selected voters show that 53% favor the Republican candidate for U.S. senator. Use these sample results to test H_0: $p = .5$ versus H_1: $p \neq .15$. Let $\alpha = .05$ and state the p-value.

7.66 A machine shop wants to institute a control chart for the length of bolts it is producing. The job order specifies that the mean length of the bolts is to be 2.7 cm. Past experience indicates that the standard deviation of the bolts manufactured by this shop is 0.2 cm. The quality-control engineer suggests that a random sample of size 15 be examined every hour to make sure the manufacturing process stays in control. Construct a two-sided 95% quality control chart for this manufacturing process.

7.67 For the situation described in Exercise 7.66 the sample means are obtained for eight successive hours with the following results: 2.72, 2.65, 2.68, 2.71, 2.73, 2.78, 2.90, and 2.95. Plot these points on the quality-control chart of Exercise 7.66 to determine if the process is out of control.

BIBLIOGRAPHY

Additional material on the topics presented in this chapter can be found in the following publications.

Conover, W. J. (1980). *Practical Nonparametric Statistics,* 2nd ed. Wiley, New York.

Iman, R. L. (1982). "Graphs for Use With the Lilliefors Test for Normal and Exponential Distributions." *The American Statistician,* **36** (2), 108–112.

8
TWO RELATED SAMPLES (MATCHED PAIRS)

PRELIMINARY REMARKS

Previous chapters have been devoted to providing the reader with an understanding of the procedures for gathering, displaying, summarizing, and analyzing sample data. Estimation and hypothesis testing presented in Chapters 6 and 7 provide a structured framework that aids the reader in the decision-making process. The remaining chapters are concerned with presenting additional procedures for estimation and hypothesis testing, procedures that have proved to be most useful in the analysis of business data.

This background should enable the reader to handle a wide variety of every-day problems involving the analysis and evaluation of data. The problem settings in the rest of this text represent data-based decision-making situations that are likely to be encountered in actual applications. It is the authors' intent to provide a clean, concise, and readable explanation of each procedure, and to do this in a manner that will allow this text to serve as a ready reference that can be read at some future date as well. Therefore the format followed in this text involves first presenting the need for a procedure to handle a particular situation, then explaining how the procedure is to be performed and analyzed along with examples. More importantly, the format includes a clear statement of the assumptions necessary for a correct analysis and interpretation of sample data. Techniques are provided for checking these assumptions, and appropriate alternative procedures are provided when assumptions are not satisfied (which occurs frequently).

8.1

THE PAIRED t-TEST

PROBLEM SETTING 8.1

The training and retraining of employees represents a large expense to businesses both large and small. Training ranges over virtually all employee educational levels and is necessitated by the changing technological advances the business must make to stay in a competitive position. For example, computers are now used by almost every business, where requirements range from bookkeeping, payroll, and inventory accounting to highly sophisticated scientific computing. Opportunities arise almost daily from groups willing to train employees in various aspects of computer usage. Other courses are designed to increase employee skills associated with particular tasks, such as typing speed, shorthand speed, writing skills including grammar and spelling, reading speed, and computing techniques.

Speed-reading courses have been popular for the last several years. Suppose an accounting firm decides to do some independent evaluation before signing up a large number of their employees to take a speed-reading course. Ten employees are randomly selected to take the course. The reading scores of these individuals, combining measurements of speed and comprehension, are recorded before and after the course to create a pair of measurements (X_i, Y_i) for each individual. Clearly it does not matter if the reading score before the course is denoted by X or Y as long as the notation is consistent throughout. Therefore, let X represent the score after completing the course and let Y represent the reading score for the same person before taking the course. The data for the two related samples are given in Figure 8.1.

Person	X_i Reading Score After Course	Y_i Reading Score Before Course	$D_i = X_i - Y_i$ Difference
1	221	211	10
2	231	216	15
3	203	191	12
4	216	224	-8
5	207	201	6
6	203	178	25
7	201	188	13
8	179	159	20
9	179	177	2
10	211	197	14

FIGURE 8.1
PAIRED DATA BEFORE AND AFTER A
SPEED-READING COURSE.

Does the speed-reading course improve the reading ability of the employees? How much improvement can be expected on the average? By considering the average amount of improvement in the reading ability of the employees, management can decide whether this training course is an effective means for increasing the efficiency of the organization.

CONTROLLING UNWANTED VARIATION

The evaluation of increased or decreased reading skills is only one of a multitude of problems that can be solved by using the techniques presented in this chapter. However, before these techniques are presented some background is needed on a very fundamental concept in statistics. That concept is the reduction or controlling of unwanted or extraneous variation in the gathering of the sample data.

Consider a study designed to determine the effectiveness of a medication used to control high blood pressure. An experiment could be set up by randomly selecting two groups of individuals. The first group is given no medication and the other group is given the medication of interest. After some period of time the average blood pressures for each group are recorded to aid in making a decision regarding the medication.

Do you see any flaws in this experimental setup? One obvious flaw is that the variation due to medication is obscured by variability from person to person. Some of the variation in the data could have been reduced by designing the experiment differently. That is, suppose that the group not given any treatment has a higher average blood pressure initially than the treatment group or that many of the individuals in the groups do not even suffer from high blood pressure. Since the original objective was to determine the effectiveness of the medication in reducing an individual's high blood pressure, an initial measurement should be taken on each individual to provide assurance that the individual suffers from high blood pressure. Then a second set of measurements should be taken after each has received the medication for a period of time. By approaching the problem in this manner, some unwanted variation in the observed measurements is controlled, as the differences observed in the pair of readings for each individual will be due primarily to the medication if it is effective. If it is not effective the only variation observed will be random variation that would normally be observed on any individual's blood pressure when taken at different time periods. This method of pairing observations is usually very effective in reducing unwanted variability in the data. These paired observations are called **matched pairs.** A procedure for analyzing such matched pairs is called the **paired *t*-test,** and is the subject of this section.

The use of matched pairs in an experiment is an important and useful method for reducing extraneous variation in the outcomes. Some examples where matched pairs are useful are as follows.

1. Two computer systems are being compared before deciding which one to install in a bank. Several transactions are selected at random from the population of all transactions for one day. Each transaction is processed first using one of the two computer systems, and then it is processed using the other computer system. The two processing times are compared in each pair to see which system is faster.

2. The same individual may be observed twice, once before a treatment and once after the treatment, to measure the effectiveness of the treatment. These two observations taken on a random sample of individuals constitute the paired observations. The "treatment" may be a training session, where the observations are job performance ratings before and after the training session. Or the "treatment" may be a lecture or memorandum on unnecessary job absenteeism, whereas the observations are the absentee rates before and after the lecture or memorandum.

3. Participants in a lotion study are asked to use one type of lotion on one side of their face and a second type of lotion on the other. After a prescribed period of time the clearness of the skin is measured on both sides of each participant's face. This method of pairing observations provides a means of controlling the variability present from one person to another, and the differences in the two readings can be attributed solely to differences between the two lotions.

THE PAIRED t-TEST

The paired t-test is used to analyze pairs of observations (X_i, Y_i), $i = 1, \ldots, n$ in a random sample to see if the mean of the X's equals the mean of the Y's. Confidence intervals for the differences in means may also be found. These procedures resemble the procedures shown in the last two chapters for a single sample of observations. The matched pairs are reduced to a single sample by taking the difference between the two observations in the matched pair, such as by subtracting one person's blood pressure reading after the medication from the person's reading before the medication is administered. These differences in readings, denoted by $D_i = X_i - Y_i$, form a single sample of observations, as in the previous two chapters, so the methods of those chapters can be applied to the differences. The t-test is repeated in this section, and restated to apply to the matched pairs problem. The paired version of the Z-test is not presented here because the paired t-test is best for normal populations, and the Wilcoxon signed-ranks test of Section 8.2 tends to have more power for nonnormal populations and therefore should be preferred in those cases.

PROCEDURE FOR TESTING HYPOTHESES ABOUT DIFFERENCES IN RELATED SAMPLES (Paired t-Test)

Data

The data $D_1, D_2, \ldots, D_n$ (where $D_i = X_i - Y_i$) are computed from pairs of measurements (X_i, Y_i) on each of the n elements in the random sample. Let

$$\overline{D} = \frac{\Sigma D_i}{n}$$

and

$$s_D = \sqrt{\frac{\Sigma D_i^2 - (\Sigma D_i)^2/n}{n - 1}} = \sqrt{\frac{\Sigma(D_i - \overline{D})^2}{n - 1}}$$

be the sample mean and sample standard deviation of the D_i, respectively.

Assumptions

1. The random variables $D_1, D_2, \ldots, D_n$ are independent of one another and are identically distributed with mean $\mu_X - \mu_Y$.

2. If $n \leq 30$, the D_i are required to be normally distributed. If $n > 30$, the normality assumption is not necessary because the central limit theorem applies to $\overline{D}$. However, in that case the Wilcoxon signed-ranks test of Section 8.2 should be preferred because of its tendency to have greater power.

Estimation

A confidence interval for $\mu_X - \mu_Y$ with a level of confidence of $100(1 - \alpha)\%$ is given as $\overline{D} \pm t_{1-\alpha/2,n-1} \, s_D/\sqrt{n}$, where $t_{1-\alpha/2,n-1}$ is the $1 - \alpha/2$ quantile of the Student's t distribution with $n - 1$ degrees of freedom as given in Table A3.

Null Hypothesis

$H_0: \mu_X = \mu_Y$. (The mean of the X measurement is the same as the mean of the Y measurement.)

Test Statistic

$$T = \frac{\overline{D}}{s_D/\sqrt{n}} = \frac{\overline{D}\sqrt{n}}{s_D} \tag{8.1}$$

The null distribution of T is the Student's t distribution with $n - 1$ degrees of freedom, whose quantiles are given in Table A3.

Decision Rule

The decision rule is based on the alternative hypothesis of interest.

(a) $H_1: \mu_X > \mu_Y$. (The X measurement tends to be higher than the Y measurement.) Reject H_0 if $T > t_{1-\alpha,n-1}$. (See Figure 8.2a.)

(b) $H_1: \mu_X < \mu_Y$. (The X measurement tends to be lower than the Y measurement.) Reject H_0 if $T < -t_{1-\alpha,n-1}$. (See Figure 8.2b.)

(c) $H_1: \mu_X \neq \mu_Y$. (The X measurement tends to be either larger or smaller than the Y measurement.) Reject H_0 if $T > t_{1-\alpha/2,n-1}$ or if $T < -t_{1-\alpha/2,n-1}$. (See Figure 8.2c.)

Note

The paired t-test can test the more general null hypothesis

$$H_0: \mu_X - \mu_Y = \mu_0$$

for some constant μ_0, by subtracting μ_0 from $\overline{D}$ before performing the preceding test. This allows testing the null hypothesis that the difference in population means is μ_0.

Care should be taken to keep the notation consistent in each application, especially with the one-tailed tests. That is, initially either measurement may be designated X or Y, but once a variable is identified as X, it must be identified as X consistently throughout the entire analysis, and the mean for that variable must be called μ_X. Otherwise, the rejection region used may inadvertently consist of values of the test statistic that tend to support H_0 rather than discredit H_0. Figure 8.2 may be helpful in preventing errors of this type from occurring.

EXAMPLE

In Problem Setting 8.1 an accounting firm sends ten employees to take a speed-reading course, and measures their reading scores after X and before Y taking

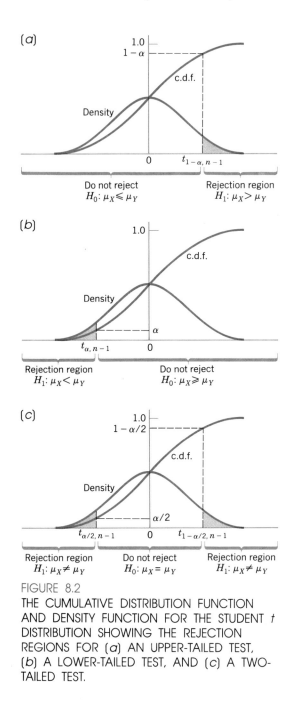

FIGURE 8.2
THE CUMULATIVE DISTRIBUTION FUNCTION
AND DENSITY FUNCTION FOR THE STUDENT t
DISTRIBUTION SHOWING THE REJECTION
REGIONS FOR (a) AN UPPER-TAILED TEST,
(b) A LOWER-TAILED TEST, AND (c) A TWO-
TAILED TEST.

the course. Those scores are given in Figure 8.1. The hypothesis to be tested
is:

$$H_0: \mu_X = \mu_Y \qquad \text{(There is no change in the mean reading score)}$$

versus

$$H_1: \mu_X > \mu_Y \qquad \text{(The mean reading score is higher after the course)}$$

For a choice of $\alpha = .01$, Table A3 is used with $10 - 1 = 9$ degrees of freedom to complete the following decision rule:

$$\text{Reject } H_0 \text{ if } T > t_{.99,9} = 2.8214$$

Sample calculations give $\overline{D} = 10.9$ and $s_D = 9.28$, since $\Sigma D_i = 109$ and $\Sigma D_i^2 = 1963$. The test statistic T is found as $T = 10.9\sqrt{10}/9.28 = 3.715$. H_0 is rejected and it is concluded that the mean reading score has increased. The *p*-value is approximately .003.

Since it is easy to make an error in the calculations, it is advisable to plot the paired data to see if the decision is reasonable. A scatterplot is useful for analyzing this type of data; one such plot appears in Figure 8.3. The 45 degree line where $X = Y$ has been added to the plot to aid in examining the results. If the speed reading course is not effective, then the pairs of points should be randomly scattered about the 45 degree line. If, however, the course is effective, then most of the points should be below the 45 degree line in the region labeled $X > Y$. Nine of the 10 pairs are in this region; hence, the decision seems justified.

A 95% confidence interval for the mean change in reading score is easily found as follows.

$$\overline{D} \pm t_{1-\alpha/2, n-1}\, s_D/\sqrt{n} = 10.9 \pm 2.2622(9.28)/\sqrt{10}$$
$$= 10.9 \pm 6.6$$
$$= 4.3 \text{ to } 17.5$$

The mean change in reading score is believed to be between 4.3 and 17.5 with 95% confidence.

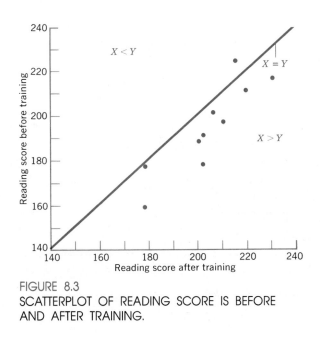

FIGURE 8.3
SCATTERPLOT OF READING SCORE IS BEFORE AND AFTER TRAINING.

CHECKING THE ASSUMPTIONS

In the previous example two important assumptions for the paired t-test have been neglected. Do you know what they are? The assumptions of independence and normality have been ignored. The assumption of independence was satisfied by randomly selecting 10 individuals for the comparison; however, the normality assumption must be checked by plotting the standardized sample values $D_i^* = (D_i - \bar{D})/s_D$ as an empirical distribution function on the Lilliefors graph. The standardized values are as follows.

D_i: 10   15   12   −8   6   25   13   20   2   14

D_i^*: −0.10 0.44 0.12 −2.04 −0.53 1.52 0.23 0.98 −0.96 0.33

The Lilliefors graph is given in Figure 8.4, and the plotted e.d.f. of the 10 observations is well within the bounds for a sample of size 10. Therefore, it is reasonable to conclude that the results and interpretation of the above test are correct.

It is impossible to overemphasize the importance of checking the D_i for normality, and that is why the Lilliefors bounds have been provided as a very easy way to do this checking. This check is necessary when the sample size is small because the test statistic T cannot be correctly compared with a Student's t random variable unless the normality assumption is satisfied. Hence,

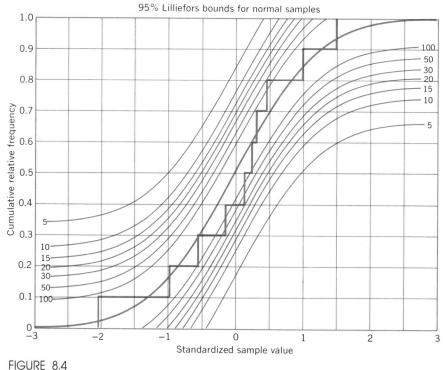

FIGURE 8.4
NORMALITY CHECK ON READING SCORES EXAMPLE.

the results of the test are correct for small sample sizes only if the normality assumption is satisfied. If the sample size is large, the central limit theorem supports the validity of the paired *t*-test even for nonnormal data, but the power of the test may be low for some types of nonnormal distributions, so the Wilcoxon signed-ranks test of Section 8.2 should be preferred in that case.

Of course, the Lilliefors test does not prove that the population is normally distributed, but it indicates whether the data could reasonably be assumed to have come from a normal distribution. The importance of checking for normality is partially offset by the low power of any test for normality when the sample sizes are small. For this reason some statisticians prefer to omit the Lilliefors test and to use the procedures in the next section whenever sample sizes are small. This is a perfectly acceptable alternative.

Many well-intentioned individuals have been led to erroneous conclusions by ignoring test assumptions. Use of a test when its assumptions are not satisfied may result in a procedure with very little power to detect the differences present in the population, or a test with an actual Type I error rate much larger or smaller than desired. The direct result of using the wrong test may be quite costly—an incorrect decision could be made regarding the merits of a new product, or the results of a marketing survey could be interpreted incorrectly. The next section considers related samples when the normality assumption is not satisfied (which frequently seems to be the case when dealing with real-world data).

EXERCISES

8.1 An office manager sends 10 randomly selected secretaries to a week-long course to improve their typing speed. The manager times their speed before and after the course to see if the course is worth the lost time and expense. State the appropriate null and alternative hypotheses and test the null hypothesis with $\alpha = .025$. (The Lilliefors test should be used to check the normality assumption because of the small sample size.) State the *p*-value. Find a 95% confidence interval for the mean increase in typing speed.

Speed after Course	Speed before Course
55	50
46	42
78	70
61	63
52	58
45	35
47	46
57	52
71	60
58	49

8.2 Twelve randomly selected students in a statistics course recorded the scores on the next page on their first and fourth exams in the course.

Score on Test 1	Score on Test 4
64	80
28	87
90	90
30	57
97	89
20	51
100	81
67	82
54	89
44	78
100	100
71	81

Test the hypothesis that there is no difference in the population mean scores for the two exams versus the alternative that the population mean is lower on the first test. Let $\alpha = .05$. The Lilliefors test verifies that the normality assumption is reasonable. State the p-value and interpret it.

8.3 Twenty randomly selected patients on a diet to lose weight had their weight recorded before starting the diet and after one month's time on the diet. The weight loss (before minus after) was recorded for each patient as follows:

$$7 \quad -6 \quad 3 \quad 1 \quad 6 \quad 4 \quad 9 \quad -5 \quad 9 \quad 7$$
$$-3 \quad 7 \quad -9 \quad 8 \quad 6 \quad -4 \quad 4 \quad 9 \quad -6 \quad 1$$

State the appropriate null and alternative hypotheses to test the effectiveness of this diet. Use a significance level of $\alpha = .05$. The Lilliefors test verifies that the assumption of normality is satisfied.

8.4 The management of a factory with large assembly lines wants to try out a new assembly technique for one of its lines. Fifteen employees are randomly selected and the number of units each assembled in one week is noted. These 15 employees are then taught the new assembly technique and the number of units each assembles using the new technique is recorded for one week. The data are given below. State the appropriate null and alternative hypotheses for testing to see if the new technique is effective in increasing worker productivity. Use a level of significance of $\alpha = .01$. Assume normality. State the p-value. Find a 90% confidence interval for the mean increase in productivity.

Worker	Number of Units Assembled Using Present Method	Number of Units Assembled Using New Technique
1	34	33
2	28	36
3	29	50
4	45	41

Worker	Number of Units Assembled Using Present Method	Number of Units Assembled Using New Technique
5	26	37
6	27	41
7	24	39
8	15	21
9	15	20
10	27	37
11	23	21
12	31	18
13	20	29
14	35	38
15	20	27

8.5 Prices for 15 randomly selected food items taken from advertisements placed in the local newspaper by two competing supermarkets are listed below. Use a level of significance of $\alpha = .05$ for a test of a hypothesis to see if there is any mean difference in the prices offered by the two supermarkets. The Lilliefors test verifies that the data satisfy the assumption of normality. State the *p*-value. (Use $\Sigma D = -.55$, $\Sigma D^2 = .3653$.)

Food item	Prices Offered by First Supermarket (Dollars)	Prices Offered by Second Supermarket (Dollars)
Chile	.98	.89
Saltines	.69	.79
Cake mix	.69	.87
Eggs	.83	.92
Catsup	.79	.82
Bologna	1.39	1.65
Orange drink	1.39	1.59
Macaroni	.30	.33
Avocados	.25	.39
Orange juice	1.45	1.39
Noodles	.59	.63
Tomato sauce	.17	.14
Margarine	.79	.89
Grapefruit	.17	.13
Crisco	2.39	1.99

8.6 An insurance adjuster has received estimates from two different repair garages for minor repairs on 20 automobiles. Assume that the differences in the individual estimates are normally distributed. State the appropriate null and alternative hypotheses to see if there is any difference in the estimates of the two garages. Let $\alpha = .05$. State the *p*-value. (Use $\Sigma D = 124$, $\Sigma D^2 = 5346$.) See data on next page.

Claim Number	Estimated by First Garage (Dollars)	Estimate by Second Garage (Dollars)
1	48	46
2	56	49
3	87	71
4	88	56
5	86	62
6	64	54
7	80	52
8	78	88
9	72	82
10	70	80
11	80	64
12	58	78
13	72	42
14	60	56
15	64	72
16	46	64
17	56	54
18	59	50
19	73	61
20	78	70

8.7 Twenty employees are asked to rate each of their fellow employees on a scale from 0 to 100 with at least one employee getting a 0 and at least one employee getting a 100. The average of these ratings is then compared with the supervisor's rating of the employees. Use a level of significance of $\alpha = .05$ to test the appropriate hypothesis to see if the two sets of ratings differ. (The Lilliefors test should be used to check the normality assumption because of the small sample size.) State the p-value. (Use $\Sigma D = -83$, $\Sigma D^2 = 7383$.)

Employee Number	Average Peer Rating	Supervisor Rating
1	1	7
2	3	0
3	10	3
4	7	11
5	7	14
6	9	14
7	9	24
8	19	7
9	27	26
10	25	54

Employee Number	Average Peer Rating	Supervisor Rating
11	36	47
12	41	29
13	46	12
14	44	72
15	57	42
16	48	96
17	72	90
18	80	84
19	83	100
20	100	75

8.8 Two individual stores of a large chain are thought to have equally attractive locations in a large city. The daily receipts for 60 randomly selected business days show store number 1 to have an average difference of $58 higher than store number 2 with a standard deviation of $200. Use these summary statistics with a level of significance of $\alpha = .05$ to see if there is any mean difference in the daily receipts of the two stores. State the p-value.

8.9 To see if there is a difference in mileage obtained using gasohol instead of unleaded gasoline, one hundred randomly selected automobiles have their mpg checked with a tank of unleaded gasoline and with a tank of gasohol. A toss of a coin is used to determine whether the automobile receives the unleaded gasoline or the gasohol first. The driver is not told which type of fuel is in the tank, but does receive instructions to keep the driving conditions as nearly constant as possible on the two tanks. Why is this procedure used?

At the end of the test the gasohol fuel shows an average mpg of 0.43 higher than the unleaded fuel with a standard deviation of 1.8 mpg for the differences. Use a level of significance of $\alpha = .01$ to test the hypothesis that there is no difference in the mpg of the two fuels. State the p-value.

8.2

THE WILCOXON SIGNED-RANKS TEST (OPTIONAL)

CHOOSING THE BEST TEST

The assumption of normality of the differences $D_i = X_i - Y_i$ is necessary for the paired t-test when the sample sizes are small. If the normality assumption is not satisfied then the procedure presented in this section should be used to test the same hypothesis $\mu_X = \mu_Y$ for matched pairs. The paired t-test is an acceptable method to use for all populations when the sample sizes are large. However, if the populations are not normal, the procedure in this section often has more power than the paired t-test even though the sample size may be large. A careful analysis of the data should include a Lilliefors test for normality to select the most appropriate test.

Tests that require the assumption of normality, or any other specific distribution, are called **parametric** tests, whereas tests that are valid for all populations are called **nonparametric** tests. The t-test is a parametric test, and the test presented in this section is a nonparametric test.

A REVISION OF THE READING SPEED EXAMPLE

Consider the reading speed example of the previous section where a new random sample produces the following differences in scores.

$$D_i: \quad 10 \quad 45 \quad 9 \quad -5 \quad 4 \quad 49 \quad 8 \quad 52 \quad -2 \quad 6$$

Sample calculations give $\overline{D} = 17.6$ and $s_D = 22.01$. The corresponding standardized values $D_i^* = (D_i - 17.6)/22.01$ are given as follows:

$$D_i^*: \quad -0.35 \quad 1.25 \quad -0.39 \quad -1.03 \quad -0.62$$
$$1.43 \quad -0.44 \quad 1.56 \quad -0.89 \quad -0.53$$

The test for normality consists of plotting the e.d.f. of these standardized values on the Lilliefors graph. Such a plot is given in Figure 8.5. This graph shows

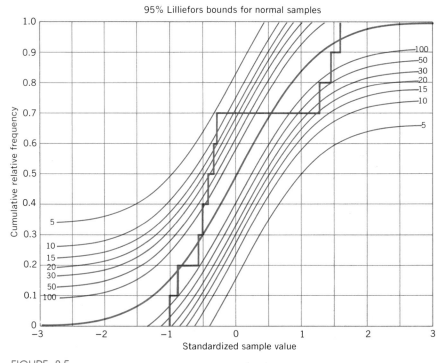

FIGURE 8.5
NORMALITY CHECK FOR REVISED READING SCORE EXAMPLE.

that the e.d.f. falls outside of the upper curve labeled with a 10; thus, the assumption of normality does not seem reasonable for these sample data. Since a basic assumption for the paired t-test has not been satisfied, the paired t-test should not be used on these data. That is, suppose the test statistic for the paired test is computed as was done in the example in the previous section, $T = 17.6\sqrt{10}/22.01 = 2.53$. If this value were compared against the tabled t value of that example, namely, $t_{.99,9} = 2.8214$, the null hypothesis would not be rejected. However, it is not correct to compare T against 2.8214, as T no longer has a Student's t distribution. Hence, the test is without meaning and the decision not to reject H_0 may be correct or it may be incorrect. One never knows for sure.

THE WILCOXON SIGNED-RANKS TEST

A procedure for testing $H_0: \mu_X = \mu_Y$ with matched pairs does not necessarily need to be based on the assumption of a normal distribution. Many statistical procedures for such a situation are valid for all kinds of populations. They are not as powerful as the paired t-test when the population is normal, but they are often more powerful than the paired t-test when the population is not normal. Recall that tests should be selected on the basis of their power when all other factors are equal. The nonparametric Wilcoxon signed-ranks test is very popular among practitioners because it is easy to use and is known to have good power for a wide variety of distributions. Although this test does not require normality of the D_i's, it is assumed that the distribution of the D_i's is symmetric. A symmetric distribution is one where the right half of the graph of the density or probability function is the mirror image of the left half.

PROCEDURE FOR TESTING HYPOTHESES ABOUT DIFFERENCES IN RELATED SAMPLES (Wilcoxon Signed-Ranks Test)

Data

The data $D_1, D_2, \ldots, D_n$ (where $D_i = X_i - Y_i$) are computed from pairs of measurements (X_i, Y_i) on each of the n elements in the random sample.

Assumptions

The random variables $D_1, D_2, \ldots, D_n$ are independent and identically distributed, and their distribution is symmetric.

Estimation

When a distribution is symmetric its mean and median are equal. Therefore a confidence interval for the mean of D is found using the procedure for finding the confidence interval for the median, given in Section 6.4, and applied to $D_1 \ldots, D_n$.

Null Hypothesis

$H_0: \mu_X = \mu_Y$ (the mean of the X measurement is the same as the mean of the Y measurement).

Test Statistic

Compute the value of each D_i. If some of the D_i equal zero, eliminate them from the analysis, and let n equal the reduced sample size, without the zeros. Assign ranks 1 to n to the absolute values of the D_i; that is, to $|D_i|$. Use average ranks in case of ties. Denote the rank of $|D_i|$ by R_i. If the original value of D_i was negative then give a negative sign to the corresponding rank R_i. Otherwise R_i will remain positive.

Let

$$\overline{R} = \frac{\Sigma R_i}{n}$$

and

$$s_R = \sqrt{\frac{\Sigma R_i^2 - (\Sigma R_i)^2 / n}{n - 1}}$$

be the usual sample mean and sample standard deviation of the R_i. The test statistic T_R is given as

$$T_R = \frac{\overline{R}\sqrt{n}}{s_R} \tag{8.2}$$

The null distribution of T_R is approximately the Student's t distribution with $n - 1$ degrees of freedom, whose quantiles are given in Table A3.

Decision Rule

The decision rule is based on the alternative hypothesis. Let α be the level of significance.

(a) H_1: $\mu_X > \mu_Y$ (the X measurement tends to be higher than the Y measurement). Reject H_0 if $T_R > t_{1-\alpha, n-1}$.

(b) H_1: $\mu_X < \mu_Y$ (the X measurement tends to be lower than the Y measurement). Reject H_0 if $T_R < -t_{1-\alpha, n-1}$.

(c) H_1: $\mu_X \neq \mu_Y$ (the X measurement tends to be either higher or lower than the Y measurement). Reject H_0 if $T_R > t_{1-\alpha/2, n-1}$ or if $T_R < -t_{1-\alpha/2, n-1}$.

The value of $t_{1-\alpha, n-1}$ is the $1 - \alpha$ quantile found in Table A3 for a Student's t random variable with $n - 1$ degrees of freedom. This provides a good approximation to the exact distribution of T_R for all sample sizes.

Note that the critical values of the Wilcoxon signed-ranks test are the same as for the paired t-test even though the test statistic is computed on ranks. The reason for this is that the t distribution provides an excellent approximation to the exact distribution of the Wilcoxon signed-ranks test statistic T_R even for small values of n. In addition, this makes the Wilcoxon signed-ranks test very easy to use because the only difference in computing the Wilcoxon signed-ranks test statistic and the paired t-test statistic is the additional step of ranking the $|D_i|$. Therefore, the authors refer to the Wilcoxon signed-ranks test as a **rank-transformation test,** since it is the result of computing the paired t-test on rank-transformed data. The rank-transformation form of the Wilcoxon signed-ranks test statistic presented here is simpler than some of the more standard forms for presentation in other textbooks, but it is equivalent to them. The Wilcoxon signed-ranks test is now demonstrated by completing the previous example where the data are nonnormal.

EXAMPLE

Recall that the revised reading scores in this section represent differences for 10 randomly selected individuals who were measured before (Y) and after (X)

taking a speed reading course. The assumption of normality for these differences was shown in Figure 8.5 not to be reasonable, as determined by the Lilliefors test. Therefore, the Wilcoxon signed-ranks test is used to test H_0 with $\alpha = .01$.

$H_0: \mu_X = \mu_Y$ (The mean reading scores are not changed by the course)

versus

$H_1: \mu_X > \mu_Y$ (The mean reading scores are higher after the course)

Person	Score After Course X_i	Score Before Course Y_i	Difference in Scores $D_i = X_i - Y_i$	Absolute Value of Difference $|D_i|$	Rank of $|D_i|$	Signed Rank R_i
1	261	251	10	10	7	7
2	292	247	45	45	8	8
3	317	308	9	9	6	6
4	253	258	-5	5	3	-3
5	271	267	4	4	2	2
6	305	256	49	49	9	9
7	238	230	8	8	5	5
8	320	268	52	52	10	10
9	267	269	-2	2	1	-1
10	281	275	6	6	4	4
						$\Sigma R = 47$

$$\overline{R} = \frac{\Sigma R_i}{n} = \frac{47}{10} = 4.7$$

$$s_R = \sqrt{\frac{385 - (47)^2/10}{9}} = 4.27$$

Test statistic $T_R = 4.7\sqrt{10}/4.27 = 3.48$.

Decision rule Reject H_0 if $T_R > t_{.99\ 9} = 2.8214$. H_0 is easily rejected. The p-value for $T_R = 3.48$ is estimated by interpolation in Table A3 to be

$$P(T_R > 3.48) = .004$$

from the Student's t distribution with nine degrees of freedom.

Recall that earlier in this section the paired t-test was applied to these same nonnormal sample data and that the value of the resulting test statistic was $T = 2.53$. When T was compared with the critical value of 2.8214, the null hypothesis was not rejected. This example serves to point out that no meaning should be attached to a paired t-test when the normality assumption is not satisfied.

To find a 90% confidence interval for the difference in the means, $\mu_X - \mu_Y$, the procedure in Section 6.4 is applied to the D's. Table A7 reveals that for $n = 10$ an 89% confidence interval for $\mu_X - \mu_Y$ extends from $D^{(3)} = 4$ to $D^{(8)} = 45$.

CHECKING THE REASONABLENESS OF THE DECISION

As in the last section it is worthwhile to make a scatterplot of the scores before and after the speed reading course to see if the conclusion of the previous example appears reasonable. Such a plot is given in Figure 8.6. An examination of Figure 8.6 shows that eight of the ten sample pairs are plotted in the region where scores after the course are higher than those before the course (i.e., $X > Y$). The remaining two scores show almost no change.

WHICH MATCHED-PAIRS TEST TO USE

A natural question to consider at this point is: Why not always use the Wilcoxon signed-ranks test? This is not a bad policy to follow. However, the choice of which test to use should be based on the following considerations. See Figure 8.7 for a summary.

1. If the normality assumption is satisfied, then the paired t-test has slightly more power than the Wilcoxon signed-ranks test. But if the normality assumption is not satisfied, the paired t-test may have considerably less power than the Wilcoxon signed-ranks test. The test that tends to have more power should be used.

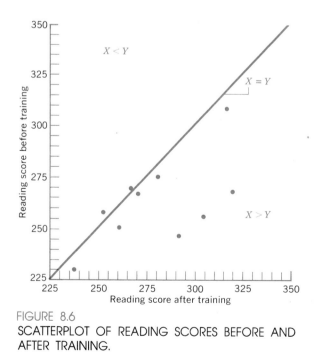

FIGURE 8.6
SCATTERPLOT OF READING SCORES BEFORE AND AFTER TRAINING.

	The differences $X - Y$ are	
	Normal	Nonnormal
$n \leq 30$	Paired t-test	Wilcoxon signed-ranks test
$n > 30$	Paired t-test	Wilcoxon signed-ranks test (Second choice: Paired t-test)

FIGURE 8.7
THE PREFERRED METHOD FOR TESTING THE HYPOTHESIS
OF EQUAL MEANS WHEN THE DATA ARE PAIRED.

2. The paired t-test is more commonly used, appearing in virtually all texts for a first course in statistics. Therefore someone reading your analysis is more likely to be familiar with a paired t-test. However, this is not a justification for ignoring the assumptions behind the t-test.

EXERCISES

8.10 Rework Exercise 8.1 using the Wilcoxon signed-ranks test. Compare the results obtained with those obtained from the paired t-test.

8.11 Change the results of the typing course reported in Exercise 8.1, so the ninth score now shows a typing score of 91 after the course, instead of 71 as reported. This more dramatic improvement increases the mean difference to 6.5. But, contrary to what one would expect, the paired t-test now has a *less* significant p-value of .04, as compared with the previous p-value, which was between .01 and .025. This larger p-value is due to the increased standard deviation caused by the new outlier in the data. What effect does this new observation have on the nonparametric analysis of these data, that was performed in Exercise 8.10? Why?

8.12 Rework Exercise 8.2 using the Wilcoxon signed-ranks test. Compare the results with those obtained from the paired t-test.

8.13 Rework Exercise 8.3 using the Wilcoxon signed-ranks test. Compare the results with those obtained from the paired t-test.

8.14 Rework Exercise 8.5 using the Wilcoxon signed-ranks test. (Use $\Sigma R = -50$, $\Sigma R^2 = 1236.5$.) Compare the results with those obtained from the paired t-test.

8.15 The city manager compares the number of new housing starts for the first eight months of this year with the comparable numbers for a nearby city of the same size. He believes that his city is growing faster than the nearby city, and wants to know if the data are sufficiently strong to support this hypothesis. What does your analysis look like?

	Jan.	Feb.	Mar.	Apr.	May	June	July	Aug.
This city	21	24	65	79	112	91	62	60
Nearby city	18	26	61	80	84	74	60	42

8.16 Rework Exercise 8.7 using the Wilcoxon signed-ranks test. Compare the results with those obtained from the paired t-test. (Use $\Sigma R = -52$, $\Sigma R^2 = 2868$.)

8.17 An insurance adjuster wants to compare estimates from two different repair garages for minor repairs on automobiles. Thirteen pairs of estimates have been randomly selected for the comparison.

Claim Number	Estimate by First Garage (Dollars)	Estimate by Second Garage (Dollars)
1	165	139
2	156	132
3	165	134
4	135	133
5	134	130
6	131	133
7	130	130
8	126	125
9	120	122
10	120	119
11	118	114
12	115	116
13	108	105

(a) State the appropriate null and alternative hypotheses to see if there is any difference in the mean estimates of the two garages. Let $\alpha = .05$, and test the null hypothesis with the Wilcoxon signed-ranks test. State the p-value.

(b) Check the differences in estimates from the two garages for normality using the Lilliefors test.

(c) Based on the results of part (b), the paired t-test should not be applied to these data; however, compute the paired t-test to test the null hypothesis in part (a) and compare it with the results of the Wilcoxon signed-ranks test.

8.18 An experiment is conducted to see if people tend to have a bias when they estimate their weight. Fifteen people were randomly selected from the employees in a large office building and were asked to estimate their exact weight. Then they were weighed on an accurate scale.

Person	Estimate	Actual	Person	Estimate	Actual
1	185	187	9	235	258
2	117	116	10	163	164
3	138	146	11	158	155
4	141	152	12	130	133
5	109	110	13	114	113
6	124	120	14	126	123
7	185	199	15	140	142
8	150	148			

Is there sufficient evidence to indicate a bias at $\alpha = .05$?

8.3

REVIEW EXERCISES

8.19 Twelve randomly selected patients have their systolic blood pressure checked before and after receiving a new medication to reduce blood pressure. The reductions in blood pressure readings (before minus after) were as follows: 11, 7, 2, 9, -7, -5, 3, 4, 13, 8, 5, and -6. Use these data to test H_0: $\mu_B = \mu_A$ versus H_1: $\mu_B > \mu_A$. Let $\alpha = .05$. (The Lilliefors test should be used to check these differences for normality because of the small sample size.)

8.20 Use the Wilcoxon signed-ranks test to test the hypothesis in Exercise 8.19.

8.21 A shoe manufacturer is field-testing the durability of a leather sole made by a new process and comparing it with the type of leather sole currently in use. Fifteen pairs of shoes are constructed with one shoe in each pair having the new type of leather sole. After six months of daily usage by 15 randomly selected supermarket employees, the shoes were examined and measured with respect to percentage of wear still remaining in the sole. The results are as follows:

Employee	New Sole	Present Sole
1	73	64
2	43	41
3	47	43
4	53	41
5	58	47
6	47	32
7	52	24
8	38	43
9	61	53
10	56	52
11	56	57
12	34	44
13	55	57
14	65	40
15	75	68

Check the assumption of normality for these paired data as required prior to a test for equality of means.

8.22 Use the appropriate test for equality of means in Exercise 8.21 to assist in deciding whether to replace the old process for making leather soles with the new process. Use $\alpha = .05$.

8.23 Make a scatterplot similar to the one in Figure 8.6 for the paired data in

Exercise 8.21 to see if the decision reached in the solution to Exercise 8.22 is justified.

8.24 In an international gymnastic competition a panel of judges rates each gymnast's performance on a scale of 0 to 10, with 10 being the best. In each of 12 performances observed, one of the judges was from the contestant's home country. Assume differences in scores behave as a random sample.

Contestant	Native Judge	Average of Foreign Judges	Contestant	Native Judge	Average of Foreign Judges
1	6.8	6.7	7	6.6	5.4
2	4.5	4.3	8	5.8	5.9
3	8.0	8.1	9	6.0	6.1
4	7.2	7.2	10	8.8	9.1
5	8.7	8.3	11	8.7	8.7
6	4.5	4.6	12	4.4	4.3

Check the assumption of normality for these paired data, as required prior to a test of equality of means.

8.25 Use the appropriate test of equality of means at $\alpha = .05$ to evaluate the data in Exercise 8.24 to determine if there is any support for the belief that judges from the contestant's home countries tend to be biased in either direction.

8.26 Make a scatterplot similar to the one in Figure 8.6 for the paired data in Exercise 8.24 to see if it supports the conclusion reached in Exercise 8.25.

8.27 Find a 95% confidence interval for any bias involved in the decision of a judge from the contestant's home country.

8.28 A company with a fleet of cars contracts for body repair work. To see if there is a tendency for the bid from one body shop to be higher or lower than a bid on the same job from another body shop, eight cars needing repairs were sent to both body shops for bids, with the following results. Assume these eight cars resemble a random sample of cars.

				Car				
	1	2	3	4	5	6	7	8
Shop A	364	112	840	610	172	83	165	216
Shop B	412	110	960	640	163	75	160	274

Use a test for equality of mean bids that does not require a preliminary test of normality, at $\alpha = .05$.

8.29 Use a scatterplot similar to the one in Figure 8.6 for the paired data in Exercise 8.28 to find out if it tends to support the conclusion reached.

8.30 An investment firm has a research staff that predicts next year's earnings per share for several hundred companies. The actual earnings per share one year later is obtained for a random sample of those companies and compared with the research staff's estimate. Numbers in parentheses

indicate losses rather than earnings. Assume that any differences between actual and predicted earnings are normally distributed.

Company	Predicted	Actual	Company	Predicted	Actual
1	$1.32	$1.04	8	0.80	(1.48)
2	4.65	2.16	9	8.85	5.10
3	0.86	1.64	10	7.21	5.40
4	8.20	3.12	11	1.74	3.48
5	5.40	4.44	12	2.25	2.33
6	2.95	(1.01)	13	3.60	2.20
7	1.64	1.58	14	3.10	3.86

Test the hypothesis that the predictions of the research staff are on target,
on the average, in the population. Use $\alpha = .05$. (Use $\Sigma D = 18.71$,
$\Sigma D^2 = 77.4087$.)

8.31 For the data in Exercise 8.30, find a 95% confidence interval for the
mean error in prediction, which is calculated as predicted minus actual
earnings.

8.32 Suppose you are leery of the normality assumption in Exercise 8.30
because of a few large errors, such as with companies, 4, 6, and 9. Use
a nonparametric test to test the hypothesis in Exercise 8.30. Compare
the p-value with the one in Exercise 8.30. (Use $\Sigma R = 67$, $\Sigma R^2 = 1015$.)

8.33 Find a nonparametric 95% confidence interval for the median difference
in Exercise 8.30.

BIBLIOGRAPHY

Additional material on the topics presented in this chapter can be found in
the following publication.

Conover, W. J. (1980). *Practical Nonparametric Statistics,* 2nd ed. Wiley, New
York.

9
ESTIMATION AND HYPOTHESIS TESTING WITH TWO INDEPENDENT SAMPLES

PRELIMINARY REMARKS

In the previous chapter the value of pairing observations was discussed. This technique of matching observations should be used whenever possible to control extraneous variation in the experimental results. However, much sample data involve two sets of measurements that cannot be paired; in fact, the samples are not related at all. Sometimes they are obtained from different populations, such as a random sample of employees over 50 years of age, and a random sample of employees under 50. Other times they are the result of randomly assigning individuals to one of two different groups, such as assigning some new employees to a classroom training session and assigning the other new employees to on-the-job training. The objective in both cases is a comparison of the two samples to see if there is a significant difference between the populations they represent.

Such samples are independent of one another, whereas the paired samples of the previous chapter are not independent. Independent samples are usually denoted by $X_1, X_2, \ldots, X_{n_x}$ for the n_x items in the random sample from one population and by $Y_1, Y_2, \ldots, Y_{n_y}$ for the n_y items in the random sample from a second population. The sample sizes n_x and n_y are not necessarily equal. Examples of settings where independent samples are used include the following.

1. **X's** The distance recorded in yards for Brand A golf balls when tested by a mechanical driver.

   Y's The distance recorded in yards for Brand B golf balls when tested by a mechanical driver.

2. **X's** Daily sales by a company for one month prior to the start of their use of TV commercials.

   Y's Daily sales by the company for one month after the start of their use of TV commercials.

3. **X's** The mileage observed on several Goodyear tires.

   Y's The mileage observed on several Firestone tires.

4. **X's** The number of sickness absentees on Monday and Friday for a period of one year.

   Y's The number of sickness absentees on Tuesday, Wednesday, and Thursday for a period of one year.

5. **X's** Achievement scores in arithmetic recorded by students taught by a teacher in an ordinary classroom setting.

   Y's Achievement scores in arithmetic recorded by students using a programmed text where the students consulted with the teacher only when they had questions.

For each of these samples the focal point of interest is a comparison of the means of respective populations to see if they are the same or different. For the preceding examples interest would center on the following comparisons.

1. Is the mean driving distance longer for one brand of golf balls?
2. Is the mean of sales after the use of TV commercials greater than the mean was before the start of the TV commercials?
3. Is the tire mileage different for Goodyear and Firestone tires?
4. Is the mean number of people calling in sick on Monday and Friday greater than the mean number on Tuesday, Wednesday, and Thursday?
5. Do students have better arithmetic skills when taught with a self-paced programming text than when taught with the conventional method of instruction?

The investigation of these questions requires a formal statement of the hypothesis of interest as well as the calculation of a test statistic. The first section of this chapter is devoted to methods that can be used with most types of populations if the sample sizes are large (>30). These methods may be used when the sample sizes are small if the populations are normal and variances are known. If the variances are unknown and the populations are normal, the methods of the second section may be used. The nonparametric method of the third section is appropriate for all populations and sample sizes, but especially for small samples from nonnormal populations because the methods of the first two sections are not valid in that situation.

9.1

LARGE SAMPLES: INFERENCES ABOUT THE DIFFERENCE BETWEEN TWO MEANS

PROBLEM SETTING 9.1

A local school board has the responsibility of providing the best possible education for the youth in the community, with a limited budget. In addition, the quality of education should be uniform for all neighborhoods in the community, regardless of the economic status or cultural background of the residents of the neighborhood.

To gain a better understanding of the differences among neighborhoods the school board conducted a survey. A listing of all families known to have school-age children was obtained and a random sample from each neighborhood was obtained. A brief but comprehensive questionnaire was drawn up and tested. Volunteer interviewers were obtained from each neighborhood, and they were carefully trained to administer this questionnaire to the families in the sample. After assuring the families of the complete confidentiality of the results, and after impressing on them that the survey results will be used only to provide a better education for their children, questions were asked regarding the family income level, the educational status of the parents, the value placed by the parents on educating their children, the amount of reading material available in the home, and the house rules regarding studying on school nights. With information such as this the school board is able to compare neighborhoods, and to tailor the educational opportunities in each neighborhood to match its inhabitants.

THE TEST FOR $\mu_X = \mu_Y$ WHEN $n_X > 30$ AND $n_Y > 30$

If the samples are large, say greater than 30 each, the central limit theorem may be used to show that $\overline{X}$, $\overline{Y}$, and $\overline{X} - \overline{Y}$ are approximately normally distributed. The fact that $\overline{X} - \overline{Y}$ is approximately normal furnishes the foundation for the validity of the test of this section.

The mean of the sampling distribution of $\overline{X} - \overline{Y}$ is $\mu_X - \mu_Y$ and the standard deviation is $\sqrt{\sigma_X^2/n_X + \sigma_Y^2/n_Y}$, where σ_X and σ_Y are the respective population standard deviations. Therefore, for large samples the approximate probability

$$P\left(-z_{1-\alpha/2} < \frac{(\overline{X} - \overline{Y}) - (\mu_X - \mu_Y)}{\sqrt{\sigma_X^2/n_X + \sigma_Y^2/n_Y}} < z_{1-\alpha/2}\right) \approx 1 - \alpha$$

obtained from Table A2 leads to the probability statement

$$P\,[(\overline{X} - \overline{Y}) - z_{1-\alpha/2}\sqrt{\sigma_X^2/n_X + \sigma_Y^2/n_Y} < \mu_X - \mu_Y$$
$$< (\overline{X} - \overline{Y}) + z_{1-\alpha/2}\sqrt{\sigma_X^2/n_X + \sigma_Y^2/n_Y}] \approx 1 - \alpha$$

which is the basis for the confidence interval for $\mu_X - \mu_Y$.

PROCEDURES FOR INFERENCES ABOUT THE DIFFERENCE IN MEANS OF TWO POPULATIONS (Large Samples, Both Sample Sizes Exceed 30) (Z-test)

Data

The data $X_1, X_2, \ldots, X_{n_X}$ and $Y_1, Y_2, \ldots, Y_{n_Y}$ represent two random samples of sizes n_X and n_Y, respectively, taken from two different populations. Denote the respective population means and variances by μ_X, σ_X, μ_Y, and σ_Y. Compute the sample means $\overline{X}$ and $\overline{Y}$, and if σ_X^2 and σ_Y^2 are unknown compute the sample variances s_X^2 and s_Y^2.

Assumptions

1. Both samples are random samples from their respective populations.

2. The two samples are independent of one another.

3. The samples are large enough so that $\overline{X}$ and $\overline{Y}$ are approximately normally distributed by the central limit theorem. Usually $n_X > 30$ and $n_Y > 30$ is sufficient for this assumption to hold true.

Confidence Interval

A $100(1 - \alpha)\%$ confidence interval for the difference $\mu_X - \mu_Y$ is given approximately by

$$\overline{X} - \overline{Y} \pm z_{1-\alpha/2}\sqrt{\sigma_X^2/n_X + \sigma_Y^2/n_Y}$$

where $z_{1-\alpha/2}$ is the $1 - \alpha/2$ quantile from Table A2. If σ_X^2 and σ_Y^2 are unknown, s_X^2 and s_Y^2 may be used instead.

Null Hypothesis

$$H_0: \mu_X = \mu_Y$$

Test Statistic

$$T = \frac{\overline{X} - \overline{Y}}{\sqrt{\sigma_X^2/n_X + \sigma_Y^2/n_Y}} \tag{9.1}$$

If σ_X^2 and σ_Y^2 are unknown, s_X^2 and s_Y^2 may be used since the sample sizes are large. The null distribution of T is approximately a standard normal for large sample sizes. It is exactly a standard normal for all sample sizes if the populations are normal and the variances are known.

Decision Rule

The decision rule is based on the alternative hypothesis of interest.

(a) H_1: $\mu_X > \mu_Y$ Reject H_0 if $T > z_{1-\alpha}$ (see Figure 9.1a).

(b) H_1: $\mu_X < \mu_Y$ Reject H_0 if $T < z_\alpha$ (see Figure 9.1b).

(c) H_1: $\mu_X \neq \mu_Y$ Reject H_0 if $T > z_{1-\alpha/2}$ or if $T < z_{\alpha/2}$ (see Figure 9.1c).

The value of z_α is the α quantile from Table A2.

Note 1

This procedure is exact for all sample sizes, if the populations are normal and σ_X^2 and σ_Y^2 are known, conditions that rarely occur in actual applications.

Note 2

This Z-test can be used to test the more general null hypothesis H_0: $\mu_X - \mu_Y = d$, for some value of d, merely by subtracting d from the numerator of the test statistic in Eq. 9.1. This permits testing the null hypothesis that μ_X is larger than μ_Y by an amount d.

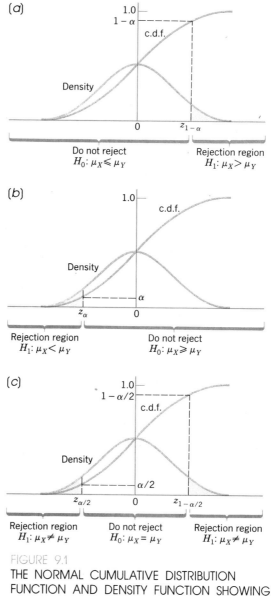

FIGURE 9.1
THE NORMAL CUMULATIVE DISTRIBUTION
FUNCTION AND DENSITY FUNCTION SHOWING
THE REJECTION REGIONS FOR LARGE
SAMPLES, USING (a) AN UPPER-TAILED TEST,
(b) A LOWER-TAILED TEST, AND (c) A TWO-
TAILED TEST.

EXAMPLE

The school board would like to compare the mean income in two different
school districts in their city. Fifty households are randomly selected in each
district with the following results.

	District 1	District 2
Sample mean	$32,070	$30,750
Sample standard deviation	$ 2,500	$ 3,000

Use these data to test $H_0: \mu_X = \mu_Y$ versus $H_1: \mu_X \neq \mu_Y$ with $\alpha = .05$. The test statistic is given by

$$T = \frac{\overline{X} - \overline{Y}}{\sqrt{s_X^2/n_X + s_Y^2/n_Y}} = \frac{32,070 - 30,750}{\sqrt{(2500)^2/50 + (3000)^2/50}} = 2.3901$$

Decision rule Reject H_0 if $T > z_{.975} = 1.9600$ or if $T < z_{.025} = -1.9600$. Reject H_0 and conclude that the mean income in the two school districts is not the same. The p-value associated with $T = 2.3901$ is found from Table A2* to be twice .0084, because this is a two-tailed test and $P(T > 2.3901) = .0084$. Thus the p-value is .0168 and the decision to reject H_0 is confirmed, since this value is less than the α-level of .05.

A 95% confidence interval for the difference in means $\mu_X - \mu_Y$ uses many of the same calculations.

$$\overline{X} - \overline{Y} \pm z_{.975}\sqrt{s_X^2/n_X + s_Y^2/n_Y} = 1320 \pm 1.9600(552.27) = 1320 \pm 1082$$

Thus a 95% confidence interval for the difference $\mu_X - \mu_Y$ is the interval from $238 to $2402. Since the interval does not contain the value zero, this indicates that the difference $\mu_X - \mu_Y$ is larger than zero or equivalently $\mu_X > \mu_Y$.

NORMAL POPULATIONS AND KNOWN VARIANCES

Suppose one or both samples have fewer than 30 observations. Then the central limit theorem cannot be used to justify the approximate normality of $\overline{X} - \overline{Y}$ and the Lilliefors test should be applied to each sample. If the Lilliefors test does not indicate lack of normality in the populations, then the procedures of this section are still valid provided the seldom-known population variances are known.

EXAMPLE

Two brands of golf balls are to be compared with respect to driving distance. The balls are tested on a mechanical driver that is known to give normally distributed distances with a standard deviation of 15 yards. Hence, the driving distance observed for Brand X and Brand Y can be thought of as normally distributed random variables with unknown means μ_X and μ_Y, each with a population variance of $15^2 = 225$. Each brand is tested based on 25 randomly selected balls. These tests give the results $\overline{X} = 275$ and $\overline{Y} = 290$. Test the hypothesis $H_0: \mu_X = \mu_Y$ versus $H_1: \mu_X \neq \mu_Y$ with $\alpha = .01$.

Test statistic

$$T = \frac{\bar{X} - \bar{Y}}{\sqrt{\sigma_X^2/n_X + \sigma_Y^2/n_Y}} = \frac{275 - 290}{\sqrt{225/25 + 225/25}} = -3.5355$$

Decision rule Reject H_0 if $T > z_{.995} = 2.5758$ or if $T < -z_{.995} = -2.5758$. Therefore, H_0 is rejected and it is concluded that the mean driving distance is not the same for the two golf balls. The p-value associated with $T = -3.5355$ is $2P(T < -3.5355)$ because this is a two-tailed test. Use of Table A2 gives the p-value of less than $2(.0005) = .001$. This small p-value indicates that the evidence is strong for rejecting H_0. A 99% confidence interval for the difference $\mu_X - \mu_Y$ in mean driving distance of the two balls is

$$\bar{X} - \bar{Y} \pm z_{.995} \sqrt{\frac{\sigma_X^2}{n_X} + \frac{\sigma_Y^2}{n_Y}} = -15 \pm 2.5758 \sqrt{\frac{225}{25} + \frac{225}{25}}$$

$$= -15 \pm 10.93$$

or from -25.93 yards to -4.07 yards. Since this interval does not contain zero, there is no reason to believe that the driving distance is the same for the two brands of golf balls.

EXERCISES

9.1 A random sample of recent graduates of a local business school revealed 36 accounting majors making an average of $28,000 a year with a standard deviation of $2100, whereas 32 general business majors were making an average of $22,000 with a standard deviation of $820. Find a 95% confidence interval for the difference in the mean income for the two majors and interpret the results.

9.2 The regional sales office for a personal computer company frequently compares the number of sales made by their salespeople each week in different cities. Past experience shows both populations to be normally distributed and the standard deviation of the number of sales in each population to be 12. Nine salespeople in Kansas City and six salespeople in St. Louis are selected at random and show the following number of sales:

Kansas City: 56   37   61   56   65   41   63   50   42

St. Louis: 46   25   46   64   34   56

State and test the appropriate null and alternative hypotheses to see if the mean number of sales differs for the two cities. Use $\alpha = .10$ and state the p-value.

9.3 Pharmaceutical companies desiring to get a share of the mild tranquilizer (like Valium and Librium) market are continually testing new compounds in their laboratories. In one of the experiments used for screening new compounds laboratory mice are placed in individual sealed beakers and the times to dormancy in minutes are recorded when testing for drugs that affect the central nervous system. The reason for this test is that an experimental drug may either speed up the metabolism rate, in which case the air supply is quickly diminished, or it may have the opposite effect, or no effect at all. A placebo was tested on 40 mice as a control group and showed a mean time to dormancy of 16.4 min with a standard deviation of 1.55 min. An experimental group of size 40 was treated with a new compound and had a mean time to dormancy of 16.9 min with a standard deviation of 1.67 min. Test the null hypothesis that mean times to dormancy are the same for both groups versus an alternative that the mean times are different with $\alpha = .01$. State the p-value.

9.4 Find a 95% confidence interval for the difference in mean times to dormancy for the two populations of mice discussed in Exercise 9.3.

9.5 A laboratory starts their experiments with mice on Monday and Wednesday of each week. To ensure that the experimental results are reliable, the weights of the mice are required to be the same on both starting days. A lab technician suspects that the mice starting on Wednesday tend to be lighter than those starting on Monday. Random samples of weights obtained on 45 mice on each day gave the following results:

$$\text{Monday: } \overline{X} = 15.88 \text{ g, } s_X = 1.25 \text{ g}$$

$$\text{Wednesday: } \overline{Y} = 15.58 \text{ g, } s_Y = 1.21 \text{ g}$$

Test the appropriate hypothesis with $\alpha = .10$. State the p-value.

9.6 In Exercise 8.9 a comparison was made on mpg obtained with unleaded gasoline and gasohol that used matched pairs. The experiment could also be performed by using two groups of drivers under similar driving conditions, one using only unleaded gasoline and the other using only gasohol. Fifty drivers in one group got 23.8 mpg on gasohol with a standard deviation of 1.7 and fifty drivers in the other group got 23.4 mpg on unleaded gasoline with a standard deviation of 1.9. Test the hypothesis that there is no difference in the mean mpg of the two fuels, with $\alpha = .01$. Comment on whether you think the matched-pairs approach or the two-independent-samples approach is the best way to conduct this experiment.

9.7 A random sample consisting of 35 McDonald's stores revealed a mean number of employees of 28.3 with a standard deviation of 3.3. A random sample of 40 Wendy's stores had a mean of 26.7 and a standard deviation of 4.9. Do Wendy's and McDonald's employ the same average number of employees nationwide? Use $\alpha = .05$ and state the p-value.

9.8 A measure of productivity has been devised for workers on an assembly line. A random sample of 40 workers in 1987 is compared with the results using a random sample of 40 workers previously obtained in 1982. The sample statistics are $\overline{X} = 115$ and $s_X = 12$ for 1987 and $\overline{Y} = 108$ and $s_Y = 22$ for 1982. Find a 95% confidence interval for the mean increase in productivity and interpret the results.

9.2

SAMPLES FROM NORMAL DISTRIBUTIONS: INFERENCES ABOUT THE DIFFERENCE BETWEEN TWO MEANS

PROBLEM SETTING 9.2

A health management firm wants to know which of two hospitals has the highest rating among ex-maternity patients. If the firm knows how ex-maternity patients rate the service, food, cleanliness, staff friendliness, etc., then they can make recommendations to their clients who are seeking maternity care. Also they can advise the hospitals on how they can improve their maternity service.

A questionnaire is devised for measuring the concerns of the ex-maternity patients. As maternity patients are preparing to check out of the hospitals they are asked to participate in the survey. Each questionnaire response is converted to a rating on a scale from 0, poor, to 100, perfect. Then the ratings are compared between the two hospitals. Suppose that the following ratings were reported:

Hospital A: 81 86 73 77 90 91 75 62 98 74
Hospital B: 89 55 59 64 37 58 35 57 65 68 42 71 69 49 67

How can the health management firm use this information to see if the population mean ratings are different for the two hospitals?

WHICH TEST SHOULD BE USED?

If the sample sizes are large enough, it is simple and convenient to use the Z-test in the previous section for samples from any population. But if the populations are normal and the variances are unknown, the tests of this section should be preferred, no matter what the sample sizes, because of their greater power. For nonnormal populations the nonparametric procedure in the next section is recommended for all sample sizes, because of its greater power in most cases.

In this section methods are given for handling two independent samples when the populations are normal and the variances are unknown. The reasonableness of the normality assumption should be checked with a goodness-of-fit test such as the Lilliefors test in each case. If the normality assumption is not satisfied for either sample, the nonparametric procedure in the next section should be used instead because it usually has more power in such cases.

SHOWING DIFFERENCES GRAPHICALLY

In Problem Setting 9.2 the health management firm is interested in knowing if the mean rating is different for the two hospitals. As a first step on a problem of this nature it is useful to plot the data. This is an application where the

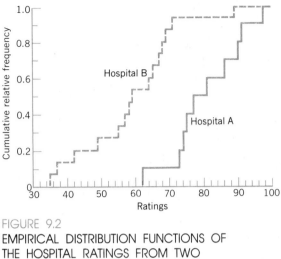

FIGURE 9.2

EMPIRICAL DISTRIBUTION FUNCTIONS OF
THE HOSPITAL RATINGS FROM TWO
INDEPENDENT SAMPLES.

e.d.f. plot is quite useful because both e.d.f.s can be put in the same figure, as shown in Figure 9.2.

What type of information is contained in Figure 9.2? First, if there were no difference in the ratings for the two hospitals, then the two e.d.f.s would be likely to cross each other several times. As it is, the Hospital A group is clearly shifted to the right of the Hospital B group. The only way such a shift can occur is for one group to have consistently higher values than the other. A comparison of the medians of the two groups, obtained easily from Figure 9.2, shows that the Hospital A group has a median of 79 and the Hospital B group has a median of 59. Hence, this lends support to the interpretation of the e.d.f. plot that one sample tends to give higher ratings than the other.

THE TWO-SAMPLE t-TEST

Two-sample problems are most commonly associated with small sample sizes and with equal but unknown variances. For this situation the small samples must be used to provide an estimate of the common but unknown variance σ^2. The sample variances s_X^2 and s_Y^2 each provide an estimate of σ^2. However, a better estimate of σ^2 can be obtained by pooling the two estimates in a weighted average called the **pooled sample variance s_p^2**:

$$s_p^2 = \frac{(n_X - 1)s_X^2 + (n_Y - 1)s_Y^2}{n_X + n_Y - 2} \tag{9.2}$$

For equal sample sizes this formula reduces to $s_p^2 = (s_X^2 + s_Y^2)/2$, which is the simple average of the two estimates. Otherwise, each of the estimates receives weights proportional to its respective degrees of freedom. As in the previous chapters, when σ^2 is estimated from small samples the resulting test and confidence interval involve using a Student's t random variable when the populations are normal. A frequently used two-sample technique, commonly called the **two-sample t-test,** is now presented and illustrated.

PROCEDURES FOR INFERENCES ABOUT THE DIFFERENCE IN MEANS OF TWO NORMAL POPULATIONS WITH EQUAL VARIANCES (Two-Sample t-Test)

Data

The data consist of two random samples $X_1, X_2, \ldots, X_{n_X}$ of size n_X and $Y_1, Y_2, \ldots, Y_{n_Y}$ of size n_Y from two populations. Denote the respective population means by μ_X and μ_Y. Compute the sample means $\overline{X}$ and $\overline{Y}$ and the sample variances s_X^2 and s_Y^2. (If the population variances σ_X^2 and σ_Y^2 are known, use the methods outlined in Section 9.1.) Also compute the *pooled sample standard deviation*:

$$s_p = \left\{ \frac{(n_X - 1)s_X^2 + (n_Y - 1)s_Y^2}{n_X + n_Y - 2} \right\}^{1/2} \tag{9.3}$$

Assumptions

1. Both samples are random samples from their respective populations.
2. The two samples are independent of one another.
3. Both populations are normal with equal variances ($\sigma_X^2 = \sigma_Y^2$).

Confidence Interval

A $100(1 - \alpha)\%$ confidence interval for the difference $\mu_X - \mu_Y$ is given by

$$(\overline{X} - \overline{Y}) \pm t_{1-\alpha/2, n_X + n_Y - 2} s_p \sqrt{\frac{1}{n_X} + \frac{1}{n_Y}}$$

where $t_{1-\alpha/2, n_X + n_Y - 2}$ is the $1 - \alpha/2$ quantile of the Student's t distribution with $n_X + n_Y - 2$ degrees of freedom, obtained from Table A3.

Null Hypothesis

$$H_0: \mu_X = \mu_Y$$

Test Statistic

$$T = \frac{\overline{X} - \overline{Y}}{s_p \sqrt{1/n_X + 1/n_Y}} \tag{9.4}$$

The null distribution of T is the Student's t distribution with $n_X + n_Y - 2$ degrees of freedom, obtained from Table A3.

Decision Rule

The decision rule depends on the alternative hypothesis of interest.

(a) $H_1: \mu_X > \mu_Y$. Reject H_0 if $T > t_{1-\alpha, n_X + n_Y - 2}$.

(b) $H_1: \mu_X < \mu_Y$. Reject H_0 if $T < -t_{1-\alpha, n_X + n_Y - 2}$.

(c) $H_1: \mu_X \neq \mu_Y$. Reject H_0 if $T > t_{1-\alpha/2, n_X + n_Y - 2}$ or $T < -t_{1-\alpha/2, n_X + n_Y - 2}$

The value of $t_{1-\alpha, n_X + n_Y - 2}$ is the $1 - \alpha$ quantile of the Student's t distribution with $n_X + n_Y - 2$ degrees of freedom, obtained from Table A3.

Note

This procedure is appropriate to use for all sample sizes when both populations are normal with equal population variances.

The hospital rating data associated with Problem Setting 9.2 are used to test H_0: $\mu_X = \mu_Y$ versus H_1: $\mu_X \neq \mu_Y$ with $\alpha = .05$, where μ_X is the population mean of Hospital A's ratings and μ_Y is the population mean for Hospital B's ratings. The Lilliefors test on each sample shows that the normality assumption is reasonable. The assumption of equal variances will be addressed later in this section. Because of the reasonableness of the assumptions the test of this section is used.

The sample data give the following results.

Hospital A: $n_X = 10$, $\overline{X} = 80.70$, $s_X^2 = 113.34$
Hospital B: $n_Y = 15$, $\overline{Y} = 59.00$, $s_Y^2 = 201.43$

$$s_p = \left\{ \frac{(10-1)113.34 + (15-1)201.43}{10+15-2} \right\}^{1/2} = \sqrt{166.96} = 12.92$$

Test statistic

$$T = \frac{80.70 - 59.00}{12.92\sqrt{1/10 + 1/15}} = \frac{21.70}{5.275} = 4.1137$$

Decision rule From Table A3, $t_{.975, 10+15-2} = 2.0687$, and so H_0 is rejected if $|T| > 2.0687$. Therefore, H_0 is soundly rejected, and it is concluded that Hospital A receives significantly higher ratings than Hospital B, as previously indicated by a comparison of the two e.d.f.s in Figure 9.2. The p-value is found by interpolation in Table A3 to be about $2(.0003) = .0006$, which indicates that these results could hardly have occurred by chance.

A 95% confidence interval for the difference in mean ratings is given by

$$\overline{X} - \overline{Y} \pm t_{.975, 23} s_p \sqrt{1/n_X + 1/n_Y} = 21.70 \pm 2.0687(5.275)$$

or from 10.79 to 32.61, which again confirms that ex-maternity patients tend to rate Hospital A higher than Hospital B, because the confidence interval does not include 0.

THE F-TEST FOR EQUALITY OF VARIANCES

In the presentation of the two-sample t-test the estimates s_X^2 and s_Y^2 were pooled together to form s_p^2 as an estimate of the common variance σ^2. However, this pooling was done under the assumption that the populations had equal variance; otherwise, the pooling is not justified and the t-test cannot be used. A procedure is now presented for testing the assumption of equal variances when two samples come from normal populations. *This test should be used only after the normality assumption has been checked and verified.* Tests are valid only when their assumptions are satisfied; the distribution of the F statistic (below) is affected by nonnormalities of the populations.

The test for equality of variances involves a distribution not yet encountered in this text, known as an F distribution. A random variable that consists of the ratio of two sample variances, s_1^2/s_2^2, has an F distribution if the two samples are independent samples from normal populations with equal population variances $\sigma_1^2 = \sigma_2^2$. Table A5 gives selected quantiles for the family of F distributions. The family of F distributions is indexed by two parameters:

PROCEDURE FOR TESTING THE EQUALITY OF VARIANCES FOR TWO NORMAL POPULATIONS (F-test)

Data

The data consist of two random samples $X_1, X_2, \ldots, X_{n_X}$ of size n_X and $Y_1, Y_2, \ldots, Y_{n_Y}$ of size n_Y from two populations. Compute the sample variances s_X^2 and s_Y^2.

Assumptions

1. Both samples are random samples from their respective populations.

2. The two samples are independent of one another.

3. Both populations are normal.

Estimation

The unknown population variances σ_X^2 and σ_Y^2 are estimated by the sample variances s_X^2 and s_Y^2, respectively.

Null Hypothesis

$$H_0: \sigma_X^2 = \sigma_Y^2$$

Alternative Hypothesis

$$H_1: \sigma_X^2 \neq \sigma_Y^2$$

Test Statistic

Let s_1^2 be the larger of the two sample variances, and let s_2^2 be the smaller of the two sample variances. The test statistic is

$$F = \frac{s_1^2}{s_2^2}$$

That is, the test statistic is the ratio of the larger of the two sample variances to the smaller of the two sample variances. The null distribution of F is the F distribution given in Table A5, with the degrees of freedom as explained below under the decision rule.

Decision Rule

If $s_X^2 > s_Y^2$, let $k_1 = n_X - 1$ and $k_2 = n_Y - 1$.

If $s_X^2 < s_Y^2$, let $k_1 = n_Y - 1$ and $k_2 = n_X - 1$.

Reject H_0 at the level of significance α if $F > F_{1-\alpha/2, k_1, k_2}$.

The value $F_{1-\alpha/2, k_1, k_2}$ is the $1 - \alpha/2$ quantile of the F distribution with k_1 and k_2 degrees of freedom given in Table A5.

Note

If the desired degrees of freedom for the F distribution cannot be found in Table A5, simply use the next smaller degrees of freedom given in the table.

k_1 = degrees of freedom in the numerator

k_2 = degrees of freedom in the denominator

The correct F distribution is selected from Table A5 by noting that the degrees of freedom parameters may be found easily from the two sample sizes

$k_1 = n_1 - 1$, where n_1 is the sample size for s_1^2

$k_2 = n_2 - 1$, where n_2 is the sample size for s_2^2

EXAMPLE

The hospital ratings in Problem Setting 9.2 are now used to test H_0: $\sigma_X^2 = \sigma_Y^2$ versus H_1: $\sigma_X^2 \neq \sigma_Y^2$ to see if the assumption of equality of variance for the two-sample t-test was justified. Let $\alpha = .10$. Recall that the assumption of normality was indicated to be reasonable in the previous example.

Test statistic The sample variances s_X^2 and s_Y^2 were previously given as 113.34 and 201.43, respectively. Since s_Y^2 is the larger of these two values, the F ratio is found as

$$F = \frac{s_Y^2}{s_X^2} = \frac{201.43}{113.34} = 1.78$$

Decision rule Since $s_X^2 < s_Y^2$, let $k_1 = n_Y - 1 = 15 - 1 = 14$ and $k_2 = n_X - 1 = 10 - 1 = 9$. The F distribution for $k_1 = 14$, $k_2 = 9$ is not given in Table A5, so the F distribution with the next smaller degrees of freedom is used, which in this case is 12 and 9. From Table A5, $F_{.95,12,9} = 3.073$ so the decision rule is to reject H_0 if $F > 3.073$. Since the observed F is less than 3.073, H_0 is not rejected. Keep in mind that this solution depends on the normality assumption being satisfied, for if it is not satisfied this test is without meaning.

If the null hypothesis of equal variances is accepted, then the two-sample t-test can be used given that the populations are normal. If the null hypothesis of equal variances is rejected and the populations are normal, then Satterthwaite's test (to follow), sometimes known as Welch's test, should be used as an approximate test. There is no exact test for equal means if the population variances are unequal and unknown.

SMALL SAMPLES AND UNEQUAL VARIANCES
(Satterthwaite's test)

An approximate procedure, called **Satterthwaite's test,** has been devised for the case where the populations are normal but the variances are unknown and unequal. If the sample sizes are large, the method of the previous section can be used, but if the sample sizes are small, some modification is required. The change suggested by Satterthwaite is to use the same test statistic used in the previous section, but instead of comparing it with the normal distribution, the t distribution is used, where the degrees of freedom in the t distribution is given by the equation

$$f = \frac{\left(\dfrac{s_X^2}{n_X} + \dfrac{s_Y^2}{n_Y}\right)^2}{\dfrac{(s_X^2/n_X)^2}{n_X - 1} + \dfrac{(s_Y^2/n_Y)^2}{n_Y - 1}} \qquad (9.5)$$

The number f will always be less than or equal to $n_X + n_Y - 2$, the degrees of freedom used in the t-test. The test statistic is not the same as in the t-test either, because a pooled estimate of the variance is no longer justified, so the test statistic in Eq. 9.1 is used.

As the sample sizes get larger, f becomes large also, and the t distribution with f degrees of freedom approaches the normal distribution. This procedure then becomes the procedure of the previous section, as one would expect to happen.

EXAMPLE

Suppose that the ratings in Problem Setting 9.2 had been collected during a five-week period in January and February, and the firm is interested in conducting a similar study during the summer months. Data for a five-week period in the summer were collected and summarized as follows.

$$\text{Hospital A: } n_X = 10, \ \overline{X} = 92.62, \ s_X^2 = 868.11$$
$$\text{Hospital B: } n_Y = 15, \ \overline{Y} = 69.72, \ s_Y^2 = 206.65$$

Both samples pass the Lilliefors test for normality. The null hypothesis of interest is H_0: $\mu_X = \mu_Y$ versus the alternative H_1: $\mu_X \neq \mu_Y$ as before. However, the assumption of equal variances should be tested first. The test for equality of variances uses the F statistic

$$F = \frac{s_1^2}{s_2^2} = \frac{s_X^2}{s_Y^2} = \frac{868.11}{206.65} = 4.201$$

The critical value $F_{.975,9,14} = 3.209$ is used for an $\alpha = .05$ test, because the degrees of freedom are 9 and 14. Since the observed F exceeds 3.209, the null hypothesis of equal variances is rejected. The p-value is less than .02.

Therefore, in the test of H_0: $\mu_X = \mu_Y$ Satterthwaite's approximation is used.

PROCEDURES FOR INFERENCES ABOUT THE DIFFERENCE IN MEANS OF TWO NORMAL POPULATIONS WITH UNEQUAL VARIANCES (Satterthwaite Test)

Data

The data consist of two random samples $X_1, X_2, \ldots, X_{n_X}$ of size n_X and $Y_1, Y_2, \ldots, Y_{n_Y}$ of size n_Y from two populations. Denote the respective population means by μ_X and μ_Y. Compute the sample means $\overline{X}$ and $\overline{Y}$ and sample variances s_X^2 and s_Y^2. (If the population variances σ_X^2 and σ_Y^2 are known use the methods outlined in Section 9.1.) Also compute the approximate degrees of freedom f given by

$$f = \frac{\left(\dfrac{s_X^2}{n_X} + \dfrac{s_Y^2}{n_Y}\right)^2}{\dfrac{(s_X^2/n_X)^2}{n_X - 1} + \dfrac{(s_Y^2/n_Y)^2}{n_Y - 1}} \tag{9.6}$$

Assumptions

1. Both samples are random samples from their respective populations.

2. The two samples are independent of one another.

3. Both populations are normal with unequal variances ($\sigma_X^2 \neq \sigma_Y^2$).

Confidence interval

An approximate $100(1 - \alpha)\%$ confidence interval for the difference $\mu_X - \mu_Y$ is given by

$$\overline{X} - \overline{Y} \pm t_{1-\alpha/2, f}\sqrt{s_X^2/n_X + s_Y^2/n_Y}$$

where $t_{1-\alpha/2, f}$ is the $1 - \alpha/2$ quantile of the Student's t distribution with f degrees of freedom, obtained from Table A3.

Null Hypothesis

$$H_0: \mu_X = \mu_Y$$

Test Statistic

$$T = \frac{\overline{X} - \overline{Y}}{\sqrt{s_X^2/n_X + s_Y^2/n_Y}} \tag{9.7}$$

The null distribution of T is approximately the Student's t distribution with f degrees of freedom.

Decision Rule

The decision rule depends on the alternative hypothesis of interest.

(a) $H_1: \mu_X > \mu_Y$. Reject H_0 if $T > t_{1-\alpha, f}$.

(b) $H_1: \mu_X < \mu_Y$. Reject H_0 if $T < -t_{1-\alpha, f}$.

(c) $H_1: \mu_X \neq \mu_Y$. Reject H_0 if $T > t_{1-\alpha/2, f}$ or if $T < -t_{1-\alpha/2, f}$.

The value $t_{1-\alpha, f}$ is the $1 - \alpha$ quantile of the Student's t distribution with f degrees of freedom. Linear interpolation may be used to find $t_{1-\alpha, f}$ in Table A3 if f is not an integer, but simply rounding f to the next smaller integer is usually sufficient to make a valid decision.

Note

This test is only approximate. It may be used for all sample sizes, but is most appropriate when either $n_X \leq 30$ or $n_Y \leq 30$.

The approximate degrees of freedom from Eq. 9.6 becomes

$$f = \frac{\left(\dfrac{868.11}{10} + \dfrac{206.65}{15}\right)^2}{\dfrac{(868.11/10)^2}{9} + \dfrac{(206.65/15)^2}{14}} = 11.9$$

The test statistic, using Eq. 9.7, becomes

$$T = \frac{92.62 - 69.72}{\sqrt{868.11/10 + 206.65/15}} = 2.2833$$

For an $\alpha = .05$ test, the null hypothesis is rejected in favor of the alternative $H_1: \mu_X \neq \mu_Y$ if T exceeds $t_{.975,\,f}$ for $f = 11.9$ degrees of freedom. From Table A3, $t_{.975,11} = 2.2010$ is used as the critical value, so the null hypothesis is rejected. The p-value is slightly less than .05, indicating a greater preference for Hospital A than for Hospital B.

EXERCISES

9.9   A college professor teaches two sections of the same subject. One section meets at 8 A.M. and the other meets at 10 A.M. with one hour separating the two classes. At examination time she gives identical exams to both classes but is concerned that the 10 A.M. section might have higher scores based on possible information received about the test from individuals in the 8 A.M. section. Test the null hypothesis $H_0: \mu_8 = \mu_{10}$ versus $H_1: \mu_8 < \mu_{10}$ with $\alpha = .05$ and state the p-value. Assume these test scores resemble two independent random samples from normal populations. (Note that for the 8 A.M. section, $n_X = 23$, $\Sigma X = 1628$, $\Sigma X^2 = 122{,}068$, and for the 10 A.M. section, $n_Y = 30$, $\Sigma Y = 2376$, $\Sigma Y^2 = 194{,}482$.) Also, plot both e.d.f.s on the same graph to see if a visual display of the data agrees with the analysis.

8 A.M. Scores			10 A.M. Scores		
98	95	94	100	98	98
91	90	89	97	96	94
87	77	76	92	92	89
75	73	70	86	85	85
67	67	66	85	82	81
65	57	56	79	79	78
55	53	45	78	77	77
43	39		74	72	67
			65	62	56
			56	49	47

9.10 Find a 95% confidence interval for the difference in the means of the two classes in Exercise 9.9.

9.11 Rework Exercise 9.1, supposing that both populations of salaries are normally distributed. How does this confidence interval compare with the confidence interval that was found in Exercise 9.1? Which confidence interval do you think is more accurate when the populations are normal?

9.12 A consumer testing service compared gas ovens to electric ovens by baking one type of bread in five ovens of each type. Assume the baking times are normally distributed. The gas ovens had an average baking time of 0.90 hours with a standard deviation of 0.09 hours and the electric ovens had an average baking time of 0.70 hours with a standard deviation of 0.16 hours. At the 5% level of significance, should the null hypothesis of identical mean baking times for the two kinds of ovens be rejected? State the p-value. Find a 95% confidence interval for the difference in mean baking times.

9.13 An appliance store decides to advertise on television to increase business. The sales for each of 10 days after the TV commercials are compared with 10 days before the start of the TV commercials. Assume that the data behave as if they were two independent random samples from normal populations. State and test the appropriate hypothesis with $\alpha = .05$. State the p-value.

Sales before Commercials		Sales after Commercials	
$564	621	$617	645
560	562	681	610
634	597	628	678
641	664	644	755
597	565	591	597

9.14 Find a 90% confidence interval for the mean increase in sales based on the data presented in Exercise 9.13.

9.15 A college senior is considering job offers for the same amount of money in two different cities and is concerned that housing costs might not be the same in each city. He obtains a copy of the Sunday paper from each city and uses a random selection procedure to choose 20 classified ads from the "Apartments for Rent" section. He finds the average rental cost in one city to be $355 with a standard deviation of $25 and the average cost in the other city to be $375 with a standard deviation of $18. Assuming that the rental costs are normally distributed, test the appropriate hypothesis with $\alpha = .05$ to see if the average rental cost differs in the two cities. State the p-value.

9.16 Two grade school classes were taught arithmetic by two different methods. In the standard group the students were taught by the teacher in an ordinary classroom situation. In the experimental group the students taught themselves using a programmed text and consulted the teacher only when they had questions. At the end of the year each group was given a standard exam with the results given below. Can it be concluded that the experimental group has a higher mean level of achievement if it is assumed that these classes were selected randomly and that the populations are normal? Let $\alpha = .01$ and state the p-value.

Scores for Standard Group	Scores for Experimental Group
72	111
75	118
77	128
80	138
104	140
110	150
125	163
	164
	169

9.3

GENERAL POPULATIONS: INFERENCES ABOUT THE DIFFERENCE BETWEEN TWO MEANS (OPTIONAL)

THE WILCOXON–MANN–WHITNEY RANK SUM TEST

The two-sample techniques of the previous section have the underlying assumption that the samples come from normal populations. If a Lilliefors test shows this assumption to be unwarranted, then the procedure presented in this section should be used.

Some statisticians prefer to use this test in all situations because the Lilliefors test does not always detect nonnormal populations, especially when sample sizes are small. In addition, this technique is not nearly as sensitive to the assumption of equal variances as the two-sample t-test, so it may be used as an approximate test when the assumption is not valid.

The name of this nonparametric procedure is the **Wilcoxon–Mann–Whitney rank sum test.** It is calculated by performing the two-sample t-test on rank-transformed data. Hence, this is also a rank-transformation test as was the Wilcoxon signed-ranks test. With this test the data are ranked from 1 to $(n_x + n_y)$. Average ranks are used in case of ties.

For example, consider the following set of 10 observations to see how the ranking is done.

Data X:	4.8	9.2	3.6	6.3	
Data Y:	7.1	6.3	11.8	10.5	8.7	6.5
Joint ranking R_X:	2	8	1	3.5	
Joint ranking R_Y:	6	3.5	10	9	7	5

The smallest observation is 3.6, so it gets rank 1. The largest observation is 11.8, so it gets the largest rank 10. The two observations tied at 6.3 each get the average rank 3.5, representing the average of the ranks 3 and 4 they would have received if they were not tied. The two-sample t-test is calculated on the ranks assigned to the X and Y samples rather than on the sample observations.

PROCEDURE FOR TESTING HYPOTHESES ABOUT THE DIFFERENCE IN MEANS OF TWO GENERAL POPULATIONS
(Wilcoxon–Mann–Whitney rank sum test)

Data

The data consist of two random samples $X_1, X_2, \ldots, X_{n_x}$ of size n_x and $Y_1, Y_2, \ldots, Y_{n_y}$ of size n_y from two populations. Denote the respective population means by μ_x and μ_y. Assign ranks 1 to $(n_x + n_y)$ to the two samples jointly, using average ranks in case of ties. Compute the sample means $\bar{R}_x$ and $\bar{R}_y$ of the ranks, and the sample variances $s_{R_x}^2$ and $s_{R_y}^2$ of the ranks for the two samples, respectively. Also compute the pooled standard deviation of the ranks

$$s_p = \left\{ \frac{(n_x - 1)s_{R_x}^2 + (n_y - 1)s_{R_y}^2}{n_x + n_y - 2} \right\}^{1/2} \tag{9.8}$$

as was done with the two-sample t-test.

Assumptions

1. Both samples are random samples from their respective populations.

2. The two samples are independent of one another.

3. The two population distribution functions are identical except for possibly different means.

Null Hypothesis

$$H_0: \mu_x = \mu_y.$$

Test Statistic

$$T_R = \frac{\bar{R}_x - \bar{R}_y}{s_p \sqrt{1/n_x + 1/n_y}} \tag{9.9}$$

The null distribution of T_R is approximately the Student's t distribution with $n_x + n_y - 2$ degrees of freedom given in Table A3.

Decision Rule

The decision rule depends on the alternative hypothesis of interest.

(a) $H_1: \mu_x > \mu_y$. Reject H_0 if $T_R > t_{1-\alpha,n_x+n_y-2}$.

(b) $H_1: \mu_x < \mu_y$. Reject H_0 if $T_R < -t_{1-\alpha,n_x+n_y-2}$.

(c) $H_1: \mu_x \neq \mu_y$. Reject H_0 if $T_R > t_{1-\alpha/2,n_x+n_y-2}$ or if $T_R < -t_{1-\alpha/2,n_x+n_y-2}$.

The value $t_{1-\alpha,n_x+n_y-2}$ is the $1 - \alpha$ quantile of the Student's t distribution with $n_x + n_y - 2$ degrees of freedom obtained from Table A3.

Note

This procedure is valid for all sample sizes and all distributions, although the Student's t distribution is merely a good approximation to the exact distribution of T_R. The approximation works well even for small sample sizes.

Of course, this procedure is exactly the same as the two-sample t-test except that it is applied to the ranks of the data rather than to the data itself. This makes the Wilcoxon–Mann–Whitney rank sum test very easy to use. This nonparametric procedure is now illustrated in a rather long example that helps to point out some of the advantages of this test.

EXAMPLE

The owners of a franchise for a fast-food chain in a large city are suspicious that the cash shortages are higher at the store under one manager than they are at another store under a different manager. For 10 randomly selected business days the following cash shortages were reported by the managers.

Manager 1 X (Dollars)		Manager 2 Y (Dollars)	
7.05	4.50	9.25	6.90
14.25	37.60	2.05	10.00
8.75	9.40	2.75	8.00
10.50	8.10	2.50	33.00
11.90	45.20	6.40	5.20

First, to get a "visual feel" for the data, an e.d.f. is plotted in Figure 9.3 for the shortages reported by each of the managers. Since the e.d.f. for Manager 1 is always to the right (indicating larger numbers) than the e.d.f. for Manager 2 in Figure 9.3, it appears that Manager 1 has higher cash shortages than Manager 2.

For the purposes of illustration only, the two-sample t-test on the original data with $\alpha = .05$ will be used to test the hypotheses

$$H_0: \mu_X = \mu_Y \quad \text{versus} \quad H_1: \mu_X > \mu_Y$$

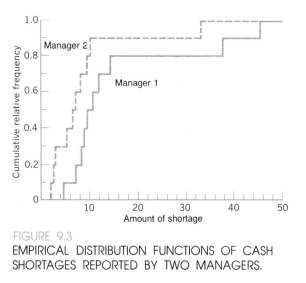

FIGURE 9.3
EMPIRICAL DISTRIBUTION FUNCTIONS OF CASH SHORTAGES REPORTED BY TWO MANAGERS.

The sample data give the following statistics.

$$\overline{X} = \$15.73 \qquad \overline{Y} = \$8.61$$
$$s_X = \$13.90 \qquad s_Y = \$9.02$$
$$s_p = \left\{ \frac{9(13.90)^2 + 9(9.02)^2}{18} \right\}^{1/2} = \sqrt{137.29} = 11.72$$

and

$$T = \frac{15.73 - 8.61}{11.72\sqrt{1/10 + 1/10}} = \frac{7.12}{5.24} = 1.36$$

with a decision rule given as:

$$\text{Reject } H_0 \text{ if } T > t_{.95,18} = 1.7341$$

For these data H_0 is not rejected despite the strong evidence suggested by the joint plot of the e.d.f.s. The assumption of normality should have been verified using a Lilliefors plot before using the two-sample t-test. The standardized sample values are as follows.

Manager 1 X*	Manager 2 Y*
−0.62	0.07
−0.11	−0.73
−0.50	−0.65
−0.38	−0.68
−0.28	−0.24
−0.81	−0.19
1.57	0.15
−0.45	−0.07
−0.55	2.71
2.12	−0.38

The Lilliefors graphs in Figures 9.4 and 9.5 make it clear that the assumption of normality is not reasonable for either population. Hence, the preceding t-test on the original data has no meaning.

Therefore the Wilcoxon–Mann–Whitney rank sum test, that is, the two-sample t-test on ranks, is used to test

H_0: the mean cash shortages are the same for both managers

versus

H_1: the mean cash shortages are higher for Manager 1 than for Manager 2.

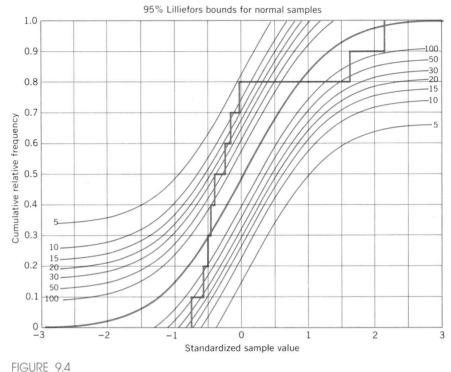

FIGURE 9.4
A LILLIEFORS TEST OF NORMALITY FOR SAMPLE 1.

The first step is to replace the sample data with a joint ranking from 1 to 20.

Manager 1 R_X		Manager 2 R_Y	
8	4	12	7
17	19	1	14
11	13	3	9
15	10	2	18
16	20	6	5

Sample statistics calculated on these ranks yield the following.

$$\bar{R}_X = 13.30 \qquad \bar{R}_Y = 7.70$$

$$s_{R_X} = 5.08 \qquad s_{R_Y} = 5.54$$

$$s_p = \left\{ \frac{9(5.08)^2 + 9(5.54)^2}{18} \right\}^{1/2} = \sqrt{28.23} = 5.314$$

Test statistic

$$T_R = \frac{13.30 - 7.70}{5.315\sqrt{1/10 + 1/10}} = \frac{5.60}{2.38} = 2.36$$

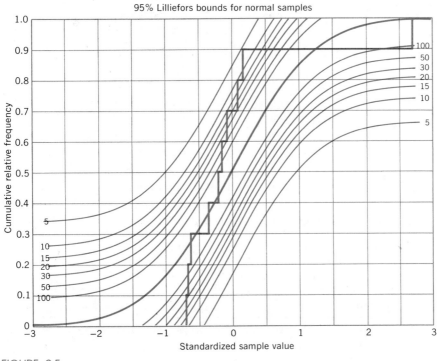

95% Lilliefors bounds for normal samples

FIGURE 9.5
A LILLIEFORS TEST OF NORMALITY FOR SAMPLE 2.

Decision rule

Reject H_0 if $T_R > t_{.95,18} = 1.7341$. Using this test H_0 is easily rejected, which agrees with the obvious interpretation of the joint plot of the e.d.f.s in Figure 9.3 for the two managers. Hence, it is safe to conclude that the cash shortages are higher for Manager 1 than they are for Manager 2. The p-value for these data is found from Table A3 to be about .017. Note that the test on ranks does better than the usual t-test in this situation because of the presence of outliers.

MANY TIES OR ORDERED CATEGORIES

The Wilcoxon–Mann–Whitney test is especially appropriate when the data for two samples have been collected into groups, and only the frequencies are given for the various group intervals. This resembles a situation where there are many ties, because all of the observations in one group may be regarded as tied. Average ranks are used to handle ties as usual, even though the number of ties is quite extensive. This application of the Wilcoxon–Mann–Whitney test is illustrated in the following example.

EXAMPLE

A random sample of faculty salaries from Big Eight schools is compared with a random sample from Big Ten schools. However, only a group summary is available for analysis.

Faculty Salaries	Frequencies	
	Big Eight	Big Ten
Less than $25,000	8	6
$25,000 to $30,000	14	13
$30,000 to $35,000	12	16
$35,000 to $45,000	6	12
Over $45,000	2	7

The null hypothesis of equal mean salaries is to be tested against the alternative $\mu_X \neq \mu_Y$ at $\alpha = .05$. Note that the interval sizes are not equal, and that the end interval "over $45,000" normally would cause a problem in calculating sample means or standard deviations. The use of ranks eliminates this problem.

Faculty Salaries (Thousands)	Frequencies f_i		Ranks That Would Be Assigned	Average Rank R_i
	Big 8	Big 10		
<$25	8	6	1–14	(1 + 14)/2 = 7.5
$25–30	14	13	15–41	(15 + 41)/2 = 28
$30–35	12	16	42–69	(42 + 69)/2 = 55.5
$35–45	6	12	70–87	(70 + 87)/2 = 78.5
≥$45	2	7	88–96	(88 + 96)/2 = 92
Total	42	54		

The average ranks are used in computing $\bar{R}$ and s in the same way class marks were used in computing the mean and standard deviation from grouped data.

$$\bar{R}_X = \frac{1}{n_X} \sum f_i R_{X_i} = \frac{1}{42} \{8(7.5) + 14(28) + 12(55.5) + 6(78.5) + 2(92)\}$$

$$= \frac{1}{42}(1773) = 42.214$$

$$\bar{R}_Y = \frac{1}{n_Y} \sum f_i R_{Y_i} = \frac{1}{54} \{6(7.5) + 13(28) + 16(55.5) + 12(78.5) + 7(92)\}$$

$$= \frac{1}{54}(2883) = 53.389$$

$$s_{R_X}^2 = \frac{1}{n_X - 1} \left\{ \sum f_i R_{X_i}^2 - \left(\sum f_i R_{X_i} \right)^2 \Big/ n_X \right\}$$

$$= \frac{1}{41} \{8(7.5)^2 + 14(28)^2 + 12(55.5)^2$$

$$+ 6(78.5)^2 + 2(92)^2 - (1733)^2/42\}$$

$$= 699.3798$$

$$s_{\bar{R}_Y}^2 = \frac{1}{n_Y - 1} \left\{ \sum f_i R_{Y_i}^2 - \left(\sum f_i R_{Y_i} \right)^2 \Big/ n_Y \right\}$$

$$= \frac{1}{53} \{ 6(7.5)^2 + 13(28)^2 + 16(55.5)^2$$

$$+ \, 12(78.5)^2 + 7(92)^2 - (2883)^2/54 \}$$

$$= 737.5157$$

The pooled standard deviation of the ranks is

$$s_p = \left\{ \frac{41(669.3798) + 53(737.5157)}{42 + 54 - 2} \right\}^{1/2} = 26.60$$

and the test statistic becomes

$$T_R = \frac{\bar{R}_X - \bar{R}_Y}{s_p \sqrt{1/n_X + 1/n_Y}} = \frac{42.214 - 53.389}{26.60 \sqrt{1/42 + 1/54}} = -2.042$$

The decision rule is to reject H_0: $\mu_X = \mu_Y$ if $T_R > 1.9855$ or if $T_R < -1.9855$ where 1.9855 is the .975 quantile from Table A3 for 94 degrees of freedom.

The statistic T_R is less than -1.9855. Therefore the null hypothesis is rejected. The p-value is between .02 and .05. That is, if there is no difference in mean salaries the chances of observing a sample difference at least as extreme as the one actually observed is between .02 and .05.

A SYNOPSIS OF THE CHAPTER

In this chapter four methods have been presented for comparing two population means on the basis of two independent samples. In some situations

Situation	Preferred Method	Also Acceptable
I. Distributions are normal		
A. Variances are unknown and equal		
1. Small samples	Section 9.2	Section 9.3
2. Large samples	Section 9.2	Sections 9.1 or 9.3
B. Variances are unknown		
1. Small samples	Section 9.2	
2. Large samples	Section 9.2	Section 9.1
C. Variances are known	Section 9.1	Section 9.2
II. Distributions are nonnormal	Section 9.3	

FIGURE 9.6
CHART FOR LOCATING THE APPROPRIATE METHOD FOR TESTING THE EQUALITY OF TWO MEANS USING TWO INDEPENDENT SAMPLES.

several of these methods may be appropriate to use, although one may be more appropriate than the others. For this reason a brief chart is given in Figure 9.6 to summarize just where to use each test.

EXERCISES

9.17 The owners of several stores belonging to a fast-food chain are getting ready to expand their business by opening another store. They are considering two potential locations, and since their business is very much dependent on the amount of traffic in the area, they hire an individual to make traffic counts for one hour at randomly selected times during the week when the store would normally be open for business. These counts are as follows:

Traffic Count at Location 1	Traffic Count at Location 2
592	622
625	644
777	664
613	853
587	608
637	635
629	885
843	668
544	649
	714
	668

(a) Use a significance level of $\alpha = .10$ with the Wilcoxon–Mann–Whitney rank sum test to test the hypothesis that there is no difference in the mean traffic counts for the two locations.

(b) Use the Lilliefors test to check each of these samples for normality to see if the methods of Section 9.2 would be appropriate and then compare the results of using the two-sample t-test with the Wilcoxon–Mann–Whitney rank sum test.

(c) Compare plots of the e.d.f.s for the two samples to see which of the conclusions in (a) or (b) seems to be correct.

(d) This data-gathering experiment could also have been done by having individuals at each location count at the same one-hour periods during the day. Comment on what might have been the advantage of using matched pairs.

9.18 The number of visitors to Carlsbad Caverns were counted for a one-week period that included the Fourth of July in 1979 and in 1980. Treat these data as random samples and use the Wilcoxon–Mann–Whitney rank sum test to see if the mean number of visitors is the same for both years.

Use $\alpha = .10$ and state the p-value.

Visitors, Week of July 4, 1979	Visitors, Week of July 4, 1980
397	314
286	257
268	278
254	252
571	613
604	646
384	253

9.19 A large department store obtained a random sample of 233 credit card applications and looked at the annual income of the applicant. These were compared with the annual income of a random sample of 260 applicants for a major bank credit card. Use the Wilcoxon–Mann–Whitney rank sum test to see if the mean income is the same for both populations of applicants. Let $\alpha = .10$ and state the p-value.

Annual Salary	Department Store Credit Card	Major Bank Credit Card
Under $5,000	19	18
$5,000 to 10,000	29	35
$10,000 to 15,000	14	16
$15,000 to 20,000	28	25
$20,000 to 25,000	40	34
$25,000 to 30,000	31	41
$30,000 to 35,000	23	36
$35,000 to 40,000	22	23
$40,000 and over	27	32

9.20 Use the Wilcoxon–Mann–Whitney rank sum test on the hospital rating data of Problem Setting 9.2 to test for equal means. Compare results with the example of Section 9.2 that used the two-sample t-test on these data. (Use $n_X = 10$, $\Sigma R_X = 192$, $\Sigma R_X^2 = 3882$, $n_Y = 15$, $\Sigma R_Y = 133$, $\Sigma R_Y^2 = 1643$.)

9.21 Use the Wilcoxon–Mann–Whitney rank sum test on the data given in Exercise 9.12 and compare results with the solution to that exercise. (Use $\Sigma R_X = 37$, $\Sigma R_X^2 = 297$, $\Sigma R_Y = 18$, $\Sigma R_Y^2 = 88$.)

9.22 Use the Wilcoxon–Mann–Whitney rank sum test on the data given in Exercise 9.9 and compare results with the solution found by using the two-sample t-test. (Use $n_X = 23$, $\Sigma R_X = 518.5$, $\Sigma R_X^2 = 17,198.75$, $n_Y = 30$, $\Sigma R_Y = 912.5$, $\Sigma R_Y^2 = 33,827.25$.)

9.23 Use the Wilcoxon–Mann–Whitney rank sum test on the data given in Exercise 9.13 and compare results with the solution found by using the two-sample t-test.

9.24 Use the Wilcoxon–Mann–Whitney rank sum test on the data given in Exercise 9.16 and compare the results with the solution found by using the two-sample t-test.

9.4

REVIEW EXERCISES

9.25 Two emission-control devices were tested on 20 cars of the same model with 10 cars randomly assigned to device A and with the other 10 cars using device B. The emission levels were checked and showed $\overline{X}_A = 1.24$ with $s_A = 0.06$, whereas $\overline{X}_B = 1.00$ with $s_B = 0.03$. Assuming that the emission levels are normally distributed test $H_0: \sigma_A = \sigma_B$ versus $H_1: \sigma_A \neq \sigma_B$ with $\alpha = .05$.

9.26 Use the appropriate test on the data in Exercise 9.25 to test $H_0: \mu_A = \mu_B$ versus $H_1: \mu_A > \mu_B$ with $\alpha = .01$.

9.27 Find a 95% confidence interval for the difference in mean emission levels in Exercise 9.25.

9.28 Two random samples of size 40 and 60 are selected to test the effectiveness of two sleeping pills. The number of hours of sleep is recorded for each individual following the administration of the sleeping pill. The results are summarized below. Find a 90% confidence interval for the true mean difference in sleeping time for the two pills.

Pill 1	Pill 2
$\overline{X} = 8.65$	$\overline{Y} = 7.15$
$s_X^2 = 9.00$	$s_Y^2 = 5.00$
$n_x = 40$	$n_y = 60$

9.29 Octane determinations have been made on a brand of gasoline at two different geographic locations. The first location was at an elevation of 5600 ft and the second location was at an elevation of 1200 ft. Test each of these two random samples for normality. (At 5600 ft, $n_x = 13$, $\Sigma X = 1073.1$, $\Sigma X^2 = 88{,}582.95$, and at 1200 ft $n_y = 16$, $\Sigma Y = 1342$, $\Sigma Y^2 = 112{,}578.06$.)

Elevation 5600:	82.1	82.1	83.1	83.0	82.8	83.0	82.1
	83.0	82.3	81.7	82.9	82.8	82.2	

Elevation 1200:	84.0	83.5	84.0	85.0	83.1	83.5	81.7
	85.4	84.1	83.0	85.8	84.0	84.2	82.2
	83.6	84.9					

9.30 Use the appropriate test for equality of means in Exercise 9.29 to determine if the octane rating is the same for this brand of gasoline at the two elevations. Let $\alpha = .05$. (Use $\Sigma R_X = 115$, $\Sigma R_X^2 = 1277$, $\Sigma R_Y = 320$, $\Sigma R_Y^2 = 7266.5$.)

9.31 Make a graph of the two empirical distribution functions for the sample data in Exercise 9.29, as was done in Figure 9.3, to see if the decision in Exercise 9.30 appears to be justified.

9.32 Use the Wilcoxon–Mann–Whitney rank sum test to see if the mean high temperature in Des Moines is greater than the mean high temperature in Spokane. Assume the data resemble two independent random samples. Let $\alpha = .05$ and state the p-value.

Des Moines:	83	91	94	89	89	96	91	92	90
Spokane:	78	82	81	77	79	81	80	81	

9.33 Make a graph of the two empirical distribution functions for the sample data in Exercise 9.32, as was done in Figure 9.3, to see if the decision made in the exercise is justified.

9.34 Although both brand A and brand B have canned peas of the same taste quality, because brand B costs less it is suspected that brand B has more water and less vegetable in its cans. Thirty randomly selected cans of each brand are opened, drained, and the vegetable residue weighed. Brand A has a mean residue of 11.2 oz with a standard deviation of .22, while brand B has a mean residue of 10.8 and a standard deviation of .24. Assume both populations are normal. Test the appropriate hypothesis at $\alpha = .05$. Find a 90% confidence interval for the difference in mean weights.

9.35 A random sample of 10 accredited business colleges was compared with a random sample of 10 nonaccredited business schools on the basis of the number of full-time faculty with doctoral degrees on their staffs.

School Number	Accredited	Nonaccredited
1	16	14
2	83	26
3	56	16
4	35	14
5	26	31
6	38	20
7	42	7
8	21	11
9	28	12
10	30	6

It is suspected that nonaccredited schools tend to have fewer faculty with doctoral degrees. Use a nonparametric procedure to analyze these data. Make a graph of the two e.d.f.s to compare the two samples. Does this graph agree with your statistical analysis? ($\Sigma R_X = 146$, $\Sigma R_X^2 = 2283.5$, $\Sigma R_Y = 64$, $\Sigma R_Y^2 = 585$.)

9.36 Assume both populations in Exercise 9.35 are normally distributed and use the appropriate test of hypothesis. Compare these results with those of Exercise 9.35. ($\Sigma X = 375$, $\Sigma X^2 = 17{,}515$, $\Sigma Y = 157$, $\Sigma Y^2 = 3035$.)

9.37 A random sample of actual sales of used cars is obtained in Los Angeles and in Detroit, to see if used cars bring more money in Los Angeles than in Detroit, relative to the blue book price listed for each type of car.

Each car sale is compared against the blue book price in order to account for some of the differences from car to car.

Los Angeles Prices		Detroit Prices	
Sale	Blue Book	Sale	Blue Book
$1850	$1800	$2750	$2840
2800	2920	2750	2840
1490	1500	3000	3165
7475	7340	2870	1945
5250	5300	6650	6880
2600	2675	5450	5595
3225	3280	8300	8340
2100	2030	4095	4220
2950	2890	2850	2980
9320	9510	2050	2200

(a) What are the hypotheses in this problem? Be precise in stating what you are testing.

(b) Check the normality assumption. What is it that needs to be normally distributed in this problem?

(c) Test the hypothesis using the appropriate test. Use $\alpha = .05$.

(d) A car dealer can transport cars from Detroit to Los Angeles for $150 each, and wants to know if this would be profitable. What changes are required in the analysis?

BIBLIOGRAPHY

Additional material on the topics presented in this chapter can be found in the following publication.

Conover, W. J. (1980). *Practical Nonparametric Statistics, 2nd ed.* Wiley, New York.

10

CONTINGENCY TABLES (Optional)

A **contingency table** is an array of counting numbers (i.e., 0, 1, 2, etc.) in matrix form, as specified below, where those numbers represent frequencies. For example, at the end of the day a stock broker may say he completed 72 transactions involving buying or selling stock, or he may wish to be more specific and say he completed 18 transactions on the New York Stock Exchange, 16 transactions on the American Stock Exchange, and 38 transactions over the counter. These transactions can be expressed in the form of a 1×3 (one-by-three) contingency table.

NYSE	ASE	OTC	Total
18	16	38	72

This contingency table has one row and three columns.

The stock broker may wish to be more specific and state whether these transactions were to buy or to sell stock for his customers. This leads to a 2×3 contingency table.

	NYSE	ASE	OTC	Total
Buy	10	10	21	41
Sell	8	6	17	31
Totals	18	16	38	72

The totals, including two **row totals,** three **column totals,** and one **grand total,** are optional but are usually included for the reader's convenience.

In general an $r \times c$ contingency table is an array of counts, or frequencies, in r rows and c columns. Such tables provide a convenient method of displaying data, especially when the data may be classified by two criteria, one represented in the rows and the other represented in the columns. The original data will usually have to be summarized so the data can be put in the form of a contingency table. For example, part of the original data for the preceding contingency tables may have looked like this.

Date	Customer Name	Transaction	Price	Exchange
Oct. 3, 1988	Murray, J.L.	Sell 300 shares CDA	$54\frac{3}{4}$	NY
Oct. 3, 1988	Pierce, H.J.	Buy 200 shares BHW	27	NY
Oct. 3, 1988	Klatt, D.L.	Buy 500 shares TT	$9\frac{3}{8}$	AS

One qualitative variable, "exchange," is used to classify each transaction into different columns. Another qualitative variable, "buy" or "sell," is used to classify each transaction further into different rows. In this instance both variables are qualitative variables, but quantitative variables may be used as well, if the range for the variable is divided into intervals. For example, the row variable could have been price instead of "buy" or "sell," where row 1 is "less than 20," and row 2 is "20 or more" in price. The numbers in the **cells** of the table (i.e., the intersection of a row and column) are the frequencies with which the respective row and column classifications occurred together in the original data.

10.1

2 × 2 CONTINGENCY TABLES

PROBLEM SETTING 10.1

A university admissions officer checks to see if the graduation rate is the same for male and female students admitted as freshmen. A random sample of students admitted as freshmen in September 1981 are classified as male or female and as to whether or not they graduated from that school within five years. The data may be presented in a contingency table with two rows and two columns.

Entering Freshmen

Sex of Student	Graduated	Did Not Graduate	Totals
Male	16	28	44
Female	18	19	37
Totals	34	47	81

PROBLEM SETTING 10.2

A production supervisor is concerned about a possible difference in the quality of production for the two shifts operating in a factory that manufactures shirts. Fifty shirts are randomly sampled from the output of the day shift, and another 50 shirts are randomly sampled from the night shift. Each shirt is carefully inspected to see if it is defective. Defective shirts are either sold as "factory seconds" at a lower price, repaired and sold at the higher price, or discarded if the defect is serious. The production supervisor summarizes the sample results in a contingency table having two rows and two columns as follows:

| | Classification of Shirt | | |
Made by	Defective	Nondefective	Totals
Day shift	3	47	50
Night shift	5	45	50
Totals	8	92	100

THE HYPOTHESIS OF INTEREST

Although there are wide variations in the types of data presented in the two problem settings, the question of interest is basically the same in both settings: Is the row classification independent of the column classification? In Problem Setting 10.1 this translates as: Of the entering freshmen students, is the proportion graduating within five years the same for male students as it is for female students? In Problem Setting 10.2 the question becomes: Is the defective proportion the same for both shifts?

Problem Settings 10.1 and 10.2 are different in one basic respect. In Problem Setting 10.1 one random sample is drawn, and the data are then classified into rows and columns. The row totals are not known prior to drawing and examining the sample of students, so they are random variables. In contrast, two random samples are drawn in Problem Setting 10.2, one for each row. The row totals are not random variables, since they represent sample sizes that are determined prior to sampling.

The same test statistic is used for both types of data, and it will be introduced later in this section. The exact distribution of the test statistic is different for the two problem settings because of the difference in the nature of the row totals. However, the exact distribution of the test statistic is almost never used because it is difficult to derive. The approximate distribution for the test statistic is the chi-square distribution, whose quantiles are given in Table A4.

INDEPENDENCE

In Chapter 4, independence between rows and columns of a frequency table was established by seeing if the joint probabilities equaled the product of the marginal probabilities. The frequency tables in Chapter 4 represented the entire population, so each frequency table was easily converted into probabilities by dividing the frequencies by the total population size.

In the problem settings of this section the frequency tables are called **contingency tables** and they represent random samples from populations, so the probabilities are not known exactly but may be estimated from the sample data. Then a measure of the degree of independence exhibited by the data is given by a comparison of the estimated joint probabilities with the product of the estimated marginal probabilities.

EXAMPLE

In Problem Setting 10.1 the frequencies in the contingency table, obtained from the random sample, may be converted to estimated probabilities by dividing each number by the total sample size 81.

Frequencies from the Random Sample

	Graduated	Did Not Graduate	Totals
Male	16	28	44
Female	18	19	37
Totals	34	47	81

Estimated Probabilities

	Graduated	Did Not Graduate	Totals
Male	.1975	.3457	.5432
Female	.2222	.2346	.4568
Totals	.4198	.5802	1.0000

The estimated joint probabilities are given in the cells, such as the joint probability of a student being female and not graduating, which is estimated as .2346. The estimated marginal probabilities are given by the numbers in the margins. The marginal probability of a student being female is estimated as .4568, and the marginal probability of a student not graduating is estimated as .5802. The product of the two estimated marginal probabilities, $(.4568)(.5802) = .2650$, is not exactly equal to the estimated joint probability .2346. Hence, one might conclude that these two variables are not independent of one another. However, one should keep in mind that this is a random sample from the population and some differences should be expected due to sampling error. Just how much sampling error can be tolerated before reaching the conclusion that these variables are, in fact, not independent is determined by the methods covered in this chapter.

SOME NOTATION

The null hypothesis of independence can be tested by using a test statistic that measures the variation between the estimated joint probability and the product of the estimated marginal probabilities, for all four cells in the contingency table. The following notation will be used.

cell (i, j) = the intersection of row i and column j.

p_{ij} = the probability of a sample observation being classified in cell (i, j).

$\hat{p}_{ij}$ = the relative frequency in cell (i, j), from an observed sample.

$p_{i.}$ = the marginal probability for row i, which is found as $\sum_j p_{ij}$.

$\hat{p}_{i.}$ = the relative frequency for row i, from an observed sample, which is found as $\sum_j \hat{p}_{ij}$.

$p_{.j}$ = the marginal probability for column j, which is found as $\sum_i p_{ij}$.

$\hat{p}_{.j}$ = the relative frequency for column j, from an observed sample, which is found as $\sum_i \hat{p}_{ij}$.

R_i = total frequency in row i.

C_j = total frequency in column j.

n = total number of observations = $\sum R_i = \sum C_j$

EXAMPLE

In the previous example the joint probabilities p_{11}, p_{12}, p_{21}, and p_{22} were unknown, as were the marginal probabilities $p_{1.}$, $p_{2.}$, $p_{.1}$, and $p_{.2}$. Their estimates are given in the table of estimated probabilities as $\hat{p}_{11} = .1975$, $\hat{p}_{12} = .3457$, $\hat{p}_{21} = .2222$, and $\hat{p}_{22} = .2346$. The estimated marginal probabilities are $\hat{p}_{1.} = .5432$, $\hat{p}_{2.} = .4568$, $\hat{p}_{.1} = .4198$, and $\hat{p}_{.2} = .5802$.

THE RATIONALE BEHIND THE TEST STATISTIC

If row classification is independent of column classification, each joint probability p_{ij} equals the product of the respective marginal probabilities $p_{i.}$ and $p_{.j}$. Since the probabilities are unknown, the independence property is judged on the basis of the estimates $\hat{p}_{ij}$, to see if they are approximately equal to the products $\hat{p}_{i.}$ times $\hat{p}_{.j}$.

A statistic that is used to measure the degree of independence exhibited by a table of sample frequencies is

$$T = \sum_{i=1}^{2} \sum_{j=1}^{2} \frac{(\hat{p}_{ij} - \hat{p}_{i.}\hat{p}_{.j})^2}{\hat{p}_{i.}\hat{p}_{.j}/n}$$

A TEST FOR INDEPENDENCE IN A 2 × 2 CONTINGENCY TABLE

Data

The data may be obtained two ways.

1. A random sample of size n is drawn from a population, and each item is classified into one of two columns of a 2 × 2 contingency table according to one criterion, and into one of two rows according to another criterion. (See Problem Setting 10.1.)

2. The first row of a 2 × 2 contingency table represents a random sample of size R_1 from one population, and the second row represents a random sample of size R_2 from another population. Each observation in each row is classified into one of the two columns. (See Problem Setting 10.2.)

In either case the data are represented in a 2 × 2 contingency table

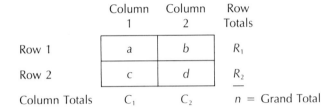

	Column 1	Column 2	Row Totals
Row 1	a	b	R_1
Row 2	c	d	R_2
Column Totals	C_1	C_2	n = Grand Total

where $a, b, c,$ and d represent the number of observations classified in each of the four cells.

Assumptions

Each observation is classified by row and/or column independently of every other observation. This assumption is satisfied if the sample is random.

Null Hypothesis

The null hypothesis is that the classification by rows is independent of the classification by columns. In symbols,

$$H_0: p_{ij} = p_{i.}p_{.j} \text{ for all four cells } (i, j)$$

Test Statistic

$$T = \frac{n(ad - bc)^2}{R_1 R_2 C_1 C_2} \tag{10.1}$$

The null distribution of T is approximately the chi-square distribution with one degree of freedom, given in Table A4.

Decision Rule

The alternative hypothesis is that the classification by rows of the sample observations is dependent on the column classification. In symbols this becomes

$$H_1: p_{ij} \neq p_{i.}p_{.j} \text{ for some cell } (i, j)$$

Reject H_0 at the α level of significance if T exceeds the $1 - \alpha$ quantile from a chi-square distribution with one degree of freedom, as given in the first row of Table A4.

which is a function of the difference between $\hat{p}_{ij}$ and the product $\hat{p}_{i.}\hat{p}_{.j}$, squared to make all of the terms positive, standardized by dividing by $\hat{p}_{i.}\hat{p}_{.j}/n$ to account for unequal cell sizes, and then summed over all of the cells. This statistic T simplifies to the form given in Eq. 10.1, which makes the computations easier.

Recall the case of the university admissions officer in Problem Setting 10.1 who wanted to know whether the probability of graduation was the same for male as for female students who were admitted as incoming freshmen. A random sample of 81 students who were incoming freshmen in September 1981 were classified according to whether they graduated within five years or not (columns 1 and 2) and whether they were male or female (rows 1 and 2). This is an example of the single random sample, classified two ways, described in part 1 of **Data.**

The results are presented in a 2 × 2 contingency table.

	Graduated	Did Not Graduate	Totals
Male	16	28	44
Female	18	19	37
Totals	34	47	81

The test statistic is

$$T = \frac{n(ad - bc)^2}{R_1 R_2 C_1 C_2} = \frac{81[16(19) - 28(18)]^2}{44(37)(34)(47)} = 1.245$$

The critical value for a 5% level test is 3.841 as found in Table A4 with one degree of freedom. Since $T = 1.245$ is less than 3.841, the null hypothesis is not rejected. The p-value is found from Table A4 by interpolation to be approximately .27. The graduation rate could be the same for male and female students. The observed apparent inequities could be due to chance fluctuation in the sample.

THE PHI COEFFICIENT

There is a measure of association that is used with 2 × 2 contingency tables, called the **phi coefficient,** or the **ϕ coefficient.** It is used as a measure of the row–column dependence in 2 × 2 contingency tables. If all of the items in row 1 are in one column (either $a = 0$ or $b = 0$), and all of the items in row 2 are in the other column (either $c = 0$ or $d = 0$), the phi coefficient will equal $+1$ when $b = 0$ and $c = 0$, or -1 when $a = 0$ and $d = 0$, representing the most extreme case of row–column dependence. If the proportions classified in each column are the same for both rows, then there is apparent independence between the rows and the columns and the phi coefficient is close to zero.

The phi coefficient can also be used in testing the hypothesis of row–column independence. The principal advantage in using ϕ instead of T presented earlier in this section for testing the same hypothesis, is that one-tailed alternative hypotheses may be considered when the test statistic is based on ϕ.

THE PHI COEFFICIENT AS A MEASURE OF ASSOCIATION IN 2 × 2 CONTINGENCY TABLES

Data

The data may be presented as counts in a 2 × 2 contingency table as follows.

	Column 1	Column 2	Row Totals
Row 1	a	b	R_1
Row 2	c	d	R_2
Column Totals	C_1	C_2	n

Phi Coefficient

$$\phi = \frac{ad - bc}{\sqrt{R_1 R_2 C_1 C_2}}$$

Properties

1. Phi is a measure of the amount of dependence between the row classification and the column classification.

2. Phi will be close to zero if there is independence between the row classification and the column classification.

3. Phi will be close to its maximum value of $+1.0$ (when $b = 0$ and $c = 0$), or its minimum value of -1.0 (when $a = 0$ and $d = 0$), if the observations in row 1 and row 2 tend to be classified in opposite columns.

THE PHI COEFFICIENT IN A ONE-SIDED TEST FOR INDEPENDENCE

The *Data, Assumptions,* and *Null Hypothesis* are the same as in "A Test for Independence in a 2 × 2 Contingency Table," presented earlier in this section.

Test Statistic

The test statistic is $\phi\sqrt{n}$, where ϕ is given as

$$\phi = \frac{ad - bc}{\sqrt{R_1 R_2 C_1 C_2}} \tag{10.2}$$

The null distribution of $\phi\sqrt{n}$ is approximately the standard normal.

Decision Rule

The use of the phi coefficient in the test statistic permits one-sided alternative hypotheses to be considered. Let $z_{1-\alpha}$ be the $1 - \alpha$ quantile from Table A2.

(a) H_1: $p_{11} > p_{1.}p_{.1}$ and $p_{22} > p_{2.}p_{.2}$. Reject H_0 if $\phi\sqrt{n} > z_{1-\alpha}$.

(b) H_1: $p_{12} > p_{1.}p_{.2}$ and $p_{21} > p_{2.}p_{.1}$. Reject H_0 if $\phi\sqrt{n} < z_{\alpha}$.

(c) H_1: $p_{ij} \neq p_{i.}p_{.j}$ for any cell (i, j). Reject H_0 if $\phi\sqrt{n} > z_{1-\alpha/2}$ or if $\phi\sqrt{n} < z_{\alpha/2}$. (This is equivalent to the previous test based on T.)

The two-tailed version of the test using ϕ is equivalent to the test using T as a test statistic.

This latter use of ϕ in a hypothesis test is to see if there is a significant association between row and column classifications. This is a different concept than the former use of ϕ, as a measure of the strength of that association.

EXAMPLE

In Problem Setting 10.2 the number of defective items in random samples from the production of the day shift and the night shift were summarized in a 2 × 2 contingency table.

	Defective	Nondefective	Totals
Day shift	3	47	50
Night shift	5	45	50
Totals	8	92	100

A measure of dependence is desired between the proportion of defectives and the shift responsible for the production. The phi coefficient is

$$\phi = \frac{ad - bc}{\sqrt{R_1 R_2 C_1 C_2}} = \frac{3(45) - 47(5)}{\sqrt{50(50)(8)(92)}} = -.074$$

which, while negative, probably represents a negligible dependence between the proportion defective and the shift responsible for the production. If prior to sampling, management suspected that the night shift tended to produce a higher proportion of defective items, because of fatigue and other factors, a one-tailed test would be set up to test the null hypothesis of independence against the alternative that p_{21} (representing the probability of the night shift producing a defective item) tends to be too large, or

$$H_1 = p_{21} > p_{2.}p_{.1}$$

The decision rule is to reject H_0 if $\phi\sqrt{n} < z_\alpha$, which is $\phi\sqrt{n} < -1.6449$ for a 5% level of significance. Since ϕ equals $-.074$, $\phi\sqrt{n}$ equals $-.074\sqrt{100} = -.74$, and the null hypothesis of independence is not rejected. The p-value is found from Table A2* to be .2296.

COMMENTS

1. The phi coefficient and the test statistic T are directly related through the relationship $T = n\phi^2$.

2. The chi-square distribution and normal distribution used in the significance tests involving T and ϕ are only approximations to the exact distribution, which is usually not practical to work with. The approximations work satisfactorily if the row and column totals are not too small. Current theory

on this subject indicates that this approximation is satisfactory if each product R_1C_1, R_1C_2, R_2C_1, and R_2C_2 is greater than n.

3. The test presented in this section can be used for testing the hypothesis that two populations have the same proportion p of successes. Let a be the number of successes in a random sample of size R_1 from population 1 and let c be the number of successes in a random sample of size R_2 from population 2. If p_1 and p_2 represent the true proportions in the two populations, then the null hypothesis is $p_1 = p_2$. To test $H_0: p_1 = p_2$ simply put the data in the form of a 2×2 contingency table

	Number of Successes	Number of Failures	Totals
First population	a	$b = R_1 - a$	R_1
Second population	c	$d = R_2 - c$	R_2
Totals	C_1	C_2	n

and use the test of this section.

EXERCISES

10.1 A random sample of 100 employees in an automobile assembly plant contains 63 men, where 48 belong to the union and 15 do not, and 37 women, where 22 belong to the union and 15 do not. Test the null hypothesis that the proportion of all employees belonging to the union is the same for men and women, against the one-sided alternative that the proportion is higher among male employees. Use a level of significance of 5% and state the p-value.

10.2 A company that manufactures after-shave lotion wants to compare two different packaging techniques. One technique uses a picture of a woman and the other does not. One hundred people were polled using the lotion with the picture of the woman; 85 of them rated the lotion as favorable over their present lotion and 15 said it was unfavorable. The package without a picture was shown to 300 people, with 210 favorable responses and 90 unfavorable. Test the hypothesis that the package does not affect opinion, that is, that the proportion of people rating the lotion favorable is the same for both packaging techniques, against the one-sided alternative that the proportion is higher for the package with the picture. Use a level of significance of $\alpha = .05$ and find the p-value.

10.3 It has been hypothesized that being left-handed is independent of eye color. Test this hypothesis against the two-sided alternative using the results of a study of 100 individuals with blue eyes and 100 individuals with brown eyes, the results of which are summarized below. Use $\alpha = .05$ and state the p-value.

	Handedness	
Color of Eyes	Left	Right
Blue	5	95
Brown	15	85

10.4 A study of inherited alcoholism by five American and Danish psychiatrists concerns whether adopted Danish male children, separated from their biological parents during early infancy, became alcoholics as adults. The results of this study, released in 1973, report that 55 of these men had one parent who was diagnosed as alcoholic. Ten of these 55 children of an alcoholic parent were found to be alcoholic. These were compared with 78 other adopted men whose biological parents had no known history of alcoholism. Four of the offspring of this nonalcoholic group were found to be alcoholic. Test the hypothesis that the probability of alcoholic children is the same for both alcoholic and nonalcoholic parents, against the appropriate one-sided alternative. Let $\alpha = .01$ and find the p-value.

10.5 The Dow Jones industrial average was examined in a random sample of 40 days where none of the days were consecutive. Each day's record noted whether the Dow was "up" compared with the previous day or not, and whether the following trading day resulted in an "up" day for the Dow or not. The results are summarized in the 2 × 2 contingency table.

| | Dow Jones Performance on the Following Day | |
First Day	Up	Down
Up	8	14
Down	9	9

Test the hypothesis that the first- and second-day performances of the Dow Jones industrial average are independent of one another against the two-sided alternative. Let $\alpha = .10$ and find the p-value.

10.6 Compute the phi coefficient for the data in Exercise 10.5.

10.7 "Free Pizza" (two for the price of one) coupons were distributed by direct mail and by handbill distribution from a local pizzeria. Of the 250 coupons sent by direct mail, 43 were redeemed. Of the 200 handbills distributed, 28 coupons were redeemed. Test the hypothesis that the proportion of persons redeeming coupons is the same, whether the coupon is received by direct mail or by handbill, against the two-sided alternative. Let $\alpha = .05$ and state the p-value.

10.8 A random sample of grade point averages was obtained for 20 graduate students with U.S. citizenship.

$$3.28 \quad 3.08 \quad 2.70 \quad 3.76 \quad 2.90$$
$$2.90 \quad 3.58 \quad 3.30 \quad 3.75 \quad 4.00$$
$$3.41 \quad 3.55 \quad 3.20 \quad 3.63 \quad 3.85$$
$$3.21 \quad 3.69 \quad 3.80 \quad 3.21 \quad 2.87$$

A random sample of grade point averages was also obtained for 20 graduate students without U.S. citizenship.

$$3.71 \quad 3.36 \quad 4.00 \quad 2.92 \quad 3.85$$
$$3.70 \quad 3.90 \quad 3.75 \quad 3.82 \quad 4.00$$

(continued)

4.00   3.20   3.50   3.21   3.05

3.85   3.46   3.92   3.59   3.91

Use a level of significance of $\alpha = .05$ to test the hypothesis that the proportion of graduate students with a grade point average above 3.50 is the same for both groups of students, against the two-sided alternative.

10.2

THE $r \times c$ CONTINGENCY TABLE

PROBLEM SETTING 10.3

Some common stocks are sold on the New York Stock Exchange. Others are listed on the American Stock Exchange. Many companies are listed on neither stock exchange and rely on other stock exchanges, such as the Pacific or Midwest stock exchanges, or over-the-counter sales for the marketing vehicle. A random sample of 59 stocks was obtained from the Standard & Poor's Stock Guide to see if there is a difference in the Standard and Poor's rating for stocks on the various exchanges.

The data, in abbreviated form, appear as the following table.

Company Name	Stock Exchange	Rating
1. Affiliated Capital	ASE	NR[a]
2. Altex Oil	Other	NR
3. American Medical Services	Other	B+
4. Application Engineering	Other	B+
5. Automated Marketing Supplies	Other	B+
...	...	...
57. Vintage Enterprises	ASE	B−
58. Webb Co.	Other	A−
59. Williams (W.W.) Co.	Other	B+

[a]NR means not rated.

The results are listed as counts in the following **3 × 9** (3 rows, 9 columns) **contingency table.**

	A+	A	A−	B+	B	B−	C	D	NR	Totals
NYSE	1	1	6	6	0	0	0	0	2	16
ASE	0	0	0	0	2	7	3	0	3	15
Other	0	0	1	10	6	3	0	0	8	28
Totals	1	1	7	16	8	10	3	0	13	59

Standard & Poor's Rating

THE CHI-SQUARE TEST FOR CONTINGENCY TABLES

Problem Setting 10.3 is an example of the more general $r \times c$ contingency table that has r rows and c columns. The hypothesis of interest is that the row and column classifications are independent of one another. As in the previous section, row and column independence implies that the joint probability p_{ij} associated with an observation being in the cell in row i and column j must equal the product of the marginal probabilities for row i, $p_{i.}$, and column j, $p_{.j}$

$$p_{ij} = p_{i.}p_{.j} \quad \text{(independence)}$$

for all the cells. If O_{ij} represents the actual cell count, the p_{ij} is estimated from the observed relative frequency

$$\hat{p}_{ij} = \frac{O_{ij}}{n}$$

for each cell. The marginal probabilities are estimated from the observed row totals R_i and column totals C_j, after division by the total sample size n:

$$\hat{p}_{i.} = \frac{R_i}{n} \qquad \hat{p}_{.j} = \frac{C_j}{n}$$

The **observed** cell count is O_{ij}. The **expected** cell count E_{ij} under the assumption of independence is obtained by multiplying the estimated probability $\hat{p}_{ij}$ of each observation being in cell (i, j) times the total number of observation n.

$$\begin{aligned}
E_{ij} &= n \cdot \hat{p}_{i.}\hat{p}_{.j} \quad \text{(By independence)} \\
&= n \cdot \frac{R_i}{n} \cdot \frac{C_j}{n} \\
&= \frac{R_i C_j}{n}
\end{aligned}$$

A statistic that is used to measure the degree of independence exhibited by a sample frequency table involves computing the difference between the observed and expected cell counts, $O_{ij} - E_{ij}$, squaring to make all terms positive, dividing by E_{ij} to standardize for the different cell frequencies, and summing over all cells:

$$T = \sum_{i=1}^{r} \sum_{j=1}^{c} \frac{(O_{ij} - E_{ij})^2}{E_{ij}}$$

This statistic T is equal to the statistic given in Eq. 10.1 for 2×2 contingency tables. If the null hypothesis is true, T may be compared with quantiles from a chi-square distribution (Table A4) with $(r - 1)(c - 1)$ degrees of freedom

CHI-SQUARE TEST OF INDEPENDENCE FOR $r \times c$ CONTINGENCY TABLES

Data

The data are classified into rows by one criterion (or variable) and into columns by another criterion (or variable). They are represented as frequency counts O_{ij} in an $r \times c$ contingency table (one with r rows and c columns), where O_{ij} equals the number of observations in cell (i, j); that is, the intersection of row i with column j.

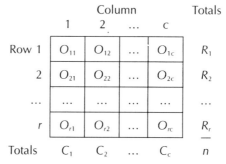

	Column 1	2	...	c	Totals
Row 1	O_{11}	O_{12}	...	O_{1c}	R_1
2	O_{21}	O_{22}	...	O_{2c}	R_2
...	...	...	...	...	...
r	O_{r1}	O_{r2}	...	O_{rc}	R_r
Totals	C_1	C_2	...	C_c	n

Assumptions

Each observation is classified independently of every other observation. This assumption is satisfied if there is one random sample being classified by two criteria (rows and columns), or if each row represents a random sample of observations being classified into columns.

Null Hypothesis

The classification by rows is independent of the classification by columns. This hypothesis requires individual interpretation for each application. In symbols

$$H_0: p_{ij} = p_{i.}p_{.j} \quad \text{for all cells } (i, j)$$

Test Statistic

Let O_{ij} be the observed count in row i, column j and let $E_{ij} = R_i C_j / n$ be the expected count in row i, column j, if the null hypothesis is true. Then one form for the test statistic is

$$T = \sum_{\substack{\text{all} \\ \text{cells}}} \frac{(O_{ij} - E_{ij})^2}{E_{ij}} \tag{10.3}$$

where $(O_{ij} - E_{ij})^2 / E_{ij}$ is computed separately for each cell, and those quantities are summed over all rc cells. An alternate form for the test statistic, which involves fewer computations, is

$$T = \sum_{\substack{\text{all} \\ \text{cells}}} \frac{O_{ij}^2}{E_{ij}} - n \tag{10.4}$$

The null distribution of T is approximately the chi-square distribution with $(r - 1)(c - 1)$ degrees of freedom, whose quantiles are given in Table A4.

Decision Rule

The alternative hypothesis is that the classification by rows of the sample observations is dependent on the column classification. In symbols this becomes

$$H_1: p_{ij} \neq p_{i.}p_{.j} \quad \text{for some cell } (i, j)$$

Reject the null hypothesis at the level of significance α if T exceeds the $1 - \alpha$ quantile of the chi-square distribution with $(r - 1)(c - 1)$ degrees of freedom. These critical values may be obtained from Table A4.

Comment

Note that the chi-square distribution is only an approximation to the true distribution of T, which is too difficult to obtain in most cases. The approximation is considered quite good unless there are many very small expected values E_{ij}. If most E_{ij} are greater than 1.0 and all E_{ij} are greater than 0.5, the approximation may still be very good, especially if the number of degrees of freedom is large. If some of the E_{ij} are less than 0.5, however, it is advisable to combine several similar rows or columns, if it is convenient, to raise those low values. This is a judgment decision that requires consideration of individual circumstances in each case.

as an approximate distribution for T. This approximation appears to be satisfactory if most of the E_{ij} are 1 or greater, and all E_{ij} are at least 0.5.

EXAMPLE

The random sample of 59 stocks in Problem Setting 10.3 was classified by exchange and by rating. The expected values E_{ij} are listed in the same cell with the observed values O_{ij} for convenience. The expected values are computed using $E_{ij} = R_i C_j / n$. For instance

$$E_{11} = \frac{R_1 C_1}{n} = \frac{16(1)}{59} = 0.27$$

and is given in the cell in row 1, column 1, below the observed count $O_{11} = 1$.

Standard & Poor's Rating

	A+	A	A−	B+	B	B−	C	D	NR	Totals
NYSE	1	1	6	6	0	0	0	0	2	16
	0.27	0.27	1.90	4.34	2.17	2.71	0.81	0	3.53	
ASE	0	0	0	0	2	7	3	0	3	15
	0.25	0.25	1.78	4.07	2.03	2.54	0.76	0	3.31	
Other	0	0	1	10	6	3	0	0	8	28
	0.47	0.47	3.32	7.59	3.80	4.75	1.42	0	6.17	
Totals	1	1	7	16	8	10	3	0	13	59

As a check on the computations, the row and column totals for the E_{ij} should be the same as listed for the O_{ij}.

The expected values under Rating D are zero, so that column should be eliminated. In general, any row or column with no observations should be dropped from the analysis.

Columns 1 and 2 both have low expected values, because of the few observed stocks with ratings A or A+. Combining those two columns makes sense in this case, since they have similar interpretations. Combining columns means adding the corresponding observed values. The new expected values are the sum of the original expected values. The revised contingency table is given below. Now all of the expected values are greater than 0.5.

<table>
<thead>
<tr><th></th><th colspan="7">Standard & Poor's Rating</th><th></th></tr>
<tr><th></th><th>A or A+</th><th>A−</th><th>B+</th><th>B</th><th>B−</th><th>C</th><th>NR</th><th>Totals</th></tr>
</thead>
<tbody>
<tr><td>NYSE</td><td>2
0.54</td><td>6
1.90</td><td>6
4.34</td><td>0
2.17</td><td>0
2.71</td><td>0
0.81</td><td>2
3.53</td><td>16</td></tr>
<tr><td>ASE</td><td>0
0.51</td><td>0
1.78</td><td>0
4.07</td><td>2
2.03</td><td>7
2.54</td><td>3
0.76</td><td>3
3.31</td><td>15</td></tr>
<tr><td>Other</td><td>0
0.95</td><td>1
3.32</td><td>10
7.59</td><td>6
3.80</td><td>3
4.75</td><td>0
1.42</td><td>8
6.17</td><td>28</td></tr>
<tr><td>Totals</td><td>2</td><td>7</td><td>16</td><td>8</td><td>10</td><td>3</td><td>13</td><td>59</td></tr>
</tbody>
</table>

With the aid of Eq. 10.3 the computations proceed as follows:

Cell	O_{ij}	E_{ij}	$(O_{ij} - E_{ij})^2/E_{ij}$
Row 1, Column 1	2	$16(2)/59 = 0.542$	$(2 - 0.542)^2/0.542 = 3.92$
Row 1, Column 2	6	$16(7)/59 = 1.898$	$(6 - 1.898)^2/1.898 = 8.86$
...	...	...	...
Row 3, Column 7	8	$28(13)/59 = 6.169$	$(8 - 6.169)^2/6.169 = 0.54$
Total	59	59	$T = 47.76$

If Eq. 10.4 had been used instead of Eq. 10.3, the calculations would look like this:

Cell	O_{ij}	E_{ij}	O_{ij}^2/E_{ij}
Row 1, Column 1	2	$16(2)/59 = 0.542$	7.37
Row 1, Column 2	6	$16(7)/59 = 1.898$	18.96
...	...	...	...
Row 3, Column 7	8	$28(13)/59 = 6.169$	10.37
Total	59	59	106.76

$$T = \sum \frac{O_{ij}^2}{E_{ij}} - n = 106.76 - 59 = 47.76$$

This is the same result as before, but with fewer computations. The null hypothesis is that the proportion receiving each rating is the same for the stocks on the NYSE, on the ASE, or not listed on either. Another way of expressing this hypothesis is that stock rating is independent of where it is listed. To test this hypothesis at the 5% level of significance the .95 quantile of a chi-square distribution with $(r - 1)(c - 1) = (3 - 1)(7 - 1) = 2(6) = 12$ degrees of freedom is obtained from Table A4. This value is 21.03. Since T easily exceeds 21.03, the null hypothesis is clearly rejected. Stocks on the different exchanges tend to have significantly different ratings. The p-value associated with $T = 47.76$ is seen from Table A4 to be less than .0001.

CLASSIFICATION USING QUANTITATIVE VARIABLES

The "natural" application of the contingency table analysis is for cases in which each observation is measured by two *qualitative* variables, as in the preceding example. However, *quantitative* variables may also be used to classify the observations into rows, or columns, or both. The range of each quantitative variable may have to be arbitrarily divided into intervals so that each interval represents one row or one column. If the quantitative variable is years of formal education the intervals may be: *did not graduate from high school, high school graduate, holds bachelor's degree, holds master's degree*, and *holds doctoral degree*. If the quantitative variable is salary, the intervals may be *less than $5000, $5000 to $10,000*, etc., with a final category that includes all salaries above a certain value.

EXAMPLE

When a company receives the customer registration cards from customers who have purchased its products, it may want to do some analysis on the data received. A memo comes down to a recent business school graduate working in marketing. It says, "Do an analysis to see if the advertising effectiveness varies according to the age of the customer." The recent graduate, having received an A in business statistics, says, "They are asking whether there is a dependence between type of advertising that led to the sale and the age of the customer." He checks his old statistics textbook and then randomly samples 200 cards from the thousands of cards in the files and summarizes the results in the following contingency table.

Reason for Selecting Product	Age of Customer (Years)					Totals
	0–20	21–30	31–40	41–50	Over 50	
Previous Ownership	10	21	28	8	6	73
Store Display	4	8	2	0	0	14
Catalog	3	4	5	0	0	12
Magazine	4	1	23	8	0	36
Newspaper	12	18	14	2	4	50
Other	5	8	2	0	0	15
	38	60	74	18	10	200

The calculations proceed as follows:

Cell	O_{ij}	E_{ij}	O_{ij}^2/E_{ij}
Row 1, Column 1	10	13.87	7.21
Row 1, Column 2	21	21.90	20.14
...	...	...	...
Row 6, Column 5	0	0.75	0
			Total 252.28

$$T = \sum \frac{O_{ij}^2}{E_{ij}} - n = 252.28 - 200 = 52.28$$

Since there are

$$(r - 1)(c - 1) = 5(4) = 20$$

degrees of freedom, the critical value for a 5% test is the .95 quantile from the chi-square distribution with 20 degrees of freedom, which is given in Table A4 as 31.41. Since T exceeds this critical value the null hypothesis of independence between the type of advertising and the age of the customer is rejected, and the conclusion may be made that different types of advertising influence the various age groups differently. The p-value is approximately .0001.

CRAMER'S CONTINGENCY COEFFICIENT

Sometimes a measure of association in a contingency table is a useful descriptive statistic to obtain. There are many different measures that are used with contingency tables. The one presented here is widely used.

EXAMPLE

A measure of association may be expressed for the relationship between advertising effectiveness and age of the customer, using the data given in the previous example. Since $r = 6$ and $c = 5$, q equals the smaller of the two, $q = 5$, and

$$\Phi = \frac{T}{n(q - 1)} = \frac{52.28}{200(4)} = .065$$

which indicates that the relationship is slight, even though it is statistically significant as shown by the test using T in the previous example.

CRAMER'S CONTINGENCY COEFFICIENT

Data

The data are counts or frequencies, in an $r \times c$ contingency table. Let q equal the smaller of r and c.

Cramer's Contingency Coefficient

$$\Phi = \frac{T}{n(q - 1)} \tag{10.5}$$

where T is the statistic defined earlier in this section by Eq. 10.3 or Eq. 10.4.

Properties

1. The contingency coefficient Φ (capital phi) is a measure of the association between the row classification and column classification in a contingency table.

2. The contingency coefficient Φ will tend to be close to zero, its minimum value, if there is independence between the row and column variables.

3. If all of the observations in each row tend to collect in one column but in a different column for each row, Φ will tend to close to its maximum value 1.0.

4. The term Φ is the $r \times c$ analog of the phi coefficient for 2×2 contingency tables. In fact Φ equals ϕ squared if $r = 2$ and $c = 2$.

Hypothesis Test

If a hypothesis test is desired using Φ, the test should be conducted using T as a test statistic, in the manner described earlier in this section. Since Φ is a function of T, any conclusions or p-values obtained using T may be applied equally to Φ.

COMMENTS

1. If the different rows represent different populations, then significance in the overall contingency table may be interpreted as: Some of the populations have distributions different from others. In cases like this, a follow-up question, "Which pairs of populations have different distributions?" is often important to answer. This question may be answered by forming smaller contingency tables with only two rows each, and analyzing each of these tables separately by using the test in this section to see which pairs of populations are significantly different. This pairwise comparison of rows should be undertaken only if the original contingency table shows a significant row–column dependence.

2. If one of the variables (e.g., rows) is quantitative, and the other variable is qualitative, it often happens that the alternative hypothesis of interest is whether some of the columns tend to yield larger observations than other columns. In this case, the rows represent ordered categories, and a multi-sample rank test is more appropriate to use because of its greater power to detect these differences if they exist. A rank test for two samples (two columns) and ordered row categories was given in Section 9.3.

3. If both variables, represented by rows and columns, are quantitative in nature, then there is a choice between the methods of this section to test for independence, and the methods of the next chapter, which are oriented toward detecting particular types of dependence, as is explained in the next chapter.

EXERCISES

10.9 The value of the test statistic T was found to be 52.28 in the second example of this section by use of Eq. 10.4. Compute the value of T using Eq. 10.3. Which equation for computing T do you think is the easier to use?

10.10 Compute Cramer's contingency coefficient for the first example in this section. How does this value compare with the value given in the final example of this section? What interpretation is associated with the value you have computed?

10.11 The placement service for a business college has 200 applicants on file, classified according to major and degree. A summary table is given as follows.

		Degree	
Major	BBA	MBA	DBA
Accounting	78	17	5
Finance	33	12	3
Other	13	31	8

Test the hypothesis that degree type is independent of major. Let $\alpha = .05$ and state the p-value.

10.12 Compute Cramer's contingency coefficient for the data in Exercise 10.11.

10.13 A personnel manager of a large company wishes to know if employee satisfaction is independent of job category. Test the hypothesis of independence based on the following data with $\alpha = .05$. State the p-value.

		Categories		
Satisfaction	I	II	III	IV
High	40	60	52	48
Medium	103	87	82	88
Low	57	53	66	64

10.14 Compute Cramer's contingency coefficient for the data in Exercise 10.13.

10.15 A political survey has been taken to compare opinion of the administration's foreign policy against the political party of the respondent.

Test the null hypothesis of independence of opinion on foreign policy and political party. Use $\alpha = .01$ and state the p-value.

| | Opinion | |
Party	Approve	Disapprove
Republican	114	53
Democrat	87	27
Other	17	8

10.16 Compute Cramer's contingency coefficient for the data in Exercise 10.15.

10.17 At the end of a course a college professor makes the following tally based on grade received and class attendance. Test the hypothesis that the grade received in the course is independent of the number of days absent. Use $\alpha = .05$ and state the p-value.

Number of Days Absent	Grade Received Pass	Fail
0–3	24	0
4–6	18	2
More than 6	3	13

10.18 A random sample of freshmen entering Texas Tech University in 1982 was examined again five years later to see how many had graduated. The students were divided into three categories, those from Lubbock County, those from other counties in Texas, and those from out of state. Test the hypothesis that graduation status is independent of where the student's home is located. Let $\alpha = .10$ and state the p-value.

Student's Home	Graduation Status Did Graduate	Did Not Graduate
Lubbock County	78	15
Other Texas Counties	92	24
Outside of Texas	38	13

10.19 Three supermarkets have a policy of advertising specials on certain days of the week to attract customers. A customer count is maintained at these supermarkets from the hours of 11 A.M. to 2 P.M. for the five weekdays. Use the summary of these counts given on the next page to test the hypothesis that the choice of supermarket is independent of the day of the week. Let $\alpha = .01$ and state the p-value.

Day of the Week

Supermarket	Monday	Tuesday	Wednesday	Thursday	Friday	Totals
Furrs	605	639	790	617	573	3224
Safeway	674	657	723	937	686	3677
A&P	564	790	477	529	501	2861
Totals	1843	2086	1990	2083	1760	9762

10.3

THE CHI-SQUARE GOODNESS-OF-FIT TEST

PROBLEM SETTING 10.4

The personnel supervisor in a plant employing over 500 people is interested in the absentee pattern of the employees. She decides to keep a record of all sick leave taken in a two-week period, with the following results.

Day on Which Sick Leave Was Taken

Monday	Tuesday	Wednesday	Thursday	Friday	Total
19	24	8	14	31	96

She is interested in testing the hypothesis that there is an equal chance of sick leave being taken on any of the five work days of the week. She believes that it is reasonable to assume that these data behave the same as a random sample would.

GOODNESS-OF-FIT TEST

This is a situation that often arises—the experimenter has in mind certain probabilities, and wants to see if the observations confirm or deny the preconceived probabilities. Any test that compares the observations with the probabilities to see how well the probabilities "fit" the observed relative frequencies, is called a **goodness-of-fit test.** One of the most widely used goodness-of-fit tests is the **chi-square goodness-of-fit test.** It is especially appropriate if the data naturally fall into a single-row contingency table; however, it may also be used with data that are grouped into intervals. The count in each interval is used to form a single-row contingency table.

The concept underlying this goodness-of-fit test is the same as the one described in earlier sections of this chapter. The observed counts O_i are compared with the expected counts E_i using a statistic that sums $(O_i - E_i)^2/E_i$ over all of the cells. In this application the expected counts E_i are not estimated from the marginal probabilities as before, but they are given directly from the individual cell probabilities p_i specified in the null hypothesis. Also only one subscript is needed because of the single-row nature of the contingency table. Thus the statistic is as follows:

$$T = \sum \frac{(O_i - E_i)^2}{E_i}$$

where $E_i = np_i$ gives the expected cell count.

THE CHI-SQUARE GOODNESS-OF-FIT TEST

Data

The data appear as counts or frequencies in a contingency table with one row. In some cases the observations may be on a continuous-valued random variable but are grouped into intervals and presented in a contingency table.

	Cells					
	1	2	3	$\cdots$	k	Total
	O_1	O_2	O_3	$\cdots$	O_k	n

Assumptions

The data represent observations from a random sample.

Null Hypothesis

The probability of an observation falling into cell 1 is p_1, cell 2 is p_2, . . . , cell k is p_k, for some specified values of p_1 to p_k.

Test Statistic

The test statistic is

$$T = \sum_{i=1}^{k} \frac{(O_i - np_i)^2}{np_i} \tag{10.6}$$

which may be also written as

$$T = \sum_{i=1}^{k} \frac{O_i^2}{np_i} - n \tag{10.7}$$

for easier computations. The null distribution of T is approximately the chi-square distribution with $k - 1$ degrees of freedom, whose quantiles are given in Table A4.

Decision Rule

The alternative hypothesis is that the true cell probabilities are different than those specified by the null hypothesis. Reject the null hypothesis at the α level of significance if T exceeds the $1 - \alpha$ quantile from the chi-square distribution with $k - 1$ degrees of freedom, where k is the number of cells. These critical values may be found in Table A4.

Comment

The chi-square distribution is only an approximation to the true distribution, which is usually too difficult to obtain. The chi-square approximation is usually fairly good, however, unless some of the values of np_i are quite small. For example, values of np_i as small as 1.0 seem to cause no difficulty. However, values less than 0.5 should be avoided by combining adjacent cells if this is reasonable to do in the experiment.

EXAMPLE

If the personnel supervisor in Problem Setting 10.4 feels that bona fide sick leave should occur with equal probability over all five work days, then she is interested in testing the null hypothesis

$$H_0: p_1 = \tfrac{1}{5}, \ p_2 = \tfrac{1}{5}, \ p_3 = \tfrac{1}{5}, \ p_4 = \tfrac{1}{5}, \ p_5 = \tfrac{1}{5}.$$

She is interested in seeing how well these probabilities fit the observations.

	Monday	Tuesday	Wednesday	Thursday	Friday	Totals
Observed	19	24	8	14	31	96
Expected (np_i)	19.2	19.2	19.2	19.2	19.2	96

The expected values are found using $np_i = 96(\tfrac{1}{5}) = 19.2$. The test statistic for the chi-square goodness of fit test is

$$
\begin{aligned}
T &= \sum_{i=1}^{5} \frac{O_i^2}{np_i} - n \\
&= \frac{(19)^2}{19.2} + \frac{(24)^2}{19.2} + \frac{(8)^2}{19.2} + \frac{(14)^2}{19.2} + \frac{(31)^2}{19.2} - 96 \\
&= \frac{2158}{19.2} - 96 = 16.40
\end{aligned}
$$

For a 5% test the .95 quantile from a chi-square distribution with $k - 1 = 4$ degrees of freedom yields a critical value of 9.488. Therefore, the null hypothesis is easily rejected, and the conclusion is that sick leave is not equally likely to be taken over all five working days. The p-value associated with $T = 16.40$ is found from Table A4 by interpolation to be about .003.

ESTIMATING PARAMETERS IN THE HYPOTHESIZED DISTRIBUTION

If some of the parameters in the hypothesized probability distribution are not specified, then they must be estimated from the data. Each such parameter estimated from the data results in a decrease of one unit in the number of degrees of freedom used in the chi-square goodness-of-fit test. The following examples illustrate parameter estimation in the binomial and exponential distributions, and the corresponding reduction in the degrees of freedom.

EXAMPLE

An operations manager is interested in building a mathematical model that will accurately describe the number of breakdowns in his production machines, and will imitate the variability of breakdowns as well. Certain assumptions of independence and equal probability of breakdown among the machines lead to a binomial distribution, where n is the number of machines (10 in this case) and p is the probability of a machine breaking down sometime during the week. The value of p is unknown and must be estimated from the data. The data are collected for 26 weeks with the following results.

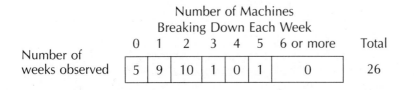

	Number of Machines Breaking Down Each Week							Total
	0	1	2	3	4	5	6 or more	
Number of weeks observed	5	9	10	1	0	1	0	26

The parameter p represents the probability of a machine breaking down in one week, so it may be estimated by dividing the total number of machines breaking down by the total number of machines times the 26 weeks.

$$\text{Estimate for } p = \frac{\text{total number of machines breaking down}}{(\text{total number of machines}) \times (\text{total number of weeks})}$$

$$= \frac{0(5) + 1(9) + 2(10) + 3(1) + 4(0) + 5(1)}{10(26)}$$

$$= .142$$

The binomial probabilities for $n = 10$ machines and $p = .142$ may be estimated from Table A1 by interpolation between $p = .10$ and $p = .15$. For this example the formula in Section 5.1 is used to find the exact probabilities $p_x = P(X = x)$. These probabilities are multiplied by the total number of weeks to get the expected number of weeks with $X = 0, 1, 2, \ldots$, breakdowns.

$$p_x = \binom{10}{x}(.142)^x(.858)^{10-x} \qquad 26\,p_x$$

	$26\,p_x$
$P(X = 0) = .2162$	5.62
$P(X = 1) = .3578$	9.30
$P(X = 2) = .2665$	6.93
$P(X = 3) = .1176$	3.06
$P(X = 4) = .0341$	0.89
$P(X = 5) = .0068$	0.18
$P(X \geq 6) = .0009$	0.02

Because of the small expected values in the last two cells, the last three cells are combined.

	Number of Machines Breaking Down Each Week					Total
	0	1	2	3	4 or more	
Observed	5	9	10	1	1	26
Expected ($26p_x$)	5.62	9.30	6.93	3.06	1.09	26

The use of Equation 10.7 gives

$$T = \frac{25}{5.62} + \frac{81}{9.30} + \frac{100}{6.93} + \frac{1}{3.06} + \frac{1}{1.09} - 26$$

$$= 28.83 - 26$$

$$= 2.83$$

The number of degrees of freedom is 3, one less than $k - 1 = 4$, because the parameter p in the binomial distribution is estimated from the data. The null hypothesis is that the number of machines breaking down each week follows a binomial distribution with $n = 10$ and unknown p. The test statistic $T = 2.83$ is compared with the critical value for 5% test, 7.815, which is the .95 quantile from a chi-square distribution with three degrees of freedom. The null hypothesis is not rejected. The binomial distribution appears to provide a reasonable model for the operations manager to work with. The p-value is about .40.

An example showing how the chi-square goodness-of-fit test may be used with continuous random variables is now given.

EXAMPLE

If accidents occur at random in a factory, it is sometimes assumed that the time between successive accidents follows an exponential distribution. As is explained in Section 5.5, the exponential distribution is a continuous distribution, so the time intervals between successive accidents must be collected into intervals if the chi-square goodness-of-fit test is to be used. Observations over a month resulted in the following times between accidents.

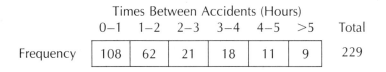

	Times Between Accidents (Hours)						
	0–1	1–2	2–3	3–4	4–5	>5	Total
Frequency	108	62	21	18	11	9	229

The exponential distribution has one parameter that must be estimated from the data using the reciprocal of the sample mean. The sample mean may be estimated by assuming that the observations in each interval all fall at the midpoint of the interval; 0.5 for the first interval, 1.5 for the second interval, and so on. The last interval includes all numbers greater than 5, so 6.0 will be used as a "midpoint," under the belief that most of the observations in that interval should be between 5 and 7. Thus

$$\overline{X} = (1\tfrac{1}{229})[0.5(108) + 1.5(62) + 2.5(21) + 3.5(18) + 4.5(11) + 6.0(9)]$$

$$= 1.60$$

The probability that an exponential random variable X will fall between any two times t_0 and t_1 is given in Section 5.5 as

$$P(t_0 < X \le t_1) = e^{-\lambda t_0} - e^{-\lambda t_1}$$

where λ is estimated by $1/\overline{X}$, and where e is a well-known constant approximately equal to 2.718. Thus the previous probability is calculated using the estimate for λ.

$$P(t_0 < X \le t_1) \approx e^{-t_0/\overline{X}} - e^{-t_1/\overline{X}}$$

as follows.

Interval	Probability p_i		np_i
0 to 1	$e^0 - e^{-1/1.6}$	= .4647	106.4
1 to 2	$e^{-1/1.6} - e^{-2/1.6}$	= .2488	57.0
2 to 3	$e^{-2/1.6} - e^{-3/1.6}$	= .1331	30.5
3 to 4	$e^{-3/1.6} - e^{-4/1.6}$	= .0713	16.3
4 to 5	$e^{-4/1.6} - e^{-5/1.6}$	= .0381	8.7
>5	$e^{-5/1.6}$	= .0439	10.1
Totals		1.0000	229.0

The test statistic is

$$T = \sum_{i=1}^{6} \frac{O_i^2}{np_i} - n$$

$$= \frac{(108)^2}{106.4} + \frac{(62)^2}{57.0} + \frac{(21)^2}{30.5} + \frac{(18)^2}{16.3} + \frac{(11)^2}{8.7} + \frac{(9)^2}{10.1} - 229$$

$$= 4.33$$

Since this value of the test statistic T is less than 9.488, the .95 quantile of a chi-square distribution with four degrees of freedom, the null hypothesis is not rejected at the 5% level. The test utilizes four degrees of freedom rather than five since λ was estimated from the sample data. The exponential distribution with $\lambda = 1/1.6 = 0.625$ seems to fit the observed time intervals between accidents. The p-value associated with $T = 4.33$ is approximately .37, as shown by Table A4.

A COMPARISON WITH THE LILLIEFORS TEST

The test presented in this section, like the Lilliefors test of Chapter 5, is a goodness-of-fit test. Therefore, it is appropriate here to compare the two tests.

Recall that the Lilliefors test is used to see if the sample data could reasonably be regarded as having come from a normal population, where neither parameter μ nor σ is specified. So the Lilliefors goodness-of-fit test, as presented, is limited. The chi-square goodness-of-fit test, however, is more versatile. It may be applied to test the goodness of fit for any distribution, discrete or continuous. It is more natural to use the Lilliefors test to test for normal distributions, however, because the data may be used directly in the Lilliefors test. The chi-square test requires grouping the data into arbitrary intervals, and in so doing it loses some "information" contained in the data. The authors recommend use of the Lilliefors test for normality because it tends to have more power and it is easier to use than the chi-square test.

EXERCISES

10.20 A bank is considering a checking account system that will pay no interest for accounts having a minimum monthly balance of less than $500, 5.25% interest for accounts with $500 to $5000, and 7.25% interest for accounts with over $5000. The bank thinks the respective proportion of accounts in each of these categories is 45%, 40%, and 15%. A random sample of 400 accounts is obtained with the results given below. Use a chi-square goodness-of-fit test to compare these sample observations with the hypothesized proportions. Use $\alpha = .05$ and state the p-value.

	Less than $500	$500 to $5000	More than $5000	Total
Number of checking accounts	165	170	65	400

10.21 An accountant for a department store knows from past experience that 23% of the store's customers pay cash for their purchases, 35% write checks, and the remaining 42% use charge cards. The accountant examines a random sample of 200 sales receipts for the week before Christmas and makes the following sales summary.

	Cash	Checks	Credit Cards	Total
Number of customers	37	47	116	200

Use a chi-square goodness-of-fit test to see if the preceding percentages fit these observations. Use $\alpha = .05$ and state the p-value. What should the accountant conclude about customer payment methods as Christmas draws near?

10.22 A recent business school graduate in marketing is asked by his boss to randomly select 200 registration cards from the thousands that have been returned by customers to the manufacturer to see if the type of store where the product was sold is in agreement with the historical percentages for these stores. Use a chi-square goodness-of-fit test to see if the sample results agree with the historical percentages. Let $\alpha = .05$ and state the p-value.

Type of Store	Historical Sales Percentage	Sample Counts
Discount	37.3	91
Department	22.8	40
Hardware	12.9	20
Appliance	13.8	22
Other	13.2	27
Total		200

10.23 To ensure that there is always adequate help on hand, the manager of a supermarket needs to schedule working hours for his employees to coincide with the number of customers expected each day of the week. The manager thinks that the percentage breakdown of customers is as given below. A customer count from cash register receipts for one randomly selected week is obtained to compare with the manager's estimate. Use a chi-square goodness-of-fit test to check the adequacy of the hypothesized percentages on these sample observations. Use $\alpha = .05$ and state the p-value.

Day of the Week	Manager's Estimate of the Percentage of Customers	Customer Count for One Week
Sunday	20	850
Monday	11	605
Tuesday	11	639
Wednesday	16	790
Thursday	11	617
Friday	11	573
Saturday	20	920
Total		4994

10.24 The following probability model is used by a fire department to describe the number of fire calls in one day. Use a chi-square goodness-of-fit test to see if the model probabilities fit the number of calls at a fire station for 100 randomly selected days. Let $\alpha = .05$ and state the p-value.

Number of fire calls:	0	1	2	3	4	5	6
Probability:	.10	.15	.15	.20	.15	.15	.10
Observed frequency:	15	20	17	21	9	10	8

10.25 An insurance company wants to compare their dollar claim settlement against the percentages given in an article by *Consumer Report*. Use a chi-square goodness-of-fit test to compare the insurance company claim settlements against the percentages given by *Consumer Report* based on 300 randomly selected claims summarized at the top of the next page. Let $\alpha = .10$ and state the p-value.

Dollar Amount of Claim Settlement	Percentage of Claims	Claims for Insurance Company
$1–$99	13.8	50
100–199	15.8	40
200–299	10.9	30
300–399	7.9	28
400–499	5.0	20
500–999	16.8	36
1,000–1,499	6.9	15
1,500–1,999	4.0	17
2,000–2,499	3.0	15
2,500–4,999	5.9	14
5,000–9,999	4.0	12
10,000–49,999	5.0	16
50,000 or more	1.0	7
Total		300

10.26 In 1970 a lottery was used to determine the order in which young men would be drafted into the armed forces. Birthdays were drawn at random, and drafting was done according to which birthdays were drawn first. A summary of these data is given below to show the number of selections in each month that were among the first half of potential draftees (i.e., draft number less than or equal to 183). If the draft were truly random, then each month should be approximately evenly split above and below draft order number 183.5. Use a chi-square goodness-of-fit test to check the randomness of the draft based on the following summaries. Let $\alpha = .01$ and state the p-value.

Month	Number of Selections in the First 183	Number Expected
January	12	15.5
February	12	14.5
March	10	15.5
April	11	15
May	14	15.5
June	14	15
July	14	15.5
August	19	15.5
September	17	15
October	13	15.5
November	21	15
December	26	15.5

10.27 A class of 44 students took an exam having 10 questions. The results are summarized on the next page. Assume these data behave as a random sample. Use a chi-square goodness-of-fit test to see if a binomial distribution can be used as a mathematical model to describe the test results as was done in the second example of this section. Use $\alpha = .05$ and state the p-value.

Number of Questions Missed	Number of Students
0	2
1	8
2	11
3	12
4	11
5 or more	0

10.4

REVIEW EXERCISES

10.28 A medication for treating the common cold is tested by giving the medication to half of a randomly selected group of people with colds. The other half receives a placebo. The summary of each patient's opinion of the treatment is given in the following table.

Treatment	Patient's Opinion Harmful	No Effect	Helpful	Totals
Medication	16	30	104	150
Placebo	20	42	88	150
Totals	36	72	192	300

Test the hypothesis that patient opinion is independent of treatment. Let $\alpha = .05$ and state the p-value.

10.29 Compute Cramer's contingency coefficient for the data in Exercise 10.28.

10.30 The number of magazines sold each hour at a corner newsstand is recorded below for several hours of business. Use a chi-square goodness-of-fit test to see how well a Poisson distribution (see Section 5.5) with $\lambda = 2$ fits these data. Let $\alpha = .01$.

Number of Magazines Sold	Frequency
0	10
1	25
2	33
3	20
4	7
5	3
6	2
More than 6	0
Total	100

10.31 To see if customers might be influenced by the colors used on discount cards for a retail store, the store mailed out 1000 cards to its customers using 250 cards for each of four color types. The assignment of color to a customer was made at random. The store then recorded the number of cards of each type that were returned and this summary is given below. Test the hypothesis that the return of a discount card is independent of the color used on the card. Let $\alpha = .05$ and state the p-value.

	Red	White	Blue	Green	Totals
		Color Used on Card			
Returned	107	106	115	127	455
Not Returned	143	144	135	123	545
Totals	250	250	250	250	1000

10.32 A production supervisor is interested in knowing if the number of breakdowns on four machines is independent of the shift using the machines. Test this hypothesis based on the following sample counts. Let $\alpha = .05$ and state the p-value.

Shift	A	B	C	D	Total
		Machine			
7–3	14	9	17	19	59
3–11	14	17	27	30	88
11–7	19	14	19	26	78
Total	47	40	63	75	225

10.33 The number of automobile fatalities for one year in a large city has been grouped by the day of the week. Use a chi-square goodness-of-fit test to see how well a uniform distribution fits these data. Let $\alpha = .01$ and state the p-value. Starting with Sunday the numbers of fatalities were 27, 20, 17, 17, 21, 28, 18. Altogether there were 148 fatalities.

10.34 New employees at an army base are given some instruction on security regulations. The instructions can be given via a videotape that the employees watch, or via a brochure that the employees read. New employees are assigned randomly to one or the other, and they are given a test on the security regulations at the completion of their indoctrination. Results for one set of employees show that of 48 who watched the videotape, 13 failed to pass the test the first time. Of 52 who read the brochure, 9 failed to pass the test. If this represents a random sample of all new employees, what do these results say about the effectiveness of the two instruction programs compared with each other?

10.35 The U.S. Army wants to know if results on a mechanical aptitude test can be used to predict success in a machinery repair course. Sixty-five applicants for the course are accepted. Forty-six applicants had scores above 110 on their mechanical aptitude tests, and 38 of them successfully completed the course. The remaining 19 applicants scored below 110, and only 6 of them successfully completed the course. Assume that these data behave as a random sample. Test the hypothesis that the mechanical aptitude test results are independent of the course results, against the one-sided alternative that high test scores tend to predict successful course performance.

10.36 A consumer group is examining the contents of breakfast cereal boxes to see if the weight of the cereal is at least as much as advertised on the box. Random samples of four different brands of cereal are examined.

| | Cereal Brand | | | |
	A	B	C	D
Total number of boxes examined	40	50	40	45
Number of boxes judged too light	6	3	8	2

Test the hypothesis that each of the four cereal manufacturers has the same probability of producing a box of cereal that contains less than the specified weight of contents.

10.37 A manufacturer of farm machinery receives bolts from six different suppliers. Random samples of 100 bolts each were drawn from shipments from each of the six suppliers, and the number of defective bolts from each was noted. These numbers of defective bolts were 6, 7, 17, 8, 10, and 4. Test the null hypothesis that the probability of receiving a defective bolt is the same from all six suppliers. Use $\alpha = .05$.

BIBLIOGRAPHY

Additional material on the topics presented in this chapter can be found in the following publication.

Conover, W. J. (1980). *Practical Nonparametric Statistics,* 2nd ed. Wiley, New York.

11

CORRELATION

In Chapter 10 the phi coefficient was used as a measure of association between the row classification and the column classification of the contingency table. In this chapter *Pearson's product-moment correlation coefficient,* often simply called the **sample correlation coefficient,** is used as a measure of association between the observations on a pair of variables X and Y. When such variables occur in pairs, a scatterplot is useful for providing information about possible relationships between X and Y that might not be easily discernible from observing the data. That is, it may be of interest to know if the data exhibit a straight-line relationship (or nearly so), or a curvilinear relationship, or perhaps no relationship at all. The correlation coefficient calculated on the sample data is quite useful in determining the strength of **linear** relationships between paired observations. Since linear relationships play an important role in statistics, particularly in the regression techniques of the next chapter, the first section of this chapter is devoted to studying the correlation coefficient computed on the sample data.

Relationships that are not linear may exist between a pair of variables. In such instances the value of Y may increase as X increases or the value of Y may decrease as X increases. Relationships like these are more generally described as **monotonic.** The strength of monotonic relationships is also of interest in statistics, and these relationships are more accurately measured by the correlation coefficient calculated on rank transformed data. The second section of this chapter considers this calculation, referred to as rank correlation or **Spearman's rho.**

11.1

INTRODUCTION TO CORRELATION

PROBLEM SETTING 11.1

Complaints have been heard from several employees in a manufacturing firm. The complaints are centered around charges of favoritism of certain employees by the manager. The manager insists that the salary raises have been justified on the basis of the employees' contribution to the company in the form of a positive attitude, general merging of the individual's goals with those of the company, and individual productivity. Of these three criteria the easiest one to quantify is productivity, although measuring productivity is still somewhat subjective.

The manager agrees to let a special committee of employees examine a random sample of employees, measure their individual productivity, and compare this with their percentage raise. A random sample of size $n = 20$ is selected and examined, with the results as given in Figure 11.1 where the 20 employees are represented by points in a scatterplot. The x-axis is the productivity measurement and the y-axis is the percentage raise received by the employee. Notice the tendency of the points to follow a straight line. Such a tendency indicates that a linear relationship exists between productivity and salary increases. However, the relationship shows some scattering around the straight line. There is a need to measure the **strength of the linear relationship**

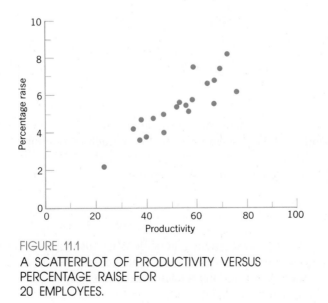

FIGURE 11.1
A SCATTERPLOT OF PRODUCTIVITY VERSUS
PERCENTAGE RAISE FOR
20 EMPLOYEES.

between the two variables. The usual statistic used for this measure is the **sample correlation coefficient.**

THE SAMPLE CORRELATION COEFFICIENT

There is a convenient statistic that can be used to measure the strength of a linearly increasing or decreasing relationship, that is, a straight-line relationship of one variable with another. The statistic is called the **sample correlation coefficient** and is denoted by r. It is a statistic that is close to zero in cases where there is no tendency for one variable to increase or decrease as the other variable increases. The correlation coefficient is always between -1 and $+1$. It equals $+1$ only if all the points in the scatterplot of the two variables are on a straight line with an upward slope (upward as viewed from left to right on the graph). It equals -1 only if all the points are on a straight line with a downward slope. The more the points scatter away from a linear relationship, the closer r comes to equaling zero. This does not imply that no relationship exists, it merely implies that the relationship is not linear.

The sample correlation coefficient results from a calculation made on a bivariate sample (X_i, Y_i), $i = 1, \ldots, n$ using Eq. 11.1.

The **sample correlation coefficient** r measures the strength of the **linear** relationship between two variables. It is computed from

SAMPLE
CORRELATION
COEFFICIENT

$$r = \frac{\sum\limits_{i=1}^{n} (X_i - \overline{X})(Y_i - \overline{Y})}{\left\{ \sum\limits_{i=1}^{n} (X_i - \overline{X})^2 \sum\limits_{i=1}^{n} (Y_i - \overline{Y})^2 \right\}^{1/2}}$$

$$= \frac{\sum\limits_{i=1}^{n} X_i Y_i - \left(\sum\limits_{i=1}^{n} X_i \right)\left(\sum\limits_{i=1}^{n} Y_i \right) \Big/ n}{\left\{ \left[\sum\limits_{i=1}^{n} X_i^2 - \left(\sum\limits_{i=1}^{n} X_i \right)^2 \Big/ n \right]\left[\sum\limits_{i=1}^{n} Y_i^2 - \left(\sum\limits_{i=1}^{n} X_i \right)^2 \Big/ n \right] \right\}^{1/2}} \quad (11.1)$$

Many inexpensive calculators are programmed to compute r with a minimum of effort, and anyone who expects to do much computing of correlation coefficients should consider investing in one of these. The steps for computing r manually are given in the following example.

EXAMPLE

The measurements of productivity and percentage raise for the random sample of 20 employees in Problem Setting 11.1 are represented in a scatterplot in Figure 11.1. A worksheet for these data is given in Figure 11.2. Use these data to calculate the sample correlation coefficient r.

Employee	Productivity X	Percentage Raise Y	XY	X²	Y²
1	47	4.2	197.4	2209	17.64
2	71	8.1	575.1	5041	65.61
3	64	6.8	435.2	4096	46.24
4	35	4.3	150.5	1225	18.49
5	43	5.0	215.0	1849	25.00
6	60	7.5	450.0	3600	56.25
7	38	4.7	178.6	1444	22.09
8	59	5.9	348.1	3481	34.81
9	67	6.9	462.3	4489	47.61
10	56	5.7	319.2	3136	32.49
11	67	5.7	381.9	4489	32.49
12	57	5.4	307.8	3249	29.16
13	69	7.5	517.5	4761	56.25
14	38	3.8	144.4	1444	14.44
15	54	5.9	318.6	2916	34.81
16	76	6.3	478.8	5776	39.69
17	53	5.7	302.1	2809	32.49
18	40	4.0	160.0	1600	16.00
19	47	5.2	244.4	2209	27.04
20	23	2.2	50.6	529	4.84
Totals	1064	110.8	6237.5	60,352	653.44

FIGURE 11.2
WORKSHEET FOR THE CALCULATION OF THE SAMPLE
CORRELATION COEFFICIENT.

Substitution of the sample calculations in Figure 11.2 into Eq. 11.1 gives

$$r = \frac{6237.5 - (1064)(110.8)/20}{\{[60352 - (1064)^2/20][653.44 - (110.8)^2/20]\}^{1/2}} = .890$$

This value of r is not equal to 1; therefore, the points do not all lie on a straight line. However, the value of r is close to 1, indicating an upward slope as viewed from left to right on the graph. Some more scatterplots are now presented that show how the pattern of points behaves for different values of r.

SCATTERPLOTS AND CORRELATION COEFFICIENTS

The variables X and Y increase together in Figures 11.3a and b (positive correlation), whereas in Figures 11.3c and d, Y decreases as X increases (negative correlation). In fact, since all pairs of points are on a straight line, Figures 11.3a and b represent the extreme in positive correlation, so $r = +1$, whereas Figures 11.3c and d represent the extreme in negative correlation, so $r = -1$.

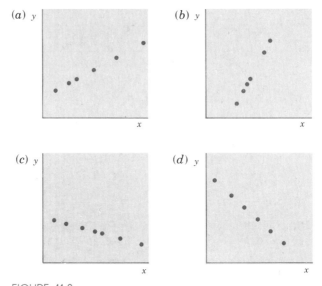

FIGURE 11.3
SAMPLE PAIRS (X_i, Y_i) THAT HAVE A PERFECT
POSITIVE OR PERFECT NEGATIVE CORRELATION.

Of course, in real-world problems, nature seldom responds with all sample pairs (X_i, Y_i) on a straight line. Rather, some scatterplots will indicate a strong linear relationship; others will not. Examples are shown in Figure 11.4 that are associated with different values of r.

BIVARIATE DISTRIBUTIONS

The scatterplots in Figure 11.4 illustrate the concept that two random variables X and Y behave jointly according to some **bivariate probability distribution.** That is, X has a probability distribution and Y also has a probability distribution; however, their joint behavior is not completely determined by their marginal behavior. When the two random variables are correlated they are no longer independent of one another. Rather, the random variable Y is influenced by the random variable X and vice versa.

If each of the marginal distributions for the random variables X and Y is normal, then the joint probability distribution of X and Y is usually given by a particular distribution known as the **bivariate normal distribution.** A graph to represent the three-dimensional nature of the bivariate normal density function is given in Figure 11.5. In Figure 11.5 lines have been added that are parallel to each of the X and Y axes such that they each intersect the elliptical mound-shaped portion of the graph. These lines illustrate the fact that the conditional distributions for the random variables X and Y are each normal.

One of the parameters of a bivariate normal distribution is the **population correlation coefficient** ρ, which is estimated by r. If ρ equals zero in a bivariate normal distribution, then the two variables are independent. In this section a procedure is given for testing whether an observed sample correlation coefficient is sufficiently large in absolute value to indicate that the population value

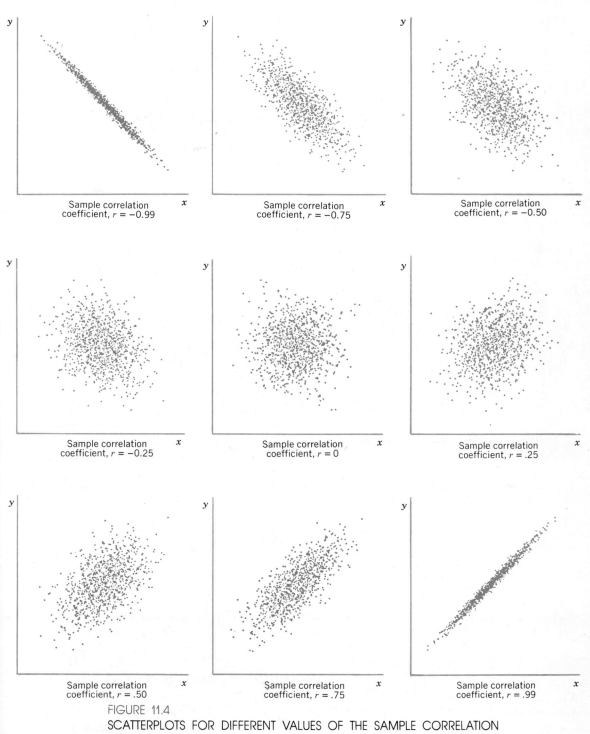

Sample correlation coefficient, $r = -0.99$

Sample correlation coefficient, $r = -0.75$

Sample correlation coefficient, $r = -0.50$

Sample correlation coefficient, $r = -0.25$

Sample correlation coefficient, $r = 0$

Sample correlation coefficient, $r = .25$

Sample correlation coefficient, $r = .50$

Sample correlation coefficient, $r = .75$

Sample correlation coefficient, $r = .99$

FIGURE 11.4
SCATTERPLOTS FOR DIFFERENT VALUES OF THE SAMPLE CORRELATION
COEFFICIENT.

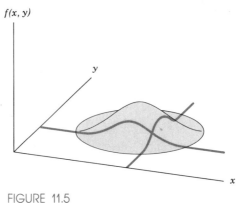

FIGURE 11.5
PLOT OF A BIVARIATE NORMAL DENSITY.

ρ is significantly different from 0. Also a method for finding a confidence interval for ρ is presented. Both procedures rely heavily on the assumption of a bivariate normal distribution for X and Y. This assumption represents the primary weakness of these procedures, as the bivariate normal distribution assumption is difficult to check.

One way to check for bivariate normality is to test for normality of the distributions of X and Y by themselves, using the Lilliefors test. If the X's are regarded as possibly having a normal distribution, then the Y's are tested for normality using the same test. Acceptance of normality for both variables *indicates the possibility* of a bivariate normal distribution, but *does not ensure* a bivariate normal distribution. However, rejection of normality by either test indicates the assumption of bivariate normality is not justified. Thus it is possible to reject bivariate normality, but not to prove it. If the assumption of bivariate normality is reasonable, the methods in this section are the best available in terms of power. If the assumption is not valid, the methods in the next section should be used. Two examples will now be given to illustrate these procedures.

EXAMPLE

The value of r was calculated to be .890 in the previous example for the measurements of productivity and percentage raise for the random sample of 20 employees in Problem Setting 11.1. Use these data to test H_0: $\rho = 0$ versus H_1: $\rho > 0$ with $\alpha = .005$.

The Lilliefors test shows the assumption of normality to be reasonable for X and Y, so the bivariate normal distribution is assumed for (X, Y).

Decision rule   Reject H_0 if $T > t_{.995,18} = 2.8784$, where $t_{.995,18}$ is obtained from Table A3 with 18 degrees of freedom. The test statistic is

$$T = .890\sqrt{\frac{20 - 2}{1 - .890^2}} = 8.2813$$

PROCEDURES FOR INFERENCES REGARDING CORRELATION
(Bivariate Normal Distribution)

Data

The data consist of n pairs (X_i, Y_i), $i = 1, \ldots, n$ from a bivariate population. Compute the sample correlation coefficient

$$r = \frac{\Sigma X_i Y_i - (\Sigma X_i)(\Sigma Y_i)/n}{\{[\Sigma X_i^2 - (\Sigma X_i)^2/n][\Sigma Y_i^2 - (\Sigma Y_i)^2/n]\}^{1/2}}$$

Assumptions

1. The sample of pairs (X_i, Y_i), $i = 1, \ldots, n$ is a random sample.
2. The joint distribution of (X, Y) is a bivariate normal distribution.

Estimation

The sample correlation coefficient provides a point estimate of the population correlation coefficient ρ. An approximate $100(1 - \alpha)\%$ confidence interval may be formed as follows.

1. Convert r to a new variable w using the table in Figure 11.6.
2. Find $w_L = w - z_{1-\alpha/2}/\sqrt{n - 3}$ and $w_U = w + z_{1-\alpha/2}/\sqrt{n - 3}$, where $z_{1-\alpha/2}$ is the $1 - \alpha/2$ quantile of a normal distribution given in Table A2.
3. Use the table in Figure 11.6 to convert w_L to r_L and w_U to r_U. The $100(1 - \alpha)\%$ confidence interval for ρ is from r_L to r_U.

Null Hypothesis

$$H_0: \rho = 0$$

(i.e., X and Y are independent).

Test Statistic

The test statistic is

$$T = r\sqrt{\frac{n - 2}{1 - r^2}}$$

The null distribution of T is the Student's t distribution with $n - 2$ degrees of freedom.

Decision Rules

The decision rule depends on the alternative hypothesis of interest and the desired level of significance α. Critical values for the test statistic are obtained by entering Table A3 with $n - 2$ degrees of freedom to obtain either the $1 - \alpha$ or $1 - \alpha/2$ quantile.

(a) H_1: $\rho > 0$. Reject H_0 if $T > t_{1-\alpha,n-2}$.
(b) H_1: $\rho < 0$. Reject H_0 if $T < -t_{1-\alpha,n-2}$.
(c) H_1: $\rho \neq 0$. Reject H_0 if $T > t_{1-\alpha/2,n-2}$ or if $T < -t_{1-\alpha/2,n-2}$.

Since 8.2813 is greater than 2.8784, H_0 is easily rejected at $\alpha = .005$. The p-value is much smaller than .0001. The positive value of the sample correlation indicates that greater salary raises are associated with higher productivity and vice versa.

r	w	r	w	r	w	r	w
0.00	0.0000	0.25	0.2554	0.50	0.5493	0.75	0.9730
0.01	0.0100	0.26	0.2661	0.51	0.5627	0.76	0.9962
0.02	0.0200	0.27	0.2769	0.52	0.5763	0.77	1.0203
0.03	0.0300	0.28	0.2877	0.53	0.5901	0.78	1.0454
0.04	0.0400	0.29	0.2986	0.54	0.6042	0.79	1.0714
0.05	0.0500	0.30	0.3095	0.55	0.6184	0.80	1.0986
0.06	0.0601	0.31	0.3205	0.56	0.6328	0.81	1.1270
0.07	0.0701	0.32	0.3316	0.57	0.6475	0.82	1.1568
0.08	0.0802	0.33	0.3428	0.58	0.6625	0.83	1.1881
0.09	0.0902	0.34	0.3541	0.59	0.6777	0.84	1.2212
0.10	0.1003	0.35	0.3654	0.60	0.6931	0.85	1.2562
0.11	0.1104	0.36	0.3769	0.61	0.7089	0.86	1.2933
0.12	0.1206	0.37	0.3884	0.62	0.7250	0.87	1.3331
0.13	0.1307	0.38	0.4001	0.63	0.7414	0.88	1.3758
0.14	0.1409	0.39	0.4118	0.64	0.7582	0.89	1.4219
0.15	0.1511	0.40	0.4236	0.65	0.7753	0.90	1.4722
0.16	0.1614	0.41	0.4356	0.66	0.7928	0.91	1.5275
0.17	0.1717	0.42	0.4477	0.67	0.8107	0.92	1.5890
0.18	0.1820	0.43	0.4599	0.68	0.8291	0.93	1.6584
0.19	0.1923	0.44	0.4722	0.69	0.8480	0.94	1.7380
0.20	0.2027	0.45	0.4847	0.70	0.8673	0.95	1.8318
0.21	0.2132	0.46	0.4973	0.71	0.8871	0.96	1.9459
0.22	0.2237	0.47	0.5101	0.72	0.9076	0.97	2.0923
0.23	0.2342	0.48	0.5230	0.73	0.9287	0.98	2.2976
0.24	0.2448	0.49	0.5361	0.74	0.9505	0.99	2.6467

FIGURE 11.6
TABLE FOR FINDING A CONFIDENCE INTERVAL FOR ρ IN BIVARIATE NORMAL POPULATIONS. *NOTE:* IF r OR w IS NEGATIVE, USE THIS TABLE, ADDING A NEGATIVE SIGN TO BOTH. ALSO, THESE VALUES OF w CAN BE FOUND FROM r USING THE RELATIONSHIP $w = \frac{1}{2} \ln[(1 + r)/(1 - r)]$; LIKEWISE, VALUES OF r CAN BE FOUND FROM w USING THE RELATIONSHIP $r = (e^{2w} - 1)/(e^{2w} + 1)$.

Confidence interval A 95% confidence interval is found by converting $r = .890$ to $w = 1.4219$ from Figure 11.6. For $n = 20$ and $z_{.975} = 1.9600$ the values of w_L and w_U become

$$w_L = 1.4219 - \frac{1.9600}{\sqrt{17}} = .9465$$

and

$$w_U = 1.4219 + \frac{1.9600}{\sqrt{17}} = 1.8973$$

which convert to $r_L = .74$ and $r_U = .96$, respectively, using Figure 11.6. The approximate 95% confidence interval for ρ is from .74 to .96.

If the value of r had been negative rather than positive in this example, that is $r = -.890$, then the confidence interval would have been from $-.96$ to $-.74$. The easiest way to find this interval is to proceed as if r were positive as in the present example and then change the signs on the final interval.

EXAMPLE

The state racing commission is interested in knowing if a significant positive correlation exists between attendance and the amount of money wagered. Assume the population distribution is bivariate normal. Use the data below for 10 racing days to test H_0: $\rho = 0$ versus H_1: $\rho > 0$ with $\alpha = .05$.

Attendance (hundreds) X	Amount Wagered (millions) Y	XY	X^2	Y^2
117	2.07	242.19	13,689	4.2849
128	2.19	280.32	16,384	4.7961
122	3.14	383.08	14,884	9.8596
119	2.26	268.94	14,161	5.1076
131	3.40	445.50	17,161	11.5600
135	2.89	390.15	18,225	8.3521
125	2.93	366.25	15,625	8.5849
120	2.66	319.20	14,400	7.0756
130	3.33	432.90	16,900	11.0889
127	3.53	448.31	16,129	12.4609
Totals 1254	28.40	3576.74	157,558	83.1706

$$r = \frac{3576.74 - (1254)(28.40)/10}{\{[157,558 - (1254)^2/10][83.1706 - (28.40)^2/10]\}^{1/2}} = .554$$

$$T = .554 \sqrt{\frac{10 - 2}{1 - .554^2}} = 1.8822$$

Decision rule From Table A3, $t_{.95,8} = 1.8595$, so H_0 will be rejected if $T > 1.8595$. Since $T = 1.8822$, H_0 is rejected. The p-value is slightly smaller than .05.

SAMPLE COVARIANCE

Another way of measuring the tendency of two variables to increase or decrease together is called the **sample covariance.** It is denoted by s_{XY} and is given in Eq. 11.2. Although the sample covariance increases or decreases as r increases or decreases, it has a definite disadvantage compared to r in that the sample covariance is not standardized, so the relative size of s_{XY} is difficult to interpret.

The **sample covariance** between X and Y in a sample of paired observations (X_1, Y_1), (X_2, Y_2), . . . , (X_n, Y_n) measures the tendency for X and Y to increase or decrease together in the sample and is computed as

SAMPLE
COVARIANCE

$$s_{XY} = \frac{1}{n-1}\left[\sum_{i=1}^{n}(X_i - \bar{X})(Y_i - \bar{Y})\right]$$

$$= \frac{1}{n-1}\left[\sum_{i=1}^{n}X_iY_i - \frac{\left(\sum_{i=1}^{n}X_i\right)\left(\sum_{i=1}^{n}Y_i\right)}{n}\right] \tag{11.2}$$

It is possible to standardize s_{XY} by dividing by the two standard deviations s_X and s_Y. In fact, it is easy to show with a little algebra that this furnishes an alternative formula for computing r

$$r = \frac{s_{XY}}{s_X s_Y} \tag{11.3}$$

that is exactly equivalent to Eq. 11.1.

EXAMPLE

For the productivity and percentage raise data in Figure 11.2 the sample covariance is

$$s_{XY} = \left(\frac{1}{19}\right)\left[6237.5 - \frac{(1064)(110.8)}{20}\right] = 18.049$$

from Eq. 11.2. The values of s_X and s_Y for these data are given as

$$s_X = \left[\frac{60352 - (1064)^2/20}{19}\right]^{1/2} = 14.044$$

$$s_Y = \left[\frac{653.44 - (110.8)^2/20}{19}\right]^{1/2} = 1.444$$

These values are substituted into Eq. 11.3 to get

$$r = \frac{18.049}{(14.044)(1.444)} = .890$$

which is exactly the same as in the first example of this section.

COMMON MISINTERPRETATIONS OF CORRELATION

This section concludes with a warning about the misinterpretation of the word *correlation* as it is used in statistics. Statisticians usually use the term *correlation* as a measure of the strength of a linear relationship between two variables. The problem arises when the word *correlation* is used by nonstatisticians to imply what is commonly referred to as a *cause and effect* relationship. That is to say, the occurrence of one event causes another event to occur, hence the events are correlated. However, *cause and effect* is never implied in the statistical use of the word *correlation*. For example, during the 1950s it was discovered that during months when consumption of soft drinks was high there were also large numbers of cases of polio, that is, a statistical correlation existed between these two sets of numbers. Was consumption of soft drinks causing polio? No! In truth, in the summer months when the weather was hot the number of polio cases and the consumption of soft drinks both naturally increased. As the weather cooled off during the rest of the year both variables decreased in value. Although there was a strong correlation between polio cases and soft drink sales, there was no reason to believe that one was causing the other.

Another common misinterpretation is that $r = 0$ indicates the lack of any relationship between X and Y. Values of r close to zero merely indicate the lack of a *linear* relationship between X and Y; other forms of relationships may still exist.

EXERCISES

11.1 If a random bivariate sample of size 20 yields the following calculations, $\Sigma XY = 3739$, $\Sigma X = 200$, $\Sigma Y = 287$, $\Sigma X^2 = 3568$, and $\Sigma Y^2 = 4956$, find the sample correlation coefficient.

11.2 If the joint distribution of (X, Y) in Exercise 11.1 can be assumed to be bivariate normal, test H_0: $\rho = 0$ versus H_1: $\rho > 0$ with a level of significance of .01.

11.3 Find a 95% confidence interval for the population correlation coefficient rho based on the calculations in Exercise 11.1.

11.4 A random bivariate sample of size 30 yields a sample correlation coefficient of $r = -.63$. Find a 95% confidence interval for the population correlation coefficient ρ, assuming the population is bivariate normal.

11.5   Thirty videotapes available for self-instruction were selected at random from a price list of several hundred videotapes. For each videotape the price (Y) and the length in minutes (X) were noted. A summary of the data yields the following results.

$$n = 30 \qquad \Sigma X = 867 \qquad \Sigma X^2 = 31{,}683$$
$$\Sigma XY = 76{,}629 \qquad \Sigma Y = 2421 \qquad \Sigma Y^2 = 203{,}297$$

Assume bivariate normality for these data and test H_0: $\rho = 0$ versus H_1: $\rho > 0$ using a level of significance of .001.

11.6   Find a 95% confidence interval for the population correlation coefficient ρ based on the sample data in Exercise 11.5.

11.7   The data for typing speeds and typing error rates for 20 applicants are given below. Assume bivariate normality in the population. Find a 95% confidence interval for the population correlation coefficient. What interpretation do you give to this interval?

Applicant Number	Typing Score	Number of Errors
1	68	8
2	72	2
3	35	9
4	91	16
5	47	9
6	52	13
7	75	12
8	63	3
9	55	0
10	65	7
11	84	0
12	45	14
13	58	9
14	61	12
15	69	2
16	22	8
17	46	5
18	55	5
19	66	13
20	71	2

11.8   Given below are the tests for length of the cotton fibers (X) and strength of cotton yarn (Y) for 10 randomly selected pieces of yarn. Test each variable for normality. What conclusion do you make about the acceptability of these data coming from a bivariate normal distribution?

($\Sigma X = 1003$, $\Sigma X^2 = 102,193$, $\Sigma Y = 798$, $\Sigma Y^2 = 64,396$, $\Sigma XY = 80,991$.)

X:	99	93	99	97	90	96	93	130	118	88
Y:	85	82	75	74	76	74	73	96	93	70

11.9 A student has completed the data summary necessary to compute the sample correlation coefficient as $\Sigma XY = 4739$, $\Sigma X = 200$, $\Sigma Y = 287$, $\Sigma X^2 = 3568$, and $\Sigma Y^2 \doteq 4956$ with $n = 20$. How can you tell that the student has made an error in the data summary calculations?

11.10 The percentage change in college football season ticket sales may be related to the percentage change in the proportion of games won during the previous season. A random sample of 12 colleges showed the following percentage changes.

Percentage Change in Season Ticket Sales	Percentage Change in Proportion of Games Won (Previous Season)
+4.3	+11
+7.0	+22
+3.0	0
−6.2	−9
−0.1	+10
−1.7	−18
+4.2	0
+6.9	+33
+2.3	+27
−10.2	−44
−8.1	−30
−1.7	−11

Use these data to test the hypothesis that sales are independent of the improvement in the won–lost record of a college team versus the alternative that they tend to increase together. Assume bivariate normality. Use $\alpha = .05$.

11.2

THE RANK CORRELATION COEFFICIENT (Optional)

PROBLEM SETTING 11.2

In peer evaluation, subordinates are asked to rank themselves and to rank each other. This can be quite enlightening for the subordinates, who are encouraged to think seriously about their own job performance relative to the job performance of their peers. It can also boost morale if the results of the ranking are used in the evaluations by supervisors. The primary purpose of such a procedure is to provide an objective independent standard against which the supervisor can check his or her own rankings. Any obvious discrepancies are given further consideration and an investigation into the cause of a discrepancy is made.

Peer rankings may be based on some kind of scaling, such as a scale from 0 to 100. It is difficult to standardize such a scale in the minds of the subordinates—a rating of 50 may represent a poor performance in one subordinate's mind, but may be interpreted as "average" by another. However, the relative ordering, "Joe, then Frank, then either Mary or Charlie," etc. is easily understood by all parties involved. So the rating scale may be used primarily as a vehicle for ranking the subordinates from "worst" to "best." Sometimes the subordinates may bypass the rating scale entirely and simply provide an ordering of their peers. Then the preceding ranking would give Joe rank 1, Frank rank 2, and Mary and Charlie a tie with an average rank of 3.5. If someone else gives Mary rank 1 and Joe rank 4, then there may be a substantial disagreement in the evaluation of job performance.

The degree of agreement or disagreement between the peer rankings and the supervisor's ranking may be summarized by a **rank correlation coefficient.** The rank correlation coefficient presented here is called **Spearman's rho.**

MONOTONIC RELATIONSHIPS

There are many situations where the bivariate data (X_i, Y_i) do not come from a bivariate normal population and where a definite relationship exists between X_i and Y_i, but this relationship is not necessarily linear. The plots in Figure 11.7 of bivariate sample data illustrate nonlinear relationships. It is clear that a strong relationship exists between X and Y in each of the plots; however, it is equally clear that a straight line will not provide an adequate description of the data. Hence, the sample correlation, which provides a measure of the strength of the **linear** relationship between two variables would not correctly describe the relationship between X and Y. Rather, relationships like these are more aptly described as **monotonic.** That is to say, a strictly **monotonic relationship** is one in which X and Y strictly increase together or one strictly increases as the other decreases, as shown in Figure 11.7. If these graphs are converted to the ranks of X versus the ranks of Y, then the relationship between the ranks is a straight line.

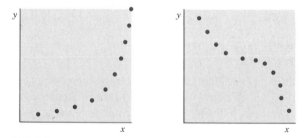

FIGURE 11.7
MONOTONIC RELATIONSHIPS

THE RANK CORRELATION COEFFICIENT

Spearman's rho is a convenient way to test the strength of the monotonic relationship between X and Y without being concerned whether or not the relationship is linear. The formula for calculating the rank correlation coefficient is obtained by calculating the sample correlation coefficient r of the previous section on the ranks of the data. The result of this calculation is given in Eq. 11.4. In the event of no ties occurring in the assignment of ranks to the data, Eq. 11.4 can be simplified to the easier-to-use Eqs. 11.5 and 11.6.

Spearman's rho r_R is the rank correlation coefficient obtained by computing the sample correlation coefficient on the ranks. It is computed from the equation

$$r_R = \frac{\Sigma R_{X_i} R_{Y_i} - C}{\sqrt{(\Sigma R_{X_i}^2 - C)(\Sigma R_{Y_i}^2 - C)}} \qquad (11.4)$$

SPEARMAN'S RHO

where R_{X_i} is the rank of X_i, R_{Y_i} is the rank of Y_i, n is the sample size and $C = n(n + 1)^2/4$. If there are no ties in the data, either of the following equations is simpler to use.

$$r_R = \frac{\Sigma R_{X_i} R_{Y_i} - C}{n(n^2 - 1)/12} \qquad (11.5)$$

$$r_R = 1 - \frac{6 \sum_{i=1}^{n} (R_{X_i} - R_{Y_i})^2}{n(n^2 - 1)} \qquad (11.6)$$

Equation 11.5 and Eq. 11.6 should never be used if there are ties.

As with the sample correlation coefficient, r_R will yield values between $+1$ and -1. Furthermore, the value of r_R can be $+1$ or -1 without r being $+1$ or -1, but the converse is not true.

A NUMERICAL ILLUSTRATION

The different values of r_R and r are now illustrated in a simple example using points from the deterministic relationship $Y = e^X$, as plotted in Figure 11.8.

X	Y
0	1.00
1	2.72
2	7.39
3	20.09
4	54.60

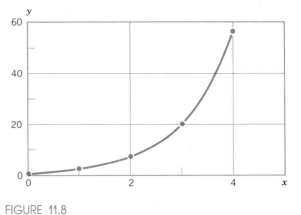

FIGURE 11.8
A SCATTERPLOT OF FIVE VALUES FROM $Y = e^x$.

First r is calculated.

X	Y	XY	X²	Y²
0	1.00	0.00	0	1.00
1	2.72	2.72	1	7.39
2	7.39	14.78	4	54.60
3	20.09	60.26	9	403.43
4	54.60	218.39	16	2980.96
Totals 10	85.79	296.15	30	3447.37

$$r = \frac{296.15 - (10)(85.79)/5}{\{[30 - (10)^2/5][3447.37 - (85.79)^2/5]\}^{1/2}} = .886$$

Now the correlation coefficient on ranks r_R is computed.

R_X	R_Y	$R_X R_Y$	R_X^2	R_Y^2
1	1	1	1	1
2	2	4	4	4
3	3	9	9	9
4	4	16	16	16
5	5	25	25	25
Totals 15	15	55	55	55

$$r_R = \frac{55 - 45}{\sqrt{(55 - 45)(55 - 45)}} = 1.000$$

Hence, r_R has achieved its maximum value of $+1.0$, indicating that the strength of the monotonic relationship between X and Y is much stronger than the strength of the linear relationship indicated by r. A plot of the five pairs of data points in Figure 11.8 shows this statement to be well founded, since the points lie on a curve and not a straight line. Hence, r_R can measure the strength of a monotonic nonlinear relationship more accurately than r.

A TEST OF INDEPENDENCE USING SPEARMAN'S RHO

Data

The data consist of n observations (X_i, Y_i), $i = 1, \ldots, n$ on the bivariate random variable (X, Y). Let R_{X_i} represent the ranks of the X's from 1 to n, and let R_{Y_i} represent ranks of the Y's. Note that the X's are ranked only among themselves, and the same is true for the Y's. Average ranks are used in case of ties.

Assumption

The sample (X_i, Y_i), $i = 1, \ldots, n$ is a random sample from some population.

Null Hypothesis

X and Y are independent.

Test Statistic

The test statistic is

$$T_R = r_R \sqrt{\frac{n - 2}{1 - r_R^2}}$$

The null distribution of T_R is approximately the Student's t distribution with $n - 2$ degrees of freedom. Exact quantiles for r_R appear in Figure 11.9 for $n \leq 18$.

Decision Rule

The decision rule depends on the alternative hypothesis of interest. Critical values for the test statistic are obtained by entering Table A3 with $n - 2$ degrees of freedom.

(a) H_1: X and Y tend to increase or decrease together (i.e., the relationship tends to be monotonically increasing). Reject H_0 if $T_R > t_{1-\alpha,n-2}$.

(b) H_1: X tends to decrease as Y increases and vice versa (i.e., relationship tends to be monotonically decreasing). Reject H_0: if $T_R < -t_{1-\alpha,n-2}$.

(c) H_1: X and Y are not independent (i.e., there is a tendency toward a monotonic relationship between them). Reject H_0 if $T_R > t_{1-\alpha/2,n-2}$ or if $T_R < -t_{1-\alpha/2,n-2}$.

Note

If $n \leq 18$ and there are no ties in the assignment of ranks, the preceding hypotheses can be tested in a similar manner by comparing r_R directly against the exact quantiles in Figure 11.9.

HYPOTHESIS TESTING

Tests can be performed for the significance of Spearman's rho just as was done with the sample correlation coefficient. Essentially the data are replaced by ranks, and the test described in the previous section is used on the ranks. If the null hypothesis is rejected, then the conclusion is that there is a tendency toward a monotonic relationship between X and Y and, therefore, X and Y are not independent. For this reason the null hypothesis is usually stated in terms of independence between X and Y, although there are some types of dependence, nonmonotonic dependence, that will not be detected by this test. A table of exact critical values is provided in Figure 11.9 for r_R for the cases where $n \leq 18$, and there are no ties in the assignment of ranks. Otherwise, the statistic T_R is compared directly against quantiles in Table A3 with $n - 2$ degrees of freedom.

EXAMPLE

Motivation for this example is provided in the problem setting of this section. Twenty subordinates are asked to rate their peers on a scale of zero to 100. They are given some standard guidelines, such as to be sure to rate at least one person as zero and at least one person as 100, with the other ratings spread out between those two extremes. The average rating is computed for each of the subordinates. This represents the rating that person has been given, averaged over all 20 evaluations including his or her own rating. These are plotted in Figure 11.10 along with the supervisor's rating for that person, to give some idea of the pattern of the relationship.

Of primary interest, however, is the agreement between the *rankings* by the peer group and the supervisor, rather than the agreement between ratings. These rankings are given in Figure 11.11 along with some preliminary calculations. The calculation of Spearman's rho and the corresponding test statistic is as follows:

$$r_R = \frac{2751 - 2205}{\{[2869 - 2205][2869 - 2205]\}^{1/2}} = .822$$

$$T_R = .822 \sqrt{\frac{20 - 2}{1 - .822^2}} = 6.1305$$

To test the null hypothesis of independence between the peer rankings and the supervisor rankings at $\alpha = .05$, T_R is compared with $t_{.95,18} = 1.7341$ from Table A3. Thus the null hypothesis of independence is easily rejected and it is concluded that supervisor rankings and peer rankings tend to agree. The p-value is seen from Table A3 to be less than .0001.

EXAMPLE

The example of the last section using the attendance-amount wagered data is reworked here as an analysis on the ranks of the data. For a value of $\alpha = .05$ the null hypothesis:

H_0: Daily racing attendance and the amount of money wagered are independent of one another

is tested versus the alternative:

H_1: Daily racing attendance and the amount of money wagered increase together monotonically.

n	.75	.90	.95	.975	.99	.995	.999
4	.400	.800					
5	.400	.700	.800	.900	.900		
6	.314	.600	.771	.829	.886	.943	
7	.286	.536	.679	.750	.857	.893	.964
8	.286	.500	.619	.714	.810	.857	.929
9	.250	.467	.583	.683	.767	.817	.900
10	.236	.442	.552	.636	.733	.782	.867
11	.227	.418	.527	.609	.700	.746	.836
12	.210	.399	.497	.580	.671	.720	.811
13	.203	.379	.478	.555	.643	.698	.786
14	.196	.363	.459	.534	.622	.675	.767
15	.186	.350	.443	.518	.600	.650	.746
16	.179	.338	.426	.500	.579	.632	.726
17	.174	.326	.412	.485	.564	.615	.708
18	.168	.315	.399	.470	.548	.598	.690

FIGURE 11.9
EXACT QUANTILES FOR THE SPEARMAN RANK
CORRELATION COEFFICIENT FOR TEST OF
INDEPENDENCE OF THE X AND Y WHEN THERE ARE
NO TIES IN THE ASSIGNMENT OF RANKS.

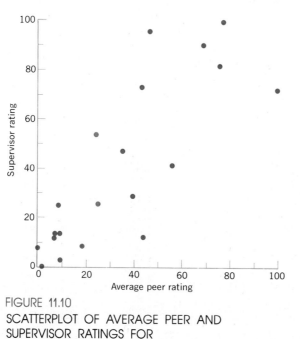

FIGURE 11.10
SCATTERPLOT OF AVERAGE PEER AND
SUPERVISOR RATINGS FOR
20 EMPLOYEES.

Average Peer Rating X	Supervisor Rating Y	R_X	R_Y	R_X^2	R_Y^2	$R_X R_Y$
1	7	1.0	3.5	1.00	12.25	3.50
2	0	2.0	1.0	4.00	1.00	2.00
7	11	3.5	5.0	12.25	25.00	17.50
7	13	3.5	7.5	12.25	56.25	26.25
9	13	5.5	7.5	30.25	56.25	41.25
9	24	5.5	9.0	30.25	81.00	49.50
10	2	7.0	2.0	49.00	4.00	14.00
19	7	8.0	3.5	64.00	12.25	28.00
24	53	9.0	14.0	81.00	196.00	126.00
26	26	10.0	10.0	100.00	100.00	100.00
36	47	11.0	13.0	121.00	169.00	143.00
40	29	12.0	11.0	144.00	121.00	132.00
43	72	13.0	15.0	169.00	225.00	195.00
45	12	14.0	6.0	196.00	36.00	84.00
47	95	15.0	19.0	225.00	361.00	285.00
56	42	16.0	12.0	256.00	144.00	192.00
70	89	17.0	18.0	289.00	324.00	306.00
78	83	18.0	17.0	324.00	289.00	306.00
80	100	19.0	20.0	361.00	400.00	380.00
99	74	20.0	16.0	400.00	256.00	320.00
		210.0	210.0	2869.00	2869.00	2751.00

$C = 20(21)^2/4 = 2205$

FIGURE 11.11
WORKSHEET FOR CALCULATING THE SAMPLE RANK
CORRELATION COEFFICIENT.

Attendance (Hundreds) (X)	R_X	Amount Wagered (Millions) (Y)	R_Y	$R_X R_Y$	R_X^2	R_Y^2
117	1	2.07	1	1	1	1
128	7	2.19	2	14	49	4
122	4	3.14	7	28	16	49
119	2	2.26	3	6	4	9
131	9	3.40	9	81	81	81
135	10	2.89	5	50	100	25
125	5	2.93	6	30	25	36
120	3	2.66	4	12	9	16
130	8	3.33	8	64	64	64
127	6	3.53	10	60	36	100
Totals	55		55	346	385	385

$$r = \frac{346 - 302.5}{\sqrt{(385 - 302.5)(385 - 302.5)}} = .527$$

Decision rule Since $n \leq 18$ and there are no ties in the assignment of ranks, the critical value can be obtained directly from Figure 11.9, for $n = 10$ as 0.552. Reject H_0 if $r_R > 0.552$. The null hypothesis is not rejected using a one-tailed test. That is, the data are not sufficient to detect a significant monotonic relationship between daily attendance and amount of money wagered. The p-value is between .05 and .10. In this case, the lack of significance may be due to the small sample size used.

If the statistic

$$T_R = .527\sqrt{\frac{10 - 2}{1 - .527^2}} = 1.7552$$

is compared against the .95 quantile from Table A3 with 8 degrees of freedom, that is $t_{.95,8} = 1.8595$, the decision would be the same. In the previous section the test statistic on the original data was $T = 1.8822$, which was significant, and the p-value was slightly smaller than .05. This illustrates the slightly greater power of the analysis on the original data whenever the data appear to be from a bivariate normal distribution.

EXERCISES

11.11 The post position in a horse race refers to the relative position of the horse from the inside rail at the beginning of the race. To see if this is related to the order in which the horses finish the race, the results were kept for one race. Find the rank correlation coefficient. Does this indicate that horses nearest the rail have an advantage in the race?

Post Position	1	2	3	4	5	6	7	8
Finishing Position	7	5	1	4	2	3	6	8

11.12 Compute Spearman's rank correlation coefficient on the data given in Exercise 11.5 and compare the result with the correlation calculated for Exercise 11.5. Test the hypothesis that the price of the tape is independent of the length in minutes, against the appropriate one-sided alternative. Let $\alpha = .01$. ($\Sigma R_X = 465$, $\Sigma R_X^2 = 9438.5$, $\Sigma R_Y = 465$, $\Sigma R_Y^2 = 9444.5$, $\Sigma R_X R_Y = 8939.5$)

11.13 Compute Spearman's rank correlation coefficient on the data given in Exercise 11.8. Test the hypothesis that the strength of cotton yarn is independent of the length of the cotton fibers, against the two-sided alternative. Let $\alpha = .05$. ($\Sigma R_X = 55$, $\Sigma R_X^2 = 384$, $\Sigma R_Y = 55$, $\Sigma R_Y^2 = 384.5$, $\Sigma R_X R_Y = 361.5$)

11.14 Nine students earned the scores listed below on their first two tests in an accounting class. Compute Spearman's rank correlation coefficient for these scores and test the hypothesis that the scores are independent of one another versus an alternative that the scores tend to agree (increase together). Let $\alpha = .005$.

Student Number	First Test	Second Test
1	90	84
2	91	83
3	82	80
4	94	90
5	92	91
6	88	89
7	89	88
8	63	62
9	86	92

11.15 Two coaches review last week's game film and independently grade each player's performance on a scale from 0 to 100. Compute Spearman's rank correlation coefficient for these scores and test the following hypothesis with $\alpha = .05$. Test

H_0: There is no association between the scores given the players by the two coaches

versus

H_1: There is an association between the two scores; that is, either the coaches tend to agree on the best and poorest performances or they tend toward opposite ratings on the players, that is, what one coach calls good the other calls bad and vice versa.

Player Number	Scores from Coach 1	Scores from Coach 2
1	64	80
2	28	87
3	90	90
4	30	57
5	97	89
6	20	51
7	100	81
8	67	82
9	54	69
10	44	78
11	100	100
12	71	81

11.16 Given below are the results of a city commission race in Manhattan, Kansas. Is there a significant correlation between the order (rank) of finish in the primary and the general elections? State and test the appropriate hypothesis. Let $\alpha = .05$.

Candidate	Votes in Primary	Votes in General
R	2251	3929
B	2073	3578
A	1993	4041
S	1489	2941
Y	1332	1802
K	1253	2525

11.17 Two noted football authorities gave the following rankings one year for the Big Eight football conference race. Is there any agreement in their rankings? State and test the appropriate hypothesis with $\alpha = .05$.

Team	Street and Smith	Minneapolis Line
Nebraska	1	2
Oklahoma	2	1
Kansas State	3	8
Colorado	4	3
Missouri	5	6
Iowa State	6	7
Kansas	7	4
Oklahoma State	8	5

11.18 Two customers are randomly selected in a supermarket and asked to smell five after-shave lotions and rank them in order of preference from 1 (best) to 5 (least desirable). Based on the results given below test the hypothesis that the two sets of rankings are independent of one another versus the alternative that they tend to agree (increase together). Let $\alpha = .05$.

	Brand of After-Shave				
	A	B	C	D	E
Customer number 1 ranking	1	4	3	2	5
Customer number 2 ranking	3	5	2	1	4

11.19 In an application of artificial intelligence the fidelity evaluation of a prototype for the selection of software packages for structural engineering problems resulted in two sets of rankings for the packages. One set of rankings is given by a machine and the other by a human consultant. Is there significant agreement between the two sets of rankings? Let $\alpha = .05$.

Program	Machine Ranking	Human Ranking
A	1	1
B	2	2
C	3	3
D	4	6
E	5	5
F	6	4

11.20 The order of finish in each of the 1986 major league baseball division championships were predicted prior to the start of the season by *Sports Illustrated*. Given below is the actual rank order of finish for each division race along with the rank order picked by *Sports Illustrated*. Calculate Spearman's rank correlation coefficient for each race. For each division race use a value of $\alpha = .05$ to test the hypothesis that the rankings are independent of one another versus the alternative that the ranks tend to agree.

AL East	Actual	Pick	AL West	Actual	Pick
Boston	1	5	California	1	6
New York	2	2	Texas	2	7
Detroit	3	3	Kansas City	3.5	1
Toronto	4	1	Oakland	3.5	2
Cleveland	5	7	Chicago	5	5
Milwaukee	6	6	Minnesota	6	3
Baltimore	7	4	Seattle	7	4
NL East	**Actual**	**Pick**	**NL West**	**Actual**	**Pick**
New York	1	1	Houston	1	5
Philadelphia	2	5	Cincinnati	2	1
St. Louis	3	2	San Francisco	3	6
Montreal	4	4	San Diego	4	3
Chicago	5	3	Los Angeles	5	2
Pittsburgh	6	6	Atlanta	6	4

11.3

REVIEW EXERCISES

11.21 Data on the heights (X) and weights (Y) of 10 randomly selected college students were given in Exercise 2.50. These data are repeated below. Test $H_0: \rho = 0$ versus $H_1: \rho > 0$. Let $\alpha = .05$ and state the p-value. Assume bivariate normality in the population. ($\Sigma X = 693$, $\Sigma Y = 1588$, $\Sigma XY = 110{,}896$, $\Sigma X^2 = 48{,}163$, $\Sigma Y^2 = 257{,}974$)

X:	64	73	71	69	66	69	75	71	63	72
Y:	121	181	156	162	142	157	208	169	127	165

11.22 Calculate Spearman's rho for the data of Exercise 11.21. Test the hypothesis that height and weight are independent of one another versus the alternative that they tend to increase together. Let $\alpha = .05$ and state the p-value.

11.23 Given below are the percentages of registered voters and the percentage turnout for 10 randomly selected cities in a nationwide election. Test each variable for normality. What conclusion do you make about the acceptability of these data coming from a bivariate normal distribution? ($\Sigma X = 672.2$, $\Sigma X^2 = 48{,}227.06$, $\Sigma Y = 558.7$, $\Sigma Y^2 = 34{,}733.27$, $\Sigma XY = 40{,}601.49$)

City	Percentage Registered	Percentage Voting
Detroit	92.0	70.0
Topeka	81.9	69.3
Philadelphia	77.6	69.8
Los Angeles	77.0	64.2
Boston	74.0	63.3
Tampa	68.8	63.6
San Francisco	68.0	64.4
Honolulu	60.0	54.7
Birmingham	39.1	13.8
Atlanta	33.8	25.6

11.24 Refer to Exercise 11.23 and test the hypothesis that the percentage of voters registered and the percentage of voters actually voting are independent of one another versus the alternative that they tend to increase together. Let $\alpha = .05$ and state the p-value.

11.25 Two judges rate 10 contestants in a beauty contest. Test the hypothesis that the ratings are independent of one another versus the alternative that the judges tend to agree on the judgment. Let $\alpha = .05$ and state the p-values.

	Contestant									
	A	B	C	D	E	F	G	H	I	J
Judge 1:	1	2	3	4	5	6	7	8	9	10
Judge 2:	2	3	1	4	6	5	9	10	8	7

11.26 Compute Spearman's rank correlation coefficient for the data given in Exercise 11.7. Test the hypothesis that typing score and number of errors are independent of one another, against the two-sided alternative. Let $\alpha = .05$ and state the p-value.

11.27 The Lilliefors test was used to check the normality assumption for the data in Exercise 11.23. Let Z_R represent the standardized values for the percentage registered, and let Z_V be the corresponding values for the percentage voting. A convenient formula for r when Z_R and Z_V are available is

$$r = \frac{\Sigma Z_R Z_V}{n - 1}$$

where the sum is over all pairs of observations. Use this formula to compute r for the data in Exercise 11.23.

11.28 In Exercise 2.30 a scatterplot was made to determine how tuition and fees are affected by the size of government appropriations. These data are repeated below. Calculate the sample correlation coefficient for these data. Assume bivariate normality and test the hypothesis H_0: $\rho = 0$ versus H_1: $\rho < 0$ with $\alpha = .01$. What is the interpretation of the results of this test? ($\Sigma X = 165.1$, $\Sigma Y = 274.9$, $\Sigma XY = 5233.23$, $\Sigma X^2 = 3783.31$, $\Sigma Y^2 = 10,001.65$)

Region	Percentage of Total Revenue From Tuition and Fees	Government Appropriations
New England	33.0	17.1
Mideast	29.1	27.2
Great Lakes	22.8	33.3
Plains	19.7	35.4
Southeast	18.0	37.9
Southwest	13.6	45.8
Rocky Mountains	17.0	36.3
Far West	11.9	41.9

11.29 Calculate Spearman's rho on the data of Exercise 11.28. Compare your answer with the sample correlation coefficient computed in that exercise.

11.30 For the countries of North and South America, the 1970 figures for educational expenditure per child and income per adult, in U.S. dollars, are given below.

(a) Draw a rank scatterplot of these data.

(b) Why is a rank scatterplot more appropriate for these data than a regular scatterplot?

(c) Find the correlation coefficient r. ($\Sigma X = 2396$, $\Sigma Y = 23,042$, $\Sigma XY = 9,296,390$, $\Sigma X^2 = 1,593,384$, $\Sigma Y^2 = 61,572,452$)

(d) Eliminate the two outliers (Canada and United States) and recompute r. What is the effect of the two outliers on r? ($\Sigma X = 647$, $\Sigma Y = 13,448$, $\Sigma XY = 674,918$, $\Sigma X^2 = 37,663$, $\Sigma Y^2 = 13,505,792$)

Country	Educational Expenditure per Child	Average Income per Adult
Argentina	71	1514
Bolivia	12	276
Brazil	19	473
Canada	760	3786
Chile	59	836
Colombia	10	547
Costa Rica	61	981
Ecuador	20	462
El Salvador	18	527
Guatemala	17	648
Honduras	18	491

Country	Educational Expenditure per Child	Average Income per Adult
Mexico	32	1074
Nicaragua	20	731
Panama	74	1179
Paraguay	11	444
Peru	29	600
United States	989	5808
Uraguay	80	778
Venezuela	96	1887

11.31 The exam scores for 34 job applicants have been recorded below along with the order in which they were handed in. Do the applicants who hand in their exams early tend to do better (or worse) than the rest of the group? Draw a scatterplot and compute r to support your answer. ($\Sigma X = 595$, $\Sigma Y = 4566$, $\Sigma XY = 79{,}495$, $\Sigma X^2 = 13{,}685$, $\Sigma Y^2 = 654{,}248$)

Order	Score	Order	Score	Order	Score	Order	Score
1	182	10	174	19	115	27	77
2	99	11	140	20	132	28	95
3	193	12	91	21	114	29	188
4	183	13	108	22	83	30	141
5	92	14	164	23	160	31	143
6	125	15	119	24	140	32	134
7	179	16	165	25	133	33	157
8	59	17	135	26	117	34	155
9	100	18	174				

11.32 Consider the following set of ranks for the nation's top 20 football teams as given at the end of a recent season by two wire services. Use Spearman's rank correlation coefficient to test the hypothesis that the two sets of rankings are independent of one another versus the alternative that the two sets of rankings tend to agree. Let $\alpha = .005$.

Football Team	AP Ranking	UPI Ranking
Georgia	1	1
Florida State	4	2
Pittsburgh	2	3
Oklahoma	5	4
Michigan	3	5
Alabama	6	6
Baylor	9	7
Notre Dame	10	8

(continued)

Football Team	AP Ranking	UPI Ranking
Nebraska	7	9
Penn State	8	10
North Carolina	11	11
UCLA	14	12
Southern California	13	13
Ohio State	15	14
Brigham Young	12	15
Washington	16	16
Mississippi State	18	17
South Carolina	17	18
SMU	19	19
Maryland	20	20

11.33 Use Spearman's rho to test the hypothesis in Exercise 11.10 and compare the p-value with the previous result.

BIBLIOGRAPHY

Additional material on the topics presented in this chapter can be found in the following publications.

Conover, W. J. (1980). *Practical Nonparametric Statistics,* 2nd ed. Wiley, New York.

Conover, W. J. and Iman, R. L. (1981). "Rank Transformations as a Bridge Between Parametric and Nonparametric Statistics." *The American Statistician,* **35** (3), 124–129.

Franklin, L. A. (1988). "The Complete Exact Null Distribution of Spearman's Rho for $n = 12(1)18$." *Journal of Statistical Computation and Simulation,* **29** (3), 255–269.

12

REGRESSION

Suppose that you are putting your house up for sale, and you must put a selling price on it. If it is priced too low, it will sell quickly but you will not get as much money as you could have. If the price is too high, the property will not sell. To arrive at a fair asking price you usually consider what you paid for the house, what has been happening to the price of houses since you bought yours, the actual selling price of other houses in your neighborhood, their square footage and lot size, your square footage and lot size, and many other variables. The actual price that your house sells for may be somewhat inconsistent with the selling prices of other houses because you may find a buyer who "just loves the rose garden," or perhaps other owners found rose garden lovers and you did not. But on the average this kind of information should prove useful in predicting how you expect the buyers are likely to react to your house.

This example involves a complex relationship between one unknown variable, the selling price of a house, and a collection of other variables. It involves the possibility of a lot of money being made or lost. It, along with many other problems of this type, has been the subject of extensive research into the best ways of expressing a mathematical relationship among the variables.

Finding a suitable mathematical relationship between the one unknown variable, called the **dependent variable,** and a group of other known quantities, called **independent variables,** is a challenging problem. One methodology for handling this type of problem is called **regression analysis.** This chapter provides an introduction to regression analysis by considering only one independent variable, along with the dependent variable. Regression analysis examines the mathematical relationship between two (or more) variables, in contrast with correlation analysis, which merely measures the strength of the relationship between two variables.

12.1

AN INTRODUCTION TO REGRESSION, AND LEAST SQUARES COMPUTATIONS FOR LINEAR REGRESSION

PROBLEM SETTING 12.1

The calculation of insurance rates is a complex process that requires highly skilled professionals. The people who do this type of analysis are called actuaries, and the body of knowledge dealing with the subject is called actuarial science. As a part of this training there are 10 challenging examinations that a trainee can take. The first exam covers undergraduate mathematics and calculus. The second exam includes topics in probability and statistics. The third, fourth, and fifth exams are concerned with various topics in mathematics, finance, and insurance. Passage of the first 5 exams qualifies one to be called an Actuarial Associate. Passage of all 10 exams results in the coveted title Fellow of the Society of Actuaries.

An undergraduate program in actuarial science is designed to help the student pass as many of the first five exams as possible before graduation. Suppose an undergraduate who is completing an actuarial program inquires of the

Graduate	Starting Salary (Y)	Exams Passed (X)	Graduate	Starting Salary (Y)	Exams Passed (X)
1	$21,900	1	15	$22,000	1
2	28,000	3	16	25,900	2
3	27,300	2	17	23,000	1
4	28,000	2	18	18,900	0
5	31,300	5	19	27,000	3
6	20,200	0	20	30,500	4
7	21,100	0	21	24,800	2
8	25,900	3	22	20,200	0
9	28,000	4	23	23,300	1
10	23,100	2	24	22,500	1
11	19,400	0	25	27,800	4
12	26,500	2	26	27,000	3
13	32,500	5	27	21,200	1
14	21,100	1	28	26,600	4

FIGURE 12.1

STARTING SALARIES AND NUMBER OF EXAMS PASSED FOR 28 RECENT GRADUATES.

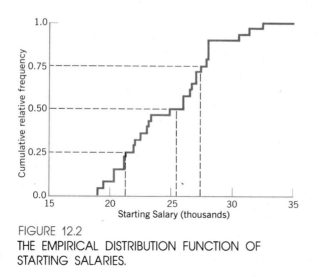

FIGURE 12.2
THE EMPIRICAL DISTRIBUTION FUNCTION OF
STARTING SALARIES.

university placement center about the salary she might expect to earn. A placement center counselor obtains a listing of recent graduates of the university
who became employed as actuaries by insurance companies. The starting
salaries and number of exams passed by the graduates are given in Figure 12.1.

To find the answer to the student's inquiry regarding the starting salary she
can expect, the counselor plots an empirical distribution function of starting
salaries, as in Figure 12.2. From this e.d.f. the sample median is easily seen
to be $25,350, and the lower and upper sample quartiles are $21,550 and
$27,550, respectively. The counselor explains to the student that she can
obtain some idea of what she might expect to earn from the fact that the middle
50% of recent graduates are in the $21,550 to $27,550 range, and the sample
median starting salary is $25,350.

At this point the student mentions that she has passed all of the first five
actuarial exams and asks the counselor if this will help him in predicting her
future starting salary. How can the counselor use this additional information
to provide the student with a more precise estimate of the starting salary she
might expect?

REGRESSION

The question posed by the student in the problem setting is similar to the
questions posed in many other real situations. That is, when the sample consists of data on a bivariate random variable (X, Y), can inferences regarding
the variable Y be improved by knowing something about the variable X? It
seems reasonable to expect that the additional information furnished by X
should improve knowledge about Y if X and Y are related in some way.

For example, the bivariate data in Figure 12.1 can be summarized by the
number of exams passed. Such a summary is given in Figure 12.3. It is fairly
obvious from Figure 12.3 that the starting salary, Y, tends to increase as the
number of exams passed, X, increases.

By examining Figure 12.3 the counselor can note that only 2 of the 28
graduates had passed all five exams as the student has done, and their starting
salaries were $31,300 and 32,500, both in the upper range of starting salaries.

0 Exams	1 Exam	2 Exams	3 Exams	4 Exams	5 Exams
$18,900	$21,100	$23,100	$25,900	$26,600	$31,300
19,400	21,200	24,800	27,000	27,800	32,500
20,200	21,900	25,900	27,000	28,000	
20,200	22,000	26,500	28,000	30,500	
21,100	22,500	27,300			
	23,000	28,000			
	23,300				
$\overline{Y}_0 = \$19,960$	$\overline{Y}_1 = \$22,143$	$\overline{Y}_2 = \$25,933$	$\overline{Y}_3 = \$26,975$	$\overline{Y}_4 = \$28,225$	$\overline{Y}_5 = \$31,900$

FIGURE 12.3
GROUP MEANS FOR SALARIES GROUPED BY NUMBER OF EXAMS PASSED.

A reasonable response to the student is that she can expect to earn about $31,900, the average of the starting salaries of the other two students who had passed five exams. This idea of predicting a value for Y by using the mean of other values of Y that have the same value for X is the underlying concept in **regression.**

SAMPLE REGRESSION CURVE

New graduates in actuarial science get a better idea of starting salaries if the number of exams passed is taken into consideration because salaries appear to be different for the different numbers of exams passed. A graduate who has passed all five exams should expect a salary closer to the average for other graduates who passed all five exams. In the scatterplot given in Figure 12.4,

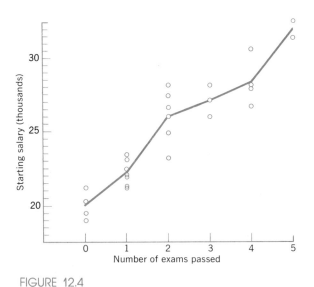

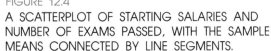

FIGURE 12.4
A SCATTERPLOT OF STARTING SALARIES AND NUMBER OF EXAMS PASSED, WITH THE SAMPLE MEANS CONNECTED BY LINE SEGMENTS.

lines connecting the average values of Y for each individual value of X are provided to give a clearer indication of the tendency for the mean of Y to depend on the observed value of X. This curve is called a **sample regression curve.**

There are many different ways of obtaining a sample regression curve. One way is simply to graph the sample means of Y for each value of X, as was done in Figure 12.4. This works well when there are several Y values for each value of X. With other types of data the usual method of obtaining a sample regression curve is to fit a straight line, or some other type of curve, to the scatterplot of bivariate data. The straight-line method is presented later in this section.

POPULATION REGRESSION CURVE

The population mean of Y for each individual value of X is called the **conditional mean of Y given X = x,** and is usually denoted by $\mu_{Y|x}$. A graph of $\mu_{Y|x}$ as a function of x is called the **population regression curve.** Suppose that the population mean starting salaries for all actuarial graduates are given as a function of the number of exams passed as follows:

$$X = 0: \mu_{Y|0} = \$21,000 \qquad X = 3: \mu_{Y|3} = \$26,000$$
$$X = 1: \mu_{Y|1} = \$23,000 \qquad X = 4: \mu_{Y|4} = \$29,500$$
$$X = 2: \mu_{Y|2} = \$24,500 \qquad X = 5: \mu_{Y|5} = \$33,000$$

A graph of these values is given in Figure 12.5, as the population regression curve. In practice, the population regression curve is seldom known and must be estimated from a sample.

THE
CONDITIONAL The **conditional mean of Y given X = x,** denoted by $\mu_{Y|x}$, represents the
MEAN OF Y   population mean of all values of Y that share that same value for X.
GIVEN X

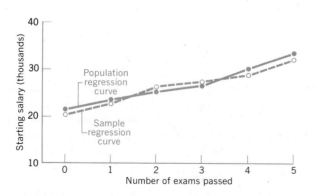

FIGURE 12.5
A GRAPH OF THE POPULATION REGRESSION
CURVE AND THE SAMPLE REGRESSION CURVE.

THE POPULATION REGRESSION CURVE	The **population regression curve** for Y is the curve that represents the mean of Y given $X = x$, $\mu_{Y|x}$, plotted as a function of x.

ESTIMATION AND PREDICTION

The specific value that is predicted by a sample regression curve for the random variable Y, for each value x of the random variable X, is usually denoted by $\hat{Y}$, and is called the **predicted value for Y, given X = x.** The value $\hat{Y}$ also serves as an estimate of the mean $\mu_{Y|x}$ of the random variable Y at a specific value $X = x$. The estimate of $\mu_{Y|x}$ is denoted by $\hat{\mu}_{Y|x}$, but is sometimes written simply as $\hat{Y}$ because the two estimates $\hat{Y}$ and $\hat{\mu}_{Y|x}$ are always equal, since they both come from the same sample regression curve with the same value of X.

EXAMPLE

The data for Problem Setting 12.1 are given in Figure 12.1, where it is seen that the observed value for (X_2, Y_2) is (3, \$28,000). The *predicted value $\hat{Y}_2$ for Y_2, given $X_2 = 3$,* is obtained from the sample mean of the four starting salaries that have $X = 3$. This is given in Figure 12.3 as

$$\hat{Y}_2 = \overline{Y}_3 = \$26{,}975$$

which would serve as an estimate of Y_2 if that value weren't already known to be \$28,000. The same number, \$26,975 also serves as an *estimate of $\mu_{Y|3}$, the population mean of Y, given $X = 3$.*

$$\hat{\mu}_{Y|3} = \overline{Y}_3 = \$26{,}975$$

This is the sample estimate of the population mean \$26,000 that was given earlier.

HOW MUCH OF THE VARIATION IN Y IS EXPLAINED BY KNOWLEDGE OF X?

If nothing were known about the number of exams passed by the undergraduate student in Problem Setting 12.1, the sample mean of all starting salaries in Figure 12.1, $\overline{Y} = \$24{,}821$, could be used to provide the student with an estimate of starting salaries. A measure of variation in starting salaries is given, of course, by the sample variance

$$s_Y^2 = \frac{1}{n-1} \sum_{i=1}^{n} (Y_i - \overline{Y})^2 = \frac{1}{27}(374{,}767{,}143)$$

$$= 13{,}880{,}265$$

The term $\Sigma(Y_i - \overline{Y})^2$ is called the **total sum of squares**. In regression this is frequently used instead of s_Y^2 to measure the total variability in the random variable Y. For the salary data the total sum of squares is given previously as 374,767,143.

TOTAL SUM OF SQUARES (TOTAL VARIATION IN Y) The total of the squared deviations between each observation Y_i and the sample mean $\overline{Y}$ is the **total sum of squares**.

$$\text{Total SS} = \Sigma(Y_i - \overline{Y})^2 = \Sigma Y_i^2 - (\Sigma Y_i)^2/n$$

Since the sample data in Figure 12.1 records both starting salary and number of exams passed for each graduate, it is natural to ask how much of the total variation in starting salaries can be explained by knowledge of the number of exams passed. Or, in general, for a set of bivariate observations on (X, Y), how much of the variation in Y can be explained by knowledge of X? First the amount of variation that is *not* explained by knowledge of X is discussed. The prediction $\hat{Y}$ for Y, given a value of X, is obtained from the sample regression curve. The unexplained variation is measured by the difference between Y and $\hat{Y}$. The sum of the squares of these differences, $\Sigma(Y_i - \hat{Y}_i)^2$ is called the **error sum of squares** and represents the variation in Y remaining after regression on X.

ERROR SUM OF SQUARES (VARIATION IN Y REMAINING AFTER REGRESSION ON X) The total of the squared deviations between each observation Y_i and its point estimate $\hat{Y}_i$ from the sample regression curve is the **error sum of squares**.

$$\text{Error SS} = \Sigma(Y_i - \hat{Y}_i)^2$$

The ratio of the *unexplained* variation to the *total* variation represents the proportion of variation in Y that is *not explained* by regression on X. Subtraction of this proportion from 1.0 gives the proportion of variation in Y that *is explained* by regression on X. The statistic used to express this proportion is called the **coefficient of determination** and is denoted by R^2. It may be written in various equivalent ways, as follows:

$$R^2 = 1 - \frac{\text{variation in } Y \text{ remaining after regression on } X}{\text{total variation in } Y}$$

$$R^2 = 1 - \frac{\text{error sum of squares}}{\text{total sum of squares}}$$

$$R^2 = 1 - \frac{\Sigma(Y_i - \hat{Y}_i)^2}{\Sigma(Y_i - \overline{Y})^2}$$

The value $R^2 = 0$ indicates that none of the variation in Y is explained by regression on X, and the value $R^2 = 1$ indicates that all of the variation of Y is explained by regression on X. The calculation of the coefficient of determination is illustrated in the following example.

COEFFICIENT OF DETERMINATION

The **coefficient of determination** R^2 is the proportion of variation in Y explained by a sample regression curve. The equation for R^2 is

$$R^2 = 1 - \frac{\Sigma(Y_i - \hat{Y}_i)^2}{\Sigma(Y_i - \overline{Y})^2} = 1 - \frac{\text{Error SS}}{\text{Total SS}} \qquad (12.1)$$

EXAMPLE

For the data given in Problem Setting 12.1 the *total sum of squares* was given earlier as

$$\Sigma(Y_i - \overline{Y})^2 = 374{,}767{,}143$$

The *error sum of squares* associated with the sample regression curve of Figure 12.4 can be computed by referring to Figure 12.3. The computations are given in Figure 12.6. The coefficient of determination is obtained using Eq. 12.1.

$$R^2 = 1 - \frac{\Sigma(Y_i - \hat{Y}_i)^2}{\Sigma(Y_i - \overline{Y})^2}$$

$$= 1 - \frac{33{,}857{,}476}{374{,}767{,}143}$$

$$= 1 - .090$$

$$= .910$$

Thus 91% of the total variation in starting salaries is accounted for by the sample regression curve in Figure 12.4. In most applications 91% would probably be regarded as a relatively large value for R^2.

LINEAR REGRESSION

The population regression curve is the curve that represents the mean of Y given $X = x$, $\mu_{Y|x}$, plotted as a function of x. When the population regression curve is a straight line the regression is said to be **linear.** The **linear regression** concept is important because scatterplots of data frequently indicate that a straight-line relationship may exist between $\mu_{Y|x}$ and x. In reality, the regression may be curved, but the assumption of a straight-line relationship may be useful as an approximation to the true but unknown regression curve. In applied statistical analyses, the first step in describing the relationship between two variables usually consists of fitting a straight line to the observed points plotted in the scatterplot.

Group	Y (Dollars)	$\hat{Y}$ (Dollars)	$(Y - \hat{Y})^2$
1	18,900	19,960	1,123,600
	19,400	19,960	313,600
	20,200	19,960	57,600
	20,200	19,960	57,600
	21,100	19,960	1,299,600
...	...	...	...
6	31,300	31,900	360,000
	32,500	31,900	360,000
$\Sigma(Y_i - \hat{Y}_i)^2$ is the error sum of squares			33,857,476

FIGURE 12.6
WORKSHEET FOR CALCULATING THE ERROR
SUM OF SQUARES.

If the population regression curve between two variables is a straight line

$$\mu_{Y|x} = \alpha + \beta x \qquad (12.2)$$

LINEAR
REGRESSION then the **regression is linear.** (Although it is customary to use α and β in linear regression, these symbols are not to be confused with probabilities of Type I and Type II errors, which is another way these same symbols are sometimes used.)

THE EQUATION FOR A STRAIGHT LINE

The general form for the equation of a straight line may be recalled from algebra as $y = a + bx$, where a is defined as the y intercept of the line and b is the slope of the line, as illustrated in Figure 12.7. That is, a is the number of units above the origin (0,0) where the straight line crosses the y-axis, and b is the change in vertical distance $(y_2 - y_1)$ between any two points on the line divided by the corresponding change in horizontal distance $(x_2 - x_1)$ between the same two points.

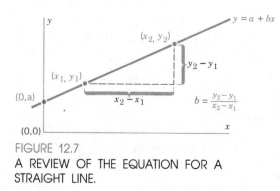

FIGURE 12.7
A REVIEW OF THE EQUATION FOR A
STRAIGHT LINE.

PRINCIPLE OF LEAST SQUARES

The problem in obtaining a *sample linear regression curve* is how to find a and b for a set of sample values (X_i, Y_i) when all points do not naturally fall on a straight line. That is, what should the criterion be for computing a and b so the line $y = a + bx$ appears to fit the data well? Since the line fitted to the sample values is generally used for predicting values of Y, simple logic would imply that there should be some agreement between values observed for Y and values predicted for Y. The criterion that has been traditionally used for this problem is to find a and b such that the error sum of squares

$$\text{Error sum of squares} = \sum_{i=1}^{n} (Y_i - \hat{Y}_i)^2$$

is minimized. Since

$$\hat{Y}_i = a + bX_i$$

in linear regression, the error sum of squares becomes

$$\text{Error SS} = \sum_{i=1}^{n} (Y_i - \hat{Y}_i)^2 = \sum_{i=1}^{n} [Y_i - (a + bX_i)]^2$$

The formulas for a and b that minimize the error sum of squares can be found using calculus, and the solution is given by Eqs. 12.3 and 12.4. This approach is known as the **principle of least squares**.

The slope and intercept needed to fit a straight line, $y = a + bx$, to a set of bivariate sample data (X_i, Y_i), $i = 1, \ldots, n$ by the **principle of least squares** are computed as follows.

LEAST SQUARES SOLUTION FOR LINEAR REGRESSION

$$\text{Slope: } b = \frac{\Sigma X_i Y_i - (\Sigma X_i)(\Sigma Y_i)/n}{\Sigma X_i^2 - (\Sigma X_i)^2/n} = \frac{\Sigma(X_i - \overline{X})(Y_i - \overline{Y})}{\Sigma(X_i - \overline{X})^2} = \frac{s_{XY}}{s_X^2} \quad (12.3)$$

$$\text{Intercept: } a = \frac{\Sigma Y_i}{n} - \frac{b\Sigma X_i}{n} = \overline{Y} - b\overline{X} \quad (12.4)$$

The equation for the **least squares regression line** is $\hat{\mu}_{Y|x} = a + bX$, sometimes written as $\hat{Y} = a + bX$.

A NUMERICAL ILLUSTRATION

The least squares calculations are now illustrated on a simple set of four bivariate observations.

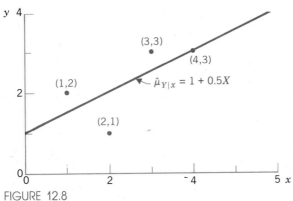

FIGURE 12.8
A SCATTERPLOT WITH THE SAMPLE
REGRESSION LINE.

X	Y	XY	X²
1	2	2	1
2	1	2	4
3	3	9	9
4	3	12	16
Totals 10	9	25	30

$$b = \frac{25 - 10(9)/4}{30 - (10)^2/4} = .5$$

$$a = \frac{9}{4} - \frac{.5(10)}{4} = 1$$

The equation for the least squares regression line is $\hat{\mu}_{Y|x} = 1 + .5X$. A scatterplot of the four points along with the least squares regression line appear in Figure 12.8. The predicted values $\hat{Y}_i$ of Y_i for each observed X are calculated by substituting each value of X into the sample regression equation as illustrated in Figure 12.9.

The principle of least squares guarantees that no other choice of values for a and b can be made that will make the error sum of squares smaller than 1.50 for these four data points.

THE COEFFICIENT OF DETERMINATION

In the case of linear regression R^2 is mathematically equal to the square of the sample correlation coefficient r, therefore the notation r^2 is customarily used to represent the coefficient of determination in linear regression.

EXAMPLE

In this example a straight line is fit to the data in Problem Setting 12.1 using the method of least squares. Preliminary calculations yield the following.

448 CHAPTER TWELVE / REGRESSION

X	Observed Y	Prediction $\hat{Y} = 1 + .5X$	Error SS $(Y - \hat{Y})^2$	Total SS $(Y - \bar{Y})^2$
1	2	1.5	.25	.0625
2	1	2.0	1.00	1.5625
3	3	2.5	.25	.5625
4	3	3.0	0.00	.5625
	9		Error SS = 1.50	Total SS = 2.7500

$\bar{Y} = \frac{9}{4} = 2.25$

FIGURE 12.9
A NUMERICAL ILLUSTRATION OF THE SUMS OF SQUARES.

$$n = 28 \quad \Sigma X = 57 \quad \Sigma X^2 = 181$$
$$\Sigma Y = 695,000 \quad \Sigma XY = 1,560,500$$

Equations 12.3 and 12.4 give the least squares coefficients a and b:

$$b = \frac{1,560,500 - 57(695,000)/28}{181 - (57)^2/28} = 2242$$

$$a = \frac{695,000}{28} + \frac{2242(57)}{28} = 20,256$$

The sample linear regression equation is therefore

$$\hat{Y} = 20,256 + 2242X$$

and it is plotted in Figure 12.10.

The interpretation of a and b in this example is as follows. The intercept a represents the estimated starting salary when $X = 0$. In this case, the estimated starting salary is $20,256 for a graduate who has passed none of the actuarial exams. The slope b represents the change in the estimated starting salary for each unit change in X. In this case the estimated starting salary increases $2242 for each additional actuarial exam that the student has passed.

The coefficient of determination represents the percent of the variability in the starting salary that is accounted for by a linear regression on X, the number of exams passed. The easiest way to find R^2 in linear regression is to find r^2, which requires calculating $\Sigma Y^2 = 17,625,660,000$, and using Eq. 11.1:

$$r^2 = \frac{[1,560,500 - 57(695,000)/28]^2}{[17,625,660,000 - (695,000)^2/28][181 - (57)^2/28]} = .872$$

Thus 87.2% of the variation in Y is accounted for by a linear regression on X. This is slightly less than the value .910 that was found earlier using the regression on the sample means. The lower value for the R^2 from linear regression means the straight line does not fit the observations as well as the regression using the sample means for each value of X.

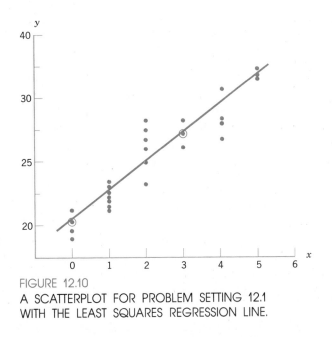

FIGURE 12.10
A SCATTERPLOT FOR PROBLEM SETTING 12.1
WITH THE LEAST SQUARES REGRESSION LINE.

LACK OF ASSUMPTIONS

The formulas given in this section for fitting a straight line to a set of bivariate data are derived using calculus and the least squares principle. They are valid for fitting a straight line to any set of bivariate data without regard to how the data were obtained or the type of population. In particular, no distributional assumptions were made and no assumptions were made about the sample being random. Those assumptions are used when testing hypotheses or when forming confidence intervals, as in the following sections. But the methods introduced thus far in this chapter do not require any assumptions and may be used wherever and whenever they are appropriate.

12.1   Quarterly changes in a local utility company's stock value for the previous three years are compared with the changes in value of Standard & Poor's 500 Stock Index (S&P Index) for the same time periods. All changes are expressed as percentage changes, and appear below and on the next page for the 12 quarters of data.

Quarter	S&P's 500 X (% change)	Utility Company Y (% change)	X^2	XY
1	−4.3	−7.0	18.49	30.10
2	−9.6	−11.4	92.16	109.44
3	−5.4	−5.3	29.16	28.62
4	−0.2	−4.3	.04	.86
5	+4.9	+1.4	24.01	6.86
6	+1.8	+1.6	3.24	2.88

7	+1.3	+4.6	1.69	5.98
8	−1.3	+0.1	1.69	−.13
9	+5.1	+7.3	26.01	37.23
10	−1.6	+1.8	2.56	−2.88
11	−9.7	−7.6	94.09	73.72
12	−12.8	−16.2	163.84	207.36
Totals	−31.8	−35.0	456.98	500.04

(a) Make a scatterplot of the data.

(b) Find the least squares regression line.

(c) Plot the sample regression line from (b) onto the scatterplot of part (a).

(d) How would you interpret the value of b in this exercise?

12.2 A real estate company specializing in sales of farms would like to know if their salespeople's sales can be predicted based on the number of years with the company. A random sample of sales is taken for 10 salespeople having years of experience ranging from 1 to 10 years and is given below.

(a) Plot the data in a scatterplot to see if a linear relationship looks reasonable for the data.

(b) Find the sample regression equation by the method of least squares and include it in the scatterplot.

(c) Find the sample correlation coefficient.

(d) Find the percentage of variation in sales accounted for by regression on the number of years with the company.

(e) Predict the sales for a salesperson with six years' experience.

Years' Experience X	Number of Sales Y	XY	X^2	Y^2
1	3	3	1	9
2	2	4	4	4
3	5	15	9	25
4	4	16	16	16
5	6	30	25	36
6	8	48	36	64
7	9	63	49	81
8	9	72	64	81
9	12	108	81	144
10	10	100	100	100
Totals 55	68	459	385	560

12.3 A random sample of recent business school graduates was classified according to grade point average (G.P.A.) and starting salary. The grade point averages were grouped into intervals and were summarized as follows.

**Starting Salaries (Thousand Dollars)
Classified by G.P.A.**

2.0–2.4	2.4–2.8	2.8–3.2	3.2–3.6	3.6–4.0
$21.4	$22.7	$23.4	$26.0	$25.3
20.7	21.5	22.3	23.3	25.1
20.1	20.9	22.1	22.8	24.3
	22.2	22.6	22.9	24.8
		24.8	25.7	
			25.5	

Graph the G.P.A. values versus starting salaries in a scatterplot. (Use the median of each G.P.A. interval to make the scatterplot, i.e., 2.2, 2.6, 3.0, 3.4, and 3.8.) Compute the total sum of squares, equal to $\Sigma(Y_i - \overline{Y})^2$ for the starting salaries.

12.4 For the sample data in Exercise 12.3 find the sample conditional mean of Y for each G.P.A. classification. Add these points to the scatterplot of Exercise 12.3 and then connect them to form a sample regression curve.

12.5 Find the error sum of squares for the sample data in Exercise 12.3.

12.6 Find the coefficient of determination for the sample data in Exercise 12.3. What is the interpretation of this value?

12.7 Find the least squares regression equation for the sample data given in Exercise 12.3. Use the median of each G.P.A. interval as the X value (i.e., 2.2, 2.6, 3.0, 3.4, and 3.8, respectively). Find the value of r^2 for these data.

12.8 Why is the value for r^2 in Exercise 12.7 not equal to the value of R^2 found in Exercise 12.6?

12.9 Predict the starting salary for a graduate with a G.P.A. of 3.45 using the least squares regression equation found in Exercise 12.7.

12.10 Daily sales (in thousands of dollars) have been recorded over a four-week period for a supermarket and are given below. Make a scatterplot of day of the week versus sales for these data and compute the total sum of squares $\Sigma(Y_i - \overline{Y})^2$ for the sales.

	Sun	Mon	Tue	Wed	Thur	Fri	Sat
	$14.5	$13.7	$12.0	$13.3	$16.5	$18.1	$20.1
	13.9	12.9	11.8	12.9	17.0	17.3	19.8
	13.8	13.6	12.3	13.8	15.8	18.5	21.4
	14.0	13.1	12.9	13.1	16.2	18.7	20.7
ΣY:	56.2	53.3	49.0	53.1	65.5	72.6	82.0
ΣY^2:	789.90	710.67	600.94	705.35	1073.33	1318.84	1682.50

Overall $\Sigma Y = 431.7$, $\Sigma Y^2 = 6881.53$

12.11 For the sample data in Exercise 12.10 find the sample conditional mean of sales for each day of the week. Add these points to the scatterplot of Exercise 12.10 and then connect them to form a sample regression curve.

12.12 Find the error sum of squares for the sample data in Exercise 12.10.

12.13 Find the coefficient of determination for the sample data in Exercise 12.10. What is the interpretation of this value?

12.14 Find the least squares regression equation for the sample data given in Exercise 12.10. Since the day of the week is a qualitative variable, it will have to be replaced by a quantitative variable by letting $X = 1$ for Sunday, $X = 2$ for Monday, and so on until $X = 7$ for Saturday. ($\Sigma X = 112$, $\Sigma X^2 = 560$, $\Sigma Y = 431.7$, $\Sigma Y^2 = 6881.53$, $\Sigma XY = 1859.3$.)

12.15 Add the graph of the least squares regression line found in Exercise 12.14 to the scatterplot found in Exercise 12.10. Do you think a straight line gives an adequate representation of these data?

12.16 Find the value of r^2 for the data given in Exercise 12.10. Why does it not equal the coefficient of determination found in Exercise 12.13?

12.2

ASSUMPTIONS FOR LINEAR REGRESSION

PROBLEM SETTING 12.2

Workers on an assembly line assemble small calculators. An engineer is interested in studying the production rate on one part of the assembly line to see how the individual worker's production varies as a function of the time spent on the assembly line. Twenty-four workers are randomly selected for observation and assigned to work on the assembly line for various lengths of time. The number of pieces completed (Y) is noted for each after a length of time (X) has elapsed. The data are given in Figure 12.11.

What sort of model may be used to describe the regression of Y on X? How can the parameters of the model be estimated? What assumptions are needed to place confidence intervals on the true model parameters? How can the estimated regression curve be used to predict values for the dependent variable Y when the independent variable X is known? These and other questions will be addressed in this and the following sections.

A MODEL FOR LINEAR REGRESSION

Although the previous section provides the mechanics of fitting a straight line to bivariate sample data, it does not address the problem of inference with respect to linear regression. To make any inferences about the population regression curve some assumptions must be satisfied regarding the population. In linear regression the following assumptions are usually made.

1. The random variables $Y_1, Y_2, \ldots, Y_n$ satisfy the linear relationship

$$Y_i = \alpha + \beta X_i + \epsilon_i, \qquad i = 1, \ldots, n \qquad (12.5)$$

where α and β are population parameters that represent, respectively, the y intercept and slope of the population regression curve, ϵ_1 through ϵ_n are

X Time Spent on Assembly Line (Hours)	Y Number of Units Assembled
0.5	36
0.5	41
0.5	33
1.0	83
1.0	61
1.0	68
1.5	116
1.5	99
1.5	91
2.0	120
2.0	135
2.0	125
2.5	170
2.5	143
2.5	151
3.0	180
3.0	205
3.0	188
3.5	212
3.5	241
3.5	208
4.0	236
4.0	261
4.0	253

FIGURE 12.11

THE NUMBER OF UNITS ASSEMBLED AND
THE TIME SPENT ON THE ASSEMBLY LINE
FOR 24 WORKERS.

random variables, and the X's may be random variables or may be fixed values.

2. The mean of ϵ_i is zero for all i.

3. The variance of ϵ_i is constant, independent of the values for X, and is denoted by σ^2 for all i.

4. The distribution of ϵ_i is normal for all i.

5. The ϵ_i are independent of each other.

Assumptions 2 through 5 are equivalent to saying that $\epsilon_1, \ldots, \epsilon_n$ is a random sample from a normal distribution with mean zero and variance σ^2. If the data come from a bivariate normal distribution, of the type discussed in Section 11.1, then all five of these assumptions are met. Bivariate normality is not a requirement, however. Many other types of data can satisfy all of the assumptions for linear regression. Each of these assumptions is discussed separately in this section.

ASSUMPTION 1: $Y = \alpha + \beta X + \epsilon$

This linear regression assumption should always be checked by making a scatterplot of the data to see if the regression curve appears to be a straight line. Then the method of least squares is used to estimate α and β, with a and b as defined in the previous section. Under the assumptions of this model, a and b are statistics and have normal distributions. The mean of a is α, and the mean of b is β, so both a and b are unbiased estimators of their respective population counterparts. This fact is useful when testing hypotheses or forming confidence intervals.

The values of X may be observations on a random variable, such as when X and Y are measurements on randomly sampled individuals from a population. For example, in a marketing survey X may be the amount spent on groceries and Y may be the corresponding amount spent on meat, and both are random variables. Or the X's may be fixed constants, such as in Problem Setting 12.2, or in a cooking experiment where X may represent predetermined oven temperature settings and Y measures the time until a 1 lb roast is finished at that temperature setting.

The ϵ's are called **population residuals** (or **population errors**) and represent the difference between Y and the population regression curve. This fact is more apparent if the model in Eq. 12.5 is rewritten as

$$\epsilon_i = Y_i - (\alpha + \beta X_i) \tag{12.6}$$

After a least squares line is fit to the data, the difference $Y_i - \hat{Y}_i$ is called a **sample residual** and is denoted by e_i. This may be written as

$$e_i = Y_i - \hat{Y}_i = Y_i - (a + bX_i) \tag{12.7}$$

The sample residuals $e_1, \ldots, e_n$, although not independent of each other, are used to study the validity of the assumptions 3, 4, and 5 concerning the population residuals.

Population residuals ϵ_i are the differences between the observed values Y_i and the conditional means $\mu_{Y|x}$

$$\epsilon_i = Y_i - \mu_{Y|x_i}$$

RESIDUALS **Sample residuals** e_i are the differences between the observed values Y_i and the estimated conditional means $\hat{\mu}_{Y|x_i}$ or the predicted values $\hat{Y}_i$

$$e_i = Y_i - \hat{\mu}_{Y|x_i} = Y_i - \hat{Y}_i$$

from a sample regression curve.

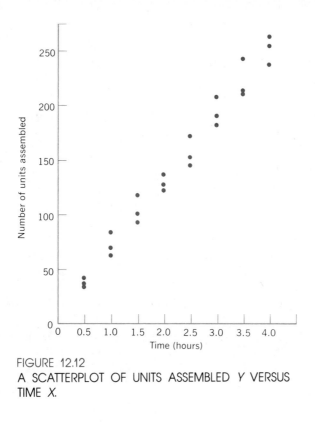

FIGURE 12.12
A SCATTERPLOT OF UNITS ASSEMBLED Y VERSUS
TIME X.

EXAMPLE

The production engineer in Problem Setting 12.2 studies the scatterplot in
Figure 12.12 for the data given in Figure 12.11 and decides to try a linear
regression model because of the nearly straight-line relationship between X
and the conditional means of Y given X.

The least squares method yields the unbiased estimates of β

$$b = \frac{\Sigma XY - \Sigma X \Sigma Y/n}{\Sigma X^2 - (\Sigma X)^2/n} = \frac{9678.5 - 54.0(3456)/24}{153 - (54.0)^2/24} = 60.4$$

and α

$$a = \frac{\Sigma Y}{n} - \frac{b \Sigma X}{n} = \frac{3456}{24} - \frac{60.4(54.0)}{24} = 8.1$$

from the calculations given in the worksheet in Figure 12.13. Note that the X
values are not observations on a random variable in this case, because the
production engineer is free to inspect each worker's production total at any
time X he chooses. This does not alter the analysis in any way.

X	Y	X^2	Y^2	XY	$\hat{Y}$	e	e*
0.5	36	0.25	1,296	18.0	38.31	−2.31	−0.21
0.5	41	0.25	1,681	20.5	38.31	2.69	0.25
0.5	33	0.25	1,089	16.5	38.31	−5.31	−0.49
1.0	83	1.00	6,889	83.0	68.50	14.50	1.35
1.0	61	1.00	3,721	61.0	68.50	−7.50	−0.70
1.0	68	1.00	4,624	68.0	68.50	−0.50	−0.05
1.5	116	2.25	13,456	174.0	98.70	17.30	1.61
1.5	99	2.25	9,801	148.5	98.70	0.30	0.03
1.5	91	2.25	8,281	136.5	98.70	−7.70	−0.72
2.0	120	4.00	14,400	240.0	128.90	−8.90	−0.83
2.0	135	4.00	18,225	270.0	128.90	6.10	0.57
2.0	125	4.00	15,625	250.0	128.90	−3.90	−0.36
2.5	170	6.25	28,900	425.0	159.10	10.90	1.01
2.5	143	6.25	20,449	357.5	159.10	−16.10	−1.49
2.5	151	6.25	22,801	377.5	159.10	−8.10	−0.75
3.0	180	9.00	32,400	540.0	189.30	−9.30	−0.86
3.0	205	9.00	42,025	615.0	189.30	15.70	1.46
3.0	188	9.00	35,344	564.0	189.30	−1.30	−0.12
3.5	212	12.25	44,944	742.0	219.50	−7.50	−0.70
3.5	241	12.25	58,081	843.5	219.50	21.50	2.00
3.5	208	12.25	43,264	728.0	219.50	−11.50	−1.07
4.0	236	16.00	55,696	944.0	249.69	−13.69	−1.27
4.0	261	16.00	68,121	1044.0	249.69	11.31	1.05
4.0	253	16.00	64,009	1012.0	249.69	3.31	0.31
Totals 54.0	3456	153.00	615,122	9678.5	3456.00	0.00	0.02

$\bar{X} = 2.25$ $\bar{Y} = 144.00$ $s_x^2 = 1.3696$ $s_y^2 = 5106.87$

FIGURE 12.13
A WORKSHEET FOR CHECKING ASSUMPTIONS IN THE LINEAR REGRESSION MODEL.

ASSUMPTION 2: $E(\epsilon) = 0$

This assumption usually causes no problem in linear regression. The sample residuals always have a sample mean equal to 0

$$\bar{e} = \frac{1}{n} \sum_{i=1}^{n} e_i = 0$$

as a by-product of the least squares method.

EXAMPLE

The sum of the sample residuals in Problem Setting 12.2 are seen in Figure 12.13 to equal 0.00, so $\bar{e}$ equals zero.

ASSUMPTION 3: EQUAL VARIANCES

This assumption says that the variance of the population residuals ϵ, and hence the variance of Y for a given X, is a constant value for all values of X. This population residual variance is denoted by σ^2 and its estimate from the sample residuals is denoted by $\hat{\sigma}^2$.

In real-world applications of regression the variance of the residuals often changes as X changes, so it is important to check this assumption of equal variances of the residuals. Violations of the equal variance assumption should not be taken lightly, as the validity of the following statistical methods may be seriously impaired. The assumption of equal variances is also called the assumption of **homogeneity of variances, or homoscedasticity.** Lack of equal variances is also called **nonhomogeneity of variances, or heteroscedasticity.**

When variances are nonhomogeneous the variation of Y often changes as X changes. Therefore, one way of testing the validity of the homoscedasticity assumption is to divide the e_i into two groups, one group associated with small values of X and the other group associated with large values of X. The sample variances s_1^2 and s_2^2 are computed on the two groups of e's separately. Then the F-test for equal variances given in Section 9.2 is used to see if the two groups of sample residuals have equal population variances. Rejection of the null hypothesis using this test is strong evidence that this assumption is not satisfied. Recall, however, that this F-test is very sensitive to the assumption of normality, so if the residuals are not normally distributed, the results of this F-test are not valid. Even if the residuals are normal, the effect of using the least squares regression line is to introduce a slight dependence among the sample residuals so the F-test becomes an approximate procedure.

The **residual variance** σ^2 is estimated from the **sample residual variance** $\hat{\sigma}^2$, with $n - 2$ as a divisor instead of $n - 1$:

$$\hat{\sigma}^2 = \frac{1}{n - 2} \Sigma e_i^2 = \frac{1}{n - 2} \Sigma(Y_i - \hat{Y}_i)^2 = \frac{\text{Error SS}}{n - 2}$$

A formula for $\hat{\sigma}^2$ that is easier to compute is given by

$$\hat{\sigma}^2 = \frac{n - 1}{n - 2} (s_Y^2 - b^2 s_X^2)$$

where s_X^2 and s_Y^2 are the sample variances of the X's and Y's, respectively, and b is the least squares estimate of the slope.

ASSUMPTION 4: NORMALITY

The sample residuals $e_1, \ldots, e_n$ may be used to check the validity of the normality assumption on the ϵ_i. The Lilliefors test may be used, but it is an approximate procedure in this case, since the e_i are not independent of each other. The approximate Lilliefors test is performed by plotting the empirical distribution function of the residuals standardized as follows:

The **population residual variance** is the variance of ϵ_i and is denoted by σ^2. The **sample residual variance** is denoted by $\hat{\sigma}^2$ and is used to estimate σ^2. It may be computed using either

$$\hat{\sigma}^2 = \frac{1}{n-2} \sum_{i=1}^{n} (Y_i - \hat{Y}_i)^2 = \frac{1}{n-2} \sum_{i=1}^{n} e_i^2 = \frac{\text{Error SS}}{n-2} \qquad (12.8)$$

RESIDUAL
VARIANCE   or the more convenient

$$\hat{\sigma}^2 = \frac{n-1}{n-2} (s_Y^2 - b^2 s_X^2) \qquad (12.9)$$

where s_Y^2 and s_X^2 are the sample variances of Y_i and X_i, respectively, and b is the least squares estimate of the slope.

$$e_i^* = \frac{e_i}{\hat{\sigma}} \qquad (12.10)$$

Note that $\bar{e} = 0$, so the only change in e_i^* from the usual z scores is that $n - 2$ is used as a divisor for $\hat{\sigma}^2$ instead of $n - 1$. If the residuals are not normal, the statistical methods in Section 12.4 should be used rather than those in Section 12.3.

ASSUMPTION 5: INDEPENDENCE

The independence of the ϵ_i may be checked by examining the method of collecting the data. If the observations (X_i, Y_i), $i = 1, \ldots, n$, are a random sample, the residuals will be independent. If the observations on X are fixed rather than random, the method of collecting observations on Y should be examined to see if the Y_i are independent of each other. Independence among the Y's implies independence among the ϵ's.

EXAMPLE

The sample residuals $e_i = Y_i - \hat{Y}_i$ for Problem Setting 12.2 are given in the worksheet in Figure 12.13. A **residual plot** of the sample residuals against X, as in Figure 12.14, is useful to detect any obvious violations of the assumptions of linear regression, homogeneity of variance, or normality. In this case, the residuals appear to be well behaved because of their apparent random scattering above and below the horizontal line at zero, but tests of the assumptions will be made to provide some confidence.

To check the assumption of normality, the sample residual variance is calculated as

$$\hat{\sigma}^2 = \frac{n-1}{n-2} (s_Y^2 - b^2 s_X^2)$$

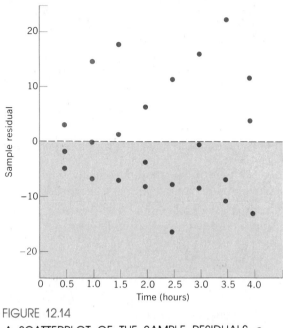

FIGURE 12.14

A SCATTERPLOT OF THE SAMPLE RESIDUALS e VERSUS TIME X.

$$= \frac{23}{22} [5106.87 - 3647.8(1.3696)]$$

$$= 116.05$$

The residuals are divided by $\hat{\sigma} = 10.77$ to get the standardized residuals e_i^*. These standardized residuals are given in the worksheet in Figure 12.13. A plot of the empirical distribution function of the e_i^* on the Lilliefors test chart (not shown here) shows that the e_i may reasonably be regarded as having come from a normal distribution.

The homogeneity of variance assumption is tested by dividing the observations into two groups of equal sizes, one group having small X values and the other group having large X values. The sample variance for the 12 e_i associated with the smallest X_i (two hours or less) is $s_1^2 = 72.357$. This is compared with the sample variance for the 12 e_i associated with X_i greater than two hours, which is $s_2^2 = 159.404$. As an approximate procedure the resulting F statistic

$$F = \frac{\text{larger } s_i^2}{\text{smaller } s_i^2} = \frac{159.404}{72.357} = 2.203$$

is compared with the .975 quantile from the F distribution with 11 degrees of freedom in both the numerator and denominator, which is 3.478 from Table A5 using interpolation. Therefore, the null hypothesis of equal variance is accepted at $\alpha = .05$. This is an indication of, but not proof of, the validity of assumption 3.

It is never possible to completely verify the validity of any of the assumptions, but checks such as these are useful to detect obvious violations of the assumptions. The data in Problem Setting 12.2 appear to satisfy all of the assumptions of the linear model. This enables the methods of the following section to be applied to make inferences in this situation.

SUMMARY

Verification of the assumptions in linear regression requires several steps.

Step 1 Make a scatterplot of the data and visually scan the data to see if a linear population regression curve seems reasonable.

Step 2 Fit a least squares regression line to the data and plot it on the scatterplot in step 1. Plot the sample residuals versus X for obvious signs of nonnormality or nonhomogeneity of variance.

Step 3 If the residual plot in step 2 suggests nonnormality of the sample residuals, use a Lilliefors test to make a final judgment.

Step 4 If the residual plot in step 2 suggests nonhomogeneity of variance, check the homogeneity of variance assumption with an F-test on the sample residuals.

Step 5 Independence of the population residuals may be verified at any stage by examining the data collection procedures to see if the Y's are mutually independent. The Y's are not mutually independent if knowledge of one of the Y values provides useful information about one or more other Y values; knowledge that is not obtained by simply knowing X.

EXERCISES

12.17 What assumptions are likely to be violated, based on the following scatterplots of the data?

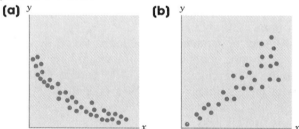

(a) y **(b)** y

12.18 What assumptions are likely to be violated based on the following residual plots?

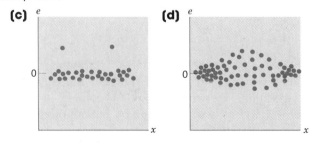

(c) e **(d)** e

12.19 The owner of an office building has noticed a linear relationship between the monthly heating bills for the office building and the monthly
heating bills for his private home. For five months these are the bills.
Assume the linear regression assumptions are true.

	Nov.	Dec.	Jan.	Feb.	Mar.
Office building (Y)	2312	2585	3482	2990	2444
Private home (X)	43	52	77	64	41

The least squares equations give $a = 1047.35$ and $b = 30.9612$.

(a) Find the sample residuals.

(b) Use these sample residuals and Eq. 12.8 to estimate the residual
variance.

(c) Use Eq. 12.9 to estimate the residual variance, and compare your
answer with (b). ($s_X^2 = 228.3$, $s_Y^2 = 226,334$)

12.20 A scatterplot of daily sales volume (Y) versus the number of customers (X) shows a possible nonhomogeneity of variance. The median number of customers is 42 in the data. For the 15 days on which the
number of customers was less than 42 the sample residual variance
was 106. For the 18 days on which the number of customers was 42
or more the sample residual variance was 362. Assume normality of
the residuals and use the F-test to test for possible heteroscedasticity.

12.21 Use the methods of this section to test the normality assumption for the
least squares residuals for the sample data given in Exercise 12.1.

12.22 Use the methods of this section to test the normality assumption for the
least squares residuals for the sample data given in Exercise 12.2.

12.23 Test the assumption of homogeneity of variance for the sample regression data given in Exercise 12.1.

12.24 Test the assumption of homogeneity of variance for the sample regression data given in Exercise 12.2.

12.3

CONFIDENCE INTERVALS AND HYPOTHESIS TESTS
WHEN THE RESIDUALS ARE NORMAL

REVIEW

In the previous section a method is presented for determining whether the
residuals

$$e_i = Y_i - \hat{Y}_i = Y_i - a - bX_i, \qquad i = 1, \ldots, n$$

should be considered normally distributed or not. The methods of this section
are useful in the case of normally distributed residuals. If the residuals are not
normally distributed, the method of the next section should be used. Although
the method in the next section is also valid for normally distributed residuals,
the methods presented in this section should be preferred in such a case,

because they are easier to use and there is a wider variety of methods available for normally distributed residuals.

Of lesser importance is the slight gain in power associated with the use of the methods requiring normality when they are appropriate. From a practical standpoint this slight gain in power is offset by the danger of using these procedures when the residuals are not normal, with the loss of power accompanying such a misapplication. Unfortunately the methods of this section are often used without any consideration given to the validity of the normality assumption. This is an unfortunate misuse of statistical methods. A check of the normality of the residuals is necessary to assure the validity of the methods of this section.

INFERENCES ABOUT SLOPE

A method is presented in this section for testing hypotheses about the slope β of the population regression line. It may be used only if the residuals are normally distributed. The hypothesis test outlined in this section uses a test statistic whose null distribution is the Student's t distribution with $n - 2$ degrees of freedom.

A confidence interval for the unknown β is easily obtainable under the assumption of normality of residuals. In general, a confidence interval for β provides an interval estimate of how much the mean of Y will increase for a given unit increase in X.

EXAMPLE

In Problem Setting 12.2 the number of pieces assembled Y was measured against the number of hours on the assembly line X. Suppose that the old standard for production had been accepted as 50 pieces per hour for each worker, not counting the initial time period when the worker was fresh and the production was known to be a little higher. The engineer has introduced some changes in the production process, and he believes that the production rate is now more than 50 pieces per hour. His belief becomes the alternative hypothesis in a hypothesis test. The slope of the regression line represents the increase in production for each hour increase in time spent on the assembly line, so the hypotheses are as follows:

$$H_0: \beta = 50$$
$$H_1: \beta > 50$$

The decision rule at $\alpha = .05$ is to reject the null hypotheses if the test statistic, given by Eq. 12.13, exceeds $t_{.95,22} = 1.7171$ from Table A3, because $n = 24$ in the problem setting.

The following calculations were obtained in the previous section:

$$\Sigma X = 54 \qquad \Sigma Y = 3456 \qquad n = 24$$
$$\Sigma X^2 = 153 \qquad \Sigma Y^2 = 615{,}122 \qquad \Sigma XY = 9678.5$$
$$\overline{X} = 2.25 \qquad \overline{Y} = 144 \qquad s_X^2 = 1.3696 \qquad s_Y^2 = 5106.87$$
$$a = 8.1 \qquad b = 60.4 \qquad \hat{\sigma}^2 = 116.05 \qquad \hat{\sigma} = 10.77$$

The only additional calculation needed here is $s_X = \sqrt{1.3696} = 1.170$. The

---

A TEST FOR SLOPE (NORMALLY DISTRIBUTED RESIDUALS)

Data

The data consist of n pairs of observations (X_i, Y_i), $i = 1, \ldots, n$. Let s_X denote the sample standard deviation of the X's.

Assumptions

1. (Linear Regression) Each Y_i is a linear function of the associated value of X_i through the equation
$$Y_i = \alpha + \beta X_i + \epsilon_i$$
where α and β are the same for all pairs (X_i, Y_i) and where ϵ_i is a random variable.

2. The ϵ_i are a random sample from a normal population with mean 0 and variance σ^2.

Estimation

The estimated slope b in Eq. 12.3 provides an unbiased **point estimate** of the true slope β. A $100(1 - \alpha)\%$ **confidence interval** for β is given by β_L and β_U where

$$\beta_L = b - t_{1-\alpha/2, n-2} \frac{\hat{\sigma}}{s_X \sqrt{n-1}} \tag{12.11}$$

$$\beta_U = b + t_{1-\alpha/2, n-2} \frac{\hat{\sigma}}{s_X \sqrt{n-1}} \tag{12.12}$$

and where $t_{1-\alpha/2, n-2}$ is the $1 - \alpha/2$ quantile of the t distribution with $n - 2$ degrees of freedom, given in Table A3. Note that $\hat{\sigma}^2$ is given by Eq. 12.9.

Null Hypothesis

$$H_0: \beta = \beta_0$$

(β_0 is some specified number).

Test Statistic

$$T = \frac{(b - \beta_0) s_X \sqrt{n-1}}{\hat{\sigma}} \tag{12.13}$$

The null distribution of T is the Student's t distribution with $n - 2$ degrees of freedom, whose upper quantiles are given in Table A3.

Decision Rule

The decision rule depends on the alternative hypothesis of interest. Critical values $t_{1-\alpha, n-2}$ for a level of significance α are given in Table A3, for $n - 2$ degrees of freedom.

(a) $H_1: \beta > \beta_0$. Reject H_0 if $T > t_{1-\alpha, n-2}$.

(b) $H_1: \beta < \beta_0$. Reject H_0 if $T < -t_{1-\alpha, n-2}$.

(c) $H_1: \beta \neq \beta_0$. Reject H_0 if $T < -t_{1-\alpha/2, n-2}$ or if $T > t_{1-\alpha/2, n-2}$.

observed value of the test statistic is

$$T = \frac{(b - \beta_0)s_x\sqrt{n-1}}{\hat{\sigma}} = \frac{(60.4 - 50)1.170\sqrt{23}}{10.77} = 5.417$$

so the null hypothesis is easily rejected. The p-value is less than .0001.

Now that the engineer has established that the marginal production rate has increased over its former value of 50 pieces per hour, he would like to estimate the new production rate with a 95% confidence interval. The lower and upper bounds for the confidence interval on β are as follows.

$$\beta_L = b - t_{.975,22}\frac{\hat{\sigma}}{s_x\sqrt{n-1}} = 60.4 - 2.0739\frac{10.77}{1.170\sqrt{23}} = 56.4$$

$$\beta_U = b + t_{.975,22}\frac{\hat{\sigma}}{s_x\sqrt{n-1}} = 60.4 + 2.0739\frac{10.77}{1.170\sqrt{23}} = 64.4$$

The engineer can be 95% confident that the current production rate is between 56.4 and 64.4 pieces per hour after the initial time period.

THE SPECIAL INTERPRETATION OF $\beta = 0$

If there is no linear relationship between two variables X and Y, then the slope of the sample regression line should be essentially equal to zero, except for chance variation due to sampling. That is, the true regression line should be horizontal if there is no linear relationship between X and Y. A horizontal regression line has the slope β equal to zero. Therefore, a test of the hypothesis that no linear relationship exists between X and Y is accomplished by testing H_0: $\beta = 0$. A test of H_0: $\beta = 0$ is equivalent to a test of H_0: $\rho = 0$, as in Section 11.1, if the population is bivariate normal.

A CONFIDENCE INTERVAL FOR $\mu_{Y|x}$

If the true regression equation were known, then the mean of Y given $X = x$ would be determined exactly from the relationship

$$\mu_{Y|x} = \alpha + \beta x$$

. However, α and β are estimated using the observed values (X_i, Y_i), $i = 1, \ldots, n$ and, hence, $\mu_{Y|x}$ is also estimated. A **confidence interval for $\mu_{Y|x}$** is based on the point estimate $\hat{\mu}_{Y|x}$ given by Eq. 12.14, and has endpoints μ_L and μ_U given by Eq. 12.15 and Eq. 12.16.

EXAMPLE

A confidence interval for the population mean number of pieces assembled in a four-hour shift is found as follows. The point estimate for $\mu_{Y|4}$ is given by Eq. 12.14:

$$\hat{\mu}_{Y|x_0} = a + bx_0 = 8.1 + 60.4(4) = 249.7$$

The lower bound for a 95% confidence interval for $\mu_{Y|x_0}$ is given by μ_L from Eq. 12.15:

$$\mu_L = \hat{\mu}_{Y|x_0} - t_{.975,22}\,\hat{\sigma}\,\sqrt{\frac{1}{n} + \frac{(x_0 - \overline{X})^2}{(n-1)s_X^2}}$$

$$= 249.7 - 2.0739(10.77)\,\sqrt{\frac{1}{24} + \frac{(4 - 2.25)^2}{23(1.3696)}} = 241.4$$

The upper bound is μ_U as given by Eq. 12.16:

$$\mu_U = \hat{\mu}_{Y|x_0} + t_{.975,22}\,\hat{\sigma}\,\sqrt{\frac{1}{n} + \frac{(x_0 - \overline{X})^2}{(n-1)s_X^2}}$$

$$= 249.7 + 2.0739(10.77)\,\sqrt{\frac{1}{24} + \frac{(4 - 2.25)^2}{23(1.3696)}} = 258.0$$

The 95% confidence interval for the mean production in a four-hour shift is from 241.4 to 258.0 pieces.

A PREDICTION INTERVAL FOR Y

Even if the true regression line were known, there would still be variation in the Y's because the Y's are normally distributed about the regression line. So

---

A CONFIDENCE INTERVAL FOR $\mu_{Y|x_0}$, THE MEAN OF Y AS PREDICTED BY A GIVEN VALUE OF X

The **Data** and **Assumptions** are the same as previously given in this section.
Estimation
A *point estimate* of $\mu_{Y|x_0}$ is provided by $\hat{\mu}_{Y|x_0}$, where

$$\hat{\mu}_{Y|x_0} = a + bx_0 \qquad (12.14)$$

and where x_0 is the given value of X. A $100(1 - \alpha)$% *confidence interval* for $\mu_{Y|x_0}$ is given by μ_L and μ_U where

$$\mu_L = \hat{\mu}_{Y|x_0} - t_{1-\alpha/2,n-2}\,\hat{\sigma}\,\sqrt{\frac{1}{n} + \frac{(x_0 - \overline{X})^2}{(n-1)s_X^2}} \qquad (12.15)$$

and

$$\mu_U = \hat{\mu}_{Y|x_0} + t_{1-\alpha/2,n-2}\,\hat{\sigma}\,\sqrt{\frac{1}{n} + \frac{(x_0 - \overline{X})^2}{(n-1)s_X^2}} \qquad (12.16)$$

The quantile $t_{1-\alpha/2,n-2}$ is obtained from Table A3, using $n - 2$ degrees of freedom, and $\hat{\sigma}^2$ is given by Eq. 12.9.

a **prediction interval for Y** considers the variation of Y about the regression line in addition to the variation of $\hat{\mu}_{Y|x}$ because the true regression line is not known. For this reason prediction intervals for Y are always much wider than confidence intervals for $\mu_{Y|x}$.

The difference between the interpretation of a confidence interval for $\mu_{Y|x}$ and a prediction interval for Y is that, for a given value of X, a confidence interval for $\mu_{Y|x}$ estimates the population mean of Y, while a prediction interval for Y provides an estimate for an individual value of Y rather than for the mean of Y. In Problem Setting 12.2 the confidence interval for $\mu_{Y|x}$ estimates the *mean* production for the population of all workers who have been on the assembly line x hours. The prediction interval gives an estimate of an *individual worker's* production for the same length of time.

A PREDICTION INTERVAL FOR Y GIVEN X

The **Data** and **Assumptions** are the same as previously given in this section.

Prediction

The predicted value $\hat{Y}$ for Y given X is

$$\hat{Y} = a + bx_0$$

where x_0 is the given value of X. A $100(1 - \alpha)\%$ prediction interval for Y is given by Y_L and Y_U where

$$Y_L = \hat{Y} - t_{1-\alpha/2, n-2}\, \hat{\sigma}\, \sqrt{1 + \frac{1}{n} + \frac{(x_0 - \overline{X})^2}{(n-1)s_X^2}} \qquad (12.17)$$

and

$$Y_U = \hat{Y} + t_{1-\alpha/2, n-2}\, \hat{\sigma}\, \sqrt{1 + \frac{1}{n} + \frac{(x_0 - \overline{X})^2}{(n-1)s_X^2}} \qquad (12.18)$$

The quantile $t_{1-\alpha/2, n-2}$ is obtained from Table A3, using $n - 2$ degrees of freedom, and $\hat{\sigma}^2$ is given by Eq. 12.9.

EXAMPLE

To estimate the productivity of an individual worker in a four-hour shift a 95% prediction interval for Y is formed using Eqs. 12.17 and 12.18. The predicted value $\hat{Y}$ for Y given X is

$$\hat{Y} = a + bx_0 = 8.1 + 60.4(4) = 249.7$$

as before. The lower limit of the prediction interval is

$$Y_L = \hat{Y} - t_{.975, 22}\, \hat{\sigma}\, \sqrt{1 + \frac{1}{n} + \frac{(x_0 - \overline{X})^2}{(n-1)s_X^2}}$$

$$= 249.7 - 2.0739(10.77)\, \sqrt{1 + \frac{1}{24} + \frac{(4 - 2.25)^2}{23(1.3696)}} = 225.9$$

while the upper limit is given by

$$Y_U = \hat{Y} + t_{.975,22}\,\hat{\sigma}\,\sqrt{1 + \frac{1}{n} + \frac{(x_0 - \overline{X})^2}{(n-1)s_X^2}}$$

$$= 249.7 + 2.0739(10.77)\sqrt{1 + \frac{1}{24} + \frac{(4 - 2.25)^2}{23(1.3696)}} = 273.5$$

The 95% prediction interval for an individual worker's output in a four-hour shift is from 225.9 to 273.5 pieces. This is a much wider interval than the 95% confidence interval for the mean production in a four-hour shift, found earlier to be from 241.4 to 258.0 pieces.

EXERCISES

12.25 Fifty employees' records are obtained in a random sample, and from each record the employee's salary Y in thousands of dollars, and years of service X are recorded. A statistical summary of the data gives the following:

$$\overline{X} = 7.3 \qquad s_X^2 = 25.6 \qquad \overline{Y} = 41.1 \qquad s_Y^2 = 48.1 \qquad s_{XY} = 12.2$$

Assume the requirements for linear regression are met. Test the null hypothesis $\beta = 0$ against the two-sided alternative using $\alpha = .05$.

12.26 Find a 90% confidence interval for β in Exercise 12.25.

12.27 Find a 90% prediction interval for an individual employee's salary in Exercise 12.25, if that employee has 10 years of service.

12.28 Obtain a 90% confidence interval for the mean salary of all employees in Exercise 12.25 with 10 years service.

12.29 Refer to the sample data given in Exercise 12.1. It is suspected that the local utility company's stock fluctuates in price a greater percentage than the Standard & Poor's Index does. Test the hypothesis $H_0: \beta = 1$ versus $H_1: \beta > 1$ using the methods of this section. Let $\alpha = .05$ and state the p-value.

12.30 Find a 95% confidence interval for β using the regression data given in Exercise 12.1.

12.31 If the S&P Index goes up 5% one quarter in Exercise 12.1, find a 95% confidence interval for the expected percentage gain in the local utility stock's value.

12.32 If the S&P Index goes up 2% one quarter in Exercise 12.1, find a 95% prediction interval for the percentage gain in the local utility stock's value.

12.33 Refer to the sample data given in Exercise 12.2. It is suspected that there is less than a one sale increase, on the average, for each additional year of experience. Test the hypothesis H_0: $\beta = 1$ versus H_1: $\beta < 1$ using the methods of this section. Let $\alpha = .05$ and state the p-value.

12.34 Find a 95% confidence interval for β using the regression data given in Exercise 12.2.

12.35 Find a 90% prediction interval for an individual saleswoman's perform-ance in Exercise 12.2, with 3 years experience.

12.36 Find a 90% confidence interval for the mean sales in Exercise 12.2, for salespeople with 3 years experience. Compare the size of this interval with the interval found in Exercise 12.35.

12.37 Refer to Exercise 12.19 and test the null hypothesis that the office heat-ing bill increases $25 for every dollar increase in the home heating bill, against the two-sided alternative at $\alpha = .05$.

12.38 Find a 95% confidence interval for the slope in Exercise 12.19.

12.39 If the home heating bill is $50 in Exercise 12.19, find a 95% prediction interval for the office heating bill.

12.40 If the home heating bill is $50 in Exercise 12.19, find a 95% confidence interval for the expected value of the office heating bill.

12.4

A NONPARAMETRIC TEST FOR SLOPE (Optional)

The test for slope presented in the previous section relies on the assumption of normality for the residuals. If that assumption is not satisfied, then the method of this section can be used. In reality the method of this section can also be used with normal data, but in such a case the methods of the previous section should be preferred due to their simplicity and versatility. That is, the previous section presented a test for slope, confidence intervals for β and $\mu_{y|x}$, and a prediction interval for Y. In this section only a test for slope is given. Other alternative procedures exist for analyzing nonnormal data, but they are not included in this text because they are somewhat cumbersome to use.

A TEST FOR SLOPE

If the wrong slope is used to fit a straight line to a set of bivariate data, the mistake will show up in a scatterplot of the data along with the erroneous regression curve. The line will tend to be above the points on one side of the graph and below the points on the other side of the graph. This means that the residuals will tend to be positive on one side of the graph and negative on the other. Thus the residuals will appear to be correlated with the x coor-dinates of the points. Spearman's rho may be used to see if the residuals are correlated with the x values. If r_R is significant, then the slope is probably not correct. This is the basis for this test on slope. Note that the residuals do not have to be normally distributed for a test using r_R to be valid.

NONPARAMETRIC TEST FOR SLOPE

Data
The data consist of n pairs of observations (X_i, Y_i), $i = 1, \ldots, n$.

Assumptions
1. (Linear Regression) Each Y_i is a linear function of the associated value of X_i through the equation

$$Y_i = \alpha + \beta X_i + \epsilon_i$$

where α and β are the same for all pairs (X_i, Y_i).

2. The ϵ_i are a random sample from a population (not necessarily normal) with mean 0 and variance σ^2.

Null Hypothesis

$$H_0: \beta = \beta_0$$

(β_0 is some specified number).

Test Statistic
The test statistic is based on Spearman's rho computed between the ranks of X_i and the ranks of the residuals assuming the slope is β_0. That is, Spearman's rho is computed as follows.

1. For each pair (X_i, Y_i) compute $U_i = Y_i - \beta_0 X_i$, $i = 1, \ldots, n$.
2. Rank the U_i from 1 to n, using average ranks in case of ties. Call these ranks R_{U_i}. (Although the U's are not the residuals because a was not used in the computation, the ranks R_{U_i} are the same as the ranks of the residuals, because subtraction of the constant a does not change the relative ranking.)
3. Rank the X_i from 1 to n, using average ranks in case of ties. Call these ranks R_{X_i}.
4. Compute Spearman's rho

$$r_R = \frac{\Sigma R_{U_i} R_{X_i} - C}{\sqrt{(\Sigma R_{U_i}^2 - C)(\Sigma R_{X_i}^2 - C)}} \qquad (12.19)$$

between the X_i and the hypothesized residuals U_i, where

$$C = \frac{n(n + 1)^2}{4}$$

For $n \leq 18$ Figure 11.9 gives the exact quantiles $r_{p,n}$ of r_R if there are no ties. Otherwise the test statistic is

$$T_R = r_R \sqrt{\frac{n - 2}{1 - r_R^2}}$$

The null distribution of T_R is approximately a Student's t distribution with $n - 2$ degrees of freedom.

Decision Rule
The decision rule depends on the alternative hypothesis of interest and the sample size. For $n \leq 18$ r_R is compared with quantiles $r_{p,n}$ from Figure 11.9 if there are no ties. Otherwise, T_R is compared with quantiles $t_{p,n-2}$ from Table A3 with $n - 2$ degrees of freedom.

(a) $H_1: \beta > \beta_0$. If $n \leq 18$ and there are no ties, reject H_0 if $r_R > r_{1-\alpha,n}$. Otherwise, reject H_0 if $T_R > t_{1-\alpha,n-2}$.

(b) H_1: $\beta < \beta_0$. If $n \leq 18$ and there are no ties, reject H_0 if $r_R < -r_{1-\alpha,n}$. Otherwise, reject H_0 if $T_R < -t_{1-\alpha,n-2}$.

(c) H_1: $\beta \neq \beta_0$. If $n \leq 18$ and there are no ties, reject H_0 if $r_R > r_{1-\alpha/2,n}$ or if $r_R < -r_{1-\alpha/2,n}$. Otherwise, reject H_0 if $T_R > t_{1-\alpha/2,n-2}$ or if $T_R < -t_{1-\alpha/2,n-2}$.

EXAMPLE

Stock prices are often very sensitive to news or rumors about strikes, contract settlements, new product lines, and changes in upper management. Such news may cause a dramatic change in stock prices in a short period of time. Figure 12.15 contains data on a computer company for 16 quarters, which includes two quarters, numbered 6 and 10, during which new product announcements had caused a flurry of stock buying. Also included is similar data on the S&P Index. It is suspected that β, the slope of the population regression line between the two variables, can no longer be assumed to equal 1.0.

Computations yield

$$s_X^2 = [352.55 - (24.1)^2/16]/15 = 21.0833$$

$$s_Y^2 = [811.16 - (29.6)^2/16]/15 = 50.4267$$

$$s_{XY} = [396.79 - 24.1(29.6)/16]/15 = 23.4803$$

Quarter	S&P Index X (% change)	Company Y (% change)	X^2	XY	Y^2	Predicted Y $\hat{Y} = a + bX$	Residual $Y - \hat{Y}$	Standardized Residual
1	-0.6	-3.2	0.36	1.92	10.24	-0.50	-2.70	-0.53
2	2.2	-1.9	4.84	-4.18	3.61	2.62	-4.52	-0.89
3	3.4	1.9	11.56	6.46	3.61	3.96	-2.06	-0.40
4	2.9	5.0	8.41	14.50	25.00	3.40	1.60	0.31
5	-3.6	-7.6	12.96	27.36	57.76	-3.84	-3.76	-0.74
6	-0.6	12.2	0.36	-7.32	148.84	-0.50	12.70	2.49
7	-5.9	-9.8	34.81	57.82	96.04	-6.40	-3.40	-0.67
8	-7.4	-8.8	54.76	65.12	77.44	-8.07	-0.73	-0.14
9	5.7	1.5	32.49	8.55	2.25	6.52	-5.02	-0.98
10	1.1	11.7	1.21	12.87	136.89	1.40	10.30	2.02
11	5.0	6.7	25.00	33.50	44.89	5.74	0.96	0.19
12	5.7	4.9	32.49	27.93	24.01	6.52	-1.62	-0.32
13	7.9	7.3	62.41	57.67	53.29	8.97	-1.67	-0.33
14	8.2	11.0	67.24	90.20	121.00	9.30	1.70	0.33
15	1.4	1.0	1.96	1.40	1.00	1.73	-0.73	-0.14
16	-1.3	-2.3	1.69	2.99	5.29	-1.28	-1.02	-0.20
Totals	24.1	29.6	352.55	396.79	811.16	29.57	0.03	0.00

FIGURE 12.15
CHANGES IN STOCK VALUE Y AND CHANGES IN VALUE OF S&P INDEX, WITH WORKSHEET

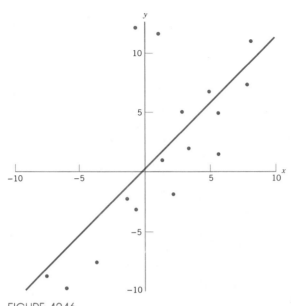

FIGURE 12.16
A SCATTERPLOT OF CHANGE IN S&P INDEX X
VERSUS CHANGES IN A COMPUTER COMPANY'S
STOCK VALUE Y, AND THE LEAST SQUARES
REGRESSION LINE.

$b = 23.4803/21.0833 = 1.1137$

$a = 29.6/16 - 1.1140(24.1/16) = 0.1725$

$\hat{\sigma}^2 = (15/14)[50.4267 - (1.1137)^2(21.0833)] = 26.0108$

A scatterplot of the data points and the least squares regression line

$$\hat{\mu}_{Y|x} = .1725 + 1.1137X$$

appear in Figure 12.16. Note the two points that lie far above the sample regression line. These two outliers suggest nonnormality of the residuals.

To test the normality assumption of the residuals, the Lilliefors test is used on the standardized residuals, as given in Figure 12.15. The empirical distribution function of the standardized residuals falls outside the Lilliefors bounds for $n = 16$ in Figure 12.17, so normality cannot be assumed for the ϵ's in this case.

To test H_0: $\beta = 1.0$ against the two-sided alternative H_1: $\beta \neq 1.0$, the rank correlation coefficient is computed between the X's and the hypothesized residuals $Y_i - (1.0)X_i$ as shown in the worksheet in Figure 12.18. Spearman's rho is calculated using Eq. 12.19,

$$r_R = \frac{1231.5 - 1156}{\sqrt{(1496 - 1156)(1495 - 1156)}} = .222$$

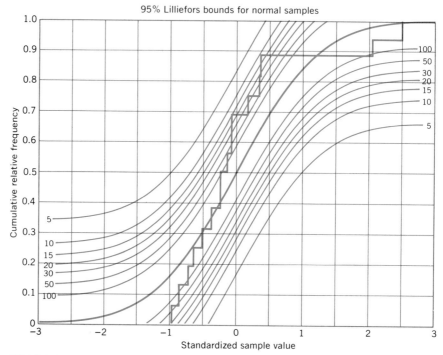

FIGURE 12.17
A LILLIEFORS TEST FOR NORMALITY OF RESIDUALS.

Because of ties in the ranks, the test statistic

$$T_R = r_R \sqrt{\frac{n-2}{1-r_R^2}} = .222 \sqrt{\frac{16-2}{1-.222^2}} = .8535$$

is compared with $t_{.975,14} = 2.1448$ from Table A3 for a two-sided test at $\alpha = .05$. Since $|T_R| < 2.1448$, the null hypothesis $\beta = 1.0$ is not rejected. The p-value is greater than .20. Therefore, it is reasonable to assume that the percentage change associated with the price per share of this company is about the same as the percentage change in the Standard & Poor's Index.

EXAMPLE

A bank is interested in determining whether there is a linear relationship between the purchase price of the new car and the annual family income of the purchaser. The bank assumes that the most recent 10 loan applications resemble a random sample of all new-car purchases. The data and the worksheet are given in Figure 12.19.

The computations are as follows:

$$s_X^2 = [6311.08 - (243.15)^2/10]/9 = 44.3209$$
$$s_Y^2 = [601.53 - (76.94)^2/10]/9 = 1.0615$$
$$s_{XY} = [1906.44 - 243.15(76.94)/10]/9 = 3.9610$$

X	Y	$U = Y - (1.0)X$	R_x	R_u	R_x^2	R_u^2	$R_x R_u$
-0.6	-3.2	-2.6	5.5	5	30.25	25	27.5
2.2	-1.9	-4.1	9.0	2	81.00	4	18.0
3.4	1.9	-1.5	11.0	6	121.00	36	66.0
2.9	5.0	2.1	10.0	13	100.00	169	130.0
-3.6	-7.6	-4.0	3.0	3	9.00	9	9.0
-0.6	12.2	12.8	5.5	16	30.25	256	88.0
-5.9	-9.8	-3.9	2.0	4	4.00	16	8.0
-7.4	-8.8	-1.4	1.0	7	1.00	49	7.0
5.7	1.5	-4.2	13.5	1	182.25	1	13.5
1.1	11.7	10.6	7.0	15	49.00	225	105.0
5.0	6.7	1.7	12.0	12	144.00	144	144.0
5.7	4.9	-0.8	13.5	9	182.25	81	121.5
7.9	7.3	-0.6	15.0	10	225.00	100	150.0
8.2	11.0	2.8	16.0	14	256.00	196	224.0
1.4	1.0	-0.4	8.0	11	64.00	121	88.0
-1.3	-2.3	-1.0	4.0	8	16.00	64	32.0
					1495.00	1496	1231.5

FIGURE 12.18
WORKSHEET FOR FINDING SPEARMAN'S RHO BETWEEN X AND U.

$$b = 3.9610/44.3209 = .0894$$
$$a = 76.94/10 - .0894(243.15/10) = 5.5210$$

A scatterplot of the data points and the least squares regression line

$$\hat{\mu}_{Y|x} = 5.5210 + 0.0894X$$

	Car Prices Y (Thousands)	Income X (Thousands)	Y^2	X^2	XY
	6.82	28.21	46.5124	795.8041	192.3922
	8.71	21.88	75.8641	478.7344	190.5748
	5.97	15.94	35.6409	254.0836	95.1618
	8.23	26.16	67.7329	684.3456	215.2968
	9.42	38.26	88.7364	1463.8276	360.4092
	7.65	26.42	58.5225	698.0164	202.1130
	6.84	18.41	46.7856	338.9281	125.9244
	7.72	16.82	59.5984	282.9124	129.8504
	8.41	23.14	70.7281	525.4596	194.6074
	7.17	27.91	51.4089	778.9681	200.1147
Totals	76.94	243.15	601.5302	6311.0799	1906.4447

FIGURE 12.19
NEW CAR PRICES AND FAMILY INCOME OF PURCHASER.

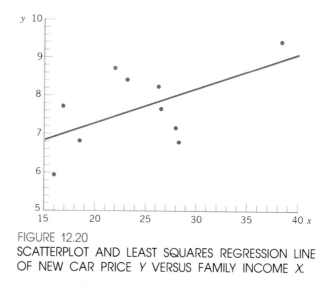

FIGURE 12.20

SCATTERPLOT AND LEAST SQUARES REGRESSION LINE OF NEW CAR PRICE Y VERSUS FAMILY INCOME X.

appear in Figure 12.20. Since the procedure of this section is equally valid for normal and nonnormal residuals, it is performed on these data for purposes of illustration, without the benefit of a Lilliefors test for normality.

To test H_0: $\beta = 0$ the rank correlation coefficient is computed between the X_i and the hypothesized residuals $Y_i - (\beta_0 X_i + a)$. The intercept a is ignored since it does not affect the ranks of the residuals in any way. Also, because $\beta_0 = 0$ in this example, the ranks of the hypothesized residuals are merely the ranks of the Y_i themselves. Therefore, the test statistic is the rank correlation coefficient computed between the X's and the Y's. The worksheet in Figure

Car Prices Y (Thousands)	Income X (Thousands)	R_y	R_x	$(R_x - R_y)^2$
6.82	28.21	2	9	49
8.71	21.88	9	4	25
5.97	15.94	1	1	0
8.23	26.16	7	6	1
9.42	38.26	10	10	0
7.65	26.42	5	7	4
6.84	18.41	3	3	0
7.72	16.82	6	2	16
8.41	23.14	8	5	9
7.17	27.91	4	8	16
				120

FIGURE 12.21

WORKSHEET FOR COMPUTING r_R BETWEEN CAR PRICES AND FAMILY INCOME.

12.21 shows there are no ties, so the simpler form for Spearman's rho, given by Eq. 11.6, is used:

$$r_R = 1 - \frac{6\Sigma(R_{X_i} - R_{Y_i})^2}{n(n^2 - 1)} = 1 - \frac{6(120)}{10(99)} = .2727$$

A comparison of this value $r_R = .2727$ with the critical value for a two-sided test at $\alpha = .05$, $r_{.975,10} = .636$, obtained from Figure 11.9, shows that the null hypothesis $\beta = 0$ is not rejected; hence, the data do not exhibit a significant linear relationship. Note that the computations in Figure 12.19 are not needed for the hypothesis test; they are used only for finding the equation of the least squares regression line.

EXERCISES

12.41 Use the nonparametric test of this section to test the null hypothesis $\beta = 25$ in Exercise 12.19, against the two-sided alternative. Use $\alpha = .10$.

12.42 Test the hypothesis $H_0: \beta = 1$ versus $H_1: \beta > 1$ using the methods of this section on the data given in Exercise 12.1. Let $\alpha = .05$ for this test. Compare your results with those obtained for Exercise 12.29.

12.43 Test the hypothesis $H_0: \beta = 1$ versus $H_1: \beta < 1$ using the methods of this section on the data given in Exercise 12.2. Let $\alpha = .05$ for this test. Compare your results with those obtained for Exercise 12.33.

12.44 Test the normality assumption for the residuals of the last example of this section.

12.5

MONOTONE REGRESSION (Optional)

PROBLEM SETTING 12.3

At the end of the year a radio station collects data for 10 of its customers, and compares their percentage change in reported sales Y for this year over the previous year with the percentage change in advertising expenditures X for those same years.

Customer:	1	2	3	4	5	6	7	8	9	10
X (advertising):	4	62	31	−11	47	88	16	−1	74	21
Y (sales):	10	33	39	−14	37	39	18	−8	45	33

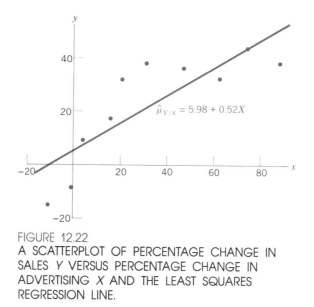

FIGURE 12.22
A SCATTERPLOT OF PERCENTAGE CHANGE IN
SALES Y VERSUS PERCENTAGE CHANGE IN
ADVERTISING X AND THE LEAST SQUARES
REGRESSION LINE.

Figure 12.22 contains a scatterplot of the data and the least squares regression line that is obtained using the techniques of the previous sections.

It is apparent from the scatterplot in Figure 12.22 that the straight line can be used to account for some of the variation in the sales, given the advertising costs, but it is also apparent that some kind of a curve, instead of a straight line, might be better to use as a sample regression function. In this section an alternative method of regression is presented for use in many situations where straight-line regression is not a reasonable assumption.

MONOTONIC REGRESSION

In Figure 12.22 the data points (X_i, Y_i) exhibit a monotonic relationship; that is, both X and Y tend to increase together. They do not increase together in a linear relationship but rather in a monotonic relationship. Many users of linear

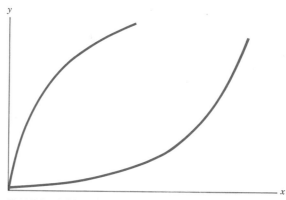

FIGURE 12.23
MONOTONIC RELATIONSHIPS.

regression techniques realize this shortcoming but proceed with linear regression on the belief that the straight line provides a "good approximation" to the curve. Other users of regression techniques try to fit a curve to the data. Although this latter approach may produce better results than fitting a straight line, it is not without drawbacks and limitations. For example, it may be possible to match the curves in Figure 12.23 by curve fitting, but it is not easy to determine what type of mathematical function to use.

There is a technique that will allow such curves to be fit accurately. It uses a linear regression computed on the rank-transformed data. The only requirement is that the data exhibit a monotonic (or nearly so) relationship; that is, as X increases the conditional mean of Y given X either increases (a **monotonic increasing** regression) or decreases (a **monotonic decreasing** regression). Linear regression is a special case of monotonic regression.

PROCEDURE FOR MONOTONIC REGRESSION

1. Replace the X_i, $i = 1, \ldots , n$ with their corresponding ranks 1 to n. Denote these ranks by R_{X_i}. Replace the Y_i, $i = 1, \ldots , n$ with their corresponding ranks 1 to n. Denote these ranks by R_{Y_i}. Use the method of average ranks in case of ties.

2. Compute the usual least squares estimates a and b on the ranks. For ranks these calculations simplify to

$$b = \frac{\Sigma R_{X_i} R_{Y_i} - n(n + 1)^2/4}{\Sigma R_{X_i}^2 - n(n + 1)^2/4} \qquad (12.20)$$

$$a = \frac{n + 1}{2} - \frac{b(n + 1)}{2} \qquad (12.21)$$

3. The regression equation is expressed in terms of ranks as

$$\hat{R}_{Y_i} = a + bR_{X_i} \qquad (12.22)$$

X	Y	R_X	R_Y	R_X^2	$R_X R_Y$
-11	-14	1	1.0	1	1.0
-1	-8	2	2.0	4	4.0
4	10	3	3.0	9	9.0
16	18	4	4.0	16	16.0
21	33	5	5.5	25	27.5
31	39	6	8.5	36	51.0
47	37	7	7.0	49	49.0
62	33	8	5.5	64	44.0
74	45	9	10.0	81	90.0
88	39	10	8.5	100	85.0
				385	376.5

FIGURE 12.24
RANKS FOR PERCENTAGE CHANGE IN SALES X AND PERCENTAGE CHANGE IN ADVERTISING Y.

EXAMPLE

As a continuation of the example given in Problem Setting 12.3, the sales and advertising data are replaced by their ranks (average ranks in case of ties) in Figure 12.24.

The regression on ranks requires calculating

$$b = \frac{376.5 - 10(11)^2/4}{385 - 10(11)^2/4} = .897$$

and

$$a = \frac{11}{2} - .897\left(\frac{11}{2}\right) = .567$$

from Eqs. 12.20 and 12.21 to get the regression equation on ranks

$$\hat{R}_{Y_i} = .567 + .897 R_{X_i} \tag{12.23}$$

from Eq. 12.22.

A scatterplot of the ranks of the X's versus the ranks of the Y's is given in Figure 12.25, along with the least squares regression equation on ranks. Note how the ranks seem to follow a linear regression even though the original points do not seem to do so.

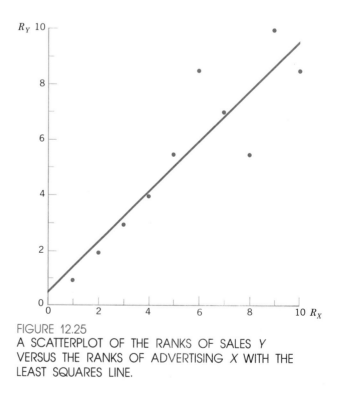

FIGURE 12.25
A SCATTERPLOT OF THE RANKS OF SALES Y
VERSUS THE RANKS OF ADVERTISING X WITH THE
LEAST SQUARES LINE.

MAKING PREDICTIONS

Thus far the regression on ranks is at least as simple as working with the original data (perhaps simpler). The only unusual part comes in making predictions, because the regression equation is expressed in terms of ranks; therefore, the ranks must be used to get a predicted value Y_0 from an original value X_0. This means that a rank must be associated with the value X_0 for which the prediction is to be made. This is done by comparing X_0 with the original observed X_i, and assigning a rank to X_0 that is commensurate with the ranks that belong to the original X's. The rank assigned to X_0 may be a noninteger. The rank for X_0 is then inserted into the least squares equation for ranks to get a predicted rank for Y_0.

---

1. Order the original X's from smallest to largest.

	Original Data	Corresponding Ranks	Ranks if There Are No Ties
Smallest	X_1	R_{X_1}	1
	X_2	R_{X_2}	2
	...	...	...
Largest	X_n	R_{X_n}	n

PREDICTING A RANK, $\hat{R}_{Y_0}$, FOR Y_0

2. If $X_0 \leq X_1$, then assign R_{X_1} to X_0.
3. If $X_0 \geq X_n$, then assign R_{X_n} to X_0.
4. If $X_0 = X_i$, then assign R_{X_i} to X_0.
5. If X_0 is between two adjacent ordered values of X, say $X_i < X_0 < X_j$, then assign R_{X_0} by linear interpolation as

$$R_{X_0} = R_{X_i} + \left(\frac{X_0 - X_i}{X_j - X_i}\right)(R_{X_j} - R_{X_i}) \qquad (12.24)$$

6. Substitute R_{X_0} into Eq. 12.22 to get a predicted rank $\hat{R}_{Y_0}$ for Y_0.

EXAMPLE

Continuing the previous example, suppose one of the radio stations' advertising customers was planning a 25% increase in their advertising expenditures. What increase in sales could they expect as determined by regression on ranks? The steps necessary to find the answer are as follows.

1. **Convert $X_0 = 25$ to a rank** Since $X_0 = 25$ lies between $X_i = 21$ (rank = 5) and $X_j = 31$ (rank = 6), the rank assigned to X_0 is between 5 and 6. It is found from Eq. 12.24 to be

$$R_{X_0} = 5 + \left(\frac{25 - 21}{31 - 21}\right)(6 - 5) = 5.4$$

2. Predict a rank for Y_0 Substitution of $R_{X_0} = 5.4$ into Eq. 12.23 gives

$$\hat{R}_{Y_0} = .567 + .897(5.4) = 5.41$$

The final step, converting the predicted *rank* for Y to a predicted *value* for Y, is deferred until the next example.

CONVERTING THE PREDICTED RANK OF Y_0 TO $\hat{Y}_0$

The final step involves converting the predicted *rank* $\hat{R}_{Y_0}$ into a predicted *value* $\hat{Y}_0$. This is done by comparing the predicted rank $\hat{R}_{Y_0}$, which may be a non-integer, with the observed ranks of the original Y's and choosing a number $\hat{Y}_0$ for $\hat{R}_{Y_0}$ that is commensurate with the original observed Y's and their corresponding ranks.

1. Order the original Y's from smallest to largest.

		Original Data	Corresponding Ranks	Ranks if There Are No Ties
Smallest		Y_1	R_{Y_1}	1
		Y_2	R_{Y_2}	2
		...	...	...
Largest		Y_n	R_{Y_n}	n

CONVERTING
$\hat{R}_{Y_0}$ TO Y_0

2. If $\hat{R}_{Y_0} \leq R_{Y_1}$, then $\hat{Y}_0 = Y_1$.

3. If $\hat{R}_{Y_0} \geq R_{Y_n}$, then $\hat{Y}_0 = Y_n$.

4. If $\hat{R}_{Y_0} = R_{Y_i}$, then $\hat{Y}_0 = Y_i$.

5. If $\hat{R}_{Y_0}$ is between two adjacent ranks assigned to the Y's, say $R_{Y_i} < \hat{R}_{Y_0} < R_{Y_j}$, then, by linear interpolation

$$\hat{Y}_0 = Y_i + \left(\frac{\hat{R}_{Y_0} - R_{Y_i}}{R_{Y_j} - R_{Y_i}} \right) (Y_j - Y_i) \qquad (12.25)$$

EXAMPLE

In continuation of the previous example, which shows how to convert a projected 25% increase in advertising expenditures into a projected increase in sales, the third and final step is now illustrated.

3. Convert $\hat{R}_{Y_0}$ into a value for $\hat{Y}_0$ The rank $\hat{R}_{Y_0} = 5.41$ obtained in step 2 lies between the observed values $R_{Y_i} = 4$ for $Y_i = 18$ and $R_{Y_j} = 5.5$ for $Y_j = 33$, as shown in Figure 12.24. Therefore, from Eq. 12.25

$$\hat{Y}_0 = 18 + \left(\frac{5.41 - 4}{5.5 - 4} \right) (33 - 18) = 32.1$$

is the predicted value of Y, given X equals 25. A 32.1% increase in sales is predicted from a 25% increase in advertising expenditures.

SOME APPLICATIONS OF REGRESSION ON RANKS

The technique of regression on ranks has a wide variety of applications and can be used to great advantage in many business settings.

1. Although sales can be expected to increase as expenditures on advertising increase, the relationship is not likely to be linear, because the marginal return of an advertising dollar added to a sparse advertising budget is likely to be greater than one that is added to an advertising campaign in a saturated market. Therefore, monotonic regression methods may be more appropriate.

2. Increasing the number of checkers in a supermarket can be expected to decrease the average waiting time for customers waiting to check out. The relationship cannot be expected to be linear, however, because the larger number of checkers will increase the amount of time in which one or more checkers will be idle. Monotonic regression would be more appropriate to analyze the relationship.

3. Increasing the amount of fertilizer can be expected to increase the yield of a crop, but the gain in yield can be expected to be less for a unit of fertilizer added to a field already rich in fertilizer than for the same unit added to a field low in fertilizer. The relationship may be monotonic but not linear.

CONCLUDING REMARKS

The method of monotone regression is analogous to the method of least squares presented in Section 12.1 in that no assumptions are made regarding distributions, independence, etc. Also no inferences are made via hypothesis tests or confidence intervals. Both methods provide a way of estimating the unknown regression curve.

One way of evaluating the two methods is to compare the sizes of the residuals obtained from the least squares regression with the sizes of the residuals from monotone regression. The usual way of measuring the overall residual size is through the **residual sum of squares,** also called the **error sum of squares:**

$$\text{Error SS} = \Sigma(Y_i - \hat{Y}_i)^2$$

A smaller error sum of squares implies a better fit to the data.

EXAMPLE

A list of all the residuals obtained by applying monotone regression to the data in Problem Setting 12.3 is useful to illustrate the computations for finding the error sum of squares in monotone regression. Also it enables the sizes of the

(1) X	(2) Y	(3) R_x	(4) R_y	(5) $\hat{R}_y$	(6) $\hat{Y}$	(7) $Y - \hat{Y}$	(8) $(Y - \hat{Y})^2$
−11	−14	1	1.0	1.46	−11.22	−2.78	7.74
−1	−8	2	2.0	2.36	−1.51	−6.49	42.13
4	10	3	3.0	3.26	12.06	−2.06	4.25
16	18	4	4.0	4.15	19.55	−1.55	2.39
21	33	5	5.5	5.05	28.52	4.48	20.11
31	39	6	8.5	5.95	34.20	4.80	23.08
47	37	7	7.0	6.85	36.59	0.41	0.17
62	33	8	5.5	7.74	37.99	−4.99	24.90
74	45	9	10.0	8.64	39.56	5.44	29.62
88	39	10	8.5	9.54	43.15	−4.15	17.18
							171.57

FIGURE 12.26
WORKSHEET FOR CALCULATING RESIDUALS FROM MONOTONE REGRESSION.

residuals using regression on ranks to be compared with the sizes of the residuals obtained using ordinary linear regression on the data.

Again the points are listed in order of increasing X (see Figure 12.26), along with the predicted rank $\hat{R}_{Y_i}$ of Y_i as determined by Eq. 12.23 for each observed X_i. This is simple to do because the actual ranks R_{X_i}, from 1 to 10, are used in the equation.

For each value of $\hat{R}_{Y_i}$ a predicted value $\hat{Y}_i$ is obtained for Column (6) as illustrated previously. The residuals are given in Column (7). The error sum of squares from the regression on ranks is easily found from Column (8) in Figure 12.26 as

$$\text{Error SS} = \sum_{i=1}^{n} (Y_i - \hat{Y}_i)^2 = 171.57 \quad \text{(Rank regression)}$$

which is considerably smaller than the error sum of squares of least squares regression on the data, obtained from Figure 12.27.

$$\text{Error SS} = \sum_{i=1}^{n} e_i^2 = 1179.94 \quad \text{(Ordinary regression)}$$

The smaller sum of squares for rank regression confirms that the monotone regression model is more appropriate for this set of data.

For use as a predictive model, monotone regression may be better or worse than the linear regression method. A smaller error sum of squares associated with the monotone regression does not necessarily imply that the monotone regression model will be better in predicting unknown values of Y, but it serves as an indication that it may be better.

Advertising X	Sales Y	Predicted Sales $\hat{Y} = 5.98 + .52X$	Residuals $e = Y - \hat{Y}$	e^2
-11	-14	0.26	-14.26	203.44
-1	-8	5.46	-13.46	181.29
4	10	8.06	1.94	3.74
16	18	14.31	3.69	13.64
21	33	16.91	16.09	258.99
31	39	22.11	16.89	285.35
47	37	30.43	6.57	43.17
62	33	38.23	-5.23	27.36
74	45	44.47	0.53	0.28
88	39	51.75	-12.75	162.66
		Totals	0.01	1179.94

FIGURE 12.27
WORKSHEET FOR CALCULATING RESIDUALS IN LINEAR
REGRESSION.

EXERCISES

12.45 A company report contains the average number of sick days per year (including time spent away from work on medical and dental appointments) for their employees with 1 to 20 years service with the company. These data are recorded as follows. Plot these data in a scatterplot. Based on this scatterplot, which assumption of the linear regression model appears inappropriate for these data?

Number of Years of Service (X)	Average Number of Sick Days (Y)
1	1.4
2	1.7
3	1.1
4	1.5
5	2.3
6	1.4
7	1.8
8	2.1
9	3.0
10	2.8
11	3.2
12	3.8
13	4.0
14	5.5
15	6.0
16	6.5

(continued)

Number of Years of Service (X)	Average Number of Sick Days (Y)
17	8.0
18	10.7
19	13.0
20	12.2

12.46 Find the least squares regression line for the data of Exercise 12.45 and add it to the scatterplot made for that exercise.

12.47 Convert the regression data in Exercise 12.45 to ranks and make a scatterplot of these ranks. How does the rank scatterplot compare with the raw data scatterplot made for Exercise 12.45?

12.48 Use the ranks of Exercise 12.47 to find the rank regression equation for the data of Exercise 12.45.

12.49 Use the rank regression equation found for Exercise 12.48 to predict the average number of sick days for employees with 16.5 years of service.

12.50 Find the error sum of squares for the least squares fit to the original data for Exercise 12.45. Also find the error sum of squares from the monotone regression fit to these same data. What conclusion is made regarding the two fits based on the error sum of squares?

12.51 Given below are the length of the cotton fibers X and tests for strength of cotton yarn Y for 10 randomly selected pieces of yarn. Plot these data in a scatterplot. Based on this scatterplot, does the linear regression model seem appropriate for these data?

X:	99	93	99	97	90	96	93	130	118	88
Y:	85	82	75	74	76	74	73	96	93	70

12.52 Find the least squares regression line for the data in Exercise 12.51 and plot it on the scatterplot obtained in that exercise. ($\Sigma X = 1003$, $\Sigma X^2 = 102{,}193$, $\Sigma Y = 798$, $\Sigma Y^2 = 64{,}396$, $\Sigma XY = 80{,}991$)

12.53 Find the rank regression equation for the data in Exercise 12.51. ($\Sigma R_X = 55$, $\Sigma R_X^2 = 384$, $\Sigma R_Y = 55$, $\Sigma R_Y^2 = 384.5$, $\Sigma R_X R_Y = 361.5$)

12.54 Use the rank regression equation found in Exercise 12.53 to predict the strength of the cotton yarn for a cotton fiber length $X = 90$.

12.55 Find the error sum of squares using linear regression in Exercise 12.51. Also find the error sum of squares using monotone regression in Exercise 12.53. What conclusion do you make regarding the two fits based on the error sum of squares?

12.6

REVIEW EXERCISES

12.56 Data on the heights X and weights Y of 10 randomly selected college students are given below. Assume the requirements for linear regression are satisfied. ($\Sigma X = 693$, $\Sigma X^2 = 48,163$, $\Sigma Y = 1588$, $\Sigma Y^2 = 257,974$, $\Sigma XY = 110,896$)

X:	64	73	71	69	66	69	75	71	63	72
Y:	121	181	156	162	142	157	208	169	127	165

Make a scatterplot of the data. Find the least squares regression line and plot it on the scatterplot.

12.57 Use the sample regression equation in Exercise 12.56 to predict the average weight for individuals with a height of 72 in. Find a 95% confidence interval for this prediction.

12.58 An experiment was set up to determine if the amount of wear on a new fabric is affected by the speed of a washing machine. Six pieces of fabric were tested at each of five machine speeds and a measurement of the amount of wear was recorded for each piece of fabric as follows.

Machine Speed (rpm)	Amount of Wear					
110	24.9	24.8	25.1	26.4	27.0	26.6
130	27.4	27.3	26.4	28.5	28.7	28.5
150	30.4	30.3	29.5	31.7	31.6	31.4
170	37.9	36.9	37.2	38.5	38.8	39.1
190	48.7	42.7	47.8	49.6	49.9	49.5

Graph machine speed versus the amount of wear in a scatterplot and calculate the total sum of squares $\Sigma(Y_i - \bar{Y})^2$ for the amount of wear. Based on this scatterplot, do you think the linear regression model is appropriate for these data? Why?

12.59 For the sample data in Exercise 12.58 find the conditional mean of Y for each machine speed. Add these points to the scatterplot of Exercise 12.58 and connect them to form a sample regression curve.

12.60 Find the error sum of squares associated with the sample regression curve in Exercise 12.59.

12.61 Find the coefficient of determination for the sample regression curve in Exercise 12.59. What is the interpretation of this value?

12.62 Find the least squares regression equation for the sample data given in Exercise 12.58. Find the value of r^2 for these data and compare it with the value found in Exercise 12.61.

12.63 Use the least squares regression equation found in Exercise 12.62 to predict the mean amount of wear on a piece of fabric washed at a machine speed of 170 rpm.

12.64 Test the normality assumption for the least squares residuals in Exercise 12.62.

12.65 Test the assumption of homogeneity of variance for the least squares residuals in Exercise 12.62.

12.66 Find a 95% confidence interval for the mean amount of wear predicted in Exercise 12.63.

12.67 Use a level of significance of $\alpha = .05$ to test the hypothesis $H_0 : \beta = 0$ versus $H_1 : \beta > 0$ for the least squares regression line found in Exercise 12.62.

12.68 Convert the regression data in Exercise 12.58 to ranks and make a scatterplot of these ranks. How does the rank scatterplot compare with the scatterplot made for the original data in Exercise 12.58?

12.69 Use the ranks of Exercise 12.68 to find the rank regression equation for the data in Exercise 12.58.

12.70 Use the nonparametric test for slope to test the hypothesis in Exercise 12.67.

12.71 Use the rank regression equation found in Exercise 12.69 to predict the amount of wear on fabric that is washed at a speed of 170 rpm. Is this closer to the linear regression estimate in Exercise 12.63 or the nonlinear regression estimate of Exercise 12.59? Which estimate do you think is the most accurate?

BIBLIOGRAPHY

Additional material on the topics presented in this chapter can be found in the following publications.

Conover, W. J. (1980). *Practical Nonparametric Statistics,* 2nd ed. Wiley, New York.

Conover, W. J. and Iman, R. L. (1981). "Rank Transformations as a Bridge between Parametric and Nonparametric Statistics." *The American Statistician,* **35**(3), 124–129.

Iman, R. L. and Conover, W. J. (1979). "The Use of the Rank Transform in Regression." *Technometrics,* **21**(4), 499–509.

13

TIME SERIES ANALYSIS (Optional)

A **time series** is a series of measurements taken at successive points in time, such as hourly temperature readings or weekly production figures. Most business and economic reports contain information in the form of time series, since sales figures, gross national product, stock prices, and many other figures are more informative when they are compared with what happened in previous time periods.

Time series data may be presented in graphical form for easy visual inspection or in tables for detailed analysis. There are several reasons for wanting to study time series data. One reason is to analyze recent or past performance to see if it is better or worse than it was expected to be. This examination is complicated by factors such as seasonal variation and long-term growth or decline in the time series as well as other less predictable factors. These factors and their influence on time series data are studied in this chapter.

A second reason for studying time series data is to predict future behavior. Of course, no crystal ball is available for gazing into the future. All that can be said is that if the pattern for future performance resembles the pattern of past performance, then some predictions can be made regarding the future performance. For the most part, future performance does not behave exactly as past performance, thus actual observations are frequently either above or below the predicted values. This behavior can then be examined to see if the cause for the difference can be discovered. This chapter presents a classic approach to time series models that has been widely used for many years.

13.1

SECULAR TREND IN TIME SERIES ANALYSIS

PROBLEM SETTING 13.1

The accurate forecasting of future sales is essential for the continued success of any business because many plans and decisions must be made based on sales forecasting. If sales were not influenced by factors such as seasonal influences, then forecasting might be accomplished using the methods of simple linear regression. However, sales in the real world seldom behave in a linear fashion because factors like Christmas or changing seasons have a pronounced effect on most businesses. This is true of the sales of soft drinks as reflected by sales recorded in Figure 13.1 for 16 quarters for the years 1976 through 1979. A graph of these data is given in Figure 13.2 where the successive quarterly sales values have been connected with straight-line segments.

TIME SERIES

Data such as those in the graph given in Figure 13.2 are taken at successive time points and will generally be recorded as a single value at any one time point. That is, only one sales value is available at the end of each quarter.

TIME SERIES   A **time series** is a sequence of observations of some quantity obtained at successive points in time.

THE MULTIPLICATIVE TIME SERIES MODEL

A traditional approach to representing time series data uses a **multiplicative time series model.** This model assumes that the product of four terms, *secular trend*, *seasonal index*, *long-term cycle*, and *irregular fluctuation*, can be used to represent the time series data.

Secular trend is the tendency for a time series to exhibit a generally increasing or decreasing pattern, such as the tendency of the Dr Pepper sales data to increase over the four-year period. Secular trend is evident in some

| | Three Months Ending | | | |
Year	March 31	June 30	Sept. 30	Dec. 31
1976	41,234	50,225	54,462	41,295
1977	44,555	59,893	68,958	53,344
1978	54,684	74,238	79,430	62,656
1979	62,691	79,865	83,637	65,569

FIGURE 13.1
QUARTERLY SALES IN THOUSANDS OF DOLLARS FOR THE DR PEPPER COMPANY.

FIGURE 13.2
QUARTERLY SALES FOR THE DR PEPPER COMPANY.

time series such as the Consumer Price Index and population of the United States, but it is not present in other time series, such as amount of rainfall or marriage rates per 100,000 people.

The **seasonal index** reflects the tendency for a time series to follow a seasonal pattern, such as in ice cream sales, amount of rainfall, or minimum daily temperature, all of which tend to increase or decrease in a predictable manner during the various times of the year. The seasonal index is not important in other time series such as the retail price of gasoline or the interest rate on corporate bonds. Many economic data, such as total unemployment, are adjusted for the seasonal effect (deseasonalized) before they are reported.

The **long-term cycle** term represents variations that are due to general conditions that exist for longer than one or two years. In economic and business data the long-term cycle represents the effect that a general period of prosperity or recession has on data such as sales, unemployment, and prices. The long-term cycle in temperature or rainfall time series data represents extended drought conditions or long-term changes in the jet stream pattern, but it may be absent in other time series, such as the retail sales of gasoline or the per capita consumption of milk.

If Y_i represents the observed value of some quantity for the ith time period in a time series, then the **multiplicative time series model** is given by

$$Y_i = T_i \times S_i \times C_i \times I_i$$

MULTIPLICATIVE TIME SERIES MODEL where

T_i is the secular trend for the ith time period
S_i is the seasonal index for the ith time period
C_i is the long-term cycle for the ith time period
I_i is the irregular fluctuation for the ith time period

The **irregular fluctuation** represents any variation not accounted for by the other terms. It is present in nearly every time series, since a model will seldom, if ever, perfectly describe all of the observations in a time series.

The multiplicative time series model simply assumes the time series can be represented as the product of the four components T_i, S_i, C_i, and I_i. Methods are given in this chapter for decomposing the time series into these four components. The secular trend is identified in this section using regression techniques. In the next section averaging methods are used to identify the seasonal indexes for the data. The cyclical and irregular components are addressed in the third section of this chapter. The following example illustrates each of those components.

EXAMPLE

The fifth quarter of data in Problem Setting 13.1 is the sales figure for the first quarter 1977 for the Dr Pepper Company. This number is

$$Y_5 = \$44,555$$

In this section the secular trend component for that quarter is found to be

$$T_5 = \$53,259$$

This is the predicted value of sales based on linear regression, if the seasonal, cyclical, and irregular factors affecting the sales are ignored.

In the next section the seasonal index is found to be

$$S_5 = .882$$

for the first quarter of each year. This value means that first-quarter sales tend to be only 88.2% as much as predicted by the secular trend. In other words, the first quarter tends to have 12% less sales, just because it is the first quarter. This is attributable in this case to less demand for soft drinks during the winter months January, February, and March.

The long-term cycle is found in Section 13.3 as

$$C_5 = .955$$

for the fifth quarter. The long-term cycle reflects general economic conditions existing at that time. In particular the decrease of 4.5%, $(1.000 - .955) \times 100\% = 4.5\%$, from the sales that would otherwise have been predicted indicates that this quarter occurs during a period of slower-than-usual economic activity.

The irregular fluctuation for quarter number five

$$I_5 = .993$$

is also derived in Section 13.3. It indicates that less than 1%, $(1.000 -$

.993) × 100% = .7% of the sales figure is attributable to factors other than secular trend, seasonal index, and long-term cycle. The actual sales figure Y_5 is reconstructed from its four components by multiplication.

$$Y_5 = T_5 \times S_5 \times C_5 \times I_5$$
$$= (\$53,259)(.882)(.955)(.993)$$
$$= \$44,547$$

which is equal to the actual sales figure $44,555 except for roundoff error caused by carrying only three decimal places in three of the factors.

IDENTIFYING THE SECULAR TREND IN TIME SERIES ANALYSIS

Perhaps the most easily identified portion of the multiplicative time series model is the **long-term trend** that is frequently apparent in time series data. Long-term trend may be a result of prolonged growth in a company, a gradual change in market behavior, or simply continually higher prices due to inflation. Trend, or secular trend as it is commonly called, can be measured by regression techniques. Figure 13.3 shows an increasing long-term trend for the Gross National Product.

SECULAR The **secular trend** in a time series describes the tendency for the time series
TREND to grow or decline over a long period of time.

All time series data may not graph as smoothly as the GNP in Figure 13.3. This is particularly true if the data are obtained at points in time that allow seasonal influences to be present, such as data obtained on sales on a daily,

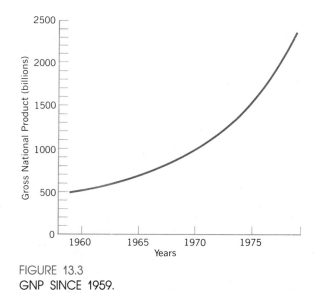

FIGURE 13.3
GNP SINCE 1959.

weekly, monthly, or quarterly basis. In Figure 13.3 the GNP is represented by a single point for each year resulting in the smooth curve of Figure 13.3. In Figure 13.2 the graph of the sales time series is not smooth because it consists of quarterly data that exhibit seasonal influences.

USING LINEAR REGRESSION TO DESCRIBE SECULAR TREND

Examination of the time series plot given in Figure 13.2 shows a long-term sales growth for the Dr Pepper Company. One way to describe this growth is by the methods of linear regression. The least squares regression equation for the Dr Pepper sales data, based on the methods of Chapter 12, is given as

$$\hat{Y} = 42134.1 + 2224.9X \tag{13.1}$$

where X is the number of the quarter for which a prediction is desired. The graph of this regression line has been added to the time series plot in Figure 13.4. The actual values for $\hat{Y}$ and the residuals $Y - \hat{Y}$ are given in Figure 13.9 in the next section. The coefficient of determination for these data is $r^2 = .605$ so that 60.5% of the variation in quarterly sales is accounted for by the linear regression equation. Clearly there is significant variation around the least squares line left unexplained by the secular trend. The other terms in the multiplicative time series model are used in later sections of this chapter to explain the remaining variation.

Linear regression is only one of several methods available for describing the secular trend in a time series. Other methods may fit curves rather than straight lines to the data or they may give more weight to recent observations rather than to earlier ones to reflect more accurately the recent trends in the observations. These are more advanced methods in time series and are not

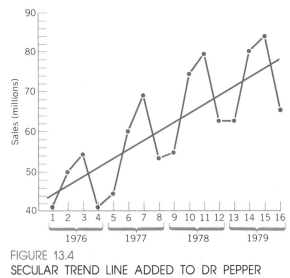

FIGURE 13.4
SECULAR TREND LINE ADDED TO DR PEPPER COMPANY SALES.

covered in this brief introduction. However, one of the simpler alternatives to linear regression is now described.

USING EXPONENTIAL GROWTH CURVES TO DESCRIBE SECULAR TREND

If you invested $1000 in a savings account at a 5% rate of interest compounded annually, at the end of the first year the account would contain $1050.

$$\$1000(1.05) = \$1050$$

At the end of the second year the account would contain $1102.50.

$$\$1102.50 = \$1050(1.05) = \$1000(1.05)^2$$

At the end of n years the account would contain $\$1000(1.05)^n$. This is an example of an **exponential growth curve.**

In more general terms, if an amount y is increased by a constant percentage each year, that amount grows according to an exponential growth curve. If the initial amount is denoted by $y_0 = a$, and the growth rate in each time period is denoted by r (corresponding to an $r \times 100\%$ increase), then the amount increases to

$$y_n = a(1 + r)^n \qquad (13.2)$$

after n time periods have elapsed.

The difference between an exponential growth curve and a linear growth curve, as previously described, is that successive terms in an exponential growth curve are obtained by multiplying the previous term by a constant, whereas successive terms in a linear growth curve are obtained by adding a constant amount to the previous term.

FITTING AN EXPONENTIAL GROWTH CURVE

The secular trend in a time series may be represented by an exponential growth curve in cases where the long-term trend in a time series grows (except for some unexplained variation) by a fixed percentage in each time period. This is a more realistic representation of many business and economic situations, such as in the savings account example just given, or in the GNP time series given in Figure 13.3. One method of fitting an exponential growth curve, such as that given in Eq. 13.2, to a set of data involves using the least squares method on the logarithms of the data. That is, taking the logarithms of both sides of Eq. 13.2 gives

$$\log y_n = \log a + n \log(1 + r) \qquad (13.3)$$

Note that this is now a linear equation of the form

$$y^* = a^* + b^*x \qquad (13.4)$$

where the variables are

$$y^* = \log y_n \tag{13.5}$$

and

$$x = n \tag{13.6}$$

The two coefficients in the equation of the straight line are given by

$$a^* = \log a \tag{13.7}$$

and

$$b^* = \log(1 + r) \tag{13.8}$$

Although any base may be used for the logarithms, the base 10 is used here for illustration. The least squares method of Chapter 12 is used to obtain a^* and b^*. The y intercept a of the exponential growth curve is estimated by solving Eq. 13.7 to get

$$\hat{a} = 10^{a^*} \tag{13.9}$$

The estimated rate of growth r is obtained from b^* by solving Eq. 13.8 for r to get

$$\hat{r} = 10^{b^*} - 1 \tag{13.10}$$

The estimated trend value $\hat{Y}$ is given by Eq. 13.2 as

$$\hat{Y}_n = \hat{a}(1 + \hat{r})^n$$

The next example provides an illustration of fitting an exponential secular trend.

EXAMPLE

The data for the Gross National Product are given in Figure 13.5, along with the log (base 10) of the data, the year number, and some intermediate calculations required for finding the equation of the least squares line that fits the logs of the data, as indicated in Eq. 13.3. The result, using more decimal places than appear in Figure 13.5, is the least squares line

$$\hat{Y}^* = 2.608 + .03443X \tag{13.11}$$

which is graphed in Figure 13.6.

    The y intercept for the exponential growth curve is estimated from Eq. 13.9 as

$$\hat{a} = 10^{2.608} = 405.51 \qquad \text{(billions of dollars)}$$

Year Number X		GNP (Billions of Dollars) Y	Log GNP Y*	X^2	XY*
1959 =	1	486.5	2.69	1	2.69
1960 =	2	506.0	2.70	4	5.41
1961 =	3	523.3	2.72	9	8.16
1962 =	4	563.8	2.75	16	11.00
1963 =	5	594.7	2.77	25	13.87
1964 =	6	635.7	2.80	36	16.82
1965 =	7	688.1	2.84	49	19.86
1966 =	8	753.0	2.88	64	23.01
1967 =	9	796.3	2.90	81	26.11
1968 =	10	868.5	2.94	100	29.39
1969 =	11	935.5	2.97	121	32.68
1970 =	12	982.4	2.99	144	35.91
1971 =	13	1063.4	3.03	169	39.35
1972 =	14	1171 1	3.07	196	42.96
1973 =	15	1306.3	3.12	225	46.74
1974 =	16	1412.9	3.15	256	50.40
1975 =	17	1528.8	3.18	289	54.13
1976 =	18	1702.2	3.23	324	58.16
1977 =	19	1899.5	3.28	361	62.29
1978 =	20	2127.6	3.33	400	66.56
1979 =	21	2368.5	3.37	441	70.86
Totals	231		62.80	3311	717.14

FIGURE 13.5
WORKSHEET FOR FITTING AN EXPONENTIAL GROWTH CURVE TO
THE GROSS NATIONAL PRODUCT.

and the growth rate is estimated from Eq. 13.10 as

$$\hat{r} = 10^{.03443} - 1 = .0825 \qquad (8.25\% \text{ annual growth})$$

This latter figure, indicating an average growth rate of over 8% per year, is usually of more interest than the y intercept. The coefficient of determination for this least squares fit is .99, indicating that 99% of the variation of the logs of the data is accounted for by the least squares line. The estimates of GNP are

$$\hat{Y} = 10^{\hat{Y}*} = 10^{2.608 + .03443X} = 405.51(1.0825)^X$$

ADJUSTING A TIME SERIES WITH INDEX NUMBERS

An alternative to fitting an exponential growth curve is first to adjust the time series by dividing each observation in the time series by an index number for that same time period. For example, it is intuitively appealing to adjust annual

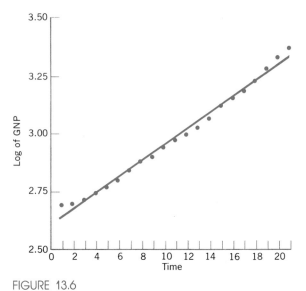

FIGURE 13.6
A GRAPH OF THE LOGARITHMS (BASE 10) OF THE
GROSS NATIONAL PRODUCT, AND THE LEAST
SQUARES LINE.

X	Y	CPI (1967 = 100)	Y/CPI
1	486.5	87.3	5.57
2	506.0	88.7	5.70
3	523.3	89.6	5.84
4	563.8	90.6	6.22
5	594.7	91.7	6.49
6	635.7	92.9	6.84
7	688.1	94.5	7.28
8	753.0	97.2	7.75
9	796.3	100.0	7.96
10	868.5	104.2	8.33
11	935.5	109.8	8.52
12	982.4	116.3	8.45
13	1063.4	121.3	8.77
14	1171.1	125.3	9.35
15	1306.3	133.1	9.81
16	1412.9	147.7	9.57
17	1528.8	161.2	9.48
18	1702.2	170.5	9.98
19	1899.5	181.5	10.47
20	2127.6	195.4	10.89
21	2368.5	217.4	10.89

FIGURE 13.7
THE GNP Y ADJUSTED FOR INFLATION
BY DIVIDING BY THE CPI.

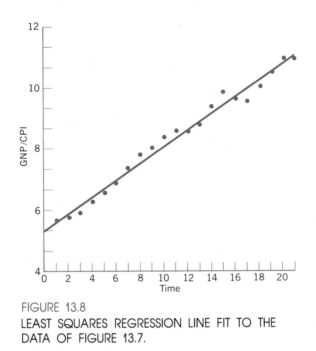

FIGURE 13.8
LEAST SQUARES REGRESSION LINE FIT TO THE
DATA OF FIGURE 13.7.

cost data, such as the Gross National Product, for inflation by dividing by some appropriate measure of inflation, such as the Consumer Price Index. The result of this division is shown in Figure 13.7. Next a straight line is fit to the adjusted figures. For the adjusted values in Figure 13.7 the following least squares regression line is obtained.

$$\hat{Y} = 5.272 + .275X \qquad (13.12)$$

where $\hat{Y}$ in this case is the predicted value of GNP/CPI. The least squares computational details are omitted. The coefficient of determination is .983, almost as high as the .99 figure obtained by working with the logarithms of the GNP and ignoring the CPI.

An examination of the graph of GNP/CPI over time, given in Figure 13.8, indicates that a linear growth curve provides a very adequate representation for these adjusted data.

EXERCISES

13.1 Quarterly net earnings in thousands of dollars are given below for the Butler Manufacturing Company for the years 1976 to 1979. Make a scatterplot for these time series data and connect the consecutive points in the scatterplot.

	Three Months Ending			
Year	March 31	June 30	Sept. 30	Dec. 31
1976	2464	4668	4692	2955
1977	2870	5980	6537	3222

(continued)

	Three Months Ending			
Year	March 31	June 30	Sept. 30	Dec. 31
1978	3552	8569	7331	3997
1979	4322	7497	8768	3464

13.2 Find the secular trend for the time series data in Exercise 13.1 by fitting a least squares regression line to the data. Add the plot of the regression line to the scatterplot of Exercise 13.1.

13.3 Quarterly revenues in millions of dollars are reported below for CBS for the years 1977 to 1979. Make a scatterplot for these time series data and connect the consecutive points in the scatterplot.

	Three Months Ending			
Year	March 31	June 30	Sept. 30	Dec. 31
1977	643.8	665.5	669.8	847.2
1978	744.2	751.4	807.4	987.1
1979	828.7	908.7	882.7	1109.6

13.4 Find the secular trend for the time series data in Exercise 13.3 by fitting a least squares regression line to the data. Add the plot of the regression line to the scatterplot of Exercise 13.3. (Let $X = 1, 2, \ldots, 12$ be the quarter number. Then $\Sigma X = 78$, $\Sigma Y = 9846.1$, $\Sigma X^2 = 650$, $\Sigma Y^2 = 8{,}291{,}296.17$, and $\Sigma XY = 68{,}721.2$.)

13.5 Net income in millions of dollars are reported below for CBS for the years 1976 to 1979. Make a scatterplot for these time series data and connect the consecutive points in the scatterplot.

	Three Months Ending			
Year	March 31	June 30	Sept. 30	Dec. 31
1976	27.8	47.5	40.8	47.9
1977	33.0	54.9	43.7	50.4
1978	33.8	59.3	48.5	56.5
1979	17.8	65.8	53.1	64.0

13.6 Find the secular trend for the time series data in Exercise 13.5 by fitting a least squares regression line to the data. Add the plot of the regression line to the scatterplot of Exercise 13.5.

13.2

SEASONAL INDEXES IN TIME SERIES ANALYSIS

SERIAL CORRELATION OF TIME SERIES RESIDUALS

If the least squares regression Eq. 13.1 were used to predict sales for the Dr Pepper Company, the predictions for any one quarter would correspond to

Quarter	Observed Y (Dollars)	Predicted Ŷ (Dollars)	Residual Y − Ŷ (Dollars)
1	41,234	44,359	− 3,125
2	50,225	46,584	3,641
3	54,462	48,809	5,653
4	41,295	51,034	− 9,739
5	44,555	53,259	− 8,704
6	59,893	55,484	4,409
7	68,958	57,709	11,249
8	53,344	59,934	− 6,590
9	54,684	62,158	− 7,474
10	74,238	64,383	9,855
11	79,430	66,608	12,822
12	62,656	68,833	− 6,177
13	62,691	71,058	− 8,367
14	79,865	73,283	6,582
15	83,637	75,508	8,129
16	65,569	77,733	− 12,164

FIGURE 13.9

RESIDUALS FROM EQ. 13.1.

points on the regression line of Figure 13.4. Note that such predictions are always too high in the first and fourth quarters of each year and are always too low in the second and third quarters of each year. This is also apparent from an examination of the residuals from Eq. 13.1 that are given in Figure 13.9 and graphed in Figure 13.10.

Another way of summarizing this result is to note that the residuals from the least squares predictions are **serially correlated.** If pairs of successive residuals are formed from the residuals in Figure 13.9, a correlation coefficient can be computed on these pairs and used to provide information for adjusting the secular trend prediction, and thus improving sales predictions.

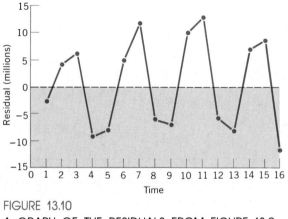

FIGURE 13.10

A GRAPH OF THE RESIDUALS FROM FIGURE 13.9.

The pairing of residuals can be done in several ways. For example, residuals 1 and 2, 2 and 3, 3 and 4, 4 and 5, etc., could be paired. Such a pairing would result in a **serial correlation of lag 1. A serial correlation of lag 2** is obtained as a result of pairing residuals 1 and 3, 2 and 4, 3 and 5, 4 and 6, etc. A **serial correlation of lag 3** results from pairing residuals 1 and 4, 2 and 5, 3 and 6, 4 and 7, etc.

SERIAL
CORRELATION
COEFFICIENT
(LAG k)

The **serial correlation coefficient of lag k** for a sequence of observations W_1, W_2, . . . , W_n measures the correlation of each observation in the series with the kth observation following it in the series. The equation is given by

$$r_k = \frac{\sum\limits_{i=1}^{n-k} W_i W_{i+k} - \left(\sum\limits_{i=1}^{n-k} W_i\right)\left(\sum\limits_{i=k+1}^{n} W_i\right) \Big/ (n-k)}{\left\{\sum\limits_{i=1}^{n-k} W_i^2 - \left(\sum\limits_{i=1}^{n-k} W_i\right)^2 \Big/ (n-k)\right\}^{1/2} \left\{\sum\limits_{i=k+1}^{n} W_i^2 - \left(\sum\limits_{i=k+1}^{n} W_i\right)^2 \Big/ (n-k)\right\}^{1/2}}$$

Figure 13.11 contains the pairing of residuals used for computing serial correlations up through lag 4. The serial correlation coefficients also appear in Figure 13.11. Note that the serial correlation of lag 0 is always equal to 1.0.

The positive serial correlation of .902 for a lag of 4 indicates a strong relationship for sales behavior between corresponding quarters from year to

Lag 1		Lag 2		Lag 3		Lag 4	
− 3125,	3641	− 3125,	5653	− 3125,	− 9739	− 3125,	− 8704
3641,	5653	3641,	− 9739	3641,	− 8704	3641,	4409
5653,	− 9739	5653,	− 8704	5653,	4409	5653,	11249
− 9739,	− 8704	− 9739,	4409	− 9739,	11249	− 9739,	− 6590
− 8704,	4409	− 8704,	11249	− 8704,	− 6590	− 8704,	− 7474
4409,	11249	4409,	− 6590	4409,	− 7474	4409,	9855
11249,	− 6590	11249,	− 7474	11249,	9855	11249,	12822
− 6590,	− 7474	− 6590,	9855	− 6590,	12822	− 6590,	− 6177
− 7474,	9855	− 7474,	12822	− 7474,	− 6177	− 7474,	− 8367
9855,	12822	9855,	− 6177	9855,	− 8367	9855,	6582
12822,	− 6177	12822,	− 8367	12822,	6582	12822,	8129
− 6177,	− 8367	− 6177,	6582	− 6177,	8129	− 6177,	− 12164
− 8367,	6582	− 8367,	8129	− 8367,	− 12164		
6582,	8129	6582,	− 12164				
8129,	− 12164						
$r_1 = -.052$		$r_2 = -.897$		$r_3 = .068$		$r_4 = .902$	

FIGURE 13.11
RESIDUAL PAIRINGS FOR CALCULATION OF SERIAL CORRELATIONS FOR PREDICTIONS OF DR PEPPER SALES.

year. That is, if sales are down in one quarter of one year, they will tend to be down in the corresponding quarter every year, or if sales are up in the first quarter of one year, they will tend to be up in the first quarter of every year. This behavior is noticeable in Figure 13.11 under the heading of *lag 4* where the members of every pair are observed to have the same sign, and it is also noticeable by a close examination of the graph of the residuals given in Figure 13.10.

Note that the regression methods of Chapter 12 rely on mutual independence of the residuals, whereas in time series data the residuals usually show a serial correlation, which indicates a lack of independence. This is the principal reason that inference methods for standard regression do not apply to time series data—methods such as a confidence interval for the slope, a confidence interval for the mean of Y, or predicted values for Y.

CORRELOGRAMS

The serial correlations in Figure 13.11 can be represented graphically in a **correlogram.** A correlogram is made by letting the horizontal axis represent the length of the lag used in calculating the serial correlation. The vertical axis represents the value of the serial correlation from −1 to 1. The points in the graph represent the serial correlation calculated for the corresponding lag on the horizontal axis. The correlogram shown in Figure 13.12 presents all of the serial correlations through lag 8. The large positive correlations for lags of 4 and 8 in the correlogram make it clear that the Dr Pepper Company sales are highly correlated for corresponding quarters from year to year.

Not all time series are as simplistically represented as are the sales for the Dr Pepper Company. For large time-series data sets where a large number of lags must be considered, the correlogram becomes a very useful tool in revealing information about the time series that might not be as obvious as in

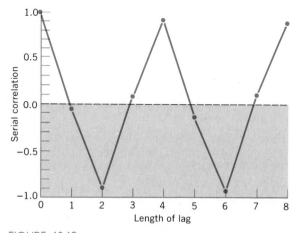

FIGURE 13.12
CORRELOGRAM FOR DR PEPPER COMPANY TIME SERIES DATA.

the Dr Pepper Company example. For larger data sets a computer should be used to produce the correlogram.

IDENTIFYING THE SEASONAL VARIATION IN A TIME SERIES

The correlogram presented in Figure 13.12 for the Dr Pepper Company time series data given in Figure 13.1 strongly identifies the presence of correlation among sales for corresponding quarters from year to year, that is, $r_4 = .902$. This suggests that the sales predictions based on the least squares trend line would be improved if an adjustment were made based on corresponding quarters of previous years. For example, if sales in the third quarter are always higher than the regression line, then predictions for the third quarter should be adjusted upward. Such adjustments can be made on a daily, weekly, monthly, or quarterly basis depending on how the time series data are recorded.

SEASONAL VARIATION **Seasonal variations** are generally short-term effects that occur on a regular basis, such as during the seasons of the year for most business data.

SEASONAL INDEXES

Seasonal variation is measured by a seasonal index, which is a number that reflects the tendency of the time series data during that time period (usually monthly or quarterly) to be higher or lower than for other time periods of equal length throughout the year. For time series data reported on a quarterly basis there will be four such seasonal indexes, whereas monthly data would have 12 indexes. Seasonal indexes are dimensionless quantities that center around 1.0. If long-term trend predictions from the least squares regression equation are seasonally adjusted by multiplying by the seasonal index, that is

$$\text{Sales forecast} = (\text{secular trend forecast}) \times (\text{seasonal index})$$

the forecasts will tend to be improved.

ESTIMATING THE SEASONAL INDEXES IN A TIME SERIES

The trend line is an idealized representation of the time series. However, for the sales data of the Dr Pepper Company in the previous section, the residuals in Figure 13.9 clearly show that the first-quarter sales are always below the least squares line. Therefore, it would be useful to have a seasonal index for the first quarter that could be used to reduce the secular trend forecast. Based on the first-quarter values in Figure 13.9 a first-quarter seasonal index may be developed by computing the ratio of observed sales to predicted sales for the first quarter of each of the four years as follows.

	1976	1977	1978	1979
Observed sales (dollars)	41,234	44,555	54,684	62,691
Predicted sales (dollars)	44,359	53,259	62,158	71,058
Ratio (observed/predicted)	.92955	.83658	.87975	.88225

Each of these ratios may be influenced to some degree by cyclical (C) and irregular (I) fluctuations. The influence of these components may be minimized somewhat by using the average of the four ratios to represent the seasonal index for the first quarter:

$$S_1 = \frac{(.930 + .837 + .880 + .882)}{4} = .882$$

The indexes for the other quarters are found in the same manner for the Dr Pepper Company sales data. Figure 13.13 contains the necessary ratios for computing all four seasonal indexes. Note that the median of these ratios for any one quarter may be preferred to the average if there are outliers present. However, the average is used in this chapter for illustration.

THE RATIO TO TREND METHOD FOR FINDING SEASONAL INDEXES

Let

Y_{ij} = the observed time series quantity for period j in year i

T_{ij} = the secular trend estimate for period j in year i

Then the **seasonal index** S_j for period j, using the **ratio to trend method** is

$$S_j = \frac{1}{n} \sum_{i=1}^{n} Y_{ij}/T_{ij}$$

where n is the number of years of observations.

The averages given at the bottom of Figure 13.13 are very nearly the values that could be used to represent the seasonal indexes. Only one additional step is needed. Since seasonal indexes are dimensionless quantities that center about 1.0, the sum of the four quarterly indexes should be $4 \times 1.0 = 4$. The sum of the averages in Figure 13.13 is $.882 + 1.100 + 1.153 + .863 = 3.998$, which is nearly equal to 4. The sum of the averages can easily be made to total 4 by multiplying each average by $4/3.998 = 1.0005$. The results, when all calculations are carried out to five decimal places, are given as follows:

$$S_1 = .88244 \qquad S_2 = 1.10064 \qquad S_3 = 1.15326 \qquad S_4 = .86365$$

RELATIONSHIP OF THE COMPUTATIONAL PROCEDURE TO THE MULTIPLICATIVE MODEL

The multiplicative model implies that the individual observations in a time series are represented as $Y_i = T_i \times S_i \times C_i \times I_i$. However, the secular trend line represents only the component $\hat{Y}_i = T_i$. Thus, the ratios computed in Figure 13.13 can each be expressed as the following ratio.

$$\frac{Y_i}{\hat{Y}_i} = \frac{T_i \times S_i \times C_i \times I_i}{T_i} = S_i \times C_i \times I_i$$

Quarter	Y (Dollars)	Ŷ (Dollars)	Y/Ŷ			
			First	Second	Third	Fourth
1	41,234	44,359	0.930			
2	50,225	46,584		1.078		
3	54,462	48,809			1.116	
4	41,295	51,034				0.809
5	44,555	53,259	0.837			
6	59,893	55,484		1.079		
7	68,958	57,709			1.195	
8	53,344	59,934				0.890
9	54,684	62,158	0.880			
10	74,238	64,383		1.153		
11	79,430	66,608			1.192	
12	62,656	68,883				0.910
13	62,691	71,058	0.882			
14	79,865	73,283		1.090		
15	83,637	75,508			1.108	
16	65,569	77,733				0.844
		Totals	3.529	4.400	4.611	3.453
		Averages	0.882	1.100	1.153	0.863

FIGURE 13.13
COMPUTATION OF SEASONAL INDEXES.

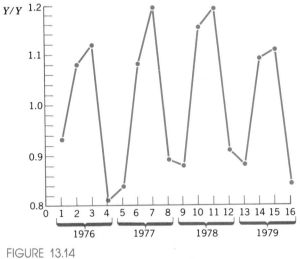

FIGURE 13.14
SCATTERPLOT OF THE RATIOS $Y/\hat{Y}$.

The term T_i divides out of this ratio meaning that the ratios given in Figure 13.13 represent $S_i \times C_i \times I_i$. These ratios are plotted in Figure 13.14.

If the scatterplot of Figure 13.14 is compared with the scatterplot of Figure 13.4, it becomes clear that the division $Y_i/\hat{Y}_i$ has removed the secular trend that was originally present in the data. Hence, the ratios in Figure 13.14 represent the product of the seasonal indexes, the long-term cyclic effects, and the irregular fluctuations in the time series. The averaging of these ratios for corresponding quarters as illustrated in Figure 13.13 tends to smooth out or eliminate the effects of both the irregular and long-term cycle components, as high values are offset by low values. Averaging this way in time series is referred to as **smoothing the data.**

The procedure presented in this section for finding the seasonal index can be used equally well to find a monthly index if the data are reported monthly. The only modification needed in Figure 13.13 would be to use 12 columns of ratios to represent each month rather than using the 4 columns used with quarterly data.

USING THE SEASONAL INDEX FOR FORECASTING

The two terms identified thus far, T_i and S_i, are the principal ones used for forecasting. For example, if a forecast were desired for Dr Pepper Company sales for the first quarter of 1980, the forecast would be made by first finding the long-term trend as

$$T_{17} = a + bx = \$42134.1 + \$2224.9(17) = \$79,958$$

and then multiplying by the seasonal index for the first quarter

$$\hat{Y}_{17} = T_{17} \times S_1 = \$79,958(.88244) = \$70,558$$

This compares with the actual sales for this quarter of $67,105 as reported by the Dr Pepper Company.

DESEASONALIZING THE DATA

Frequently economic and business data are adjusted to remove the seasonal effect. This is accomplished quite simply by dividing the original time series observation by the seasonal index. For example, the actual sales for the first quarter of 1980 are $67,105 for the Dr Pepper Company. This is deseasonalized by dividing by the first quarter seasonal index $S_1 = .88244$ to get $76,045 as the seasonally adjusted sales figure. This may be compared with the previous quarter's sales figure $65,569 after a seasonal adjustment, by dividing it by the fourth-quarter seasonal index $S_4 = .86365$, which gives $75,921. When both quarters are seasonally adjusted, the sales are seen to reflect only a slight increase from one quarter to the next.

Exercises 13.7 through 13.10 refer to the time series data based on quarterly net earnings for the Butler Manufacturing Company as given in Exercise 13.1.

13.7 Make a correlogram for the time series data of Exercise 13.1 using lags of 1, 2, 3, and 4 for the serial correlation coefficient computed on the secular trend residuals $Y_i - \hat{Y}_i$.

13.8 Compute the seasonal indexes for the quarterly net earnings of the Butler Manufacturing Company.

13.9 Make a scatterplot of the ratio $Y_i/\hat{Y}_i$ similar to Figure 13.14 for the quarterly net earnings of the Butler Manufacturing Company.

13.10 Predict the net earnings for the Butler Manufacturing Company for the first quarter of 1980 using the long-term secular trend and the seasonal index. The actual earnings for this quarter were $4790. How does your answer compare with the actual earnings?

Exercises 13.11 through 13.14 refer to the time series data based on quarterly revenues for CBS as given in Exercise 13.3.

13.11 Make a correlogram for the time series data of Exercise 13.3 using lags of 1, 2, 3, and 4 for the serial correlation coefficient computed on the secular trend residuals $Y_i - \hat{Y}_i$. ($r_1 = -.415, r_2 = -.186, r_3 = -.382$) Find r_4 given the residuals

1977:	4.890	−6.428	−35.146	109.236
1978:	−26.782	−52.599	−29.617	117.065
1979:	−74.353	−27.371	−86.389	107.494

13.12 Compute the seasonal indexes for quarterly revenues of CBS, given that the ratio to trend is as follows.

1977:	1.008	.990	.950	1.148
1978:	.965	.935	.965	1.135
1979:	.918	.971	.911	1.107

13.13 Make a scatterplot of the ratio $Y_i/\hat{Y}_i$ similar to Figure 13.14 for the quarterly revenues of CBS.

13.14 Predict the revenue for CBS for the first quarter of 1980 using the long-term secular trend and the seasonal index. The actual revenue for this quarter was $968.9. How does your answer compare with the actual revenue?

Exercises 13.15 through 13.18 refer to the time series data based on net income for CBS as given in Exercise 13.5.

13.15 Make a correlogram for the time series of Exercise 13.5 using lags of 1, 2, 3 and 4 for the serial correlation coefficient computed on the secular trend residuals $Y_i - \hat{Y}_i$.

13.16 Compute the seasonal indexes for the quarterly net revenues of CBS.

13.17 Make a scatterplot of the ratio $Y_i/\hat{Y}_i$ similar to Figure 13.14 for the quarterly net revenues of CBS.

13.18 Predict the net revenue for CBS for the first quarter of 1980 using the long-term secular trend and the seasonal index. The actual net revenue for this quarter was $13.1. How does your answer compare with the actual net revenue?

13.3

THE CYCLICAL AND IRREGULAR COMPONENTS IN A TIME SERIES

USE OF THE MULTIPLICATIVE MODEL

The cyclical and irregular components of a time series can be isolated through use of the multiplicative model, as was done in the previous section to obtain the seasonal index. The previous two sections have identified two components of the time series, T_i and S_i. Since the time series is represented as a multiplicative model, division by the term $T_i \times S_i$ isolates the term $C_i \times I_i$:

$$\frac{Y_i}{T_i \times S_i} = \frac{T_i \times S_i \times C_i \times I_i}{T_i \times S_i} = C_i \times I_i$$

Figure 13.15 contains the results of this division for the Dr Pepper Company sales data given in Figure 13.9.

The values $C_i \times I_i$ from Figure 13.15 are plotted in Figure 13.16. Comparison of Figure 13.16 with Figure 13.2 shows that both the long-term trend

Quarter	(1) Y (Dollars)	(2) $\hat{Y} = T$ (Dollars)	(3) S	(4) $T \times S$ (Dollars)	(1) ÷ (4) $C \times I$
1	41,234	44,359	0.88244	39,144	1.05338
2	50,225	46,584	1.10064	51,272	0.97958
3	54,462	48,809	1.15326	56,290	0.96753
4	41,295	51,034	0.86365	44,076	0.93692
5	44,555	53,259	0.88244	46,998	0.94802
6	59,893	55,484	1.10064	61,068	0.98077
7	68,958	57,709	1.15326	66,553	1.03613
8	53,344	59,934	0.86365	51,762	1.03057
9	54,684	62,158	0.88244	54,851	0.99695
10	74,238	64,383	1.10064	70,863	1.04763
11	79,430	66,608	1.15326	76,817	1.03402
12	62,656	68,833	0.86365	59,448	1.05396
13	62,691	71,058	0.88244	62,705	0.99978
14	79,865	73,283	1.10064	80,658	0.99016
15	83,637	75,508	1.15326	87,081	0.96045
16	65,569	77,733	0.86365	67,134	0.97669

FIGURE 13.15
FINDING THE TERM $C \times I$ IN THE MULTIPLICATIVE MODEL.

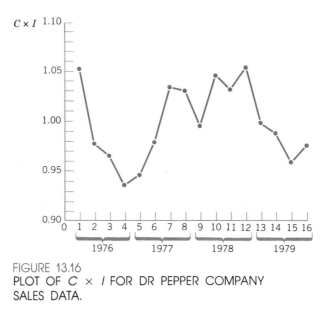

FIGURE 13.16
PLOT OF $C \times I$ FOR DR PEPPER COMPANY
SALES DATA.

and seasonal effect have been eliminated. What remains in Figure 13.16 is
the presence of a long-term cyclical effect that bottoms out at the end of 1976
and reaches a peak toward the end of 1978. However, the curve is somewhat
jagged in appearance. This is because it still contains the irregular component
I_i. The irregular component will now be separated from the time series by use
of a smoothing technique called a **moving average.**

COMPUTING THE MOVING AVERAGE

The moving average is a commonly used tool in time series analysis for smoothing
out short-term fluctuations in the time series. A moving average is obtained
by computing arithmetic averages over successive observations in the time
series. The moving average will always have a **length** associated with it, which
represents the number of terms being averaged. For example, suppose a time
series consists of 10 observations $Y_1, Y_2, \ldots, Y_{10}$ and it is desired to compute
a **moving average of length 3** on these observations. The following averages
would be found.

First average: $(Y_1 + Y_2 + Y_3)/3$
Second average: $(Y_2 + Y_3 + Y_4)/3$
Third average: $(Y_3 + Y_4 + Y_5)/3$
Fourth average: $(Y_4 + Y_5 + Y_6)/3$
Fifth average: $(Y_5 + Y_6 + Y_7)/3$
Sixth average: $(Y_6 + Y_7 + Y_8)/3$
Seventh average: $(Y_7 + Y_8 + Y_9)/3$
Eighth average: $(Y_8 + Y_9 + Y_{10})/3$

The first average is computed on the first three observations in the series. The second average is computed by dropping off the first element of the series and including the fourth element in its place. Likewise the third average drops off the second element and picks up the fifth observation. The average continues to use three observations in this manner until the end of the series is reached.

THE MOVING
AVERAGE

A **moving average** is a technique used to smooth out short-term fluctuations in a time series. A moving average of length k is obtained by computing repeated averages on k successive observations in the time series starting with the first k observations in the time series and continuing until the last moving average is computed on the last k observations in the time series. If a time series consists of n observations $Y_1, Y_2, \ldots, Y_n$ the following averages would be computed.

First computation: $(Y_1 + Y_2 + \cdots + Y_k)/k$

Second computation: $(Y_2 + Y_3 + \cdots + Y_{k+1})/k$ $\left.\begin{array}{l} \\ \\ \\ \\ \end{array}\right\}$ $n - k + 1$ computations

$\cdots$

Last computation: $(Y_{n-k+1} + Y_{n-k+2} + \cdots + Y_n)/k$

The moving average results in a smoothing of data points involved, but the number of smoothed observations is less than the original number of observations. In general, if the length of a moving average is k and the number of observations in the time series is n, the number of averages computed will be $n - k + 1$. For the preceding example this formula gives $10 - 3 + 1 = 8$ moving averages. If a moving average of length 4 had been used, then $10 - 4 + 1 = 7$ moving averages would be computed.

OBTAINING THE CYCLICAL COMPONENT FROM $C_i \times I_i$

The moving average technique can be used to smooth out the irregular component from $C_i \times I_i$ as given in Figure 13.15. A moving average of length three will be sufficient to smooth out the irregular component. The results of these computations are given in Figure 13.17. The first moving average in Figure 13.17 is computed as $(1.053 + .980 + .968)/3 = 1.000$, whereas the second moving average is found as $(.980 + .968 + .937)/3 = .961$, and so on until all $16 - 3 + 1 = 14$ moving averages are computed. Note that the first moving average in Figure 13.17 is placed on line number two, since it represents the average of the first, second, and third quarters.

A plot of the C_i from Column (2) of Figure 13.17 is given in Figure 13.18. When Figure 13.18 is compared with Figure 13.16 it becomes clear that the irregular component has been smoothed from the series. A moving average of length 4 would have provided an even smoother curve. Note that the plot in Figure 13.18 starts with Quarter 2 and ends with Quarter 15 since two points were absorbed in calculating the moving average. Also, the cycle in the time series starts to increase in the fourth quarter and continues to do so until the eleventh quarter, after which a decrease is noted through the fifteenth quarter.

Quarter	(1) C × I	(2) Moving Average (C)	(1) ÷ (2) I
1	1.053		
2	0.980	1.000	0.979
3	0.968	0.961	1.006
4	0.937	0.951	0.985
5	0.948	0.955	0.992
6	0.981	0.988	0.992
7	1.036	1.016	1.020
8	1.031	1.021	1.009
9	0.997	1.025	0.973
10	1.048	1.026	1.021
11	1.034	1.045	0.989
12	1.054	1.029	1.024
13	1.000	1.015	0.985
14	0.990	0.983	1.007
15	0.960	0.976	0.984
16	0.977		

FIGURE 13.17
MOVING AVERAGES FOR CYCLICAL AND
IRREGULAR COMPONENTS.

This pattern was not nearly as clear in Figure 13.16 where the irregular component was still present. Many years of data are necessary to properly identify a long-term cycle.

The most recent cyclical component in a time series can be used to improve a forecast. In the previous section the trend and the seasonal index were used to predict a value for Dr Pepper sales. The trend value for the first quarter of

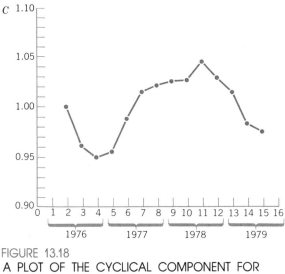

FIGURE 13.18
A PLOT OF THE CYCLICAL COMPONENT FOR
DR PEPPER SALES DATA.

1980, $x = 17$, was $Y_{17} = \$79,958$. The seasonal index for the first quarter was $S_1 = .88244$, and the product gave an estimate of sales as \$70,558.

$$\hat{Y} = T_{17} \times S_1 = \$79,958 \times .88244 = \$70,558$$

This estimate can be refined by multiplying by the most recent indicator of the long-term cycle, $C_{15} = .976$ in this case. This assumes that the long-term cycle will not change much in only two quarters.

$$\hat{Y} = T_{17} \times S_1 \times C_{15} = \$79,958 \times .88244 \times .976 = \$68,865$$

The resulting forecast, \$68,865, is closer to the sales actually observed for the first quarter of 1980, which was \$67,105.

IDENTIFYING THE IRREGULAR (RESIDUAL) COMPONENT

The irregular component of the time series represents what is left over after the T_i, S_i, and C_i components have been removed from the time series. Hence, it is frequently referred to as the **residual component.** The irregular component can be identified as follows by using the multiplicative model, since three of the components of the time series have been identified.

$$\frac{Y_i}{T_i \times S_i \times C_i} = \frac{T_i \times S_i \times C_i \times I_i}{T_i \times S_i \times C_i} = I_i$$

Or, an equivalent method is to do the following division

$$\frac{C_i \times I_i}{C_i} = I_i$$

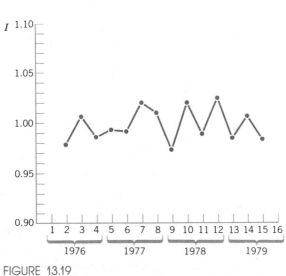

FIGURE 13.19
A PLOT OF THE IRREGULAR COMPONENT FOR DR PEPPER SALES DATA.

where the terms $C_i \times I_i$ and C_i have been identified in Figure 13.17 in Column (1) and Column (2), respectively. This division provides the irregular component and appears in Figure 13.17. A graph of the irregular component appears in Figure 13.19 and shows only random fluctuation about 1.0; that is, there is no longer any indication of trend, seasonal influence, or long-term cycle.

EXERCISES

Exercises 13.19 through 13.21 refer to the time series data based on quarterly net earnings for the Butler Manufacturing Company as given in Exercise 13.1.

13.19 Make a plot of $C_i \times I_i$ similar to Figure 13.16 for the quarterly net earnings of the Butler Manufacturing Company.

13.20 Find the cyclical component for the quarterly net earnings of the Butler Manufacturing Company. Make a plot of this cyclical component similar to the one given in Figure 13.18.

13.21 Find the irregular component for the quarterly net earnings of the Butler Manufacturing Company. Make a plot of this irregular component similar to the one given in Figure 13.19.

Exercises 13.22 through 13.24 refer to the time series data based on quarterly revenues for CBS as given in Exercise 13.3.

13.22 Make a plot of $C_i \times I_i$ similar to Figure 13.16 for the quarterly revenues of CBS.

13.23 Find the cyclical component for the quarterly revenues of CBS. Make a plot of this cyclical component similar to the one given in Figure 13.18.

13.24 Find the irregular component for the quarterly revenues of CBS. Make a plot of this irregular component similar to the one given in Figure 13.19.

Exercises 13.25 though 13.27 refer to the time series data based on net income for CBS as given in Exercise 13.5.

13.25 Make a plot of $C_i \times I_i$ similar to Figure 13.16 for the quarterly net income for CBS.

13.26 Find the cyclical component for the quarterly net income of CBS. Make a plot of this cyclical component similar to the one given in Figure 13.18.

13.27 Find the irregular component for the quarterly net income of CBS. Make a plot of this irregular component similar to the one given in Figure 13.19.

13.4

REVIEW EXERCISES

13.28 Given below are the monthly sales data in thousands of dollars for a retail store during a recent four-year period. Plot these data on a graph and discuss any information that appears from an inspection of the graph.

	J	F	M	A	M	J	J	A	S	O	N	D
1977	46	40	48	49	53	49	34	43	62	62	84	134
1978	33	33	42	53	58	55	34	48	62	70	84	144
1979	35	38	53	65	58	58	39	48	62	65	86	142
1980	33	36	39	60	56	56	38	48	63	68	87	150

13.29 Use linear regression to obtain the secular trend for the data in Exercise 13.28. Plot the results onto the plot of the data to see if the results agree with the data.

13.30 Use the data in Exercise 13.28 to obtain seasonal indexes for January and February. Plot a graph of the seasonal components. (The seasonal indexes for March through December are .784, .957, .949, .909, .600, .769, 1.020, 1.075, 1.375, 2.276, and the mean unadjusted seasonal component is 1.000.)

13.31 Use the results of Exercises 13.29 and 13.30, and a moving average of length 9, to get the first long-term cycle and irregular components of the multiplicative time series model for the data in Exercise 13.28. The remaining components are:

Long-term cycle
1.10	1.09	1.07	1.06	1.05	1.04	1.03	1.03
1.02	1.01	1.00	1.00	1.02	1.03	1.04	1.03
1.02	1.02	1.04	1.03	1.02	1.02	1.01	1.00
1.00	1.00	0.99	0.95	0.94	0.91	0.89	0.88
0.88	0.88	0.87	0.86	0.88	0.88	0.90	

Irregular
0.94	0.99	0.98	1.07	1.02	1.08	1.05	0.89	0.91
	0.95	0.98	1.07	1.03	0.95	1.03	1.00	1.08	1.00
	1.01	0.87	0.95	1.08	1.08	0.98	1.01	1.02	0.99
	0.99	1.00	1.06	1.06	0.87	0.97	0.85	1.07	1.01
	1.03	1.04	1.00						

Graph the long-term cycle component and interpret the results. Also graph the irregular components and see if these values appear to be randomly distributed.

13.32 A linear model for gasoline consumption has been fit to data for the past six years with the following results for secular trend

$$\hat{Y}_i = 4.634 + .0863X$$

where X is the number of the quarter, beginning with $X = 1$ for the first quarter, 1975. Use this equation to predict gasoline consumption for the first quarter of 1981.

13.33 The data referred to in Exercise 13.32 were used to obtain the following seasonal indexes for the four quarters in each year:

$$S_1 = .884 \qquad S_2 = .986 \qquad S_3 = 1.306 \qquad S_4 = .824$$

Use the appropriate seasonal index to adjust the prediction in Exercise 13.32 for seasonal effects.

13.34 Suppose the long-term cycle prediction for the first quarter of 1981 is $C_{25} = 1.144$ as obtained from government economists. Use this value to further refine the predicted gasoline consumption figure from Exercise 13.33.

13.35 The numbers of new housing starts for one town in the years 1970 to 1981 are given as follows.

1970	331	**1976**	445
1971	354	**1977**	472
1972	350	**1978**	490
1973	396	**1979**	538
1974	402	**1980**	533
1975	428	**1981**	568

Fit an exponential growth curve to these data. Plot it along with the data points to see if the fit seems reasonable. Estimate the annual growth rate from these 12 years of data.

13.36 The numbers of bankruptcy procedures initiated in a recent six-year period are given below, by quarters. Plot these data on a graph. Interpret the results of the graph with respect to the type of model that appears to be appropriate.

	Quarter			
Year	**1**	**2**	**3**	**4**
1	31	26	24	30
2	34	29	25	33
3	36	32	27	38
4	40	38	33	39
5	54	37	34	42
6	58	43	38	48

13.37 Use the exponential growth curve to identify the secular trend for the data in Exercise 13.36. Plot the fitted curve along with the data to see if the fit seems reasonable.

13.38 Use the results of Exercise 13.37 to obtain seasonal index values for the data in Exercise 13.36. Plot the deseasonalized data along with the fitted growth curve to see if there is a pattern in the residuals.

13.39 Find the long-term cycle and irregular components for the data in Exercise 13.36, using the results of Exercises 13.37 and 13.38. Plot the long-term cycle values and interpret the results. Plot the irregular fluctuations and see if some pattern still appears to be present.

FORMULATING GENERAL LINEAR MODELS FOR FITTING DATA

Chapters 12 and 13 are concerned with fitting mathematical equations, called models, to sets of bivariate data. This chapter is concerned with formulating models for multivariate data, that is, data where each observation has several variables associated with it. Examples of multivariate data include employee personnel records, which show sex, age, salary, length of employment, and the like, for each employee, or quarterly business reports, which include sales, profit, expenses, taxes, interest paid, and the like, for each quarter. Strictly speaking, bivariate data are simple forms of multivariate data, but usually the term **multivariate** implies more than just two variables for each unit in a sample.

MULTIVARIATE **Multivariate** sample data consist of values of two or more variables measured
DATA on each unit in a sample.

In contrast to the **simple linear model,** which is based on the straight-line equation $y = a + bx$, the models in this chapter are called **general linear models** and are based on equations of the form

$$y = \beta_0 + \beta_1 x_1 + \beta_2 x_2 + \cdots + \beta_k x_k$$

where $x_1, x_2, \ldots, x_k$ may be different variables or may be functions of other

x's. This model is called a general linear model because the individual terms in the model are added together. The individual terms themselves may represent variables raised to various powers, such as x^2, x^3, or perhaps products of variables, such as $x_1 x_2$.

The **general linear model** is a mathematical equation of the form

GENERAL
LINEAR MODEL

$$y = \beta_0 + \beta_1 x_1 + \beta_2 x_2 + \cdots + \beta_k x_k$$

that is used to describe a relationship between one dependent variable Y and k independent variables $X_1, \ldots, X_k$.

The fitting of a general linear model to a set of data involves finding coefficients to use in place of the β's that appear in the model. The coefficients that are obtained are denoted by $\hat{\beta}_0$ for β_0, $\hat{\beta}_1$ for β_1, and so on. The method of least squares is used in this text to obtain the estimates $\hat{\beta}$, but there are other methods that could be used. However, most computer programs use the least squares method, and it is by far the most widespread method used today for fitting models to data.

Since the least squares method is a suitable procedure for fitting models to data, the next step that follows naturally is to make some statistical inferences regarding the adequacy of the model, or to find a statistical justification for including individual terms into the model. Such inference procedures require that the data satisfy additional requirements, such as being a random sample from a normal population. Discussion of such inference procedures is deferred to the next chapter. This chapter is concerned only with the formulation of a model to be fit to a set of data, which may or may not be a random sample from any particular distribution. Therefore, the methods of this chapter may be applied to any set of data.

14.1

MULTIVARIATE MODELS

PROBLEM SETTING 14.1

A pharmaceutical firm has set up an experiment in which patients with a common type of pain are treated with a new analgesic. The analgesic is to be applied to each patient in one of the following dosage levels in grams: 2, 5, 7, or 10. For each patient the time in minutes is recorded until a noticeable relief in the pain is detected. The patients selected for the study are equally divided on the basis of sex (female = 0, male = 1) and also fall very loosely into low, medium, and high blood pressure groups. The blood pressure groups represent approximately the .25, .50, and .75 quantiles of the distribution of all blood pressure readings. Additionally, the age and weight of each patient is recorded. At the conclusion of the experiment the firm has recorded the

Minutes to Relief Y	Dosage in Grams X_1	Age of Patient X_2	Weight of Patient X_3	Sex Code of Patient X_4	Blood Pressure Quantile X_5
25	2	31	119	0	.25
43	2	19	125	0	.50
55	2	30	125	0	.75
47	2	27	152	1	.25
43	2	41	173	1	.50
57	2	25	146	1	.75
26	5	40	132	0	.25
27	5	35	145	0	.50
25	5	44	103	0	.75
29	5	39	141	1	.25
22	5	64	181	1	.50
29	5	31	188	1	.75
13	7	34	108	0	.25
11	7	29	120	0	.50
14	7	55	113	0	.75
20	7	14	139	1	.25
20	7	32	141	1	.50
30	7	33	204	1	.75
13	10	66	113	0	.25
8	10	20	150	0	.50
3	10	45	126	0	.75
27	10	22	144	1	.25
26	10	27	157	1	.50
5	10	53	155	1	.75

FIGURE 14.1
A MULTIVARIATE DATA SET RESULTING FROM THE TESTING OF
A NEW ANALGESIC.

multivariate data given in Figure 14.1. The firm would like to know if the time
(Y) until a noticeable relief in pain is detected can be predicted based on the
dosage level of the analgesic (X_1), and the age (X_2), weight (X_3), sex (X_4), and
blood pressure quantile (X_5) of the patient.

FITTING THE GENERAL LINEAR MODEL

The pharmaceutical firm in Problem Setting 14.1 would like to find a general
linear model to describe the relationship between the dependent variable Y
and the independent variables X_1, X_2, X_3, X_4, and X_5. That is, they would like
to fit a model of the following form:

$$y = \beta_0 + \beta_1 x_1 + \beta_2 x_2 + \beta_3 x_3 + \beta_4 x_4 + \beta_5 x_5 \qquad (14.1)$$

The model in Eq. 14.1 is fit using the principle of least squares to estimate the
parameters as was done in Chapter 12. That is, the coefficients are determined

SOURCE	DF	SUM OF SQUARES	MEAN SQUARE	F VALUE	PR > F
MODEL	5	3711.6174	742.3235	10.82	.0001
ERROR	18	1234.8826	68.6046		
TOTAL	23	4946.5000		R SQUARE	
				.7504	

PARAMETER	ESTIMATE	T FOR H0: PARAMETER = 0	PR > |T|	STD ERROR OF ESTIMATE
INTERCEPT	51.7790	3.87	.0011	13.3717
DOSAGE	− 3.7408	− 6.31	.0001	0.5927
AGE	− 0.2174	− 1.63	.1207	0.1335
WEIGHT	− 0.0249	− 0.24	.8126	0.1035
SEX	7.8589	1.54	.1416	5.1125
BDPRESSQ	7.5342	0.85	.4080	8.8936

FIGURE 14.2
COMPUTER PRINTOUT FOR A LEAST SQUARES FIT TO THE MODEL IN EQ. 14.1.

such that the sum of squares between the differences of the observed Y_i and the predicted Y_i is minimized. Since there are more parameters to be estimated in fitting the model in Eq. 14.1 than for a straight line the calculations can become very tedious without the aid of a computer. Hence, such calculations are usually done on a computer.

INDEPENDENT VARIABLES   The variables on the right-hand side of the equals sign in a general linear model are called **independent variables.** The single variable on the left-hand side of the equals sign is called the **dependent variable.** This terminology does not imply independence or dependence in the statistical sense.

There are many packaged computer programs available for performing the calculations involved in statistical analyses. These programs are continually being updated to enlarge the number of different types of statistical analyses they will handle, and to make them easier for the statistician to use. Two of the more popular and convenient-to-use programs are SAS® and Minitab®. Supplements describing these two programs are available for this text.

SAS® was used in this text, but the computer output has been modified slightly in format here; it will not look exactly like the reader's SAS® output. Most packaged programs available today present the same details in their output as SAS® produces, with slight cosmetic variations. Learning about SAS® computer output will prepare you to use most of the other packaged statistical programs as well.

INTERPRETING THE COMPUTER PRINTOUT

There are many components in the computer output and each will be explained as it is encountered in the analysis. The computer output for a least

MODEL	X_i	$\hat{\beta}_0$	$\hat{\beta}_1$	MODEL R-SQUARE
1	DOSAGE	49.3382	-3.9314	.6374
2	AGE	39.8915	-0.3965	.1335
3	WEIGHT	8.7334	0.1201	.0448
4	SEX	21.9167	7.6667	.0713
5	BDPRESQ	23.5000	4.5000	.0041

FIGURE 14.3
RESULTS OF FIVE SEPARATE SIMPLE LINEAR REGRESSIONS.

squares fit of the model in Eq. 14.1 for the data given in Figure 14.1 appears in Figure 14.2. This output contains the numerical values for $\hat{\beta}_0, \hat{\beta}_1, \ldots,$ $\hat{\beta}_5$. These values are found in Figure 14.2 under the heading **ESTIMATE,** since they are estimates of the population values based on sample data. From Figure 14.2 the estimated regression equation is as follows,

$$\hat{Y} = 51.7790 - 3.7408X_1 - 0.2174X_2 - 0.0249X_3$$
$$+ 7.8589X_4 + 7.5342X_5 \qquad (14.2)$$

This equation can be used to predict the time to relief for different combinations of the independent variables X_1 to X_5. For example, the predicted time to relief for a 25-year-old female weighing 120 pounds with a blood pressure quantile of .25 and receiving a 5-gram dosage is found as

$$\hat{Y} = 51.7790 - 3.7408(5) - 0.2174(25) - 0.0249(120)$$
$$+ 7.8589(0) + 7.5342(.25)$$
$$= 26.5356$$

which rounds to 27 minutes.

Elsewhere in Figure 14.2 the model **R-SQUARE** value is given as .7504. This R^2 was defined in Chapter 12 as the *coefficient of determination, $R^2 =$* $1 -$ Error SS/Total SS. When the principle of least squares is used to estimate the parameters, then R^2 can also be written as $R^2 =$ Model SS/Total SS, where Model SS is the **model sum of squares** defined below. In Figure 14.2 R^2 is found as the ratio of the model sum of squares (3711.6174) to the total sum of squares (4946.5000). It indicates that collectively the five independent variables explain approximately 75% of the variation in the observed time to relief of the patients in the experiment.

THE MODEL SUM OF SQUARES

The total of the squared deviations between the values $\hat{Y}_i$ predicted for the dependent variable by a general linear model and the sample mean $\overline{Y}$ of the dependent variable is the **model sum of squares.**

$$\text{Model SS} = \Sigma(\hat{Y}_i - \overline{Y})^2$$

When the principle of least squares is used to fit the model, then

$$\text{Model SS} + \text{Error SS} = \text{Total SS}$$

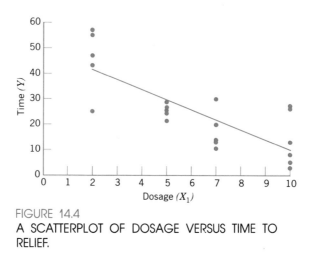

FIGURE 14.4
A SCATTERPLOT OF DOSAGE VERSUS TIME TO RELIEF.

COMPARISON WITH SIMPLE LINEAR REGRESSION

The fitted general linear model given in Eq. 14.2 expresses the simultaneous role of all five independent variables in influencing the time to relief. It is worthwhile to compare the role of the individual X's in the full model of Eq. 14.1 and their roles in simple linear regressions. Figure 14.3 displays the results of five separate simple linear regressions obtained by fitting each X_i versus Y using the techniques of Chapter 12.

Note that the largest R-square value of .6374 is obtained for Model 1 using the dosage by itself. Hence, the R^2 of .7504 from the full model in Eq. 14.2 indicates that collectively the five X's explain more variation in Y than any individual X_i.

How does the role of each X_i in the full model compare with the role of the individual X's in Figure 14.2? One way to answer this question is to make a scatterplot for each of the five X's versus Y and add a straight line to each of these scatterplots based on the coefficients in Figure 14.3. These scatterplots are given in Figures 14.4 to 14.8.

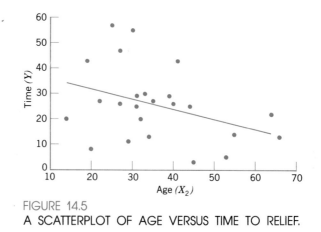

FIGURE 14.5
A SCATTERPLOT OF AGE VERSUS TIME TO RELIEF.

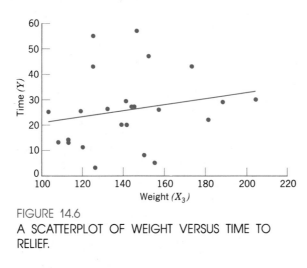

FIGURE 14.6
A SCATTERPLOT OF WEIGHT VERSUS TIME TO
RELIEF.

Figure 14.4 shows that time to relief decreases rapidly as dosage increases. This result seems reasonable over the range of doses considered in the experiment. The coefficient -3.9314 in Figure 14.3 is similar to -3.7408 found in Figure 14.2. A decrease in time to relief also occurs as age increases as shown in Figure 14.5. The coefficients -0.3965 and -0.2174 from Figures 14.3 and 14.2, respectively, are in agreement as to the effect of age on time to relief. Time to relief increases as weight increases as indicated in Figure 14.6 and confirmed by the coefficient 0.1201 in Figure 14.3. However, the coefficient for weight in Figure 14.2 is slightly negative (-0.0249), indicating that increased weight has the effect of decreasing time to relief. This disagreement may be due to an inadequacy in one or both of the models. This question is explored in more detail in the next subsection. In Figure 14.1 sex was coded as 0 for female and 1 for male; thus, the line in Figure 14.7 indicates a slightly longer time to relief for males than for females. The coefficients for sex in both Figures 14.2 and 14.3 are in agreement for this variable. There is no special significance attached to the coding for sex. In fact if the coding were inter-

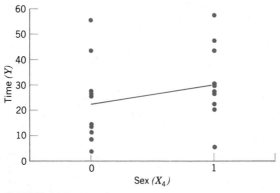

FIGURE 14.7
A SCATTERPLOT OF SEX CODE VERSUS TIME TO
RELIEF.

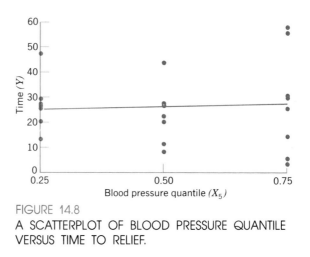

FIGURE 14.8

A SCATTERPLOT OF BLOOD PRESSURE QUANTILE
VERSUS TIME TO RELIEF.

changed, the dots above 0 and 1 in Figure 14.7 would merely change places, in which case the straight line would slope downward rather than upward. However, the interpretation would remain the same as before. The line added to Figure 14.8 is nearly horizontal, indicating that the blood pressure quantile may not be directly tied to time to relief.

IMPROVING THE FITTED MODEL

The fitted general linear model given in Eq. 14.2 accounted for 75.04% of the variation in Y. Sometimes it is possible to improve the fitted model by including other terms in the model. Typically, these terms include squares of the independent variables (X_i^2) or interactions (X_iX_j). For a five-variable model there are five squared terms to choose from and 10 interaction terms. These terms should not ordinarily be added to the model unless there is some reasonable

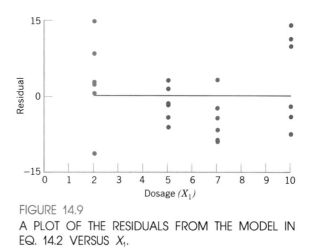

FIGURE 14.9

A PLOT OF THE RESIDUALS FROM THE MODEL IN
EQ. 14.2 VERSUS X_1.

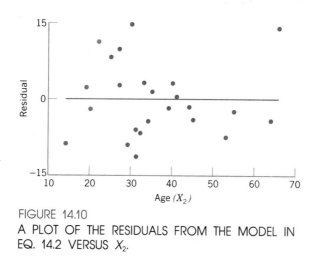

FIGURE 14.10

A PLOT OF THE RESIDUALS FROM THE MODEL IN EQ. 14.2 VERSUS X_2.

justification that doing so will result in an improved fit. One indication of whether or not an improvement will occur in a fitted model can be found in an examination of plots of the residuals from the fitted model versus each independent variable. The residual plots based on the model in Equation 14.2 for the variables X_1 to X_5 appear in Figures 14.9 to 14.13. These plots can provide useful clues as to possible improvements in the fitting of the model. For example, if a model adequately incorporates a variable, then residual plots will exhibit a random scattering about zero with no apparent pattern present. This would seem to be the case in Figures 14.10 to 14.13 for X_2 to X_5, respectively.

The residual plot for X_1 in Figure 14.9 is a different matter, as the residuals at dosages of 5 and 7 grams are more closely scattered about zero than at the other two dosage levels. Also, the residuals appear to demonstrate some curvature. There could be more than one reason for this pattern, but one possibility is that time to relief has a quadratic response with respect to dosage

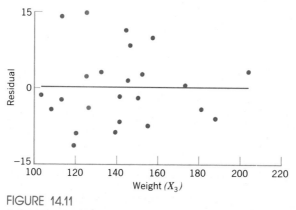

FIGURE 14.11

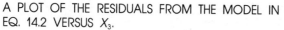

A PLOT OF THE RESIDUALS FROM THE MODEL IN EQ. 14.2 VERSUS X_3.

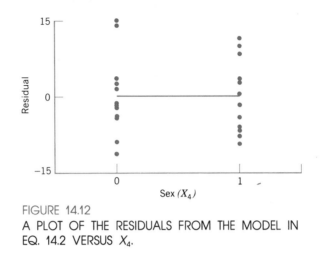

FIGURE 14.12
A PLOT OF THE RESIDUALS FROM THE MODEL IN
EQ. 14.2 VERSUS X_4.

level. Thus, the first attempt at improving the model is to add a quadratic term, X_1^2. The model to be fit is of the form,

$$y = \beta_0 + \beta_1 x_1 + \beta_2 x_2 + \beta_3 x_3 + \beta_4 x_4 + \beta_5 x_5 + \beta_6 x_1^2 \qquad (14.3)$$

The computer printout for this model appears in Figure 14.14. The new estimated regression equation is as follows,

$$\hat{Y} = 60.9836 - 8.9562 X_1 - 0.1803 X_2 - 0.0096 X_1$$
$$+ 7.4187 X_4 + 6.7064 X_5 + 0.4321 X_1^2 \qquad (14.4)$$

The estimated regression equation in Eq. 14.4 has an R^2 of .7999 compared to the R^2 of .7504 for the five-variable model. Thus, the fitted model of Eq. 14.4 explains more of the variation in Y than its predecessor given in Eq. 14.2. It may be possible to improve the fit substantially through some additional examination of the residual plots.

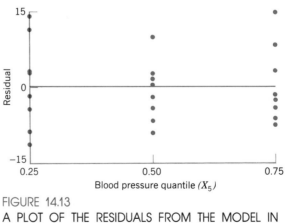

FIGURE 14.13
A PLOT OF THE RESIDUALS FROM THE MODEL IN
EQ. 14.2 VERSUS X_5.

SOURCE	DF	SUM OF SQUARES	MEAN SQUARE	F VALUE	PR > F
MODEL	6	3956.9066	659.4844	11.33	.0001
ERROR	17	989.5934	58.2114		
TOTAL	23	4946.5000		R SQUARE	
				.7999	

PARAMETER	ESTIMATE	T FOR H0: PARAMETER = 0	PR > |T|	STD ERROR OF ESTIMATE
INTERCEPT	60.9836	4.65	.0002	13.1081
DOSAGE	−8.9562	−3.45	.0031	2.5987
AGE	−0.1803	−1.45	.1650	0.1243
WEIGHT	−0.0096	−0.10	.9213	0.0956
SEX	7.4187	1.57	.1340	4.7142
BDPRESSQ	6.7064	0.82	.4249	8.2022
DOSAGE**2	0.4321	2.05	.0558	0.2105

FIGURE 14.14

COMPUTER PRINTOUT FOR A LEAST SQUARES FIT TO THE MODEL IN EQ. 14.3.

INTERACTION TERMS

It may be that some of the five variables are acting jointly in their relationship with time to relief. Such relationships or *interactions,* can be included in the model as cross products X_iX_j. For a five-variable model, there are 10 possible pairs of interactions. Their potential usefulness can frequently be detected by graphical methods. This approach is demonstrated with the term X_1X_5. The residual plots in Figures 14.9 and 14.13 are reconstructed in Figures 14.15 and 14.16, respectively. The points in Figure 14.15 represent the residual versus the dosage X_1, but each point has also been identified by X_5, the blood

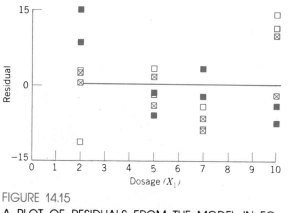

FIGURE 14.15

A PLOT OF RESIDUALS FROM THE MODEL IN EQ. 11.2 VERSUS X_1. (X_5 VALUES. □ = .25 QUANTILE; ⊠ = .50 QUANTILE; ■ = .75 QUANTILE.)

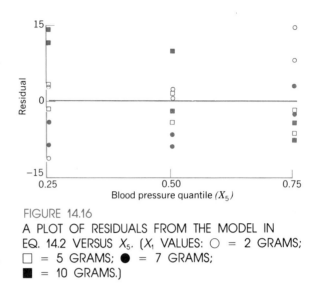

FIGURE 14.16
A PLOT OF RESIDUALS FROM THE MODEL IN
EQ. 14.2 VERSUS X_5. (X_1 VALUES: ○ = 2 GRAMS;
□ = 5 GRAMS; ● = 7 GRAMS;
■ = 10 GRAMS.)

pressure quantile present when the residual was produced. Thus, it can be seen that the lowest blood pressure quantile of .25 produces one large negative residual at a dosage of 2 grams, and two positive residuals at a dosage of 10 grams. In contrast, the residuals for a blood pressure quantile of .75 are both positive at 2 grams and both negative at 10 grams.

Likewise, the points in Figure 14.16 represent the residual versus X_5, the blood pressure quantile. In addition each point has been identified by X_1, the level of dosage used when the residual was produced. The residual for a

SOURCE	DF	SUM OF SQUARES	MEAN SQUARE	F VALUE	PR > F
MODEL	7	4465.9723	637.9960	21.24	.0001
ERROR	16	480.5277	30.0330		
TOTAL	23	4946.5000		R SQUARE	
				.9029	

PARAMETER	ESTIMATE	T FOR H0: PARAMETER = 0	PR > |T|	STD ERROR OF ESTIMATE
INTERCEPT	31.2819	2.64	.0179	11.8614
DOSAGE	−5.2787	−2.55	.0214	2.0693
AGE	−0.1289	−1.43	.1718	0.0901
WEIGHT	0.0370	0.53	.6021	0.0696
SEX	5.8735	1.72	.1040	3.4069
BDPRESSQ	52.3380	4.17	.0007	12.5521
DOSAGE**2	0.4524	2.99	.0086	0.1513
DOS*BPQ	−7.9148	−4.12	.0008	1.9224

FIGURE 14.17
COMPUTER PRINTOUT FOR A LEAST SQUARES FIT TO THE MODEL IN EQ.
14.5.

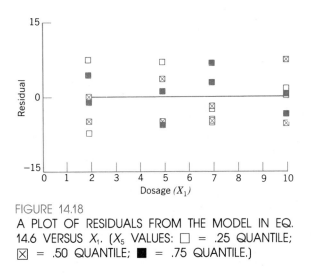

FIGURE 14.18
A PLOT OF RESIDUALS FROM THE MODEL IN EQ.
14.6 VERSUS X_1. (X_5 VALUES: □ = .25 QUANTILE;
⊠ = .50 QUANTILE; ■ = .75 QUANTILE.)

dosage of 2 grams can be seen to increase as the blood pressure quantile increases, while the residual for a dosage of 10 grams decreases as blood pressure increases. The pattern of the residuals in these figures clearly indicates the presence of an interaction involving X_1 and X_5. The last model to be fit with the present set of data includes the term X_1X_5 and is of the form,

$$y = \beta_0 + \beta_1 x_1 + \beta_2 x_2 + \beta_3 x_3 + \beta_4 x_4 + \beta_5 x_5 + \beta_6 x_1^2 + \beta_7 x_1 x_5 \quad (14.5)$$

The computer printout for this model is given in Figure 14.17.

The estimated regression equation obtained from the computer printout in Figure 14.17 is as follows,

$$\hat{Y} = 31.2819 - 5.2787X_1 - 0.1289X_2 + 0.0370X_3 + 5.8735X_4$$
$$+ 52.3380X_5 + 0.4524X_1^2 - 7.9148X_1X_5 \quad (14.6)$$

This estimated regression equation has a corresponding R^2 value of .9029, representing a substantial increase over the previous model.

The residuals for the estimated regression equation in Eq. 14.6 have been plotted versus X_1 in Figure 14.18. When Figures 14.15 and 14.18 are compared, it is clear that the magnitude of the residuals has decreased considerably, and that the curvature present in Figure 14.15 is not present in Figure 14.18. Rather, the residuals are now spread in a reasonably uniform pattern for all dosage levels. Finally, the interaction effect previously present is no longer apparent in Figure 14.18. The same statements hold for residuals plotted versus X_5 in Figures 14.16 and 14.19.

FURTHER MODEL REFINEMENT AND THE DANGER OF OVERFITTING

The intent of the previous discussion was to show how a simple regression diagnostic such as plotting residuals can lead to an improvement in the fitted

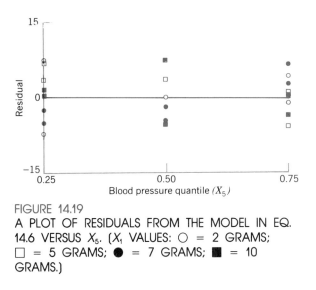

FIGURE 14.19
A PLOT OF RESIDUALS FROM THE MODEL IN EQ.
14.6 VERSUS X_5. (X_1 VALUES: ○ = 2 GRAMS;
□ = 5 GRAMS; ● = 7 GRAMS; ■ = 10
GRAMS.)

model. However, the reader should be aware that there is an underlying danger of overfitting the data by concentrating too closely on increasing the R-square value. The problem with concentrating on R-square is that models may become distorted so that they fit the observed sample data points well, but may provide a poor approximation to the underlying population model that is generating those points. This is commonly referred to as **overfitting** the data. Such models are typically poor predictors of future observations. As an example of this, consider the four data points plotted in the graph in Figure 14.20. These four data points exhibit reasonably well a straight-line relationship, but have been fit by a general linear model of the form $y = \beta_0 + \beta_1 x + \beta_2 x^2 + \beta_3 x^3$, whose graph also appears in Figure 14.20. This graph intersects each of the four data points, and thus has $R^2 = 1.0$, but clearly would not provide useful predictions of Y for most other values of X if the true regression equation is a straight line.

There are many other more sophisticated techniques that can be used to improve the fit of a model. Many of these techniques have associated tests of significance. Tests of significance in relationship to model building are covered in the next chapter.

FIGURE 14.20
EXAMPLE OF A
MODEL OVERFITTING
THE DATA.

14.1 A general linear model is to be fit to a set of data containing six independent variables.

(a) Write out the general linear model for the data set in the form given in Eq. 14.1.

(b) How many parameters will be estimated in fitting the model in part (a)?

(c) What procedure is commonly used to provide the estimates in part (b)?

(d) If the fitted model produces a MODEL sum of squares 1236.38 and a TOTAL sum of squares 1745.75, what is the associated R-SQUARE value?

(e) What is the interpretation of the R-SQUARE value in part (d)?

14.2 Use the estimated regression equation given in Eq. 14.2 to predict the time to relief for a 45-year-old male weighing 150 pounds with a blood pressure quantile of .75 and receiving a 10-gram dose.

14.3 Use the five fitted models given in Figure 14.3 to make five individual predictions based on the information given in Exercise 14.2. How do these five answers compare with the single answer obtained in Exercise 14.2?

14.4 If all other factors remain the same in Eq. 14.2, how much does the estimated time to relief change when

(a) the dosage is increased by one gram?

(b) age decreases by five years?

(c) weight increases by 20 pounds?

(d) the patient is male rather than female?

(e) the blood pressure quantile decreases from .75 to .25?

14.5 A truck dispatcher obtained 20 observations on the following variables in order to predict the number of driver hours required:

$$Y = \text{driver time in hours}$$
$$X_1 = \text{distance of route in miles}$$
$$X_2 = \text{weight of load in tons}$$
$$X_3 = \text{number of deliveries to be made}$$
$$X_4 = \text{historical route speed in miles per hour}$$

Use the computer printout on the next page from the resulting regression analysis to answer the questions below.

(a) What is the estimated regression equation?

(b) What is the predicted driving time when the distance is 125 miles, the load is 4 tons, there are 5 deliveries to be made, and the historical speed for the route is 45 mph?

(c) What percent of the variability in driving time is explained by these four independent variables?

SOURCE	DF	SUM OF SQUARES	MEAN SQUARE	F VALUE	PR > F
MODEL	4	47.3825	11.8456	25.51	.0001
ERROR	15	6.9655	.4644		
TOTAL	19	54.3480		R SQUARE	
				.9337	

PARAMETER	ESTIMATE	T FOR H0: PARAMETER = 0	PR > |T|	STD ERROR OF ESTIMATE
INTERCEPT	1.5045	2.21	.0002	0.6814
DISTANCE	0.0297	6.02	.0001	0.0049
WEIGHT	0.2503	1.30	.2124	0.1922
DELIVERIES	0.4436	5.29	.0001	0.0838
SPEED	−0.0380	−1.48	.0001	0.0255

Exercises 14.6 to 14.8 refer to the set of data below, which was given in McClave and Benson (1985). In these data Y = the auction price of grandfather clocks, X_1 = age of the clock being auctioned, and X_2 = the number of individuals bidding on the clock being auctioned.

Clock Number	X_1	X_2	Y	Residuals Model 1	Residuals Model 2	Residuals Model 3
1	108	6	729	175.3	31.4	78.6
2	108	14	1055	−185.2	−16.8	−12.5
3	111	7	785	107.3	10.7	23.0
4	111	15	1175	−189.2	−4.5	35.1
5	113	9	946	71.2	45.3	14.8
6	115	7	744	15.4	−70.1	−58.7
7	115	12	1080	−77.7	−13.4	−43.8
8	117	11	1024	−73.4	−43.8	−81.5
9	117	13	1152	−117.0	−32.7	−45.0
10	126	10	1336	209.8	202.0	166.6
11	127	13	1235	−161.4	−127.2	−131.6
12	127	7	845	−36.5	−88.6	−79.7
13	132	10	1253	50.4	35.8	2.2
14	137	8	1147	52.3	29.4	15.0
15	137	15	1713	17.6	4.6	73.4
16	137	9	1297	116.5	95.0	67.6
17	143	6	854	−145.4	−146.8	−112.5
18	150	9	1522	176.0	156.8	131.0
19	153	6	1092	−34.8	4.6	35.2
20	156	12	1822	142.1	54.3	49.5

Clock Number	X_1	X_2	Y	Residuals Model 1	Residuals Model 2	Residuals Model 3
21	156	6	1047	−118.0	−66.4	−36.9
22	159	9	1483	22.3	4.8	−19.8
23	162	11	1884	213.5	134.4	117.3
24	168	7	1262	−141.7	−79.9	−79.4
25	170	14	2131	101.2	−121.5	−57.9
26	175	8	1545	−33.6	−0.4	−16.2
27	179	9	1792	76.6	62.7	40.6
28	182	11	1979	53.8	−73.6	−81.5
29	182	8	1550	−117.8	−74.2	−90.3
30	184	10	2041	176.1	103.5	85.2
31	187	8	1593	−138.5	−87.4	−103.8
32	194	5	1356	−207.2	71.9	115.8

14.6 A regression model of the form

$$\text{Model 1: } y = \beta_0 + \beta_1 x_1 + \beta_2 x_2$$

was fit to the preceding set of 32 observations and produced the residuals in the column labeled Model 1.

(a) Make a scatterplot of X_1 versus these residuals. Use the plot symbol "●" when the corresponding value of $X_2 \leq 9$ (18 points). When $X_2 \geq 10$, use the plot symbol "○" (14 points).

(b) Make a scatterplot of X_2 versus these residuals. Use the plot symbol "●" when the corresponding value of $X_1 \leq 139$ (16 points). When $X_1 \geq 140$ use the plot symbol "○" (16 points). (Note that the data have been sorted on X_1 so the first 16 points are plotted with the symbol "●" and the second set of 16 points with the symbol "○".)

(c) In the scatterplot in part (a) the symbol "●" represents those values with $X_2 \leq 9$. What behavior do the residuals with the symbol "●" exhibit as X_1 increases?

(d) In the scatterplot in part (a) the symbol "○" represents those values with $X_2 \geq 10$. What behavior do the residuals with the symbol "○" exhibit as X_1 increases?

(e) In the scatterplot in part (b) the symbol "●" represents those values with $X_1 \leq 139$. What behavior do the residuals with the symbol "●" exhibit as X_2 increases?

(f) In the scatterplot in part (b) the symbol "○" represents those values with $X_1 \geq 140$. What behavior do the residuals with the symbol "○" exhibit as X_2 increases?

(g) On the basis of what you have observed in parts (a) to (f), what recommendations, if any, would you make to the modeler with respect to improving the fitted model? Be sure to compare the plots on (a) and (b) with Figures 14.15 and 14.16.

14.7 The regression model of the previous exercise was expanded to the form

$$\text{Model 2: } y = \beta_0 + \beta_1 x_1 + \beta_2 x_2 + \beta_3 x_1 x_2$$

so that the interaction term $x_1 x_2$ could be included. The residuals corresponding to this interaction model appear in the column labeled Model 2.

(a) Make a scatterplot of X_1 versus these residuals. Use the symbol "●" as a plot symbol when the corresponding value of $X_2 \leq 9$ (18 points). When $X_2 \geq 10$ use the symbol "○" as the plot symbol (14 points).

(b) Make a scatterplot of X_2 versus these residuals. Use the symbol "●" as a plot symbol when the corresponding value of $X_1 \leq 139$ (16 points). When $X_1 \geq 140$ use the symbol "○" as the plot symbol (16 points).

(c) Is the presence of interaction evident in these plots?

(d) Do the residual plots on parts (a) and (b) suggest the addition of any other term to the model that may improve the fit?

14.8 The regression model of the previous exercise was expanded to the form

$$\text{Model 3: } y = \beta_0 + \beta_1 x_1 + \beta_2 x_2 + \beta_3 x_1 x_2 + \beta_4 x_2^2$$

so that the quadratic term x_2^2 could be included. The residuals corresponding to this interaction model appear in the column labeled Model 3.

(a) Make a scatterplot of X_1 versus these residuals. Use the symbol "●" as a plot symbol when the corresponding value of $X_2 \leq 9$ (18 points). When $X_2 \geq 10$ use the symbol "○" as the plot symbol (14 points).

(b) Make a scatterplot of X_2 versus these residuals. Use the symbol "●" as a plot symbol when the corresponding value of $X_1 \leq 139$ (16 points). When $X_1 \geq 140$ use the symbol "○" as the plot symbol (16 points).

(c) Is either the presence of interaction or curvature indicated in either of the plots in parts (a) or (b), or do the residuals appear to be randomly scattered about zero?

14.2

MULTIVARIATE MODELS WITH ONE QUALITATIVE VARIABLE

PROBLEM SETTING 14.2

A real estate appraiser in Albuquerque, New Mexico, would like to develop a model for predicting the selling price of homes in the Albuquerque area. She obtains information on 24 recent sales. This information includes sales price, square footage of the house, and type of house exterior, as summarized in Figure 14.21. An examination of Figure 14.21 shows information on both qualitative and quantitative variables. Since the real estate appraiser's experience indicates that the type of exterior influences the value of a house, she

Sale Price (Y)	Square Footage (X)	Type of Exterior
$ 53,400	1600	Wood
59,450	1800	Wood
48,100	1400	Wood
39,900	1100	Wood
82,850	1900	Wood
86,500	2000	Wood
81,650	1850	Wood
85,300	1950	Wood
71,650	1800	Brick
70,250	1750	Brick
65,550	1600	Brick
69,100	1700	Brick
104,150	2050	Brick
117,700	2400	Brick
114,150	2300	Brick
112,500	2250	Brick
58,250	1500	Adobe
64,250	1680	Adobe
55,550	1400	Adobe
49,400	1200	Adobe
112,050	2400	Adobe
108,600	2300	Adobe
95,900	1950	Adobe
101,700	2100	Adobe

FIGURE 14.21
DATA ON 24 RECENT HOUSE SALES
IN ALBUQUERQUE.

wonders whether or not the information contained in the qualitative variable can be incorporated into a model for predicting the selling price of a house.

THE SIMPLE LINEAR REGRESSION MODEL

Perhaps a linear regression with square footage (X) as the independent variable and selling price (Y) as the dependent variable is adequate for developing a model to predict the selling price of a house for the real estate appraiser in the problem setting. This type of model has the following form:

$$y = \beta_0 + \beta_1 x \qquad (14.7)$$

The least squares fit to the data for this model gives the prediction equation

$$\hat{Y} = -37,730.39 + 63.97X \qquad (14.8)$$

This model suggests that additional square footage contributes to the selling price of the house at the rate of about $64 per square foot. The R^2 value

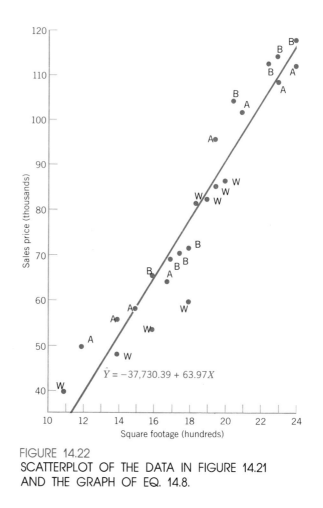

FIGURE 14.22
SCATTERPLOT OF THE DATA IN FIGURE 14.21
AND THE GRAPH OF EQ. 14.8.

associated with Eq. 14.8 is .9222, so 92.2% of the variation in selling price is accounted for by this model. A plot of selling price versus square footage is given in Figure 14.22. The graph of Equation 14.8 also appears in Figure 14.22 and shows that a straight-line fit provides a reasonable description of the data.

CHANGING THE INTERCEPT BY THE USE OF A DUMMY VARIABLE

Although the quantitative data in Figure 14.21 accounts for 92.2% of the variation in selling price through Eq. 14.8, it may be possible to incorporate some of the information contained in the qualitative variable and obtain an even better fit to the data. For example, the first eight entries in Figure 14.21 correspond to houses with wood exteriors, whereas the last 16 entries correspond to houses with exteriors other than wood. If the type of exterior influences the selling price, then it may be reasonable to fit the model of Eq. 14.7 to the first eight pairs of data values only and then fit it again to the last 16 pairs of data values. This would mean that the two estimated prediction equations could have different intercepts, or different slopes, or both. Although these two fits could be obtained, it is easier to obtain both prediction equations simultaneously through the use of **dummy variables.** Furthermore, this method allows all of the data to be used to estimate one common slope while estimating two separate intercepts, or to estimate one common intercept with two separate slopes, if these variations are desired for a particular application.

DUMMY VARIABLE A **dummy variable** is a variable that assumes only the value 0 or 1. It is used to indicate the absence or presence, respectively, of a particular qualitative characteristic of the observation.

A dummy variable is used to quantify the qualitative variable. For example, suppose a dummy variable is defined to represent the type of exterior in Figure 14.21 as follows

$$D = 0 \quad \text{when the exterior is wood}$$
$$D = 1 \quad \text{when the exterior is other than wood}$$

This means that the qualitative information from Figure 14.21 is represented quantitatively as in Figure 14.23 and the model used is of the form

$$y = \beta_0 + \beta_1 x + \beta_2 D \qquad (14.9)$$

When $D = 0$, the model in Eq. 14.9 reduces to

$$y = \beta_0 + \beta_1 x \qquad (14.10)$$

and when $D = 1$ the model in Eq. 14.9 can be rewritten as

$$y = (\beta_0 + \beta_2) + \beta_1 x \qquad (14.11)$$

The only difference between the models in Eqs. 14.10 and 14.11 involves the intercepts, since the slope β_1 remains the same in both. The computer printout for the least squares fit of the model of Eq. 14.9 to the data in Figure 14.23 is given in Figure 14.24 and provides the following prediction equation

$$\hat{Y} = -37{,}869.17 + 61.77X + 6250.88D \qquad (14.12)$$

When $D = 0$ (corresponding to a house with a wood exterior) Eq. 14.12 reduces to

$$\hat{Y} = -37{,}869.17 + 61.77X \qquad (14.13)$$

and when $D = 1$ (corresponding to a house with an exterior other than wood) Eq. 14.12 simplifies to

$$\hat{Y} = -31{,}618.29 + 61.77X \qquad (14.14)$$

Sale Price (Y)	Square Footage (X)	Type of Exterior (D)
$ 53,400	1600	0
59,450	1800	0
48,100	1400	0
39,900	1100	0
82,850	1900	0
86,500	2000	0
81,650	1850	0
85,300	1950	0
71,650	1800	1
70,250	1750	1
65,550	1600	1
69,100	1700	1
104,150	2050	1
117,700	2400	1
114,150	2300	1
112,500	2250	1
58,250	1500	1
64,250	1680	1
55,550	1400	1
49,400	1200	1
112,050	2400	1
108,600	2300	1
95,900	1950	1
101,700	2100	1

FIGURE 14.23

DATA FROM FIGURE 14.21 REWRITTEN WITH A DUMMY
VARIABLE USED TO QUANTIFY THE QUALITATIVE VARIABLE.

SOURCE	DF	SUM OF SQUARES	MEAN SQUARE	F VALUE	PR > F
MODEL	2	1.253E10	6.264E09	155.48	.0001
ERROR	21	8.460E08	4.029E07		
TOTAL	23	1.337E10		R SQUARE	
				.9367	

PARAMETER	ESTIMATE	T FOR H0: PARAMETER = 0	PR > |T|	STD ERROR OF ESTIMATE
INTERCEPT	− 37869.17	− 5.55	.0001	6824.125
X	61.77	16.29	.0001	3.791
D	6250.88	2.19	.0397	2849.850

FIGURE 14.24

COMPUTER PRINTOUT FOR THE MODEL IN EQ. 14.9.

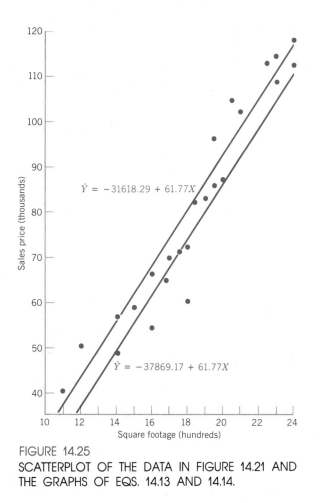

FIGURE 14.25
SCATTERPLOT OF THE DATA IN FIGURE 14.21 AND
THE GRAPHS OF EQS. 14.13 AND 14.14.

Note that the slope, and therefore the increase of sales price of $61.77 for an increase of one square foot, is the same for both models because of the way the model was specified in Eq. 14.9. However, the different intercepts may be interpreted as saying that a house with a wood exterior costs less than a house with an exterior other than wood. The prediction equation of Eq. 14.12 has an associated R^2 value of .9367 so the amount of variation in selling price has increased slightly from 92.2% for the model of Eq. 14.7 to 93.7% for the model of Eq. 14.9. The graphs of Eq. 14.13 and 14.14 appear as parallel lines in Figure 14.25 along with a scatterplot of the pairs of quantitative variables.

Note that the use of a dummy variable enables all of the points to be used in estimating the slope, whereas the estimates of the two intercepts are obtained from the two separate data sets for wood and nonwood exteriors. This method provides more flexibility than one that treats the two data sets separately as suggested previously.

Scientific notation is used in Figure 14.24 to represent some very large numbers that have been calculated for the sums of squares. For example, the scientific notation for the Error SS in Figure 14.24 is 8.460E08, which is used to represent $8.460 \times 10^8 = 846,000,000$. Thus the numbers following E represent the exponent of 10 or the number of places the decimal should be moved. Very small numbers can be written with negative exponents such as $.0007113 = 7.113\text{E-}04$.

CHANGING THE SLOPE BY THE USE OF A DUMMY VARIABLE

In Section 14.1 an interaction term was introduced in an attempt to improve the fit to a set of data. If an interaction term involves a quantitative variable and a dummy variable, such as Dx, the result is a change in the slope of the regression line. For example, consider the following model

$$y = \beta_0 + \beta_1 x + \beta_2 Dx \qquad (14.15)$$

When $D = 0$, Eq. 14.15 can be written as

$$y = \beta_0 + \beta_1 x \qquad (14.16)$$

and when $D = 1$ Eq. 14.15 simplifies to

$$y = \beta_0 + (\beta_1 + \beta_2)x \qquad (14.17)$$

The only difference between Eqs. 14.16 and 14.17 is in the coefficient of x, which of course represents the slope of the line. If the interaction term is included with the model in Eq. 14.9, then both the slope and intercept are changed for the regression lines for different types of house exteriors. Such a model is written as

$$y = \beta_0 + \beta_1 x + \beta_2 D + \beta_3 Dx \qquad (14.18)$$

The computer printout for the least squares fit to the model in Eq. 14.18 is given in Figure 14.26 and gives the prediction equation as

$$\hat{Y} = -27{,}975.59 + 55.95X - 7215.92D + 7.702DX \qquad (14.19)$$

with an R^2 value of .9390, which is almost the same as with Eq. 14.12. When $D = 0$ (corresponding to a house with a wood exterior), Eq. 14.19 reduces to

$$\hat{Y} = -27{,}975.59 + 55.95X \qquad (14.20)$$

and for $D = 1$ (corresponding to a house with an exterior other than wood), Eq. 14.19 simplifies to

$$\hat{Y} = -35{,}191.51 + 63.65X \qquad (14.21)$$

This indicates a difference in the value of an additional square foot of house, the value being only about \$56 for houses with wood exteriors, as compared with about \$64 for other types of exteriors.

SOURCE	DF	SUM OF SQUARES	MEAN SQUARE	F VALUE	PR > F
MODEL	3	1.256E10	4.186E09	102.69	.0001
ERROR	20	8.153E08	4.077E07		
TOTAL	23	1.337E10		R SQUARE	
				.9390	

PARAMETER	ESTIMATE	T FOR H0: PARAMETER = 0	PR > |T|	STD ERROR OF ESTIMATE
INTERCEPT	− 27975.593	− 2.10	.0484	13307.562
X	55.953	7.25	.0001	7.715
D	− 7215.919	− 0.46	.6524	15780.380
D*X	7.702	0.87	.3958	8.875

FIGURE 14.26
COMPUTER PRINTOUT FOR THE MODEL IN EQ. 14.18.

Equation 14.20 is the same equation as would be obtained from a least squares fit of the model in Eq. 14.7 to the first eight data points (wood exterior) of Figure 14.21. Also, Eq. 14.21 corresponds to a least squares fit of the model in Eq. 14.7 to the last 16 data points (brick or adobe exterior) of Figure 14.21. Thus, the equivalent of two separate straight-line fits can be obtained from one fit using the general linear model approach with a dummy variable. The graphs of the prediction Eqs. 14.20 and 14.21 are left as an exercise for the reader.

USE OF TWO DUMMY VARIABLES

An examination of Figure 14.21 shows that only part of the information on the type of house exterior has been utilized. That is, rather than wood versus nonwood exterior, the distinction is really wood, brick, or adobe. This information can be included in a model using two dummy variables D_1 and D_2 as follows:

$$D_1 = 0, D_2 = 0 \quad \text{when exterior is wood}$$
$$D_1 = 1, D_2 = 0 \quad \text{when exterior is brick}$$
$$D_1 = 0, D_2 = 1 \quad \text{when exterior is adobe}$$

The information in Figure 14.21 would be rewritten as in Figure 14.27 where the qualitative variable representing type of exterior has been quantified. The model to be fit to the data in Figure 14.27 has the form:

$$y = \beta_0 + \beta_1 x + \beta_2 D_1 + \beta_3 D_2 + \beta_4 D_1 x + \beta_5 D_2 x \qquad (14.22)$$

The model in Eq. 14.22 provides for a different slope and intercept for each type of house exterior. For example, with a wood exterior ($D_1 = 0, D_2 = 0$),

Sale Price (Y)	Square Footage (X)	Brick (D_1)	Adobe (D_2)
$ 53,400	1600	0	0
59,450	1800	0	0
48,100	1400	0	0
39,900	1100	0	0
82,850	1900	0	0
86,500	2000	0	0
81,650	1850	0	0
85,300	1950	0	0
71,650	1800	1	0
70,250	1750	1	0
65,550	1600	1	0
69,100	1700	1	0
104,150	2050	1	0
117,700	2400	1	0
114,150	2300	1	0
112,500	2250	1	0
58,250	1500	0	1
64,250	1680	0	1
55,550	1400	0	1
49,400	1200	0	1
112,050	2400	0	1
108,600	2300	0	1
95,900	1950	0	1
101,700	2100	0	1

FIGURE 14.27
DATA FROM FIGURE 14.21 REWRITTEN, USING TWO DUMMY
VARIABLES TO QUANTIFY THE QUALITATIVE VARIABLE.

the model in Eq. 14.22 becomes

$$y = \beta_0 + \beta_1 x \tag{14.23}$$

For a brick exterior ($D_1 = 1$, $D_2 = 0$), the model is

$$y = (\beta_0 + \beta_2) + (\beta_1 + \beta_4)x \tag{14.24}$$

Finally, the adobe exterior ($D_1 = 0$, $D_2 = 1$) has the form

$$y = (\beta_0 + \beta_3) + (\beta_1 + \beta_5)x \tag{14.25}$$

Note that there is no need to include the product $D_1 D_2$ in the model, since this product is always zero.

The computer printout for the least squares fit to the model of Eq. 14.22 is

SOURCE	DF	SUM OF SQUARES	MEAN SQUARE	F VALUE	PR > F
MODEL	5	1.267E10	2.534E09	64.73	.0001
ERROR	18	7.046E08	3.915E07		
TOTAL	23	1.337E10		R SQUARE	
				.9473	

PARAMETER	ESTIMATE	T FOR H0: PARAMETER = 0	PR > |T|	STD ERROR OF ESTIMATE
INTERCEPT	− 27975.593	− 2.15	.0458	13040.274
X	55.953	7.40	.0001	7.560
D1	− 28669.800	− 1.42	.1719	20152.423
D2	2143.977	0.13	.8978	16461.068
D1*X	18.383	1.71	.1051	10.772
D2*X	2.709	0.29	.7739	9.287

FIGURE 14.28
COMPUTER PRINTOUT FOR THE MODEL IN EQ. 14.22.

given in Figure 14.28 and provides the following equation:

$$\hat{Y} = -27,975.59 + 55.95X - 28,669.80D_1 + 2143.98D_2$$
$$+ 18.38D_1X + 2.709D_2X \tag{14.26}$$

The R^2 value for Eq. 14.26 is .9473, which represents a slight increase over the model in Eq. 14.18.

The three regression equations for wood, brick, and adobe exteriors corresponding to Eqs. 14.23, 14.24, and 14.25 are, respectively,

$$\hat{Y} = -27,975.59 + 55.95X \tag{14.27}$$
$$\hat{Y} = -56,645.39 + 74.34X \tag{14.28}$$
$$\hat{Y} = -25,831.62 + 58.66X \tag{14.29}$$

An examination of these equations shows Eqs. 14.27 and 14.29 to be nearly the same, with Eq. 14.28 differing greatly from the other two in both slope and intercept. Also note that Eq. 14.27 for wood exterior is the same as Eq. 14.20.

In general, if a qualitative variable has several different possible values, several dummy variables need to be introduced into the model where the number of dummy variables is one less than the number of possible values of the qualitative variable.

The general linear model approach provides a great deal of flexibility in model building because information on a qualitative variable is easily incorporated into the analysis through the use of dummy variables. This approach also allows several fits to be obtained from a single analysis. In the next section the methods of this section are extended to include two qualitative variables.

EXERCISES

14.9 Make a scatterplot similar to the one in Figure 14.25 for the quantitative data given in Figure 14.21. Add the graphs of the prediction Eqs. 14.20 and 14.21 to the scatterplot. How do the graphs of these two lines compare to those in Figure 14.25 in terms of slopes and intercepts?

14.10 Find the least squares fit of the model in Eq. 14.7 for the first set of eight points in Figure 14.21, the second set of eight points, and the last set of eight points. Show that your results correspond to Eqs. 14.27, 14.28, and 14.29, respectively. (First eight points: $\Sigma X = 13,600$, $\Sigma Y = 537,150$, $\Sigma XY = 951,482,500$, $\Sigma X^2 = 23,805,000$; second eight points: $\Sigma X = 1^-,850$, $\Sigma Y = 725,050$, $\Sigma XY = 1,485,915,000$, $\Sigma X^2 = 32,067,500$; third eight points: $\Sigma X = 14,530$, $\Sigma Y = 645,700$, $\Sigma XY = 1,251,640,000$, $\Sigma X^2 = 27,734,900$.)

14.11 Construct graphs of Eqs. 14.27, 14.28, and 14.29 on the same set of axes for values of X ranging from 1000 to 2500. These three straight-line graphs correspond, respectively, to wood, brick, and adobe exteriors. What does the intersecting of these lines imply about the interaction of square footage and type of exterior in terms of influencing selling price?

14.12 For 25 recent sales of condominiums the selling price, age in years, square footage, and the number of bedrooms and type of parking for cars was recorded as follows.

Selling Price (Y)	Age (X_1)	Square Footage (X_2)	Number of Bedrooms	Type of Parking
$44,000	0	880	1	Open
50,000	1	1000	3	Open
52,000	0	970	3	Covered
32,000	16	900	2	Open
32,000	4	830	3	Open
50,000	8	1110	2	Open
51,000	0	970	1	Covered
38,000	11	940	1	Covered
43,000	3	890	1	Covered
53,000	5	1090	2	Open
30,000	8	730	1	Covered
53,000	8	1200	3	Open
49,000	7	1080	1	Covered
59,000	1	1100	2	Covered
35,000	3	760	1	Open
52,000	1	1000	2	Open
55,000	2	1120	3	Open
50,000	7	1120	3	Open
69,000	7	1410	2	Covered
43,000	0	830	1	Covered
44,000	0	880	1	Open

Selling Price (Y)	Age (X_1)	Square Footage (X_2)	Number of Bedrooms	Type of Parking
57,000	19	1460	3	Covered
45,000	12	1030	2	Covered
37,000	3	770	1	Covered
48,000	1	910	2	Open

Write a model that incorporates age and square footage as quantitative variables and type of parking as a qualitative variable, using a dummy variable. Do not include interaction terms.

14.13 For the data in Exercise 14.12 write a model that incorporates square footage as a quantitative variable and number of bedrooms as a qualitative variable. Do not include interaction terms.

14.14 The data in Exercise 14.12 were used to fit the model

$$y = \beta_0 + \beta_1 x_2 + \beta_2 D_1 + \beta_3 D_2$$

where

$$D_1 = 1 \quad \text{if 2 bedrooms, 0 otherwise}$$

$$D_2 = 1 \quad \text{if 3 bedrooms, 0 otherwise}$$

Write the overall prediction equation indicated by the following computer printout of results.

SOURCE	DF	SUM OF SQUARES	MEAN SQUARE	F VALUE	PR > F
MODEL	3	1.496E09	4.988E08	17.97	.0001
ERROR	21	5.830E08	2.776E07		
TOTAL	24	2.079E09		R SQUARE	
				.7196	

PARAMETER	ESTIMATE	T FOR H0: PARAMETER = 0	PR > |T|	STD ERROR OF ESTIMATE
INTERCEPT	3241.781	0.49	.6273	6578.165
X2	43.709	6.00	.0001	7.289
D1	1043.905	0.36	.7204	2877.973
D2	−1464.868	−0.48	.6392	3079.036

14.15 Using the computer printout in Exercise 14.14 write the three individual prediction equations for one, two, and three bedroom condominiums. Use these equations to find the residuals for the first eight sales listed in Exercise 14.12.

14.16 Write the model for the data in Exercise 14.12 that incorporates the quantitative variable X_2, the number of bedrooms as a qualitative variable, and allows three different intercepts and slopes.

14.17 The seasonal effect in time series data can be treated as a qualitative variable that takes four possible values, winter, spring, summer, and fall. Therefore, three dummy variables can be used to model time series data. For the Dr Pepper Company, sales data are fit to the model

$$y = \beta_0 + \beta_1 t + \beta_2 D_1 + \beta_3 D_2 + \beta_4 D_3$$

where t = time (quarters); D_1 = 1 if summer, 0 otherwise; D_2 = 1 if fall, 0 otherwise; D_3 = 1 if winter, 0 otherwise. The Dr Pepper sales data, in thousands of dollars, are as follows.

	March 31	**June 30**	**Sept. 30**	**Dec. 31**
1976	41,234	50,225	54,462	41,295
1977	44,555	59,893	68,958	53,344
1978	54,684	74,238	79,430	62,656
1979	62,691	79,865	83,637	65,569

Write the full prediction equation as indicated by the computer print-out. Use it to find the residuals for all four sales values in the first year, that is t = 1, 2, 3, and 4.

SOURCE	DF	SUM OF SQUARES	MEAN SQUARE	F VALUE	PR > F
MODEL	4	2.683E09	6.708E08	75.42	.0001
ERROR	11	9.783E07	8.894E06		
TOTAL	15	2.781E09		R SQUARE	
				.9648	

PARAMETER	ESTIMATE	T FOR H0: PARAMETER = 0	PR > |T|	STD ERROR OF ESTIMATE
INTERCEPT	35133.05	18.55	.0001	1893.487
T	2236.85	13.42	.0001	116.712
D1	13027.40	6.16	.0001	2115.341
D2	16357.05	7.66	.0001	2134.958
D3	−1785.55	−0.82	.4275	2167.259

14.3
MULTIVARIATE MODELS WITH TWO QUALITATIVE VARIABLES

INCLUDING TWO QUALITATIVE VARIABLES IN THE GENERAL LINEAR MODEL

In Problem Setting 14.2 an Albuquerque, New Mexico, real estate appraiser expressed a desire to develop a model for predicting the selling price of houses in the Albuquerque area. The data included information on a qualitative variable that described the type of house exterior. A model without the qualitative variable, Eq. 14.7, explained 92.2% of the variation in selling price, whereas a model that included the qualitative variable, Eq. 14.22, provided essentially

Sale Price (Y)	Square Footage (X)	Brick (D_1)	Adobe (D_2)	Location (D_3)
$ 53,400	1600	0	0	0
59,450	1800	0	0	0
48,100	1400	0	0	0
39,900	1100	0	0	0
82,850	1900	0	0	1
86,500	2000	0	0	1
81,650	1850	0	0	1
85,300	1950	0	0	1
71,650	1800	1	0	0
70,250	1750	1	0	0
65,550	1600	1	0	0
69,100	1700	1	0	0
104,150	2050	1	0	1
117,700	2400	1	0	1
114,150	2300	1	0	1
112,500	2250	1	0	1
58,250	1500	0	1	0
64,250	1680	0	1	0
55,550	1400	0	1	0
49,400	1200	0	1	0
112,050	2400	0	1	1
108,600	2300	0	1	1
95,900	1950	0	1	1
101,700	2100	0	1	1

FIGURE 14.29
HOUSE SALES DATA WITH TWO QUALITATIVE VARIABLES QUANTIFIED
WITH DUMMY VARIABLES.

three regression equations (Eqs. 14.27, 14.28, and 14.29). These equations explained 94.7% of the variation in selling price.

Upon further examination of the sales information, the real estate appraiser notes that she has additional information that gives the location of the houses. In particular she can tell whether the house was in an old development of the city (older than 15 years) or in a newer development. This additional information is given in Figure 14.29 where the location in the city has been quantified with the use of a dummy variable defined as

$$D_3 = 0 \quad \text{house is in an old development}$$
$$D_3 = 1 \quad \text{house is in a new development}$$

Figure 14.29 is an extension of Figure 14.27 where the dummy variables D_1 and D_2 were given.

If the dummy variable D_3 is added to the model of Eq. 14.22, the following

SOURCE	DF	SUM OF SQUARES	MEAN SQUARE	F VALUE	PR > F
MODEL	6	1.337E10	2.228E09	6929.01	.0001
ERROR	17	5.467E06	3.216E05		
TOTAL	23	1.337E10		R SQUARE	
				.9996	

PARAMETER	ESTIMATE	T FOR H0: PARAMETER = 0	PR > |T|	STD ERROR OF ESTIMATE
INTERCEPT	10235.314	7.12	.0001	1438.188
X	26.993	29.19	.0001	0.925
D1	−7273.923	−3.86	.0012	1883.243
D2	−2880.589	−1.93	.0710	1495.813
D1∗X	11.694	11.85	.0001	0.987
D2∗X	7.329	8.65	.0001	0.848
D3	22041.660	46.63	.0001	472.703

FIGURE 14.30
COMPUTER PRINTOUT FOR THE MODEL IN EQ. 14.30.

model is formed

$$y = \beta_0 + \beta_1 x + \beta_2 D_1 + \beta_3 D_2 + \beta_4 D_1 x + \beta_5 D_2 x + \beta_6 D_3 \quad (14.30)$$

The model in Eq. 14.30 reduces to the model in Eq. 14.22 when $D_3 = 0$ and, therefore, will again produce the same models given in Eqs. 14.23, 14.24, and 14.25. However, when $D_3 = 1$ the intercept in each of these three equations is increased by β_6. The computer printout for the least squares fit to the model in Eq. 14.30 is given in Figure 14.30, and gives the following prediction equation:

$$\hat{Y} = 10{,}235.31 + 26.99X - 7273.92D_1 - 2880.59D_2$$
$$+ 11.69D_1X + 7.33D_2X + 22{,}041.66D_3 \quad (14.31)$$

Equation 14.31 has an associated R^2 value of .9996, which is very close to the maximum of 1.0000, so the inclusion of location has allowed for almost all of the variation in selling price to be explained.

With the almost perfect fit obtained by the model in Eq. 14.30, there does not seem to be any need to increase the R^2 value. However, the following development shows that there are other reasons for wanting an accurate model, other than just a high R^2.

INTERACTION WITH TWO QUALITATIVE VARIABLES

The six regression equations that are derived from Eq. 14.30 have six unique intercepts but only three unique slopes. The slope can be adjusted as in Section 14.2 by including the term $D_3 x$ in the model of Eq. 14.30. The result is six unique slopes to accompany the six intercepts corresponding to the six combinations of location and exterior.

Two other interaction terms could also be added to the model of Eq. 14.30. These are D_1D_3 and D_2D_3, which represent interaction between qualitative variables. That is, it is possible that the exterior of the house and the location of the house act together in a way that is not additive to affect the selling price. Thus the model used is

$$y = \beta_0 + \beta_1 x + \beta_2 D_1 + \beta_3 D_2 + \beta_4 D_1 x + \beta_5 D_2 x + \beta_6 D_3$$
$$+ \beta_7 D_3 x + \beta_8 D_1 D_3 + \beta_9 D_2 D_3 \tag{14.32}$$

The computer printout for the model in Eq. 14.32 using a least squares fit is given in Figure 14.31, and gives the prediction equation

$$\hat{Y} = 9234.66 + 27.78X + 4282.47D_1 + 3790.52D_2$$
$$+ 4.697D_1 X + 2.556D_2 X + 10{,}210.37D_3$$
$$+ 5.792D_3 X + 2286.69D_1 D_3 + 2293.30D_2 D_3 \tag{14.33}$$

with an R^2 value of .9999. This shows the price of brick homes in general increased \$4282.47, regardless of the neighborhood. In the previous model the negative coefficient on D_1 gave the appearance of decreasing the price of the house if it was brick, which is contrary to common experience. By allowing for different slopes in different neighborhoods this difficulty is eliminated. The same comments can be made about the treatment of adobe houses in the two models.

The effect of using this model rather than the previous model may be explained as follows. In the previous model the intercept is increased by an

SOURCE	DF	SUM OF SQUARES	MEAN SQUARE	F VALUE	PR > F
MODEL	9	1.337E10	1.486E09	16290.78	.0001
ERROR	14	1.277E06	9.121E04		
TOTAL	23	1.337E10		R SQUARE	
				.9999	

PARAMETER	ESTIMATE	T FOR H0: PARAMETER = 0	PR > |T|	STD ERROR OF ESTIMATE
INTERCEPT	9234.657	10.76	.0001	857.879
X	27.782	48.52	.0001	0.573
D1	4282.472	1.85	.0853	2312.966
D2	3790.519	2.69	.0177	1410.570
D1*X	4.697	3.42	.0041	1.373
D2*X	2.556	2.67	.0184	0.958
D3	10210.366	5.21	.0001	1958.724
D3*X	5.792	5.72	.0001	1.013
D1*D3	2286.688	3.36	.0047	681.055
D2*D3	2293.296	3.75	.0021	611.002

FIGURE 14.31
COMPUTER PRINTOUT FOR THE MODEL IN EQ. 14.32.

amount $\beta_6 = 22{,}041.66$ for all houses in new developments, whether those houses are made out of wood, brick, or adobe. In the new model the intercept for wood houses is increased by the new amount for $\beta_6 = 10{,}210.37$. The intercept for brick houses is increased by the amount $\beta_6 + \beta_8 = 12{,}497.05$, and the intercept for adobe houses is increased by the amount $\beta_6 + \beta_9 = 12{,}503.66$.

Another way of looking at the difference in intercepts caused by adding the interaction terms D_1D_3 and D_2D_3 to the model is through graphs of the intercept values. In Figure 14.32a the six intercepts resulting from Eq. 14.31 are plotted, with lines connecting the intercepts for models in similar neighborhoods. The intercepts for new neighborhoods are uniformly a distance β_6 above the intercepts for old neighborhoods. On the other hand, the intercepts arising from Eq. 14.33 plotted in Figure 14.32b do not have that equidistant relationship. The lines connecting intercepts in similar neighborhoods are no longer a uniform distance apart, for the reasons just explained. The intercepts in Figure 14.32b exhibit the interaction present between the type of exterior and the neighborhood. Note that the extent of the interaction is slight.

Although the effect of the interaction between the qualitative variables on the intercepts is shown in Figure 14.32a and b, part of the different appearance

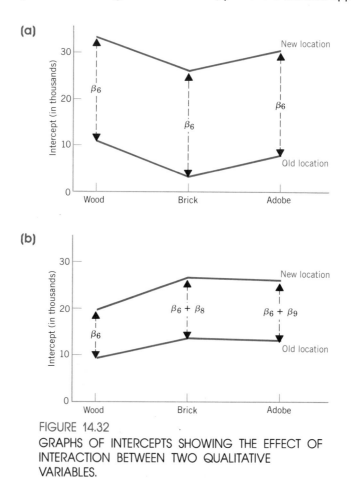

FIGURE 14.32

GRAPHS OF INTERCEPTS SHOWING THE EFFECT OF INTERACTION BETWEEN TWO QUALITATIVE VARIABLES.

in the graphs in those two figures is due to the other interaction term D_3x, which is present in the latter model but not in the former. However, the six slopes in the latter model, given by Eq. 14.33, still have the same equidistant property shared by the intercepts in Eq. 14.31. That is, the slopes for the models in the new neighborhood are each an amount $\beta_7 = 5.79$ greater than the corresponding slopes for wood, brick, and adobe homes in the old neighborhood. This equidistant property can be eliminated by adding two more terms to the model, the three-way interaction terms D_1D_3x and D_2D_3x. This result is deferred to the review problems.

THE SIX MODELS

Six models can be derived from the model in Eq. 14.32 and corresponding prediction equations can be derived from Eq. 14.33. These six pairs of equations depend on the combination of dummy variables used to indicate type of house exterior and location of the house. Either the dummy variables can be substituted directly into Eq. 14.33 along with the value of X (square footage) to predict the selling price, or the simplified versions given below can be used. Both give the same predictions.

1. $D_1 = 0$, $D_2 = 0$, $D_3 = 0$ (wood exterior in an old development).

$\qquad$ Model: $y = \beta_0 + \beta_1 x$

$\qquad$ Prediction equation: $\hat{Y} = 9234.66 + 27.78X$ $\qquad$ (14.34)

2. $D_1 = 0$, $D_2 = 0$, $D_3 = 1$ (wood exterior in a new development).

$\qquad$ Model: $y = (\beta_0 + \beta_6) + (\beta_1 + \beta_7)x$

$\qquad$ Prediction equation: $\hat{Y} = 19{,}445.02 + 33.57X$ $\qquad$ (14.35)

3. $D_1 = 1$, $D_2 = 0$, $D_3 = 0$ (brick exterior in an old development).

$\qquad$ Model: $y = (\beta_0 + \beta_2) + (\beta_1 + \beta_4)x$

$\qquad$ Prediction equation: $\hat{Y} = 13{,}517.13 + 32.48X$ $\qquad$ (14.36)

4. $D_1 = 1$, $D_2 = 0$, $D_3 = 1$ (brick exterior in a new development).

$\qquad$ Model: $y = (\beta_0 + \beta_2 + \beta_6 + \beta_8) + (\beta_1 + \beta_4 + \beta_7)x$

$\qquad$ Prediction equation: $\hat{Y} = 26{,}014.18 + 38.27X$ $\qquad$ (14.37)

5. $D_1 = 0$, $D_2 = 1$, $D_3 = 0$ (adobe exterior in an old development).

$\qquad$ Model: $y = (\beta_0 + \beta_3) + (\beta_1 + \beta_5)x$

$\qquad$ Prediction equation: $\hat{Y} = 13{,}025.18 + 30.34X$ $\qquad$ (14.38)

6. $D_1 = 0$, $D_2 = 1$, $D_3 = 1$ (adobe exterior in a new development).

$\qquad$ Model: $y = (\beta_0 + \beta_3 + \beta_6 + \beta_9) + (\beta_1 + \beta_5 + \beta_7)x$

$\qquad$ Prediction equation: $\hat{Y} = 25{,}528.84 + 36.13X$ $\qquad$ (14.39)

The graphs of these six prediction equations are given in Figure 14.33. These graphs show that the set of points that once appeared to be reasonably fit by a straight line in Figure 14.22 are much more accurately described by six

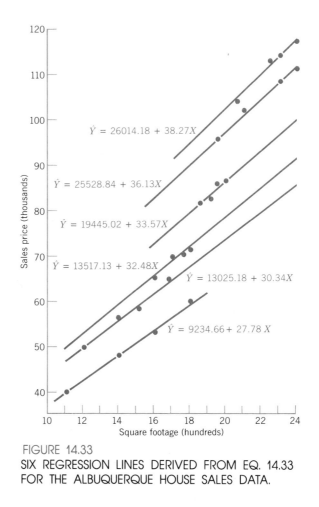

FIGURE 14.33
SIX REGRESSION LINES DERIVED FROM EQ. 14.33
FOR THE ALBUQUERQUE HOUSE SALES DATA.

different lines in sets of four points corresponding to the combination of house exterior and location.

Much more important than the slight increase in R^2 is the interpretation of the new models. The additional square footage adds anywhere from \$27.78 to \$38.27 per square foot to the price of the house, depending on the neighborhood and the type of construction. This is more accurate than the inflated figures in the previous section, which failed to separate out the effect of the neighborhood on the selling price. Also, as mentioned earlier in this section, the additional value of brick or adobe construction is brought to light in this model, whereas the earlier model in this section failed to separate that facet from other concepts, such as unequal prices per square foot.

LACK OF ASSUMPTIONS

This chapter presents different procedures for formulating general linear models for a set of sample data. These include straight-line models, quadratic models, models with interaction terms, and models with one or more dummy variables. These models can be used with any set of data where the objective is to improve a fit to the data, because they all were fit based on least squares calculations that have no underlying distributional assumptions. However, it is quite likely that terms will be included in a general linear model that do not make a "significant" improvement in the model. Judging the "significance" of a variable becomes quite important when there are many variables to choose from. It is best to base such judgments on statistical tests of hypotheses that have underlying distribution assumptions. Therefore, the inference aspect of model building is considered in the next chapter.

EXERCISES

14.18 Refer to Eq. 14.31 and write the six prediction equations corresponding to the six possible sets of values to which the dummy variables D_1, D_2, and D_3 can be assigned.

14.19 For the condominium sale price data in Exercise 14.12, a model involving both number of bedrooms ($D_1 = 1$ for two bedrooms, $D_2 = 1$ for three bedrooms) and type of parking ($D_3 = 1$ for covered) as qualitative variables is fit using the least squares method. The result is given in the following computer printout. Write the complete prediction equation suggested in the printout. Then, write individual prediction equations for the six situations covered by the dummy variables.

SOURCE	DF	SUM OF SQUARES	MEAN SQUARE	F VALUE	PR > F
MODEL	4	1.497E09	3.742E08	12.85	.0001
ERROR	20	5.825E08	2.913E07		
TOTAL	24	2.079E09		R SQUARE	
				.7199	

PARAMETER	ESTIMATE	T FOR H0: PARAMETER = 0	PR > |T|	STD ERROR OF ESTIMATE
INTERCEPT	3377.239	0.50	.6256	6815.839
X2	43.287	5.33	.0001	8.125
D1	1234.551	0.38	.7110	3284.399
D2	−1231.349	−0.34	.7372	3618.493
D3	332.545	0.13	.8966	2526.497

14.20 In an attempt to further refine the model in Exercise 14.19, interaction terms are included to account for possibly different values of additional square footage in the six situations. Write the overall prediction equation suggested by the following computer printout. Then write the individual prediction equations for the six different situations and make comparisons with the results in Exercise 14.19.

SOURCE	DF	SUM OF SQUARES	MEAN SQUARE	F VALUE	PR > F
MODEL	7	1.591E09	2.273E08	7.91	.0003
ERROR	17	4.882E08	2.872E07		
TOTAL	24	2.079E09		R SQUARE	
				.7652	

PARAMETER	ESTIMATE	T FOR H0: PARAMETER = 0	PR > |T|	STD ERROR OF ESTIMATE
INTERCEPT	−10723.481	−0.60	.5562	17862.152
X2	58.925	2.95	.0089	19.952
D1	−9341.631	−0.46	.6531	20421.457
D2	17146.730	0.89	.3833	19159.694
D3	11760.776	0.86	.4025	13696.019
D1*X2	8.476	0.40	.6929	21.098
D2*X2	−18.657	−0.93	.3636	19.983
D3*X2	−12.158	−0.88	.3887	13.745

14.21 Use the prediction equation in Exercise 14.19 to estimate the net effect on sale price of having two bedrooms instead of one. Use the same prediction equation to estimate the effect of having covered parking. Do these results seem reasonable to you?

14.22 Use the prediction equation in Exercise 14.20 to estimate the net effect on sale price of having two bedrooms instead of one. Use the same prediction equation to estimate the effect of having covered parking. Do these results seem reasonable to you? Which model, that of Exercise 14.19 or 14.20, seems more reasonable on this basis?

14.23 In a further extension of the models of Exercises 14.19 and 14.20, terms expressing interaction between the two qualitative variables are added. What is the effect of including such terms in the model? Write the complete prediction equation, using the following computer printout as a guide. Also write the six individual prediction equations and compare them with the equations in Exercises 14.19 and 14.20. Are there any interesting differences in the coefficients caused by adding these interaction terms to the model?

SOURCE	DF	SUM OF SQUARES	MEAN SQUARE	F VALUE	PR > F
MODEL	9	1.668E09	1.853E08	6.75	.0007
ERROR	15	4.118E08	2.745E07		
TOTAL	24	2.079E09		R SQUARE	
				.8020	(continued)

PARAMETER	ESTIMATE	T FOR H0: PARAMETER = 0	PR > |T|	STD ERROR OF ESTIMATE
INTERCEPT	− 26510.225	− 1.33	.2019	19863.898
X2	80.369	3.44	.0037	23.372
D1	− 4326.440	− 0.21	.8400	21057.808
D2	23093.305	1.20	.2474	19186.352
D3	25500.539	1.62	.1257	15724.596
D1*X2	− 2.688	− 0.12	.9081	22.899
D2*X2	− 31.587	− 1.49	.1557	21.133
D3*X2	− 32.371	− 1.79	.0938	18.095
D1*D3	9536.966	1.34	.1999	7112.989
D2*D3	12476.558	1.59	.1317	7824.515

14.24 The addition of two more terms to the model of Exercise 14.23 results in the following computer printout. Write the prediction equation as indicated by the printout. Also write the six individual prediction equations.

SOURCE	DF	SUM OF SQUARES	MEAN SQUARE	F VALUE	PR > F
MODEL	11	1.698E09	1.543E08	5.26	.0031
ERROR	13	3.818E08	2.937E07		
TOTAL	24	2.079E09		R SQUARE .8164	

PARAMETER	ESTIMATE	T FOR H0: PARAMETER = 0	PR > |T|	STD ERROR OF ESTIMATE
INTERCEPT	− 22000.000	− 0.47	.6444	46563.516
X2	75.000	1.36	.1982	55.307
D1	9319.749	0.17	.8663	54260.782
D2	9133.782	0.18	.8597	50660.750
D3	20473.863	0.42	.6849	49329.255
D1*X2	− 15.439	− 0.25	.8068	61.855
D2*X2	− 17.252	− 0.30	.7724	58.414
D3*X2	− 26.420	− 0.45	.6575	58.229
D1*D3	− 13598.828	− 0.22	.8270	60984.399
D2*D3	34494.396	0.61	.5530	56634.887
D1*D3*X2	20.648	0.31	.7635	67.206
D2*D3*X2	− 21.124	− 0.33	.7434	63.154

14.25 The computer printout in Exercise 14.24 indicates an increase in selling price of the condominium of $20,473.86 if the parking is covered, according to the coefficient of D_3. Yet the data in Exercise 14.12 do not show the condominiums with covered parking as being that much more expensive than the other condominiums. How do you account for the apparent paradox?

14.4

REVIEW EXERCISES

14.26 The housing sale price data in Section 14.3 can be modeled using the full prediction equation with the three-way interaction terms D_1D_3x and D_2D_3x added to Eq. 14.32. The result of a least squares fit is given in the following computer printout. Write the full prediction equation as indicated by the printout. Write the six individual prediction equations and compare them with Eqs. 14.34 to 14.39.

SOURCE	DF	SUM OF SQUARES	MEAN SQUARE	F VALUE	PR > F
MODEL	11	1.337E10	1.216E09	13142.52	.0001
ERROR	12	1.110E06	9.250E04		
TOTAL	23	1.337E10		R SQUARE	
				.9999	

PARAMETER	ESTIMATE	T FOR H0: PARAMETER = 0	PR > |T|	STD ERROR OF ESTIMATE
INTERCEPT	9264.019	10.52	.0001	880.608
X	27.762	47.21	.0001	0.588
D1	7764.553	2.14	.0539	3633.178
D2	3158.425	2.04	.0643	1550.535
D3	9360.981	1.76	.1035	5312.362
D1*X	2.667	1.25	.2362	2.139
D2*X	2.993	2.83	.0151	1.056
D3*X	6.238	2.24	.0447	2.783
D1*D3	− 1928.014	− 0.28	.7853	6919.061
D2*D3	4646.062	0.80	.4378	5788.345
D1*D3*X	2.295	0.63	.5425	3.660
D2*D3*X	− 1.275	− 0.42	.6829	3.045

In general, what is the purpose of adding those two terms to this model? In this particular case, is there a substantial change in the equations? Could you have predicted this result from the high R^2 associated with Eq. 14.33?

14.27 *Fortune* magazine lists the "second 500" companies in sales each year in addition to the top 500. A random sample of 25 companies from *Fortune*'s second 500 is obtained to model net income as a function of sales and assets. The data are as follows (in millions of dollars).

Net Income (Y)	Sales (X_1)	Assets (X_2)
2.88	155.71	104.5
25.46	186.43	284.5
2.71	108.73	44.5
9.61	129.95	86.2

Net Income (Y)	Sales (X_1)	Assets (X_2)
16.16	209.77	137.1
4.13	130.56	72.3
5.15	137.63	78.5
6.21	144.40	97.0
5.29	147.12	87.6
0.78	145.11	87.8
19.16	209.31	134.6
14.44	161.49	210.7
1.79	127.32	88.6
2.70	125.33	68.0
−4.88	112.49	69.6
9.64	157.63	79.9
11.43	162.27	159.7
−1.46	126.30	101.7
7.41	115.88	97.2
23.72	230.64	177.9
18.14	243.80	169.2
18.67	349.62	168.6
7.48	148.60	68.1
8.79	206.68	180.7
20.97	291.50	195.9

Plot net income versus sales in a graph. Does there appear to be either a linear or quadratic relationship between the two variables? Is there an outlier in the data?

14.28 The data in Exercise 14.27 are fit to the model $y = \beta_0 + \beta_1 x_1$ using the method of least squares, with the results given in the following computer printout. Write the fitted prediction equation as indicated by the printout.

SOURCE	DF	SUM OF SQUARES	MEAN SQUARE	F VALUE	PR > F
MODEL	1	908.992	908.992	30.40	.0001
ERROR	23	687.836	29.906		
TOTAL	24	1596.828		R SQUARE	
				.5692	

PARAMETER	ESTIMATE	T FOR H0: PARAMETER = 0	PR > |T|	STD ERROR OF ESTIMATE
INTERCEPT	−8.336	−2.45	.0225	3.4073
X1	0.104	5.51	.0001	0.0189

14.29 The data in Exercise 14.27 are fit to the model

$$y = \beta_0 + \beta_1 x_1 + \beta_2 x_1^2 + \beta_3 x_2 + \beta_4 x_2^2 + \beta_5 x_1 x_2$$

using the least squares method as indicated in the following computer printout. Write the prediction equation indicated by the printout. Use this equation to find residuals for the first five data points listed in Exercise 14.27.

SOURCE	DF	SUM OF SQUARES	MEAN SQUARE	F VALUE	PR > F
MODEL	5	1295.548	259.110	16.34	.0001
ERROR	19	301.280	15.857		
TOTAL	24	1596.828		R SQUARE	
				.8113	

PARAMETER	ESTIMATE	T FOR H0: PARAMETER = 0	PR > |T|	STD ERROR OF ESTIMATE
INTERCEPT	−28.316	−2.31	.0323	12.257
X1	0.341	2.74	.0130	0.125
X1*X1	−4.833E-04	−1.70	.1052	2.841E-04
X2	−5.613E-02	−0.48	.6334	0.116
X2*X2	4.660E-04	1.52	.1445	3.061E-04
X1*X2	−2.770E-04	−0.37	.7130	7.417E-04

14.30 The condominium sales data in Exercise 14.12 are fit to the model

$$y = \beta_0 + \beta_1 x_2 + \beta_2 D_1 + \beta_3 D_2 + \beta_4 D_1 x_2 + \beta_5 D_2 x_2$$

using the method of least squares, as indicated in the following computer printout, where $D_1 = 1$ for two bedrooms, and $D_2 = 1$ for three bedrooms. Write the prediction equation as indicated by the computer printout. Give an interpretation to each of the five coefficients β_1 to β_5.

SOURCE	DF	SUM OF SQUARES	MEAN SQUARE	F VALUE	PR > F
MODEL	5	1.569E09	3..137E08	11.67	.0001
ERROR	19	5.108E08	2.688E07		
TOTAL	24	2.079E09		R SQUARE	
				.7543	

PARAMETER	ESTIMATE	T FOR H0: PARAMETER = 0	PR > |T|	STD ERROR OF ESTIMATE
INTERCEPT	−1719.288	−0.12	.9050	14212.103
X2	49.392	3.05	.0065	16.171
D1	−8254.317	−0.43	.6753	19401.894
D2	17138.463	0.93	.3652	18473.090
D1*X2	7.659	0.38	.7099	20.280
D2*X2	−18.085	−0.94	.3611	19.324

14.31 Construct a model for the data in Exercise 14.12 that uses both quantitative variables [age (X_1) and square footage (X_2)] and both qualitative variables [number of bedrooms $(D_1$ and $D_2)$ and type of parking (D_3)]. Do not include any interaction terms. How many different situations are represented by this model? Write the individual model for each different situation.

14.32 Add the interaction terms necessary to allow for different slopes in each individual model in Exercise 14.31. Write the individual model for each different situation.

14.33 Add the remaining interaction terms to the model in Exercise 14.32 to remove the constant additive effect of covered parking on both the intercepts and the slopes. Write the individual models for each different situation.

BIBLIOGRAPHY

Additional material on the topics presented in this chapter can be found in the following publications

Daniel, C. and Wood, F. S. (1980). *Fitting Equations to Data,* 2nd ed. Wiley, New York.

McClave, J. T. and Benson, P. G. (1985). *Statistics for Business and Economics,* 3rd ed. Dellen-Macmillan, San Francisco.

15

MULTIPLE LINEAR REGRESSION

PRELIMINARY REMARKS

In the previous chapter, techniques were presented for formulating general linear models. These models were fit using the method of least squares. The calculation of least squares coefficients does not require any underlying distributional assumptions, but merely provides a mechanical technique for fitting a desired general linear model of the form

$$y = \beta_0 + \beta_1 x_1 + \beta_2 x_2 + \cdots + \beta_k x_k$$

to a set of sample data. Thus, the possibility exists that some of the terms $\beta_i x_i$ included in the model may not make a significant additional contribution (i.e., in the presence of other x_i) toward explaining the variation in the dependent variable y. **Multiple linear regression** provides the framework for determining whether a variable x_i makes a significant contribution by testing the null hypothesis H_0: $\beta_i = 0$ against H_1: $\beta_i \neq 0$ for $i = 1, \ldots, k$. Unless H_0 is rejected for some coefficient β_i, there is no convincing evidence that the corresponding variable x_i makes a significant contribution to the model. Then the term $\beta_i x_i$ can be deleted from the model, thus simplifying it.

The testing of such hypotheses relies on underlying distributional assumptions similar to those presented in Chapter 12 for simple linear regression. In the first section of this chapter the procedure for testing hypotheses such as

<table>
<tr><td>

MULTIPLE
REGRESSION
EQUATION

</td><td>

If the mean of a random variable Y is a function of other variables $x_1, x_2, \ldots,$ x_k, the equation that expresses that functional relationship

$$E(Y) = f(x_1, x_2, \ldots, x_k)$$

is a **multiple regression equation.**

</td></tr>
</table>

<table>
<tr><td>

MULTIPLE
LINEAR
REGRESSION

</td><td>

If the multiple regression equation is in terms of a general linear model, such as

$$E(Y) = \beta_0 + \beta_1 x_1 + \beta_2 x_2 + \cdots + \beta_k x_k$$

then the regression is a **multiple linear regression.**

</td></tr>
</table>

those previously given is described. The second section addresses the questions of how to make predictions from the regression equation and how to provide confidence intervals for such predictions.

The third section discusses how to select a regression model in the presence of a large number of independent variables. Statisticians have devised many techniques for identifying a useful subset of a large number of independent variables for use in a regression equation. The subset selection technique presented in this chapter is one of the more widely used procedures and is referred to as stepwise regression because, with it, a model is built up one step at a time until all "significant" variables have been included.

15.1
TESTS OF HYPOTHESES
IN MULTIPLE REGRESSION

PROBLEM SETTING 15.1

An industrial consulting firm has been hired to survey a large research laboratory to analyze employee job satisfaction. The laboratory employees include clerical workers, secretaries, laboratory assistants, professional staff, and management, with the educational degrees ranging from high school to advanced graduate. The consulting firm randomly selects 50 individuals for its survey and has each of them provide information on the following qualitative and quantitative variables (while their identity remains unknown).

X_1 Age.

X_2 Sex (coded as 0 for female and 1 for male).

X_3 Salary.

X_4 Number of years with the laboratory.

X_5 Number of years of related job experience.

X_6 Education level (in years).

X_7 Percentage of total income that their salary represents.

X_8 Management position (coded as 0 for no and 1 for yes).

Additionally, each employee responds to a number of questions that are designed to measure job satisfaction. The responses to these questions are used to produce an employee job satisfaction score for each employee on a scale from 1 (highly dissatisfied) to 15 (highly satisfied). The results of the survey appear in Figure 15.1. The consulting firm would like to determine what relationship exists between the variables X_1 to X_8 and degree of job satisfaction (Y), that would allow job satisfaction to be predicted. The firm would also like to report to the laboratory which of the variables are related to employee job satisfaction.

THE NEED FOR RESTRICTING THE NUMBER OF TERMS IN A MODEL

In the previous chapter, models were considered for describing a relationship among the variables dosage, age, weight, sex, blood pressure, and time to relief. Another problem setting considered a model for predicting the selling

Number	X_1	X_2	X_3	X_4	X_5	X_6	X_7	X_8	Y
1	23	1	35980	5	0	17	93	0	9.9
2	31	0	20420	11	0	14	99	0	7.8
3	64	1	59090	30	5	16	56	1	11.8
4	46	1	34480	12	7	16	94	0	10.4
5	34	1	23980	12	1	15	94	0	9.3
6	39	0	41560	7	2	17	93	0	8.8
7	31	1	18100	4	6	15	100	0	6.1
8	19	0	18510	1	0	15	100	0	7.0
9	33	0	13160	2	1	14	100	0	4.3
10	26	0	15410	2	1	14	92	0	5.1
11	62	1	57260	27	6	16	79	0	12.9
12	18	0	10310	1	0	13	95	0	5.2
13	21	1	13360	4	0	13	99	0	9.6
14	60	1	37770	16	7	16	98	0	8.6
15	26	1	21190	7	0	16	93	0	5.5
16	25	0	30740	1	0	16	83	0	6.2
17	18	0	10450	1	0	12	100	0	3.8
18	40	1	59860	6	2	19	90	1	11.7
19	30	1	10940	4	2	13	100	0	4.9
20	20	1	19770	2	0	15	97	0	11.2
21	61	0	33210	22	0	15	95	0	8.5
22	22	0	19930	2	2	16	92	0	5.4

Number	X_1	X_2	X_3	X_4	X_5	X_6	X_7	X_8	Y
23	35	1	46620	5	0	17	98	0	10.1
24	18	0	19690	1	0	13	97	0	9.9
25	40	1	38600	9	1	16	92	0	5.6
26	22	0	11370	2	1	12	91	0	5.7
27	43	0	41480	7	1	16	99	0	10.6
28	27	0	27580	2	0	16	100	0	9.3
29	70	0	50550	6	9	16	99	0	10.2
30	37	1	55520	16	0	19	93	1	6.8
31	32	0	12000	5	3	12	99	0	6.2
32	30	0	19910	9	0	13	94	0	10.2
33	36	1	41750	5	2	18	93	0	6.4
34	69	0	21300	21	2	13	99	0	8.7
35	24	0	12540	2	0	13	99	0	5.1
36	29	0	17220	4	0	13	100	0	8.4
37	36	0	17480	9	0	12	91	0	6.8
38	31	0	19400	5	2	14	99	0	10.1
39	20	1	18480	2	0	15	99	0	8.3
40	46	1	35990	11	8	14	92	0	11.2
41	20	1	17710	3	0	14	97	0	9.0
42	46	1	49560	7	5	18	71	0	8.6
43	62	0	30660	7	12	16	69	0	9.7
44	37	1	40960	14	0	16	92	0	11.9
45	34	0	15210	4	3	15	99	0	4.3
46	48	1	43930	22	0	16	92	0	10.5
47	34	1	39640	12	0	16	99	0	8.9
48	26	0	10700	5	2	12	93	0	5.6
49	29	1	25700	13	0	13	97	0	13.1
50	43	1	18900	18	2	13	95	0	9.3

FIGURE 15.1
RESULTS OF A SURVEY OF 50 LABORATORY EMPLOYEES
REGARDING JOB SATISFACTION.

price of a house based on both qualitative and quantitative information. Frequently, model builders are confronted with many more independent variables (perhaps hundreds). In cases where the number of independent variables is large the complexity of models containing linear and quadratic terms (including interactions) becomes quite unwieldy. In fact, the number of terms in a full quadratic model with k independent variables is $(k + 1)(k + 2)/2$. As the reader can verify, this expression produces quite large values even when k is quite small. Hence, a systematic procedure is needed that will in some sense produce a model containing only the "important" variables. That is, in many model-building situations data are typically available on several independent variables. The list of independent variables frequently includes variables that turn out to have negligible influence on the dependent variable. It is up to the model builder to determine which variables are important enough to be included in the model and which variables are not important enough to be included.

This section presents tests of hypotheses as a tool to aid the modeler in determining the importance of variables. However, decisions on whether to include or exclude individual variables should not be made solely on the basis of hypothesis tests. Any additional information available to the model builder that is more convincing than a hypothesis test should be utilized.

THE MODEL FOR MULTIPLE REGRESSION

In Section 12.2 a model was presented for simple linear regression (one independent variable) along with the assumptions for such a regression. Those assumptions can be extended to cover the case of multiple linear regression (several independent variables) where the underlying general linear model is of the form

$$Y_i = \beta_0 + \beta_1 X_{1i} + \beta_2 X_{2i} + \cdots + \beta_k X_{ki} + \epsilon_i, \qquad i = 1, \ldots, n \quad (15.1)$$

The parameter β_0 is the intercept and the remaining parameters β_1 to β_k are the slopes associated with each independent variable. They measure the change in Y associated with a change in each X. The ϵ_i are assumed to be independent random variables, each with the same normal distribution with mean 0 and variance σ^2. The X's may be random variables, dummy variables, or fixed values. Throughout this chapter the job satisfaction data in Problem Setting 15.1 are assumed to satisfy these conditions.

HYPOTHESIS TESTING IN MULTIPLE REGRESSION

It is common modeling practice to include enough terms in the initial version of the model so that all independent variables are incorporated. The reason for this is that the modeler does not know a priori which independent variables are important. Consequently, variables frequently get incorporated in a model that do not contribute significantly toward explaining the variation in the dependent variable Y and therefore probably should not have been included in the model. If a variable X_j is not making a significant contribution toward explaining the variation in the dependent variable, then the least squares procedure automatically adjusts by making the estimated coefficient $\hat{\beta}_j$ relatively close to zero, as measured in terms of its standard error. Thus, the hypothesis of interest is $H_0: \beta_j = 0$ versus $H_1: \beta_j \neq 0$. The test of this hypothesis determines whether the estimated coefficient $\hat{\beta}_j$ is "significantly different from zero." If H_0 is rejected, then the corresponding variable X_j should be included in the model. Otherwise there is no convincing evidence that X_j makes a significant contribution in explaining the variation of Y and therefore X_j can be eliminated from the model. Of course, this does not preclude the modeler from including the variable X_j in the model based on prior information regarding the importance of the variable.

The test of H_0 is based on the statistic

$$T = \frac{\hat{\beta}_j}{s_{\hat{\beta}_j}} \qquad (15.2)$$

where $\hat{\beta}_j$ is the least squares estimate of the coefficient β_j of X_j in the general linear model and $s_{\hat{\beta}_j}$ is the estimated standard error of the estimate $\hat{\beta}_j$. The statistic T is compared with quantiles of the Student's t distribution with degrees of freedom equal to $n - k - 1$ (i.e., the error degrees of freedom given in the computer printout of the regression analysis). Fortunately, the work in calculating the test statistic T is part of the computer printout.

Because there are many terms in the model, there are also many tests that may be performed, one on each term, and each has its own level of significance. Therefore, the overall probability of a Type I error may be quite large. For this reason, it is a sound statistical procedure to precede these individual tests with one overall test of the model and to follow with individual tests only if the overall test of the model is significant.

AN OVERALL TEST OF THE MODEL

It is useful to see if the model, considering all of the parameters simultaneously, accounts for a significant amount of the variation in Y, before checking the individual terms in the model one by one. This overall test is easy to do with most types of computer programs. The statistic is an F statistic that may be compared with the F tables to see if it is large enough to be significant. Most computer packages not only compute the F statistic but they also make the comparison with the F distribution and print out the p-value, which is the probability of getting an F value as large or larger than the observed value when H_0 is assumed to be true.

For example, in Figure 15.2 the F statistic for testing simultaneously that all β's except β_0 are equal to zero,

$$H_0: \beta_1 = \cdots = \beta_8 = 0$$

SOURCE	DF	SUM OF SQUARES	MEAN SQUARE	F VALUE	PR > F
MODEL	8	150.91	18.86	5.39	.0001
ERROR	41	143.57	3.50		
TOTAL	49	294.49			
			R SQUARE		
			.5125		

PARAMETER	ESTIMATE	T FOR H0: PARAMETER = 0	PR > |T|	STD ERROR OF ESTIMATE
INTERCEPT	12.728	2.28	.0279	5.584
X1	−6.461 E-02	−1.13	.2653	5.722 E-02
X2	6.076 E-01	0.80	.4293	7.611 E-01
X3	1.961 E-04	3.94	.0003	4.979 E-05
X4	9.191 E-02	1.00	.3244	9.214 E-02
X5	1.535 E-01	0.86	.3973	1.795 E-01
X6	−7.678 E-01	−2.48	.0173	3.094 E-01
X7	2.711 E-02	0.63	.5294	4.274 E-02
X8	−1.894	−1.34	.1864	1.410

FIGURE 15.2
COMPUTER PRINTOUT FOR THE MODEL IN EQ. 15.3.

in the model

$$E(Y) = \beta_0 + \beta_1 x_1 + \beta_2 x_2 + \beta_3 x_3 + \beta_4 x_4 + \beta_5 x_5$$
$$+ \beta_6 x_6 + \beta_7 x_7 + \beta_8 x_8 \qquad (15.3)$$

is given as $F = 5.39$, which has a p-value of .0001. The actual p-value may be less than .0001, but the printout shows .0001 as the p-value in those cases. This indicates that H_0 should be rejected at $\alpha = .05$, and it is safe to conclude that *one or more* of the terms in the model given by Eq. 15.3 is significant. Since H_0 for the overall model has been rejected, it is appropriate to look at the terms in the model one by one using separate tests of significance to see which terms should be included and which terms should be excluded from the model, except for the intercept. The intercept β_0 is automatically included in all regression models.

INTERPRETING THE COMPUTER OUTPUT

The F value in the computer printout (**F VALUE**) measures the agreement of the data with the null hypothesis that all of the β's except β_0 are equal to zero. It is the ratio of the two mean squares in the **MEAN SQUARE** column of the printout,

$$F = \frac{\text{MODEL MEAN SQUARE}}{\text{ERROR MEAN SQUARE}} \qquad (15.4)$$

where the numerator and denominator degrees of freedom are the model degrees of freedom (**MODEL DF**) and the error degrees of freedom (**ERROR DF**), respectively, and are found in the **DF** column. The p-value associated with the observed value for F is printed in the **PR > F** column. This is the same p-value that the reader may obtain from the F tables in the back of this text.

The model degrees of freedom (MODEL DF) in the DF column is always the number of parameters in the model, not counting the intercept. The total degrees of freedom (**TOTAL DF**) is one less than the total number of observations, and the error degrees of freedom (ERROR DF) is the difference between the two.

$$\text{ERROR DF} = \text{TOTAL DF} - \text{MODEL DF}$$
$$\text{MODEL DF} = k$$
$$\text{ERROR DF} = n - k - 1$$
$$\text{TOTAL DF} = n - 1$$

The model sum of squares measures the variation in Y *accounted for* by the model. The error sum of squares measures the variation in Y *not accounted for* by the model, and the total sum of squares measures the total variation in Y.

$$\text{MODEL SUM OF SQUARES} = \Sigma(\hat{Y}_i - \overline{Y})^2$$

$$\text{ERROR SUM OF SQUARES} = \Sigma(Y_i - \hat{Y}_i)^2$$

$$\text{TOTAL SUM OF SQUARES} = \Sigma(Y_i - \overline{Y})^2$$

The total sum of squares is the sum of the other two sums of squares. The mean squares are obtained by dividing the sums of squares by their respective degrees of freedom. R SQUARE is the coefficient of determination, and, as before, it is the proportion of variation in Y accounted for by the fitted regression model.

$$\text{R SQUARE} = \frac{\text{MODEL SS}}{\text{TOTAL SS}} = 1 - \frac{\text{ERROR SS}}{\text{TOTAL SS}} \qquad (15.5)$$

The column headed by **ESTIMATE** in the computer printout gives the values of $\hat{\beta}_j$; the column headed by **STD ERROR OF ESTIMATE** provides the corresponding values of $s_{\hat{\beta}_j}$. The values of the test statistic corresponding to Eq. 15.2 are given in the column headed by **T FOR H0**; and finally the p-values corresponding to the tests of $H_0: \beta_j = 0$ versus $H_1: \beta_j \neq 0$ appear in the column headed by **PR > |T|**. Thus, if a level of significance of .05 is used, any p-value greater than .05 would indicate that the corresponding variable is a candidate for elimination from the general linear model.

EXAMPLE

The computer printout in Figure 15.2 is used to illustrate how to interpret the computer output from a least squares fit to the data in Figure 15.1. It is also used to illustrate how to test the individual terms in the multiple linear regression model given by Eq. 15.3 to see if any variables appear to be nonsignificant.

Since there are eight parameters β_i in Eq. 15.3, other than the intercept β_0, k equals 8; hence, the model has 8 degrees of freedom. The total degrees of freedom (49) is one less than the total number of observations (50) from the data in Figure 15.1. The difference between these two degrees of freedom $(49 - 8 = 41)$ is the error degrees of freedom. These degrees of freedom are divided into the respective sums of squares to get the mean squares:

$$\text{MODEL MEAN SQUARE} = \frac{150.91}{8} = 18.86$$

$$\text{ERROR MEAN SQUARE} = \frac{143.57}{41} = 3.50$$

The ratio of these two mean squares gives the F statistic used in an overall test of the model, as previously discussed:

$$F = \frac{\text{MODEL MEAN SQUARE}}{\text{ERROR MEAN SQUARE}} = \frac{18.86}{3.50} = 5.39$$

The computer-calculated p-value for this F statistic is given as .0001, but it also may be found from the F tables (Table A5) with 8 and 41 degrees of freedom. The value for R^2 is found from the sum of the squares as

$$\text{R SQUARE} = \frac{\text{MODEL SUM OF SQUARES}}{\text{TOTAL SUM OF SQUARES}} = \frac{150.91}{294.49} = .5125$$

or as

$$\text{R SQUARE} = 1 - \frac{\text{ERROR SUM OF SQUARES}}{\text{TOTAL SUM OF SQUARES}} = 1 - \frac{143.57}{294.49} = .5125$$

which are equivalent forms for R^2, since the total sum of squares is the sum of the other sums of squares.

The values of T in the T FOR H0: PARAMETER = 0 column at the bottom of Figure 15.2 are obtained by dividing the parameter estimate in the ESTIMATE column by the corresponding value in the STD ERROR OF ESTIMATE column. Each T value may be compared with the tables of the Student's t distribution with $n - k - 1 = 41$ degrees of freedom to find the p-value. The large p-values for the estimated coefficients of X_1, X_2, X_4, X_5, X_7, and X_8 in this case indicate that those six terms are candidates for elimination from the model.

A PARTIAL TEST OF THE MODEL

Earlier in this section a method was shown for testing individual terms in the model by testing the hypothesis

$$H_0: \beta_j = 0$$

for a specified value of j using a t statistic. Then a method for testing the overall model was given, which used an F statistic to test the hypothesis

$$H_0: \beta_1 = 0, \beta_2 = 0, \ldots, \beta_k = 0$$

which involves all of the terms in the general linear model. There exists a third statistical procedure that lies between these two tests. It allows several terms in the model to be tested simultaneously. It is useful when the experimenter thinks that some terms belong in the model, but is unsure about a group of other terms and wants a test that will enable a decision to be made about those other terms simultaneously.

Specifically, if β_1 through β_r are coefficients of terms that are to remain in the model, and β_{r+1} through β_k are coefficients that are to be tested simultaneously for possible elimination, then the null hypothesis of interest is

$$H_0: \beta_{r+1} = 0, \beta_{r+2} = 0, \ldots, \beta_k = 0$$

To test this hypothesis the full model

$$E(Y) = \beta_0 + \beta_1 x_1 + \cdots + \beta_k x_k \qquad (15.6)$$

is fit to the data as before. The error sum of squares for the full model is denoted by SSE_{FM} and the error degrees of freedom is denoted by DF_{FM}. Then the partial model of interest

$$E(Y) = \beta_0 + \beta_1 x_1 + \cdots + \beta_r x_r, \qquad r < k \qquad (15.7)$$

is fit to the data using the least squares procedure. The error sum of squares from this partial model is denoted by SSE_{PM}, and the error degrees of freedom is denoted by DF_{PM}. The F statistic for testing this hypothesis is

$$F = \frac{(SSE_{PM} - SSE_{FM})/(DF_{PM} - DF_{FM})}{SSE_{FM}/DF_{FM}} \qquad (15.8)$$

which is compared with quantiles from the F distribution in Table A5, with $(DF_{PM} - DF_{FM})$ as the numerator degrees of freedom, and DF_{FM} as the denominator degrees of freedom.

Note that the error sum of squares SSE_{FM} for the full model will always be less than or equal to the error sum of squares SSE_{PM} for the partial model, because the full model contains all of the terms from the partial model plus $(k - r)$ additional terms. Thus the numerator of this F statistic measures the reduction in the error sum of squares resulting from the inclusion of $(k - r)$ additional terms in the model. This reduction in the error sum of squares is divided by the number $(k - r)$ of additional terms. The denominator of the F statistic is the mean squared error for the full model, the same as before.

Also note that the F statistic given before in Eq. 15.4 is really only a special case of this F statistic, since the previous test is actually a test of the full model against the model

$$E(Y) = \beta_0 \qquad (15.9)$$

which is a partial model with $r = 0$. The error sum of squares for this partial model is the same as the total sum of squares, since there are no regression variables in the model. Thus the F statistic becomes

$$\begin{aligned} F &= \frac{(\text{TOTAL SS} - \text{ERROR SS})/[(n - 1) - (n - k - 1)]}{\text{ERROR SUM OF SQUARES}/(n - k - 1)} \\ &= \frac{\text{MODEL SUM OF SQUARES}/k}{\text{ERROR SUM OF SQUARES}/(n - k - 1)} \\ &= \frac{\text{MODEL MEAN SQUARE}}{\text{ERROR MEAN SQUARE}} \end{aligned}$$

which shows that Eq. 15.8 leads to Eq. 15.4.

EXAMPLE

Suppose the personnel director at the research laboratory in Problem Setting 15.1 feels that sex (X_2), salary (X_3), education level (X_6), and management position (X_8) are variables that are relevant to job satisfaction, but does not know whether any of the remaining variables belong in the model. She would like a statistical test, at $\alpha = .05$, of the hypothesis

$$H_0: \beta_1 = 0, \beta_4 = 0, \beta_5 = 0, \beta_7 = 0$$

Since the variables X_1, X_4, X_5, and X_7 are not included, the partial model is written as

$$E(Y) = \beta_0 + \beta_2 x_2 + \beta_3 x_3 + \beta_6 x_6 + \beta_8 x_8 \qquad (15.10)$$

Note that the partial model in Eq. 15.10 includes four terms; hence, $r = 4$, whereas $k = 8$ for the full model given by Eq. 15.3. The error sum of squares is obtained by fitting the model in Eq. 15.10 to the data in Figure 15.1 using the method of least squares. The computer printout of the results of the least squares fit is given in Figure 15.3. The error sum of squares and degrees of freedom for the model in Eq. 15.10 are obtained from Figure 15.3 as

$$SSE_{PM} = 148.50 \qquad \text{and} \qquad DF_{PM} = 45$$

whereas the corresponding figures for the full model are

$$SSE_{FM} = 143.57 \qquad \text{and} \qquad DF_{FM} = 41$$

from Figure 15.2. The F statistic in Eq. 15.8 becomes

$$F = \frac{(SSE_{PM} - SSE_{FM})/(DF_{PM} - DF_{FM})}{SSE_{FM}/DF_{FM}}$$

$$= \frac{(148.50 - 143.57)/(45 - 41)}{143.57/41}$$

$$= 0.35$$

A comparison with the F quantiles in Table A5 shows the p-value to be much larger than .25. The numerator and denominator degrees of freedom are 4 and 41, respectively.

The decision in this case is not to reject the null hypothesis, and conclude that there is insufficient statistical evidence to indicate that X_1, X_4, X_5, or X_7 should be included in the model. Note also that on the basis of the statistical evidence available, one might question the value of including X_8 in the model, since the p-value associated with the estimated coefficient of X_8 is .1974, which is fairly large. However, there may be other, nonstatistical, reasons for wanting to include management position as a variable in the model.

SOURCE	DF	SUM OF SQUARES	MEAN SQUARE	F VALUE	PR > F
MODEL	4	145.99	36.50	11.06	.0001
ERROR	45	148.50	3.30		
TOTAL	49	294.49		R SQUARE	
				.4957	

PARAMETER	ESTIMATE	T FOR H0: PARAMETER = 0	PR > |T|	STD ERROR OF ESTIMATE
INTERCEPT	14.867			
X2	1.141	1.96	.0558	5.808 E-01
X3	1.774 E-04	5.32	.0001	3.332 E-05
X6	−8.088 E-01	−3.28	.0020	2.465 E-01
X8	−1.665	−1.31	.1974	1.273

FIGURE 15.3
COMPUTER PRINTOUT FOR THE MODEL IN EQ. 15.10.

CONFIDENCE INTERVAL FOR β

With the computer printout, such as those in Figures 15.2 and 15.3, it is very easy to find a confidence interval for each coefficient β_j. These confidence intervals are found just as they were in simple linear regression. Recall that the confidence interval is formed around a point estimate, provided by $\hat{\beta}_j$, by adding and subtracting a constant times the estimated standard error of $\hat{\beta}_j$. The constant is obtained from the Student's t distribution (Table A3) with $n - k - 1$ degrees of freedom (error degrees of freedom), and the estimated standard error of $\hat{\beta}_j$ is provided by the computer printout. The confidence interval for β_j is interpreted just the same as all confidence intervals. That is, there is $100(1 - \alpha)\%$ confidence that the interval provided by Eq. 15.11 contains the true value β_j.

A $100(1 - \alpha)\%$ **confidence interval for β_j** is provided by

CONFIDENCE
INTERVAL
$$\hat{\beta}_j \pm t_{1-\alpha/2, n-k-1} s_{\hat{\beta}_j} \tag{15.11}$$
FOR β_j where $\hat{\beta}_j$ is the least squares estimate of β_j, $t_{1-\alpha/2, n-k-1}$ is the $1 - \alpha/2$ quantile of the Student's t distribution with $n - k - 1$ degrees of freedom, and $s_{\hat{\beta}_j}$ is the estimated standard error of $\hat{\beta}_j$.

EXAMPLE

To find a 95% confidence interval for β_2 in the model provided by Eq. 15.10, the point estimate 1.141 and the estimated standard error .5808 from Figure

15.3 are combined with the .975 quantile 2.0141 from Table A3 with 45 degrees of freedom:

$$\hat{\beta}_2 \pm t_{.975,45} s_{\hat{\beta}_2} = 1.141 \pm 2.0141(.5808)$$
$$= 1.141 \pm 1.170$$
$$= -.029 \text{ to } 2.311$$

Since X_2 is a dummy variable representing sex, 0 for females and 1 for males, this confidence interval provides an interval estimate of the amount by which the mean job satisfaction score for males differs from that for females.

EXERCISES

15.1 Write the least squares equation for the model given by Eq. 15.10 as provided by the computer printout in Figure 15.3. Interpret the estimated coefficients of X_2, X_3, X_6, and X_8 in terms of the job satisfaction scores.

15.2 Write the least squares equation for the model given by Eq. 15.3 as provided by the computer printout in Figure 15.2. Why are the coefficients of X_2, X_3, X_6, and X_8 different than those found in Exercise 15.1?

15.3 Find a 95% confidence interval for the coefficient of x_6 in the model of Eq. 15.10 using the computer printout in Figure 15.3. Interpret this confidence interval. Does this agree with your preconceived ideas regarding level of education?

15.4 Find a 95% confidence interval for the coefficient of x_6 in the model of Eq. 15.3 using the computer printout in Figure 15.2. Why is this confidence interval different than the one found in Exercise 15.3?

15.5 A real estate appraiser is using multiple regression to get a base figure to work with for the market value of a house. Several nonquantifiable factors will then be considered before obtaining a final appraised value. To find the base figure three independent variables are considered: X_1 = age of house, X_2 = square footage, X_3 = 0 if wooden exterior or 1 if brick. The results from 45 recent house sales are used to fit the model. A partial printout of the least squares analysis is given below. Fill in the components (a) through (o).

SOURCE	DF	SUM OF SQUARES	MEAN SQUARE	F VALUE	PR > F
MODEL	(a)	2.882 E04	(e)	(g)	(h)
ERROR	(b)	(d)	(f)		
TOTAL	(c)	4.063 E04		R SQUARE	
				(i)	

PARAMETER	ESTIMATE	T FOR H0: PARAMETER = 0	PR > |T|	STD ERROR OF ESTIMATE
INTERCEPT	2.088 E03			
X1	−9.357 E02	−2.66	(l)	(o)
X2	4.581 E01	(k)	(m)	1.656 E01
X3	(j)	1.88	(n)	3.190 E03

15.6 An income tax investigator is establishing a quick procedure by which an individual's income tax liability may be approximated using a few easily obtained variables. A sample of 100 tax returns from the previous year is examined and three relevant variables are obtained from each return, in addition to the income tax paid. The computer printout of a least squares fit to the data is given in part below. Fill in the missing components (a) through (o).

SOURCE	DF	SUM OF SQUARES	MEAN SQUARE	F VALUE	PR > F
MODEL	(a)	(d)	(e)	(g)	(h)
ERROR	(b)	1.012E06	(f)		
TOTAL	(c)	3.882 E06		R SQUARE	
				(i)	

PARAMETER	ESTIMATE	T FOR H0: PARAMETER = 0	PR > |T|	STD ERROR OF ESTIMATE
INTERCEPT	−8.416 E-01			
X1	3.608 E-03	2.03	(m)	(o)
X2	(j)	(l)	.0500	7.3751 E-03
X3	(k)	1.79	(n)	4.8831 E-02

15.7 The top 500 companies in sales are listed each year in *Fortune* magazine and are called "Fortune's top 500." A random sample of 27 of these companies one year yielded the following values for sales, assets, and net income, all in millions of dollars:

Net Income (Y)	Sales (X_1)	Assets (X_2)
7.52	808.8	674.9
18.68	727.3	355.6
11.80	491.6	366.8
etc.		

The linear model $y = \beta_0 + \beta_1 x_1$ is fit to the sample data with a resulting R^2 of .966. Find the error sum of squares and the error degrees of freedom for this model. *Hint:* The total sum of squares is given in the next exercise as 1.104 E06.

15.8 Assume that the sample of the Fortune 500 companies given in Exercise 15.7 satisfies the multiple linear regression model. Test the null hypothesis that the terms involving x_1^2, x_2, x_2^2, and $x_1 x_2$ do not belong in the model

$$E(Y) = \beta_0 + \beta_1 x_1 + \beta_2 x_1^2 + \beta_3 x_2 + \beta_4 x_2^2 + \beta_5 x_1 x_2$$

by comparing the results of Exercise 15.7 with the least squares fit given in the following computer printout. Use $\alpha = .05$. Interpret your results.

SOURCE	DF	SUM OF SQUARES	MEAN SQUARE	F VALUE	PR > F
MODEL	5	1.096 E06	2.192 E05	574.93	.0001
ERROR	21	8.006 E03	3.812 E02		
TOTAL	26	1.104 E06		R SQUARE	
				.9927	

15.9 Assume that the housing sale price data of Section 14.3 satisfy the multiple linear regression model. Compare the results given in Exercise 14.26 with those in Figure 14.30 to test the null hypothesis that the coefficients of the last five terms in the model

$$\begin{aligned} E(Y) = \beta_0 &+ \beta_1 x + \beta_2 D_1 + \beta_3 D_2 + \beta_4 D_3 + \beta_5 D_1 x + \beta_6 D_2 x \\ &+ \beta_7 D_3 x + \beta_8 D_1 D_3 + \beta_9 D_2 D_3 \\ &+ \beta_{10} D_1 D_3 x + \beta_{11} D_2 D_3 x \end{aligned}$$

are equal to zero. Use $\alpha = .05$ and interpret your results.

15.10 Test the null hypothesis that the coefficients β_4, β_5, β_6, and β_7 in the model for condominium sale prices

$$\begin{aligned} E(Y) = \beta_0 &+ \beta_1 x_2 + \beta_2 D_1 + \beta_3 D_2 + \beta_4 D_3 + \beta_5 D_1 x_2 + \beta_6 D_2 x_2 \\ &+ \beta_7 D_3 x_2 \end{aligned}$$

are all equal to zero. Use these computer printouts and use $\alpha = .05$. Interpret your results.

SOURCE	DF	SUM OF SQUARES	MEAN SQUARE	F VALUE	PR > F
MODEL	3	1.496 E09	4.988 E08	17.97	.0001
ERROR	21	5.830 E08	2.776 E07		
TOTAL	24	2.079 E09		R SQUARE	
				.7196	

PARAMETER	ESTIMATE	T FOR H0: PARAMETER = 0	PR > |T|	STD ERROR OF ESTIMATE
INTERCEPT	3241.781	0.49	.6273	6578.165
X2	43.709	6.00	.0001	7.289
D1	1043.905	0.36	.7204	2877.973
D2	−1464.868	−0.48	.6392	3079.036

SOURCE	DF	SUM OF SQUARES	MEAN SQUARE	F VALUE	PR > F
MODEL	7	1.591 E09	2.273 E08	7.91	.0003
ERROR	17	4.882 E08	2.872 E07		
TOTAL	24	2.079 E09		R SQUARE	
				.7652	

PARAMETER	ESTIMATE	T FOR H0: PARAMETER = 0	PR > |T|	STD ERROR OF ESTIMATE
INTERCEPT	− 10723.481	−0.60	.5562	17862.152
X2	58.925	2.95	.0089	19.952
D1	− 9341.631	−0.46	.6531	20421.457
D2	17146.730	0.89	.3833	19159.694
D3	11760.776	0.86	.4025	13696.019
D1*X2	8.476	0.40	.6929	21.098
D2*X2	− 18.657	−0.93	.3636	19.983
D3*X2	− 12.158	−0.88	.3887	13.745

15.2
THE BACKWARD ELIMINATION PROCEDURE, PREDICTION INTERVALS, AND CONFIDENCE INTERVALS FOR $\mu_{Y|x}$

THE BACKWARD ELIMINATION PROCEDURE

Statisticians have devised several methods for determining the best subset from a list of possible independent regression variables. One method (explained in the next section) proceeds by adding one significant variable at a time to the model. This method is commonly called a *forward procedure*. The other method (discussed here) starts with all variables in the model and eliminates the nonsignificant variables one at a time. This is commonly referred to as a **backward elimination procedure**.

Based on the p-values for the t tests in Figure 15.2 for the job satisfaction data in the previous section, six of the eight independent variables are candidates for elimination from the model given in Eq. 15.3. The backward elimination procedure proceeds one step at a time by deleting the variable at each step with the largest p-value as long as that p-value exceeds the level of significance α. Thus in Figure 15.2, reprinted below, if a level of significance of .05 is used, the variable X_7 would be dropped at step 1 because it has the largest p-value .5294, which is greater than .05. The result of a least squares fit to the data, after eliminating X_7 from the model, is given in Figure 15.4. Note that the coefficients and p-values associated with the remaining variables in Figure 15.4 are all changed from what they were in Figure 15.2 when X_7 was present.

Figures 15.4 to 15.9 contain the computer printouts for the remaining steps

SOURCE	DF	SUM OF SQUARES	MEAN SQUARE	F VALUE	PR > F
MODEL	8	150.91	18.86	5.39	.0001
ERROR	41	143.57	3.50		
TOTAL	49	294.49		R SQUARE	
				.5125	

PARAMETER	ESTIMATE	T FOR H0: PARAMETER = 0	PR > |T|	STD ERROR OF ESTIMATE
INTERCEPT	12.728	2.28	.0279	5.584
X1	-6.461 E-02	-1.13	.2653	5.722 E-02
X2	6.076 E-01	0.80	.4293	7.611 E-01
X3	1.961 E-04	3.94	.0003	4.979 E-05
X4	9.191 E-02	1.00	.3244	9.214 E-02
X5	1.535 E-01	0.86	.3973	1.795 E-01
X6	-7.678 E-01	-2.48	.0173	3.094 E-01
X7	2.711 E-02	0.63	.5294	4.274 E-02
X8	-1.894	-1.34	.1864	1.410

FIGURE 15.2
COMPUTER PRINTOUT FOR THE MODEL IN EQ. 15.3.

in the backward elimination procedure. At step 2 the variable X_5 is eliminated, since it has a p-value of .5267. The variable X_4 is eliminated at step 3 because of its p-value .5659. This process continues until step 6, where it can be seen that the p-values for the remaining two variables are both less than .05. Thus no more variables are eliminated from the model and the final model contains only the variables X_3 and X_6. In reference to Problem Setting 15.1 these are the variables that the consulting firm would report to the laboratory as being related to job satisfaction. It should be noted at this point that other types of subset selection procedures would not necessarily give the same set of variables, although there would generally be good agreement. The two variables that remain in the model, X_3 and X_6, were the only ones not designated as candidates for elimination from the model at step 1 in the example. However, frequently there is considerable disagreement between the variables in the final model and those designated as candidates for elimination at step 1. Since the elimination of one variable changes the p-value associated with the remaining variables, it makes sense to eliminate only one variable at a time, especially when a computer is available to take care of all of the computations.

THE PROBABILITY OF A TYPE I ERROR

At each step in the backward elimination procedure, a hypothesis test is performed at some stated level of significance α. That value for α supposedly represents the probability of making a Type I error at each step. However, it represents the true probability of a Type I error only on the *first* step and only then if the distributional assumptions regarding the ϵ_i are met. After the first step the tests become *conditional tests,* conditional on the event that at least one variable has been removed from the model, so that the α value is no

SOURCE	DF	SUM OF SQUARES	MEAN SQUARE	F VALUE	PR > F
MODEL	7	149.51	21.36	6.19	.0001
ERROR	42	144.98	3.45		
TOTAL	49	294.49		R SQUARE	
				.5077	

PARAMETER	ESTIMATE	T FOR H0: PARAMETER = 0	PR > |T|	STD ERROR OF ESTIMATE
INTERCEPT	15.313			
X1	-5.319 E-02	-0.99	.3295	5.392 E-02
X2	7.246 E-01	0.99	.3286	7.331 E-01
X3	1.902 E-04	3.92	.0003	4.856 E-05
X4	7.182 E-02	0.84	.4079	8.591 E-02
X5	1.009 E-01	0.64	.5267	1.580 E-01
X6	-7.726 E-01	-2.52	.0158	3.071 E-01
X8	-2.072	-1.51	.1384	1.372

FIGURE 15.4
STEP 1 OF THE BACKWARD ELIMINATION PROCEDURE. VARIABLE X_7 WAS REMOVED.

longer an accurate representation of the true probability of a Type I error. Also, the fact that several hypothesis tests are performed on the same set of data tends to increase the chances of making a Type I error somewhere in the process. Therefore, care should be taken when interpreting α in any procedure such as this one, or the one presented in the next section, where multiple tests are made on the same data set.

SOURCE	DF	SUM OF SQUARES	MEAN SQUARE	F VALUE	PR > F
MODEL	6	148.10	24.68	7.25	.0001
ERROR	43	146.39	3.40		
TOTAL	49	294.49		R SQUARE	
				.5029	

PARAMETER	ESTIMATE	T FOR H0: PARAMETER = 0	PR > |T|	STD ERROR OF ESTIMATE
INTERCEPT	15.080			
X1	-2.689 E-02	-0.78	.4406	3.455 E-02
X2	9.040 E-01	1.34	.1859	6.725 E-01
X3	1.874 E-04	3.90	.0003	4.802 E-05
X4	4.056 E-02	0.58	.5659	7.010 E-02
X6	-7.907 E-01	-2.60	.0126	3.037 E-01
X8	-1.986	-1.47	.1502	1.356

FIGURE 15.5
STEP 2 OF THE BACKWARD ELIMINATION PROCEDURE. VARIABLE X_5 WAS REMOVED.

SOURCE	DF	SUM OF SQUARES	MEAN SQUARE	F VALUE	PR > F
MODEL	5	146.96	29.39	8.77	.0001
ERROR	44	147.53	3.35		
TOTAL	49	294.49		R SQUARE	
				.4990	

PARAMETER	ESTIMATE	T FOR H0: PARAMETER = 0	PR > |T|	STD ERROR OF ESTIMATE
INTERCEPT	15.811			
X1	− 1.442 E-02	− 0.54	.5932	2.680 E-02
X2	1.078	1.81	.0777	5.968 E-01
X3	1.944 E-04	4.22	.0001	4.611 E-05
X6	− 8.673 E-01	− 3.20	.0026	2.712 E-01
X8	− 1.804	− 1.38	.1750	1.308

FIGURE 15.6
STEP 3 OF THE BACKWARD ELIMINATION PROCEDURE. VARIABLE X_4
WAS REMOVED.

The proper interpretation of α in this procedure and the one in the next section is that the probability of making a Type I error on the test at the first step is truly given by α if the distributional assumptions on the ϵ_i are true. The probability of making a Type I error on the second step or subsequent steps is unknown. For those steps the nominal value of α used in the test is merely a convenient yardstick against which the observed values may be measured, to see whether additional terms should be excluded from the model. The overall probability of making a Type I error at least once in this multiple

SOURCE	DF	SUM OF SQUARES	MEAN SQUARE	F VALUE	PR > F
MODEL	4	145.99	36.50	11.06	.0001
ERROR	45	148.50	3.30		
TOTAL	49	294.49		R SQUARE	
				.4957	

PARAMETER	ESTIMATE	T FOR H0: PARAMETER = 0	PR > |T|	STD ERROR OF ESTIMATE
INTERCEPT	14.867			
X2	− 1.141	1.96	.0558	5.808 E-01
X3	1.774 E-04	5.32	.0001	3.332 E-05
X6	− 8.088 E-01	− 3.28	.0020	2.465 E-01
X8	− 1.665	− 1.31	.1974	1.273

FIGURE 15.7
STEP 4 OF THE BACKWARD ELIMINATION PROCEDURE. VARIABLE X_1
WAS REMOVED.

SOURCE	DF	SUM OF SQUARES	MEAN SQUARE	F VALUE	PR > F
MODEL	3	140.34	46.78	13.96	.0001
ERROR	46	154.14	3.35		
TOTAL	49	294.49		R SQUARE	
				.4766	

PARAMETER	ESTIMATE	T FOR H0: PARAMETER = 0	PR > |T|	STD ERROR OF ESTIMATE
INTERCEPT	15.165			
X2	1.125	1.92	.0608	5.851 E-01
X3	1.631 E-04	5.14	.0001	3.173 E-05
X6	−8.081 E-01	−3.25	.0021	2.484 E-01

FIGURE 15.8
STEP 5 OF THE BACKWARD ELIMINATION PROCEDURE. VARIABLE X_8 WAS REMOVED.

decision-making process is always larger than α, since the probability associated with the first step alone is α and subsequent steps provide additional opportunities for making a Type I error. This compounding of the probability of a Type I error, and the lack of an overall measure for that probability, is not a serious problem as long as the analyst realizes that such a situation exists.

PREDICTION INTERVAL

Regression analysis serves two primary purposes. One is to identify important relationships between the dependent variable and one or more independent variables. The other is to use a regression model to predict unknown values

SOURCE	DF	SUM OF SQUARES	MEAN SQUARE	F VALUE	PR > F
MODEL	2	127.96	63.98	18.06	.0001
ERROR	47	166.52	3.54		
TOTAL	49	294.49		R SQUARE	
				.4345	

PARAMETER	ESTIMATE	T FOR H0: PARAMETER = 0	PR > |T|	STD ERROR OF ESTIMATE
INTERCEPT	14.464			
X3	1.740 E-04	5.42	.0001	3.211 E-05
X6	−7.437 E-01	−2.94	.0051	2.530 E-01

FIGURE 15.9
STEP 6 OF THE BACKWARD ELIMINATION PROCEDURE. VARIABLE X_2 WAS REMOVED.

of Y for given values of the independent variables. A high value of R^2 is more important for this latter purpose, because a high R^2 generally indicates a good fit to the data, and a better predictive capability for the model.

In Chapter 12 prediction intervals were formed in simple linear regression by expanding the point estimate $\hat{Y}$ to an interval estimate by adding and subtracting $t_{1-\alpha/2,n-k-1}$ (the $1-\alpha/2$ quantile from a Student's t distribution) times the standard error of $\hat{Y}$:

$$\text{Prediction interval: } \hat{Y} \pm t_{1-\alpha/2,n-k-1}s_{\hat{Y}} \qquad (15.12)$$

In multiple regression the procedure is the same, except that the standard error of $\hat{Y}$ is much more difficult to find. Therefore, the entire procedure is usually handled in a statistical packages such as SAS® or Minitab®. All that is necessary to obtain a prediction interval is a model that provides a prediction.

PREDICTION INTERVAL

A $100(1-\alpha)\%$ **prediction interval** for Y may be obtained from a multiple linear regression model by using Eq. 15.12, where $\hat{Y}$ is obtained by substituting specific values of the independent variables into the least squares fitted model, $t_{1-\alpha/2,n-k-1}$ is the $1-\alpha/2$ quantile from Table A3 for the Student's t distribution with $n-k-1$ degrees of freedom, and $s_{\hat{Y}}$ is the estimated standard error of $\hat{Y}$.

EXAMPLE

The model obtained from the backward elimination procedure in Figure 15.9 is used to obtain

$$\hat{Y} = 14.464 + .000174X_3 - .7437X_6 \qquad (15.13)$$

Suppose that a prediction regarding job satisfaction rating is to be made for an employee of the laboratory. The employee is 35 years old with a salary of $36,000, has been with the company 11 years, has a master's degree, and is in a nonmanagerial position. It is also known that the employee is male, has two years additional professional experience, and his salary constitutes 95% of his total income. Of these independent variables the only ones required by Eq. 15.13 are salary (X_3) and education level (X_6). The values $X_3 = 36,000$ and $X_6 = 17$ are substituted into Eq. 15.13 to get

$$\hat{Y} = 8.085$$

Thus the point estimate of the employee's job satisfaction score is $\hat{Y} = 8.085$. How much uncertainty is associated with this point estimate? Clearly some sort of a prediction interval would be more informative than a single-point estimate. The correct quantile from the Student's t distribution is multiplied by the standard error of $\hat{Y}$ to form the prediction interval. The SAS® statistical package gives a prediction interval for Y in the form

LOWER 95% CL INDIVIDUAL	UPPER 95% CL INDIVIDUAL
4.192	11.977

which says that there is 95% confidence that an individual with those characteristics has a job satisfaction score between 4.192 and 11.977.

This prediction interval is quite wide when compared to the range of values observed for the job satisfaction scores in the sample presented in Figure 15.1. Thus it is difficult to state with any sense of precision what this individual's satisfaction will be with his job. The size of the prediction interval depends on the amount of variability shown by the individuals in the sample in addition to how well the model fits the data. For the model in Eq. 15.13 the R^2 value was only .4345 as shown in Figure 15.9; thus, predictions associated with this model will have a large degree of uncertainty.

CONFIDENCE INTERVAL FOR THE MEAN

If the true regression equation were known, the specific values of the independent variables could be substituted into it to get the exact value of $\mu_{Y|x}$, which is the mean of Y for those specific values of the independent variables. However, the true regression equation is always unknown. It is estimated on the basis of the sample data. Therefore, the mean of Y for the given values of the independent variables is always estimated.

A point estimate for $\mu_{Y|x}$ is obtained by substituting the specific values of the independent variables into the least squares estimate of the regression equation. This point estimate $\hat{\mu}_{Y|x}$ is identical to the point estimate $\hat{Y}$ previously found for use in prediction intervals. However, the standard error of $\hat{\mu}_{Y|x}$ is always smaller than the standard error of $\hat{Y}$, so the confidence interval for the mean of Y, $\mu_{Y|x}$, is always smaller than the prediction interval for an individual value of Y.

As in the case of simple linear regression the confidence interval for $\mu_{Y|x}$ is obtained by adding and subtracting the product of the standard error of $\hat{\mu}_{Y|x}$ times a quantile from Student's t distribution, from the point estimate $\hat{\mu}_{Y|x}$ for

---

CONFIDENCE INTERVAL FOR $\mu_{Y|x}$ A $100(1 - \alpha)\%$ **confidence interval for the mean of Y** given specific values of the independent variables is given by Eq. 15.14. The point estimate $\hat{\mu}_{Y|x}$ is obtained by substituting the specific values of the independent variables into the least squares estimate of the regression model; $t_{1-\alpha/2, n-k-1}$ is the $1 - \alpha/2$ quantile of the Student's t distribution with $n - k - 1$ degrees of freedom obtained from Table A3; and $s_{\hat{\mu}_{Y|x}}$ is the estimated standard error of $\hat{\mu}_{Y|x}$.

$\mu_{Y|x}$:

$$\text{Confidence interval for } \mu_{Y|x}: \quad \hat{\mu}_{Y|x} \pm t_{1-\alpha/2,n-k-1}s_{\hat{\mu}_{Y|x}} \qquad (15.14)$$

The standard error of $\hat{\mu}_{Y|x}$ is much more difficult to calculate in multiple regression than in simple linear regression; consequently, all of the calculations are usually handled via the computer.

EXAMPLE

The mean of Y is estimated by substituting the values $X_3 = 36{,}000$ and $X_6 = 17$ into Eq. 15.13 to get

$$\hat{\mu}_{Y|x} = 14.464 + .000174(36{,}000) - .7437(17) = 8.085 \quad (15.15)$$

Note that this point estimate is the same as the point estimate $\hat{Y}$ for a predicted value of Y. However, the width of the confidence interval for $\mu_{Y|x}$ is less than the width of the prediction interval in the previous example.

For the given values of X_3 and X_6 the SAS® statistical package gives a confidence interval for $\mu_{Y|x}$ as follows:

LOWER 95% CL FOR MEAN	UPPER 95% CL FOR MEAN
7.184	8.986

There is a confidence of 95% that the true mean job satisfaction score for all employees with the given values for the X's lies between the values 7.184 and 8.986.

EXERCISES

15.11 Use Eq. 15.13 to predict the job satisfaction rating for a 45-year-old male employee in a nonmanagement position who makes a salary of $13,500, has been with the company 17 years with no other professional experience, has 13 years of education, and whose salary constitutes 100% of his total income.

15.12 The 95% prediction interval for the individual's job satisfaction rating in Exercise 15.11 is from 3.28 to 11.01, whereas the 95% confidence interval for the mean response of such individuals is from 6.36 to 7.93. What are the interpretations of these intervals?

15.13 (a) Judging from the model suggested by the printout in Figure 15.9, does higher pay tend to be associated with greater job satisfaction?

(b) Does more education tend to be associated with greater job satisfaction?

(c) If one person has two years of education more than another person, how much more should the first person be making in salary in order to have the same job satisfaction as the other person?

15.14 (a) Use Eq. 15.13 to predict the mean job satisfaction score for all persons with a salary of $36,000 and a master's degree (17 years of education).

(b) How many degrees of freedom are associated with the t distribution used in finding the prediction interval?

(c) What is the value of the quantile that would be used in finding a 95% prediction interval for Y?

(d) What is the value of the t distribution quantile that would be used in finding a 95% confidence interval for $\mu_{Y|x}$?

(e) For a 95% prediction interval for Y from 4.19 to 11.98, find the standard error of $\hat{Y}$.

(f) For a 95% confidence interval for $\mu_{Y|x}$ from 7.18 to 8.99, find the standard error of $\hat{\mu}_{Y|x}$.

15.15 The computer printout in this exercise represents the least squares fit of the model

$$y = \beta_0 + \beta_1 x_1 + \beta_2 x_2 + \beta_3 x_1 x_2$$

to the data in Exercise 14.12. Which of the variables should be eliminated first in backward regression? What are the assumptions that need to be made for this hypothesis test to have a valid interpretation?

SOURCE	DF	SUM OF SQUARES	MEAN SQUARE	F VALUE	PR > F
MODEL	3	1.994 E09	6.646 E08	163.06	.0001
ERROR	21	8.559 E07	4.076 E06		
TOTAL	24	2.079 E09		R SQUARE	
				.9588	

PARAMETER	ESTIMATE	T FOR H0: PARAMETER = 0	PR > |T|	STD ERROR OF ESTIMATE
INTERCEPT	−7491.816	−1.92	.0682	3897.445
X1	−469.320	−1.17	.2564	402.262
X2	59.443	14.80	.0001	4.015
X1*X2	−0.490	−1.31	.2037	0.374

15.16 Given in this exercise is more computer printout for the data in Exercise 14.12, this time fitting the model

$$y = \beta_0 + \beta_1 x_2 + \beta_2 x_2^2 + \beta_3 x_1 x_2$$

SOURCE	DF	SUM OF SQUARES	MEAN SQUARE	F VALUE	PR > F
MODEL	3	1.990 E09	6.634 E08	156.20	.0001
ERROR	21	8.919 E07	4.247 E06		
TOTAL	24	2.079 E09		R SQUARE	
				.9571	

PARAMETER	ESTIMATE	T FOR H0: PARAMETER = 0	PR > |T|	STD ERROR OF ESTIMATE
INTERCEPT	−3135.453	−0.27	.7909	11673.001
X2	47.675	2.13	.0453	22.399
X2*X2	0.007	0.68	.5052	0.011
X1*X2	−0.944	−10.15	.0001	0.093

Which of the variables should be eliminated first in backward regression? Note that elimination of this variable results in the same model as the one obtained in Exercise 15.15.

15.17 The computer output for Exercise 14.17 is reprinted in this exercise. Which of the variables should be eliminated first in backward regression? What is the interpretation of the resulting model?

SOURCE	DF	SUM OF SQUARES	MEAN SQUARE	F VALUE	PR > F
MODEL	4	2.683 E09	6.708 E08	75.42	.0001
ERROR	11	9.783 E07	8.894 E06		
TOTAL	15	2.781 E09		R SQUARE .9648	

PARAMETER	ESTIMATE	T FOR H0: PARAMETER = 0	PR > |T|	STD ERROR OF ESTIMATE
INTERCEPT	35133.05	18.55	.0001	1893.487
T	2236.85	13.42	.0001	116.712
D1	13027.40	6.16	.0001	2115.341
D2	16357.05	7.66	.0001	2134.958
D3	−1785.55	−0.82	.4275	2167.259

15.18 List all of the variables in Exercise 14.19 that are not significant at $\alpha = .05$. What would be the interpretation of a model after those variables are eliminated? Does the inclusion of additional terms in Exercises 14.20, 14.23, and 14.24 seem to result in an improvement in the model, in view of the associated p-values?

15.19 If the stocks in Exercise 15.7 were randomly selected from the Fortune 500, and if the other assumptions stated in this section are met, then a 95% confidence interval for the mean income of companies with sales of $950.2 million and assets of 688.3 million can be found. A new MBA on the job used the following model

$$y = \beta_0 + \beta_1 x_1 + \beta_2 x_1^2 + \beta_3 x_2 + \beta_4 x_2^2 + \beta_5 x_1 x_2$$

and obtained an interval for the mean of $30.175 million to $56.052 million as the 95% confidence interval. An old-timer did some calculations on an old desk calculator, and in less than 2 minutes stated flatly that the confidence interval was incorrect. Is the old-timer correct?

SOURCE	DF	SUM OF SQUARES	MEAN SQUARE	F VALUE	PR > F
MODEL	5	1.096 E06	2.192 E05	574.93	.0001
ERROR	21	8.006 E03	3.812 E02		
TOTAL	26	1.104 E06		R SQUARE	
				.9927	

PARAMETER	ESTIMATE	T FOR H0: PARAMETER = 0	PR > |T|	STD ERROR OF ESTIMATE
INTERCEPT	1.758	0.25	.8033	6.968
X1	3.168 E-02	1.88	.0743	1.687 E-02
X1*X1	− 1.978 E-05	− 2.63	.0157	7.530 E-06
X2	2.125 E-02	1.12	.2735	1.890 E-02
X2*X2	− 3.414 E-05	− 3.43	.0025	9.962 E-06
X1*X2	5.298 E-05	3.22	.0041	1.646 E-05

15.3

STEPWISE REGRESSION

PROBLEM SETTING 15.2

A governmental health service agency wants to make an epidemiological study to see if mortality rates in different regions of the country can be predicted. Studies of this type generally do not involve well-designed or controlled experiments and the data usually consist of measurements on various weather, socioeconomic, and pollution variables. In Figure 15.10 such measurements are given on the following variables from a random sample of 30 locations throughout the country.

X_1 = mean annual precipitation (in.)

X_2 = mean January temperature (°F)

X_3 = mean July temperature (°F)

X_4 = median school years completed for those over age 25

X_5 = percentage of urbanized area population that is nonwhite

X_6 = relative pollution potential of sulfur dioxide

Y = total age-adjusted mortality rate from all causes, expressed as deaths per 100,000 population

In a setting of this type the number of potentially important predictor variables can easily become unmanageable. Hence, the problem becomes one of systematically sorting through a large list of variables to develop a useful model

Observation Number	X_1	X_2	X_3	X_4	X_5	X_6	Y
1	32.2	21.2	70.9	11.1	1.9	94.6	942
2	17.6	37.0	69.3	12.3	1.1	13.1	852
3	37.6	36.9	79.3	10.1	10.0	9.8	924
4	31.7	36.4	72.3	11.0	2.0	28.4	928
5	33.0	34.5	73.1	10.2	5.4	243.0	949
6	36.0	32.2	73.2	11.2	0.9	1.4	901
7	39..7	36.7	71.3	10.7	1.7	11.0	938
8	42.8	35.4	78.8	11.3	2.4	18.5	901
9	40.0	25.5	75.9	10.7	4.0	56.4	942
10	40.1	31.2	72.2	9.8	5.6	87.6	1010
11	29.0	32.2	69.8	11.7	1.1	89.7	885
12	42.9	34.5	75.7	10.5	35.1	78.7	1039
13	37.1	41.2	77.2	12.1	15.0	129.0	988
14	54.6	34.5	75.5	9.0	4.0	74.1	981
15	37.3	38.4	74.4	11.3	3.6	54.7	944
16	33.0	36.4	74.8	9.8	22.2	36.3	998
17	33.7	28.7	69.3	11.8	0.8	1.9	892
18	39.3	42.7	76.4	9.9	2.9	3.9	918
19	52.6	38.1	74.1	10.9	24.5	1.9	960
20	28.5	34.1	71.0	11.4	1.9	16.9	908
21	31.8	25.9	73.8	10.8	3.7	21.6	913
22	40.5	51.7	76.8	9.1	29.6	84.7	1002
23	45.1	30.4	81.6	10.3	5.0	2.2	952
24	40.9	45.4	77.7	12.1	30.4	31.5	975
25	35.2	38.8	70.4	11.7	10.7	5.4	956
26	20.2	48.1	74.8	12.0	7.8	239.0	913
27	33.9	31.1	76.0	11.6	23.9	4.3	959
28	48.3	52.5	76.1	11.0	14.9	2.0	901
29	23.9	42.7	75.2	11.5	5.1	3.1	842
30	41.2	30.8	69.9	9.8	4.3	266.0	946

FIGURE 15.10

MORTALITY DATA AND SIX POTENTIAL PREDICTOR VARIABLES FROM 30 RANDOMLY SELECTED LOCATIONS THROUGHOUT THE COUNTRY.

for predicting mortality. This section addresses this problem under the heading of stepwise regression. There are two reasons why the backward elimination procedure may not be appropriate when the number of variables is large. First, if the number of variables exceeds the number of observations, then a backward elimination procedure is mathematically impossible. Second, when a variable is dropped from the model with the backward elimination procedure, it does not have a chance to reenter the model at a subsequent step. However, the stepwise regression procedure explained in this section does allow such reentry.

Number	Variable	Correlation	Number	Variable	Correlation
1	X_1	.5326	6	X_6	.2362
2	X_2	.0062	7	$X_1 X_2$	.3816
3	X_3	.2754	8	$X_1 X_3$	.5258
4	X_4	−.4924	9	$X_1 X_6$	.3429
5	X_5	.6445	10	$X_4 X_5$	.6150

FIGURE 15.11
SIMPLE CORRELATIONS OF THE INDEPENDENT VARIABLES WITH THE DEPENDENT VARIABLE.

STEP 1 IN THE STEPWISE REGRESSION PROCEDURE

The stepwise regression procedure starts by scanning the list of independent variables and finding the variable that has the highest simple correlation with the dependent variable. This list includes not only the original list of independent variables but also any additional variables that are created as functions of these, such as X_i^2 or $X_i X_j$. To illustrate the first step, suppose in addition to X_1 through X_6 in Problem Setting 15.2 the following interaction terms are considered: $X_1 X_2$, $X_1 X_3$, $X_1 X_6$, and $X_4 X_5$. The first two interaction terms represent the nonadditivity of the effects of precipitation with temperature, and the other two interaction terms represent other pairs of variables that are believed to affect the mortality rates in a nonadditive manner. Thus, there are 10 variables to choose from at the first step in the stepwise procedure.

The simple correlations of each of these 10 variables with Y are given in Figure 15.11. The simple correlations in Figure 15.11 are used to determine the variable having the greatest absolute correlation with the dependent variable. In this case, that variable is X_5 with a correlation of .6445. Thus, the mathematical model at step 1 in the stepwise analysis is

$$y = \beta_0 + \beta_1 x_5 \tag{15.16}$$

The computer printout for this model is given in Figure 15.12, which gives the least squares prediction equation as

$$\hat{Y} = 911.81 + 2.858 X_5 \tag{15.17}$$

with an R^2 value of $(.6445)^2 = .4154$. This means that 41.54% of the variation in mortality rates in the different cities is explained by the percentage of the urbanized area population that is nonwhite.

Note at step 1 the stepwise model consists only of a simple linear regression model and, in such a case, the percentage of variation in the dependent variable explained by regression is found as the square of the simple correlation coefficient. That is, $(.6445)^2 = .4154$. Also, note that the p-value associated with X_5 in Figure 15.12 is .0001. This p-value must be less than some level of significance specified as part of the input to the program by the user.

SOURCE	DF	SUM OF SQUARES	MEAN SQUARE	F VALUE	PR > F
MODEL	1	24497.80	24497.80	19.90	.0001
ERROR	28	34477.17	1231.33		
TOTAL	29	58974.97			

R SQUARE
.4154

PARAMETER	ESTIMATE	T FOR H0: PARAMETER = 0	PR > |T|	STD ERROR OF ESTIMATE
INTERCEPT	911.812			
X5	2.858	4.46	.0001	0.641

FIGURE 15.12
COMPUTER PRINTOUT FOR STEP 1 OF THE STEPWISE REGRESSION PROCEDURE.
VARIABLE X_5 ENTERED THE REGRESSION EQUATION.

In this illustration that level of significance was set at .05. In the event that the p-value was greater than .05 no variables would be included in the model and the analysis would end.

STEP 2 IN THE STEPWISE REGRESSION PROCEDURE

At the conclusion of a successful first step (meaning that a significant variable was found) a search is made for a two-variable model. This is accomplished by considering models containing X_5 in turn with each of the remaining variables. In this case, nine such two-variable models are considered. As each variable is paired with the variable selected at the first step, the test of H_0: $\beta_j = 0$ versus H_1: $\beta_j \neq 0$ is made for the new variable added to the model. The new variable with the smallest p-value noted from each of these tests is paired with the variable at the first step to produce a two-variable model provided the p-value is smaller than that value specified by the user as input

SOURCE	DF	SUM OF SQUARES	MEAN SQUARE	F VALUE	PR > F
MODEL	2	35388.09	17694.04	20.25	.0001
ERROR	27	23586.88	873.59		
TOTAL	29	58974.97			

R SQUARE
.6001

PARAMETER	ESTIMATE	T FOR H0: PARAMETER = 0	PR > |T|	STD ERROR OF ESTIMATE
INTERCEPT	1152.170			
X4	−21.905	−3.53	.0015	6.204
X5	2.666	4.91	.0001	0.543

FIGURE 15.13
COMPUTER PRINTOUT FOR STEP 2 OF THE STEPWISE REGRESSION PROCEDURE.
VARIABLE X_4 ENTERED THE REGRESSION EQUATION.

to the program. If the p-value is not small enough, the analysis concludes at this point producing a model with only one variable. The computer printout for step 2 is given in Figure 15.13.

The stepwise procedure produces a two-variable mathematical model of the form:

$$y = \beta_0 + \beta_1 x_4 + \beta_2 x_5 \qquad (15.18)$$

The least squares prediction equation is

$$\hat{Y} = 1152.17 - 21.91 X_4 + 2.666 X_5 \qquad (15.19)$$

with an R^2 value of .6001. Note that the p-values for X_4 and X_5 are .0015 and .0001, respectively. The p-value for X_5 does not appear to have changed, but in general the p-values may change each time a variable is added to or deleted from the model.

STEP 3 IN THE STEPWISE REGRESSION PROCEDURE

Following the conclusion of a successful second step in the stepwise procedure a search is made of the variables not yet included in the model in order to produce a three-variable model. This part of the analysis is identical to step 2, with the smallest p-value associated with tests of H_0: $\beta_j = 0$ versus H_1: $\beta_j \neq 0$ used as the criterion to determine which of the remaining variables (if any) will be added to the model.

Each time a new variable is added to the model tests of H_0: $\beta_j = 0$ versus H_1: $\beta_j \neq 0$ are made for those variables already in the model in the same manner as was done in the backward elimination procedure. If any of these tests indicate that the corresponding variables are no longer significant, then those variables are deleted from the model. Therefore, the usual procedure is to specify two levels of significance α_1 and α_2 as input to the stepwise program. The value of α_1 is used to screen variables for inclusion into the model, and α_2 is used with the tests for those variables already included in the model. The proper operation of a stepwise model requires $\alpha_2 \geq \alpha_1$ to prevent an infinite loop where a variable is added to the model and then deleted. In this example, α_2 was .10.

The remaining steps in the stepwise procedure are similar to step 3. The computer printout for step 3 is given in Figure 15.14 and indicates that the variables X_2, X_4, and X_5 explain 66.31% of the variation in mortality rates. For this example, no more significant variables were found after step 3; hence, step 3 was also the final step. Also, no p-values were greater than .10 so no variables were eliminated from the model.

A COMMENT ON CONDITIONAL TESTS AT STEP 2

Note that the test of H_0: $\beta_j = 0$ versus H_1: $\beta_j \neq 0$ as just explained for step 3 could also take place in step 2. However, this could reduce the model to a one-variable model, which would have a smaller R^2 than the original one-

SOURCE	DF	SUM OF SQUARES	MEAN SQUARE	F VALUE	PR > F
MODEL	3	39104.05	13034.68	17.06	.0001
ERROR	26	19870.92	764.27		
TOTAL	29	58974.97			
			R SQUARE		
			.6631		

PARAMETER	ESTIMATE	T FOR H0: PARAMETER = 0	PR > |T|	STD ERROR OF ESTIMATE
INTERCEPT	1196.165			
X2	−1.732	−2.21	.0365	0.785
X4	−20.649	−3.54	.0015	5.831
X5	3.195	5.69	.0001	0.561

FIGURE 15.14
COMPUTER PRINTOUT FOR STEP 3 OF THE STEPWISE REGRESSION PROCEDURE.
VARIABLE X_2 ENTERED THE REGRESSION EQUATION.

variable model. For this reason conditional tests at step 2 are not usually conducted.

A COMPARISON OF THE STEPWISE AND BACKWARD ELIMINATION PROCEDURES

Both the backward elimination procedure of Section 15.1 and the stepwise procedure presented in this section represent methods used by statisticians to find a subset of independent variables that are useful in predicting the dependent variable. Since the two procedures are different, they cannot be relied on to produce identical subsets of variables. However, there will generally be

SOURCE	DF	SUM OF SQUARES	MEAN SQUARE	F VALUE	PR > F
MODEL	4	41260.62	10315.16	14.56	.0001
ERROR	25	17714.34	708.57		
TOTAL	29	58974.97			
			R SQUARE		
			.6996		

PARAMETER	ESTIMATE	T FOR H0: PARAMETER = 0	PR > |T|	STD ERROR OF ESTIMATE
INTERCEPT	1105.682			
X1	2.524	2.39	.0245	1.054
X4	−20.554	−3.05	.0054	6.741
X1*X2	−4.887 E-02	−2.44	.0221	2.002 E-02
X4*X5	2.881 E-01	5.45	.0001	5.290 E-02

FIGURE 15.15
BACKWARD ELIMINATION RESULTS FOR PROBLEM SETTING 15.2.

good agreement between the two subsets. As an illustration, the backward elimination procedure is applied to the 10 variables used with the stepwise procedure in the mortality rate data of Problem Setting 15.2. The final (seventh) step of the backward elimination procedure is given in Figure 15.15. A level of significance of .05 was used at each step in the analysis.

The stepwise procedure has previously identified the subset of significant variables as X_2, X_4, and X_5. Figure 15.15 shows that the backward elimination procedure produces the subset containing X_1, X_4, X_1X_2, and X_4X_5. Although both procedures have selected X_2, X_4, and X_5 for inclusion in the model, these variables are used differently in each of the models. Additionally, the backward elimination procedure has included X_1 in the final model. This points out that the backward elimination and the stepwise procedure will not necessarily lead to the same subset of variables.

As a final comment, although both of these procedures are attempting to identify only significant variables, there is no assurance that all important variables have been identified or that the models contain only important variables. Both procedures use sample data and depend on multiple tests of the hypotheses being performed on the same set of data. A different set of data may produce different results. The results could also be influenced by outliers or nonlinearity in the data. In either of these cases a transformation such as logs or ranks prior to the regression analysis often proves to be quite helpful in identifying the important variables.

EXAMPLE

Use the results of the stepwise analysis and the backward elimination procedure to predict the mortality rate and to provide a 95% confidence interval for the mean mortality rate for a location with the following properties: $X_1 = 43.3$, $X_2 = 30.2$, $X_3 = 76.5$, $X_4 = 11.0$, $X_5 = 18.8$, $X_6 = 166.0$. From the stepwise analysis in Figure 15.14

$$\hat{Y} = 1196.165 - 1.732(30.2) - 20.649(11.0) + 3.195(18.8)$$
$$= 976.79$$

The 95% confidence limits for the mean mortality rate are found by use of the computer to be from 956.39 to 997.20. From the backward elimination in Figure 15.15

$$\hat{Y} = 1105.682 + 2.524(43.3) - 20.554(11.0) - .04887(43.3)(30.2)$$
$$\quad + .2881(11.0)(18.8)$$
$$= 984.55$$

The 95% confidence limits for the mean mortality rate are found by use of the computer to be from 962.19 to 1006.94. Since the respective figures are within 1% of each other, the results from the two models are in good agreement.

15.20 Use the computer printouts in Figures 15.14 and 15.15 to write the prediction equations corresponding, respectively, to the stepwise and backward elimination procedures for Problem Setting 15.2.

15.21 Use each of the prediction equations in Exercise 15.20 to predict the mortality rates for a city with $X_1 = 35.3$, $X_2 = 48.1$, $X_3 = 81.0$, $X_4 = 10.8$, $X_5 = 7.0$, and $X_6 = 75.2$. Why do these two predictions differ?

15.22 The 95% prediction interval for the mortality rate in the city in Exercise 15.21 based on the stepwise model is from 850.85 to 973.59, whereas the 95% confidence interval for the mean mortality rate of cities of this type is from 889.05 to 935.40. What are the interpretations of these intervals?

15.23 A regression analysis procedure is used on the condominium sales data given below:

Selling Price (Y)	X_1	X_2	D_1	D_2	D_3	X_3	X_4
$44,000	0	880	0	0	0	1	6
50,000	1	1000	0	1	0	2	7
52,000	0	970	0	1	1	1	14
32,000	16	900	1	0	0	6	1
32,000	4	830	0	1	0	1	3
50,000	8	1100	1	0	0	3	2
51,000	0	970	0	0	1	1	4
38,000	11	940	0	0	1	2	1
43,000	3	890	0	0	1	1	7
53,000	5	1090	1	0	0	1	3
30,000	8	730	0	0	1	4	1
53,000	8	1200	0	1	0	7	2
49,000	7	1080	0	0	1	3	2
59,000	1	1100	1	0	1	1	8
35,000	3	760	0	0	0	1	5
52,000	1	1000	1	0	0	2	4
55,000	2	1120	0	1	0	1	6
50,000	7	1120	0	1	0	3	3
69,000	7	1410	1	0	1	2	4
43,000	0	830	0	0	1	3	8
44,000	0	880	0	0	0	1	7
57,000	19	1460	0	1	1	1	1
45,000	12	1030	1	0	1	14	1
37,000	3	770	0	0	1	8	2
48,000	1	910	1	0	0	1	4

The variables are as follows:

$X_1 = $ age in years

$X_2 = $ square footage

D_1 = 1 if two bedroom, 0 otherwise

D_2 = 1 if three bedroom, 0 otherwise

D_3 = 1 if covered parking, 0 otherwise

X_3 = the floor number for main entrance

X_4 = miles to center of town

Additional terms considered in the model are X_1^2, X_2^2, X_3^2, X_4^2, the two-way interaction terms X_1X_2, X_1X_3, X_1X_4, X_2X_3, X_2X_4, X_3X_4, D_1X_2, D_2X_2, D_3X_2, D_1D_3, D_2D_3, and the three-way interaction terms $D_1D_3X_2$, $D_2D_3X_2$. Note that there are 24 variables plus the intercept, so there are 25 parameters being fit by only 25 points. Why will backward regression not work in this problem?

15.24 What is the purpose of including the following variables in the model described in Exercise 15.23?

(a) X_1^2, X_2^2, X_3^2, X_4^2

(b) D_1X_2, D_2X_2

(c) D_3X_2

(d) D_1D_3, D_2D_3

(e) $D_1D_3X_2$, $D_2D_3X_2$

15.25 Stepwise regression on the model in Exercise 15.23, with $\alpha_1 = .05$ and $\alpha_2 = .10$, results in the following computer printout for the first step.

SOURCE	DF	SUM OF SQUARES	MEAN SQUARE	F VALUE	PR > F
MODEL	1	1.473 E09	(b)	55.87	.0001
ERROR	23	6.064 E08	(c)		
TOTAL	24	(a)		R SQUARE	
				.7084	

PARAMETER	ESTIMATE	T FOR H0: PARAMETER = 0	PR > |T|	STD ERROR OF ESTIMATE
INTERCEPT	3737.205			
X2	43.137	7.47	.0001	5.771

(a) What number belongs in the position designated by (a) in the computer printout?

(b) What number belongs in (b)?

(c) What number belongs in (c)?

(d) Which of the 24 variables listed in Exercise 15.23 has the highest correlation with Y?

(e) What is the correlation coefficient between Y and the variable identified in part (d)?

(f) How would you interpret the coefficient 43.137 given in the computer printout?

15.26 The second step in the stepwise regression on the model in Exercise 15.23 results in the following computer printout.

SOURCE	DF	SUM OF SQUARES	MEAN SQUARE	F VALUE	PR > F
MODEL	2	1.988 E09	9.941 E08	239.97	.0001
ERROR	22	9.114 E07	4.143 E06		
TOTAL	24	2.079 E09		R SQUARE	
				(a)	

PARAMETER	ESTIMATE	T FOR H0: PARAMETER = 0	PR > |T|	STD ERROR OF ESTIMATE
INTERCEPT	−10834.152			
X2	62.730	(b)	.0001	2.884
X1*X2	−0.915	(c)	.0001	0.082

(a) What should R^2 be in the computer printout?

(b) What are the values for T that belong in (b) and (c)?

(c) Which variable was added to the model?

(d) Taking into consideration the variables now in the model, and their coefficients given by the computer printout, give an interpretation to the net effect of adding the new variable into the model.

15.27 The third step in the stepwise regression on the model in Exercise 15.23 results in the following computer printout.

SOURCE	DF	SUM OF SQUARES	MEAN SQUARE	F VALUE	PR > F
MODEL	(a)	2.015 E09	6.716 E08	218.43	.0001
ERROR	(b)	(d)	3.075 E06		
TOTAL	(c)	(e)		R SQUARE	
				.9689	

PARAMETER	ESTIMATE	T FOR H0: PARAMETER = 0	PR > |T|	STD ERROR OF ESTIMATE
INTERCEPT	−12437.967			
X2	65.059	24.94	.0001	2.608
D2	−2455.459	−2.94	.0078	835.304
X1*X2	−0.922	−13.03	.0001	7.075 E-02

(a) Which degrees of freedom belong in (a), (b), and (c)?

(b) Fill in the missing sums of squares (d) and (e).

(c) How would you interpret the new variable just added to the model, and its coefficient?

(d) Why were no variables deleted from the model at this step?

15.28 Step 4 in the stepwise regression procedure on the model in Exercise 15.23 results in the following computer printout. Fill in the missing blanks (a) through (r). *Hint*: Proceed in alphabetical order, and refer to the previous exercises.

SOURCE	DF	SUM OF SQUARES	MEAN SQUARE	F VALUE	PR > F
MODEL	(a)	2.046 E09	(g)	(i)	(j)
ERROR	(b)	(e)	(h)		
TOTAL	(c)	(d)		R SQUARE	
				(f)	

PARAMETER	ESTIMATE	T FOR H0: PARAMETER = 0	PR > |T|	STD ERROR OF ESTIMATE
INTERCEPT	−11618.450			
X2	64.685	(k)	(o)	1.913
D2	−3644.513	(l)	(p)	669.601
D2*D3*X2	4.327	(m)	(q)	0.989
X1*X2	−1.020	(n)	(r)	5.643 E-02

15.29 The fifth and final step in the stepwise regression on the model in Exercise 15.23 results in the following computer printout. Write out the equation for the resulting fitted model. Give a variable-by-variable interpretation of this model, explaining what each variable means and how each affects the selling price of the condominium.

SOURCE	DF	SUM OF SQUARES	MEAN SQUARE	F VALUE	PR > F
MODEL	5	2.059 E09	4.118 E08	386.34	.0001
ERROR	19	2.025 E07	1.066 E06		
TOTAL	24	2.079 E09		R SQUARE	
				.9903	

PARAMETER	ESTIMATE	T FOR H0: PARAMETER = 0	PR > |T|	STD ERROR OF ESTIMATE
INTERCEPT	−13507.748			
X3	286.798	3.46	.0026	82.931
X2	66.223	41.38	.0001	1.600
D2	−3765.273	−6.98	.0001	539.317
D2*D3*X2	5.446	6.34	.0001	0.858
X1*X2	−1.118	−20.89	.0001	5.350 E-02

15.30 In the stepwise regression of Exercises 15.25 to 15.29, the procedure stopped after entering the variable X_3 into the model, because no other variables were significant at α_1 = .05, where α_1 is the level of significance for the inclusion of variables into the model. If someone does not know the size of α_1, what clues can he or she get about the size of α_1 from the computer printouts of Exercises 15.25 to 15.29?

15.4

REVIEW EXERCISES

15.31 Use Eq. 15.13 to predict the job satisfaction rating for a 55-year-old female manager who makes a salary of $58,000, has been with the company for 23 years with 5 years of related professional experience, has 19 years of education, and whose salary constitutes 80% of her total income.

15.32 The 95% prediction interval for the individual's job satisfaction rating in Exercise 15.31 is from 6.41 to 14.44, whereas the 95% confidence interval for the mean response of such individuals is from 9.01 to 11.76. What are the interpretations of these intervals?

15.33 The computer printout in Figure 15.2 results when the full model with all eight variables is used to fit the job satisfaction data. Using the full model, the 95% prediction interval for the person described in Exercise 15.31 is from 4.04 to 14.19, and the 95% confidence interval for the mean of all persons with those characteristics is from 5.73 to 12.50. Are these intervals, obtained from the model with all eight variables, better or worse than the corresponding intervals obtained from the model with only two variables, as given in Exercise 15.32?

15.34 The stepwise regression procedure on the job satisfaction data consists of two steps. The first step introduces the variable X_3 and the second step brings in the variable X_6, so the stepwise regression results in the same model as given by Figure 15.9 for backward regression. The first step in the stepwise procedure results in the following computer printout. Fill in the missing values (a) through (l).

SOURCE	DF	SUM OF SQUARES	MEAN SQUARE	F VALUE	PR > F
MODEL	(a)	97.354	(f)	(i)	(j)
ERROR	(b)	(d)	(g)		
TOTAL	(c)	(e)		R SQUARE	
				(h)	

PARAMETER	ESTIMATE	T FOR H0: PARAMETER = 0	PR > |T|	STD ERROR OF ESTIMATE
INTERCEPT	5.568			
X3	9.679 E-05	(k)	(l)	1.988 E-05

15.35 A stepwise regression was run on the job satisfaction data of Figure 15.1. The model included not only the variables X_1 to X_8 but also several squares and cross products, for a total of 24 variables. The result is given in the following computer printout. Notice that one of the standard errors is zero when rounded off to the eighth decimal place. Use the estimate of the coefficient and the value of T to find the real value of that standard error.

SOURCE	DF	SUM OF SQUARES	MEAN SQUARE	F VALUE	PR > F
MODEL	4	165.681	41.420	14.47	.0001
ERROR	45	128.804	2.862		
TOTAL	49	294.485		R SQUARE .5626	

PARAMETER	ESTIMATE	T FOR H0: PARAMETER = 0	PR > |T|	STD ERROR OF ESTIMATE
INTERCEPT	14.116			
X2	1.108	2.05	.0462	0.541
X3	4.361 E-04	4.53	.0001	9.633 E-05
X6	−0.993	−4.18	.0001	0.238
X3*X3	−3.869 E-09	−2.98	.0047	0.000 E-05

15.36 Write the equation for the model indicated by the computer printout in Exercise 15.35. Use it to obtain point estimates of job satisfaction ratings for the following employees.

(a) A male employee with a master's degree (17 years of education) making $36,000 a year.

(b) A female employee with one year of college, making $13,500 a year.

(c) A male employee with a doctorate (19 years of education) earning $58,000 a year.

15.37 Use each of the prediction equations in Exercise 15.20 to predict the mortality rates for a city with $X_1 = 20.4$, $X_2 = 38.6$, $X_3 = 76.2$, $X_4 = 11.5$, $X_5 = 9.9$, and $X_6 = 5.6$. Why do these two predictions differ?

15.38 The 95% prediction interval for the mortality rate in the city in Exercise 15.37 based on the stepwise model is from 865.18 to 981.79, whereas the 95% confidence interval for the mean mortality rate of cities of this type is from 910.43 to 936.54. What are the interpretations of these intervals?

15.39 Assume that the sample of the second 500 companies listed by *Fortune* magazine, as given in Exercise 14.27, satisfies the multiple linear regression model. Test the null hypothesis that the coefficients of X_1^2, X_2, X_2^2, and $X_1 X_2$ all equal zero in the model

$$Y = \beta_0 + \beta_1 X_1 + \beta_2 X_1^2 + \beta_3 X_2 + \beta_4 X_2^2 + \beta_5 X_1 X_2 + \epsilon$$

by comparing the following computer printouts. Use $\alpha = .05$. Interpret your results.

SOURCE	DF	SUM OF SQUARES	MEAN SQUARE	F VALUE	PR > F
MODEL	1	908.992	908.992	30.40	.0001
ERROR	23	687.836	29.906		
TOTAL	24	1596.828		R SQUARE .5692	

PARAMETER	ESTIMATE	T FOR HO: PARAMETER = 0	PR > |T|	STD ERROR OF ESTIMATE
INTERCEPT	−8.336	−2.45	.0225	3.4073
X1	0.104	5.51	.0001	0.0189

SOURCE	DF	SUM OF SQUARES	MEAN SQUARE	F VALUE	PR > F
MODEL	5	1295.548	259.110	16.34	.0001
ERROR	19	301.280	15.857		
TOTAL	24	1596.828		R SQUARE .8113	

PARAMETER	ESTIMATE	T FOR HO: PARAMETER = 0	PR > |T|	STD ERROR OF ESTIMATE
INTERCEPT	−28.316	−2.31	.0323	12.257
X1	0.341	2.74	.0130	0.125
X1*X1	−4.833E-04	−1.70	.1052	2.841E-04
X2	−5.613E-02	−0.48	.6334	0.116
X2*X2	4.660E-04	1.52	.1445	3.061E-04
X1*X2	−2.770E-04	−0.37	.7130	7.417E-04

15.40 Test the null hypothesis that the coefficients of $D_3 X$, $D_1 D_3$, and $D_2 D_3$ all equal zero in the housing sale price data of Section 14.3. Use the computer printout in Figures 14.30 and 14.31, and use $\alpha = .05$. Assume that the multiple linear regression model assumptions are satisfied. Interpret your results.

15.41 Assume that the condominium sale price data of Exercise 14.12 follow the multiple linear regression model. Compare the results in Exercises 14.19 and 14.23 to test the null hypothesis that the coefficients of all of the interaction terms in the model

$$Y = \beta_0 + \beta_1 X_2 + \beta_2 D_1 + \beta_3 D_2 + \beta_4 D_3 + \beta_5 D_1 X_2$$
$$+ \beta_6 D_2 X_2 + \beta_7 D_3 X_2 + \beta_8 D_1 D_3 + \beta_9 D_2 D_3 + \epsilon$$

equal zero. Use $\alpha = .05$. Interpret your results.

15.42 Find a 95% confidence interval for the coefficient of X_6 in the model estimated by Eq. 15.13. Interpret your results.

15.43 Find a 95% confidence interval for β_4 in the model

$$Y = \beta_0 + \beta_1 X_2 + \beta_2 X_3 + \beta_3 X_6 + \beta_4 X_8 + \epsilon$$

as it applies to the job satisfaction data of Problem Setting 15.1. Interpret your results.

BIBLIOGRAPHY

Additional material on the topics presented in this chapter can be found in the following publications.

Draper, N. R. and Smith, H. (1981). *Applied Regression Analysis,* 2nd ed. Wiley, New York.

Iman, R. L. and Conover, W. J. (1979). "The Use of the Rank Transform in Regression" *Technometrics,* **21** (4), 499–509.

16
ANALYSIS OF VARIANCE
FOR ONE-FACTOR EXPERIMENTS

PRELIMINARY REMARKS

The concept of an experimental design was introduced in Chapter 8 where matched pairs were used as a method of controlling unwanted variation in experimental results. There it was pointed out that the design, or set of plans, of an experiment is very important in obtaining results that are sensitive to the purposes of the experiment. The usage of matched pairs of data in an experimental design enables one to detect differences between population means that are not possible to detect when independent samples are obtained. Since it is not always possible to collect data in matched pairs, methods of analysis for independent samples are also needed. These were given in Chapter 9 where comparisons of means were made using two independent samples in another type of experimental design. These experimental designs are but two of many existing experimental designs that cover a variety of experimental requirements. This chapter is the first of three chapters that consider some of the more commonly used experimental designs that have proven useful in a variety of business settings.

EXPERIMENTAL DESIGN   The **experimental design** is the set of plans and instructions by which the data in an experiment are collected.

It is very important that a good experimental design be established prior to the collection of the data. In this way the experimenter is assured that a method of analysis is available for the data. If the data are collected prior to the consideration of an experimental design, often it is not possible to extract the desired information from the data, either because the data were collected improperly, or because the data were collected in a manner that was inconsistent with any known method of analysis.

The computations associated with the experimental design in this chapter are referred to as a **one-factor** (or **one-way**) **analysis of variance.** *One-factor* refers to the fact that the data are classified into groups according to one criterion, or factor. *Analysis of variance* refers to the fact that the analysis consists of an examination and identification of the sources of variation present in the sample data. For example, suppose the purpose of an experiment is to determine the average mpg of three types of small pickup truck. A test is set up using six Toyota, five Nissan, and four Mazda pickups. Each vehicle is driven over the same 300-mile course at 55 mph with the following mpg results recorded:

Toyota	Nissan	Mazda
27.1	25.3	23.1
27.5	26.5	24.3
27.0	26.4	23.4
26.9	26.8	24.2
27.7	26.5	$23.75 \ \bar{x}$
27.3		

The single *factor* used for separating the data into groups is the make of the truck. A first step in examining these sample data is to ask what is causing the variation in the results; that is, only two of the numbers are identical, so there must be some explanation for the observed variability. One obvious response to this query is that some of the variation is due to the fact that different varieties of pickups produce different average mpg readings because of different engineering. For example, in these sample data the four Mazda readings are less than all the results recorded for either Toyota or Nissan. But even among pickups of the same brand, mileage varies. Many things could account for this. The six Toyota observations were obtained from six different Toyota pickups that may have been driven by different drivers. Perhaps the tire inflations differed from vehicle to vehicle, or maybe the carburetors were adjusted differently, or the weather conditions (windy, rainy) may have differed when the six tests were made, and so on.

The preceding illustration points out that the experimental results contain several sources of variability. A good experimental design enables the primary sources of variability to be identified and the amount of variability due to each source to be separated out of the total variability in the data. The remaining variability in the data is attributed to randomness, and this source of variation is called **error.** A good experimental design reduces the error variation as much as possible so that the factors of interest, such as the different brands of pickup trucks, can be examined more accurately. The analysis associated with an experimental design attempts to identify all of the sources contributing to the variability in the results and to measure the amount of variability due to each source. One of the simplest and most frequently used experimental designs is the **completely randomized design,** which is considered in this chapter. Subsequent chapters consider other designs.

16.1
AN OVERVIEW OF COMPLETELY RANDOMIZED DESIGNS

PROBLEM SETTING 16.1

An air quality engineer has the job of monitoring the amount of contaminants in the output of smokestacks from factories in a large city. At randomly selected times smokestacks of factories associated with particular industries are observed and measurements of the amount of contaminants are made. One week, three mills are selected for monitoring, and the amount of contaminant is measured at five randomly selected times at each factory. The results are given in Figure 16.1. The air quality engineer must make comparisons among the factories to see if they produce different amounts of contaminants.

PROBLEM SETTING 16.2

A consumer product evaluation group is interested in comparing the mean life in minutes of four types of batteries commonly used with children's toys. A random sample of each of the four battery types is selected and put on a continuous test where the time required for the energy output of the battery to fall below a predetermined acceptable level is measured. The test results are given in Figure 16.2.

The evaluation group would like to make statements about the relative merits of the battery types as measured by average lifetime.

PROBLEM SETTING 16.3

As input to a consumer price index, the price of a representative market basket of food is obtained at each of 10 randomly selected grocery stores in each of five cities. The prices observed in each of these stores is recorded in Figure 16.3.

Factory A	Factory B	Factory C
46.3	48.6	45.1
43.7	52.3	46.7
51.2	50.9	41.8
49.6	53.6	40.4
48.8	55.7	42.6

FIGURE 16.1
MEASUREMENTS ON THE AMOUNT
OF CONTAMINANT AT EACH OF
THREE FACTORIES.

Battery 1	Battery 2	Battery 3	Battery 4
43	45	45	45
47	48	43	48
48	49	41	55
45	46	41	47
46	52	38	58
42	45	46	50
46	44	45	46
45	47	41	53
49		43	56
		41	

FIGURE 16.2
RESULTS OF LIFE TESTS ON FOUR
BATTERY TYPES.

The prices in Figure 16.3 will be used to determine an index for food costs across the country as well as for each city. Additionally, the prices can be used to make comparisons about the relative cost of living in each of the five cities.

THE COMPLETELY RANDOMIZED DESIGN

Each of the problem settings (16.1, 16.2, and 16.3) involves sample data obtained as independent random samples. This is the same situation encountered in Chapter 9 where two independent random samples were used to make inferences about the means of the corresponding populations. In this chapter the case of two or more independent random samples is considered to make inferences about the respective population means. Hence, the methods of this chapter represent a generalization of the methods in Chapter 9.

St. Louis	San Francisco	Omaha	Washington, D.C.	Dallas
$73.34	$76.29	$70.28	$75.81	$74.61
71.57	74.78	74.25	78.19	73.55
77.46	77.41	70.51	77.94	76.51
73.86	75.83	74.59	76.42	75.11
74.10	79.25	72.58	77.46	73.41
71.66	75.65	71.91	75.29	74.24
75.75	74.12	71.58	78.82	76.87
76.82	78.56	73.13	77.71	75.34
73.88	75.90	72.18	76.12	74.66
72.49	77.03	74.43	78.88	75.46

FIGURE 16.3
PRICES OF A REPRESENTATIVE MARKET BASKET OF FOOD IN
EACH OF FIVE CITIES.

A design where the data consist of independent random samples from each of k populations is an **independent sampling design,** but is usually referred to as a **completely randomized design.** Strictly speaking, a completely randomized design is one in which a group of experimental units, such as people, are randomly assigned to treatments, so that each experimental unit is assigned to one of the k treatments. Because the matching of experimental units with treatments is done in a completely random manner, this is called a completely randomized design. The statistical analysis is the same whether experimental units are assigned at random to k treatments, or whether random samples are taken from k populations; thus, both designs are treated as completely randomized designs in this text.

INDEPENDENT
SAMPLING
DESIGN (THE
COMPLETELY
RANDOMIZED
DESIGN)

An **independent sampling design** is one in which independent random samples are obtained from each of k populations. This design is often referred to as a **completely randomized (CR) design.**

The main purpose of a completely randomized design is to compare the means of the populations from which the sample data are obtained. Hence, for k populations the null hypothesis of interest is of the form, $H_0: \mu_1 = \mu_2 = \cdots = \mu_k$ (i.e., all k populations have the same mean). Procedures for testing this hypothesis are given in Sections 16.2 and 16.4. If H_0 is rejected, then more specific information is required with respect to the individual population means. For example, in Problem Setting 16.1, the null hypothesis takes the form $H_0: \mu_A = \mu_B = \mu_C$ (i.e., the mean amount of contaminant is the same for all factories) and if H_0 is rejected, then it is desirable to know how the means μ_A, μ_B, and μ_C differ from one another. This is the subject of Section 16.3.

The remainder of this section is devoted to a graphical presentation of the sample data from a completely randomized design. This graphical presentation is meant to aid the reader in further understanding the problem of interest as well as to provide a check on the conclusions of Sections 16.2 and 16.3.

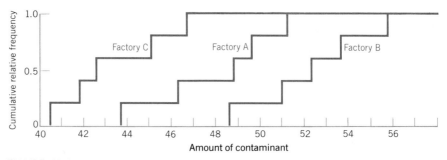

FIGURE 16.4
EMPIRICAL DISTRIBUTION FUNCTIONS FOR THE SMOKESTACK DATA FROM FIGURE 16.1.

A GRAPHICAL COMPARISON OF
SAMPLE DATA FROM A CR DESIGN

The empirical distribution functions associated with the two-sample data of Chapter 9 were used in Figure 9.2 to demonstrate the fact that the mean rating of one hospital was higher than the mean rating for a second hospital as determined by two randomly selected groups of ex-maternity patients. Of course, the graph did not prove in and of itself that such a difference existed in the two hospitals, but rather it provided a pictorial representation that supported the conclusions of the analysis in Chapter 9. The e.d.f. also provides a convenient method to use with data from a completely randomized design. For example, consider the smokestack data of Problem Setting 16.1, for which a graphical display is given in Figure 16.4.

Figure 16.4 indicates that it would be reasonable to think in terms of rejecting the null hypothesis H_0: $\mu_A = \mu_B = \mu_C$ because there appears to be a clear separation of the amount of contaminants released by each factory. Additionally, the comparison of the means would seem to indicate that $\mu_C < \mu_A < \mu_B$. That is, contamination from Factory C is lower than the other two factories, with contamination from Factory B higher than the other two factories. These tentative conclusions are examined in detail in Sections 16.2 and 16.3.

Next consider the e.d.f.s for the battery-life test data given in Figure 16.3. These e.d.f.s appear in Figure 16.5. Figure 16.5 does not make the decision regarding the null hypothesis H_0: $\mu_1 = \mu_2 = \mu_3 = \mu_4$ (i.e., equal mean battery lives) as clear-cut as was the case with Figure 16.4 for the smokestack data. It does seem safe to say from observing Figure 16.5 that the data do not support H_0. However, statements about the relative ordering of the population means are more difficult to make, since the e.d.f.s for populations 1 and 2 appear to be very nearly the same and both are in between the other two. Probably the best guess at this point would be $\mu_3 < \mu_1 = \mu_2 < \mu_4$, which merely puts into symbols what the previous statement said in words. Again, these tentative conclusions will be examined in detail in the next two sections.

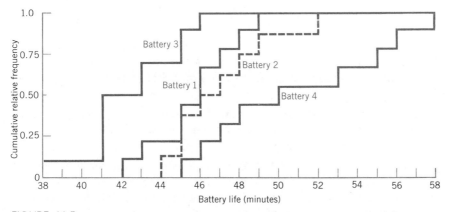

FIGURE 16.5
EMPIRICAL DISTRIBUTION FUNCTIONS FOR THE BATTERY-LIFE TEST DATA
FROM FIGURE 16.2.

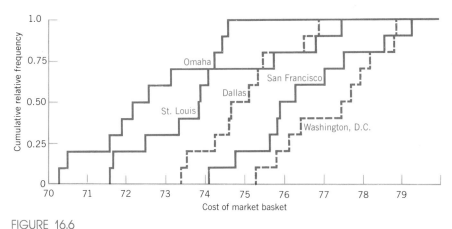

FIGURE 16.6
EMPIRICAL DISTRIBUTION FUNCTIONS FOR THE MARKET BASKET COSTS FROM
FIGURE 16.3.

As a last example of e.d.f.s, consider the market basket costs given in Figure
16.3. These e.d.f.s appear in Figure 16.6 and indicate that the null hypothesis

$$H_0: \mu_{SL} = \mu_{SF} = \mu_O = \mu_W = \mu_D$$

(i.e., equal mean market basket costs) could probably be rejected due to
separation of the e.d.f.s. However, statements regarding the means become
more difficult. It would appear that the costs are higher in San Francisco and
Washington, D.C., than in the other three cities. Also the cost in Omaha
appears to be less than that in the other cities, whereas the e.d.f.s for St. Louis
and Dallas overlap somewhat. Hence, it may be reasonable to think in terms
of the relative ordering of the population means as $\mu_O < \mu_{SL} = \mu_D < \mu_{SF} = \mu_W$. As with the previous examples, these tentative conclusions are investigated in the next two sections.

EXERCISES

16.1 Refer to the sample mpg data given in the preliminary remarks of this
chapter and graph the e.d.f.s for each model of pickup truck.

16.2 Make a tentative guess about the acceptability of the null hypothesis
$H_0: \mu_T = \mu_N = \mu_M$ from the e.d.f.s in Exercise 16.1. If your decision
is to reject H_0, then make another tentative guess about the relative
ordering of the population means.

16.3 To compare four different methods of growing corn, a large number of
plots were used where one of the methods of growing corn was randomly
assigned to each plot of land. The number of bushels per acre was
recorded for each plot of land with the following results. Graph the
e.d.f.s for these sample data.

Method 1	Method 2	Method 3	Method 4
83	91	101	78
91	90	100	82
94	81	91	81
89	83	93	77
89	84	96	79
96	83	95	81
91	88	94	80
92	91		81
90	89		
	84		

16.4 Make a tentative guess about the acceptability of the null hypothesis H_0: $\mu_1 = \mu_2 = \mu_3 = \mu_4$ from the e.d.f.s in Exercise 16.3. If your decision is to reject H_0, then make another tentative guess about the relative ordering of the population means.

16.5 A random sample of the number of units of production each hour was obtained for each of five workers with the results as follows. Graph the e.d.f.s for these sample data.

Worker 1	Worker 2	Worker 3	Worker 4	Worker 5
23	28	19	31	29
27	23	21	25	23
34	31	29	37	32
24	18	17	28	21

16.6 Make a tentative guess about the acceptability of the null hypothesis H_0: $\mu_1 = \mu_2 = \mu_3 = \mu_4 = \mu_5$ from the e.d.f.s in Exercise 16.5. If your decision is to reject H_0, then make another tentative guess about the relative ordering of the population means.

16.7 The tar content (in milligrams) is measured on randomly selected samples of each of four brands of cigarettes. The test measurements are as follows. Graph the e.d.f.s for these sample data.

Brand 1	Brand 2	Brand 3	Brand 4
16.4	17.7	16.1	16.6
15.5	17.1	17.9	19.6
15.4	17.2	18.1	17.3
15.8	17.3	16.5	17.9
15.1	17.1	16.8	16.6
16.8	17.3	16.1	16.9
16.8	16.5	17.2	16.2
	17.5	16.7	17.6
		17.8	17.4
			15.5

16.8 Make a tentative guess about the acceptability of the null hypothesis H_0: $\mu_1 = \mu_2 = \mu_3 = \mu_4$ from the e.d.f.s in Exercise 16.7. If your decision is to reject H_0, then make another tentative guess about the relative ordering of the population means.

16.2
THE ANALYSIS OF VARIANCE FOR THE COMPLETELY RANDOMIZED DESIGN

Although the graphs of the e.d.f.s of the previous section aid intuition with respect to testing the null hypothesis of equality of means of k populations, that is, H_0: $\mu_1 = \mu_2 = \cdots = \mu_k$, the graphical procedure should not be relied upon entirely to make a decision regarding the acceptance or rejection of H_0 because it is subjective. Instead, a testing procedure that removes the subjectivity such as the two-sample t-tests of Chapter 9 for testing the equality of two population means (H_0: $\mu_1 = \mu_2$) is needed. Such a test procedure is given in this section and is referred to as a *one-way analysis of variance*. The term one-way means only one factor is used to classify the sample data.

IDENTIFYING THE SOURCES OF VARIATION IN THE SAMPLE DATA

As alluded to in the preliminary remarks of this chapter, an *analysis of variance* is simply an analysis of the sources of variation present in the sample data. The mpg data presented in the preliminary remarks are used for demonstration. The sample mean for all 15 data points is $\overline{X} = 26.0$ and the total sum of squares

$$\sum_{i=1}^{15} (X_i - \overline{X})^2 = 32.940$$

represents the total variation present in the sample data. That is, if all 15 data points were exactly the same number, there would be no variation in the data and the total sum of squares would be equal to zero; otherwise, the total sum of squares will always be greater than zero.

If the null hypothesis H_0: $\mu_T = \mu_N = \mu_M$ *is true,* then all three model types tend to yield the same average miles per gallon (mpg), and the only source of variation in the data is the variation naturally observed from one vehicle to another *within* a given model type. For the mpg data, the average mpg observed for Toyota is $\overline{X}_T = 27.25$ and the variation *within* the Toyota vehicles is given as

$$\sum_{i=1}^{6} (X_{iT} - 27.25)^2 = 0.475$$

Likewise the sample means for Nissan and Mazda are 26.30 and 23.75, respectively, and the corresponding measures of variation *within* each of these model types are

$$\sum_{i=1}^{5} (X_{iN} - 26.30)^2 = 1.340$$

and

$$\sum_{i=1}^{4} (X_{iM} - 23.75)^2 = 1.050$$

The sum of these three sources of variation is $0.475 + 1.340 + 1.050 = 2.865$.

If the null hypothesis H_0: $\mu_T = \mu_N = \mu_M$ *is not true*, then some of the variation present in the sample data is due to differences *between* model types. This *between* variation is measured by determining how much each of the three sample means (27.25, 26.30, and 23.75) differs from the overall mean (26.0) when each difference is weighted by the respective sample size. The calculation is

$$6(27.25 - 26.0)^2 + 5(26.30 - 26.0)^2 + 4(23.75 - 26.0)^2 = 30.075$$

An interesting result can now be observed. If the variation within each model type is added to the variation between model types the total variation is obtained; that is,

$$2.865 + 30.075 = 32.940$$

A more general way of stating this result is that the total sum of squares has been partitioned (divided) into two parts that are generally referred to as the treatment (between) sum of squares and the error (within) sum of squares.

The total of the squared deviations between each observation X_{ij} and the mean $\bar{X}$ of all of the observations is the **total sum of squares.** In symbols, if

X_{ij} = the *i*th observation in the *j*th population or treatment,

N = the total number of observations, and

TOTAL SUM OF SQUARES $\bar{X} = \dfrac{1}{N} \sum_{i,j} X_{ij}$ = overall sample mean

then the total sum of squares is given as

$$\text{Total SS} = \sum_{i,j} (X_{ij} - \bar{X})^2 = \sum_{i,j} X_{ij}^2 - \frac{\left(\sum_{i,j} X_{ij} \right)^2}{N} \tag{16.1}$$

---

total sum of squares = treatment sum of squares + error sum of squares

The total sum of squares may be computed using Eq. 16.1, and the treatment sum of squares may be computed using Eq. 16.2. The error sum of squares is easily found as the difference of the total sum of squares and the treatment sum of squares.

The partitioning shows that most of the total sum of squares in this example is obtained from the treatment (between) sum of squares. However, this type of variation is large only when the null hypothesis is not true. Thus, the result of the imbalance observed in the partitioning may be a strong indication that the null hypothesis is not true.

THE ONE-WAY ANALYSIS OF VARIANCE

The importance of partitioning the total sum of squares is that it leads to the development of a test statistic. For example, it has been pointed out in the mpg data that almost all of the total variation (32.940) is due to differences between model types (30.075), which is perhaps strong evidence that H_0 is not true. On the other hand, if each of the model types has yielded a sample mean exactly equal to the overall mean (26.0), the sum of the squares between model types (SST) would have been zero,

$$SST = 6(26.0 - 26.0)^2 + 5(26.0 - 26.0)^2 + 4(26.0 - 26.0)^2 = 0$$

Hence, the contribution to the total variation, Total SS, from model-type differences would have been zero and H_0 would probably be true. If SST is large, relative to the Total SS or to the Error SS, the statistical evidence indicates that the population means are not equal. This corresponds to large values of the F statistic given by Eq. 16.4. The formal testing procedure for H_0 is called a **one-way analysis of variance.**

The sum of squared deviations between each sample mean $\overline{X}_j$ and the overall mean $\overline{X}$, weighted by the number of observations in each sample, is called the **treatment sum of squares (SST).** In symbols, if

n_j = the sample size from the jth population, and

TREATMENT
SUM OF
SQUARES

$$\overline{X}_j = \frac{1}{n_j} \sum_{i=1}^{n_j} X_{ij}$$

= sample mean of the observations from the jth population,

then the treatment sum of squares is given as

$$SST = \sum_{j=1}^{k} n_j (\overline{X}_j - \overline{X})^2 = \sum_{j=1}^{k} \frac{\left(\sum_{i=1}^{n_j} X_{ij}\right)^2}{n_j} - \frac{\left(\sum_{i,j} X_{ij}\right)^2}{N} \qquad (16.2)$$

ERROR SUM OF SQUARES

The sum of squared deviations between each observation X_{ij} and the sample mean $\bar{X}_j$ for the population from which X_{ij} was obtained is called the **error sum of squares:**

$$\text{Error SS} = \sum_{j=1}^{k} \sum_{i=1}^{n_j} (X_{ij} - \bar{X}_j)^2 = \text{Total SS} - \text{SST} \tag{16.3}$$

PROCEDURE FOR INFERENCE ABOUT THE DIFFERENCE IN MEANS OF k NORMAL POPULATIONS WITH EQUAL VARIANCES (One-Way Analysis of Variance)

Data

The data consist of k independent samples of sizes $n_1, n_2, \ldots, n_k$ from each of k populations. The respective population means are unknown and are denoted by $\mu_1, \mu_2, \ldots, \mu_k$. The total sample size is N.

Assumptions

1. All samples are random samples from their respective populations.
2. The samples are all independent of one another.
3. All populations are normal with equal variances ($\sigma_1^2 = \sigma_2^2 = \cdots = \sigma_k^2$).

Null Hypothesis

$$H_0: \mu_1 = \mu_2 = \cdots = \mu_k$$

Alternative Hypothesis

H_1: At least two population means are unequal ($\mu_i \neq \mu_j$).

Test Statistic

The test statistic is

$$F = \frac{\text{MST}}{\text{MSE}} \tag{16.4}$$

where MST = $\text{SST}/(k - 1)$ and MSE = $\text{Error SS}/(N - k)$, with SST and Error SS defined by Eqs. 16.2 and 16.3, respectively. The null distribution of F is the F distribution with $(k - 1)$ and $(N - k)$ degrees of freedom, whose quantiles are given in Table A5.

Decision Rule

H_0 is rejected and it is concluded that at least two population means are not equal to one another if $F > F_{1-\alpha, k-1, N-k}$, where $F_{1-\alpha, k-1, N-k}$ is the $1 - \alpha$ quantile from Table A5 with $k - 1$ and $N - k$ degrees of freedom.

Note

If the desired degrees of freedom for the F distribution cannot be found in Table A5, simply use the next smaller degrees of freedom given in the table.

ANALYSIS OF VARIANCE TABLE (CR DESIGN)

The sums of squares, degrees of freedom (DF), and mean squares may be summarized in an analysis of variance table, along with the F statistic and its p-value, as follows:

SOURCE	DF	SUM OF SQUARES	MEAN SQUARE	F VALUE	PR > F
TREATMENT	$k - 1$	SST	MST $= $ SST$/(k - 1)$	MST/MSE	
ERROR	$N - k$	Error SS	MSE $= $ Error SS$/(N - k)$		
TOTAL	$N - 1$	Total SS			

A WORD ABOUT THE TEST ASSUMPTIONS

The first two assumptions for the one-way analysis of variance are usually within the control of the experimenter and may be verified by examining the protocol by which the data were obtained. However, the assumption of normality should be tested by using the Lilliefors test on each set of sample data. If the sample sizes are small, nonnormality is difficult to detect. The authors recommend that the Kruskal–Wallis test in Section 16.4 be used with small sample sizes whenever the experimenter doubts that the populations are normal, even if a goodness-of-fit test fails to detect the nonnormality.

The assumption of equal variances should be checked only after a normality assumption has been checked and verified. Good tests for equality of variance involve considerable computation and in general have low power. See the references at the end of this chapter for a recent survey article on tests for equality of variances. If the equality of variance assumption is suspect, then the procedure in Section 16.4 is recommended, because it is less sensitive to unequal variances. The one-way analysis of variance for testing equality of population means is now demonstrated with three examples.

EXAMPLE

The smokestack data given in Problem Setting 16.1 are now used to test H_0: $\mu_A = \mu_B = \mu_C$ (i.e., equal mean levels of contaminants for all factories) at $\alpha = .05$. Calculations with the sample data show the test assumptions to be satisfied and reveal the following.

$$\sum_i X_{iA} = 239.6, \qquad \sum_i X_{iB} = 261.1, \qquad \sum_i X_{iC} = 216.6$$

$$\sum_{i,j} X_{ij} = 717.3, \qquad \sum_{i,j} X_{ij}^2 = 34{,}588.99$$

From Eq. 16.1 the total sum of squares is found as

$$\text{Total SS} = 34{,}588.99 - \frac{(717.3)^2}{15} = 287.704$$

Equation 16.2 yields the treatment sum of squares as

$$SST = \frac{(239.6)^2}{5} + \frac{(261.1)^2}{5} + \frac{(216.6)^2}{5} - \frac{(717.3)^2}{15}$$

$$= 34{,}499.386 - 34{,}301.286$$

$$= 198.100$$

And finally from Eq. 16.3 the error sum squares is

$$\text{Error SS} = \text{Total SS} - SST$$

$$= 287.704 - 198.100$$

$$= 89.604$$

From Eq. 16.4,

$$MST = 198.100/(3 - 1) = 99.05$$

and

$$MSE = 89.604/(15 - 3) = 7.467,$$

from which the F ratio is given as

$$F = \frac{99.05}{7.467} = 13.27$$

This value of F is much larger than the critical value for a level of significance of $\alpha = .05$, $F_{.95,2,12} = 3.885$. The results of these calculations are conveniently summarized in an analysis of variance table.

SOURCE	DF	SUM OF SQUARES	MEAN SQUARE	F VALUE	PR > F
TREATMENT	2	198.100	99.050	13.27	.0009
ERROR	12	89.604	7.467		
TOTAL	14	287.704			

The p-value is less than .05, so H_0 is rejected. It is concluded that the mean amount of contaminants is not the same for all factories. This conclusion is in agreement with tentative conclusions of Section 16.1 based on the e.d.f.s graphed in Figure 16.4

THE MODEL

The assumptions in the one-way analysis of variance imply a particular mathematical model for the variables involved. This model resembles in some ways the multiple regression model of Chapter 14. In this case the model is usually written in the form

$$X_{ij} = \mu_j + \epsilon_{ij} \qquad j = 1, \ldots, k; \quad i = 1, \ldots, n_j$$

where the ϵ_{ij} are independent, identically distributed normal random variables, with mean zero. Then the mean of X_{ij} is simply μ_j, the mean for the population from which X_{ij} is obtained.

THE RELATIONSHIP WITH THE GENERAL LINEAR MODEL

The general linear model of Chapters 14 and 15 can be extended to include the model for the one-way analysis of variance by using dummy variables. The form for the model is

$$X_i = \beta_0 + \sum_{j=1}^{k-1} \beta_j D_j + \epsilon_i \qquad i = 1, \ldots, N$$

where

$$D_j = 1, \text{ if } X_i \text{ is from population } j, j = 1, \ldots, k - 1;$$
$$= 0, \text{ otherwise}$$

At first glance the two models may appear to be different, but actually they are identical. There is a slight change in notation from one model to the other, in that β's are used in the general linear model instead of μ's, but the general concept of the two models is the same. For an observation from population j the mean is either μ_j or $\beta_0 + \beta_j$, depending on which notation is being used. The error term in both cases is normal with mean zero, so the observations are normal with mean μ_j or $\beta_0 + \beta_j$. The β_j simply represent the differences in the means.

The null hypothesis H_0: $\mu_1 = \mu_2 = \cdots = \mu_k$ is equivalent to the null hypothesis H_0: $\beta_1 = \cdots = \beta_{k-1} = 0$. The analysis of variance procedure for the former hypothesis is algebraically identical to the multiple regression method of Section 15.1 for testing the latter hypothesis. The F statistics obtained in both tests will be identical. The choice of which procedure to use is based on factors such as convenience and personal preference.

USE OF THE COMPUTER WITH THE ONE-WAY ANALYSIS OF VARIANCE

Although the calculations in the previous example were relatively easy, the analysis of other CR designs involving more treatments and more observations will be more time-consuming. For this reason, computer programs such as SAS® and Minitab® are used to do the analysis on a CR design. Additionally, the computer program provides other output that will be useful in subsequent sections.

EXAMPLE

The battery-life test data given in Problem Setting 16.2 will now be analyzed to test H_0: $\mu_1 = \mu_2 = \mu_3 = \mu_4$. The calculations will first be done by hand, and then the computer printout will be presented as a check on the calculations. The sample data calculations provide the following:

$$\sum_i X_{i1} = 411, \qquad \sum_i X_{i2} = 376, \quad \sum_i X_{i3} = 424, \qquad \sum_i X_{i4} = 458$$

$$\sum_{i,j} X_{ij} = 1669 \qquad \sum_{i,j} X_{ij}^2 = 78,049$$

From Eq. 16.1

$$\text{Total SS} = 78,049 - \frac{(1669)^2}{36} = 672.306$$

Equation 16.2 yields

$$\text{SST} = \frac{(411)^2}{9} + \frac{(376)^2}{8} + \frac{(424)^2}{10} + \frac{(458)^2}{9} - \frac{(1669)^2}{36}$$
$$= 349.017$$

Equation 16.3 gives

$$\text{Error SS} = 672.306 - 349.017$$
$$= 323.289$$

From Eq. 16.4,

$$\text{MST} = 349.017/(4 - 1) = 116.339$$

and

$$\text{MSE} = 323.289/(36 - 4) = 10.103,$$

from which the F ratio is given as

$$F = 116.339/10.103 = 11.52.$$

This value of F exceeds the critical value for a level of significance of $\alpha = .05$, $F_{.95,3,32} \approx 2.922$, obtained from Table A5 with 3 and 30 degrees of freedom (30 degrees of freedom is used since 32 degrees of freedom is not given in Table A5). Therefore, H_0 is rejected, and it is concluded that the different battery types do indeed differ with respect to their mean lives. The p-value is less than .001, according to Table A5. This conclusion is consistent with the tentative guess made in Section 16.1 based on e.d.f.s displayed in Figure 16.5. The computer printout for this analysis is as follows.

SOURCE	DF	SUM OF SQUARES	MEAN SQUARE	F VALUE	PR > F
TREATMENT	3	349.017	116.339	11.52	.0001
ERROR	32	323.289	10.103		
TOTAL	35	672.306			

EXAMPLE

The sample data on the cost of a market basket of food in five cities will now be analyzed to test H_0: $\mu_{SL} = \mu_{SF} = \mu_0 = \mu_W = \mu_D$ (i.e., the mean market basket price is the same for all cities). This analysis is done using only the computer printout.

SOURCE	DF	SUM OF SQUARES	MEAN SQUARE	F VALUE	PR > F
TREATMENT	4	141.514	35.379	14.73	.0001
ERROR	45	108.078	2.402		
TOTAL	49	249.592			

Since the p-value is .0001, the null hypothesis is soundly rejected at $\alpha = .05$, and it is concluded that the mean market basket prices are different in the five cities. However, it should be pointed out that the analysis thus far does not reveal which of the five cities are different. Again, this conclusion agrees with the tentative conclusion based on the e.d.f.s appearing in Figure 16.6.

This section has considered testing the equality of the means from k normal populations. Remaining unanswered is the question regarding the relative ordering of the population means when the null hypothesis is rejected. This question is answered in the next section.

EXERCISES

16.9   Test the null hypothesis H_0: $\mu_T = \mu_N = \mu_M$ for the mpg sample data of the pickup trucks given in the preliminary remarks of this chapter. Let $\alpha = .05$ and state the p-value. (If a computer is used to analyze these data, then check the calculations by hand using Eqs. 16.1 to 16.4.) Assume that the assumptions of the test are satisfied. ($\Sigma X = 390$, $\Sigma X^2 = 10{,}172.94$, $\Sigma X_{iT} = 163.5$, $\Sigma X_{iN} = 131.5$, $\Sigma X_{iM} = 95.0$.)

16.10 Compare your conclusion in Exercise 16.9 with your first tentative guess made in Exercise 16.2.

16.11 Test the null hypothesis H_0: $\mu_1 = \mu_2 = \mu_3 = \mu_4$ for the four methods of growing corn using the sample data given in Exercise 16.3. Let $\alpha = .05$ and state the p-value. What assumptions are you making regarding the samples?

16.12 Compare your conclusion in Exercise 16.11 with your first tentative guess made in Exercise 16.4.

16.13 Test the null hypothesis H_0: $\mu_1 = \mu_2 = \mu_3 = \mu_4 = \mu_5$ for the productivity of five workers using the sample data given in Exercise 16.5. Let $\alpha = .05$ and state the p-value. Assume that the assumptions of the test are satisfied. The computer output is as follows.

SOURCE	DF	SUM OF SQUARES	MEAN SQUARE	F VALUE	PR > F
TREATMENT	4	161.4	40.375	1.47	.2610
ERROR	15	412.5	27.500		
TOTAL	19	574.0			

16.14 Compare your conclusion in Exercise 16.13 with your first tentative guess made in Exercise 16.6.

16.15 Test the null hypothesis H_0: $\mu_1 = \mu_2 = \mu_3 = \mu_4$ for mean tar content for four brands of cigarettes using the sample data given in Exercise 16.7. What assumptions are you making regarding the samples? The computer output is as follows.

SOURCE	DF	SUM OF SQUARES	MEAN SQUARE	F VALUE	PR > F
TREATMENT	3	7.625	2.542	3.89	.0184
ERROR	30	19.603	0.653		
TOTAL	33	27.227			

16.16 Compare your conclusion in Exercise 16.15 with your first tentative guess made in Exercise 16.8.

16.17 Check the computer computations given in the third example of this section by hand using Eqs. 16.1 to 16.3.

16.18 Use the model $X_{ij} = \mu_j + \epsilon_{ij}$ to describe the data from the mpg experiment referred to in the preliminary remarks of this chapter. Write a separate model for each brand of pickup truck. Interpret the parameters on each model.

16.19 Use the general linear model given in this section to describe the data from the mpg experiment referred to in the preliminary remarks of this chapter. Give an interpretation for each dummy variable and each parameter that appears in the model.

16.3
COMPARING POPULATION MEANS
IN A COMPLETELY RANDOMIZED DESIGN

The one-way analysis of variance was presented in the previous section as a method of testing for the equality of k population means, that is, $H_0: \mu_1 = \mu_2 = \cdots = \mu_k$. If the null hypothesis is rejected, a question still remains about the correct relative ordering of the population means. That question is addressed in this section. For example, suppose a null hypothesis concerning only three populations ($H_0: \mu_1 = \mu_2 = \mu_3$) is rejected. A decision must now be made about the relative ordering of the population means. In this case, these orderings could include the following.

$$\mu_1 < \mu_2 < \mu_3 \qquad \mu_1 < \mu_2 = \mu_3$$
$$\mu_1 < \mu_3 < \mu_2 \qquad \mu_2 < \mu_1 = \mu_3$$
$$\mu_2 < \mu_1 < \mu_3 \qquad \mu_3 < \mu_1 = \mu_2$$
$$\mu_2 < \mu_3 < \mu_1 \qquad \mu_1 = \mu_2 < \mu_3$$
$$\mu_3 < \mu_1 < \mu_2 \qquad \mu_1 = \mu_3 < \mu_2$$
$$\mu_3 < \mu_2 < \mu_1 \qquad \mu_2 = \mu_3 < \mu_1$$

With four or more populations even more comparisons are possible. Fortunately an orderly procedure exists for making the comparisons of the population means. This is an area in statistics known as **multiple comparisons.**

FISHER'S LEAST SIGNIFICANT DIFFERENCE

There are many multiple comparisons procedures in statistics, and much research effort has been devoted to the examination of multiple comparisons techniques over the years. One procedure that continually ranks among the best in studies by various researchers, and is preferred by the authors, is also one of the oldest. The procedure was developed by the famed British statistician R. A. Fisher (1890–1962) and is referred to as **Fisher's Least Significant Difference (LSD).** The LSD procedure, used only after the null hypothesis has been rejected, provides a "measuring stick" for comparing the amount of separation necessary between any two sample means (i.e., $|\overline{X}_i - \overline{X}_j|$) before a significant difference can be declared to exist between the corresponding population means. The LSD multiple comparisons procedure is now stated formally.

Note that with three populations it is mathematically possible for $|\overline{X}_1 - \overline{X}_3|$ to exceed LSD_α, while neither $|\overline{X}_1 - \overline{X}_2|$ nor $|\overline{X}_2 - \overline{X}_3|$ exceeds LSD_α, and this situation often occurs in practice. In this instance, the conclusion is simply that $\mu_1 < \mu_3$, but there is not enough statistical evidence to show that either $\mu_1 < \mu_2$ or $\mu_2 < \mu_3$. The LSD multiple comparisons procedure is now demonstrated with three examples.

If the null hypothesis H_0: $\mu_1 = \mu_2 = \cdots = \mu_k$ has been rejected, the population means μ_i and μ_j are declared to be significantly different at a level of significance α if

$$|\overline{X}_i - \overline{X}_j| > LSD_\alpha$$

where

FISHER'S LEAST
SIGNIFICANT
DIFFERENCE
PROCEDURE

$$LSD_\alpha = t_{1-\alpha/2, N-k}\sqrt{MSE}\sqrt{\frac{1}{n_i} + \frac{1}{n_j}}$$

and where

$\overline{X}_i$ and $\overline{X}_j$ are the sample means being compared,

MSE is the mean square error defined in Section 16.2,

n_i and n_j are the respective sample sizes,

$t_{1-\alpha/2, N-k}$ is the $1 - \alpha/2$ quantile from Student's t distribution, with $N - k$ degrees of freedom, obtained from Table A3.

EXAMPLE

The smokestack data of Problem Setting 16.1 were used to test H_0: $\mu_A = \mu_B = \mu_C$ in an example in Section 16.2. The null hypothesis was rejected in that example. The sample calculations show the following results:

$$\overline{X}_A = 47.92 \qquad \overline{X}_B = 52.22 \qquad \overline{X}_C = 43.32$$
$$n_A = 5 \qquad n_B = 5 \qquad n_C = 5$$
$$MSE = 7.467$$

For a level of significance of $\alpha = .05$, the Student's t value from Table A3 is $t_{.975,12} = 2.1788$, and the $LSD_{.05}$ value for comparing all population means (since all sample sizes are equal) is given as

$$LSD_{.05} = 2.1788\sqrt{7.467}\sqrt{\frac{1}{5} + \frac{1}{5}} = 3.77$$

It is easier to make comparisons if the sample means are ordered from smallest to largest as follows.

Population	C	A	B
Sample mean:	43.32	47.92	52.22

The value of $|\overline{X}_C - \overline{X}_A| = 4.60$ exceeds the $LSD_{.05}$ value of 3.77, so μ_C, and μ_A are declared to be significantly different with $\mu_C < \mu_A$. Additionally the value of $|\bar{X}_A - \bar{X}_B| = 4.30$ exceeds the $LSD_{.05}$ value, so that μ_A and μ_B are declared to be significantly different with $\mu_A < \mu_B$. It obviously follows that $\mu_C < \mu_B$, so the correct relative ordering of sample means is

$$\mu_C < \mu_A < \mu_B$$

The conclusion is that the mean amount of contaminant produced by the three factories is different for all three and that their means are in the respective order shown. This result is in agreement with the tentative guess made in Section 16.1 regarding the ordering of the population means based on the e.d.f.s appearing in Figure 16.4.

EXAMPLE

The battery-life test data given in Problem Setting 16.2 were used in an example in Section 16.2 to test $H_0: \mu_1 = \mu_2 = \mu_3 = \mu_4$. The null hypothesis was rejected in that example. Calculations based on the sample data give the following results:

$$\bar{X}_1 = 45.667 \quad \bar{X}_2 = 47.000 \quad \bar{X}_3 = 42.400 \quad \bar{X}_4 = 50.889$$

$$n_1 = 9 \quad\quad n_2 = 8 \quad\quad n_3 = 10 \quad\quad n_4 = 9$$

$$MSE = 10.103$$

For a level of significance of $\alpha = .05$ the Student's t value from Table A3 is $t_{.975,32} = 2.0369$. Since the sample sizes are unequal, a separate $LSD_{.05}$ value must be computed for each possible pair of sample sizes. That is, $LSD_{.05}$ values are needed for the following pairs of sample sizes (8,9), (8,10), (9,9), and (9,10).

$$LSD_{.05,(8,9)} = 2.0369\sqrt{10.103}\sqrt{\frac{1}{8} + \frac{1}{9}} = 3.146$$

$$LSD_{.05,(8,10)} = 2.0369\sqrt{10.103}\sqrt{\frac{1}{8} + \frac{1}{10}} = 3.071$$

$$LSD_{.05,(9,9)} = 2.0369\sqrt{10.103}\sqrt{\frac{1}{9} + \frac{1}{9}} = 3.052$$

$$LSD_{.05,(9,10)} = 2.0369\sqrt{10.103}\sqrt{\frac{1}{9} + \frac{1}{10}} = 2.975$$

Again it is easier to make comparisons if the sample means are ordered from smallest to largest:

Population:	3	1	2	4
Sample mean:	42.400	45.667	47.000	50.889

The value of $|\bar{X}_3 - \bar{X}_1| = 3.267$ is greater than $\text{LSD}_{.05,(9,10)} = 2.975$ and so μ_1 and μ_3 are declared to be significantly different, with $\mu_3 < \mu_1$. Comparisons of $|\bar{X}_3 - \bar{X}_2| = 4.600$ and $|\bar{X}_3 - \bar{X}_4| = 8.489$ with $\text{LSD}_{.05,(8,10)} = 3.071$ and $\text{LSD}_{.05,(9,10)} = 2.975$, respectively, also show that μ_3 is less than both μ_2 and μ_4. Next, proceeding left to right with the ordered sample means, the value $|\bar{X}_1 - \bar{X}_2| = 1.333$ is less than $\text{LSD}_{.05,(8,9)} = 3.146$; hence, no significant difference is declared between μ_1 and μ_2. The value $|\bar{X}_1 - \bar{X}_4| = 5.222$ is greater than $\text{LSD}_{.05,(9,9)} = 3.052$, and so μ_1 and μ_4 are declared to be significantly different, with $\mu_1 < \mu_4$. The last comparison has $|\bar{X}_2 - \bar{X}_4| = 3.889$ larger than $\text{LSD}_{.05,(8,9)} = 3.146$, and so μ_2 and μ_4 are declared to be significantly different, with $\mu_2 < \mu_4$. Putting all of the parts together results in the following relative ordering of population means.

$$\mu_3 < \mu_1 = \mu_2 < \mu_4$$

This represents the conclusion concerning which battery types have mean lives different from or the same as other types. This result is in agreement with the tentative guess made in Section 16.1 about the relative ordering on the e.d.f.s in Figure 16.5.

EXAMPLE

The cost of a market basket of food was sampled for five cities in Problem Setting 16.3. The null hypothesis $H_0: \mu_{SL} = \mu_{SF} = \mu_O = \mu_W = \mu_D$ was tested in an example in Section 16.2 and was rejected. Sample data calculations give the following results.

$$\bar{X}_{SL} = 74.093 \qquad n_{SL} = 10 \qquad \text{MSE} = 2.402$$
$$\bar{X}_{SF} = 76.482 \qquad n_{SF} = 10$$
$$\bar{X}_O = 72.544 \qquad n_O = 10$$
$$\bar{X}_W = 77.264 \qquad n_W = 10$$
$$\bar{X}_D = 74.976 \qquad n_D = 10$$

For a level of significance of $\alpha = .05$ the Student's t value from Table A3 is $t_{.975,45} = 2.0141$. The $\text{LSD}_{.05}$ value for comparing all population means (since all sample sizes are equal) is given as

$$\text{LSD}_{.05} = 2.0141\sqrt{2.402}\sqrt{\frac{1}{10} + \frac{1}{10}} = 1.396$$

The sample means are ordered from smallest to largest to make comparisons easier.

Population:	Omaha	St. Louis	Dallas	San Francisco	Washington
Sample mean:	72.544	74.093	74.976	76.482	77.264

The value $|\bar{X}_O - \bar{X}_{SL}| = 1.549$ exceeds the $\mathrm{LSD}_{.05}$ value of 1.396; therefore, μ_O and μ_{SL} are declared to be significantly different, with $\mu_O < \mu_{SL}$. Obviously μ_O will also be declared less than all other population means. Moving from left to right across the sample means, the value $|\bar{X}_{SL} - \bar{X}_D| = 0.883$ is less than the $\mathrm{LSD}_{.05}$ value, so μ_{SL} and μ_D cannot be declared significantly different. Next, $|\bar{X}_{SL} - \bar{X}_{SF}| = 2.389$ is compared against the $\mathrm{LSD}_{.05}$ value, and it is concluded that $\mu_{SL} < \mu_{SF}$. Also, it is clear that $\mu_{SL} < \mu_W$. The value $|\bar{X}_D - \bar{X}_{SF}| = 1.506$ is found to be greater than the $\mathrm{LSD}_{.05}$ value, so it is concluded that μ_D and μ_{SF} are significantly different, with $\mu_D < \mu_{SF}$. Again, it follows immediately that $\mu_D < \mu_W$. The last comparison shows the value $|\bar{X}_{SF} - \bar{X}_W| = 0.782$ to be less than the $\mathrm{LSD}_{.05}$ value, so that μ_{SF} and μ_W cannot be declared significantly different. In summary, this analysis concludes that the relative ordering of the population mean market basket prices is as follows:

$$\mu_O < \mu_{SL} = \mu_D < \mu_{SF} = \mu_W$$

This result is in agreement with the tentative conclusion given in Section 16.1 based on the e.d.f.s in Figure 16.6.

OTHER MULTIPLE COMPARISON PROCEDURES

Fisher's LSD multiple comparisons procedure is only one of many procedures available for making several comparisons among population means, based on the observed differences between sample means. Some of these procedures are intended for usage only after an overall test, like the F-test in this case, declares that a difference exists. Other procedures may be used without any prior overall test; they detect overall differences as well as differences between pairs of means, and some even allow combinations of means as groups to be tested against combinations of other means. Much has been written about the advantages and disadvantages of the various multiple comparisons procedures. However, the LSD procedure given here has relatively good power to detect differences in means when the differences are present, and still retains protection against a Type 1 error by virtue of the F-test, which must be significant before this procedure is to be used. Therefore, it is recommended for making multiple comparisons.

EXERCISES

16.20 Apply the LSD multiple comparisons procedure with $\alpha = .05$ to the analysis of variance problem in Exercise 16.9 to see which makes of pickup truck differ in their mean gas mileage.

16.21 Compare your conclusion in Exercise 16.20 with your second tentative guess made in Exercise 16.2.

16.22 Apply the LSD multiple comparisons procedure with $\alpha = .05$ to the analysis of variance problem in Exercise 16.11 to see which methods of growing corn tend to produce higher yields.

16.23 Compare your conclusion in Exercise 16.22 with your second tentative guess made in Exercise 16.14.

16.24 Apply the LSD multiple comparisons procedure with $\alpha = .05$ to the analysis of variance problem in Exercise 16.13 to see which workers produce more than other workers.

16.25 Compare your conclusion in Exercise 16.24 with your second tentative guess made in Exercise 16.6.

16.26 Find out which cigarettes have lower tar content than other brands by applying the LSD multiple comparisons procedure with $\alpha = .05$ to the analysis of variance problem in Exercise 16.15.

16.27 Compare your conclusion in Exercise 16.26 with your second tentative guess made in Exercise 16.8.

16.4

A COMPARISON OF MEANS FOR GENERAL POPULATIONS (OPTIONAL)

THE KRUSKAL–WALLIS TEST

The one-way analysis of variance procedure presented in Section 16.2 has the underlying assumption of normality for testing the equality of means of k populations. If a Lilliefors test shows this not to be a reasonable assumption, then the procedure in this section should be preferred. Additionally, the technique presented in this section is not nearly as sensitive to the assumption of equal variances as is the F-test in the one-way analysis of variance. This nonparametric procedure is called the **Kruskal–Wallis test** and is calculated by performing the one-way analysis on rank transformed data. Therefore, this test, like the Wilcoxon signed-ranks test and the Wilcoxon–Mann–Whitney test previously presented, is a rank transformation test. An overall ranking is used with this test; that is, the sample data from k independent samples are replaced by their ranks from the rank 1 for the smallest observation to the rank N for the largest observation, where N is the total number of observations as before. The notation is the same as in earlier sections of this chapter, with $R_{X_{ij}}$ used to denote the rank of the observation X_{ij}.

Obviously the Kruskal–Wallis test is the same as the one-way analysis of variance, except that it is applied to the ranks of the data rather than to the data. This means that existing computer programs for the one-way analysis of variance can be used to produce the Kruskal–Wallis test after the data are replaced by their ranks. This makes the Kruskal–Wallis test very easy to use, since computer packages such as SAS® and Minitab® will also rank transform the data. The Kruskal–Wallis test will now be demonstrated with an example.

EXAMPLE

The smokestack data given in Problem Setting 16.1 are used to demonstrate the Kruskal–Wallis test. The data are repeated here with the ranks appearing in parentheses.

PROCEDURE FOR TESTING HYPOTHESES ABOUT THE DIFFERENCE IN THE MEANS OF k GENERAL POPULATIONS
(Kruskal–Wallis Test)

Data

The data consist of k samples of sizes $n_1, n_2, \ldots, n_k$ from each of k populations. The population means are denoted by $\mu_1, \mu_2, \ldots, \mu_k$. Assign ranks 1 to N, where $N = n_1 + n_2 + \cdots + n_k$, to the k samples jointly, assigning average ranks to ties. All of the computations are made on the ranks as numbers, rather than on the data as in the one-way analysis of variance. That is, all of the ranks in each sample are added and divided by the sample size to get the sample means of the ranks, $\bar{R}_1, \bar{R}_2, \ldots, \bar{R}_k$. The sums of squares are computed using Eqs. 16.1, 16.2, and 16.3, but the computations are made on the ranks, not on the original data.

Assumptions

1. All samples are random samples from their respective populations.
2. All k samples are independent of one another.
3. Either the k population distributions are identical, or else some of the populations have larger means than other populations.

Null Hypothesis

$$H_0: \mu_1 = \mu_2 = \cdots = \mu_k$$

Alternative Hypothesis

H_1: At least two of the populations have unequal means ($\mu_i \neq \mu_j$).

Test Statistic

The test statistic is

$$F_R = \frac{MST}{MSE}$$

where MST and MSE are defined following Eq. 16.4, except that they are computed on the ranks of the data rather than on the original data. The null distribution of F_R is approximately the F distribution with $k - 1$ and $N - k$ degrees of freedom, as given in Table A5.

Decision Rule

Reject H_0 if $F_R > F_{1-\alpha, k-1, N-k}$, where $F_{1-\alpha, k-1, N-k}$ is the $1 - \alpha$ quantile from Table A5 with $k - 1$ and $N - k$ degrees of freedom.

Note

If the desired degrees of freedom for the F distribution cannot be found in Table A5, simply use the next smaller degrees of freedom given in the table.

Multiple Comparisons

The multiple comparisons LSD procedure proceeds as in Section 16.3, only with all computations done on the ranks of the data, including using $\bar{R}_1, \bar{R}_2, \ldots, \bar{R}_k$ rather than the sample means $\bar{X}_1, \bar{X}_2, \ldots, \bar{X}_k$. As with the analysis on the original data the multiple comparisons procedure should be used only when H_0 is rejected.

Factory A	Factory B	Factory C
46.3(6)	48.6(8)	45.1(5)
43.7(4)	52.3(13)	46.7(7)
51.2(12)	50.9(11)	41.8(2)
49.6(10)	53.6(14)	40.4(1)
48.8(9)	55.7(15)	42.6(3)

Calculations on the ranks yield the following

$$\sum_i R_{X_{iA}} = 41 \qquad \sum_i R_{X_{iB}} = 61 \qquad \sum_i R_{X_{iC}} = 18$$

$$\sum_{i,j} R_{X_{ij}} = 120 \qquad \sum_{i,j} R_{X_{ij}}^2 = 1240$$

Equation 16.1 gives the total sum of squares as

$$\text{Total SS} = 1240 - \frac{(120)^2}{15} = 280$$

Equation 16.2 yields the treatment sum of squares as

$$SST = \frac{(41)^2}{5} + \frac{(61)^2}{5} + \frac{(18)^2}{5} - \frac{(120)^2}{15}$$

$$= 185.20$$

Equation 16.3 uses subtraction to produce the error sum of squares as

$$\text{Error SS} = 280 - 185.20 = 94.80$$

From Eq. 16.4, MST $= 185.20/(3 - 1) = 92.6$, MSE $= 94.80/(15 - 3) = 7.90$, and the F ratio is computed as

$$F_R = \frac{92.6}{7.90} = 11.72$$

The value of F_R is slightly less than the value of $F = 13.27$ found from the analysis on the original data in Section 16.2 but still leads to a sound rejection of the null hypothesis of equal means at $\alpha = .05$ when it is compared with 3.885, the .95 quantile of the F distribution from Table A5, with 2 and 12 degrees of freedom. The computer printout summary for this analysis on ranks appears as follows.

SOURCE	DF	SUM OF SQUARES	MEAN SQUARE	F VALUE	PR > F
TREATMENT	2	185.2	92.6	11.72	.0015
ERROR	12	94.8	7.9		
TOTAL	14	280.0			

For multiple comparisons at a level of significance of $\alpha = .05$, the value $t_{.975,12} = 2.1788$ from Table A3 is used, and the $LSD_{.05}$ value for all comparisons (since all sample sizes are equal) is given as

$$LSD_{.05} = 2.1788\sqrt{7.90}\sqrt{\frac{1}{5} + \frac{1}{5}} = 3.87$$

The comparisons are made on rank means, which when ordered from smallest to largest appear as:

Population	C	A	B
Rank mean	3.6	8.2	12.2

Since the difference between all pairs of means exceeds the $LSD_{.05}$ value of 3.87, all mean levels of contaminant are declared to be significantly different from one another for the three factories, which is in agreement with the conclusion in Section 16.3.

THE EFFECT OF OUTLIERS ON THE F-TEST

In the previous example the results of the analysis using the Kruskal–Wallis test agree well with the results of the analysis using the one-way analysis of variance with the original data. Both the normality assumption and the equal variance assumption are satisfied, as the reader can verify. When the assumptions of the analysis of variance procedure are satisfied, the results of the two tests tend to be in agreement with each other. But when the analysis of variance assumptions are not met, the Kruskal–Wallis test tends to have more power, especially when outliers are present in the data. This is illustrated by changing the last example slightly. The largest sample observation is the value $X^{(15)} = 55.7$, which occurs with Factory B. What happens to the Kruskal–Wallis test as this observation gets larger? Clearly the Kruskal–Wallis test is unaffected by such a change because no matter how large $X^{(15)}$ becomes, it will still be assigned a rank of 15 and the results of the Kruskal–Wallis test remain the same.

However, the F-test on the original data does not remain the same as $X^{(15)}$ increases. Consider the e.d.f.s in Figure 16.4 for these data. What effect does increasing $X^{(15)}$ have on this graph? Quite simply the last step for Factory B at 55.7 is moved to the right, apparently increasing the distance between Factories A and B (at least, as measured by their sample means, this would appear to be the case). However, increasing $X^{(15)}$ causes adverse problems for the F-test on the original data. To illustrate this point, consider Figure 16.7, which shows what happens to the F-test on the original data as $X^{(15)}$ is increased in steps of 10 from 55.7 to 95.7.

Figure 16.7 shows that SST increases as $X^{(15)}$ increases. However, SST does not increase at the same rate as either the Total SS or the Error SS. Since the degrees of freedom are fixed, the F ratio is forced to become smaller. In fact, the last two values of F in Figure 16.7 are no longer significant at the .05 level

$X^{(15)}$	Total SS	SST	Error SS	F
55.7	287.70	198.10	89.60	13.27
65.7	538.64	299.43	239.21	7.51
75.7	976.24	427.43	548.81	4.67
85.7	1600.50	538.83	1061.67	3.05
95.7	2411.44	720.16	1691.28	2.55

FIGURE 16.7
A MODIFICATION OF PROBLEM SETTING 16.1
TO SHOW THE ADVERSE EFFECT ON THE
F-TEST WHEN ASSUMPTIONS ARE VIOLATED.

of significance. Increasing the value of $X^{(15)}$ causes problems with both the normality assumption and the equal variance assumption, and this example should serve to point out that these assumptions cannot be casually ignored. That is, when outliers are present in the data, nonnormality of the distributions is indicated; therefore, the nonparametric Kruskal–Wallis test should be preferred because the assumptions of the F-test are not satisfied and because the power of the F-test is likely to be diminished by the presence of outliers.

EXERCISES

16.28 Use the Kruskal–Wallis test to test the null hypothesis $H_0: \mu_T = \mu_N = \mu_M$ for the mpg sample data of pickup trucks in the Preliminary Remarks of this chapter. Compare your answer with the one obtained in Exercise 16.9.

16.29 Apply the LSD multiple comparisons procedure on the rank transformed data in Exercise 16.28. Compare your answer with the one obtained in Exercise 16.20.

16.30 Use the Kruskal–Wallis test to test the null hypothesis $H_0: \mu_1 = \mu_2 = \mu_3 = \mu_4$ for the four methods of growing corn using the sample data given in Exercise 16.3. Compare your answer with the one obtained in Exercise 16.11.

16.31 Apply the LSD multiple comparisons procedure on the rank-transformed data in Exercise 16.30. Compare your answer with the one obtained in Exercise 16.22.

16.32 Use the Kruskal–Wallis test to test the null hypothesis $H_0: \mu_1 = \mu_2 = \mu_3 = \mu_4 = \mu_5$ for the productivity of five workers using the sample data given in Exercise 16.5. Compare your answer with the one obtained in Exercise 16.13. The computer output is as follows.

SOURCE	DF	SUM OF SQUARES	MEAN SQUARE	F VALUE	PR > F
TREATMENT	4	171.5	42.875	1.31	.3096
ERROR	15	489.5	32.63		
TOTAL	19	661.0			

16.33 Apply the LSD multiple comparisons procedure on the rank transformed data in Exercise 16.32. Compare your answer with the one obtained in Exercise 16.24.

16.34 Use the Kruskal–Wallis test to test the null hypothesis $H_0: \mu_1 = \mu_2 = \mu_3 = \mu_4$ for the mean tar content for four brands of cigarettes using the sample data in Exercise 16.7. Compare your answer with the one obtained in Exercise 16.15. The computer output is as follows.

SOURCE	DF	SUM OF SQUARES	MEAN SQUARE	F VALUE	PR > F
TREATMENT	3	922.025	307.342	3.94	.0176
ERROR	30	2442.975	78.099		
TOTAL	33	3265.000			

16.35 Transform the data in Exercise 16.7 to ranks and apply the LSD multiple comparisons procedure. Compare your answer with the one obtained in Exercise 16.26.

16.5

REVIEW EXERCISES

16.36 Use the Kruskal–Wallis test with the data given in Problem Setting 16.2. Compare your results with those obtained in the example of Section 16.2 that used the one-way analysis of variance on the original data.

16.37 Use the LSD multiple comparisons procedure on the rank-transformed data given in Problem Setting 16.2. Compare your results with those obtained in the example of Section 16.3 that applied the LSD procedure to the original data.

16.38 Use the Kruskal–Wallis test with the data given in Problem Setting 16.3. Compare your results with those obtained in the example of Section 16.2 that used the one-way analysis of variance on the original data.

16.39 Use the LSD multiple comparisons procedure on the rank transformed data given in Problem Setting 16.3. Compare your results with those obtained in the example of Section 16.3 that applied the LSD procedure to the original data.

16.40 An experiment has been designed to assess the effect of alcoholic consumption on keypunch operators. Twenty keypunch operators were randomly selected and, in turn, randomly assigned to receive either 0, 1, 2, or 3, ounces of liquor one hour prior to keypunching the same data set. The number of keypunch errors was then noted for each operator and recorded below. Graph the e.d.f.s for these data.

0 Ounces	1 Ounce	2 Ounces	3 Ounces
0	1	5	7
1	2	8	11
1	3	8	12
2	3	9	13
3	4	10	15

16.41 Make a tentative guess about the acceptability of the null hypothesis H_0: $\mu_0 = \mu_1 = \mu_2 = \mu_3$ from the e.d.f.s in Exercise 16.40. If your decision is to reject H_0, then make another tentative guess about the relative ordering of the population means.

16.42 Test the null hypothesis H_0: $\mu_0 = \mu_1 = \mu_2 = \mu_3$ for the sample data given in Exercise 16.40. Let $\alpha = .05$ and state the p-value. Assume that the analysis of variance assumptions are met.

16.43 Compare your answer in Exercise 16.42 with your first tentative guess in Exercise 16.41.

16.44 Apply the LSD multiple comparisons procedure with $\alpha = .05$ to the analysis-of-variance problem in Exercise 16.42.

16.45 Compare your conclusion in Exercise 16.44 with your second tentative guess made in Exercise 16.41.

16.46 A department examines their last 17 credit card sales and records the amount charged (nearest dollar) for each of three types of credit cards. Graph the e.d.f.s for these charges.

TYPE OF CREDIT CARD

Store	VISA	MasterCard
$56	$ 80	$ 73
20	51	56
37	40	123
28	72	56
	132	37
	60	44
		40

16.47 Make a tentative guess about the acceptability of the null hypothesis H_0: $\mu_S = \mu_V = \mu_{MC}$ from the e.d.f.s in Exercise 16.46. If your decision is to reject H_0, then make another tentative guess about the relative ordering of the population means.

16.48 Test the null hypothesis H_0: $\mu_S = \mu_V = \mu_{MC}$ for the sample data given in Exercise 16.46. Let $\alpha = .05$ and state the p-value. Notice the outliers present in the data. What test should be used with these data?

16.49 Compare your answer in Exercise 16.48 with your first tentative guess in Exercise 16.47.

16.50 Apply the appropriate LSD multiple comparisons procedure with $\alpha = .05$ to the analysis of variance problem in Exercise 16.48.

16.51 Compare your conclusion in Exercise 16.50 with your second tentative guess made in Exercise 16.47.

BIBLIOGRAPHY

Additional material on the topics presented in this chapter can be found in the following publications.

Carmer, S. G. and Swanson, M. R. (1973). "An Evaluation of Ten Pairwise Multiple Comparisons Procedures by Monte Carlo Methods." *Journal of the American Statistical Association*, **68**(341), 66–74.

Conover, W. J. (1980). *Practical Nonparametric Statistics*, 2nd ed. Wiley, New York.

Conover, W. J. and Iman, R. L. (1981). "Rank Transformations as a Bridge between Parametric and Nonparametric Statistics." *The American Statistician*, **35**(3), 124–133.

ANALYSIS OF VARIANCE FOR TWO-FACTOR EXPERIMENTS

The experimental design presented in this chapter is an extension of the matched pairs design introduced in Chapter 8 where a pair of measurements on each of several different individuals (experimental units) was used as a method of controlling unwanted variation in the experimental results. Such an experiment can be used to compare the effect of one treatment (such as a diet) versus a different treatment (a different diet) or nontreatment (no diet) on individuals who are matched according to similar characteristics. These similar individuals are referred to as an experimental **block.** In this chapter the block concept is extended to cover k treatments. Examples of experiments using blocks include the following.

Three different brands of word processing software (treatments) are tested by each of six operators (blocks).

Four different real estate appraisers (treatments) make an appraisal on the same 10 pieces of property (blocks).

Ten leadership traits (treatments) are ranked in order of importance by each of seven managers (blocks).

Weekly receipts for each of three restaurants (treatments) are compared on the basis of five randomly selected weeks (blocks).

The number of units produced by each of five workers (treatments) is observed for four randomly selected weeks (blocks).

Three brands of pickups (treatments) are mileage tested by having six drivers (blocks) drive each vehicle.

Four different types of grass (treatments) are ranked by each of 12 different homeowners (blocks).

The rate of return for each of five stocks (treatments) is recorded at the end of each of nine randomly selected three-month periods (blocks).

As with matched pairs, blocks are used to account for some of the variation in the data collected in an experiment. Therefore, the analysis takes into account the presence of blocks. The result is a test more likely to detect treatment differences when they actually exist.

CHECKING THE ASSUMPTIONS

The assumption of normality is no less important in this chapter and the next than it was in previous chapters. However, good tests for normality essentially do not exist for the experimental designs about to be introduced. The experimenter should visually check for obvious signs of nonnormality, such as discrete valued data or the presence of outliers. In the absence of obvious nonnormality, the usual practice is to use the procedures in Sections 17.1 and 17.2, which are based on the normality assumption. The experimenter who suspects that the distributions are nonnormal should use the method in Section 17.3.

17.1
THE ANALYSIS OF VARIANCE FOR THE RANDOMIZED COMPLETE BLOCK DESIGN

PROBLEM SETTING 17.1

A purchasing agent must purchase word processing software for a new branch office of the company. The agent would like to purchase the software that will yield the best production rate from the operators. The purchasing agent consults with a company office manager and together they decide to set up a test where the length of time (in minutes to input a standard document) will be measured. Six operators not familiar with any of the software are randomly selected. Each operator is given an equal amount of training time on each type of software. Then the length of time to input a standard document is measured for each operator on each of the three candidate brands of software. The test results are recorded in Figure 17.1. Based on these results, is there any reason to prefer one brand of word processing software over another?

THE RANDOMIZED COMPLETE BLOCK DESIGN

In Problem Setting 17.1 the purchasing agent wants to secure the word processing software for which the operators will have the best production rate. The method of evaluation consists of measuring the length of time required for each of six randomly selected operators to input a standard document with

	Brand of Software		
Operator	1	2	3
1	42	45	45
2	37	36	40
3	53	56	55
4	68	73	75
5	48	45	47
6	36	39	40

FIGURE 17.1

TEST RESULTS OF INPUT TIMES (MINUTES) FOR THREE BRANDS OF WORD PROCESSING SOFTWARE.

each brand of software. In this test setup the software package is the treatment of interest and the operators act as blocks since each operator tries all three brands of software. This is an example of a **randomized complete block design.**

THE RANDOMIZED COMPLETE BLOCK DESIGN	A **randomized complete block** (RCB) design is one in which measurements are recorded for each of k treatments in each of b homogeneous (matched) blocks.

AN INAPPROPRIATE ANALYSIS

As a first step in understanding the analysis required for the data given in Problem Setting 17.1 suppose the fact that each operator had tried each software package is ignored and that the experiment is treated as a completely

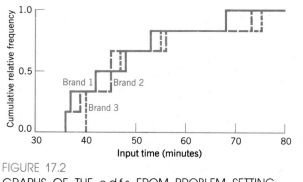

FIGURE 17.2

GRAPHS OF THE e.d.f.s FROM PROBLEM SETTING 17.1 WHEN THE EXPERIMENT IS TREATED AS A CR DESIGN.

SOURCE	DF	SUM OF SQUARES	MEAN SQUARE	F VALUE	PR > F
TREATMENT	2	27.11	13.56	0.08	.9232
ERROR	15	2532.67	168.84		
TOTAL	17	2559.78			

FIGURE 17.3
COMPUTER PRINTOUT FOR THE CR ANALYSIS ON PROBLEM SETTING 17.1.

randomized design as in the previous chapter with six observations on each brand of software. This would mean that 18 operators would be needed (6 for each brand of software). A graphical display of the data when treated as a completely randomized design is given in Figure 17.2 in the useful format of e.d.f.s. This figure shows a great deal of overlap for the input times using each of the three brands of software and this indicates no differences among them. A one-way analysis of variance shown in the computer printout in Figure 17.3 confirms this conclusion. The small F value leads to the conclusion that the null hypothesis of equal mean input times for the three software packages should not be rejected.

A GRAPHICAL COMPARISON OF SAMPLE DATA FROM AN RCB DESIGN

As mentioned, the preceding analysis is really not appropriate since the operators act as blocks or homogeneous units. This means that the input times are influenced by the operator. For example, the three slowest (longest) times were all made by operator number 4, and the three next slowest times were all made by operator number 3. The six fastest (shortest) times were all made by operators 2 and 6.

Much of the variability in input rates can be accounted for by considering the presence of blocks in the experimental design. Removing the variability attributed to blocks leads to a much clearer graphical display. This is easily done by converting the scores in each block to z scores by subtracting the block mean and dividing by the block standard deviation (i.e., the sample mean and standard deviation for the observations within each block). By converting each block to z scores, each block will then have a mean of zero and the only effect remaining will be whatever is due to the different treatments.

The mean and standard deviation of the three times for operator number 1 are 44 and 1.73, respectively. The z scores are

$$z_1 = \frac{42 - 44}{1.73} = -1.16, \ z_2 = \frac{45 - 44}{1.73} = .58, \text{ and } z_3 = \frac{45 - 44}{1.73} = .58$$

In a similar manner the block mean and standard deviation can be found for each of the other blocks and the times in each block can be converted to z scores. These z scores appear in Figure 17.4.

Operator	Brand of Software		
	1	2	3
1	−1.16	0.58	0.58
2	−0.32	−0.80	1.12
3	−1.09	0.87	0.22
4	−1.11	0.28	0.83
5	0.87	−1.09	0.22
6	−1.12	0.32	0.80

FIGURE 17.4
TEST RESULTS FROM PROBLEM
SETTING 17.1 CONVERTED TO z
SCORES WITHIN EACH BLOCK.

The next step is to graph the e.d.f.s of the z scores. This has been done in Figure 17.5 where quite a different situation is indicated than Figure 17.2 would lead one to believe. The total overlap in Figure 17.2 has been replaced by a clear separation, which indicates that some differences exist among the types of word processing packages.

THE TWO-WAY ANALYSIS OF VARIANCE

Two factors are used to classify the data in a **two-way analysis of variance.** One factor (treatments) classifies the data in columns, and a second factor (blocks) classifies the data into rows. In the two-way analysis of variance the same hypothesis of equal treatment means used with the completely randomized design is still of interest; that is, $H_0: \mu_1 = \mu_2 = \cdots = \mu_k$. However, due to the presence of blocks some of the calculations are changed from the one-way analysis of variance. In the completely randomized design the total sum of squares (Total SS) was partitioned into a sum of squares from treatments (SST) and an error sum of squares (Error SS). For the randomized complete

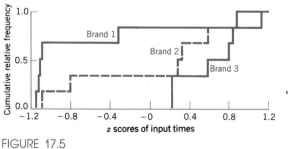

FIGURE 17.5
GRAPHS OF THE e.d.f.s BASED ON THE z SCORES
GIVEN IN FIGURE 17.4.

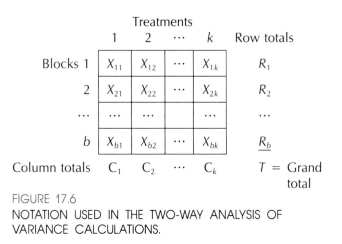

FIGURE 17.6

NOTATION USED IN THE TWO-WAY ANALYSIS OF VARIANCE CALCULATIONS.

block analysis the Error SS is further divided to account for the difference due to blocks. This is summarized as follows for the two designs.

$$\text{CR design: Total SS} = \text{SST} + \text{Error SS}_{CR}$$

$$\text{RCB design: Total SS} = \text{SST} + \text{SSB} + \text{Error SS}_{RCB}$$

The subscripts CR and RCB on the error term are usually not used and were added here to help clarify the fact that the two Error SS terms are not the same number. This means that the Total SS and SST calculations proceed as before and the only new calculation is for the sum of squares due to blocks (SSB). The new Error SS will be found by subtraction as before. To keep the notation straight for the calculations it is helpful to use a diagram such as the one given in Figure 17.6. There the number of treatments is denoted by k as in the completely randomized design. The number of blocks is denoted by b (for blocks).

Following the notation in Figure 17.6 the sum of squares calculations for the RCB design, proceed as follows where $CF = T^2/bk$ is called the **correction factor,** and $\bar{X}_i$ and $\bar{X}_j$ are the block (row) and treatment (column) sample means, respectively. The overall sample mean is $\bar{X} = T/bk$.

$$\text{Total SS} = \sum_i \sum_j (X_{ij} - \bar{X})^2 = \sum_i \sum_j X_{ij}^2 - CF \qquad (17.1)$$

$$\text{SST} = b \sum_{j=1}^{k} (\bar{X}_j - \bar{X})^2 = \frac{1}{b} \sum_{j=1}^{k} C_j^2 - CF \qquad (17.2)$$

$$\text{SSB} = k \sum_{i=1}^{b} (\bar{X}_i - \bar{X})^2 = \frac{1}{k} \sum_{i=1}^{b} R_i^2 - CF \qquad (17.3)$$

$$\text{Error SS}_{RCB} = \sum_i \sum_j (X_{ij} - \bar{X}_i - \bar{X}_j + \bar{X})^2$$

$$= \text{Total SS} - \text{SST} - \text{SSB} \qquad (17.4)$$

The results of these calculations and the relevant statistics are conveniently summarized in an **analysis of variance** table.

ANALYSIS OF VARIANCE TABLE (RCB DESIGN)

The sums of squares, degrees of freedom, mean squares, and related statistics for a two-way analysis of variance are conveniently summarized in an analysis of variance table such as the following.

SOURCE	DF	SUM OF SQUARES	MEAN SQUARE	F VALUE	PR > F
TREATMENT	$k - 1$	SST	$MST = SST/(k - 1)$	MST/MSE	
BLOCK	$b - 1$	SSB	$MSB = SSB/(b - 1)$	MSB/MSE	
ERROR	$(b - 1)(k - 1)$	Error SS	$MSE = \text{Error SS}/(b - 1)(k - 1)$		
TOTAL	$bk - 1$	Total SS			

THE MODEL

The simplest way to write the model for the analysis of the randomized complete block design is as

$$X_{ij} = \mu_j + \beta_i + \epsilon_{ij}, \qquad i = 1, \ldots, b; \quad j = 1, \ldots, k \qquad (17.5)$$

where

X_{ij} = the random variable observed in block i and treatment j

μ_j = the population mean for treatment j

β_i = the change in the mean of X_{ij} caused by being in block i

ϵ_{ij} = independent normal random variables with the same mean 0 and the same variance σ^2

When it is written in this form, the role of the block effects β_i is easy to see. The block effects change the mean of X_{ij} from what it would be if there were no blocks in the experimental design. The mean of X_{11} for example, is $\mu_1 + \beta_1$, whereas the mean of X_{21} is $\mu_1 + \beta_2$, which may be different from the mean of X_{11} even though both are in the same treatment, treatment 1, because they are in different blocks. Another assumption implicit in the model is that the blocks have an additive effect; the difference between the means in block 1 and block 2 is a constant $(\beta_1 - \beta_2)$ no matter which treatment is being considered.

EXAMPLE

Motivation for this example is provided by Problem Setting 17.1. Test H_0: $\mu_1 = \mu_2 = \mu_3$ with $\alpha = .05$. The calculations will be done by hand and then output from a computer program will be given. Sample data calculations give the following results.

$C_1 = 284$	$C_2 = 294$	$C_3 = 302$
$\overline{X}_1 = 47.33$	$\overline{X}_2 = 49.00$	$\overline{X}_3 = 50.33$
$R_1 = 132$	$R_2 = 113$	$R_3 = 164$
$R_4 = 216$	$R_5 = 140$	$R_6 = 115$

$$T = 880 \qquad \Sigma\Sigma X_{ij}^2 = 45{,}582 \qquad k = 3 \qquad b = 6$$

$$CF = \frac{880^2}{(3)(6)} = 43{,}022.22$$

PROCEDURE FOR INFERENCE ABOUT THE DIFFERENCE IN MEANS OF k NORMAL POPULATIONS WITH EQUAL VARIANCES (Two-Way Analysis of Variance)

Data

The data occur in b homogeneous blocks, each of which contains k observations, one observation from each of k populations. Denote the observations by X_{ij}, $i = 1, \ldots, k; j = 1, \ldots, b$. The k population means are denoted by μ_1, $\mu_2, \ldots, \mu_k$, respectively.

Assumptions

1. The random variables being observed follow the model

$$X_{ij} = \mu_i + \beta_i + \epsilon_{ij}$$

where

μ_i = treatment mean

β_i = change in the mean of X_{ij} due to block i ($\Sigma\beta_i = 0$)

ϵ_{ij} = independent and identically distributed random variables

2. The ϵ_{ij} are normal with mean 0 and equal variances σ^2.

Null Hypothesis

$$H_0: \mu_1 = \mu_2 = \cdots = \mu_k$$

Alternative Hypothesis

H_1: The treatment means are not all equal (some $\mu_i \neq \mu_j$).

Test Statistic

$$F = \frac{MST}{MSE} \qquad (17.6)$$

where $MST = SST/(k - 1)$ and $MSE = \text{Error } SS/(k - 1)(b - 1)$ with SST and Error SS calculated from Eqs. 17.2 and 17.4, respectively. The null distribution of F is the F distribution with $(k - 1)$ and $(k - 1)(b - 1)$ degrees of freedom.

Decision Rule

H_0 is rejected if $F > F_{1-\alpha, \ k-1, \ (b-1)(k-1)}$, where $F_{1-\alpha, \ k-1, \ (b-1)(k-1)}$ is the $1 - \alpha$ quantile from Table A5 with $k - 1$ and $(k - 1)(b - 1)$ degrees of freedom.

Note:

If the desired degrees of freedom for the F distribution cannot be found in Table A5, simply use the next smaller degrees of freedom given in the table.

Multiple Comparisons

If the null hypothesis is *rejected*, the population means μ_i and μ_j are declared significantly different at a level of significance α if

$$|\overline{X}_i - \overline{X}_j| > LSD_\alpha$$

where $\overline{X}_i$ and $\overline{X}_j$ are the sample means for treatments i and j, and

$$LSD_\alpha = t_{1-\alpha/2, \ (b-1)(k-1)} \sqrt{MSE} \sqrt{\frac{2}{b}} \qquad (17.7)$$

MSE is defined with Eq. 17.6 and $t_{1-\alpha/2, \ (b-1)(k-1)}$ is the $1 - \alpha/2$ quantile from Student's t distribution in Table A3 with $(b - 1)(k - 1)$ degrees of freedom.

The total sum of squares is obtained from Eq. 17.1

$$\text{Total SS} = 45,582 - 43,022.22 = 2559.78$$

and the treatment sum of squares from Eq. 17.2,

$$\text{SST} = \tfrac{1}{6}(284^2 + 294^2 + 302^2) - 43,022.22 = 27.11$$

Equation 17.3 yields the sum of squares due to differences among blocks

$$\text{SSB} = \tfrac{1}{3}(132^2 + 113^2 + 164^2 + 216^2 + 140^2 + 115^2) - 43,022.22$$
$$= 2501.11$$

Subtraction gives the error sum of squares,

$$\text{Error SS} = 2559.78 - 27.11 - 2501.11 = 31.56$$

as explained in Eq. 17.4. The mean squares are found by dividing the sums of squares by their respective degrees of freedom:

$$\text{MST} = \frac{\text{SST}}{(3 - 1)} = \frac{27.11}{2} = 13.56$$

$$\text{MSE} = \frac{\text{SSE}}{(6 - 1)(3 - 1)} = \frac{31.56}{10} = 3.16$$

From Eq. 17.6, the test statistic is $F = 13.56/3.16 = 4.30$. Since the decision rule is to reject H_0 if $F > F_{.95,2,10} = 4.103$, H_0 is rejected and the conclusion is that the mean input time is different for the three software packages. The p-value is slightly less than .05. A computer printout of the analysis of these data appears as follows.

SOURCE	DF	SUM OF SQUARES	MEAN SQUARE	F VALUE	PR > F
PACKAGE	2	27.11	13.56	4.30	.0450
OPERATOR	5	2501.11	500.22	158.52	.0001
ERROR	10	31.56	3.16		
TOTAL	17	2559.78			

Since H_0 was rejected, multiple comparisons can be made to determine which of the three software packages have different mean input times. If a level of significance of .05 is used, the LSD value is

$$\text{LSD}_{.05} = t_{.975,10} \sqrt{\text{MSE}} \sqrt{\frac{2}{b}}$$

$$= 2.2281 \sqrt{3.16} \sqrt{\frac{2}{6}}$$

$$= 2.29$$

Absolute Difference of Sample Means	Conclusion about Population Means
$\lvert \overline{X}_1 - \overline{X}_2 \rvert = 1.67$	$\mu_1 = \mu_2$
$\lvert \overline{X}_1 - \overline{X}_3 \rvert = 3.00$	$\mu_1 < \mu_3$
$\lvert \overline{X}_2 - \overline{X}_3 \rvert = 1.33$	$\mu_2 = \mu_3$

The data do not indicate a significant difference between software packages 1 and 2, nor between packages 2 and 3, but the slight differences present are additive, so that the difference between software packages 1 and 3 is significant.

One way of representing the presence or absence of such differences graphically is to underline groups of means that are not considered different. If two means do not share the same line beneath them, they are considered different:

$$\underline{\mu_1 \quad \mu_2} \quad \mu_3$$

Thus the data are not sufficiently strong to indicate a difference between software package 2 and the other word processing software packages, but they are sufficient to declare a difference between software packages 1 and 3.

An interesting aspect of the analysis in the preceding example is the complete turnaround of the conclusion from that demonstrated in Figure 17.3 where the analysis was treated as a one-way layout. A comparison of the two analyses shows the same values for SST and Total SS; however, the value of Error SS has decreased from 2532.67 to 31.56 and the value of MSE has decreased from 168.84 to 3.16. Since this latter value is the divisor in the F ratio and the numerator is fixed at 13.56, the test is now more sensitive to treatment differences. This shows the value of accounting for variation in the data due to differences from block to block.

THE RELATIONSHIP WITH THE GENERAL LINEAR MODEL

The randomized complete block design model of Eq. 17.5

$$X_{ij} = \mu_j + \beta_i + \epsilon_{ij} \tag{17.8}$$

is a special case of the general linear model (GLM) introduced in Chapters 14 and 15. Thus the analysis of the data could have been made using the computer programs for the general linear model. When performing the calculations without a computer, however, the procedures shown in this section are much simpler for the RCB design than the procedures for the general linear model, which are not given in this text.

To see how the two models, RCB and GLM, relate, consider the GLM with two qualitative variables and no interaction as described in Section 14.3.

$$X_{ij} = \beta_0 + \underbrace{\beta_1 D_1 + \cdots + \beta_{k-1} D_{k-1}}_{\text{Treatments}}$$

$$+ \underbrace{\beta_k D_k + \cdots + \beta_{k+b-2} D_{k+b-2}}_{\text{Blocks}} + \epsilon_{ij} \tag{17.9}$$

The first qualitative variable represents the k treatments. It appears at k levels, and therefore is represented by the $(k - 1)$ dummy variables D_1 through D_{k-1} and the intercept term β_0. The second qualitative variable represents the b blocks and appears at b levels. The $(b - 1)$ dummy variables D_k through D_{k+b-2} are used with β_0 to represent the different blocks.

In other words, β_0 is the mean for the observations in the first block and first treatment. The change in this mean caused by an observation being in a different treatment is given by one of the coefficients β_1 through β_{k-1}, depending on which treatment the observation is in. Similarly the change in the mean caused by the observation being in a block other than the first block is given by one of the coefficients β_k through β_{k+b-2}.

The analysis given in this section may be used only when there is exactly one observation for each block–treatment combination. The general linear model analysis has no such restriction, and therefore many practitioners use the general linear model to analyze all data from experiments in designs such as the CR or RCB designs.

EXERCISES

17.1 Weekly receipts (in thousands) from three restaurants are given below for five randomly selected weeks. Plot the e.d.f.s for these data as in Figure 17.2. Assume the analysis of variance requirements are satisfied, and perform a one-way analysis of variance on these data to test H_0: $\mu_1 = \mu_2 = \mu_3$ with $\alpha = .05$. ($\Sigma X = 293.6$, $\Sigma X^2 = 6383.82$, $C_1 = 83.5$, $C_2 = 101$, $C_3 = 109.1$.)

	Restaurant		
Week	1	2	3
1	14.6	19.4	17.3
2	19.9	22.7	24.6
3	17.9	20.3	28.9
4	4.8	11.1	12.7
5	26.3	27.5	25.6

17.2 Standardize the data within each week in Exercise 17.1 and graph the e.d.f.s of z scores as was done in Figure 17.5. ($\overline{X}$ and s for week 1: 17.1, 2.41; week 2: 22.4, 2.36; week 3: 22.37, 5.78; week 4: 9.53, 4.18; week 5: 26.47, 0.96.)

17.3 Assume that the assumptions are satisfied and test the null hypothesis H_0: $\mu_1 = \mu_2 = \mu_3$ using a two-way analysis of variance on the data given in Exercise 17.1. Let $\alpha = .05$. Compare your answer with the one-way analysis of variance in Exercise 17.1. ($R_1 = 51.3$, $R_2 = 67.2$, $R_3 = 67.1$, $R_4 = 28.6$, $R_5 = 79.4$.)

17.4 Apply the LSD multiple comparisons procedure with $\alpha = .05$ to the analysis of variance problem in Exercise 17.3.

17.5 Use Eq. 17.5 to write a model describing the experiment in Exercise 17.1. Interpret the parameters appearing in the model.

17.6 Use Eq. 17.9 to write a general linear model to describe the data in

Exercise 17.1. Interpret the variables and parameters appearing in the model.

17.7 The number of units produced by each of four workers is given below for four randomly selected weeks. Test the null hypothesis that the mean productivity is the same for all workers, that is, $H_0: \mu_1 = \mu_2 = \mu_3 = \mu_4$ with $\alpha = .05$.

	Workers			
Week	1	2	3	4
1	20	16	14	18
2	10	5	6	7
3	17	16	14	13
4	14	12	10	15

17.8 Apply the LSD multiple comparisons procedure with $\alpha = .05$ to the analysis of variance problem in Exercise 17.7.

17.9 The rate of return on an investment in several common stocks over a period of time is calculated by taking the market price of each stock at the end of the time period plus the amount of any dividends that were paid during the time period and dividing by the price of the stock at the beginning of the time period. The rate of return for five stocks for four randomly selected quarters is given below. Test the null hypothesis $H_0: \mu_A = \mu_B = \mu_C = \mu_D = \mu_E$ with $\alpha = .05$ using the computer printout.

	Stock				
Quarter	A	B	C	D	E
1	1.022	1.018	1.031	1.009	1.018
2	0.996	0.998	1.021	0.981	0.992
3	1.064	1.073	1.020	1.051	1.061
4	0.993	1.004	1.010	0.998	0.987

SOURCE	DF	SUM OF SQUARES	MEAN SQUARE	F VALUE	PR > F
STOCK	4	4.503 E-04	1.126 E-04	0.51	.7312
QUARTER	3	1.041 E-02	3.471 E-03	15.66	.0002
ERROR	12	2.660 E-03	2.217 E-04		
TOTAL	19	1.352 E-02			

17.10 Apply the LSD multiple comparisons procedure with $\alpha = .05$ to the analysis of variance problem in Exercise 17.9.

17.11 Use Eq. 17.5 to write a model describing the experiment in Exercise 17.9. Interpret the parameters appearing in the model.

17.12 Use Eq. 17.9 to write a general linear model to describe the data in Exercise 17.9. Interpret the variables and parameters appearing in the model.

17.2

INTERACTION IN TWO-FACTOR EXPERIMENTS

Suppose in Problem Setting 17.1 the purchasing agent and office manager obtained three independent test measurements on the time of input for each operator with each word processing package. These data are reported in Figure 17.7.

SHOWING POPULATION DIFFERENCES WITH A GRAPH

Since comparisons of population means are still of interest, it is natural to ask what effect these repeated measurements have on conclusions from the test. An obvious response is that with more observations the sample e.d.f.s should tend to provide better estimates of the population c.d.f.s than those in Figure

Operator	Brand of Software		
	1	2	3
1	42	45	45
	43	43	44
	39	45	47
2	37	36	40
	36	37	41
	36	34	42
3	53	56	55
	56	55	57
	54	58	55
4	68	73	75
	70	71	78
	65	76	75
5	48	45	47
	44	46	49
	47	45	47
6	36	39	40
	34	39	42
	37	37	38

FIGURE 17.7
RESULTS OF REPEATED MEASUREMENTS FOR SIX OPERATORS USING THREE DIFFERENT WORD PROCESSING PACKAGES.

Operator	Brand of Software		
	1	2	3
1	−0.73	0.58	0.58
	−0.29	−0.29	0.15
	−2.04	0.58	1.45
2	−0.25	−0.62	0.87
	−0.62	−0.25	1.24
	−0.62	−1.36	1.61
3	−1.62	0.37	−0.29
	0.37	−0.29	1.03
	−0.96	1.69	−0.29
4	−1.04	0.16	0.64
	−0.56	−0.32	1.35
	−1.75	0.88	0.64
5	0.98	−0.91	0.35
	−1.54	−0.28	1.61
	0.35	−0.91	0.35
6	−0.85	0.43	0.85
	−1.71	0.43	1.71
	−0.43	−0.43	0.00

FIGURE 17.8
REPEATED MEASUREMENTS FROM
FIGURE 17.7 CONVERTED TO z
SCORES WITHIN EACH BLOCK.

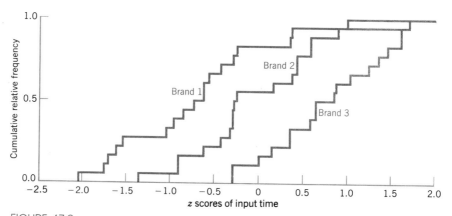

FIGURE 17.9
GRAPHS OF THE e.d.f.s BASED ON THE z SCORES GIVEN IN FIGURE 17.8.

17.5. To obtain the e.d.f.s the observations within each block (operator) are converted to z scores as in the previous section, only now there are nine rather than three observations in each block. The z scores are given in Figure 17.8 and the corresponding e.d.f.s are graphed in Figure 17.9. Figure 17.9 shows a clear separation of the e.d.f.s and indicates the likely ordering of the mean population input times of the three brands of word processing software to be $\mu_1 < \mu_2 < \mu_3$.

THE RANDOMIZED COMPLETE BLOCK MODEL WITH INTERACTION

Another natural question to ask with respect to the repeated observations is: Have these repeated measurements introduced any additional variation in the sample data? The answer is yes. Before, there was only one observation for each operator with each word processing package; now there are three. These three observations will not usually be exactly equal to each other because the input times for the operators will naturally be different at some times than at others. Thus, some additional variation is introduced. At the same time, however, more measurements generally mean a better design, one that is more apt to pick up differences among the treatment means.

  The additional observations also permit a more general model to be used, one that does not require that the block effects be constant from treatment to treatment as was done in the previous section. That is, in the previous section the mean of X_{ij} was changed by an amount β_i because X_{ij} was located in block i, no matter which treatment was used. In the example, if the mean input time for a particular operator was 3 minutes faster for one word processing package, it was assumed to be 3 minutes faster for all word processing packages. In this section the change in the mean is allowed to vary from treatment to treatment, because having several observations in each block–treatment combination permits a statistical analysis on a more complicated, and more accurate, model. In this new model the mean input time for a particular operator may be faster or slower, depending on which word processing package the operator is using. This nonadditivity of block effects is expressed by adding a term γ_{ij} (Greek letter lower case gamma) to the model given in Eq. 17.5 to get

$$X_{ijm} = \mu_j + \beta_i + \gamma_{ij} + \epsilon_{ijm} \qquad (17.10)$$

where the γ_{ij} represents the **interaction** (nonadditivity) between block i and treatment j. The other terms in this model have the same meanings they had in the previous section, with additional subscript $m = 1, 2, \ldots, n$ representing the multiple observations in each block-treatment combination.

INTERACTION
  BETWEEN The **interaction** between blocks and treatments is the nonadditive effect
BLOCK AND exhibited by each block–treatment mean.
  TREATMENT

  With repeated measurements available a mean can be calculated for each cell in Figure 17.7 (i.e., for each operator and machine combination) and a sum of squares (SSI) based on the variation of these means can be found. This is known as the sum of squares due to interaction. This new sum of squares results from partitioning the error sum of squares from the RCB design. The relationship of the sums of squares for the CR design, the RCB design, and the RCB design with repeated measurements is given as follows:

CR design: Total SS = SST + Error SS_{CR}

RCB design: Total SS = SST + SSB + Error SS_{RCB}

RCB design: Total SS = SST + SSB + SSI + Error $SS_{RCB \text{ (interaction)}}$

(repeated measurements)

Figure 17.10 has been constructed to aid the reader in understanding the notation used in calculating these sums of squares. Note that the number of observations in each block–treatment combination is the same, n. In many practical applications the number of observations may vary from one block–treatment combination to another, so that the equations in this section may not be used. In those instances the analysis is easily handled by any one of

Treatments

	1	2	...	k	Block totals
Blocks 1	X_{111} X_{112} ... X_{11n} $\overline{S_{11}} = $ Total	X_{121} X_{122} ... X_{12n} $\overline{S_{12}} = $ Total	...	X_{1k1} X_{1k2} ... X_{1kn} $\overline{S_{1k}} = $ Total	R_1
2	X_{211} X_{212} ... X_{21n} $\overline{S_{21}} = $ Total	X_{221} X_{222} ... X_{22n} $\overline{S_{22}} = $ Total	...	X_{2k1} X_{2k2} ... X_{2kn} $\overline{S_{2k}} = $ Total	R_2
...	...	...	...	...	...
b	X_{b11} X_{b12} ... X_{b1n} $\overline{S_{b1}} = $ Total	X_{b21} X_{b22} ... X_{b2n} $\overline{S_{b2}} = $ Total	...	X_{bk1} X_{bk2} ... X_{bkn} $\overline{S_{bk}} = $ Total	R_b
Column totals	C_1	C_2	...	C_k	T = Grand total

FIGURE 17.10

NOTATION USED IN TWO-WAY ANALYSIS OF VARIANCE CALCULATIONS WITH REPEATED MEASUREMENTS.

many computer programs, and the computer output is interpreted in the same way as when all numbers of observations are equal. Unequal numbers of observations can also be analyzed easily by using the general linear model explained at the end of this section. The numbers of observations are equal in the examples of this section for simplicity in handling the calculations.

SUM OF SQUARES CALCULATIONS WITH REPEATED MEASUREMENTS

Following the notation in Figure 17.10 the sum of squares calculations for the RCB design with repeated measurements proceed as follows where $\overline{X}_i$ represents the block (row) sample means R_i/kn; $\overline{X}_j$ represents the treatment (column) sample mean C_j/bn; $\overline{X}_{ij}$ represents the block–treatment (cell) sample mean S_{ij}/n; and CF $= T^2/bkn$ represents the correction factor. The overall mean is given by $\overline{X} = T/bkn$.

$$\text{Total SS} = \sum_i \sum_j \sum_m (X_{ijm} - \overline{X})^2 = \sum_i \sum_j \sum_m X_{ijm}^2 - \text{CF} \quad (17.11)$$

$$\text{SST} = bn \sum_{j=1}^{k} (\overline{X}_j - \overline{X})^2 = \frac{1}{nb} \sum_{j=1}^{k} C_j^2 - \text{CF} \quad (17.12)$$

$$\text{SSB} = kn \sum_{i=1}^{b} (\overline{X}_i - \overline{X})^2 = \frac{1}{nk} \sum_{i=1}^{b} R_i^2 - \text{CF} \quad (17.13)$$

$$\text{SSI} = n \sum_i \sum_j (\overline{X}_{ij} - \overline{X}_i - \overline{X}_j + \overline{X})^2$$

$$= \frac{1}{n} \sum_i \sum_j S_{ij}^2 - \frac{1}{nk} \sum_{i=1}^{b} R_i^2 - \frac{1}{nb} \sum_{j=1}^{k} C_j^2 + \text{CF} \quad (17.14)$$

$$\text{Error SS} = \sum_i \sum_j \sum_m (X_{ijm} - \overline{X}_{ij})^2$$

$$= \text{Total SS} - \text{SST} - \text{SSB} - \text{SSI} \quad (17.15)$$

In the statement of the formal testing procedure that follows, two tests are presented. The first F-test listed in the procedure is used for testing the equality of population means for the treatments, as has been done previously. The second test listed is for interaction. The calculations are conveniently summarized in an analysis of variance table.

ANALYSIS OF VARIANCE TABLE (RCB DESIGN WITH INTERACTION)

The sums of squares, degrees of freedom, mean squares, and related statistics for a two-way analysis of variance with interaction are conveniently summarized in an analysis of variance table such as the following:

SOURCE	DF	SUM OF SQUARES	MEAN SQUARE	F VALUE	PR > F
TREATMENT	$k-1$	SST	MST $= \text{SST}/(k-1)$	MST/MSE	
BLOCK	$b-1$	SSB	MSB $= \text{SSB}/(b-1)$	MSB/MSE	
INTERACTION	$(b-1)(k-1)$	SSI	MSI $= \text{SSI}/(b-1)(k-1)$	MSI/MSE	
ERROR	$bk(n-1)$	Error SS	MSE $= \text{Error SS}/bk(n-1)$		
TOTAL	$nbk-1$	Total SS			

PROCEDURE FOR INFERENCE ABOUT THE DIFFERENCE IN MEANS OF k NORMAL POPULATIONS WITH EQUAL VARIANCES
(Two-Way Analysis of Variance with Interaction)

Data

The data occur in b homogeneous blocks, each of which contains kn observations X_{ijm}, $i = 1, \ldots, b$; $j = 1, \ldots, k$; $m = 1, \ldots, n$. The observations originate from each of k populations whose respective means are denoted by μ_1, $\mu_2, \ldots, \mu_k$ and result by taking n repeated measurements on each treatment and block combination.

Assumptions

1. The random variables X_{ijk} being observed follow the model

$$X_{ijm} = \mu_j + \beta_i + \gamma_{ij} + \epsilon_{ijm}$$

where

μ_j = mean for treatment j

β_i = difference in treatment mean due to the observation being in block i

$$\Sigma \beta_i = 0$$

γ_{ij} = the nonadditivity of block i with treatment j

$$\sum_i \gamma_{ij} = 0 \text{ (the sum of the interactions in each block is zero)}$$

$$\sum_j \gamma_{ij} = 0 \text{ (the sum of the interactions for each treatment is zero)}$$

ϵ_{ijm} = independent and identically distributed random variables

2. The ϵ_{ijm} are normally distributed with zero mean and equal variances σ^2.

Null Hypotheses

(a) H_0: $\mu_1 = \mu_2 = \cdots = \mu_k$ (the treatment means are all equal).

(b) H_0: $\gamma_{ij} = 0$ for all i and j (there is no interaction between block and treatment).

Alternative Hypotheses

(a) H_1: The treatment means are not all equal (some $\mu_i \neq \mu_j$).

(b) H_1: There is some interaction present between block and treatment (some $\gamma_{ij} \neq 0$).

Test Statistics

(a)

$$F = \frac{MST}{MSE} \tag{17.16}$$

The null distribution of F is the F distribution with $(k - 1)$ and $bk(n - 1)$ degrees of freedom.

(b)

$$F_1 = \frac{MSI}{MSE} \tag{17.17}$$

The null distribution of F_1 is the F distribution with $(b - 1)(k - 1)$ and $bk(n - 1)$ degrees of freedom.

In these test statistics,

$$MST = \frac{SST}{(k - 1)}$$

$$MSI = \frac{SSI}{(b - 1)(k - 1)}$$

$$MSE = \frac{Error\ SS}{bk(n - 1)}$$

with SST, SSI, and Error SS calculated from Eqs. 17.12, 17.14, and 17.15, respectively.

Decision Rules

(a) H_0 is rejected if $F > F_{1-\alpha,k-1,bk(n-1)}$, where $F_{1-\alpha,k-1,bk(n-1)}$ is the $1 - \alpha$ quantile from Table A5 with $k - 1$ and $bk(n - 1)$ degrees of freedom.

(b) H_0 is rejected if $F_1 > F_{1-\alpha,(b-1)(k-1),bk(n-1)}$, where $F_{1-\alpha,(b-1)(k-1),bk(n-1)}$ is the $1 - \alpha$ quantile from Table A5 with $(b - 1)(k - 1)$ and $bk(n - 1)$ degrees of freedom.

Note

If the desired degrees of freedom for the F distribution cannot be found in Table A5, simply use the next smaller degrees of freedom given in the table.

Multiple Comparisons

If the null hypothesis in part (a) is *rejected*, multiple comparisons between pairs of treatments may be made. The population means μ_i and μ_j are declared significantly different at a level of significance α if

$$|\bar{X}_i - \bar{X}_j| > LSD_\alpha$$

where $\bar{X}_i$ and $\bar{X}_j$ are the sample means for treatments i and j,

$$LSD_\alpha = t_{1-\alpha/2,bk(n-1)} \sqrt{MSE} \sqrt{\frac{2}{bn}} \qquad (17.18)$$

and where $t_{1-\alpha/2,bk(n-1)}$ is the $1 - \alpha/2$ quantile from Student's t distribution in Table A3 with $bk(n - 1)$ degrees of freedom.

EXAMPLE

The sample data of Figure 17.7 are used to test $H_0: \mu_1 = \mu_2 = \mu_3$ (the mean input time in minutes is the same for all three word processing packages). A level of significance of $\alpha = .05$ is used. Sample data calculations give the following results:

$$T = 2642 \qquad \Sigma\Sigma\Sigma X_{ijm}^2 = 137,418 \qquad k = 3 \qquad b = 6 \qquad n = 3$$

$$CF = \frac{(2642)^2}{6(3)(3)} = 129,262.296$$

$R_1 = 393$	$R_2 = 339$	$R_3 = 499$
$R_4 = 651$	$R_5 = 418$	$R_6 = 342$
$C_1 = 845$	$C_2 = 880$	$C_3 = 917$
$S_{11} = 124$	$S_{12} = 133$	$S_{13} = 136$
$S_{21} = 109$	$S_{22} = 107$	$S_{23} = 123$
$S_{31} = 163$	$S_{32} = 169$	$S_{33} = 167$
$S_{41} = 203$	$S_{42} = 220$	$S_{43} = 228$
$S_{51} = 139$	$S_{52} = 136$	$S_{53} = 143$
$S_{61} = 107$	$S_{62} = 115$	$S_{63} = 120$

The total sum of squares is obtained from Eq. 17.11

$$\text{Total SS} = 137{,}418.000 - 129{,}262.296 = 8155.704$$

whereas Eq. 17.12 gives the treatment sum of squares

$$SST = \frac{1}{3(6)} \{845^2 + 880^2 + 917^2\} - CF$$

$$= 129{,}406.333 - 129{,}262.296 = 144.037$$

From Eq. 17.13 the block sum of squares is

$$SSB = \frac{1}{3(3)} \{393^2 + 339^2 + 499^2$$

$$+ 651^2 + 418^2 + 342^2\} - CF$$

$$= 137{,}095.556 - 129{,}262.296 = 7833.259$$

and the interaction sum of squares is obtained from Eq. 17.14

$$SSI = \frac{1}{3} \{124^2 + 133^2 + \cdots + 120^2\} - 129{,}406.333$$

$$- 137{,}095.556 + CF$$

$$= 137{,}324.000 - 129{,}406.333 - 137{,}095.556 + 129{,}262.296$$

$$= 84.407$$

The error sum of squares is obtained using subtraction as indicated by Eq. 17.15

$$\text{Error SS} = 8155.704 - 144.037 - 7833.259 - 84.407$$

$$= 94.000$$

The F statistic for testing for treatment differences is obtained as follows:

$$MST = \frac{144.037}{(3-1)} = 72.019$$

$$MSE = \frac{94.000}{6(3)(3-1)} = 2.611$$

and

$$F = \frac{72.019}{2.611} = 27.58$$

The decision rule is to reject H_0 if $F > F_{.95,2,36} \approx 3.316$ (using 2 and 30 degrees of freedom in Table A5). Therefore, H_0 is soundly rejected with a p-value of less than .001, and the conclusion is that the mean input time is different among the three brands of word processing packages. A computer printout for this analysis is given as part of the next example.

Since H_0 was rejected, multiple comparisons can be made. The sample means are $\bar{X}_1 = 46.944$, $\bar{X}_2 = 48.889$, and $\bar{X}_3 = 50.944$. If a level of significance of .05 is used, the LSD value is

$$\text{LSD}_{.05} = t_{.975,36} \sqrt{\text{MSE}} \sqrt{\frac{2}{(6)(3)}}$$

$$= 2.0281 \sqrt{2.611} \sqrt{\frac{2}{18}}$$

$$= 1.092$$

Absolute Difference of Sample Means	Conclusion about Population Means
$\lvert \bar{X}_1 - \bar{X}_2 \rvert = 1.944$	$\mu_1 < \mu_2$
$\lvert \bar{X}_1 - \bar{X}_3 \rvert = 4.000$	$\mu_1 < \mu_3$
$\lvert \bar{X}_2 - \bar{X}_3 \rvert = 2.056$	$\mu_2 < \mu_3$

Thus significant differences appear to exist between the mean input times in all pairs of word processing packages.

THE CONCEPT OF INTERACTION

Thus far this section has dealt only with the two-way analysis of variance when repeated measurements are obtained. However, the repeated measurements allow for the calculation of means for each treatment and block combination, which provides the opportunity to see whether the treatment and blocks *interact* with one another. The concept of *interaction* is most easily understood through the use of graphical techniques. Such graphical techniques are based

	Brand of Software		
Operator	1	2	3
1	41.3	44.3	45.3
2	36.3	35.7	41.0
3	54.3	56.3	55.7
4	67.7	73.3	76.0
5	46.3	45.3	47.7
6	35.7	38.3	40.0

FIGURE 17.11
MEANS FOR EACH
COMBINATION OF SOFTWARE
AND OPERATOR BASED ON
THE SCORES IN FIGURE 17.7.

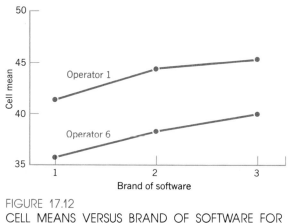

FIGURE 17.12
CELL MEANS VERSUS BRAND OF SOFTWARE FOR
OPERATORS 1 AND 6.

on cell means computed from the scores in Figure 17.7. These cell means are
given in Figure 17.11. They show that it is not possible to identify one software
package as always being the best, since the mean input time using one word
processing package is better for some operators and worse for other operators
when compared with other packages.

First consider the plot of the cell means in Figure 17.11 for only operators
1 and 6 versus brand of word processing package as given in Figure 17.12.
This figure presents a clear comparison of the test results on mean input times.
That is, the lowest mean input time is with brand number 1 and the highest
mean input time is with brand number 3. In general, the mean input times for
these two operators tend to increase by the same amount as they change from
one brand to another. This fact is made evident in Figure 17.12 by the parallel
line segments. On the other hand, consider Figure 17.13, which adds operator
2 to the graph given in Figure 17.12. This figure demonstrates interaction or
nonadditivity, as these line segments are not all parallel.

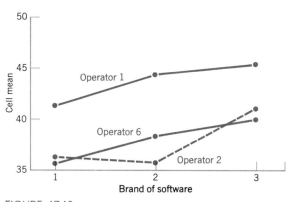

FIGURE 17.13
CELL MEANS VERSUS BRAND OF SOFTWARE FOR
OPERATORS 1, 2, AND 6.

For the sample data in Figure 17.7 test the following hypothesis with $\alpha = .05$:

H_0: There is no interaction between operator and brand of word processing package, that is, $\gamma_{ij} = 0$ for all cells (i, j)

The sums of squares obtained in the previous example of this section furnish the following mean squares

$$MST = \frac{84.407}{(6 - 1)(3 - 1)} = 8.441$$

$$MSE = \frac{94.000}{6(3)(3 - 1)} = 2.611.$$

From Eq. 17.17 the F statistic for testing interaction is

$$F_1 = \frac{8.441}{2.611} = 3.23$$

The decision rule is to reject H_0 if $F_1 > F_{.95,10,36} \approx 2.165$. Therefore, H_0 is rejected with a p-value of less than .01, and it is concluded that there is a significant interaction between operator and brand of word processing software. That is, the relative ease with which word processing software can be operated depends to a significant extent on the person who is using the software.

A computer analysis of these data provides the following printout, where the term PAC*OPER refers to the interaction between word processing packages and operators.

SOURCE	DF	SUM OF SQUARES	MEAN SQUARE	F VALUE	PR > F
PACKAGE	2	144.037	72.019	27.58	.0001
OPERATOR	5	7833.259	1566.652	599.99	.0001
PAC*OPER	10	84.407	8.441	3.23	.0046
ERROR	36	94.000	2.611		
TOTAL	53	8155.704			

RELATIONSHIP WITH THE GENERAL LINEAR MODEL

Just as the models used in the experimental design of the previous sections were shown to be special cases of the general linear model, the model in this section can be expressed as a general linear model. That is, the model given by Eq. 17.10 for the analysis of the RCB design with interaction may be written in general linear model form. However, since the notation is so cumbersome, the method is illustrated using the data given in Figure 17.7 as an example. The three word processing packages are represented by two dummy variables D_1 and D_2, and the six operators are represented by five dummy variables D_3 through D_7. The interaction term γ_{ij} in Eq. 17.10 is represented by interaction terms involving D_1 with each of D_3 through D_7, and D_2 with each of D_3 through D_7. Thus the model becomes

$$X_{ijm} - \underbrace{\beta_0 + \beta_1 D_1 + \beta_2 D_2}_{\text{software packages}} + \underbrace{\beta_3 D_3 + \beta_4 D_4 + \beta_5 D_5 + \beta_6 D_6 + \beta_7 D_7}_{\text{operators}}$$

$$+ \underbrace{\beta_8 D_1 D_3 + \beta_9 D_1 D_4 + \beta_{10} D_1 D_5 + \beta_{11} D_1 D_6 + \beta_{12} D_1 D_7}_{\text{interaction of brand 2 with operators}}$$

$$+ \underbrace{\beta_{13} D_2 D_3 + \beta_{14} D_2 D_4 + \beta_{15} D_2 D_5 + \beta_{16} D_2 D_6 + \beta_{17} D_2 D_7}_{\text{interaction of brand 3 with operators}} + \epsilon_{ijm} \quad (17.19)$$

One advantage of the model form in Eq. 17.10 becomes apparent; it is much simpler to write, and therefore easier to understand. The main advantage of the general linear model approach is still one of general applicability. That is, Eq. 17.19 applies even though there may be unequal numbers of observations in each block–treatment combination, whereas the computations for the two-way analysis of variance described in this section may be used only if the number of observations is the same for all block–treatment combinations.

EXERCISES

17.13 An experiment to investigate the effects of three types of herbicide and four levels of nitrogen on the yield of wheat produced the sample data given below. Convert these data to z scores within each herbicide and plot the e.d.f.s of wheat yield for each level of nitrogen. ($\overline{X}$ and s for herbicide 1: 39.5, 3.295; herbicide 2: 39.5, 5.707; herbicide 3: 39.0, 4.036)

| | Nitrogen Level | | | |
Herbicide	1	2	3	4
1	35	38	41	45
	37	38	39	43
2	31	39	44	47
	33	37	40	45
3	38	34	39	46
	38	36	37	44

17.14 Assume that the necessary assumptions are met and use the sample data in Exercise 17.13 to test the null hypothesis H_0: $\mu_1 = \mu_2 = \mu_3 = \mu_4$ (the mean wheat yield is the same for all levels of nitrogen). Let $\alpha = .05$. ($\Sigma X = 944$, $\Sigma X^2 = 37,550$, $\Sigma C^2 = 224,728$, $\Sigma R^2 = 297,056$, $\Sigma S^2 = 75,048$.)

17.15 Apply the LSD multiple comparisons procedure with $\alpha = .05$ to the analysis of variance problem in Exercise 17.14. ($\overline{X}_1 = 35.3$, $\overline{X}_2 = 37$, $\overline{X}_3 = 40$, $\overline{X}_4 = 45$.)

17.16 Assume that the necessary assumptions are met and use the sample data in Exercise 17.13 to test the null hypothesis H_0: There is no interaction between type of herbicide and level of nitrogen. Let $\alpha = .05$.

17.17 Three types of unleaded gasoline were tested with two different types of carburetors. The mpg is recorded below for each test. Convert these data to z scores for each carburetor type and plot the e.d.f.s of mpg for each gasoline type.

Brand of Gasoline

Carburetor type	A	B	C
1	19	17	20
	22	18	21
	20	17	19
2	24	23	25
	23	22	23
	23	20	22

17.18 Use the sample data in Exercise 17.17 to test the null hypothesis H_0: $\mu_A = \mu_B = \mu_C$ (the mean mpg is the same for all three brands of gasoline). Let $\alpha = .05$. What assumptions are you making? Do they appear to be reasonable?

17.19 Apply the LSD multiple comparisons procedure with $\alpha = .05$ to the analysis of variance problem in Exercise 17.18.

17.20 Use the sample data in Exercise 17.17 to test the null hypothesis H_0: there is no interaction between type of carburetor and brand of gasoline. Let $\alpha = .05$. What assumptions are you making? Do they appear to be reasonable?

17.21 In an experiment to test the drying ability of various types of clothes dryers, similar batches of clothing are weighed when dry, again after wetting, and again after drying for 10 minutes. From these data the percentage of moisture remaining in the clothes is easily obtained. These values are recorded below. Convert these data to z scores for each appliance type and plot the e.d.f.s of percent of moisture for each brand of dryer. ($\overline{X}$ and s for gas: 13.33, 3.559; electric: 12.67, 3.222.)

Brand of Dryer

Type	A	B	C
Gas	17	11	12
	13	14	9
	21	10	13
	19	9	11
	15	11	15

(continued)

Brand of Dryer

Type	A	B	C
Electric	11	12	15
	8	16	11
	15	9	18
	10	11	12
	11	12	19

17.22 Use the sample data in Exercise 17.21 to test the null hypothesis, H_0: $\mu_A = \mu_B = \mu_C$ (the mean amount of moisture removed is the same for all brands of dryers). Let $\alpha = .05$. What assumptions are you making? The computer printout is as follows.

SOURCE	DF	SUM OF SQUARES	MEAN SQUARE	F VALUE	PR > F
BRAND	2	35.00	17.50	2.39	.1134
TYPE	1	3.33	3.33	0.45	.5066
BRAND*TYPE	2	111.67	55.83	7.61	.0028
ERROR	24	176.00	7.33		
TOTAL	29	326.00			

17.23 Apply the LSD multiple comparisons procedure with $\alpha = .05$ to the analysis of variance problem in Exercise 17.22.

17.24 Use the sample data in Exercise 17.21 to test the null hypothesis H_0: there is no interaction between dryer type and brand of dryer. Let $\alpha = .05$. What assumptions are you making?

17.25 Write the model given in Eq. 17.10 to describe the data given in Exercise 17.21. Interpret each parameter in the model.

17.26 Write the general linear model that corresponds to the data given in Exercise 17.21. Interpret each parameter and each dummy variable in the model.

17.3

ANALYSIS OF TWO-FACTOR EXPERIMENTS FOR GENERAL POPULATIONS (OPTIONAL)

PROBLEM SETTING 17.2

At a management leadership seminar the entire group is asked to break up into smaller groups and have each individual rank 10 leadership traits from 1 (least important in a leader) to 10 (most important in a leader). One group of seven managers reported the rankings in Figure 17.14. How should the data in Figure 17.14 be analyzed to see if the managers have a tendency to agree on which leadership traits are most important? That is, do some leadership traits tend to get higher scores than other leadership traits?

	Leadership Trait									
Manager	A	B	C	D	E	F	G	H	I	J
1	1	3	2	9	10	5	6	8	4	7
2	2	1	7	10	9	8	3	5	4	6
3	7	1	9	6	10	4	5	2	3	8
4	7	2	5	4	9	6	10	3	1	8
5	4	5	8	6	10	3	7	2	1	9
6	2	1	7	6	9	4	10	5	3	8
7	3	7	8	6	9	2	5	4	1	10

FIGURE 17.14
RANKINGS OF TEN LEADERSHIP TRAITS BY EACH OF
SEVEN MANAGERS.

PROBLEM SETTING 17.3

Three real estate appraisers are each asked to make an appraisal of the same
five pieces of property. The result of their appraisals (in thousands of dollars)
is given in Figure 17.15. Do some real estate appraisers tend to give higher
evaluations than others?

THE FRIEDMAN TEST

Problem Setting 17.2 is another example of a randomized complete block
design; the managers act as blocks and the leadership traits are the treatments.
However, since the data are discrete, consisting of the ranks 1 to 10 within
each block, the assumption of normality stated in Section 17.1 for analyzing
randomized complete block designs is clearly not satisfied. Problem Setting
17.3 is also an example of a randomized complete block design and, if the
assumptions are satisfied, the methods of Section 17.1 would be appropriate

	Real Estate Appraiser		
Property	A	B	C
1	35.6	34.3	37.3
2	47.4	43.8	47.9
3	60.0	61.4	64.5
4	36.9	36.8	38.0
5	45.5	46.2	46.6

FIGURE 17.15
APPRAISALS OF FIVE PIECES
OF PROPERTY BY EACH OF
THREE REAL ESTATE
APPRAISERS.

for analyzing the sample data. However, if the assumptions are not satisfied, then the nonparametric **Friedman test** may be preferred. The Friedman test is one more example of a rank-transformation procedure.

To apply the Friedman test the original data must be replaced with their corresponding *ranks* from 1 to *k within each block*. In Problem Setting 17.2 the original data as shown in Figure 17.14 are already in this form, so no replacement is needed. To apply the Friedman test to Problem Setting 17.3, it is necessary to replace the observations within each block in Figure 17.15 by their corresponding ranks from 1 to 3. Such a ranking is given in Figure 17.16 and makes it clear that real estate appraiser C always gives the highest appraisal. The Friedman test can be applied to RCB design where the original data consist of ranks assigned within blocks or to data that must be replaced by ranks within blocks.

The Friedman test statistic is obtained by applying Eq. 17.6 directly to the ranks such as appear in Figures 17.14 and 17.16. However, since ranks are used, Eqs. 17.1 through 17.4 for sums of squares are greatly simplified as follows. Let R_{ij} represent the rank assigned to treatment j in block i. Then the sum of squares of all observations is denoted as

$$A = \sum_{i=1}^{b} \sum_{j=1}^{k} R_{ij}^2 \qquad (17.20)$$

The sum of all ranks is $T = bk(k + 1)/2$, so the total sum of squares is

$$\text{Total SS} = A - \frac{T^2}{bk} = A - \frac{bk(k + 1)^2}{4}.$$

If there are no ties in the assignment of ranks in any block, then A simplifies to

$$A = \frac{bk(k + 1)(2k + 1)}{6} \qquad (17.21)$$

Property	Real Estate Appraiser		
	A	B	C
1	2	1	3
2	2	1	3
3	1	2	3
4	2	1	3
5	1	2	3

FIGURE 17.16
REAL ESTATE
APPRAISALS OF
PROBLEM SETTING
17.3 CONVERTED TO
RANKS WITHIN
BLOCKS.

and the total sum of squares simplifies to $bk(k + 1)(k - 1)/12$.

Let C_j represent the total of the jth column of ranks (see Figure 17.6). Then a sum of squares based on column totals is

$$B = \frac{1}{b} \sum_{j=1}^{k} C_j^2 \tag{17.22}$$

The treatment sum of squares is

$$SST = B - \frac{bk(k + 1)^2}{4}$$

and the error sum of squares is $A - B$. The block sum of squares is always zero in the Friedman test. A formal statement of the Friedman test is now given.

EXAMPLE

Use the data in Problem Setting 17.2 and a level of significance of $\alpha = .05$ to test

H_0: there is no difference in the perceived importance of the various leadership traits.

Since there are no ties in the rankings in Figure 17.14, Eq. 17.21 is used to find

$$A = \frac{7(10)(10 + 1)(20 + 1)}{6} = 2695$$

The column totals are computed from Figure 17.14 and are used with Eq. 17.22 to find

$$B = \frac{1}{7} [26^2 + 20^2 + 46^2 + 47^2 + 66^2 + 32^2 + 46^2$$
$$+ 29^2 + 17^2 + 56^2]$$
$$= 2451.86$$

The test statistic is found from Eq. 17.23 as

$$F_R = \frac{(7 - 1)[2451.86 - 7(10)(11)^2/4]}{2695 - 2451.86}$$
$$= 8.25$$

Since $F_{.95,9,54} \approx 2.073$, H_0 is rejected, and it is concluded that the mean preference on the leadership traits is not the same for all traits. The p-value is less than .001.

ANALYSIS OF VARIANCE FOR RANDOMIZED COMPLETE BLOCK DESIGNS WITH GENERAL POPULATIONS (Friedman Test)

Data

The data consist of b blocks, each of which contains k observations X_{ij}, $i = 1, \ldots, k$. These observations appear either as ranks, R_{ij}, from 1 to k within each block, or are replaced with their ranks 1 to k within each block.

Assumptions

1. The observations within each block are independent of the observations within every other block.

2. Within each block the observations may be ranked according to some criterion of interest.

Null Hypothesis

H_0: each ranking of the random variables within a block is equally likely (i.e., the treatments have identical means).

Alternative Hypothesis

H_1: the treatment means are not all equal.

Test Statistic

$$F_R = \frac{(b - 1)[B - bk(k + 1)^2/4]}{A - B} \qquad (17.23)$$

where A and B are given by Eqs. 17.20 and 17.22. In the case of identical treatment rankings within each of the blocks, the statistic F_R is undefined because $A = B$. In this case, the null hypothesis should be rejected with a p-value exactly equal to $(1/k!)^{b-1}$. The null distribution of F_R is approximately the F distribution with $(k - 1)$ and $(k - 1)(b - 1)$ degrees of freedom.

Decision Rule

Reject H_0 if $F_R > F_{1-\alpha,k-1,(b-1)(k-1)}$, where $F_{1-\alpha,k-1,(b-1)(k-1)}$ is the $1 - \alpha$ quantile from Table A5 with $k - 1$ and $(b - 1)(k - 1)$ degrees of freedom.

Note

If the desired degrees of freedom for the F distribution cannot be found in Table A5, simply use the next smaller degrees of freedom given in the table.

Multiple Comparisons

If H_0 is rejected, then the means of treatments i and j are declared to be significantly different if

$$|\bar{R}_i - \bar{R}_j| > \text{LSD}_\alpha \qquad (17.24)$$

where $\bar{R}_i$ and $\bar{R}_j$ are the mean ranks for treatments i and j, and

$$\text{LSD}_\alpha = t_{1-\alpha/2,(b-1)(k-1)}\sqrt{\text{MSE}}\,\sqrt{\frac{2}{b}}$$

$$= t_{1-\alpha/2,(b-1)(k-1)}\sqrt{\frac{2(A - B)}{b(b - 1)(k - 1)}}$$

with A and B as defined in Eqs. 17.20 and 17.22, respectively. Recall that $t_{1-\alpha/2,(b-1)(k-1)}$ is the $1 - \alpha/2$ quantile from Table A3 with $(b - 1)(k - 1)$ degrees of freedom.

Because H_0 was rejected, multiple comparisons can be made on the rankings of leadership traits.

$$LSD_{.05} = t_{.975,54} \sqrt{\frac{2(2695 - 2451.86)}{7(7 - 1)(10 - 1)}}$$

$$= 2.0049(1.134)$$

$$= 2.27$$

The $\bar{R}_i$ are listed from smallest to largest with the number of the leadership trait noted. Lines are drawn under those $\bar{R}_i$ values that the multiple comparisons procedure would group together. Thus leadership trait 5 is perceived as more important than all other leadership traits except for trait number 10, and leadership trait number 10 is considered more important than traits numbered 9, 2, 1, 8, and 6. Similar statements may be made about all pairs of traits based on this multiple comparisons analysis.

Trait:	9	2	1	8	6	3	7	4	10	5
$\bar{R}_i$:	2.43	2.86	3.71	4.14	4.57	6.57	6.57	6.71	8.00	9.43

EXAMPLE

For Problem Setting 17.3 test the null hypothesis,

H_0: there is no difference in the mean evaluation by the three real estate appraisers.

Let $\alpha = .05$. The block ranks for these data appear in Figure 17.16. Equation 17.21 is used to find A, since there are no ties in the rankings.

$$A = \frac{5(3)(4)(7)}{6} = 70$$

From Eq. 17.22,

$$B = \frac{1}{5}[8^2 + 7^2 + 15^2] = 67.60$$

and from Eq. 17.23,

$$F_R = \frac{(5 - 1)[67.60 - 5(3)(4)^2/4]}{70 - 67.60} = 12.67$$

because $F_{.95,2,8} = 4.459$, H_0 is rejected, and it is concluded that the mean appraisal figures given by the different real estate appraisers are not equal. The p-value is less than .01.

Since H_0 was rejected, multiple comparisons are in order.

$$\text{LSD}_{.05} = t_{.975,8} \sqrt{\frac{2(70 - 67.60)}{5(4)(2)}} = 0.80$$

Appraiser:	B	A	C
$\overline{R}_i$:	1.4	1.6	3.0

This means that real estate appraisers A and B tend to give the same evaluations, and real estate appraiser C gives significantly higher appraisals than the other two.

THE EFFECT OF OUTLIERS

If the regular two-way analysis of variance is calculated on the original data in the preceding example, the result is $F = 5.25$. This is less than the Friedman test statistic, but is still significant at the .05 level of significance. However, as with the Kruskal–Wallis test in Section 16.4 an interesting situation can be created by changing one of the original observations in Figure 17.15. Specifically, consider changing the largest observation 64.5 in steps of 5.0 to 84.5. The Friedman test statistic is unaffected by these changes, but the F-test on the original data is greatly affected by them, as is illustrated in Figure 17.17.

The result in Figure 17.17 is contrary to what one might expect. That is, if one of appraiser C's appraisals is quite large, this is even stronger evidence of the tendency for appraiser C to appraise higher than the other two. But as the appraisal gets higher, the F statistic actually decreases in value until it is no longer significant. This points out once again the harmful effect that one or a few outliers can have on the power of a test based on the original data, and the lack of such an effect on a test based on ranks.

The Friedman test is easily extended to cover the situation in Section 17.2 where repeated measurements are present in a randomized complete block

$X^{(15)}$	F
64.5	5.25
69.5	3.55
74.5	2.55
79.5	2.08
84.5	1.83

FIGURE 17.17
A MODIFICATION OF PROBLEM SETTING 17.3 TO SHOW THE ADVERSE EFFECT ON THE F-TEST OF AN OUTLIER IN THE SAMPLE DATA.

design. However, such an extension is not covered in this text, and the reader is referred to texts on nonparametric statistics.

17.27 A survey was taken of all seven hospitals in a large city to obtain the number of babies born over a 12-month period.

Hospital	Season			
	Winter	Spring	Summer	Fall
A	92	112	94	77
B	9	11	10	12
C	98	109	92	81
D	19	26	19	18
E	21	22	23	24
F	58	71	51	62
G	42	49	44	41

The time period was divided into four seasons to test the null hypothesis that the birth rate is constant over all four seasons. Use the Friedman test to test the null hypothesis with $\alpha = .05$. ($A = 209.5$, $\Sigma C_i^2 = 1304.5$.)

17.28 Apply the LSD multiple comparisons procedure with $\alpha = .05$ to the analysis of variance problem in Exercise 17.27. ($C_1 = 13.5$, $C_2 = 25$, $C_3 = 16.5$, $C_4 = 15$.)

17.29 Twelve randomly selected students are involved in a learning experiment. Four lists of words are made up by the experimenter. Each list contains 20 words and each student is given five minutes to study it. Each student is then tested on his or her ability to remember the words. This procedure is repeated for all four lists for each student, the order of the lists being rotated from one student to the next. Use a Friedman test to see if some lists are easier to learn than others. Let $\alpha = .05$. The examination scores are as follows (20 is perfect).

Student	List			
	1	2	3	4
1	18	14	16	20
2	7	6	5	10
3	13	14	16	17
4	15	10	12	14
5	12	11	12	18
6	11	9	9	16
7	15	16	10	14
8	10	8	11	16
9	14	12	13	15
10	9	9	9	10
11	8	6	9	14
12	10	11	13	16

17.30 Apply the LSD multiple comparisons procedure with $\alpha = .05$ to the analysis of variance problem in Exercise 17.29.

17.31 Twelve homeowners are randomly selected and asked to plant four different types of grass in their yards and then to rank the grasses from 1 (best) to 4 (least preferred). Use a Friedman test with $\alpha = .05$ to see if differences exist among the grass types. The computer printout is given below.

	Grass Type			
Homeowner	A	B	C	D
1	4	3	2	1
2	4	2	3	1
3	3	1.5	1.5	4
4	3	1	2	4
5	4	2	1	3
6	2	2	2	4
7	1	3	2	4
8	2	4	1	3
9	3.5	1	2	3.5
10	4	1	3	2
11	4	2	3	1
12	3.5	1	2	3.5

SOURCE	DF	SUM OF SQUARES	MEAN SQUARE	F VALUE	PR > F
GRASS	3	12.71	4.24	3.19	.0362
HOMEOWNER	11	0.00	0.00	0.00	1.0000
ERROR	33	43.79	1.33		
TOTAL	47	56.50			

17.32 Apply the LSD multiple comparisons procedure with $\alpha = .05$ to the analysis of variance problem in Exercise 17.31.

17.4

REVIEW EXERCISES

17.33 Perform a two-way analysis of variance on the sample data given in Exercise 17.27 to test the null hypothesis H_0: $\mu_W = \mu_{Sp} = \mu_{Su} = \mu_F$. Compare your results with those obtained from the Friedman test in Exercise 17.27. ($\Sigma X = 1387$, $\Sigma X^2 = 99{,}113$, $\Sigma R_i^2 = 391{,}153$, $\Sigma C_j^2 = 485{,}035$.)

17.34 Apply the LSD multiple comparisons procedure with $\alpha = .05$ to the analysis of variance problem in Exercise 17.33 and compare results with those obtained in Exercise 17.28. ($C_1 = 339$, $C_2 = 400$, $C_3 = 333$, $C_4 = 315$.)

17.35 Perform a two-way analysis of variance on the sample data given in Exercise 17.29 to test the null hypothesis H_0: $\mu_1 = \mu_2 = \mu_3 = \mu_4$. Compare your results with those obtained from the Friedman test in Exercise 17.29.

17.36 Apply the LSD multiple comparisons procedure with $\alpha = .05$ to the analysis of variance problem in Exercise 17.35 and compare results with those obtained in Exercise 17.30.

17.37 Perform the Friedman test on the data given in Problem Setting 17.1 and compare results with the example given in Section 17.1.

17.38 Apply the LSD multiple comparisons procedure with $\alpha = .05$ to the analysis of variance problem in Exercise 17.37 and compare results with the example given in Section 17.1.

17.39 Perform the Friedman test on the data given in Exercise 17.1 and compare results with those in Exercise 17.3. ($A = 70$, $B = 64.8$.)

17.40 Apply the LSD multiple comparisons procedure with $\alpha = .05$ to the analysis in Exercise 17.39 and compare results with Exercise 17.4.

17.41 Perform the Friedman test on the data given in Exercise 17.7 and compare results with that exercise.

17.42 Apply the LSD multiple comparisons procedure with $\alpha = .05$ to the analysis in Exercise 17.41 and compare results with Exercise 17.8.

17.43 Perform the Friedman test on the data given in Exercise 17.9 and compare results with that exercise. Use the following computer printout.

SOURCE	DF	SUM OF SQUARES	MEAN SQUARE	F VALUE	PR > F
STOCK	4	16.625	4.156	2.18	.1330
QUARTER	3	0.000	0.000	0.00	1.0000
ERROR	12	22.875	1.906		
TOTAL	19	39.500			

17.44 Apply the LSD multiple comparisons procedure with $\alpha = .05$ to the analysis in Exercise 17.43 and compare results with Exercise 17.10.

17.45 Three measurements are made on each of three workers for the number of units they produce on each of three different machines. These data are recorded below. Use these sample data to test whether the production rate is the same for all three operators. Let $\alpha = .05$. Assume that the assumptions of the model hold true for this case.

Operator

Machine	1	2	3
A	44	42	36
	38	36	34
	40	38	40

(continued)

Operator

Machine	1	2	3
B	42	46	38
	38	48	36
	36	42	42
C	34	40	38
	34	44	42
	30	38	46

17.46 Apply the LSD multiple comparisons procedure with $\alpha = .05$ to the analysis of variance problem in Exercise 17.45.

17.47 Convert the sample data in Exercise 17.45 to z scores within each block and plot the e.d.f.s of production for each worker. How do these e.d.f.s compare with your conclusions in Exercises 17.45 and 17.46?

17.48 Use the sample data in Exercise 17.45 to test the null hypothesis, H_0: there is no interaction between worker and machine type. Let $\alpha = .05$.

17.49 Use the scores in Exercise 17.45 to compute means for each worker–machine combination. Plot these means in a graph similar to Figure 17.13 and use this graph as a basis for checking the results in Exercise 17.48.

17.50 The owners of a fast-food chain in a large city have decided to promote one of three store managers to general manager. They set up an experiment where each of the three candidates rotates as manager among four of their stores (none of which is currently run by these three managers). The amount of sales at each store is recorded for three days for each manager. These sales data are recorded below (in thousands). Use these data to see whether the sales differ from manager to manager. Let $\alpha = .05$. The computer printout is as follows. What assumptions are you making?

SOURCE	DF	SUM OF SQUARES	MEAN SQUARE	F VALUE	PR > F
MANAGER	2	1.356 E-01	6.778 E-02	1.08	.3541
STORE	3	22.143	7.381	118.10	.0001
MAN*STORE	6	1.691	0.2819	4.51	.0034
ERROR	24	1.500	0.0625		
TOTAL	35	25.470			

Manager

Store	1	2	3
1	$2.7	$3.4	$2.2
	2.4	2.9	3.0
	2.9	3.0	2.6

(continued)

Manager

Store	1	2	3
2	1.5	1.1	1.0
	1.8	1.3	1.0
	1.6	1.0	0.6
3	1.8	1.5	2.1
	2.2	2.1	1.6
	2.1	2.0	1.8
4	0.3	0.5	1.0
	0.7	0.5	0.7
	0.7	0.6	1.3

17.51 Apply the LSD multiple comparisons procedure with $\alpha = .05$ to the analysis of variance problem in Exercise 17.50.

17.52 Convert the sample data in Exercise 17.50 to z scores within each block and plot the e.d.f.s of sales for each manager. How do these e.d.f.s compare with your conclusions in Exercises 17.50 and 17.51?

17.53 Use the sample data in Exercise 17.50 to test the null hypothesis, H_0: there is no interaction between manager and store. Let $\alpha = .05$.

17.54 Use the scores in Exercise 17.50 to compute means for each manager–store combination. Plot these means in a graph similar to Figure 17.13 and use this graph as a basis for checking the results in Exercise 17.53.

BIBLIOGRAPHY

Additional material on the topics presented in this chapter can be found in the following publications.

Conover, W. J. (1980). *Practical Nonparametric Statistics,* 2nd ed. Wiley, New York.

Conover, W. J. and Iman, R. L. (1981). "Rank Transformations as a Bridge Between Parametric and Nonparametric Statistics." *The American Statistician,* **35**(3), 124–133.

18
OTHER USEFUL TOPICS
IN EXPERIMENTAL DESIGN

PRELIMINARY REMARKS

The previous two chapters considered two fundamental experimental designs, namely the completely randomized design and the randomized complete block design. The main objective of an experimental design is to enable the experimenter to detect small differences between treatments or populations, using a minimum amount of cost or time. Since each situation varies in the type of data that may be collected and the cost of collecting the data, different types of experimental designs are needed. A thorough study of available experimental designs would be very extensive. Thus, the previous two chapters and the present chapter provide only a brief survey of the subject.

The completely randomized design in Chapter 16 is sometimes called a **one-factor** design, because the observations are divided into different categories by only one criterion—treatments (or populations). The different treatments represent different **levels** of the factor. Chapter 17, however, presented a **two-factor** design. The observations were divided into groups by one factor, treatments, and also by a second factor, blocks. This was the randomized complete block design. In this chapter a **three-factor** design is presented. These designs are simple examples of a class of designs called **factorial designs.**

When the observed response variable in an experimental design has an associated quantitative measurement available with it (e.g., with observed test scores of students, an IQ score might also be available), the analysis comes under the heading of **analysis of covariance.** An analysis of covariance combines both regression and analysis of variance and is considered in the second section of this chapter. The last section of this chapter provides alternative procedures for use with general populations. Since all of the procedures presented in this chapter require more calculations than the analyses in the previous two chapters, it is recommended that a computer be used in the analysis.

18.1

ANALYSIS OF VARIANCE FOR THREE-FACTOR EXPERIMENTS

PROBLEM SETTING 18.1

A marketing analyst desires to gauge public response to a new product prior to the start of its national advertising campaign. The response is gauged by recording the volume of sales noted in a test marketing. The marketing analyst decides to set up an experiment to compare three different prices on the product in three different chains of stores for two weeks. The volume of sales is recorded for two days of each week on each price in each store. The sales volumes are given in Figure 18.1.

What type of information should the analysis of the sample data in Figure 18.1 provide for the marketing analyst? That is, should a comparison of the

		Store		
Week	Price	S_1	S_2	S_3
1	P_1	22	31	17
		24	30	17
	P_2	25	36	18
		24	38	15
	P_3	29	36	22
		28	34	21
		S_1	S_2	S_3
2	P_1	24	30	15
		21	35	18
	P_2	27	34	16
		24	36	19
	P_3	26	37	24
		31	42	21

FIGURE 18.1
VOLUME OF SALES RECORDED AT EACH OF THREE STORES AT EACH OF THREE PRICES FOR EACH OF TWO WEEKS.

mean volume of sales from store to store, or perhaps from price to price, be made? Interaction might exist between price and store, and this might be of interest. Probably of lesser interest would be a comparison between weeks, unless some other factor such as the amount of advertising had been different in one of the weeks. The answers to these questions are considered in this section.

USE OF GRAPHICAL TECHNIQUES

Graphs of e.d.f.s based on z scores can be used to check for treatment differences and interaction in three-factor experiments, as was done in the previous two chapters. However, care must be taken to keep the bookkeeping straight in computing the z scores. The basic rule to follow in computing z scores for one factor is to compute the z scores based on only those observations where the remaining two factors do not change. For example, the six observations comprising the first two rows in Figure 18.1 were obtained with the week fixed at Level 1 and price at Level 1. Hence, the z scores computed on these six observations are used for comparing *stores*. Likewise, z scores would be computed based on the second set of six observations in Figure 18.1 where week is fixed at level 1 and price at level 2. Continuing on down through Figure 18.1 in this manner and using additional groups of size 6, all observations are converted to z scores (see Figure 18.2), which are then used to plot e.d.f.s for the three stores (see Figure 18.3).

		Store		
Week	Price	S_1	S_2	S_3
1	P_1	-0.25	1.23	-1.07
		0.08	1.07	-1.07
	P_2	-0.11	1.07	-0.86
		-0.21	1.29	-1.18
	P_3	0.11	1.26	-1.04
		-0.05	0.93	-1.20
		S_1	S_2	S_3
2	P_1	0.02	0.82	-1.17
		-0.38	1.48	-0.78
	P_2	0.13	1.00	-1.25
		-0.25	1.25	-0.88
	P_3	-0.52	0.85	-0.76
		0.10	1.46	-1.13

FIGURE 18.2
z SCORES FOR CONSTRUCTING e.d.f.s TO COMPARE THE VOLUME OF SALES AT THREE STORES.

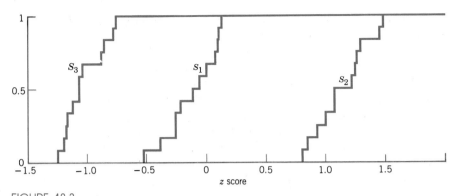

FIGURE 18.3
EMPIRICAL DISTRIBUTION FUNCTIONS OF THE z SCORES FOR THE THREE
STORES IN PROBLEM SETTING 18.1.

In a similar manner, z scores for *price* comparisons are computed by working only with groups of size six where both week and store are fixed at given levels. The ż scores for *week* differences are obtained by working with only four observations at a time where both store and price are fixed at given levels.

SUMS OF SQUARES IN THE THREE-FACTOR ANALYSIS
OF VARIANCE

The data in Figure 18.1 were obtained on the basis of three factors: (1) store chain, (2) price of product, and (3) week of sale. The sources of variation have been increased over the two-factor experiment by the inclusion of a third factor. That is, it is necessary to compute a sum of squares for each of the following sources of variation.

1. Total sum of squares.
2. Sum of squares due to store differences.
3. Sum of squares due to price differences.
4. Sum of squares due to week differences.
5. Sum of squares due to interaction between store and price.
6. Sum of squares due to interaction between store and week.
7. Sum of squares due to interaction between price and week.
8. Sum of squares due to interaction among store, price, and week (i.e., a three-way interaction).
9. Sum of squares due to error.

The calculation equations for these sums of squares are similar to those stated in the previous two chapters; however, the amount of hand calculations is sufficient to encourage the use of a computer on three-factor problems. Therefore, the calculation formulas are not given here.

PROCEDURE FOR INFERENCE ABOUT THE DIFFERENCE IN MEANS
OF k NORMAL POPULATIONS
(Three-Way Analysis of Variance with Interaction)

Data

The data may be classified into categories according to three different factors, A, B, and C. Each factor occurs at several levels, a levels for factor A, b levels for factor B, and c levels for factor C, so there are abc different categories into which the data are classified. There are n observations in each category.

Assumptions

1. The random variables being observed follow the model

$$X_{ijkm} = \mu + \tau_i + \beta_j + \gamma_k + (\tau\beta)_{ij} + (\tau\gamma)_{ik} + (\beta\gamma)_{jk} + (\tau\beta\gamma)_{ijk} + \epsilon_{ijkm}$$

where

μ = the overall mean

τ_i = the difference in mean due to factor A ($\Sigma\tau_i = 0$)

β_j = the difference in mean due to factor B ($\Sigma\beta_j = 0$)

γ_k = the difference in mean due to factor C ($\Sigma\gamma_k = 0$)

$(\tau\beta)_{ij}$ = the difference in mean due to interaction between factors A and B

$$\left(\sum_i (\tau\beta)_{ij} = 0 \text{ for each } j \text{ and } \sum_j (\tau\beta)_{ij} = 0 \text{ for each } i\right)$$

$(\tau\gamma)_{ik}$ = the difference in mean due to interaction between factors A and C

$$\left(\sum_i (\tau\gamma)_{ik} = 0 \text{ for each } k \text{ and } \sum_k (\tau\gamma)_{ik} = 0 \text{ for each } i\right)$$

$(\beta\gamma)_{jk}$ = the difference in mean due to interaction between factors B and C

$$\left(\sum_j (\beta\gamma)_{jk} = 0 \text{ for each } k \text{ and } \sum_k (\beta\gamma)_{jk} = 0 \text{ for each } j\right)$$

$(\tau\beta\gamma)_{ijk}$ = the difference in mean due to the three-way interaction among factors

$$A, B, \text{ and } C \left(\sum_i (\tau\beta\gamma)_{ijk} = 0, \sum_j (\tau\beta\gamma)_{ijk} = 0 \text{ and } \sum_k (\tau\beta\gamma)_{ijk} = 0\right)$$

ϵ_{ijkm} = independent and identically distributed random variables

2. The ϵ_{ijkm} are normally distributed with zero mean and equal variance σ^2.

Hypotheses

(a) H_0: $\tau_1 = \tau_2 = \cdots = \tau_a = 0$ (factor A means are all equal)
H_1: some $\tau_i \neq 0$ (the levels of factor A have some unequal means)

(b) H_0: $\beta_1 = \beta_2 = \cdots = \beta_b = 0$ (factor B means are all equal)
H_1: some $\beta_j \neq 0$ (the levels of factor B have some unequal means)

(c) H_0: $\gamma_1 = \gamma_2 = \cdots = \gamma_c = 0$ (factor C means are all equal)
H_1: some $\gamma_k \neq 0$ (the levels of factor C have some unequal means)

(d) H_0: $(\tau\beta)_{ij} = 0$ for all (i, j) (there is no two-way interaction between factors A and B)
H_1: $(\tau\beta)_{ij} \neq 0$ for some (i, j) (the effects of factors A and B are nonadditive)

(e) H_0: $(\tau\gamma)_{ik} = 0$ for all (i, k) (there is no two-way interaction between factors A and C)
H_1: $(\tau\gamma)_{ik} \neq 0$ for some (i, k) (the effects of factors A and C are nonadditive)

(f) H_0: $(\beta\gamma)_{jk} = 0$ for all (j, k) (there is no two-way interaction between factors B and C)

H_1: $(\beta\gamma)_{jk} = 0$ for some (j, k) (the effects of factors B and C are nonadditive)

(g) H_0: $(\tau\beta\gamma)_{ijk} = 0$ for all (i, j, k) (there is no three-way interaction among factors A, B, and C)

H_1: $(\tau\beta\gamma)_{ijk} \neq 0$ for some (i, j, k) (the nonadditivity among factors A, B, and C are not adequately explained by two-way interactions)

Test Statistics

(See the analysis of variance table in Figure 18.4.) The test statistics have the F distribution given in Table A5 when the null hypothesis is true.

Decision Rules

H_0 is rejected if the test statistic exceeds the appropriate critical value from Table A5. If the desired degrees of freedom cannot be found in Table A5, simply use the next smaller degrees of freedom given in the table.

Multiple Comparisons

If the null hypothesis in parts (a), (b), or (c) is rejected, the respective factor levels may be compared pairwise using the following procedure. The factor levels τ_i and τ_j are declared unequal if the sample mean of all observations under the ith level of factor A differs from the corresponding sample mean for all observations under the jth level of factor A by more than LSD_α, where

$$LSD_\alpha = t_{1-\alpha/2, abc(n-1)}\sqrt{MSE}\ \sqrt{\frac{2}{bcn}}$$

where $t_{1-\alpha/2, abc(n-1)}$ is the $1 - \alpha/2$ quantile of the Student's t distribution with $abc(n-1)$ degrees of freedom given in Table A3. Factor B means are compared to

$$LSD_\alpha = t_{1-\alpha/2, abc(n-1)}\sqrt{MSE}\ \sqrt{\frac{2}{acn}}$$

and factor C means are compared to

$$LSD_\alpha = t_{1-\alpha/2, abc(n-1)}\sqrt{MSE}\ \sqrt{\frac{2}{abn}}$$

HYPOTHESES OF INTEREST IN THE THREE-FACTOR ANALYSIS OF VARIANCE

Several hypotheses can be tested in three-factor experiments as follows.

1. H_0: there is no difference in sales volume from store to store.

2. H_0: there is no difference in sales volume from price to price.

3. H_0: there is no difference in sales volume from week to week.

4. H_0: there is no interaction between store and price.

5. H_0: there is no interaction between store and week.

6. H_0: there is no interaction between price and week.

7. H_0: there is no interaction among store, price, and week.

The test of each of these hypotheses requires the calculation of a separate

ANALYSIS OF VARIANCE TABLE (THREE-FACTOR EXPERIMENT)

The sum of squares, degrees of freedom, mean squares, and related statistics for a three-way analysis of variance are usually summarized in an analysis of variance table such as the following.

SOURCE	DF	SUM OF SQUARES	MEAN SQUARE	F VALUE	PR > F
FACTOR A	$a - 1$	SSA	$MSA = \dfrac{SSA}{a - 1}$	MSA/MSE	
FACTOR B	$b - 1$	SSB	$MSB = \dfrac{SSB}{b - 1}$	MSB/MSE	
FACTOR C	$c - 1$	SSC	$MSC = \dfrac{SSC}{c - 1}$	MSC/MSE	
A*B	$(a - 1)(b - 1)$	SS(A*B)	$MS(A*B) = \dfrac{SS(A*B)}{(a - 1)(b - 1)}$	MS(A*B)/MSE	
A*C	$(a - 1)(c - 1)$	SS(A*C)	$MS(A*C) = \dfrac{SS(A*C)}{(a - 1)(c - 1)}$	MS(A*C)/MSE	
B*C	$(b - 1)(c - 1)$	SS(B*C)	$MS(B*C) = \dfrac{SS(B*C)}{(b - 1)(c - 1)}$	MS(B*C)/MSE	
A*B*C	$(a - 1)(b - 1)(c - 1)$	SS(A*B*C)	$MS(A*B*C) = \dfrac{SS(A*B*C)}{(a - 1)(b - 1)(c - 1)}$	MS(A*B*C)/MSE	
ERROR	$abc(n - 1)$	Error SS	$MSE = \dfrac{Error\ SS}{abc(n - 1)}$		
TOTAL	$abcn - 1$	Total SS			

FIGURE 18.4
THE ANALYSIS OF VARIANCE TABLE FOR A THREE-FACTOR EXPERIMENT.

F statistic. However, in a particular application all seven hypotheses might not be of interest as in the following example where only the first four hypotheses are tested.

EXAMPLE

The computer printout for the set of three-factor sample data given in Figure 18.1 is as follows. Note that calculations were made only for the three factors and the store–price interaction, in accordance with the desires of the marketing analyst.

SOURCE	DF	SUM OF SQUARES	MEAN SQUARE	F VALUE	PR > F
STORE	2	1614.89	807.44	206.72	.0001
PRICE	2	188.72	94.36	24.16	.0001
WEEK	1	4.69	4.69	1.20	.2830
STORE*PRICE	4	21.78	5.44	1.39	.2636
ERROR	26	101.56	3.91		
TOTAL	35	1931.64			

The hypotheses of interest are as follows.

1. H_0: there is no difference in sales volume from store to store. From the printout the F value for testing this hypothesis is 206.72 with a p-value of .0001, so H_0 is soundly rejected. For multiple comparisons the MSE = 3.91 with 26 degrees of freedom is used as follows, since there are 12 observations on each store:

$$\text{LSD}_{.05} = t_{.975,26}\sqrt{\text{MSE}}\ \sqrt{\frac{2}{12}}$$

$$= 2.0555\sqrt{3.91}\ \sqrt{\frac{2}{12}}$$

$$= 1.66$$

The sample means for the stores are

Store	3	1	2
Sample mean:	18.58	25.42	34.92

Since all pairs of these means differ by more than 1.66, all population means are declared to be significantly different, that is, $\mu_3 < \mu_1 < \mu_2$.

2. H_0: there is no difference in sales volume from price to price. From the printout the F value for testing this hypothesis is 24.16 with a p-value of .0001; hence, H_0 is soundly rejected. The LSD value for multiple com-

parisons is

$$LSD_{.05} = t_{.975,26}\sqrt{MSE}\;\sqrt{\frac{2}{12}}$$

$$= 2.0555\sqrt{3.91}\;\sqrt{\frac{2}{12}}$$

$$= 1.66$$

which is the same as the previous LSD value, since 12 observations were made on each price. The sample means for price are

Price	1	2	3
Sample mean:	23.67	26.00	29.25

As before, all population means are declared to be significantly different with $\mu_1 < \mu_2 < \mu_3$.

3. H_0: there is no difference in sales volume from week to week. From the printout the F value for testing this hypothesis is 1.20 with a p-value of .2830, so H_0 is not rejected. Since there are only two weeks, multiple comparisons would not be necessary even if H_0 had been rejected.

4. H_0: there is no interaction between store and price. From the printout the F value for testing this hypothesis is 1.39 with a p-value of .2636; hence, H_0 is not rejected.

   In summary, the mean volume of sales is different from store to store, and is sensitive to price differences, but not to differences from week to week. There appears to be additivity in mean volume of sales with respect to different stores and different prices (that is, there is no interaction).

RELATIONSHIP WITH THE GENERAL LINEAR MODEL

The model used in the three-way analysis of variance can be expressed in the form of a general linear model. The model has three qualitative variables—one for each factor. Pairwise interactions between dummy variables are used to represent the two-way interactions, and products of three dummy variables are used to represent the three-way interaction. The general linear model has many more terms in it than the linear model given in this section; consequently, it is not as easy to work with. However, use of the general linear model approach to analyze the data in this section will give the same results as are obtained with an analysis of variance. If the number of observations is not the same for all combinations of factors, then the general linear model computer packages should be used, because many analysis of variance computer packages are not designed to handle unequal numbers of observations.

EXERCISES

18.1 Compute the z scores in Figure 18.1 for Week 1, Price P_1, to verify that the numbers in the first row of cells in Figure 18.2 are indeed correct.

18.2 In order to convert the data in Figure 18.1 to z scores for comparing the volume of sales for the three different prices, the sales figures need to be grouped differently than they were for Figure 18.2. Show how the groups need to be formed, so that group means and standard deviations can be found.

18.3 The z scores for comparing prices are given as follows:

P_1: -1.25, -0.50, -1.01, -1.33, -0.50, -0.50, -0.44, -1.33, -1.44, -0.17, -1.16, -0.25

P_2: -0.13, -0.50, 0.59, 1.23, -0.13, -1.25, 0.44, -0.44, -0.42, 0.08, -0.86, 0.05

P_3: 1.38, 1.00, 0.59, -0.05, 1.38, 1.00, 0.15, 1.62, 0.34, 1.61, 1.56, 0.65

Plot the e.d.f.s of these z scores to compare sales volume at the three different price levels.

18.4 Prior to the marketing of plywood manufactured by a new process, breaking strength tests are conducted and provide the results recorded below. These test results are classified according to type of glue used, type of wood used, and width of plywood. Use the computer printout on the following page to test the following hypotheses at $\alpha = .05$.

(a) H_0: there is no difference in the mean breaking strengths for the three types of glue.

(b) H_0: there is no difference in the mean breaking strengths for the two types of wood.

(c) H_0: there is no difference in the mean breaking strengths for the two widths of plywood.

(d) H_0: there is no interaction between glue and wood.

(e) H_0: there is no interaction between glue and width.

(f) H_0: there is no interaction between wood and width.

(g) H_0: there is no interaction among glue, wood, and width.

		Glue Type		
Width	Wood Type	1	2	3
$\frac{3}{8}$ in.	Pine	373	382	444
		359	405	462
		357	360	454
		368	368	453
	Fir	373	352	382
		357	335	385
		370	321	378
		368	315	402
		1	2	3
$\frac{1}{2}$ in.	Pine	436	444	409
		438	439	415
		408	434	422
		429	439	415

(CONTINUED)

		Glue Type		
Width	Wood Type	1	2	3
	Fir	352	397	459
		355	428	460
		335	406	463
		356	415	454

SOURCE	DF	SUM OF SQUARES	MEAN SQUARE	F VALUE	PR > F
GLUE	2	22926.13	11463.06	94.88	.0001
WOOD	1	10063.02	10063.02	83.29	.0001
WIDTH	1	16317.19	16317.19	135.06	.0001
GLUE*WOOD	2	1855.04	927.52	7.68	.0017
GLUE*WIDTH	2	6852.88	3426.44	28.36	.0001
WOOD*WIDTH	1	825.02	825.02	6.83	.0130
GLUE*WO*WI	2	18311.29	9155.65	75.78	.0001
ERROR	36	4349.25	120.81		
TOTAL	47	81499.81			

18.5 Apply the LSD multiple comparisons procedure with $\alpha = .05$ to the means associated with the tests of the first three hypotheses in Exercise 18.4.

18.6 To improve on the collection of delinquent accounts, a large department store randomly selects 12 delinquent accounts that have previously been in arrears and 12 delinquent accounts that have not previously been in arrears. Half of each of these sets of 12 are sent only a second notice; the other half receive a "tough letter" with the notice. Further, these billings are divided again so that half of the billings contain a postage paid return envelope and the other half contain only a return envelope. The store notes the percentage paid on each of these 24 accounts over the next 30 days. Use the computer printout to test the following hypotheses at $\alpha = .05$.

(a) H_0: there is no difference in the mean percentage paid for accounts that have previously been in arrears and those that have not been in arrears.

(b) H_0: there is no difference in the mean percentage paid for those billings with return postage paid and those without return postage paid.

(c) H_0: there is no difference in the mean percentage paid for those notices with the accompanying tough letter and those without the tough letter.

Billing	Return Postage	Account Previously in Arrears	
		Yes	No
Second Notice Only	Paid	10	100
		60	40
		80	60
	Not Paid	5	80
		0	20
		25	50
		Yes	No
Second Notice with Tough Letter	Paid	50	100
		100	100
		50	75
	Not Paid	25	100
		40	50
		80	80

SOURCE	DF	SUM OF SQUARES	MEAN SQUARE	F VALUE	PR > F
ACCT	1	4537.50	4537.50	6.27	.0235
POST	1	3037.50	3037.50	4.20	.0573
BILL	1	4266.67	4266.67	5.89	.0274
ACCT*POST	1	266.67	266.67	0.37	.5524
ACCT*BILL	1	4.17	4.17	0.01	.9405
POST*BILL	1	204.17	204.17	0.28	.6027
ACCT*POST*BILL	1	150.00	150.00	0.21	.6551
ERROR	16	11583.33	723.96		
TOTAL	23	24050.00			

18.7 Apply the LSD multiple comparisons procedure with $\alpha = .05$ to the test of the main factors in Exercise 18.6.

18.2

THE ANALYSIS OF COVARIANCE

PROBLEM SETTING 18.2

The owners of four stores belonging to a fast-food chain have obtained some sample data on the number of seconds required to serve their customers. These data are recorded in Figure 18.5.

Store 1 (Sec)	Store 2 (Sec)	Store 3 (Sec)	Store 4 (Sec)
156	38	145	56
127	40	137	115
58	33	98	108
127	51	156	128
41	101	112	30
107	37	86	28
174	50		88
146	110		

FIGURE 18.5
RESULTS OF RANDOM SAMPLES OF TIME IN SECONDS
REQUIRED TO SERVE CUSTOMERS AT FOUR STORES.

The primary question of interest to the owners is whether or not the mean service time is the same for all stores; hence, the data appear very much as in the completely randomized design of Chapter 16. However, there is additional information available with these data in the form of the amount of the purchase by each customer. These amounts are reported in Figure 18.6 along with the corresponding time to service from Figure 18.5. What should be the role, if any, of this additional information in the analysis of the data in Figure 18.5 with respect to mean service time?

BACKGROUND ON THE ANALYSIS OF COVARIANCE

Problem Setting 18.2 outlines an analysis of variance concerning one variable (service time) in the presence of a covariate (amount of purchase). The procedure for analyzing data such as these is called an **analysis of covariance,** which is really an analysis of variance and regression analysis combined. The need for an analysis of covariance will now be demonstrated using Problem Setting 18.2.

Store 1		Store 2		Store 3		Store 4	
Seconds	Amount ($)	Seconds	Amount ($)	Seconds	Amount ($)	Seconds	Amount ($)
156	4.73	38	1.27	145	4.36	56	0.89
127	5.23	40	2.45	137	3.52	115	1.56
58	1.84	33	0.87	98	2.89	108	1.33
127	2.89	51	1.45	156	4.76	128	2.31
41	0.37	101	1.67	112	1.84	30	0.63
107	3.66	37	0.63	86	2.22	28	0.63
174	4.75	50	0.92			88	1.87
146	4.28	110	2.33				

FIGURE 18.6
CORRESPONDING AMOUNTS OF PURCHASES FOR THE TIMES TO SERVICE GIVEN IN
FIGURE 18.5.

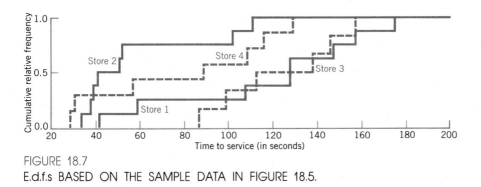

FIGURE 18.7
E.d.f.s BASED ON THE SAMPLE DATA IN FIGURE 18.5.

If the presence of the covariate (amount of purchase) is ignored for the time being, the data in Figure 18.5 can be analyzed by itself. First consider the sample mean service times for each store.

Store	1	2	3	4
Sample mean:	117.0	57.5	122.3	79.0

These sample means indicate that the mean service time at Store 2 is well below the other stores and that Store 1 and Store 3 are well above the other two stores, with Store 4 somewhere in the middle. A graph of the e.d.f.s for the data of Figure 18.5 is given in Figure 18.7 and is consistent with these observations. Additionally, a computer printout for a one-way analysis of variance for the data of Figure 18.5 is given in Figure 18.8 and confirms the observation that there is a significant difference in the mean service times from store to store. Upon completion of the LSD multiple comparisons procedure it would seem on the surface that all is well with the analysis of the data in Figure 18.5. However, consider Figure 18.6 and calculate the corresponding average amount of purchase for each store.

Store	1	2	3	4
Sample mean:	$3.47	$1.45	$3.27	$1.32

This calculation shows that Stores 1 and 3 have the largest average purchases, and previous calculations show these same two stores also have the largest

SOURCE	DF	SUM OF SQUARES	MEAN SQUARE	F VALUE	PR > F
TREATMENT	3	21217.91	7072.64	4.96	.0078
ERROR	25	35653.33	1426.13		
TOTAL	28	56871.24			

FIGURE 18.8
A COMPUTER PRINTOUT SHOWING A SIGNIFICANT DIFFERENCE IN THE MEAN SERVICE TIMES FROM FIGURE 18.5.

average service time. Does it seem reasonable to conclude that the time of service to a customer is influenced by the size of the customer's order? Probably so. In fact, the simple correlation between the pairs of observations in Figure 18.6 is .848, indicating a strong linear relationship between the two variables. The analysis of covariance is useful with data of this type because it addresses the question of differences in mean service time while taking into account the role of the covariate (amount of purchase).

THE MODELS FOR THE ANALYSIS OF COVARIANCE IN THE COMPLETELY RANDOMIZED DESIGN

The hypothesis of interest in the analysis of covariance is essentially the same as in the one-way analysis of variance for the completely randomized design with k treatments. However, in the analysis of covariance the test is conditional on a linear regression equation, with the test statistic using two error sum of squares. One error sum of squares (denoted by SSE_{PM}) comes from the simple regression of Y on X and corresponds to the partial model

$$Y_{ij} = \beta_0 + \beta_k X_{ij} + \epsilon_{ij} \tag{18.1}$$

which has intercept β_0. The second error sum of squares (denoted by SSE_{FM}) corresponds to the full regression model

$$Y_{ij} = \beta_0 + \beta_1 D_1 + \beta_2 D_2 + \cdots + \beta_{k-1} D_{k-1} + \beta_k X_{ij} + \epsilon_{ij} \tag{18.2}$$

where the D_i are dummy variables that change the intercept of the regression line. The role of the dummy variables in this model can be explained by writing the regression model from Eq. 18.2 for the analysis of covariance with the data in Figure 18.6. First the dummy variables are added to the data set given in Figure 18.6 to produce the data set given in Figure 18.9.

Figure 18.9 makes it clear that the dummy variables are used to identify the population (store) from which the sample was obtained. For example when $D_1 = D_2 = D_3 = 0$ the observations are from Store 1. Observations from Store 2 are designated with $D_1 = 1$, $D_2 = D_3 = 0$, and for Stores 3 and 4 the designations $D_1 = 0$, $D_2 = 1$, $D_3 = 0$ and $D_1 = D_2 = 0$, $D_3 = 1$ are used, respectively. The number of dummy variables is $k - 1$ (one less than the number of populations). The model corresponding to Eq. 18.2 is

$$Y_{ij} = \beta_0 + \beta_1 D_1 + \beta_2 D_2 + \beta_3 D_3 + \beta_4 X_{ij} + \epsilon_{ij} \tag{18.3}$$

The model in Eq. 18.3 really expresses four separate regressions. That is, if an observation is from Store 1, then Eq. 18.3 reduces to

$$Y_{i1} = \beta_0 + \beta_4 X_{i1} + \epsilon_{i1}$$

and for Store 2

$$Y_{i2} = (\beta_0 + \beta_1) + \beta_4 X_{i2} + \epsilon_{i2}$$

	D_1	D_2	D_3	Amount X($)	Time Y
Store 1	0	0	0	4.73	156
	0	0	0	5.23	127
	0	0	0	1.84	58
	0	0	0	2.89	127
	0	0	0	0.37	41
	0	0	0	3.66	107
	0	0	0	4.75	174
	0	0	0	4.28	146
Store 2	1	0	0	1.27	38
	1	0	0	2.45	40
	1	0	0	0.87	33
	1	0	0	1.45	51
	1	0	0	1.67	101
	1	0	0	0.63	37
	1	0	0	0.92	50
	1	0	0	2.33	110
Store 3	0	1	0	4.36	145
	0	1	0	3.52	137
	0	1	0	2.89	98
	0	1	0	4.76	156
	0	1	0	1.84	112
	0	1	0	2.22	86
Store 4	0	0	1	0.89	56
	0	0	1	1.56	115
	0	0	1	1.33	108
	0	0	1	2.31	128
	0	0	1	0.63	30
	0	0	1	0.63	28
	0	0	1	1.87	88

FIGURE 18.9
DUMMY VARIABLES ADDED TO THE DATA
IN FIGURE 18.6 FOR USE WITH THE
MODEL IN EQ. 18.2.

and for Store 3

$$Y_{i3} = (\beta_0 + \beta_2) + \beta_4 X_{i3} + \epsilon_{i3}$$

and, finally, for Store 4

$$Y_{i4} = (\beta_0 + \beta_3) + \beta_4 X_{i4} + \epsilon_{i4}$$

It is clear that these models all have the same slope β_4 (this is a basic assumption in an analysis of covariance), but possibly have different intercepts and the different intercepts correspond to different mean levels in the response variable Y. Hence, in an analysis of covariance the test for equality of intercepts (given equal slopes) is equivalent to testing for equality of population means after adjusting for the covariate. The test procedure for the analysis of covariance is now outlined. Note that it is really only an application of the procedure given in Section 15.1 for testing whether some of the parameters equal zero in a general linear model.

THE ANALYSIS OF COVARIANCE FOR USE WITH THE COMPLETELY RANDOMIZED DESIGN AND ONE COVARIATE

Data

The data consist of n_j pairs of observations (X_{ij}, Y_{ij}), $i = 1, \ldots, n_j$, from each of k populations, $j = 1, \ldots, k$.

Assumptions

1. Each Y_{ij} is a linear function of the associated value of X_{ij} through Eq. 18.2 where ϵ_{ij} is a random variable.

2. The ϵ_{ij} are a random sample from a normal population with mean 0 and variance σ^2.

Null Hypothesis

H_0: $\beta_1 = \beta_2 = \cdots = \beta_{k-1} = 0$ (i.e., the regression lines have equal intercepts given the slopes are equal). This is equivalent to testing H_0: $\mu_1 = \mu_2 = \cdots = \mu_k$ after adjusting for the covariate X.

Alternative Hypothesis

H_1: $\beta_i \neq \beta_j$ for some $i, j \leq k - 1$.

Test Statistic

$$F = \frac{(SSE_{PM} - SSE_{FM})/(k - 1)}{SSE_{FM}/(N - 1 - k)} \qquad (18.4)$$

where SSE_{PM} and SSE_{FM} are the error sums of squares associated with the models in Eqs. 18.1 and 18.2, respectively, and $N = \Sigma n_j$. Note that the degrees of freedom in the numerator is always the numerical difference between the two error degrees of freedom. The null distribution of F is the F distribution with $k - 1$ and $N - 1 - k$ degrees of freedom.

Decision Rule

Reject H_0 and conclude that the population means are different if $F > F_{1-\alpha, k-1, N-1-k}$, where $F_{1-\alpha, k-1, N-1-k}$ is the $1 - \alpha$ quantile from Table A5 with $k - 1$ and $N - 1 - k$ degrees of freedom.

Note

If the desired degrees of freedom for the F distribution cannot be found in Table A5, simply use the next smaller degrees of freedom given in the table.

The analysis of covariance is now illustrated on the sample data given in Problem Setting 18.2.

Use the sample data in Problem Setting 18.2 to test for equality of mean service times using an analysis of covariance. Let $\alpha = .05$.

First the value of SSE_{PM} corresponding to the model in Eq. 18.1 is obtained. The computer printout for this analysis is as follows.

SOURCE	DF	SUM OF SQUARES	MEAN SQUARE	F VALUE	PR > F
MODEL	1	40871.26	40871.26	68.97	.0001
ERROR	27	15999.99	592.59		
TOTAL	28	56871.24		R SQUARE .7187	

PARAMETER	ESTIMATE	T FOR H0: PARAMETER = 0	PR > |T|	STD ERROR OF ESTIMATE
INTERCEPT	31.691	3.68	.0010	8.607
X	25.884	8.30	.0001	3.117

From this analysis the value of SSE_{PM} is seen to be 15,999.99. Next the value of SSE_{FM}, corresponding to the model in Eq. 18.3, is obtained from the computer printout for that model.

SOURCE	DF	SUM OF SQUARES	MEAN SQUARE	F VALUE	PR > F
MODEL	4	43582.79	10895.70	19.68	.0001
ERROR	24	13288.45	553.69		
TOTAL	28	56871.24		R SQUARE .7663	

PARAMETER	ESTIMATE	T FOR H0: PARAMETER = 0	PR > |T|	STD ERROR OF ESTIMATE
INTERCEPT	25.3027	1.52	.1418	16.65
AMOUNT	26.4353	6.36	.0001	4.16
D1	-6.1008	-0.42	.6768	14.46
D2	10.7195	0.84	.4083	12.74
D3	18.8783	1.25	.2237	15.11

From this analysis the value of SSE_{FM} is seen to be 13,288.45. Next the test statistic is calculated from Eq. 18.4 as

$$F = \frac{(15{,}999.99 - 13{,}288.45)/(4 - 1)}{13{,}288.45/(29 - 1 - 4)} = 1.63$$

Decision Rule

From Table A5 $F_{.95,3,24} = 3.009$, so H_0 is not rejected. The p-value is between .10 and .25. Therefore, there is insufficient evidence to conclude that mean service time, when adjusted for the amount of the purchase, differs from store to store.

A COMPARISON OF THE ANALYSIS OF COVARIANCE RESIDUALS

The conclusion in the previous example may come as a surprise, since it is a direct contradiction of the conclusion given for the one-way analysis of variance in Figure 18.8. However, what the analysis of covariance correctly points out is that if the sample data in Figure 18.6 are adjusted for the amount of the purchase, then there is no difference in the mean service times. The simple one-way analysis of variance does not provide for such an adjustment.

One way to see the effect of the analysis of covariance adjustment is to consider the residuals associated with the simple linear model

$$\hat{Y} = 31.691 + 25.884X \qquad (18.5)$$

which is obtained from the first computer printout in the previous example for the model in Eq. 18.1. The residuals from this estimated regression equation are given in Figure 18.10. A one-way analysis of variance computed on these residuals is an alternative method for conducting an analysis of covariance in the example.

Next the e.d.f.s of the residuals are plotted for each store. These graphs are given in Figure 18.11. When the e.d.f.s in Figure 18.11 are compared with the e.d.f.s in Figure 18.7, it is apparent that the analysis of covariance has

Store 1	Store 2	Store 3	Store 4
1.9	−26.6	0.5	1.3
−40.1	−55.1	14.2	42.9
−21.3	−21.2	−8.5	41.9
20.5	−18.2	1.1	36.5
−0.3	26.1	32.7	−18.0
−19.4	−11.0	−3.2	−20.0
19.4	−5.5		7.9
3.5	18.0		

FIGURE 18.10
RESIDUALS FROM EQ. 18.5 WHEN APPLIED TO THE DATA IN FIGURE 18.6.

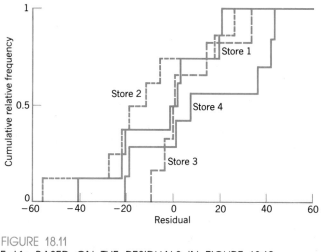

FIGURE 18.11
E.d.f.s BASED ON THE RESIDUALS IN FIGURE 18.10.

adjusted the sample service times for the amount of the purchase. The overlap of the e.d.f.s in Figure 18.11 makes it clear that there really is not any difference in the mean service times for the four stores.

WHEN SHOULD AN ANALYSIS OF COVARIANCE BE USED

An analysis of covariance applies specifically if both qualitative and quantitative supplementary data are available, and the objective of the analysis is to see if one of the qualitative variables is contributing a significant amount to the model. It is like an analysis of variance, except that additional information in the form of a quantitative variable is available on the variable being analyzed. In this section, only one independent variable is involved in the model, but it is easy to see how this can be extended to consider information on several independent variables and several additional factors in the experimental design.

EXERCISES

18.8 The productivity rates (Y) of four randomly selected workers were sampled for each of five different machines. These data are reported here along with the number of years (X) that each worker has been with the company. Analyze these data using an analysis of covariance to see if the production rate is the same for all machines. The computer printouts are given on the next page. Let $\alpha = .05$.

Machine 1		Machine 2		Machine 3		Machine 4		Machine 5	
Rate	Years	Rate	Years	Rate	Years	Rate	Years	Rate	Years
21	2	24	3	29	7	33	11	37	18
21	2	23	3	28	7	29	7	34	13
19	1	23	2	25	5	28	6	31	8
17	1	18	1	23	4	27	5	31	9

SOURCE	DF	SUM OF SQUARES	MEAN SQUARE	F VALUE	PR > F
MODEL	1	531.47	531.47	172.43	.0001
ERROR	18	55.48	3.08		
TOTAL	19	586.95		R SQUARE	
				.9055	

PARAMETER	ESTIMATE	T FOR H0: PARAMETER = 0	PR > |T|	STD ERROR OF ESTIMATE
INTERCEPT	19.248	29.61	.0001	0.650
X	1.183	13.13	.0001	0.090

SOURCE	DF	SUM OF SQUARES	MEAN SQUARE	F VALUE	PR > F
MODEL	5	557.81	111.56	53.59	.0001
ERROR	14	29.14	2.08		
TOTAL	19	586.95		R SQUARE	
				.9504	

PARAMETER	ESTIMATE	T FOR H0: PARAMETER = 0	PR > |T|	STD ERROR OF ESTIMATE
INTERCEPT	18.181	24.07	.0001	0.755
YEARS	0.879	5.89	.0001	0.149
D1	1.840	1.79	.0946	1.026
D2	3.013	2.51	.0251	1.202
D3	4.694	3.52	.0034	1.334
D4	4.517	2.41	.0301	1.871

18.9 Perform a one-way analysis of variance on the production rates for each machine in Exercise 18.8 and compare results with the analysis of covariance in that exercise. Let $\alpha = .05$. The computer printout is as follows:

SOURCE	DF	SUM OF SQUARES	MEAN SQUARE	F VALUE	PR > F
TREATMENT	4	485.70	121.43	17.99	.0001
ERROR	15	101.25	6.75		
TOTAL	19	586.95			

18.10 Plot the e.d.f.s for the sample production rates for each machine in Exercise 18.8 as was done in Figure 18.7. Compare these e.d.f.s with the e.d.f.s based on the following analysis of covariance residuals as was done in Figure 18.11.

Machine 1 $-0.61, -0.61, -1.43, -3.43$

Machine 2 $1.20, 0.20, 1.39, -2.43$ (CONTINUED)

Machine 3 1.47, 0.47, −0.16, −0.98

Machine 4 0.74, 1.47, 1.65, 1.84

Machine 5 −3.54, −0.63, 2.29, 1.11

18.11 A comparison of four diets records the number of pounds lost by each patient (Y) and the initial weight of each patient (X). These data are recorded here. Use an analysis of covariance to analyze these data for differences in the mean number of pounds lost for each method of diet. The computer printouts are given also. Let $\alpha = .05$.

Diet 1		Diet 2		Diet 3		Diet 4	
Y	X	Y	X	Y	X	Y	X
20	215	14	230	24	206	31	251
17	220	21	219	22	239	22	223
10	180	10	221	25	188	20	250
11	206	16	211	25	240	25	243
23	217	10	230	20	210	18	192
19	201	15	220	19	222	26	239

SOURCE	DF	SUM OF SQUARES	MEAN SQUARE	F VALUE	PR > F
MODEL	1	110.95	110.95	3.96	.0591
ERROR	22	616.01	28.00		
TOTAL	23	726.96		R SQUARE	
				.1526	

PARAMETER	ESTIMATE	T FOR H0: PARAMETER = 0	PR > |T|	STD ERROR OF ESTIMATE
INTERCEPT	−6.420	−0.50	.6253	12.962
X	0.117	1.99	.0591	0.059

SOURCE	DF	SUM OF SQUARES	MEAN SQUARE	F VALUE	PR > F
MODEL	4	407.03	101.76	6.04	.0026
ERROR	19	319.93	16.84		
TOTAL	23	726.96		R SQUARE	
				.5599	

PARAMETER	ESTIMATE	T FOR H0: PARAMETER = 0	PR > |T|	STD ERROR OF ESTIMATE
INTERCEPT	−0.612	−0.05	.9567	11.123
YEARS	0.084	1.57	.1326	0.053
D1	−3.616	−1.44	.1653	2.506
D2	4.913	2.01	.0585	2.440
D3	4.783	1.73	.0990	2.758

18.12 Perform a one-way analysis of variance on the weight losses for each diet in Exercise 18.11 and compare results with the analysis of covariance in the exercise. Let $\alpha = .05$. The computer printout follows:

SOURCE	DF	SUM OF SQUARES	MEAN SQUARE	F VALUE	PR > F
TREATMENT	3	365.46	121.82	6.74	.0025
ERROR	20	361.50	18.08		
TOTAL	23	726.96			

18.13 In Exercise 16.3 the yield in bushels per acre (Y) was reported for each of four methods of growing corn. These data are repeated here along with the number of inches of irrigation water (X) used. Analyze these data with an analysis of covariance on the mean yields for each method of growing corn and compare your results with those obtained in Exercise 16.11 where a one-way analysis of variance was used. See Exercise 18.20 for another way of analyzing these same data. Let $\alpha = .05$.

Method 1		Method 2		Method 3		Method 4	
Y	X	Y	X	Y	X	Y	X
83	22	91	27	101	50	78	20
91	27	90	27	100	46	82	22
94	31	81	22	91	28	81	22
89	26	83	22	93	30	77	20
89	26	84	23	96	34	79	21
96	34	83	22	95	32	81	21
91	28	88	25	94	31	80	21
92	29	91	27			81	21
90	27	89	26				
		84	23				

SOURCE	DF	SUM OF SQUARES	MEAN SQUARE	F VALUE	PR > F
MODEL	1	1112.17	1112.17	141.59	.0001
ERROR	32	251.36	7.85		
TOTAL	33	1363.53		R SQUARE	
				.8157	

PARAMETER	ESTIMATE	T FOR H0: PARAMETER = 0	PR > |T|	STD ERROR OF ESTIMATE
INTERCEPT	64.685	32.21	.0001	2.008
X	0.864	11.90	.0001	0.073

SOURCE	DF	SUM OF SQUARES	MEAN SQUARE	F VALUE	PR > F
MODEL	4	1252.02	313.00	81.40	.0001
ERROR	29	111.51	3.85		
TOTAL	33	1363.53		R SQUARE	
				.9182	

PARAMETER	ESTIMATE	T FOR H0: PARAMETER = 0	PR > |T|	STD ERROR OF ESTIMATE
INTERCEPT	73.328	31.17	.0001	2.353
WATER	0.620	7.62	.0001	0.081
D1	−2.061	−2.19	.0369	0.942
D2	0.148	0.12	.9017	1.187
D3	−6.477	−5.88	.0001	1.101

. 18.3
PROCEDURES FOR USE WITH GENERAL POPULATIONS (OPTIONAL)

THE THREE-FACTOR ANALYSIS FOR GENERAL POPULATIONS

Since the three-factor analysis of variance of Section 18.1 is just an extension of the two-factor analysis, the same assumptions apply to both analyses. However, there is no corresponding nonparametric three-factor procedure such as the Friedman test for two factors. But there is the possibility of using the methods of Section 18.1 to analyze the ranks of data instead of analyzing the original data when normality assumptions are not satisfied. This is a procedure that has been investigated by the authors and others in a number of papers related to the analysis of experimental designs (including the two-factor analysis of variance). The following example demonstrates the results of performing the three-factor analysis on the ranks of the data.

EXAMPLE

The example in Section 18.1 will be reworked after rank-transforming the data. The data in Figure 18.1 are first replaced by their corresponding ranks from 1 to 36, using average ranks where ties are present. These ranks are given in Figure 18.12. The computer printout for the analysis on the ranks is given on the next page.

The results of the analysis on ranks are in agreement with the analysis on the original data in Section 18.1 for all four tests of hypotheses and multiple comparisons. Thus, they are not repeated here. However, if outliers were present in the data, the results probably would not be the same, and in that case the analysis on ranks would be preferred.

Week	Price	Store		
		S_1	S_2	S_3
1	P_1	12.5	26.5	4.5
		16.0	24.5	4.5
	P_2	19.0	32.0	6.5
		16.0	35.0	1.5
	P_3	23.0	32.0	12.5
		22.0	28.5	10.0
2	P_1	16.0	24.5	1.5
		10.0	30.0	6.5
	P_2	21.0	28.5	3.0
		16.0	32.0	8.0
	P_3	20.0	34.0	16.0
		26.5	36.0	10.0

FIGURE 18.12
RANKS OF THE DATA GIVEN IN
FIGURE 18.1.

SOURCE	DF	SUM OF SQUARES	MEAN SQUARE	F VALUE	PR > F
STORE	2	3245.38	1622.69	213.69	.0001
PRICE	2	365.79	182.90	24.09	.0001
WEEK	1	4.69	4.69	0.62	.4388
STORE*PRICE	4	54.21	13.55	1.78	.1622
ERROR	26	197.43	7.59		
TOTAL	35	3867.50			

Store 1		Store 2		Store 3		Store 4	
R_Y	R_X	R_Y	R_X	R_Y	R_X	R_Y	R_X
27.5	26	5	8	25	25	10	6
21.5	29	6	19	24	22	20	11
11	13.5	3	5	14	20.5	17	9
21.5	20.5	9	10	27.5	28	23	17
7	1	15	12	19	13.5	2	3
16	23	4	3	12	16	1	3
29	27	8	7			13	15
26	24	18	18				

FIGURE 18.13
RANKS OF THE SAMPLE ANALYSIS OF
COVARIANCE DATA GIVEN IN FIGURE 18.5.

ANALYSIS OF COVARIANCE FOR GENERAL POPULATIONS

If the normality assumption cannot be satisfied for the analysis of covariance, then replace each of the quantitative variables with its corresponding rank and use the general linear model approach with dummy variables as presented in Eqs. 18.1 and 18.2 of the previous section. The test statistic is based on the rank analogue of Eq. 18.4. The rank-transform procedure is now stated formally.

ANALYSIS OF COVARIANCE FOR GENERAL POPULATIONS

Data

The data consist of n_j pairs of observations (X_{ij}, Y_{ij}), $i = 1, \ldots, n_j$, from each of k populations, $j = 1, 2, \ldots, k$. Replace the X_{ij} with their corresponding ranks from 1 to N. Replace the Y_{ij} with their corresponding ranks from 1 to N.

Use the method of average ranks in case of ties. $\left(N = \sum_{j=1}^{k} n_j \right)$

Assumptions

The k samples (X_{ij}, Y_{ij}), $j = 1, \ldots, k$ are random samples from each of k populations.

Null Hypothesis

The conditional means of Y given X are equal for all populations.

Alternative Hypothesis

The conditional means of Y given X are not equal for some populations.

Test Statistic

$$F = \frac{(SSE_{PM} - SSE_{FM})/(k - 1)}{SSE_{FM}/(N - 1 - k)} \qquad (18.6)$$

where SSE_{PM} and SSE_{FM} are the error sums of squares associated, respectively, with the models in Eqs. 18.1 and 18.2 based on rank-transformed data. The null distribution of the test statistic is approximately the F distribution with $(k - 1)$ and $(N - 1 - k)$ degrees of freedom.

Decision Rule

As an approximate test, reject H_0 and conclude that the conditional distributions of Y on X are not identical if $F > F_{1-\alpha, k-1, N-1-k}$, where $F_{1-\alpha, k-1, N-1-k}$ is the $1 - \alpha$ quantile from Table A5 with $(k - 1)$ and $(N - 1 - k)$ degrees of freedom.

Note

If the desired degrees of freedom for the F distribution cannot be found in Table A5, simply use the next smaller degrees of freedom given in the table.

SOURCE	DF	SUM OF SQUARES	MEAN SQUARE	F VALUE	PR > F
MODEL	1	1429.53	1429.53	64.39	.0001
ERROR	27	599.47	22.20		
TOTAL	28	2029.00			

FIGURE 18.14
COMPUTER PRINTOUT FOR THE MODEL IN EQ. 18.1 APPLIED TO RANKS.

SOURCE	DF	SUM OF SQUARES	MEAN SQUARE	F VALUE	PR > F
MODEL	4	1516.39	379.10	17.75	.0001
ERROR	24	512.61	21.36		
TOTAL	28	2029.00		R SQUARE	
				.7474	

PARAMETER	ESTIMATE	T FOR H0: PARAMETER = 0	PR > |T|	STD ERROR OF ESTIMATE
INTERCEPT	3.276	1.02	.3200	3.226
AMOUNT	0.813	5.99	.0001	0.136
D1	−3.107	−1.15	.2607	2.697
D2	0.042	0.02	.9868	2.496
D3	1.579	0.55	.5841	2.845

FIGURE 18.15
COMPUTER PRINTOUT FOR THE MODEL IN EQ. 18.3 APPLIED TO RANKS.

EXAMPLE

The example of the previous section will be reworked using the rank-transformation procedure. First the data in Figure 18.6 are replaced by their corresponding ranks where Y is the time and X is the amount. These ranks are given in Figure 18.13. The printout for the model in Eq. 18.1 applied to ranks is given in Figure 18.14. This output is used to find $SSE_{PM} = 599.47$. Next the value of SSE_{FM} corresponding to the model in Eq. 18.3 applied ranks is obtained from the computer prin'   ... uiat model in Figure 18.15. From the analysis in Figure 18.15 the value of SSE_{FM} is seen to be 512.61. Then the test statistic from Eq. 18.6 is calculated as

$$F = \frac{(599.47 - 512.61)/(4 - 1)}{512.61/(29 - 1 - 4)} = 1.36$$

From Table A5, $F_{.95,3,24} = 3.009$, so H_0 is not rejected. The p-value is greater than .25. Therefore, as with the analysis of the original data in the previous section, there is insufficient evidence to conclude that the mean service time, when adjusted for the amount of purchase, differs from store to store.

EXERCISES

18.14 Rework Exercise 18.4 after replacing the data with ranks and compare your results with those obtained on that exercise. The computer output is as follows.

SOURCE	DF	SUM OF SQUARES	MEAN SQUARE	F VALUE	PR > F
GLUE	2	2882.63	1441.31	131.61	.0001
WOOD	1	1102.08	1102.08	100.63	.0001
WIDTH	1	1564.08	1564.08	142.82	.0001
GLUE*WOOD	2	251.04	125.52	11.46	.0001
GLUE*WIDTH	2	754.54	377.27	34.45	.0001
WOOD*WIDTH	1	60.75	60.75	5.55	.0241
GLUE*WO*WI	2	2194.63	1097.31	100.20	.0001
ERROR	36	394.25	10.95		
TOTAL	47	9204.00			

18.15 Apply the LSD multiple comparisons procedure with $\alpha = .05$ to the means associated with the tests of the first three hypotheses in Exercise 18.14 and compare results with those obtained in Exercise 18.5.

18.16 Rework Exercise 18.6 after replacing the data with ranks and compare your results with those obtained in that exercise. The computer output is as follows.

SOURCE	DF	SUM OF SQUARES	MEAN SQUARE	F VALUE	PR > F
ACCT	1	192.67	192.67	5.21	.0364
POST	1	140.17	140.17	3.79	.0692
BILL	1	176.04	176.04	4.76	.0443
ACCT*POST	1	13.50	13.50	0.37	.5541
ACCT*BILL	1	0.38	0.38	0.01	.9210
POST*BILL	1	9.38	9.38	0.25	.6214
ACCT*POST*BILL	1	5.04	5.04	0.14	.7167
ERROR	16	591.33	36.96		
TOTAL	23	1128.50			

18.17 Apply the LSD multiple comparisons procedure with $\alpha = .05$ to the analysis in Exercise 18.16 and compare results with those obtained in Exercise 18.7.

18.18 Use the rank transform procedure on the sample data in Exercise 18.8 and compare results with the analysis of covariance in that exercise. The computer output appears below and on the next page.

SOURCE	DF	SUM OF SQUARES	MEAN SQUARE	F VALUE	PR > F
MODEL	1	644.07	644.07	684.96	.0001
ERROR	18	16.93	0.94		
TOTAL	19	661.00			

SOURCE	DF	SUM OF SQUARES	MEAN SQUARE	F VALUE	PR > F
MODEL	5	650.14	130.03	167.61	.0001
ERROR	14	10.86	0.78		
TOTAL	19	661.00		R SQUARE .9836	

PARAMETER	ESTIMATE	T FOR H0: PARAMETER = 0	PR > |T|	STD ERROR OF ESTIMATE
INTERCEPT	−0.294	−0.54	.5998	0.547
YEARS	1.012	10.93	.0001	0.093
D1	0.975	1.50	.1557	0.650
D2	−0.729	−0.73	.4757	0.995
D3	0.499	0.44	.6644	1.126
D4	0.070	0.05	.9632	1.481

18.19 Use the rank-transform procedure on the sample data in Exercise 18.11 and compare results with the analysis of covariance in that exercise. The computer output is as follows.

SOURCE	DF	SUM OF SQUARES	MEAN SQUARE	F VALUE	PR > F
MODEL	1	135.05	135.05	2.95	.1000
ERROR	22	1007.95	45.82		
TOTAL	23	1143.00			

SOURCE	DF	SUM OF SQUARES	MEAN SQUARE	F VALUE	PR > F
MODEL	4	620.12	155.03	5.63	.0037
ERROR	19	522.88	27.52		
TOTAL	23	1143.00		R SQUARE .5425	

PARAMETER	ESTIMATE	T FOR H0: PARAMETER = 0	PR > |T|	STD ERROR OF ESTIMATE
INTERCEPT	7.706	3.07	.0063	2.513
WEIGHT	0.221	1.18	.2543	0.188
D1	−4.499	−1.38	.1835	3.259
D2	6.554	2.08	.0514	3.153
D3	6.091	1.67	.1122	3.657

18.20 Use the rank-transform procedure to analyze the data in Exercise 18.13 and compare results with the analysis of covariance in that exercise.

How do you explain the vastly different conclusions for these two analyses (a scatterplot of the data in Exercise 18.13 may aid in the explanation)? The computer output is as follows.

SOURCE	DF	SUM OF SQUARES	MEAN SQUARE	F VALUE	PR > F
MODEL	1	3194.63	3194.63	1797.57	.0001
ERROR	32	56.87	1.78		
TOTAL	33	3251.50			

SOURCE	DF	SUM OF SQUARES	MEAN SQUARE	F VALUE	PR > F
MODEL	4	3196.16	799.04	418.71	.0001
ERROR	29	55.34	1.91		
TOTAL	33	3251.50		R SQUARE .9830	

PARAMETER	ESTIMATE	T FOR H0: PARAMETER = 0	PR > |T|	STD ERROR OF ESTIMATE
INTERCEPT	0.576	0.47	.6412	1.224
YEARS	0.976	18.75	.0001	0.052
D1	0.034	0.05	.9633	0.725
D2	−0.078	−0.10	.9237	0.811
D3	−0.644	−0.58	.5632	1.102

18.4

REVIEW EXERCISES

18.21 Three factors are considered in a study of employee productivity where productivity is measured in pieces produced per hour. One factor is the piece rate, that is, the price paid to the worker for each piece produced; another factor is the shift worked; and the last factor is the years of experience of the worker. Four randomly selected observations are made on each worker and are recorded here. Use the computer printout to analyze these data and test the following hypotheses. Let $\alpha = .05$.

(a) H_0: there is no difference in worker productivity for the three piece rates.

(b) H_0: there is no difference in worker productivity for the three shifts.

(c) H_0: there is no difference in worker productivity for the two levels of experience.

(d) H_0: there is no interaction between piece rate and shift worked.

(e) H_0: there is no interaction between piece rate and years of experience.

Years Experience	Shift Worked	Piece Rate 1	2	3
Less than one	Swing	9	9	11
		8	8	9
		9	9	9
		8	8	9
	Night	9	9	9
		9	10	10
		11	10	9
		8	9	9
	Day	8	9	10
		7	8	8
		8	8	9
		9	9	8

Years of Experience	Shift Worked	Piece Rate 1	2	3
More than one	Swing	8	9	10
		10	11	10
		6	10	12
		7	12	12
	Night	12	11	12
		10	10	12
		9	12	12
		11	13	13
	Day	8	10	13
		13	14	14
		12	12	14
		12	14	13

SOURCE	DF	SUM OF SQUARES	MEAN SQUARE	F VALUE	PR > F
PIECE	2	27.69	13.85	10.31	.0002
SHIFT	2	19.53	9.76	7.27	.0016
YEARS	1	98.00	98.00	72.99	.0001
PIECE*SHIFT	4	3.22	0.81	0.60	.6642
PIECE*YEARS	2	11.08	5.54	4.13	.0215
SHIFT*YEARS	2	29.08	14.54	10.83	.0001
PIECE*SHIFT*YR	4	2.83	0.71	0.53	.7159
ERROR	54	72.50	1.34		
TOTAL	71	263.94			

(f) H_0: there is no interaction between shift worked and years of experience.

(g) H_0: there is no interaction among piece rate, shift worked, and years of experience.

18.22 Apply the LSD multiple comparisons procedure with $\alpha = .05$ to the means associated with the tests of the first three hypotheses in Exercise 18.21.

18.23 Rework Exercise 18.21 after replacing the data with ranks and compare your results with those obtained on that exercise. The computer printout is as follows.

SOURCE	DF	SUM OF SQUARES	MEAN SQUARE	F VALUE	PR > F
PIECE	2	3328.40	1664.20	10.18	.0002
SHIFT	2	1934.02	967.01	5.92	.0048
YEARS	1	11577.35	11577.35	70.83	.0001
PIECE*SHIFT	4	539.77	134.94	0.83	.5147
PIECE*YEARS	2	842.38	421.19	2.58	.0853
SHIFT*YEARS	2	2511.55	1255.77	7.68	.0012
PIECE*SHIFT*YR	4	445.91	111.48	0.68	.6074
ERROR	54	8826.13	163.45		
TOTAL	71	30005.50			

18.24 Apply the LSD multiple comparisons procedure with $\alpha = .05$ to the analysis in Exercise 18.23 and compare the results with those obtained in Exercise 18.22.

18.25 Which observations in Exercise 18.21 need to be grouped together in order to convert the data to z scores for comparing worker productivity based on piece rates? Plot these z scores, which are given below, in e.d.f.s to compare worker productivity at each piece rate.

P_1 0.20, −1.00,   0.20, −1.00, −0.43, −0.43,   2.14, −1.71, −0.53, −1.79, −0.53,   0.74, −0.89,   0.13, −1.91, −1.40, 0.47, −1.14, −1.95, −0.34, −2.41,   0.32, −0.23, −0.23

P_2 0.20, −1.00,   0.20, −1.00, −0.43,   0.86,   0.86, −0.43, 0.74, −0.53, −0.53,   0.74, −0.38,   0.64,   0.13,   1.15, −0.34, −1.14,   0.47,   1.28, −1.32,   0.86, −0.23,   0.86

P_3 −2.60,   0.20,   0.20,   0.20, −0.43,   0.86, −0.43, −0.43, 2.00, −0.53,   0.74, −0.53,   0.13,   0.13,   1.15,   1.15, 0.47,   0.47,   0.47,   1.28,   0.32,   0.86,   0.86,   0.32

18.26 Starting salaries (in thousands of dollars) are recorded on the next page for recent university graduates from three universities. Use the computer printout to analyze these data and test the following hypotheses.

(a) H_0: there is no difference in starting salaries for the three universities.

(b) H_0: there is no difference in starting salaries for the three fields of study.

(c) H_0: there is no difference in starting salaries for students with and without honors.

(d) H_0: there is no interaction between university attended and major field of study.

		University		
Honors	Major Field	1	2	3
Yes	Education	16.2	15.0	14.4
		16.5	15.6	14.1
		16.2	16.2	12.0
	Accounting	19.8	19.8	17.4
		20.4	19.8	19.8
		20.4	18.6	17.7
	Chemical engineering	22.2	21.0	20.4
		22.5	21.6	20.1
		22.2	22.2	18.0

		University		
Honors	Major Field	1	2	3
No	Education	15.3	15.3	15.3
		15.3	14.4	14.7
		15.6	14.1	15.0
	Accounting	18.0	14.7	14.4
		19.2	14.4	12.3
		17.7	14.7	12.0
	Chemical engineering	20.4	21.6	14.7
		20.7	21.0	15.0
		21.0	21.3	16.5

SOURCE	DF	SUM OF SQUARES	MEAN SQUARE	F VALUE	PR > F
UNIV	2	89.90	44.95	23.84	.0001
MAJOR	2	232.24	116.12	61.59	.0001
HONORS	1	57.04	57.04	30.26	.0001
UNIV*MAJOR	4	23.29	5.82	3.09	.0252
ERROR	44	82.95	1.89		
TOTAL	53	485.43			

18.27 Apply the LSD multiple comparisons procedure with $\alpha = .05$ to the first three tests of hypothesis in Exercise 18.26.

18.28 Rework Exercise 18.26 after replacing the data with ranks and compare your results with those obtained in that exercise. The computer printout is as follows.

SOURCE	DF	SUM OF SQUARES	MEAN SQUARE	F VALUE	PR > F
UNIV	2	2672.58	1336.29	26.26	.0001
MAJOR	2	6075.03	3037.51	59.69	.0001
HONORS	1	1514.74	1514.74	29.76	.0001
UNIV*MAJOR	4	579.39	144.85	2.85	.0350
ERROR	44	2239.26	50.89		
TOTAL	53	13081.00			

18.29 Apply the LSD multiple comparisons procedure with $\alpha = .05$ to the first three tests of hypotheses in Exercise 18.28 and compare results with those obtained in Exercise 18.27.

18.30 Three different methods of training are to be compared. The subjects selected for the experiment are all given a common aptitude test and the score (X) is recorded. After the training, all subjects are given a common achievement test over the training material and the score (Y) is noted. Use an analysis of covariance to analyze the sample data given here to see whether differences exist in the mean achievement test scores for the three methods. The computer printout is given below and at the top of the next page.

Method 1		Method 2		Method 3	
Y	X	Y	X	Y	X
79	64	92	80	80	65
93	86	95	83	94	61
93	81	97	89	85	92
77	77	79	76	78	64
68	65	84	93	76	65
92	67	76	81	89	92
		73	68	86	76
		68	64		

SOURCE	DF	SUM OF SQUARES	MEAN SQUARE	F VALUE	PR > F
MODEL	1	389.67	389.67	6.03	.0239
ERROR	19	1227.57	64.61		
TOTAL	20	1617.24			

SOURCE	DF	SUM OF SQUARES	MEAN SQUARE	F VALUE	PR > F
MODEL	3	444.54	148.18	2.15	.1318
ERROR	17	1172.70	68.98		
TOTAL	20	1617.24		R SQUARE	
				.2749	

PARAMETER	ESTIMATE	T FOR H0: PARAMETER = 0	PR > T	STD ERROR OF ESTIMATE
INTERCEPT	50.716	3.76	.0015	13.471
SCORE	0.449	2.53	.0217	0.178
D1	−3.325	−0.72	.4803	4.607
D2	0.226	0.05	.9615	4.621

18.31   Use the rank-transform procedure to analyze the sample data in Exercise 18.30 and compare results with the analysis of covariance on the data.

SOURCE	DF	SUM OF SQUARES	MEAN SQUARE	F VALUE	PR > F
MODEL	1	140.51	140.51	4.26	.0530
ERROR	19	626.99	33.00		
TOTAL	20	767.50			

SOURCE	DF	SUM OF SQUARES	MEAN SQUARE	F VALUE	PR > F
MODEL	3	164.90	54.97	1.55	.2379
ERROR	17	602.60	35.45		
TOTAL	20	767.50		R SQUARE	
				.2149	

PARAMETER	ESTIMATE	T FOR H0: PARAMETER = 0	PR > T	STD ERROR OF ESTIMATE
INTERCEPT	6.235	1.88	.0776	3.320
SCORE	0.481	2.14	.0468	0.224
D1	−1.921	−0.58	.5667	3.287
D2	0.623	0.19	.8532	3.316

18.32   Compute a one-way analysis of variance on the achievement test scores in Exercise 18.30 and compare your results with the analysis of covariance in that exercise. The computer output is as follows.

SOURCE	DF	SUM OF SQUARES	MEAN SQUARE	F VALUE	PR > F
MODEL	2	3.90	1.95	0.02	.9785
ERROR	18	1613.33	89.63		
TOTAL	20	1617.24			

18.33 Three different types of packaging are used on a new product. Each type of packaging is randomly assigned to five stores. The number of units sold (Y) is recorded in each store along with the number of square feet of display space (X) used to display the item. These data are recorded below. Use the rank transformation to analyze these data. The computer output for the ranks is given below.

Package 1		Package 2		Package 3	
Y	X	Y	X	Y	X
16	10	44	8	17	1
60	4	67	11	28	7
82	15	87	2	105	15
126	18	100	5	149	18
137	20	142	21	160	13

SOURCE	DF	SUM OF SQUARES	MEAN SQUARE	F VALUE	PR > F
MODEL	1	132.13	132.13	11.62	.0047
ERROR	13	147.87	11.37		
TOTAL	14	280.00			

SOURCE	DF	SUM OF SQUARES	MEAN SQUARE	F VALUE	PR > F
MODEL	3	157.05	52.35	4.68	.0242
ERROR	11	122.95	11.18		
TOTAL	14	280.00		R SQUARE .5609	

PARAMETER	ESTIMATE	T FOR H0: PARAMETER = 0	PR > T	STD ERROR OF ESTIMATE
INTERCEPT	0.091	0.04	.9710	2.446
SPACE	0.756	3.67	.0037	0.206
D1	2.615	1.20	.2538	2.172
D2	2.961	1.38	.1952	2.147

18.34 Use an analysis of covariance to analyze the data in Exercise 18.33 and compare results with the rank-transformation procedure used in that exercise.

SOURCE	DF	SUM OF SQUARES	MEAN SQUARE	F VALUE	PR > F
MODEL	1	16352.57	16352.50	12.68	.0035
ERROR	13	16769.43	1289.96		
TOTAL	14	33122.00			

SOURCE	DF	SUM OF SQUARES	MEAN SQUARE	F VALUE	PR > F
MODEL	3	18243.77	6081.26	4.50	.0272
ERROR	11	14878.23	1352.57		
TOTAL	14	33122.00			

R SQUARE
.5508

PARAMETER	ESTIMATE	T FOR H0: PARAMETER = 0	PR > T	STD ERROR OF ESTIMATE
INTERCEPT	8.371	0.32	.7577	26.462
SPACE	5.659	3.66	.0038	1.547
D1	26.436	1.10	.2955	24.069
D2	22.313	0.95	.3648	23.605

BIBLIOGRAPHY

Additional material on the topics presented in this chapter can be found in the following publications.

Conover, W. J. (1980). *Practical Nonparametric Statistics,* 2nd ed. Wiley, New York.

Conover, W. J. and Iman, R. L. (1981). "Rank Transformations as a Bridge Between Parametric and Nonparametric Statistics." *The American Statistician,* **35**(3), 124–133.

DECISION THEORY UNDER UNCERTAINTY (Optional)

Thus far in this text much effort has been expended regarding hypothesis testing where decisions are made either to reject or not reject a null hypothesis. The last two chapters of this text are devoted to the topic of decision making, but the decisions made in these chapters are not placed in the usual framework of statistical inference associated with hypothesis testing. Instead, the decision-making approaches presented in these two chapters are much more intuitive and frequently involve only basic concepts of probability. This chapter concerns only decisions made in the face of information that is currently available (prior information), whereas the next chapter is concerned with the incorporation of sample information into the decision-making process.

There are three basic categories associated with **decision theory.**

DECISION MAKING UNDER CERTAINTY

In decisions of this type there are usually several courses of action available to the decision maker, and the decision maker knows the result of each course of action. If the number of courses of action is small, then the problem of making the optimal decision may be quite easy. But as the number of courses of action increases, a computer program may be needed. This type of problem is solved by using various methods such as linear programming, which is not covered in this text, and is frequently concerned with the allocation of available manpower and resources to maximize profits. For example, in which order should a delivery van make its deliveries to minimize the total distance traveled? Or on a larger scale, how should a metropolitan mass transit system schedule their bus drivers to meet the demanded bus schedule with the fewest man-hours, under the constraints that no driver drives more than four hours without a one-hour break, no driver works more than nine hours a day, and the man-hours extend from the time the drivers check in at the central bus depot to the time they check out at the same location?

DECISION MAKING UNDER UNCERTAINTY

As in decision making under certainty, decision making under uncertainty entails several courses of action; however, each action may lead to several possible states of nature that are governed by chance and are therefore uncertain. This is the type of decision making considered in this text. It covers, for example, whether to book a particular entertainer for a local concert that may make money if many tickets are sold or may lose money otherwise. It also includes whether to open a branch store in a new location, whether to purchase a new line of products for later sale, or whether to drill for oil.

DECISION MAKING UNDER CONFLICT (GAME THEORY)

Decision making under conflict is similar to decision making under uncertainty except that the outcomes are governed by actions of "opponents" or "competitors" instead of by chance. This type of decision making is referred to as game theory and is not covered in this text. It includes everything from a simple game of chess to a strategy to follow at an important conference.

19.1

INTRODUCTION TO DECISION THEORY

PROBLEM SETTING 19.1

An investor with $100,000 to invest knows that he can invest all $100,000 in certificates of deposit (CDs) at 15% and earn $15,000 in one year, but after taxes his profit would be $9000. This investor could also invest the entire amount at 12% interest in municipal bonds and make a profit of $12,000 in one year, since interest earned from municipal bonds is tax exempt. However, while the amount of interest is guaranteed to be $12,000, the actual value of the bonds varies with the prime interest rate set by the Federal Reserve. For example, an increase of three percentage points may decrease the value of the bonds by $10,000 and his net profit would be $12,000 − $10,000 = $2000 if he sells the bonds at the end of the year. A tax adjustment of $1600 on this loss leaves a profit of $2000 + $1600 = $3600. On the other hand, a decrease of three percentage points in the prime rate may increase the value of the bonds by $10,000. The investor will have to pay $1600 in taxes on this $10,000 increase if he sells the bonds, thus leaving a profit of $12,000 +

Change in Prime Interest Rate

Investor's Action	−3%	0%	+3%
100% in CDs	$ 9,000	$ 9,000	$9,000
50% in CDs and 50% in municipal bonds	$14,700	$10,500	$6,300
100% in municipal bonds	$20,400	$12,000	$3,600

FIGURE 19.1
PAYOFF POSSIBILITIES FOR AN INVESTMENT OF $100,000.

(10,000 − $1600) = $20,400. Using another strategy, the investor could split his investment evenly between CDs and municipal bonds.

Of course, the prime interest rate is not restricted to just these three changes, and the investor could alter the splitting of the investment into any ratio. However, for simplicity only three possible changes in the prime interest rate are considered along with three investment strategies. The after-taxes position of the investor is given in Figure 19.1 for each combination of decision and change in the prime interest rate. What is the best decision for the investor?

THE PAYOFF TABLE

The summary table in Figure 19.1 shows the profit made by the investor for each combination of action by the investor and change in the prime interest rate. A table of this type is generally referred to as a **payoff table** because the entries represent the payoff for each combination of action by the investor and change in prime interest rate. The change in the prime interest rate is something that the investor has no control over and, hence, introduces the uncertainty into the decision-making process. The uncertainty categories are usually denoted as the **state of nature** to further emphasize that the investor has no control over it. The general form of the payoff table is given in Figure 19.2 where the table entries O_{ij} represent the outcome of the ith action with the jth state of nature. Of course, these payoffs could be negative representing a loss on an investment.

State of Nature

Action	s_1	s_2	$\cdots$	s_k
a_1	O_{11}	O_{12}	$\cdots$	O_{1k}
a_2	O_{21}	O_{22}	$\cdots$	O_{2k}
$\cdots$	$\cdots$	$\cdots$	$\cdots$	$\cdots$
a_m	O_{m1}	O_{m2}	$\cdots$	O_{mk}

FIGURE 19.2
GENERAL FORM OF A PAYOFF TABLE.

State of Nature
Payoffs

Action	s_1	s_2	s_3
a_1	$-\$1000$	$\$2000$	$\$3500$
a_2	$-\$2000$	$\$1500$	$\$4500$
a_3	$-\$1500$	$\$1750$	$\$3000$

FIGURE 19.3
A PAYOFF TABLE WITH AN INADMISSIBLE
ACTION.

INADMISSIBLE ACTIONS

Consider the payoff table in Figure 19.3. For each state of nature the payoffs associated with action a_1 are always higher than those corresponding payoffs associated with action a_3. Thus action a_1 is said to dominate action a_3, and a_3 is referred to as an **inadmissible** action. Inadmissible actions should always be eliminated from a payoff table (or opportunity loss table) as they represent actions that should never be taken.

INADMISSIBLE
ACTION

An action a_j is **inadmissible** if there is another action a_i that is better. That is the payoff for action a_i is greater than or equal to the payoff for action a_j for each state of nature, and the payoff for action a_i is greater than the payoff for action a_j for at least one state of nature.

THE DECISION TREE

Another way of organizing the outcomes for combinations of actions and states of nature is through the use of a decision tree. Decision trees are better suited

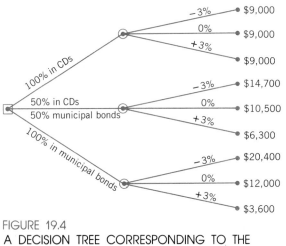

FIGURE 19.4
A DECISION TREE CORRESPONDING TO THE
PAYOFF TABLE IN FIGURE 19.1.

to problems where the number of actions and states of nature are both small or when several decisions are to be made sequentially. A decision tree is laid out in the same order in which the decision and outcome are realized. Branches in a decision tree represent several possible actions or states of nature at each point in time. Furthermore, for clarity the branches representing actions emanate from a junction denoted with a square, whereas branches representing various states of nature emanate from a circle. This is a good device for communicating with a manager.

A decision tree corresponding to the payoff table in Figure 19.1 is given in Figure 19.4. The decision tree is read from left to right with a branch for each state of nature at the end of each action branch. Thus the uppermost branch in Figure 19.4 represents investing all $100,000 in certificates of deposit and then having the prime interest rate decrease by three percentage points.

MAXIMAX AND MAXIMIN CRITERIA

Both the payoff table in Figure 19.1 and the decision tree in Figure 19.4 are convenient methods of summarizing the decision problem at hand. However, a decision must be made with respect to the best action to take. There are numerous criteria for making such a decision, and many are presented throughout the rest of this chapter and in the next. The first of these criteria is called **maximax.** The maximax criterion approach ignores all information in the payoff table except for the maximum payoff and optimistically hopes for the best state of nature to occur. This might be used when there is a lot of emotional satisfaction involved in obtaining the greatest possible payoff. However, this criterion is not as popular as some of the others introduced later in this section.

The **maximax criterion** is as follows:

MAXIMAX **1.** Find the maximum payoff for each action over all states of nature in a
CRITERION   payoff table.
2. Find the maximum of the maximum payoffs in step 1 and choose the action corresponding to this maximum of the maximums.

EXAMPLE

Find the decision based on the maximax criterion for the payoff table in Figure 19.1.

The maximum payoff for the action that invests all monies in CDs is $9000 (which occurs with each state of nature). The maximum payoff for the action that splits the investments 50–50 is $14,700 for the state of nature where the prime rate decreases by three percentage points. The maximum payoff for the action that has all monies in municipal bonds is $20,400, occurring with a decrease of three percentage points in the prime rate. Thus, the maximums for each of the three actions are, respectively, $9000, $14,700, and $20,400. The maximum of these values is $20,400 corresponding to the action that invests all monies in municipal bonds. Thus this is the decision that corresponds to the maximax criterion.

A second criterion is called the **maximin criterion.** The maximin criterion reflects more of a conservative approach than maximax because it looks for the decision that will maximize the payoff in the event that the least favorable state of nature occurs. An older person investing her life savings might prefer this approach.

The **maximin criterion** is as follows:

MAXIMIN
CRITERION

1. Find the minimum payoff for each action over all states of nature in a payoff table.

2. Find the maximum of the minimum payoffs in step 1 and choose the action corresponding to this maximum of the minimums.

EXAMPLE

Find the decision based on the maximin criterion for the payoff table in Figure 19.1.

The minimum payoff for the action that invests all monies in CDs is $9000 (occurring at all states of nature). The minimum payoff for the action to split the investments is $6300, which occurs if there is an increase of three percentage points in the prime rate. The minimum payoff for the action to invest all monies in municipal bonds is $3600, occurring when the prime rate increases three percentage points.

The three minimums are, respectively, $9000, $6300, and $3600 for the three actions. The maximum of these three values is $9000, corresponding to an investment of all monies in CDs. Thus, this is the decision that corresponds to the maximin criterion.

OPPORTUNITY LOSS

If the investor decides to invest all monies in CDs and the prime interest rate decreases by three percentage points, then the investor has lost the opportunity to make an additional profit of $20,400 − $9000 = $11,400, which would have been his had he decided to invest all monies in municipal bonds. This difference between what the investor (decision maker) could have made and what he actually made is sometimes referred to as regret, but more commonly as an **opportunity loss,** as it represents a lost opportunity to gain a larger payoff.

OPPORTUNITY
LOSS

The **opportunity loss** represents the difference between the maximum payoff a decision maker could obtain for a given state of nature and the payoff associated with the action taken by the decision maker.

A payoff table is easily converted to an opportunity loss table by finding the maximum payoff in a given column and subtracting all entries in the column from this maximum value. The opportunity loss table may be used as the basis for making a decision also. For example, the decision maker may prefer to make a decision that may not be optimal but will never be too far from being optimal. That is, the decision maker may choose the action that minimizes the maximum opportunity loss. This is called the **minimax** procedure, and is illustrated in the next example.

EXAMPLE

Convert the payoff table in Figure 19.1 to an opportunity loss table.

The entries in the first column of the payoff table in Figure 19.1 are $9000, $14,700, and $20,400, with the maximum value being $20,400. Each of these column entries is subtracted from $20,400 giving

$$\$20,400 - \$9000 = \$11,400$$
$$\$20,400 - \$14,700 = \$5700$$
$$\$20,400 - \$20,400 = \$0$$

as the respective column entries in the opportunity loss table. In like manner all entries in the second column of the payoff table of Figure 19.1 are subtracted from the maximum value of $12,000 in that column, resulting in the values

$$\$12,000 - \$9000 = \$3000$$
$$\$12,000 - \$10,500 = \$1500$$
$$\$12,000 - \$12,000 = \$0$$

for the second column of the opportunity loss table. The final column of the opportunity loss table is completed by subtracting all values in the third column of the payoff table from $9000, resulting in

$$\$9000 - \$9000 = \$0$$
$$\$9000 - \$6300 = \$2700$$
$$\$9000 - \$3600 = \$5400$$

	State of Nature Change in Prime Interest Rate		
Action	−3%	0%	3%
100% in CDs	$11,400	$3,000	0
50% in CDs 50% in municipal bonds	$ 5,700	$1,500	$2,700
100% in municipal bonds	0	0	$5,400

FIGURE 19.5
OPPORTUNITY LOSS TABLE CORRESPONDING TO THE PAYOFF TABLE OF FIGURE 19.1.

The completed opportunity loss table appears in Figure 19.5. A decision based on minimizing the maximum opportunity loss would be the decision to invest 100% in municipal bonds. The worst opportunity loss in this case is $5400, which occurs if the prime interest rate increases 3%. Other actions could result in opportunity losses of $5700 and $11,400, both occurring if the prime interest rate drops 3%.

EXERCISES

19.1 The following payoff table shows the units of profit for each state of nature of sales combined with the possible advertising strategy that may be used. Find the decision based on the maximax criterion.

Advertising Strategy	Strength of Sales		
	Good (s_1)	Average (s_2)	Poor (s_3)
a_1	27	8	-8
a_2	23	18	-4
a_3	24	18	-2
a_4	15	10	0

19.2 Find the decision based on the maximin criterion for the payoff table in Exercise 19.1.

19.3 Are there any inadmissible actions associated with the payoff table in Exercise 19.1?

19.4 After eliminating any inadmissible actions from the payoff table in Exercise 19.1, convert it to an opportunity loss table.

19.5 Formulate a decision tree for the payoff table in Exercise 19.1.

19.6 The buyer for an appliance store must decide how many air conditioners to purchase for sale during the coming summer months. The store manager feels that the number of air conditioners sold will be 30, 40, or 50. Each air conditioner costs the store $240 and will be sold for $400. Any air conditioners left over at the end of the summer will be cleared out at a price of $200. Construct the payoff table for this decision problem if the air conditioners are purchased in lots of size 10.

19.7 Find the decision based on the maximax criterion for the decision problem in Exercise 19.6.

19.8 Find the decision based on the maximin criterion for the decision problem in Exercise 19.6.

19.9 Are there any inadmissible actions associated with the decision problem in Exercise 19.6?

19.10 Convert the payoff table in Exercise 19.6 to an opportunity loss table.

19.2
ESTIMATING THE UNCERTAINTY IN THE STATES OF NATURE

SUBJECTIVE PROBABILITY

In the previous section procedures were presented for helping the decision maker organize information into a payoff table for a decision problem with regard to possible actions and states of nature whose specific occurrence is uncertain. Although it is true that the decision maker cannot control the different states of nature, it is possible that he or she has some knowledge about the likelihood of the occurrence of each state of nature. This likelihood of occurrence is often based on a **subjective probability** using the terminology of Section 4.1 where subjective probabilities were referred to as probabilities obtained from the opinions of one or more people.

The concept of subjective probability is different from the two types of probability that are used in the rest of this text. One type is by assumption, such as in hypothesis testing when the probability of a customer selecting Brand X is assumed to be .5 (when only two brands are involved) under the null hypothesis of "no preference for Brand X." Another type is by observing relative frequencies over a long period of time to estimate a probability, such as to estimate the probability of a manufactured part being defective. In this chapter a probability is needed for an event that may never have happened under similar circumstances and may never happen again. Examples include estimating the probability that a highly recruited possible employee will accept a job offer once it is made, or estimating the probability that the commercial loan interest rates will rise in the next six months. In these cases expert opinion is the most valuable source of information. But the subjective probability is just that, subjective, and the decision maker should keep this fact in mind.

The information contained in the subjective probabilities can be used to aid in making the decision. For example, suppose the investor in Problem Setting 19.1 believes that there is about a 90% chance that the prime rate will decrease by three percentage points with only about a 5% chance of occurrence for each of the other two states of nature summarized in Figure 19.1. Then the investor will strongly consider putting all monies in municipal bonds where the payoff is the highest when the prime rate decreases by three percentage points.

A critical part of using subjective probabilities is the decision maker's ability to provide good estimates of the probabilities. Such estimates would likely be based on historical data and the decision maker's subject-matter expertise. If correctly used, the information on the probabilities of the occurrence of different states of nature should provide a better basis for making a decision than the simplistic maximax, maximin, or minimax procedures of the previous section.

EXPECTED PAYOFF

If subjective probabilities have been assigned to each state of nature, then one way of making a decision is to choose the action with the **maximum expected payoff.** This criterion is useful when the amounts involved are not large relative to total assets, or when repetitive decisions are being made. Examples of repetitive decisions include routine buying and selling of stock for an investment portfolio, replenishing inventory in a clothing store where fashions change continually, or selecting new employees after interviews with several applicants. The method for finding the maximum expected payoff is now outlined.

MAXIMUM EXPECTED PAYOFF

The **maximum expected payoff** is found as follows.

1. Let $r(a_i, s_j)$ be the payoff associated with action a_i and state of nature s_j.
2. Find the expected payoff $E[r(a_i, s)]$ for each action a_i over all states of nature s using the formula

$$E[r(a_i, s)] = \sum_{j=1}^{k} r(a_i, s_j) \cdot P(s_j)$$

where $P(s_j)$ is the subjective probability assigned to state of nature s_j, $j = 1, \ldots, k$. Note that $\Sigma P(s_j) = 1$.

3. The maximum of the expected values in step 2 over all possible actions a_i, $i = 1, \ldots, m$, is designated as the maximum expected payoff.

The following example illustrates how the maximum expected payoff is obtained.

EXAMPLE

Suppose the investor in Problem Setting 19.1 has been able to assign the following subjective probabilities to each state of nature.

s_1 = prime rate decreases by three percentage points　　$P(s_1) = .1$

s_2 = prime rate remains unchanged　　$P(s_2) = .6$

s_3 = prime rate increases by three percentage points　　$P(s_3) = .3$

The payoff table of Figure 19.1 is repeated here for ready reference. Use this table and the subjective probabilities given by the investor to find the maximum expected payoff.

	State of Nature (Payoff)		
Action	s_1 (.1)	s_2 (.6)	s_3 (.3)
a_1	\$9,000	\$9,000	\$9,000
a_2	\$14,700	\$10,500	\$6,300
a_3	\$20,400	\$12,000	\$3,600

The payoffs for action a_1 (all monies are invested in CDs) are $9000, $9000, and $9000 over all three states of nature. The expected payoff for a_1 is found as

$$E[r(a_1, s)] = \$9000(.1) + \$9000(.6) + \$9000(.3)$$
$$= \$9000$$

Note the payoff for action a_1 in this problem is always fixed at $9000, so the expected payoff cannot be affected by the assignment of subjective probabilities.

For action a_2 (the investment is split equally between CDs and municipal bonds) the payoffs are $14,700, $10,500, and $6300 over the three states of nature, so the expected payoff is

$$E[r(a_2, s) = \$14,700(.1) + \$10,500(.6) + \$6300(.3)$$
$$= \$9660$$

Note that the expected payoff is higher for action a_2 than it is for action a_1. The last action a_3 (invest all monies in municipal bonds) has payoffs of $20,400, $12,000, and $3600, respectively, for the three states of nature. The expected payoff for action a_3 is

$$E[r(a_3, s)] = \$20,400(.1) + \$12,000(.6) + \$3600(.3)$$
$$= \$10,320$$

The maximum of the three expected payoffs is $10,320 associated with action a_3; thus, the decision according to the maximum expected payoff criterion is to invest all monies in municipal bonds (a_3). Note this turns out to be the same as the maximax decision.

EXPECTED OPPORTUNITY LOSS

The subjective probabilities can also be used with an opportunity loss table, such as given in Figure 19.5 of the previous section. However, since oppor-

---

The **minimum expected opportunity loss** is found as follows:

MINIMUM EXPECTED OPPORTUNITY LOSS

1. Let $OL(a_i, s_j)$ be the opportunity loss associated with action a_i and state of nature s_j.

2. Find the expected opportunity loss $E[OL(a_i, s)]$ for each action a_i as

$$E[OL(a_i, s)] = \sum_{j=1}^{k} OL(a_i, s_j) \cdot P(s_j)$$

where $P(s_j)$ is the probability assigned to state of nature s_j, $j = 1, \ldots, k$.

3. The minimum of the expected values in step 2 over all possible actions a_i, $i = 1, \ldots, m$, is designated as the minimum expected opportunity loss.

---

tunity loss represents a lost opportunity to gain a larger payoff, the decision associated with **expected opportunity loss** would be the one that minimizes the expected opportunity loss. The procedure for finding the **minimum expected opportunity loss** is given on previous page.

The calculation of the minimum expected opportunity loss is illustrated in the next example.

EXAMPLE

The opportunity loss table from Figure 19.5 is repeated here for easy reference. Use this table and the subjective probabilities of the previous example to find the decision corresponding to the minimum opportunity loss criterion.

State of Nature
(Opportunity Losses)

Action	s_1 (.1)	s_2 (.6)	s_3 (.3)
a_1	$11,400	$3,000	0
a_2	$5,700	$1,500	$2,700
a_3	0	0	$5,400

The opportunity losses for action a_1 are $11,400, $3000, and 0 for the three states of nature. The expected opportunity loss for action a_1 is

$$E[OL(a_1, s)] = \$11,400(.1) + \$3000(.6) + 0(.3)$$
$$= \$2940$$

For action a_2 the opportunity losses are $5700, $1500, and $2700 over all states of nature, and the expected opportunity loss for action a_2 is

$$E[OL(a_2, s)] = \$5700(.1) + \$1500(.6) + \$2700(.3)$$
$$= \$2280$$

Note that action a_2 has a smaller expected opportunity loss than action a_1, and in the previous example action a_2 also had a larger expected payoff than action a_1. Thus, the assignment of these particular subjective probabilities makes action a_2 preferable to action a_1 according to either the maximum expected payoff criterion or the minimum expected opportunity loss criterion. The last action a_3 has opportunity losses of 0, 0, and $5400 over all states of nature with a corresponding expected opportunity loss of

$$E[OL(a_3, s)] = 0(.1) + 0(.6) + \$5400(.3)$$
$$= \$1620$$

Thus, the three expected opportunity losses for actions a_1, a_2, and a_3 are, respectively, $2940, $2280, and $1620. The minimum of these expected

State of Nature
(Payoff)

Action	s_1	s_2	s_3
a_1	15	10	20
a_2	27	8	-8
a_3	18	24	-2

FIGURE 19.6
PAYOFF TABLE.

values is $1620, so the decision corresponding to the minimum expected opportunity loss is to invest all monies in municipal bonds (a_3). This is the same decision as that associated with the maximum expected payoff. In fact, it can be shown that the two procedures *always* result in the same decision.

THE SENSITIVITY OF THE DECISION TO THE
SELECTION OF SUBJECTIVE PROBABILITIES

The decision resulting from use of maximum expected payoff (or equivalently minimum expected opportunity loss) may be sensitive to the actual values assigned as subjective probabilities. For example, consider the hypothetical payoff table given in Figure 19.6. The assignment of the subjective probabilities $P(s_1) = .6$, $P(s_2) = .2$, and $P(s_3) = .2$ leads to the decision to take action a_2 as indicated in Figure 19.7. However, use of other subjective probabilities could easily change the result as evidenced in Figure 19.7, where the assignment of $P(s_1) = .2$, $P(s_2) = .6$, and $P(s_3) = .2$ under the maximum expected payoff criterion leads to the decision to take action a_3. Likewise in Figure 19.7, the assignment of $P(s_1) = .2$, $P(s_2) = .2$, and $P(s_3) = .6$ leads to the decision to take action a_1.

If the decision changes when the subjective probabilities change only slightly, then the decision is said to be sensitive to the choice of subjective probabilities, and great care should be taken to make the subjective proba-

Subjective Probabilities			Expected Payoffs			Maximum Expected Payoff Decision
$P(s_1)$	$P(s_2)$	$P(s_3)$	$E[r(a_1, s)]$	$E[r(a_2, s)]$	$E[r(a_3, s)]$	
.6	.2	.2	15.0	16.2	15.2	a_2
.2	.6	.2	13.0	8.6	17.6	a_3
.2	.2	.6	17.0	2.2	7.2	a_1

FIGURE 19.7
MAXIMUM EXPECTED PAYOFF DECISIONS ASSOCIATED WITH DIFFERENT CHOICES OF SUBJECTIVE PROBABILITIES FOR THE PAYOFF TABLE IN FIGURE 19.6.

bilities as accurate as possible. On the other hand, if the decision is not sensitive to the choice of subjective probabilities, uncertainty in the subjective probabilities is not such a critical factor. Any application of decision theory should be accompanied by a sensitivity analysis of the type just illustrated.

EXERCISES

19.11 Perform the necessary calculations to find the expected payoffs in Figure 19.7.

19.12 Suppose the investor in Problem Setting 19.1 chose the following subjective probabilities $P(s_1) = .01$, $P(s_2) = .60$, and $P(s_3) = .39$. What is the maximum expected payoff for this choice of probabilities and what is the corresponding decision?

19.13 Use the subjective probabilities in Exercise 19.12 to find the minimum expected opportunity loss for the investor in Problem Setting 19.1. How does the decision according to the minimum expected opportunity loss compare with the decision for the maximum expected payoff found in Exercise 19.12?

19.14 Use the subjective probabilities $P(s_1) = .1$, $P(s_2) = .1$, and $P(s_3) = .8$ to find the maximum expected payoff decision for the payoff table given in Exercise 19.1. (Be sure to see Exercise 19.3.)

19.15 Use the subjective probabilities $P(s_1) = .7$, $P(s_2) = .2$, and $P(s_3) = .1$ to find the minimum expected opportunity loss decision for the opportunity loss table given in Exercise 19.4.

19.16 Suppose the appliance dealer in Exercise 19.6 knows that in the last 10 years 30 air conditioners have been sold in 5 of those years, 40 air conditioners have been sold in 3 of those years, and 50 air conditioners have been sold in 2 of those years. Based on this information, what subjective probabilities would you assign to the sale of 30, 40, and 50 air conditioners?

19.17 Suppose the appliance dealer in Exercise 19.6 feels that the probabilities of selling 30, 40, and 50 air conditioners are .5, .3, and .2, respectively. Use these subjective probabilities to find the number of air conditioners the dealer should order according to the maximum expected payoff criteria.

19.18 Use the subjective probabilities assigned by the appliance dealer in Exercise 19.17 to find the minimum expected opportunity loss decision for the opportunity loss table of Exercise 19.10.

19.19 A fuel-conscious consumer is considering buying one of four economy model cars. He estimates that 80% of his mileage will be put on in city driving and the rest in highway mileage. The city and highway EPA mileage estimates are summarized in the payoff table below for the four models. Find the expected payoff associated with the decision to purchase each of these models. What is the optimal decision based on the maximum expected payoff?

	Mileage Estimates (mpg)	
Model	City	Highway
A	21	27
B	18	24
C	19	34
D	22	26

19.20 Convert the payoff matrix in Exercise 19.19 to an opportunity loss matrix.

19.21 For the opportunity loss matrix in Exercise 19.20, find the decision based on the minimum expected opportunity loss. Compare this with the decision in Exercise 19.19.

19.22 The consumer in Exercise 19.19 decides to buy a car for his business also. The expected mileage for business travel is about 70% on the road and 30% in the city. Does this affect the decision on which car to buy?

19.3

USING UTILITY AS THE PAYOFF

PROBLEM SETTING 19.2

The attorneys representing a large oil company believe that the plaintiff is virtually assured of winning an award of $1 million in a lawsuit filed against the company. Therefore, the attorneys advise the oil company to settle out of court and to offer the plaintiff a choice of a $1 million settlement outright or, in its place, part ownership in a lease on which the oil company plans to drill. If oil is found, the plaintiff can sell her shares back to the oil company for $5 million and if no oil is found, the lease will be worthless. The oil company assesses the probability of finding oil on the lease at .25 and the probability of a dry hole at .75. What decision should the plaintiff make regarding this out-of-court settlement?

ATTITUDES TOWARD RISK

The plaintiff in the problem setting is faced with the decision either to accept $1 million with certainty or to take a risk and face the possibility of ending up with nothing in an attempt to gain an additional $4 million. The attitude toward risk was not considered in the previous section where the decision was based on the action having the maximum expected payoff. Although maximum expected payoff may be a desirable objective, it may carry more of a risk than the decision maker is willing to take and, thus, the decision maker may choose an action other than the one giving the maximum expected payoff simply because of the lower risk.

For example, suppose you are offered a one-time choice of receiving $1 as a gift or playing a game where you receive $2 with probability .6 and $0 with probability .4. With the gift the expected payoff is $1, whereas with the game the expected payoff is $2(.6) + $0(.4) = $1.20. Clearly playing the game has the higher expected payoff, and you could easily choose to play the game because the amounts involved are not large enough to cause much concern. On the other hand, suppose only the stakes of the one-time game are changed so that you either receive a gift of $1 million or play the game where you receive $2 million with probability .6 or get $0 with probability .4. The expected payoff for not playing the game is $1 million, and the expected payoff for playing the game is $2 million(.6) + $0(.4) = $1.2 million. Playing the game still has the highest expected payoff, but the decision to play the game is not taken as lightly this time. For you can have $1 million for certain, which will be quite useful to you or, if you play the game, you have a 40% chance of going home empty-handed and a 60% chance of having $2 million. There may be a great deal you can do with $1 million, but you might not feel that the additional advantage of having $2 million is worth the risk necessary to try to obtain it.

In Problem Setting 19.2 the expected payoff for settling out of court and not taking the option on the lease is $1 million and the expected payoff with the lease is $5 million(.25) + $0(.75) = $1.25 million. Thus, even though taking the lease has the higher expected payoff, the plaintiff might not believe that the added risk is worth taking, for a chance to gain an additional $4 million.

The preceding discussion has centered on one's attitude toward risk. There are a number of factors that could affect this attitude, and present financial position would probably be one of them. For most individuals and small companies the gain from having a gift of $1 million would be preferable to a 40% chance of gaining nothing. However, a large company with corresponding large assets might prefer to play the game because of the higher expected payoff. This would correspond to a management more inclined toward risky ventures. Choosing not to play the game reflects more of a conservative outlook.

THE UTILITY FUNCTION

A term commonly used in decision theory to characterize one's willingness to take certain risks is **utility.** A graph of utility is obtained from a **utility function,** which is simply a function that provides for a quantification of how much relative value a person places on several alternatives, on a scale from 0 to 1. Thus, the willingness of an individual to take a risk (or play a game) may be obtained from the utility function for the individual if that function is known. The value 0 is assigned to the utility of the smallest payoff in the payoff matrix, which presumably is the payoff that has the least utility to the decision maker. The greatest utility is assumed to belong to the largest payoff in the payoff matrix; hence, it is given a utility of 1.0. All other possible payoffs lie between the smallest and largest payoffs; hence, they have utilities between 0 and 1.

To determine the utility of an individual payoff A, the decision maker is asked to state a preference for one of the two options:

1. Receiving the amount A with certainty.
2. Playing a game of chance, in which the maximum possible payoff is received with probability p, and the minimum possible payoff is received with probability $1 - p$, where p is given as some number.

If the decision maker prefers Option 1, the game of chance is made more attractive by increasing the value of p, and the decision maker is again asked to choose an option. If the decision maker prefers Option 2, the value of p is lowered to make that option less attractive and, again, the decision maker is given a choice. Only when the decision maker is indifferent to the two options, in other words, prefers neither one to the other, does the process end, and the utility for the amount A is determined to be the final value for p. This utility is expressed as

$$U(A) = p$$

By repeating this process for each payoff in the payoff matrix, the complete list of utilities is obtained for that individual. A utility function may be expressed graphically or as a mathematical function, as is done in the exercises.

EXAMPLE

For Problem Setting 19.2 the minimum payoff is $0 and the maximum payoff is $5 million. These utilities are expressed as

$$U(\$0) = 0$$
$$U(\$5,000,000) = 1$$

These endpoints are graphed in Figure 19.8. The list of utilities is completed by considering the only other possible payoff, $1 million. By asking the plaintiff to consider the probability p for striking oil to be .5, .3, or other values, a

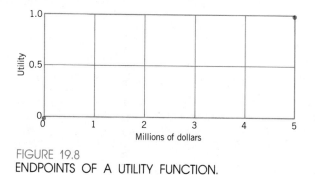

FIGURE 19.8
ENDPOINTS OF A UTILITY FUNCTION.

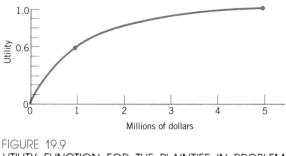

FIGURE 19.9
UTILITY FUNCTION FOR THE PLAINTIFF IN PROBLEM
SETTING 19.2.

probability is obtained for which the plaintiff is indifferent to receiving $1 million with certainty, or taking a chance on striking oil. This becomes her utility for $1 million, or $U(\$1,000,000)$. If the plaintiff is unwilling to take the oil lease with the probability of striking oil (and thus a $5 million payoff) at .25, it follows that the plaintiff associates a utility greater than .25 with the $1 million or $U(\$1,000,000) > .25$. Suppose the value of p is gradually increased until it reaches .60, where the plaintiff states that she is indifferent to taking the million dollar settlement or the lease (and thus the risk). Since $p = .60$ is the probability of striking oil for which the plaintiff is willing either to take the lease or to take the settlement, the plaintiff's utility for $1 million is .60. The point with coordinates ($1,000,000,.60) is added to the plot of Figure 19.8 and a smooth curve drawn in to complete the utility function for the plaintiff. This utility function is given in Figure 19.9.

EXPECTED UTILITY

The payoffs in a payoff table can be replaced by the utilities associated with each payoff. Then the payoff table becomes a **utility table.** The utility table is useful as a device for determining the preferred action, where the criterion for making a decision regarding which action to take is called the **maximum expected utility** criterion.

The payoff table for the decision problem in Problem Setting 19.2 is given in Figure 19.10. The expected payoffs resulting from this table have previously been calculated as $1 million and $1.25 million for not taking and taking the

	State of Nature (Payoff)	
Action	Oil is found ($p = .25$)	Hole is dry ($1 - p = .75$)
Do not take lease	$1,000,000	$1,000,000
Take lease	$5,000,000	0

FIGURE 19.10
PAYOFF TABLE FOR PROBLEM SETTING 19.2.

State of Nature
(Utility)

Action	Oil is found ($p = .25$)	Hole is dry ($1 - p = .75$)
Do not take lease	.60	.60
Take lease	1	0

FIGURE 19.11
UTILITY TABLE FOR PLAINTIFF BASED ON FIGURE 19.9.

lease, respectively. Therefore, the maximum expected payoff action is to take the lease.

An alternative to the maximum expected payoff criterion is to use the utility function to construct a utility table and, from it, find the maximum expected utility. The utility table for the plaintiff in Problem Setting 19.2 is given in Figure 19.11. The calculation of the expected utility from the utility table yields

$$E[U(a_1)] = .25(.60) + .75(.60)$$
$$= .60$$

for not taking the lease and

$$E[U(a_2)] = .25(1) + .75(0)$$
$$= .25$$

for taking the lease. The maximum of these values is .60, so the action corresponding to maximum expected utility is not to take the lease, which is in contradiction to the maximum expected payoff criterion. However, this method may explain more accurately the reasoning used by the plaintiff.

TYPES OF UTILITY FUNCTIONS

The plaintiff in Problem Setting 19.2 by not taking the action with the maximum expected payoff has demonstrated a conservative approach to risk, otherwise known as **risk aversion.** The utility function of risk averse individuals is concave downward as in Figure 19.9. By contrast, if the person assigns a

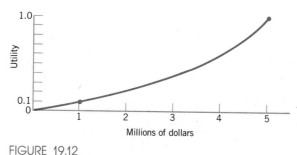

FIGURE 19.12
UTILITY FUNCTION FOR A RISK SEEKER.

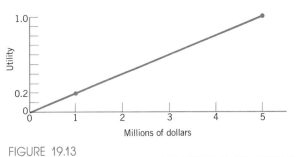

FIGURE 19.13
UTILITY FUNCTION FOR A RISK NEUTRAL INDIVIDUAL.

utility of .1 to $1 million, instead of .6 as before, then the form of the utility function changes. An individual who assigns a utility of .1 to $1 million under these conditions is referred to as a **risk seeker.** The utility function for a risk seeker is given in Figure 19.12 where it appears as convex downward (or concave upward). An individual may be a risk seeker if the larger payoff may be unusually useful, perhaps enabling the individual to pay off a particularly annoying debt, or may carry with it a glamorous fringe benefit, such as being the owner of an oil lease.

If the person assigns a utility of .2 to $1 million, then the utility function is a straight line, as in Figure 19.13, and the individual is called **risk neutral.** With such a utility function, the maximum expected utility criterion is equivalent to the maximum expected payoff criterion with regard to the preferred action.

EXERCISES

19.23 The following utility function has been developed for an individual who owns a rental condominium in a ski resort area. Let x represent the number of days the condo is rented in one season, which can range from 21 to 100 days.

$$U(x) = \frac{\ln(x - 20)}{4.382}, \qquad 21 < x < 100$$

Make a graph of this utility function and use it to determine the individual's attitude toward risk.

19.24 Suppose you have the condominium and the utility function in Exercise 19.23 and you are offered a choice of leasing for the entire season for the equivalent of 40 days' rent, or to put it into the rental pool with a probability .5 of renting 30 days and a probability .5 of renting 60 days. Construct a utility table for this decision problem.

19.25 Use the utility table in Exercise 19.24 to determine the action corresponding to the maximum expected utility.

19.26 Construct a payoff table for the decision problem in Exercise 19.24 and use it to find the action corresponding to the maximum expected payoff. How does this action compare with the one selected in Exercise 19.25?

19.27 Use the following utility table to determine the decision associated with maximum expected utility. Probabilities for the states of nature are given in parentheses.

State of Nature

Action	s_1 (.2)	s_2 (.5)	s_3 (.3)
a_1	.4	.4	.4
a_2	.3	.5	.6
a_3	0	.1	.7

19.28 Suppose your utility function is given in Figure 19.12 and you are offered an opportunity to receive $1 million with certainty or invest and receive $0 with probability .8 or $5 million with probability .2. Construct a utility table for this decision problem.

19.29 Use the utility table constructed in Exercise 19.28 to find the decision corresponding to maximum expected utility.

19.30 Suppose you are offered a chance to win $12 with probability $\frac{1}{6}$ or lose $3 with probability $\frac{5}{6}$. Would this be a fair game? What is the maximum amount of money you would be willing to pay to avoid having to play this game?

19.4

REVIEW EXERCISES

19.31 Convert the payoff table in Figure 19.3 to an opportunity loss table after eliminating the inadmissible action.

19.32 Find the decision based on the maximax criterion for the payoff table in Figure 19.3.

19.33 Find the decision based on the maximin criterion for the payoff table in Figure 19.3.

19.34 If the subjective probabilities $P(s_1) = .3$, $P(s_2) = .4$, and $P(s_3) = .3$ are assigned to the states of nature in Figure 19.3, find the decision corresponding to the maximum expected payoff criteria. How does this decision compare with those in Exercises 19.32 and 19.33? (Be sure any inadmissible actions are eliminated from Figure 19.3.)

19.35 Use the subjective probabilities in Exercise 19.34 to find the decision corresponding to the minimum expected opportunity loss criteria for Figure 19.3. (See Exercise 19.31.)

19.36 The following utility function has been developed for a professor who works as a private consultant in the summer. Let x be the number of days employed as a consultant. This may range from 10 to 50 days.

$$U(x) = \frac{(x - 10)^2}{1600}, \qquad 10 < x < 50$$

Make a graph of this utility function and use it to determine the individual's attitude toward risk.

19.37 Suppose you are the professor in Exercise 19.36 and your utility function corresponds to the one in Exercise 19.36. You are offered a choice of a contract for 30 days consulting or an opportunity to free-lance, which has a probability .6 of getting 20 days' work and a probability .4 of getting 40 days' work. Construct a utility table for this decision problem.

19.38 Use the utility table in Exercise 19.37 to determine the action corresponding to maximum expected utility.

19.39 Construct a payoff table for the decision problem in Exercise 19.37 and use it to find the action corresponding to the maximum expected payoff. How does this action compare with the one selected in Exercise 19.38?

19.40 Are there any inadmissible actions in the following payoff table? The table entries represent the savings in the number of hours required to complete project s_j if the suggested change a_i is made in the management system.

State of Nature

Action	s_1 (.3)	s_2 (.4)	s_3 (.2)	s_4 (.1)
a_1	15	60	-20	16
a_2	-25	14	-10	10
a_3	30	5	40	-5
a_4	20	-15	9	38
a_5	50	20	25	12

19.41 What is the decision based on the maximax criterion for the payoff table in Exercise 19.40?

19.42 What is the decision based on the maximin criterion for the payoff table in Exercise 19.40?

19.43 Find the maximum expected value decision for the payoff table in Exercise 19.40. Probabilities are given in parentheses with the states of nature.

19.44 Convert the payoff table in Exercise 19.40 to an opportunity loss table and find the minimum expected opportunity loss decision. Verify that it is the same as the decision in Exercise 19.43.

19.45 A production manager is considering offering one of three types of contracts to his employees.

Contract 1: Employees receive $7.80 per hour.

Contract 2: Employees receive $5.00 per hour plus 10 cents per item produced.

Contract 3: Employees receive 36 cents per item produced.

Calculate the labor cost per item for each combination of contract and rate of production and complete the following table.

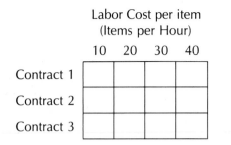

Labor Cost per item
(Items per Hour)

	10	20	30	40
Contract 1				
Contract 2				
Contract 3				

19.46 Find the decision based on the minimax criterion for the cost table in Exercise 19.45. Who benefits from this decision, the employer or employee?

19.47 Find the decision based on the maximin criterion for the cost table in Exercise 19.45. Who benefits from this decision, the employer or employee?

19.48 Past records show that 20% of the employees of the company in Exercise 19.45 produce 10 items per hour; 40% produce 20 items per hour; 30% produce 30 items per hour; and 10% produce 40 items per hour. Use these relative frequencies to determine the expected labor cost per item for each contract. Based on these expected costs, which contract should the production manager favor? Which contract should the employees prefer?

19.49 The employees are interested in maximizing the hourly pay. Construct a payoff matrix for Exercise 19.45, where the payoff is the hourly wage under the various contracts.

19.50 Which contract maximizes the expected payoff for the employees? How does this compare with the result of Exercise 19.48?

BIBLIOGRAPHY

Additional material on the topics presented in this chapter can be found in the following publications.

Aitchison, J. (1970). *Choice Against Chance: An Introduction to Statistical Decision Theory.* Addison-Wesley, Reading, Mass.

Chernoff, H. and Moses, L.E. (1959). *Elementary Decision Theory.* Wiley, New York.

Lindley, D.V. (1965). *Introduction to Probability and Statistics from a Bayesian Viewpoint.* Cambridge Univ. Press, London/New York.

Raiffa, H. (1968). *Decision Analysis.* Addison-Wesley, Reading, Mass.

Raiffa, H. and Schlaifer, R. (1961). *Applied Statistical Decision Theory.* Harvard Business School, Boston, Mass.

Schlaifer, R. (1959). *Probability and Statistics for Business Decisions.* McGraw-Hill, New York.

Wald, A. (1950). *Statistical Decision Functions.* Wiley, New York.

20
DECISION THEORY
WITH SAMPLE INFORMATION
(Optional)

The decision-making procedures presented in Chapter 19 were based on prior or subjective judgment. That is to say, there was no attempt to incorporate information obtained from sampling into the decision-making process. In this chapter sample information is incorporated. The first section of this chapter shows how such sample information can be used to revise the original state-of-nature probabilities. The probabilities before being revised are called **prior probabilities.** The revised prior probabilities are referred to as **posterior probabilities** to identify these probabilities as being calculated after the sample is obtained rather than prior to obtaining the sample. For example, the prior probability of being low bidder on a city street paving project might be .3 based on past history, but the posterior probability may be .6 after learning that a major competitor did not bid on the project.

The second section considers the potential value of removing all uncertainty from the decision-making process. In such a setting the decision maker is said to have **perfect information.** However, since perfect information does not usually exist in practice, the concept is presented as a means of determining the maximum amount that a decision maker would be willing to pay for sample information that would remove uncertainty from the decision-making process. For example, perfect information may be the knowledge that 18% of the population is listening to the TV show you are sponsoring. How much are you willing to pay for a survey to estimate this audience percentage? Of course, the more you pay, the larger the survey and the better the estimate.

In the final section of this chapter techniques are presented for determining, prior to actually obtaining a sample, the effect of various outcomes of a sample on the decision. If a sample were obtained, the information gained from it could be used to calculate posterior probabilities. However, since the analysis in the final section takes place before the sample is obtained, it is referred to as a **preposterior analysis.** The preposterior analysis is used to assign a dollar value to the sample that is based on the potential impact of the sample on the decision.

20.1
USING SAMPLE INFORMATION TO REVISE PRIOR PROBABILITIES

PROBLEM SETTING 20.1

An oil company is trying to make a decision whether or not to drill at a particular site. Past experience shows that there is probability .3 of striking oil and a .7 probability of a dry hole. The oil company can gain additional information by collecting seismic data. Past records show that when oil has been found, a geologic dome has been present 60% of the time. Moreover, a geologic dome has been present in only 20% of the dry holes. The geologic records can be summarized in terms of conditional probabilities as follows.

$$P(\text{dome}|\text{oil}) = .6$$

and

$$P(\text{dome}|\text{dry hole}) = .2$$

If the seismic records show the presence of a geologic dome, how should the management of the oil company revise the probability of striking oil?

REVISION OF PRIOR PROBABILITIES

The original probabilities assigned to the states of nature are called **prior probabilities.** They are probabilities that may have been obtained by a rigorous process, such as by looking at past records to get accurate relative frequencies, or by a less rigorous process, for instance, by a subjective evaluation of the situation by a manager or group of managers. Sometimes additional information becomes available that is applicable for a particular decision. This additional information may affect the probabilities of the various states occurring. Therefore, it should be used, if at all possible, to aid the manager in making a better decision for that particular situation. If this additional information is used to revise the prior probabilities, the new revised probabilities are called **posterior probabilities.** These are conditional probabilities, where the given condition is the new information now available. The reader may wish to review the concepts of joint and conditional probabilities from Section 4.1.

PRIOR
PROBABILITIES

The probabilities originally assigned to the states of nature are called **prior probabilities,** to contrast them with a revised set of probabilities that may be obtained later.

POSTERIOR
PROBABILITIES

If the prior probabilities are revised in accordance with additional information, the revised probabilities, which are conditional probabilities given the information, are called **posterior probabilities** or **probabilities of causes.**

The posterior probabilities are easily obtained from a joint probability table constructed through the use of Eq. 4.1, which is repeated here.

$$P(A|B) = \frac{P(A \text{ and } B)}{P(B)} \tag{20.1}$$

The prior probability is denoted by $P(B)$, and the additional information is in the form of $P(A|B)$. The problem is to find the posterior probability $P(B|A)$.

Equation 20.1 can be written as

$$P(A \text{ and } B) = P(A|B)P(B) \tag{20.2}$$

The ultimate objective is to adjust the prior probability $P(B)$ to take into consideration the given event A; hence, the adjustment gives the posterior probability $P(B|A)$. It is accomplished by rewriting Eq. 20.1, with events A and B interchanged

$$P(B|A) = \frac{P(A \text{ and } B)}{P(A)}$$

Since $P(A \text{ and } B)$ is given by Eq. 20.2, all that remains is to find $P(A)$. However, $P(A)$ can be found by using Eq. 20.2 to find $P(A \text{ and } B_i)$ for all states of nature B_i, and then using the relationship:

$$P(A) = \sum_i P(A \text{ and } B_i)$$

EXAMPLE

A question relevant to the problem posed in Problem Setting 20.1 is whether or not the seismic data should be used. One way to evaluate the situation is to consider the case where the presence of a geologic dome always indicates oil. To ignore the seismic data in such circumstances would be foolish. However, it would also be inappropriate to ignore the seismic data, even when the presence of a dome is not a perfect indicator of oil. The question becomes one of how to incorporate the geologic information into a revision of the known probabilities of finding oil. The known probabilities of finding oil are referred to as *prior probabilities*, whereas those probabilities resulting from the revision or reweighting are called *posterior probabilities*. To construct the joint probability table the following calculations are made using Eq. 20.2 with the information given.

$$P(dome \text{ and } oil) = P(dome|oil)P(oil) = .6(.3) = .18$$

This becomes the first entry of the joint probability table, given in completed form in Figure 20.1.

The other entries in the 2 × 2 table in Figure 20.1 are found in a similar manner as follows.

$$P(dome \text{ and } dry \text{ } hole) = P(dome|dry \text{ } hole)P(dry \text{ } hole) = .2(.7) = .14$$

and

$$P(no \text{ } dome \text{ and } oil) = P(no \text{ } dome|oil)P(oil) = .4(.3) = .12$$

The last table entry is

$$P(no \text{ } dome \text{ and } dry \text{ } hole) = P(no \text{ } dome|dry \text{ } hole)P(dry \text{ } hole)$$
$$= .8(.7) = .56$$

The marginal totals in Figure 20.1 are found by summing the appropriate table entries. In the event that the seismic data show the presence of a geologic dome, it is desirable to use this information to revise the prior probabilities assigned to finding oil and to finding a dry hole. That is, the following conditional probabilities are desired.

$$P(oil|dome) \quad \text{and} \quad P(dry \text{ } hole|dome)$$

These posterior probabilities are found by using only the first row of the table in Figure 20.1 and Eq. 20.1. These calculations are as follows:

$$P(oil|dome) = \frac{P(oil \text{ and } dome)}{P(dome)} = \frac{.18}{.32} = .56$$

and

$$P(dry \text{ } hole|dome) = \frac{P(dry \text{ } hole \text{ and } dome)}{P(dome)} = \frac{.14}{.32} = .44$$

	Oil	Dry hole	Totals
Dome	.18	.14	.32 = P(dome)
No dome	.12	.56	.68 = P(no dome)
Totals	.30	.70	1.00

FIGURE 20.1
JOINT PROBABILITY TABLE FOR PROBLEM SETTING 20.1.

> Thus, the management of the oil company in Problem Setting 20.1 must revise the prior probabilities of .3 and .7 to .56 and .44, respectively, for finding oil and finding a dry hole, to reflect the information that a dome is present.

BAYES' THEOREM

The procedure just explained for calculating posterior probabilities is an application of a theorem attributed to an English Presbyterian minister and mathematician, Thomas Bayes (1702–1761). The formal statement of **Bayes' theorem** is given in Eq. 20.3. The posterior probabilities in Eq. 20.3 are calculated by referring to a joint probability table like the one given in Figure 20.1 and dividing each entry in a particular row (or column) by the total of that row (or column).

The posterior probability of the occurrence of the state of nature s_i given the event B, is given by **Bayes' theorem** as

HOW TO FIND POSTERIOR PROBABILITIES USING BAYES' THEOREM

$$P(s_i|B) = \frac{P(B|s_i)P(s_i)}{P(B)} \qquad (20.3)$$

where $P(s_i)$ is the prior probability of the occurrence of the state of nature s_i, $P(B|s_i)$ is the conditional probability of the event B given the state of nature s_i, and $P(B)$ is usually found from the equation

$$P(B) = \sum_i P(B|s_i)P(s_i) = \sum_i P(B \text{ and } s_i)$$

AN APPLICATION OF BAYES' THEOREM

In Problem Setting 19.1 an investor was confronted with a decision about investing his money entirely in certificates of deposit (a_1), in a mixture of certificates of deposit and municipal bonds (a_2), or all in municipal bonds (a_3). The payoff table in Figure 19.1 provides the payoffs for each of these three actions over three possible changes in the prime interest rate (PIR): a decrease of three percentage points (s_1), no change (s_2), and an increase of three percentage points (s_3). In Section 19.2 the investor used his judgment to assign subjective prior probabilities to the occurrence of each state of nature as $P(s_1) = .1$, $P(s_2) = .6$, and $P(s_3) = .3$. Based on these prior probabilities, the investor was able to determine that action a_3 had the maximum expected payoff of $10,320.

Suppose the investor decides to seek some professional advice on the likelihood of change in the PIR, and he hires the research services of an investment firm to make such a determination. The firm reports that the PIR is likely to increase three percentage points in the next year. Since this forecast differs from the investor's feeling about the likely behavior of the PIR, he naturally wants to know the firm's record on forecasting changes in the PIR. The firm indicates that in the past when the PIR increased, their predictions were correct 75% of the time. However, they also indicate that in the past when the PIR remained stable, they had incorrectly forecast an increase 30% of the time, and when the PIR decreased, they had incorrectly forecast an increase 15% of the time. In terms of conditional probabilities the firm's track record is summarized as follows.

P(firm had forecast increase in PIR|PIR increased) $= .75$

P(firm had forecast increase in PIR|PIR remained stable) $= .30$

P(firm had forecast increase in PIR|PIR decreased) $= .15$

The joint probability table for this decision problem is calculated by using Eq. 20.2, as was done earlier in this section for Problem Setting 20.1. The joint probability table is given in Figure 20.2. Since the firm forecasts an increase in the PIR, attention is directed only at the first row in the joint probability table in Figure 20.2. The posterior probabilities for each state of nature are obtained by dividing each respective table entry in the first row by the total of the first row, .42. The calculations yield the following results:

$$P(s_1|\text{firm had forecast an increase in PIR}) = \frac{.15(.1)}{.42} = .0357$$

$$P(s_2|\text{firm had forecast an increase in PIR}) = \frac{.30(.6)}{.42} = .4286$$

$$P(s_3|\text{firm had forecast an increase in PIR}) = \frac{.75(.3)}{.42} = .5357$$

The investor can now calculate his maximum expected payoff using the posterior probabilities. These calculations proceed as in Section 19.2, and are as follows:

$E[r(a_1, s)] = \$9000(.0357)\ \ \ + \$9000(.4286)\ \ \ + \$9000(.5357) = \$9000$

$E[r(a_2, s)] = \$14{,}700(.0357) + \$10{,}500(.4286) + \$6300(.5357) = \$8400$

$E[r(a_3, s)] = \$20{,}400(.0357) + \$12{,}000(.4286) + \$3600(.5357) = \$7800$

Thus, the maximum expected payoff using posterior probabilities is $9000, and the corresponding decision is to invest all monies in CDs.

This decision differs from the one made using the subjective prior probabilities of the investor. The previous decision was to invest all monies in

	s_1	s_2	s_3	Totals
Forecast increase	.15(.1)	.30(.6)	.75(.3)	.42
Forecast no increase	.85(.1)	.70(.6)	.25(.3)	.58
Totals	.1	.6	.3	1.00

FIGURE 20.2
JOINT PROBABILITY TABLE FOR THE DECISION PROBLEM
RELATED TO CHANGES IN THE PRIME INTEREST RATE.

municipal bonds, and it was based on subjective probabilities that indicated the most likely state of nature to be no change in the prime rate. The investment firm predicts an increase in the prime rate. While they are not correct all of the time in their predictions, the revised probabilities still indicate the increase in the prime rate to be the most likely state of nature, with a posterior probability of .5357. Since the price of municipal bonds tends to decrease when the prime rate goes up, it seems reasonable to expect a different decision, which in this case is to invest all monies in CDs.

EXERCISES

20.1   Refer to Figure 20.1 and calculate the posterior probabilities for the case where the seismic data show no geologic dome. How do these posterior probabilities compare with those worked out in this section for the case where the seismic data show a geologic dome?

20.2   Refer to Figure 20.2 and calculate the posterior probabilities for the case where the firm did not forecast an increase in the prime interest rate. How do these posterior probabilities compare with those worked out in this section for the case where the investment firm forecast an increase in the prime interest rate?

20.3   Find the posterior probabilities associated with the decision problem in Problem Setting 20.1 when the oil company assesses the prior probabilities of striking oil or having a dry hole to both be equal to .5.

20.4   Find the posterior probabilities associated with Problem Setting 20.1 for the case where the seismic data show a geologic dome 75% of the time when oil has been found, and a geologic dome has been present in only 10% of the dry holes.

20.5   Find the posterior probabilities associated with Problem Setting 20.1 for the case where the seismic data show a geologic dome 50% of the time when oil has been found, and a geologic dome has been present in 50% of the dry holes. Compare these posterior probabilities with the prior probabilities in the problem setting and explain the results of the comparison.

20.6   Refer to Figure 20.2 and calculate the posterior probabilities for the case when the firm forecasts an increase in the prime interest rate when the prior probabilities have been assigned as $P(s_1) = .5$, $P(s_2) = .3$, and $P(s_3) = .2$.

20.7   Determine the decision associated with the maximum expected payoff based on the posterior probabilities from Exercise 20.6.

20.8   Employees in a factory produce 10, 20, 30, or 40 items per hour with respective probabilities of .2, .4, .3, and .1. Each worker's hourly production rate is matched with the score made on an aptitude test when they were hired. Of the workers who produce 10 items per hour, 20% scored high and 80% scored low on the aptitude test. That is,

$$P(\text{high score}|\text{production rate is 10}) = .2$$
$$P(\text{low score}|\text{production rate is 10}) = .8$$

Test scores for other production rates are summarized as follows:

$$P(\text{high score}|\text{production rate is } 20) = .3$$
$$P(\text{low score}|\text{production rate is } 20) = .7$$
$$P(\text{high score}|\text{production rate is } 30) = .6$$
$$P(\text{low score}|\text{production rate is } 30) = .4$$
$$P(\text{high score}|\text{production rate is } 40) = .9$$
$$P(\text{low score}|\text{production rate is } 40) = .1$$

Use this information to construct a joint probability table. Find the posterior probabilities for the various production rates given that a worker scores high.

20.9 The production manager in the factory of Exercise 20.8 is considering offering one of three types of contracts to his employees.

Contract 1 Employees receive $7.80 per hour.

Contract 2 Employees receive $5.00 per hour plus 10 cents per item produced.

Contract 3 Employees receive 36 cents per item produced.

Use the posterior probabilities in Exercise 20.8 to find the expected cost per item for each contract, for the workers with high scores on the aptitude test. Which contract is most favorable from the point of view of the manager for this case?

20.10 Use the joint probability table in Exercise 20.8 to determine the posterior probabilities for a worker who scores low on the aptitude test.

20.11 Use the posterior probabilities found in Exercise 20.10 to find the expected cost per item for each contract in Exercise 20.9 for the employees who score low. Which contract is most favorable from the point of view of the manager for this case?

20.2

THE EXPECTED VALUE OF PERFECT INFORMATION

PERFECT INFORMATION

In Problem Setting 19.1 the investor is faced with making a decision when the states of nature are unknown. If the decision maker had access to a crystal ball that would somehow predict the state of nature perfectly, then he could always make the correct decision and would maximize profit. This is referred to as having **perfect information.** Of course, in practice perfect information does not exist and so it is used only as a frame of reference to determine the value of having perfect information. To illustrate the use of perfect information, consider the payoff table given in Figure 19.1, which is repeated in Figure 20.3. If the investor knew for certain that the PIR was going to decrease by

	Change in Prime Interest Rate		
Action	− 3%	0%	3%
Buy all CDs	$ 9000	$ 9000	$9000
Split investment	$14,700	$10,500	$6300
Buy all municipal bonds	$20,400	$12,000	$3600

FIGURE 20.3
PAYOFF TABLE FOR PROBLEM SETTING 19.1.

three percentage points (s_1), then the optimal action to maximize the payoff is to invest all monies in municipal bonds (a_3) with a payoff of $20,400. Likewise if the investor knew for certain that the PIR was not going to change (s_2), then the maximum payoff is again associated with buying only municipal bonds. However, the payoff in this case is $12,000. In the event that the PIR increases by three percentage points (s_3), the maximum payoff is $9000 when all monies are invested in CDs. These payoffs are summarized in Figure 20.4 along with the prior subjective probabilities from Section 19.2. The information in Figure 20.4 can be used to find the **expected payoff under perfect information** as

$$\$20,400(.1) + \$12,000(.6) + \$9000(.3) = \$11,940$$

THE EXPECTED VALUE OF PERFECT INFORMATION

Of course, perfect information usually does not exist in practice. Even so, the expected payoff under perfect information is of use in making comparisons on the expected value of having perfect information. That is, the amount $11,940 represents the expected payoff under perfect information, whereas the maximum expected payoff of $10,320, found in Section 19.2, is calculated in the presence of uncertainty (i.e., less than perfect information). The difference between these two amounts $11,940 − $10,320 = $1620 is the **expected value of perfect information.**

THE EXPECTED VALUE OF PERFECT INFORMATION   The **expected value of perfect information** is the difference between the expected payoff under perfect information and the maximum expected payoff under uncertainty.

State of Nature	Maximum Payoff	Prior Probabilities
s_1	$20,400	.1
s_2	$12,000	.6
s_3	$ 9,000	.3

FIGURE 20.4
SUMMARY OF PAYOFFS WITH PERFECT INFORMATION.

Since $1620 represents the increase in the expected payoff with perfect information over not having perfect information, it may be interpreted as the maximum amount of money the investor would be willing to pay to obtain perfect information. Another way to think of the $1620 is as the cost of uncertainty. That is, if the uncertainty could be eliminated, then the expected payoff would increase by $1620.

The expected value of perfect information can also be found by using an opportunity loss table. It always equals the minimum expected opportunity loss. To demonstrate this idea, consider the opportunity loss table associated with Problem Setting 19.1. This table was originally given in Figure 19.5 and is repeated in Figure 20.5. The minimum expected opportunity loss associated with the table in Figure 20.5 was found in Section 19.2 as $1620. This amount agrees with the previous calculation and, in general, this is an alternative method of finding the expected value of perfect information.

THE EXPECTED VALUE OF PERFECT INFORMATION USING POSTERIOR PROBABILITIES

In the previous section posterior probabilities were calculated based on information obtained from an investment firm. A subsequent analysis was used to determine whether the investment strategy would be altered based on these revised probabilities. Posterior probabilities can also be used to determine the value of perfect information. The procedure is the same as the one just demonstrated with prior probabilities, except that posterior probabilities are used in place of prior probabilities.

The expected payoff associated with Figure 20.4 and the minimum expected opportunity loss associated with Figure 20.5 were both determined by using the prior probabilities given in Section 19.2 as $P(s_1) = .1$, $P(s_2) = .6$, and $P(s_3) = .3$. However, the joint probability distribution in Figure 20.2 allowed these prior probabilities to be converted to the following posterior probabilities, .0357, .4286, and .5357, respectively. Recall that these posterior probabilities are conditional probabilities, based on the investment firm forecasting an increase in the PIR. The posterior probabilities appear with the maximum payoffs in Figure 20.6 and are used to find the revised expected payoff under perfect information as

$$\$20,400(.0357) + \$12,000(.4286) + \$9000(.5357) = \$10,693$$

Action	Change in Prime Interest Rate		
	−3%	0%	3%
Buy all CDs	$11,400	$3000	0
Split investment	$ 5700	$1500	$2700
Buy all municipal bonds	0	0	$5400

FIGURE 20.5
OPPORTUNITY LOSS TABLE FOR PROBLEM SETTING 19.1.

State of Nature	Maximum Payoff ($)	Posterior Probability
s_1	20,400	.0357
s_2	12,000	.4286
s_3	9,000	.5357

FIGURE 20.6
SUMMARY OF PAYOFFS WITH PERFECT INFORMATION.

The maximum expected payoff using posterior probabilities was given in the previous section as $9000. It follows that the expected value of perfect information is $10,693 − $9000 = $1693 for the case where the investment firm forecast an increase in the PIR. This amount is slightly more than the value $1620 found earlier using prior probabilities. One reason for this change is that the prior and posterior probabilities are considerably different. Another reason is that the decision associated with the maximum expected payoff changes from the prior analysis to the posterior analysis.

Frequently the expected value of perfect information decreases in the presence of the additional information that leads to the calculation of posterior probabilities. This would indicate that the price the investor is willing to pay for perfect information is not as great as it was previously. Of course, the posterior analysis can also be used with opportunity loss and will yield the same value of $1693. The next example applies the concepts of this section.

EXAMPLE

A general road contractor receives a bonus for each day of early completion of the project before the final contract deadline. He has a choice of using three different subcontractors. The ability of each of the subcontractors to finish the job early is influenced by the type of weather conditions. The general contractor estimates the number of days that each subcontractor will be able to finish early for each type of weather condition. This summary is given in Figure 20.7 along with the general contractor's estimate of the probabilities for each weather condition.

Type of Weather Condition
(Number of Days Ahead of Schedule)

	Clear (.4)	Light Rain (.3)	Heavy Rain (.1)	Rain with Heavy Wind (.2)
Subcontractor 1	14	10	9	6
Subcontractor 2	12	9	7	8
Subcontractor 3	16	11	6	5

FIGURE 20.7
PAYOFF TABLE WITH PRIOR PROBABILITIES GIVEN IN PARENTHESES.

(a) Draw a decision tree representing the possible decisions and states of nature. (A decision tree for this example appears in Figure 20.8.)

(b) Find the maximum expected payoff and corresponding decision:

$$E[r(a_1, s)] = 14(.4) + 10(.3) + 9(.1) + 6(.2) = 10.7$$

$$E[r(a_2, s)] = 12(.4) + 9(.3) + 7(.1) + 8(.2) = 9.8$$

$$E[r(a_3, s)] = 16(.4) + 11(.3) + 6(.1) + 5(.2) = 11.3$$

The maximum of these expected payoffs is 11.3 so that the maximum expected payoff decision is to use subcontractor 3.

(c) Find the expected payoff under perfect information. The maximum payoff for each state of nature and the prior probabilities are as follows:

State of Nature	Maximum Payoff	Prior Probability
Clear	16	.4
Light rain	11	.3
Heavy rain	9	.1
Rain with heavy wind	8	.2

The expected payoff under perfect information is calculated from these values as

$$16(.4) + 11(.3) + 9(.1) + 8(.2) = 12.2$$

(d) Find the expected value of perfect information and interpret its meaning. The expected value of perfect information represents the maximum

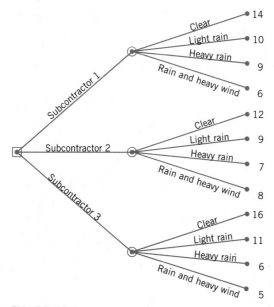

FIGURE 20.8

A DECISION TREE SUMMARIZING PAYOFFS FOR DIFFERENT COMBINATIONS OF CHOICE OF SUBCONTRACTOR AND TYPE OF WEATHER CONDITION.

amount the contractor would be willing to pay for perfect information. It is the difference between the answers in parts (c) and (b) or $12.2 - 11.3 = 0.9$ days. This leads directly to the amount of the bonus the contractor would gain in the presence of perfect information.

(e) Convert the payoff table in Figure 20.7 to an opportunity loss table and then find the minimum expected opportunity loss. Compare this value with the expected value of perfect information found on part (d). The opportunity loss table is as follows.

Type of Weather Condition
(Opportunity Loss, Days)

	Clear (.4)	Light Rain (.3)	Heavy Rain (.1)	Rain with Heavy Wind (.2)
Subcontractor 1	2	1	0	2
Subcontractor 2	4	2	2	0
Subcontractor 3	0	0	3	3

$$E[OL(a_1, s)] = 2(.4) + 1(.3) + 0(.1) + 2(.2) = 1.5$$
$$E[OL(a_2, s)] = 4(.4) + 2(.3) + 2(.1) + 0(.2) = 2.4$$
$$E[OL(a_3, s)] = 0(.4) + 0(.3) + 3(.1) + 3(.2) = 0.9$$

The minimum of these values is 0.9, and therefore represents the minimum expected opportunity loss. This value is identical to the expected value of perfect information.

(f) The long-range forecast is for clear weather. The posterior probabilities under this additional information are given as .686, .171, .086, and .057 for the states of nature s_1 to s_4, respectively. Calculate the maximum expected payoff using posterior probabilities:

$$E[r(a_1, s)] = 14(.686) + 10(.171) + 9(.086) + 6(.057) = 12.4$$
$$E[r(a_2, s)] = 12(.686) + 9(.171) + 7(.086) + 8(.057) = 10.8$$
$$E[r(a_3, s)] = 16(.686) + 11(.171) + 6(.086) + 5(.057) = 13.7$$

The maximum of these values is 13.7 corresponding to the selection of subcontractor 3.

(g) Find the expected payoff under perfect information using posterior probabilities. The maximum payoff for each state of nature and the posterior probabilities are as follows:

State of Nature	Maximum Payoff	Prior Probability
Clear	16	.686
Light rain	11	.171
Heavy rain	9	.086
Rain with heavy wind	8	.057

The expected payoff under perfect information is calculated from these values as

$$16(.686) + 11(.171) + 9(.086) + 8(.057) = 14.1$$

(h) Find the expected value of perfect information using posterior probabilities and comment on the comparison with the expected value of perfect information found in part (d). The expected value of perfect information is found as the difference between the answers for parts (g) and (f), or $14.1 - 13.7 = 0.4$. The expected value of perfect information based on posterior probabilities is less than the value based on prior probabilities. Thus, the cost of uncertainty has decreased.

(i) Find the minimum expected opportunity loss based on posterior probabilities and compare it with the expected value of perfect information found in part (h). Using the opportunity loss table from part (e) and the posterior probabilities given in part (f), the expected opportunity losses are calculated as

$$E[OL(a_1, s)] = 2(.686) + 1(.171) + 0(.086) + 2(.057) = 1.7$$
$$E[OL(a_2, s)] = 4(.686) + 2(.171) + 2(.086) + 0(.057) = 3.3$$
$$E[OL(a_3, s)] = 0(.686) + 0(.171) + 3(.086) + 3(.057) = 0.4$$

The minimum of these expected opportunity losses is 0.4, which is identical to the expected value of perfect information found in part (h).

EXERCISES

The entries in the following payoff table represent the savings in the number of hours required to complete project s_j if the suggested change a_i is made in the management system. The probabilities of getting each type of project are given in parentheses. Use this payoff table in reference to the requirements of Exercises 20.12 to 20.19.

Type of Project Received
(Savings in Hours)

		$s_1(.3)$	$s_2(.4)$	$s_3(.2)$	$s_4(.1)$
	a_1	15	60	-20	16
Suggested Changes	a_2	30	5	40	-5
to Management System	a_3	20	-15	9	38
	a_4	50	20	25	12

20.12 Draw a decision tree representing the possible decisions and states of nature.

20.13 Find the expected payoff under perfect information for the given payoff table.

20.14 Find the expected value of perfect information and interpret its meaning for the given payoff table. See the answer to Exercise 19.43 for the maximum expected payoff.

20.15 Compare the minimum expected opportunity loss with the expected value of perfect information found in Exercise 20.14. See the answer to Exercise 19.44 for the minimum expected opportunity loss.

20.16 Use the posterior probabilities .2, .3, .3, and .2 for the states of nature s_1 to s_4, respectively, for the given payoff table and calculate the maximum expected payoff using posterior probabilities.

20.17 Find the expected payoff under perfect information for the given payoff table using the posterior probabilities given in Exercise 20.16.

20.18 Find the expected value of perfect information for the given payoff table using the posterior probabilities given in Exercise 20.16. Comment on the comparison with the expected value of perfect information found in Exercise 20.14.

20.19 Find the minimum expected opportunity loss for the given payoff table based on the posterior probabilities given in Exercise 20.16. Compare this value with the expected value of perfect information found in Exercise 20.18.

The following information is used in Exercises 20.20 to 20.22. A business executive has agreed to meet an associate at a motel in Dallas, Texas. He needs to stay overnight, but he cannot remember which of two motels they agreed on. Motel 1 costs $50 and motel 2 costs $54. If he stays at the wrong motel, he will have to pay $10 taxi fare back and forth to meet his associate. He is about 80% sure that motel 2 is the correct motel.

20.20 Draw a decision tree representing the possible decisions, states of nature, and costs.

20.21 Construct a payoff table of costs and find the correct choice of motel based on the minimum expected cost.

20.22 The business executive can telephone his associate's secretary to find out for sure where his associate will be staying. The long-distance telephone call will cost $3. Is it cost effective?

20.3
THE EXPECTED VALUE
OF SAMPLE INFORMATION

PROBLEM SETTING 20.2

Manufacturers routinely examine warranty claims to see if changes are needed in the manufacturing process. If only a small percentage of claims are made, then it is less costly for the manufacturer to replace the item than it is to change the manufacturing process. However, if a large number of claims are made, then it may be cost effective to make the change. The payoff table in Figure 20.9 illustrates the situation for one such manufacturer, who makes large

Percentage Defective

Action	1%	5%	10%
Do not change process (a_1)	−100	−500	−1000
Change process (a_2)	−400	−400	−400
Prior probabilities	.85	.10	.05

FIGURE 20.9
PAYOFF TABLE FOR PROBLEM SETTING 20.2.

valves for industrial use. The payoffs represent the amount of money gained (or lost) for each action under each state of nature. The cost of changing the process is $400, independent of the percentage defective. The probabilities represent the manufacturer's best estimate of the probabilities associated with various percentages of defectives based on past data. A decision tree is given in Figure 20.10.

The percentage defective may change from time to time, so it may be worthwhile occasionally to draw a random sample of valves and submit them to extensive tests and thorough inspection to see if they will prove defective. However, the sample is expensive to examine and the manufacturer would like to know beforehand (i.e., prior to obtaining the sample) if the expected value of the sample information will offset the cost of obtaining the sample.

THE PRIOR ANALYSIS

The valve manufacturer in Problem Setting 20.2 would like to reduce the uncertainty associated with the percentage of manufactured items that are defective, because an unacceptably large percentage of defectives will require

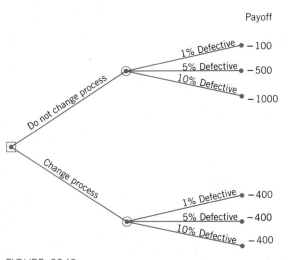

FIGURE 20.10
A DECISION TREE FOR USE WITH PROBLEM SETTING 20.2.

some costly changes in the manufacturing process. From the first row in the payoff table of Figure 20.9 it can be seen that as the percentage of defective items increases, it becomes more costly for the manufacturer if no changes are made in the manufacturing process. On the other hand, changing the manufacturing process is expensive and does not become cost effective, as is evidenced by comparing the two rows of the payoff table, unless the percentage of defective parts is high. The manufacturer's prior subjective probabilities are given in Figure 20.9 and can be used to determine the action with the maximum expected payoff:

$$E[r(a_1, s)] = .85(-100) + .10(-500) + .05(-1000) = -185$$
$$E[r(a_2, s)] = .85(-400) + .10(-400) + .05(-400) \quad = -400$$

In this case, the decision is not to change the manufacturing process, as this action has the maximum expected payoff (or minimum expected cost for this decision problem). This decision is based entirely on **prior** subjective judgment and does not involve sample information. If sample information were available, it could be used to revise the prior probabilities and to determine the value of the **posterior** probabilities. The posterior probabilities could be used to calculate the expected payoffs as in Section 20.1, and this might alter the decision from the prior analysis.

THE PREPOSTERIOR ANALYSIS

A decision to obtain sample information is not made automatically, as the sample may be expensive and/or time-consuming to obtain. Hence, this section presents a middle-ground approach and determines the value of sample information *prior to actually obtaining a sample*. That is to say, posterior probabilities are calculated for each possible outcome of a random sample of size n. These posterior probabilities are used to calculate the expected payoff associated with each possible sample outcome to see whether the sample can provide cost-effective information. Since this analysis is done previous to obtaining a sample and calculating posterior probabilities based on the sample information, it is usually called a **preposterior analysis.**

FINDING
POSTERIOR
PROBABILITIES
GIVEN SAMPLE
OUTCOMES

Let the possible outcomes of the sample be denoted by $B_1, B_2, \ldots, B_k$. The posterior probabilities $P(s_j|B_i)$ are found using Eq. 20.3, which is repeated here for convenience:

$$P(s_j|B_i) = \frac{P(B_i|s_j)P(s_j)}{P(B_i)}$$

where $P(B_i)$ is found by using $P(B_i) = \sum_j P(B_i|s_j)P(s_j)$.

Of course, sample information is not as valuable as perfect information; thus, the expected value of the sample information will always be less than the expected value of perfect information. The latter was defined in the previous section. In Problem Setting 20.2, the expected payoff under perfect information is found as

$$-100(.85) - 400(.10) - 400(.05) = -145$$

which represents a savings of 40 units over taking action a_1. Since the expected value of perfect information is 40 units, the expected value of sample information is worth somewhere between 0 and 40 units. Its precise value is computed in the remainder of this section.

STEPS IN
CONDUCTING
A
PREPOSTERIOR
ANALYSIS

1. Construct a joint probability table, with the states of nature s_j as columns, and the sample outcomes B_i as rows. The entries in the table are the joint probabilities $P(B_i \text{ and } s_j)$.

2. Find the posterior probabilities $P(s_j|B_i)$ for each sample outcome B_i.

3. Find the maximum expected payoff for each B_i for the various actions in the payoff matrix, using the posterior probabilities obtained in step 2.

4. Multiply each maximum expected payoff in step 3 by the corresponding marginal probability $P(B_i)$ from step 1, and sum to find the expected payoff over all sample outcomes.

COMPUTING POSTERIOR PROBABILITIES FOR EACH POSSIBLE SAMPLE OUTCOME

In Problem Setting 20.2 the manufacturer has assigned prior subjective probabilities to the percentage of defective items as .85, .10, and .05, respectively, for 1% defective, 5% defective, and 10% defective, thus reflecting a strong belief that the percentage of defective items is small. If the manufacturer examines a random sample of valves and finds the number of defective valves is small, then this sample outcome tends to support the manufacturer's belief. On the other hand, if most of the valves in the sample prove to be defective, then the manufacturer's prior probabilities may be in need of revision. For example, suppose a random sample of three valves is examined. The possible number of defective valves is 0, 1, 2, or 3. The outcome of 0 and possibly 1 would indicate a small percentage of defective valves and tends to support the manufacturer's belief, whereas outcomes of 2 or 3 would lend support to a need to revise the prior probabilities.

The revision of the prior probabilities involves first finding the joint probabilities using Eq. 20.2, as was done in Section 20.2. The binomial probability function of Chapter 5 is needed to calculate the conditional probability in Eq. 20.2. The calculations will be demonstrated by using a sample of size 3, although the procedure easily extends to larger sample sizes. For larger sample sizes it is helpful to use a computer to do the calculations.

Suppose a sample of size 3 shows no defective valves, that is, $X = 0$. The conditional probability that is needed is

$P(\text{sample shows no defectives}|\text{proportion of defectives is .01})$

or

$$P(X = 0|s_1)$$

From Eq. 5.1 this probability is calculated as

$$\binom{3}{0}(.01)^0(.99)^3 = .9703$$

Next, Eq. 20.2 is used to find the joint probability

$$P(X = 0|s_1)P(s_1) = .9703(.85) = P(X = 0 \text{ and } s_1)$$

This value becomes the first entry in the first row of the joint probability table in Figure 20.11. The second entry is found as

P(sample shows no defective | proportion of defectives is .05)

or

$$P(X = 0|s_2) = \binom{3}{0}(.05)^0(.95)^3 = .8574$$

then

$$P(X = 0|s_2)P(s_2) = .8574(.10) = P(X = 0 \text{ and } s_2)$$

The last entry in the first row is found as

P(sample shows no defectives | proportion of defectives is .10)

or

$$P(X = 0|s_3) = \binom{3}{0}(.10)^0(.90)^3 = .7290$$

Sample Outcome	Percentage Defective			Totals
	1%	5%	10%	
$X = 0$	.9703(.85)	.8574(.10)	.7290(.05)	.9469
$X = 1$	.0294(.85)	.1354(.10)	.2430(.05)	.0507
$X = 2$	.0003(.85)	.0071(.10)	.0270(.05)	.0023
$X = 3$	.0000(.85)	.0001(.10)	.0010(.05)	.0001
Totals	.85	.10	.05	1.0000

FIGURE 20.11
JOINT PROBABILITY TABLE FOR CALCULATING POSTERIOR
PROBABILITIES BASED ON A RANDOM SAMPLE OF SIZE 3.

then

$$P(X = 0|s_3)P(s_3) = .7290(.05) = P(X = 0 \text{ and } s_3)$$

The posterior probabilities for the sample outcome $X = 0$ are found as before by dividing each entry in the first row by the row total .9469 to obtain .8710, .0905, and .0385. These posterior values are very much in line with the manufacturer's prior probabilities, thus indicating that the sample outcome is in agreement with the manufacturer's preconceived feeling about the manufacturing process. The expected payoffs for these posterior probabilities are found as

$$E[r(a_1, s)] = .8710(-100) + .0905(-500) + .0385(-1000) = -171$$

and

$$E[r(a_2, s)] = .8710(-400) + .0905(-400) + .0385(-400) = -400$$

Therefore, the manufacturer's decision remains the same as that found by using the prior subjective probabilities, and no change is made in the manufacturing process.

The second row in the joint probability table in Figure 20.11 corresponds to exactly one defective item in a random sample of size 3. The entry-by-entry calculations for the second row are as follows:

Entry 1 $P(X = 1|s_1) = \binom{3}{1}(.01)^1(.99)^2 = .0294$

$\qquad P(X = 1|s_1)P(s_1) = .0294(.85) = P(X = 1 \text{ and } s_1)$

Entry 2 $P(X = 1|s_2) = \binom{3}{1}(.05)^1(.95)^2 = .1354$

$\qquad P(X = 1|s_2)P(s_2) = .1354(.10) = P(X = 1 \text{ and } s_2)$

Entry 3 $P(X = 1|s_3) = \binom{3}{1}(.10)^1(.90)^2 = .2430$

$\qquad P(X = 1|s_3)P(s_3) = .2430(.05) = P(X = 1 \text{ and } s_3)$

The posterior probabilities for $X = 1$ are found by dividing the entries in the second row by the total of the second row, .0507. These posterior probabilities are .4931, .2671, and .2397, respectively, indicating some disagreement with the manufacturer's subjective prior probabilities. The expected payoffs based on these posterior probabilities are calculated as

$$E[r(a_1, s)] = .4931(-100) + .2671(-500) + .2397(-1000) = -423$$

and

$$E[r(a_2, s)] = .4931(-400) + .2671(-400) + .2397(-400) = -400$$

Thus, the decision according to maximum expected payoff is to change the manufacturing process.

The entries in the third and fourth rows of the joint probability table are found in a similar manner. A summary of the posterior probabilities corresponding to each possible sample outcome, $X = 0, 1, 2,$ and 3, is given in Figure 20.12. The reader should be aware that the calculation of the posterior probabilities in Figure 20.12 requires carrying more than the four decimal place accuracy indicated in Figure 20.11.

THE EXPECTED VALUE OF SAMPLE INFORMATION

The material presented thus far in this section has shown how to process sample information, but a determination of the worth of sample information has not yet been made. Of course, such a determination is made prior to obtaining a sample and, as such, represents an integral part of any preposterior analysis.

To find the **expected value of sample information** perform the following steps:

THE EXPECTED VALUE OF SAMPLE INFORMATION

1. Find the maximum expected payoff based on prior subjective probabilities.
2. Find the expected payoff over all sample outcomes using a preposterior analysis.
3. Subtract the value in step 1 from the value in step 2. The result is the expected value of sample information.

To find the expected value of sample information it is helpful to have a summary of the maximum expected payoffs based on each sample outcome from the preposterior analysis. This summary appears in Figure 20.13. The marginal probabilities in Figure 20.13 are marginal probabilities given for each sample outcome in the joint probability table in Figure 20.11. The maximum expected payoffs are worked out for each sample outcome as was done earlier in this section for the outcomes $X = 0$ and $X = 1$. The maximum expected payoff, called the expected payoff over all sample outcomes, is computed from the values in Figure 20.13:

$$.9469(-171) + .0507(-400) + .0023(-400) + .0001(-400) = -183$$

| | Percentage Defective | | |
Sample Outcome	1%	5%	10%
$X = 0$	.8710	.0905	.0385
$X = 1$	.4931	.2671	.2397
$X = 2$	.1091	.3078	.5832
$X = 3$	.0134	.1973	.7893

FIGURE 20.12
POSTERIOR PROBABILITIES CORRESPONDING TO EACH POSSIBLE SAMPLE OUTCOME FROM A RANDOM SAMPLE OF SIZE $n = 3$.

Sample Outcome	Best Action	Maximum Expected Payoff	Marginal Probabilities of X
X = 0	a₁	−171	.9469
X = 1	a₂	−400	.0507
X = 2	a₂	−400	.0023
X = 3	a₂	−400	.0001

FIGURE 20.13
SUMMARY OF THE PREPOSTERIOR ANALYSIS FOR USE IN COMPUTING
THE EXPECTED VALUE OF SAMPLE INFORMATION.

This value is greater than the maximum expected payoff of -185 found earlier in this chapter based on prior information. The difference between these values, $-183 - (-185) = 2$, is the expected value of sample information. The interpretation of this value is that the manufacturer could afford to pay up to 2 units for sample information.

Of course, this does not mean that each sample is worth 2 units to the manufacturer. Some samples will be worth more and some less, depending on the particular sample values. However, prior to drawing the sample the *expected* value of the sample information in this case is 2 units. If each of the three valves in the sample costs .5 units to inspect, the sample will pay for itself in the long run. However, if each of the three valves costs 1 unit to inspect, drawing a sample is not cost effective.

In practice, these computations will usually be done on a computer. Then instead of merely examining a sample of size 3, many other sample sizes will be examined. For each sample size the cost of the sample may be subtracted from the expected value of sample information to get the net expected value of the sample. In this way the optimal sample size, in terms of maximum expected payoff, can be determined prior to sampling. The calculations are somewhat involved so they are not demonstrated here, although no new principles are used in that type of analysis.

APPLYING DECISION THEORY IN THE BUSINESS WORLD

Chapters 19 and 20 provide only a brief introduction to the area known formally as decision theory. The examples and exercises are purposely kept very simple to illustrate the principles involved. Actual applications of decision theory in the business world will differ from what you have studied in this text in the following ways.

1. **The situations will be more complex** There will be more possible actions, more states of nature possible, and there will be additional actions and states of nature following the initial actions and states of nature. Decision trees will be utilized much more, just to keep the sequences of actions, states of nature, then more actions, more states of nature, and so forth, clear in everybody's mind. Complex situations may be solved piece by piece, from right to left as represented on a decision tree, using the methods in Chapters 19 and 20.

2. **Prior distributions may be inaccurate** Prior probabilities are difficult to obtain, partly because the people who need to supply the prior probabilities do not understand what probabilities represent, and partly because many people are reluctant to put a number (a prior probability) on the possibility of a state of nature occurring, especially when there is very little firm information on which to base the number. The numbers obtained as prior probabilities may be quite inaccurate. And as if this were not enough, the choice of an optimal decision may be very sensitive to the prior probabilities. Thus a sensitivity analysis is needed, that is, an analysis is needed to determine how much the prior probabilities can change without affecting the payoff a significant amount. Every application of decision theory should be followed by an analysis of the sensitivity of the decision to the various elements that went into making the decision, if those elements have an aura of doubt surrounding them.

3. **Payoffs are difficult to determine** The entire analysis depends on the payoffs in the initial payoff matrix. These payoffs are often very difficult to determine in a manner consistent with the manager's objectives. Sometimes nonquantifiable concepts, such as the president's pet project, are very important to consider and are very difficult to incorporate into a payoff matrix.

4. **Some states of nature may be forgotten** Sometimes an action is taken and a state of nature develops that was not even considered by the decision maker. This is even more likely to happen if decision theory is not used, however. The formal structure of decision theory outlined in these chapters encourages more formal thought and discussion on each decision, and that alone has a tendency to improve the decision.

5. **Decision theory does not make decisions, managers do** One mistake commonly made by young, well-educated junior executives is to rely too much on models and theory for making a decision, and not to consider sufficiently factors such as company tradition, employer preferences, external image, and reputation.

AUSTIN'S LAW Models don't make decisions—managers do.

EXERCISES

20.23 The distributor of a new product has made the following payoff table based on the amount spent on advertising and the percentage of the market the product currently has. Determine the decision associated with the maximum expected payoff.

Advertising Budget	Percentage of Market		
	5%	10%	15%
Small	−20	10	80
Large	−100	−20	150
Prior probabilities	.2	.5	.3

20.24 Suppose the distributor in Exercise 20.23 is considering taking a random sample of four customers to determine the likelihood that customers will buy the product. Let X be a random variable denoting the number of customers in the sample who say they will buy the product. As part of a preposterior analysis, form a joint probability table similar to the one in Figure 20.11 for each sample outcome $X = 0, 1, 2, 3,$ and 4. Table A1 may be used to find $P(X = x)$ for $p = .05, .10,$ and .15 as required in this problem.

20.25 Use the joint probability table in Exercise 20.24 to find the posterior probabilities associated with each sample outcome. Be sure to carry enough decimal places in the calculations to ensure accuracy of the results.

20.26 Use the posterior probabilities found in Exercise 20.25 to determine the maximum expected payoff and corresponding decision associated with each sample outcome.

20.27 Use the results found in Exercise 20.26 to determine the expected payoff over all sample outcomes.

20.28 Find the expected value of sample information for the decision problem in Exercise 20.23 based on a random sample of size 4 and interpret the results of this calculation.

20.29 A small business is concerned that some of the $100 bills it receives in the normal course of business may be counterfeit. The store may accept or refuse each bill, with the following payoffs. These payoffs are based on profit, loss of customer's good will, and loss of cash. If the manager feels that 5% of the bills are counterfeit, determine the decision associated with the maximum expected payoff.

	Good	Counterfeit
Accept	0	− 100
Refuse	− 10	0

20.30 If the manager in Exercise 20.29 believes that 10% of the bills may be counterfeit, determine the decision associated with the maximum expected payoff. Form a new payoff matrix with the two States of Nature 5% and 10% counterfeit, and with the two actions Accept and Refuse. Let the payoffs be the expected payoffs considered in this exercise and the previous one.

20.31 The manager in Exercise 20.29 is considering taking five randomly selected bills to the bank for examination to see if they are counterfeit. Let X be a random variable denoting the number of counterfeit bills in the sample. Assume that the store manager feels there is a 50–50 chance of the percentage of counterfeit bills being 5% or 10%. Form a joint probability table, similar to the one given in Figure 20.11, for each sample outcome $X = 0, 1, 2, 3, 4,$ and 5. The binomial probabilities required to work this problem may be found from Table A1 for $p = .05$ and $p = .10$.

20.32 Use the joint probability table in Exercise 20.31 to find posterior probabilities associated with each sample outcome. Be sure to carry enough decimal places in your calculations to ensure accurate results.

20.33 Use the posterior probabilities found in Exercise 20.32 to determine the maximum expected payoff and corresponding decision associated with each sample outcome.

20.34 Use the results found in Exercise 20.33 to determine the expected payoff over all sample outcomes.

20.35 Find the expected value of sample information for the decision problem in Exercise 20.29, based on a random sample size 5, and interpret the results of this calculation.

20.4

REVIEW EXERCISES

20.36 A company planning to market a new product has established prior subjective probabilities of .60, .25, and .15 for high sales (20% of the market), moderate sales (10% of the market), and low sales (5% of the market), respectively. The company has also established the payoff table given here for the sales of the product based on the quality of material used to manufacture the product. Find the action associated with the maximum expected payoff.

	Percentage of Market		
Quality of Material	20%	10%	5%
Low	100	20	− 40
Medium	90	40	− 50
High	80	60	20
Prior probabilities	.60	.25	.15

20.37 Suppose the company in Exercise 20.36 is considering hiring a firm to conduct a survey of potential customers to assess the sales of the product. The firm's record shows that in the past they have correctly predicted high sales 80% of the time, but have incorrectly forecast high sales 40% of the time when sales were moderate and have also forecast high sales 20% of the time when sales were low. Suppose the firm forecasts high sales. Use this information to establish posterior probabilities for the sales of the new product.

20.38 Use the posterior probabilities found in Exercise 20.37 to find the action associated with the maximum expected payoff. How does this decision compare with the one made in Exercise 20.36?

20.39 Find the expected payoff under perfect information using the prior subjective probabilities given in Exercise 20.36 for the payoff table in that exercise.

20.40 Find the expected value of perfect information and interpret its meaning for the payoff table given in Exercise 20.36.

20.41 Find the minimum expected opportunity loss for the payoff table in Exercise 20.36 and compare this value with the expected value of perfect information found in Exercise 20.40.

20.42 Find the expected payoff under perfect information for the payoff table given in Exercise 20.36 using the posterior probabilities found in Exercise 20.37.

20.43 Find the expected value of perfect information for the payoff table given in Exercise 20.36 using the posterior probabilities found in Exercise 20.37. Comment on the comparison of this value with the expected value of perfect information found in Exercise 20.40 that was based on prior probabilities.

20.44 Suppose the company in Exercise 20.36 is considering taking a random sample of size 3 to determine the potential sales of the new product. Let X be a random variable denoting the number of individuals in the sample who say they will buy the new product. As part of a preposterior analysis, form a joint probability table similar to the one in Figure 20.11 for each sample outcome $X = 0, 1, 2,$ and 3. The binomial probabilities required to work this problem may be obtained from Table A1.

20.45 Use the joint probability table in Exercise 20.44 to find the posterior probabilities associated with each sample outcome. Be sure to carry enough decimal places in the calculation to ensure accuracy of the results.

20.46 Use the posterior probabilities found in Exercise 20.45 to determine the maximum expected payoff and corresponding decision associated with each sample outcome.

20.47 Use the results found in Exercise 20.46 to determine the expected payoff over all sample outcomes.

20.48 Find the expected value of sample information for the decision problem in Exercise 20.47 based on a random sample of size 3 and interpret the results of this calculation.

BIBLIOGRAPHY

Additional material on the topics presented in this chapter can be found in the following publications.

Aitchison, J. (1970). *Choice Against Chance: An Introduction to Statistical Decision Theory.* Addison-Wesley, Reading, Mass.

Chernoff, H. and Moses, L. E. (1959). *Elementary Decision Theory.* Wiley, New York.

Lindley, D. V. (1965). *Introduction to Probability and Statistics from a Bayesian Viewpoint.* Cambridge Univ. Press, London/New York.

Raiffa, H. (1968). *Decision Analysis*. Addison-Wesley, Reading, Mass.

Raiffa, H. and Schlaifer, R. (1961). *Applied Statistical Decision Theory*. Harvard Business School, Boston, Mass.

Schlaifer, R. (1959). *Probability and Statistics for Business Decisions*. McGraw-Hill, New York.

Wald, A. (1950). *Statistical Decision Functions*. Wiley, New York.

APPENDIX TABLES

TABLE A1
BINOMIAL DISTRIBUTION

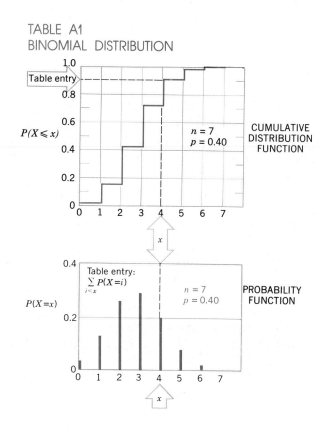

Source: Table generated by W. J. Conover and R. L. Iman.

TABLE A1
BINOMIAL DISTRIBUTION

n	x	p = .05	.10	.15	.20	.25	.30	.35	.40	.45
1	0	.9500	.9000	.8500	.8000	.7500	.7000	.6500	.6000	.5500
	1	1.0000	1.0000	1.0000	1.0000	1.0000	1.0000	1.0000	1.0000	1.0000
2	0	.9025	.8100	.7225	.6400	.5625	.4900	.4225	.3600	.3025
	1	.9975	.9900	.9775	.9600	.9375	.9100	.8775	.8400	.7975
	2	1.0000	1.0000	1.0000	1.0000	1.0000	1.0000	1.0000	1.0000	1.0000
3	0	.8574	.7290	.6141	.5120	.4219	.3430	.2746	.2160	.1664
	1	.9928	.9720	.9392	.8960	.8438	.7840	.7182	.6480	.5748
	2	.9999	.9990	.9966	.9920	.9844	.9730	.9571	.9360	.9089
	3	1.0000	1.0000	1.0000	1.0000	1.0000	1.0000	1.0000	1.0000	1.0000
4	0	.8145	.6561	.5220	.4096	.3164	.2401	.1785	.1296	.0915
	1	.9860	.9477	.8905	.8192	.7383	.6517	.5630	.4752	.3910
	2	.9995	.9963	.9880	.9728	.9492	.9163	.8735	.8208	.7585
	3	1.0000	.9999	.9995	.9984	.9961	.9919	.9850	.9744	.9590
	4	1.0000	1.0000	1.0000	1.0000	1.0000	1.0000	1.0000	1.0000	1.0000
5	0	.7738	.5905	.4437	.3277	.2373	.1681	.1160	.0778	.0503
	1	.9774	.9185	.8352	.7373	.6328	.5282	.4284	.3370	.2562
	2	.9988	.9914	.9734	.9421	.8965	.8369	.7648	.6826	.5931
	3	1.0000	.9995	.9978	.9933	.9844	.9692	.9460	.9130	.8688
	4	1.0000	1.0000	.9999	.9997	.9990	.9976	.9947	.9898	.9815
	5	1.0000	1.0000	1.0000	1.0000	1.0000	1.0000	1.0000	1.0000	1.0000
6	0	.7351	.5314	.3771	.2621	.1780	.1176	.0754	.0467	.0277
	1	.9672	.8857	.7765	.6554	.5339	.4202	.3191	.2333	.1636
	2	.9978	.9842	.9527	.9011	.8306	.7443	.6471	.5443	.4415
	3	.9999	.9987	.9941	.9830	.9624	.9295	.8826	.8208	.7447
	4	1.0000	.9999	.9996	.9984	.9954	.9891	.9777	.9590	.9308
	5	1.0000	1.0000	1.0000	.9999	.9998	.9993	.9982	.9959	.9917
	6	1.0000	1.0000	1.0000	1.0000	1.0000	1.0000	1.0000	1.0000	1.0000
7	0	.6983	.4783	.3206	.2097	.1335	.0824	.0490	.0280	.0152
	1	.9556	.8503	.7166	.5767	.4449	.3294	.2338	.1586	.1024
	2	.9962	.9743	.9262	.8520	.7564	.6471	.5323	.4199	.3164
	3	.9998	.9973	.9879	.9667	.9294	.8740	.8002	.7102	.6083
	4	1.0000	.9998	.9988	.9953	.9871	.9712	.9444	.9037	.8471
	5	1.0000	1.0000	.9999	.9996	.9987	.9962	.9910	.9812	.9643
	6	1.0000	1.0000	1.0000	1.0000	.9999	.9998	.9994	.9984	.9963
	7	1.0000	1.0000	1.0000	1.0000	1.0000	1.0000	1.0000	1.0000	1.0000

TABLE A1
BINOMIAL DISTRIBUTION (CONTINUED)

n	x	p = .50	.55	.60	.65	.70	.75	.80	.85	.90	.95
1	0	.5000	.4500	.4000	.3500	.3000	.2500	.2000	.1500	.1000	.0500
	1	1.0000	1.0000	1.0000	1.0000	1.0000	1.0000	1.0000	1.0000	1.0000	1.0000
2	0	.2500	.2025	.1600	.1225	.0900	.0625	.0400	.0225	.0100	.0025
	1	.7500	.6975	.6400	.5775	.5100	.4375	.3600	.2775	.1900	.0975
	2	1.0000	1.0000	1.0000	1.0000	1.0000	1.0000	1.0000	1.0000	1.0000	1.0000
3	0	.1250	.0911	.0640	.0429	.0270	.0156	.0080	.0034	.0010	.0001
	1	.5000	.4252	.3520	.2818	.2160	.1562	.1040	.0608	.0280	.0072
	2	.8750	.8336	.7840	.7254	.6570	.5781	.4880	.3859	.2710	.1426
	3	1.0000	1.0000	1.0000	1.0000	1.0000	1.0000	1.0000	1.0000	1.0000	1.0000
4	0	.0625	.0410	.0256	.0150	.0081	.0039	.0016	.0005	.000i	.0000
	1	.3125	.2415	.1792	.1265	.0837	.0508	.0272	.0120	.0037	.0005
	2	.6875	.6090	.5248	.4370	.3483	.2617	.1808	.1095	.0523	.0140
	3	.9375	.9085	.8704	.8215	.7599	.6836	.5904	.4780	.3439	.1855
	4	1.0000	1.0000	1.0000	1.0000	1.0000	1.0000	1.0000	1.0000	1.0000	1.0000
5	0	.0312	.0185	.0102	.0053	.0024	.0010	.0003	.0001	.0000	.0000
	1	.1875	.1312	.0870	.0540	.0308	.0156	.0067	.0022	.0005	.0000
	2	.5000	.4069	.3174	.2352	.1631	.1035	.0579	.0266	.0086	.0012
	3	.8125	.7438	.6630	.5716	.4718	.3672	.2627	.1648	.0815	.0226
	4	.9688	.9497	.9222	.8840	.8319	.7627	.6723	.5563	.4095	.2262
	5	1.0000	1.0000	1.0000	1.0000	1.0000	1.0000	1.0000	1.0000	1.0000	1.0000
6	0	.0156	.0083	.0041	.0018	.0007	.0002	.0001	.0000	.0000	.0000
	1	.1094	.0692	.0410	.0223	.0109	.0046	.0016	.0004	.0001	.0000
	2	.3438	.2553	.1792	.1174	.0705	.0376	.0170	.0059	.0013	.0001
	3	.6562	.5585	.4557	.3529	.2557	.1694	.0989	.0473	.0158	.0022
	4	.8906	.8364	.7667	.6809	.5798	.4661	.3446	.2235	.1143	.0328
	5	.9844	.9723	.9533	.9246	.8824	.8220	.7379	.6229	.4686	.2649
	6	1.0000	1.0000	1.0000	1.0000	1.0000	1.0000	1.0000	1.0000	1.0000	1.0000
7	0	.0078	.0037	.0016	.0006	.0002	.0001	.0000	.0000	.0000	.0000
	1	.0625	.0357	.0188	.0090	.0038	.0013	.0004	.0001	.0000	.0000
	2	.2266	.1529	.0963	.0556	.0288	.0129	.0047	.0012	.0002	.0000
	3	.5000	.3917	.2898	.1998	.1260	.0706	.0333	.0121	.0027	.0002
	4	.7734	.6836	.5801	.4677	.3529	.2436	.1480	.0738	.0257	.0038
	5	.9375	.8976	.8414	.7662	.6706	.5551	.4233	.2834	.1497	.0444
	6	.9922	.9848	.9720	.9510	.9176	.8665	.7903	.6794	.5217	.3017
	7	1.0000	1.0000	1.0000	1.0000	1.0000	1.0000	1.0000	1.0000	1.0000	1.0000

TABLE A1
BINOMIAL DISTRIBUTION (CONTINUED)

n	x	p = .05	.10	.15	.20	.25	.30	.35	.40	.45
8	0	.6634	.4305	.2725	.1678	.1001	.0576	.0319	.0168	.0084
	1	.9428	.8131	.6572	.5033	.3671	.2553	.1691	.1064	.0632
	2	.9942	.9619	.8948	.7969	.6785	.5518	.4278	.3154	.2201
	3	.9996	.9950	.9786	.9437	.8862	.8059	.7064	.5941	.4770
	4	1.0000	.9996	.9971	.9896	.9727	.9420	.8939	.8263	.7396
	5	1.0000	1.0000	.9998	.9988	.9958	.9887	.9747	.9502	.9115
	6	1.0000	1.0000	1.0000	.9999	.9996	.9987	.9964	.9915	.9819
	7	1.0000	1.0000	1.0000	1.0000	1.0000	.9999	.9998	.9993	.9983
	8	1.0000	1.0000	1.0000	1.0000	1.0000	1.0000	1.0000	1.0000	1.0000
9	0	.6302	.3874	.2316	.1342	.0751	.0404	.0207	.0101	.0046
	1	.9288	.7748	.5995	.4362	.3003	.1960	.1211	.0705	.0385
	2	.9916	.9470	.8591	.7382	.6007	.4628	.3373	.2318	.1495
	3	.9994	.9917	.9661	.9144	.8343	.7297	.6089	.4826	.3614
	4	1.0000	.9991	.9944	.9804	.9511	.9012	.8283	.7334	.6214
	5	1.0000	.9999	.9994	.9969	.9900	.9747	.9464	.9006	.8342
	6	1.0000	1.0000	1.0000	.9997	.9987	.9957	.9888	.9750	.9502
	7	1.0000	1.0000	1.0000	1.0000	.9999	.9996	.9986	.9962	.9909
	8	1.0000	1.0000	1.0000	1.0000	1.0000	1.0000	.9999	.9997	.9992
	9	1.0000	1.0000	1.0000	1.0000	1.0000	1.0000	1.0000	1.0000	1.0000
10	0	.5987	.3487	.1969	.1074	.0563	.0282	.0135	.0060	.0025
	1	.9139	.7361	.5443	.3758	.2440	.1493	.0860	.0464	.0233
	2	.9885	.9298	.8202	.6778	.5256	.3828	.2616	.1673	.0996
	3	.9990	.9872	.9500	.8791	.7759	.6496	.5138	.3823	.2660
	4	.9999	.9984	.9901	.9672	.9219	.8497	.7515	.6331	.5044
	5	1.0000	.9999	.9986	.9936	.9803	.9527	.9051	.8338	.7384
	6	1.0000	1.0000	.9999	.9991	.9965	.9894	.9740	.9452	.8980
	7	1.0000	1.0000	1.0000	.9999	.9996	.9984	.9952	.9877	.9726
	8	1.0000	1.0000	1.0000	1.0000	1.0000	.9999	.9995	.9983	.9955
	9	1.0000	1.0000	1.0000	1.0000	1.0000	1.0000	1.0000	.9999	.9997
	10	1.0000	1.0000	1.0000	1.0000	1.0000	1.0000	1.0000	1.0000	1.0000
11	0	.5688	.3138	.1673	.0859	.0422	.0198	.0088	.0036	.0014
	1	.8981	.6974	.4922	.3221	.1971	.1130	.0606	.0302	.0139
	2	.9848	.9104	.7788	.6174	.4552	.3127	.2001	.1189	.0652
	3	.9984	.9815	.9306	.8389	.7133	.5696	.4256	.2963	.1911
	4	.9999	.9972	.9841	.9496	.8854	.7897	.6683	.5328	.3971
	5	1.0000	.9997	.9973	.9883	.9657	.9218	.8513	.7535	.6331
	6	1.0000	1.0000	.9997	.9980	.9924	.9784	.9499	.9006	.8262
	7	1.0000	1.0000	1.0000	.9998	.9988	.9957	.9878	.9707	.9390
	8	1.0000	1.0000	1.0000	1.0000	.9999	.9994	.9980	.9941	.9852
	9	1.0000	1.0000	1.0000	1.0000	1.0000	1.0000	.9998	.9993	.9978
	10	1.0000	1.0000	1.0000	1.0000	1.0000	1.0000	1.0000	1.0000	.9998
	11	1.0000	1.0000	1.0000	1.0000	1.0000	1.0000	1.0000	1.0000	1.0000

TABLE A1
BINOMIAL DISTRIBUTION (CONTINUED)

n	x	p = .50	.55	.60	.65	.70	.75	.80	.85	.90	.95
8	0	.0039	.0017	.0007	.0002	.0001	.0000	.0000	.0000	.0000	.0000
	1	.0352	.0181	.0085	.0036	.0013	.0004	.0001	.0000	.0000	.0000
	2	.1445	.0885	.0498	.0253	.0113	.0042	.0012	.0002	.0000	.0000
	3	.3633	.2604	.1737	.1061	.0580	.0273	.0104	.0029	.0004	.0000
	4	.6367	.5230	.4059	.2936	.1941	.1138	.0563	.0214	.0050	.0004
	5	.8555	.7799	.6846	.5722	.4482	.3215	.2031	.1052	.0381	.0058
	6	.9648	.9368	.8936	.8309	.7447	.6329	.4967	.3428	.1869	.0572
	7	.9961	.9916	.9832	.9681	.9424	.8999	.8322	.7275	.5695	.3366
	8	1.0000	1.0000	1.0000	1.0000	1.0000	1.0000	1.0000	1.0000	1.0000	1.0000
9	0	.0020	.0008	.0003	.0001	.0000	.0000	.0000	.0000	.0000	.0000
	1	.0195	.0091	.0038	.0014	.0004	.0001	.0000	.0000	.0000	.0000
	2	.0898	.0498	.0250	.0112	.0043	.0013	.0003	.0000	.0000	.0000
	3	.2539	.1658	.0994	.0536	.0253	.0100	.0031	.0006	.0001	.0000
	4	.5000	.3786	.2666	.1717	.0988	.0489	.0196	.0056	.0009	.0000
	5	.7461	.6386	.5174	.3911	.2703	.1657	.0856	.0339	.0083	.0006
	6	.9102	.8505	.7682	.6627	.5372	.3993	.2618	.1409	.0530	.0084
	7	.9805	.9615	.9295	.8789	.8040	.6997	.5638	.4005	.2252	.0712
	8	.9980	.9954	.9899	.9793	.9596	.9249	.8658	.7684	.6126	.3698
	9	1.0000	1.0000	1.0000	1.0000	1.0000	1.0000	1.0000	1.0000	1.0000	1.0000
10	0	.0010	.0003	.0001	.0000	.0000	.0000	.0000	.0000	.0000	.0000
	1	.0107	.0045	.0017	.0005	.0001	.0000	.0000	.0000	.0000	.0000
	2	.0547	.0274	.0123	.0048	.0016	.0004	.0001	.0000	.0000	.0000
	3	.1719	.1020	.0548	.0260	.0106	.0035	.0009	.0001	.0000	.0000
	4	.3770	.2616	.1662	.0949	.0473	.0197	.0064	.0014	.0001	.0000
	5	.6230	.4956	.3669	.2485	.1503	.0781	.0328	.0099	.0016	.0001
	6	.8281	.7340	.6177	.4862	.3504	.2241	.1209	.0500	.0128	.0010
	7	.9453	.9004	.8327	.7384	.6172	.4744	.3222	.1798	.0702	.0115
	8	.9893	.9767	.9536	.9140	.8507	.7560	.6242	.4557	.2639	.0861
	9	.9990	.9975	.9940	.9865	.9718	.9437	.8926	.8031	.6513	.4013
	10	1.0000	1.0000	1.0000	1.0000	1.0000	1.0000	1.0000	1.0000	1.0000	1.0000
11	0	.0005	.0002	.0000	.0000	.0000	.0000	.0000	.0000	.0000	.0000
	1	.0059	.0022	.0007	.0002	.0000	.0000	.0000	.0000	.0000	.0000
	2	.0327	.0148	.0059	.0020	.0006	.0001	.0000	.0000	.0000	.0000
	3	.1133	.0610	.0293	.0122	.0043	.0012	.0002	.0000	.0000	.0000
	4	.2744	.1738	.0994	.0501	.0216	.0076	.0020	.0003	.0000	.0000
	5	.5000	.3669	.2465	.1487	.0782	.0343	.0117	.0027	.0003	.0000
	6	.7256	.6029	.4672	.3317	.2103	.1146	.0504	.0159	.0028	.0001
	7	.8867	.8089	.7037	.5744	.4304	.2867	.1611	.0694	.0185	.0016
	8	.9673	.9348	.8811	.7999	.6873	.5448	.3826	.2212	.0896	.0152
	9	.9941	.9861	.9698	.9394	.8870	.8029	.6779	.5078	.3026	.1019
	10	.9995	.9986	.9964	.9912	.9802	.9578	.9141	.8327	.6862	.4312
	11	1.0000	1.0000	1.0000	1.0000	1.0000	1.0000	1.0000	1.0000	1.0000	1.0000

TABLE A1
BINOMIAL DISTRIBUTION (CONTINUED)

n	x	p = .05	.10	.15	.20	.25	.30	.35	.40	.45
12	0	.5404	.2824	.1422	.0687	.0317	.0138	.0057	.0022	.0008
	1	.8816	.6590	.4435	.2749	.1584	.0850	.0424	.0196	.0083
	2	.9804	.8891	.7358	.5583	.3907	.2528	.1513	.0834	.0421
	3	.9978	.9744	.9078	.7946	.6488	.4925	.3467	.2253	.1345
	4	.9998	.9957	.9761	.9274	.8424	.7237	.5833	.4382	.3044
	5	1.0000	.9995	.9954	.9806	.9456	.8822	.7873	.6652	.5269
	6	1.0000	.9999	.9993	.9961	.9857	.9614	.9154	.8418	.7393
	7	1.0000	1.0000	.9999	.9994	.9972	.9905	.9745	.9427	.8883
	8	1.0000	1.0000	1.0000	.9999	.9996	.9983	.9944	.9847	.9644
	9	1.0000	1.0000	1.0000	1.0000	1.0000	.9998	.9992	.9972	.9921
	10	1.0000	1.0000	1.0000	1.0000	1.0000	1.0000	.9999	.9997	.9989
	11	1.0000	1.0000	1.0000	1.0000	1.0000	1.0000	1.0000	1.0000	.9999
	12	1.0000	1.0000	1.0000	1.0000	1.0000	1.0000	1.0000	1.0000	1.0000
13	0	.5133	.2542	.1209	.0550	.0238	.0097	.0037	.0013	.0004
	1	.8646	.6213	.3983	.2336	.1267	.0637	.0296	.0126	.0049
	2	.9755	.8661	.6920	.5017	.3326	.2025	.1132	.0579	.0269
	3	.9969	.9658	.8820	.7473	.5843	.4206	.2783	.1686	.0929
	4	.9997	.9935	.9658	.9009	.7940	.6543	.5005	.3530	.2279
	5	1.0000	.9991	.9925	.9700	.9198	.8346	.7159	.5744	.4268
	6	1.0000	.9999	.9987	.9930	.9757	.9376	.8705	.7712	.6437
	7	1.0000	1.0000	.9998	.9988	.9944	.9818	.9538	.9023	.8212
	8	1.0000	1.0000	1.0000	.9998	.9990	.9960	.9874	.9679	.9302
	9	1.0000	1.0000	1.0000	1.0000	.9999	.9993	.9975	.9922	.9797
	10	1.0000	1.0000	1.0000	1.0000	1.0000	.9999	.9997	.9987	.9959
	11	1.0000	1.0000	1.0000	1.0000	1.0000	1.0000	1.0000	.9999	.9995
	12	1.0000	1.0000	1.0000	1.0000	1.0000	1.0000	1.0000	1.0000	1.0000
	13	1.0000	1.0000	1.0000	1.0000	1.0000	1.0000	1.0000	1.0000	1.0000
14	0	.4877	.2288	.1028	.0440	.0178	.0068	.0024	.0008	.0002
	1	.8470	.5846	.3567	.1979	.1010	.0475	.0205	.0081	.0029
	2	.9699	.8416	.6479	.4481	.2811	.1608	.0839	.0398	.0170
	3	.9958	.9559	.8535	.6982	.5213	.3552	.2205	.1243	.0632
	4	.9996	.9908	.9533	.8702	.7415	.5842	.4227	.2793	.1672
	5	1.0000	.9985	.9885	.9561	.8883	.7805	.6405	.4859	.3373
	6	1.0000	.9998	.9978	.9884	.9617	.9067	.8164	.6925	.5461
	7	1.0000	1.0000	.9997	.9976	.9897	.9685	.9247	.8499	.7414
	8	1.0000	1.0000	1.0000	.9996	.9978	.9917	.9757	.9417	.8811
	9	1.0000	1.0000	1.0000	1.0000	.9997	.9983	.9940	.9825	.9574
	10	1.0000	1.0000	1.0000	1.0000	1.0000	.9998	.9989	.9961	.9886
	11	1.0000	1.0000	1.0000	1.0000	1.0000	1.0000	.9999	.9994	.9978
	12	1.0000	1.0000	1.0000	1.0000	1.0000	1.0000	1.0000	.9999	.9997
	13	1.0000	1.0000	1.0000	1.0000	1.0000	1.0000	1.0000	1.0000	1.0000
	14	1.0000	1.0000	1.0000	1.0000	1.0000	1.0000	1.0000	1.0000	1.0000

TABLE A1
BINOMIAL DISTRIBUTION (CONTINUED)

n	x	p = .50	.55	.60	.65	.70	.75	.80	.85	.90	.95
12	0	.0002	.0001	.0000	.0000	.0000	.0000	.0000	.0000	.0000	.0000
	1	.0032	.0011	.0003	.0001	.0000	.0000	.0000	.0000	.0000	.0000
	2	.0193	.0079	.0028	.0008	.0002	.0000	.0000	.0000	.0000	.0000
	3	.0730	.0356	.0153	.0056	.0017	.0004	.0001	.0000	.0000	.0000
	4	.1938	.1117	.0573	.0255	.0095	.0028	.0006	.0001	.0000	.0000
	5	.3872	.2607	.1582	.0846	.0386	.0143	.0039	.0007	.0001	.0000
	6	.6128	.4731	.3348	.2127	.1178	.0544	.0194	.0046	.0005	.0000
	7	.8062	.6956	.5618	.4167	.2763	.1576	.0726	.0239	.0043	.0002
	8	.9270	.8655	.7747	.6533	.5075	.3512	.2054	.0922	.0256	.0022
	9	.9807	.9579	.9166	.8487	.7472	.6093	.4417	.2642	.1109	.0196
	10	.9968	.9917	.9804	.9576	.9150	.8416	.7251	.5565	.3410	.1184
	11	.9998	.9992	.9978	.9943	.9862	.9683	.9313	.8578	.7176	.4596
	12	1.0000	1.0000	1.0000	1.0000	1.0000	1.0000	1.0000	1.0000	1.0000	1.0000
13	0	.0001	.0000	.0000	.0000	.0000	.0000	.0000	.0000	.0000	.0000
	1	.0017	.0005	.0001	.0000	.0000	.0000	.0000	.0000	.0000	.0000
	2	.0112	.0041	.0013	.0003	.0001	.0000	.0000	.0000	.0000	.0000
	3	.0461	.0203	.0078	.0025	.0007	.0001	.0000	.0000	.0000	.0000
	4	.1334	.0698	.0321	.0126	.0040	.0010	.0002	.0000	.0000	.0000
	5	.2905	.1788	.0977	.0462	.0182	.0056	.0012	.0002	.0000	.0000
	6	.5000	.3563	.2288	.1295	.0624	.0243	.0070	.0013	.0001	.0000
	7	.7095	.5732	.4256	.2841	.1654	.0802	.0300	.0075	.0009	.0000
	8	.8666	.7721	.6470	.4995	.3457	.2060	.0991	.0342	.0065	.0003
	9	.9539	.9071	.8314	.7217	.5794	.4157	.2527	.1180	.0342	.0031
	10	.9888	.9731	.9421	.8868	.7975	.6674	.4983	.3080	.1339	.0245
	11	.9983	.9951	.9874	.9704	.9363	.8733	.7664	.6017	.3787	.1354
	12	.9999	.9996	.9987	.9963	.9903	.9762	.9450	.8791	.7458	.4867
	13	1.0000	1.0000	1.0000	1.0000	1.0000	1.0000	1.0000	1.0000	1.0000	1.0000
14	0	.0001	.0000	.0000	.0000	.0000	.0000	.0000	.0000	.0000	.0000
	1	.0009	.0003	.0001	.0000	.0000	.0000	.0000	.0000	.0000	.0000
	2	.0065	.0022	.0006	.0001	.0000	.0000	.0000	.0000	.0000	.0000
	3	.0287	.0114	.0039	.0011	.0002	.0000	.0000	.0000	.0000	.0000
	4	.0898	.0426	.0175	.0060	.0017	.0003	.0000	.0000	.0000	.0000
	5	.2120	.1189	.0583	.0243	.0083	.0022	.0004	.0000	.0000	.0000
	6	.3953	.2586	.1501	.0753	.0315	.0103	.0024	.0003	.0000	.0000
	7	.6047	.4539	.3075	.1836	.0933	.0383	.0116	.0022	.0002	.0000
	8	.7880	.6627	.5141	.3595	.2195	.1117	.0439	.0115	.0015	.0000
	9	.9102	.8328	.7207	.5773	.4158	.2585	.1298	.0467	.0092	.0004
	10	.9713	.9368	.8757	.7795	.6448	.4787	.3018	.1465	.0441	.0042
	11	.9935	.9830	.9602	.9161	.8392	.7189	.5519	.3521	.1584	.0301
	12	.9991	.9971	.9919	.9795	.9525	.8990	.8021	.6433	.4154	.1530
	13	.9999	.9998	.9992	.9976	.9932	.9822	.9560	.8972	.7712	.5123
	14	1.0000	1.0000	1.0000	1.0000	1.0000	1.0000	1.0000	1.0000	1.0000	1.0000

TABLE A1
BINOMIAL DISTRIBUTION (CONTINUED)

n	x	p = .05	.10	.15	.20	.25	.30	.35	.40	.45
15	**0**	.4633	.2059	.0874	.0352	.0134	.0047	.0016	.0005	.0001
	1	.8290	.5490	.3186	.1671	.0802	.0353	.0142	.0052	.0017
	2	.9638	.8159	.6042	.3980	.2361	.1268	.0617	.0271	.0107
	3	.9945	.9444	.8227	.6482	.4613	.2969	.1727	.0905	.0424
	4	.9994	.9873	.9383	.8358	.6865	.5155	.3519	.2173	.1204
	5	.9999	.9978	.9832	.9389	.8516	.7216	.5643	.4032	.2608
	6	1.0000	.9997	.9964	.9819	.9434	.8689	.7548	.6098	.4522
	7	1.0000	1.0000	.9994	.9958	.9827	.9500	.8868	.7869	.6535
	8	1.0000	1.0000	.9999	.9992	.9958	.9848	.9578	.9050	.8182
	9	1.0000	1.0000	1.0000	.9999	.9992	.9963	.9876	.9662	.9231
	10	1.0000	1.0000	1.0000	1.0000	.9999	.9993	.9972	.9907	.9745
	11	1.0000	1.0000	1.0000	1.0000	1.0000	.9999	.9995	.9981	.9937
	12	1.0000	1.0000	1.0000	1.0000	1.0000	1.0000	.9999	.9997	.9989
	13	1.0000	1.0000	1.0000	1.0000	1.0000	1.0000	1.0000	1.0000	.9999
	14	1.0000	1.0000	1.0000	1.0000	1.0000	1.0000	1.0000	1.0000	1.0000
	15	1.0000	1.0000	1.0000	1.0000	1.0000	1.0000	1.0000	1.0000	1.0000
16	**0**	.4401	.1853	.0743	.0281	.0100	.0033	.0010	.0003	.0001
	1	.8108	.5147	.2839	.1407	.0635	.0261	.0098	.0033	.0010
	2	.9571	.7892	.5614	.3518	.1971	.0994	.0451	.0183	.0066
	3	.9930	.9316	.7899	.5981	.4050	.2459	.1339	.0651	.0281
	4	.9991	.9830	.9209	.7982	.6302	.4499	.2892	.1666	.0853
	5	.9999	.9967	.9765	.9183	.8103	.6598	.4900	.3288	.1976
	6	1.0000	.9995	.9944	.9733	.9204	.8247	.6881	.5272	.3660
	7	1.0000	.9999	.9989	.9930	.9729	.9256	.8406	.7161	.5629
	8	1.0000	1.0000	.9998	.9985	.9925	.9743	.9329	.8577	.7441
	9	1.0000	1.0000	1.0000	.9998	.9984	.9929	.9771	.9417	.8759
	10	1.0000	1.0000	1.0000	1.0000	.9997	.9984	.9938	.9809	.9514
	11	1.0000	1.0000	1.0000	1.0000	1.0000	.9997	.9987	.9951	.9851
	12	1.0000	1.0000	1.0000	1.0000	1.0000	1.0000	.9998	.9991	.9965
	13	1.0000	1.0000	1.0000	1.0000	1.0000	1.0000	1.0000	.9999	.9994
	14	1.0000	1.0000	1.0000	1.0000	1.0000	1.0000	1.0000	1.0000	.9999
	15	1.0000	1.0000	1.0000	1.0000	1.0000	1.0000	1.0000	1.0000	1.0000
	16	1.0000	1.0000	1.0000	1.0000	1.0000	1.0000	1.0000	1.0000	1.0000

TABLE A1
BINOMIAL DISTRIBUTION (CONTINUED)

n	x	p = .50	.55	.60	.65	.70	.75	.80	.85	.90	.95
15	0	.0000	.0000	.0000	.0000	.0000	.0000	.0000	.0000	.0000	.0000
	1	.0005	.0001	.0000	.0000	.0000	.0000	.0000	.0000	.0000	.0000
	2	.0037	.0011	.0003	.0001	.0000	.0000	.0000	.0000	.0000	.0000
	3	.0176	.0063	.0019	.0005	.0001	.0000	.0000	.0000	.0000	.0000
	4	.0592	.0255	.0093	.0028	.0007	.0001	.0000	.0000	.0000	.0000
	5	.1509	.0769	.0338	.0124	.0037	.0008	.0001	.0000	.0000	.0000
	6	.3036	.1818	.0950	.0422	.0152	.0042	.0008	.0001	.0000	.0000
	7	.5000	.3465	.2131	.1132	.0500	.0173	.0042	.0006	.0000	.0000
	8	.6964	.5478	.3902	.2452	.1311	.0566	.0181	.0036	.0003	.0000
	9	.8491	.7392	.5968	.4357	.2784	.1484	.0611	.0168	.0022	.0001
	10	.9408	.8796	.7827	.6481	.4845	.3135	.1642	.0617	.0127	.0006
	11	.9824	.9576	.9095	.8273	.7031	.5387	.3518	.1773	.0556	.0055
	12	.9963	.9893	.9729	.9383	.8732	.7639	.6020	.3958	.1841	.0362
	13	.9995	.9983	.9948	.9858	.9647	.9198	.8329	.6814	.4510	.1710
	14	1.0000	.9999	.9995	.9984	.9953	.9866	.9648	.9126	.7941	.5367
	15	1.0000	1.0000	1.0000	1.0000	1.0000	1.0000	1.0000	1.0000	1.0000	1.0000
16	0	.0000	.0000	.0000	.0000	.0000	.0000	.0000	.0000	.0000	.0000
	1	.0003	.0001	.0000	.0000	.0000	.0000	.0000	.0000	.0000	.0000
	2	.0021	.0006	.0001	.0000	.0000	.0000	.0000	.0000	.0000	.0000
	3	.0106	.0035	.0009	.0002	.0000	.0000	.0000	.0000	.0000	.0000
	4	.0384	.0149	.0049	.0013	.0003	.0000	.0000	.0000	.0000	.0000
	5	.1051	.0486	.0191	.0062	.0016	.0003	.0000	.0000	.0000	.0000
	6	.2272	.1241	.0583	.0229	.0071	.0016	.0002	.0000	.0000	.0000
	7	.4018	.2559	.1423	.0671	.0257	.0075	.0015	.0002	.0000	.0000
	8	.5982	.4371	.2839	.1594	.0744	.0271	.0070	.0011	.0001	.0000
	9	.7728	.6340	.4728	.3119	.1753	.0796	.0267	.0056	.0005	.0000
	10	.8949	.8024	.6712	.5100	.3402	.1897	.0817	.0235	.0033	.0001
	11	.9616	.9147	.8334	.7108	.5501	.3698	.2018	.0791	.0170	.0009
	12	.9894	.9719	.9349	.8661	.7541	.5950	.4019	.2101	.0684	.0070
	13	.9979	.9934	.9817	.9549	.9006	.8029	.6482	.4386	.2108	.0429
	14	.9997	.9990	.9967	.9902	.9739	.9365	.8593	.7161	.4853	.1892
	15	1.0000	.9999	.9997	.9990	.9967	.9900	.9719	.9257	.8147	.5599
	16	1.0000	1.0000	1.0000	1.0000	1.0000	1.0000	1.0000	1.0000	1.0000	1.0000

TABLE A1
BINOMIAL DISTRIBUTION (CONTINUED)

n	x	p = .05	.10	.15	.20	.25	.30	.35	.40	.45
17	0	.4181	.1668	.0631	.0225	.0075	.0023	.0007	.0002	.0000
	1	.7922	.4818	.2525	.1182	.0501	.0193	.0067	.0021	.0006
	2	.9497	.7618	.5198	.3096	.1637	.0774	.0327	.0123	.0041
	3	.9912	.9174	.7556	.5489	.3530	.2019	.1028	.0464	.0184
	4	.9988	.9779	.9013	.7582	.5739	.3887	.2348	.1260	.0596
	5	.9999	.9953	.9681	.8943	.7653	.5968	.4197	.2639	.1471
	6	1.0000	.9992	.9917	.9623	.8929	.7752	.6188	.4478	.2902
	7	1.0000	.9999	.9983	.9891	.9598	.8954	.7872	.6405	.4743
	8	1.0000	1.0000	.9997	.9974	.9876	.9597	.9006	.8011	.6626
	9	1.0000	1.0000	1.0000	.9995	.9969	.9873	.9617	.9081	.8166
	10	1.0000	1.0000	1.0000	.9999	.9994	.9968	.9880	.9652	.9174
	11	1.0000	1.0000	1.0000	1.0000	.9999	.9993	.9970	.9894	.9699
	12	1.0000	1.0000	1.0000	1.0000	1.0000	.9999	.9994	.9975	.9914
	13	1.0000	1.0000	1.0000	1.0000	1.0000	1.0000	.9999	.9995	.9981
	14	1.0000	1.0000	1.0000	1.0000	1.0000	1.0000	1.0000	.9999	.9997
	15	1.0000	1.0000	1.0000	1.0000	1.0000	1.0000	1.0000	1.0000	1.0000
	16	1.0000	1.0000	1.0000	1.0000	1.0000	1.0000	1.0000	1.0000	1.0000
	17	1.0000	1.0000	1.0000	1.0000	1.0000	1.0000	1.0000	1.0000	1.0000
18	0	.3972	.1501	.0536	.0180	.0056	.0016	.0004	.0001	.0000
	1	.7735	.4503	.2241	.0991	.0395	.0142	.0046	.0013	.0003
	2	.9419	.7338	.4797	.2713	.1353	.0600	.0236	.0082	.0025
	3	.9891	.9018	.7202	.5010	.3057	.1646	.0783	.0328	.0120
	4	.9985	.9718	.8794	.7164	.5187	.3327	.1886	.0942	.0411
	5	.9998	.9936	.9581	.8671	.7175	.5344	.3550	.2088	.1077
	6	1.0000	.9988	.9882	.9487	.8610	.7217	.5491	.3743	.2258
	7	1.0000	.9998	.9973	.9837	.9431	.8593	.7283	.5634	.3915
	8	1.0000	1.0000	.9995	.9957	.9807	.9404	.8609	.7368	.5778
	9	1.0000	1.0000	.9999	.9991	.9946	.9790	.9403	.8653	.7473
	10	1.0000	1.0000	1.0000	.9998	.9988	.9939	.9788	.9424	.8720
	11	1.0000	1.0000	1.0000	1.0000	.9998	.9986	.9938	.9797	.9463
	12	1.0000	1.0000	1.0000	1.0000	1.0000	.9997	.9986	.9942	.9817
	13	1.0000	1.0000	1.0000	1.0000	1.0000	1.0000	.9997	.9987	.9951
	14	1.0000	1.0000	1.0000	1.0000	1.0000	1.0000	1.0000	.9998	.9990
	15	1.0000	1.0000	1.0000	1.0000	1.0000	1.0000	1.0000	1.0000	.9999
	16	1.0000	1.0000	1.0000	1.0000	1.0000	1.0000	1.0000	1.0000	1.0000
	17	1.0000	1.0000	1.0000	1.0000	1.0000	1.0000	1.0000	1.0000	1.0000
	18	1.0000	1.0000	1.0000	1.0000	1.0000	1.0000	1.0000	1.0000	1.0000

TABLE A1
BINOMIAL DISTRIBUTION (CONTINUED)

n	x	p = .50	.55	.60	.65	.70	.75	.80	.85	.90	.95
17	0	.0000	.0000	.0000	.0000	.0000	.0000	.0000	.0000	.0000	.0000
	1	.0001	.0000	.0000	.0000	.0000	.0000	.0000	.0000	.0000	.0000
	2	.0012	.0003	.0001	.0000	.0000	.0000	.0000	.0000	.0000	.0000
	3	.0064	.0019	.0005	.0001	.0000	.0000	.0000	.0000	.0000	.0000
	4	.0245	.0086	.0025	.0006	.0001	.0000	.0000	.0000	.0000	.0000
	5	.0717	.0301	.0106	.0030	.0007	.0001	.0000	.0000	.0000	.0000
	6	.1662	.0826	.0348	.0120	.0032	.0006	.0001	.0000	.0000	.0000
	7	.3145	.1834	.0919	.0383	.0127	.0031	.0005	.0000	.0000	.0000
	8	.5000	.3374	.1989	.0994	.0403	.0124	.0026	.0003	.0000	.0000
	9	.6855	.5257	.3595	.2128	.1046	.0402	.0109	.0017	.0001	.0000
	10	.8338	.7098	.5522	.3812	.2248	.1071	.0377	.0083	.0008	.0000
	11	.9283	.8529	.7361	.5803	.4032	.2347	.1057	.0319	.0047	.0001
	12	.9755	.9404	.8740	.7652	.6113	.4261	.2418	.0987	.0221	.0012
	13	.9936	.9816	.9536	.8972	.7981	.6470	.4511	.2444	.0826	.0088
	14	.9988	.9959	.9877	.9673	.9226	.8363	.6904	.4802	.2382	.0503
	15	.9999	.9994	.9979	.9933	.9807	.9499	.8818	.7475	.5182	.2078
	16	1.0000	1.0000	.9998	.9993	.9977	.9925	.9775	.9369	.8332	.5819
	17	1.0000	1.0000	1.0000	1.0000	1.0000	1.0000	1.0000	1.0000	1.0000	1.0000
18	0	.0000	.0000	.0000	.0000	.0000	.0000	.0000	.0000	.0000	.0000
	1	.0001	.0000	.0000	.0000	.0000	.0000	.0000	.0000	.0000	.0000
	2	.0007	.0001	.0000	.0000	.0000	.0000	.0000	.0000	.0000	.0000
	3	.0038	.0010	.0002	.0000	.0000	.0000	.0000	.0000	.0000	.0000
	4	.0154	.0049	.0013	.0003	.0000	.0000	.0000	.0000	.0000	.0000
	5	.0481	.0183	.0058	.0014	.0003	.0000	.0000	.0000	.0000	.0000
	6	.1189	.0537	.0203	.0062	.0014	.0002	.0000	.0000	.0000	.0000
	7	.2403	.1280	.0576	.0212	.0061	.0012	.0002	.0000	.0000	.0000
	8	.4073	.2527	.1347	.0597	.0210	.0054	.0009	.0001	.0000	.0000
	9	.5927	.4222	.2632	.1391	.0596	.0193	.0043	.0005	.0000	.0000
	10	.7597	.6085	.4366	.2717	.1407	.0569	.0163	.0027	.0002	.0000
	11	.8811	.7742	.6257	.4509	.2783	.1390	.0513	.0118	.0012	.0000
	12	.9519	.8923	.7912	.6450	.4656	.2825	.1329	.0419	.0064	.0002
	13	.9846	.9589	.9058	.8114	.6673	.4813	.2836	.1206	.0282	.0015
	14	.9962	.9880	.9672	.9217	.8354	.6943	.4990	.2798	.0982	.0109
	15	.9993	.9975	.9918	.9764	.9400	.8647	.7287	.5203	.2662	.0581
	16	.9999	.9997	.9987	.9954	.9858	.9605	.9009	.7759	.5497	.2265
	17	1.0000	1.0000	.9999	.9996	.9984	.9944	.9820	.9464	.8499	.6028
	18	1.0000	1.0000	1.0000	1.0000	1.0000	1.0000	1.0000	1.0000	1.0000	1.0000

TABLE A1
BINOMIAL DISTRIBUTION (CONTINUED)

n	x	p = .05	.10	.15	.20	.25	.30	.35	.40	.45
19	0	.3774	.1351	.0456	.0144	.0042	.0011	.0003	.0001	.0000
	1	.7547	.4203	.1985	.0829	.0310	.0104	.0031	.0008	.0002
	2	.9335	.7054	.4413	.2369	.1113	.0462	.0170	.0055	.0015
	3	.9868	.8850	.6841	.4551	.2631	.1332	.0591	.0230	.0077
	4	.9980	.9648	.8556	.6733	.4654	.2822	.1500	.0696	.0280
	5	.9998	.9914	.9463	.8369	.6678	.4739	.2968	.1629	.0777
	6	1.0000	.9983	.9837	.9324	.8251	.6655	.4812	.3081	.1727
	7	1.0000	.9997	.9959	.9767	.9225	.8180	.6656	.4878	.3169
	8	1.0000	1.0000	.9992	.9933	.9713	.9161	.8145	.6675	.4940
	9	1.0000	1.0000	.9999	.9984	.9911	.9674	.9125	.8139	.6710
	10	1.0000	1.0000	1.0000	.9997	.9977	.9895	.9653	.9115	.8159
	11	1.0000	1.0000	1.0000	1.0000	.9995	.9972	.9886	.9648	.9129
	12	1.0000	1.0000	1.0000	1.0000	.9999	.9994	.9969	.9884	.9658
	13	1.0000	1.0000	1.0000	1.0000	1.0000	.9999	.9993	.9969	.9891
	14	1.0000	1.0000	1.0000	1.0000	1.0000	1.0000	.9999	.9994	.9972
	15	1.0000	1.0000	1.0000	1.0000	1.0000	1.0000	1.0000	.9999	.9995
	16	1.0000	1.0000	1.0000	1.0000	1.0000	1.0000	1.0000	1.0000	.9999
	17	1.0000	1.0000	1.0000	1.0000	1.0000	1.0000	1.0000	1.0000	1.0000
	18	1.0000	1.0000	1.0000	1.0000	1.0000	1.0000	1.0000	1.0000	1.0000
	19	1.0000	1.0000	1.0000	1.0000	1.0000	1.0000	1.0000	1.0000	1.0000
20	0	.3585	.1216	.0388	.0115	.0032	.0008	.0002	.0000	.0000
	1	.7358	.3917	.1756	.0692	.0243	.0076	.0021	.0005	.0001
	2	.9245	.6769	.4049	.2061	.0913	.0355	.0121	.0036	.0009
	3	.9841	.8670	.6477	.4114	.2252	.1071	.0444	.0160	.0049
	4	.9974	.9568	.8298	.6296	.4148	.2375	.1182	.0510	.0189
	5	.9997	.9887	.9327	.8042	.6172	.4164	.2454	.1256	.0553
	6	1.0000	.9976	.9781	.9133	.7858	.6080	.4166	.2500	.1299
	7	1.0000	.9996	.9941	.9679	.8982	.7723	.6010	.4159	.2520
	8	1.0000	.9999	.9987	.9900	.9591	.8867	.7624	.5956	.4143
	9	1.0000	1.0000	.9998	.9974	.9861	.9520	.8782	.7553	.5914
	10	1.0000	1.0000	1.0000	.9994	.9961	.9829	.9468	.8725	.7507
	11	1.0000	1.0000	1.0000	.9999	.9991	.9949	.9804	.9435	.8692
	12	1.0000	1.0000	1.0000	1.0000	.9998	.9987	.9940	.9790	.9420
	13	1.0000	1.0000	1.0000	1.0000	1.0000	.9997	.9985	.9935	.9786
	14	1.0000	1.0000	1.0000	1.0000	1.0000	1.0000	.9997	.9984	.9936
	15	1.0000	1.0000	1.0000	1.0000	1.0000	1.0000	1.0000	.9997	.9985
	16	1.0000	1.0000	1.0000	1.0000	1.0000	1.0000	1.0000	1.0000	.9997
	17	1.0000	1.0000	1.0000	1.0000	1.0000	1.0000	1.0000	1.0000	1.0000
	18	1.0000	1.0000	1.0000	1.0000	1.0000	1.0000	1.0000	1.0000	1.0000
	19	1.0000	1.0000	1.0000	1.0000	1.0000	1.0000	1.0000	1.0000	1.0000
	20	1.0000	1.0000	1.0000	1.0000	1.0000	1.0000	1.0000	1.0000	1.0000

TABLE A1
BINOMIAL DISTRIBUTION (CONTINUED)

n	x	p = .50	.55	.60	.65	.70	.75	.80	.85	.90	.95
19	0	.0000	.0000	.0000	.0000	.0000	.0000	.0000	.0000	.0000	.0000
	1	.0000	.0000	.0000	.0000	.0000	.0000	.0000	.0000	.0000	.0000
	2	.0004	.0001	.0000	.0000	.0000	.0000	.0000	.0000	.0000	.0000
	3	.0022	.0005	.0001	.0000	.0000	.0000	.0000	.0000	.0000	.0000
	4	.0096	.0028	.0006	.0001	.0000	.0000	.0000	.0000	.0000	.0000
	5	.0318	.0109	.0031	.0007	.0001	.0000	.0000	.0000	.0000	.0000
	6	.0835	.0342	.0116	.0031	.0006	.0001	.0000	.0000	.0000	.0000
	7	.1796	.0871	.0352	.0114	.0028	.0005	.0000	.0000	.0000	.0000
	8	.3238	.1841	.0885	.0347	.0105	.0023	.0003	.0000	.0000	.0000
	9	.5000	.3290	.1861	.0875	.0326	.0089	.0016	.0001	.0000	.0000
	10	.6762	.5060	.3325	.1855	.0839	.0287	.0067	.0008	.0000	.0000
	11	.8204	.6831	.5122	.3344	.1820	.0775	.0233	.0041	.0003	.0000
	12	.9165	.8273	.6919	.5188	.3345	.1749	.0676	.0163	.0017	.0000
	13	.9682	.9223	.8371	.7032	.5261	.3322	.1631	.0537	.0086	.0002
	14	.9904	.9720	.9304	.8500	.7178	.5346	.3267	.1444	.0352	.0020
	15	.9978	.9923	.9770	.9409	.8668	.7369	.5449	.3159	.1150	.0132
	16	.9996	.9985	.9945	.9830	.9538	.8887	.7631	.5587	.2946	.0665
	17	1.0000	.9998	.9992	.9969	.9896	.9690	.9171	.8015	.5797	.2453
	18	1.0000	1.0000	.9999	.9997	.9989	.9958	.9856	.9544	.8649	.6226
	19	1.0000	1.0000	1.0000	1.0000	1.0000	1.0000	1.0000	1.0000	1.0000	1.0000
20	0	.0000	.0000	.0000	.0000	.0000	.0000	.0000	.0000	.0000	.0000
	1	.0000	.0000	.0000	.0000	.0000	.0000	.0000	.0000	.0000	.0000
	2	.0002	.0000	.0000	.0000	.0000	.0000	.0000	.0000	.0000	.0000
	3	.0013	.0003	.0000	.0000	.0000	.0000	.0000	.0000	.0000	.0000
	4	.0059	.0015	.0003	.0000	.0000	.0000	.0000	.0000	.0000	.0000
	5	.0207	.0064	.0016	.0003	.0000	.0000	.0000	.0000	.0000	.0000
	6	.0577	.0214	.0065	.0015	.0003	.0000	.0000	.0000	.0000	.0000
	7	.1316	.0580	.0210	.0060	.0013	.0002	.0000	.0000	.0000	.0000
	8	.2517	.1308	.0565	.0196	.0051	.0009	.0001	.0000	.0000	.0000
	9	.4119	.2493	.1275	.0532	.0171	.0039	.0006	.0000	.0000	.0000
	10	.5881	.4086	.2447	.1218	.0480	.0139	.0026	.0002	.0000	.0000
	11	.7483	.5857	.4044	.2376	.1133	.0409	.0100	.0013	.0001	.0000
	12	.8684	.7480	.5841	.3990	.2277	.1018	.0321	.0059	.0004	.0000
	13	.9423	.8701	.7500	.5834	.3920	.2142	.0867	.0219	.0024	.0000
	14	.9793	.9447	.8744	.7546	.5836	.3828	.1958	.0673	.0113	.0003
	15	.9941	.9811	.9490	.8818	.7625	.5852	.3704	.1702	.0432	.0026
	16	.9987	.9951	.9840	.9556	.8929	.7748	.5886	.3523	.1330	.0159
	17	.9998	.9991	.9964	.9879	.9645	.9087	.7939	.5951	.3231	.0755
	18	1.0000	.9999	.9995	.9979	.9924	.9757	.9308	.8244	.6083	.2642
	19	1.0000	1.0000	1.0000	.9998	.9992	.9968	.9885	.9612	.8784	.6415
	20	1.0000	1.0000	1.0000	1.0000	1.0000	1.0000	1.0000	1.0000	1.0000	1.0000

TABLE A1
BINOMIAL DISTRIBUTION (CONTINUED)

n	x	p = .05	.10	.15	.20	.25	.30	.35	.40	.45
21	0	.3406	.1094	.0329	.0092	.0024	.0006	.0001	.0000	.0000
	1	.7170	.3647	.1550	.0576	.0190	.0056	.0014	.0003	.0001
	2	.9151	.6484	.3705	.1787	.0745	.0271	.0086	.0024	.0006
	3	.9811	.8480	.6113	.3704	.1917	.0856	.0331	.0110	.0031
	4	.9968	.9478	.8025	.5860	.3674	.1984	.0924	.0370	.0126
	5	.9996	.9856	.9173	.7693	.5666	.3627	.2009	.0957	.0389
	6	1.0000	.9967	.9713	.8915	.7436	.5505	.3567	.2002	.0964
	7	1.0000	.9994	.9917	.9569	.8701	.7230	.5365	.3495	.1971
	8	1.0000	.9999	.9980	.9856	.9439	.8523	.7059	.5237	.3413
	9	1.0000	1.0000	.9996	.9959	.9794	.9324	.8377	.6914	.5117
	10	1.0000	1.0000	.9999	.9990	.9936	.9736	.9228	.8256	.6790
	11	1.0000	1.0000	1.0000	.9998	.9983	.9913	.9687	.9151	.8159
	12	1.0000	1.0000	1.0000	1.0000	.9996	.9976	.9892	.9648	.9092
	13	1.0000	1.0000	1.0000	1.0000	.9999	.9994	.9969	.9877	.9621
	14	1.0000	1.0000	1.0000	1.0000	1.0000	.9999	.9993	.9964	.9868
	15	1.0000	1.0000	1.0000	1.0000	1.0000	1.0000	.9999	.9992	.9963
	16	1.0000	1.0000	1.0000	1.0000	1.0000	1.0000	1.0000	.9998	.9992
	17	1.0000	1.0000	1.0000	1.0000	1.0000	1.0000	1.0000	1.0000	.9999
	18	1.0000	1.0000	1.0000	1.0000	1.0000	1.0000	1.0000	1.0000	1.0000
	19	1.0000	1.0000	1.0000	1.0000	1.0000	1.0000	1.0000	1.0000	1.0000
	20	1.0000	1.0000	1.0000	1.0000	1.0000	1.0000	1.0000	1.0000	1.0000
	21	1.0000	1.0000	1.0000	1.0000	1.0000	1.0000	1.0000	1.0000	1.0000
22	0	.3235	.0985	.0280	.0074	.0018	.0004	.0001	.0000	.0000
	1	.6982	.3392	.1367	.0480	.0149	.0041	.0010	.0002	.0000
	2	.9052	.6200	.3382	.1545	.0606	.0207	.0061	.0016	.0003
	3	.9778	.8281	.5752	.3320	.1624	.0681	.0245	.0076	.0020
	4	.9960	.9379	.7738	.5429	.3235	.1645	.0716	.0266	.0083
	5	.9994	.9818	.9001	.7326	.5168	.3134	.1629	.0722	.0271
	6	.9999	.9956	.9632	.8670	.6994	.4942	.3022	.1584	.0705
	7	1.0000	.9991	.9886	.9439	.8385	.6713	.4736	.2898	.1518
	8	1.0000	.9999	.9970	.9799	.9254	.8135	.6466	.4540	.2764
	9	1.0000	1.0000	.9993	.9939	.9705	.9084	.7916	.6244	.4350
	10	1.0000	1.0000	.9999	.9984	.9900	.9613	.8930	.7720	.6037
	11	1.0000	1.0000	1.0000	.9997	.9971	.9860	.9526	.8793	.7543
	12	1.0000	1.0000	1.0000	.9999	.9993	.9957	.9820	.9449	.8672
	13	1.0000	1.0000	1.0000	1.0000	.9999	.9989	.9942	.9785	.9383
	14	1.0000	1.0000	1.0000	1.0000	1.0000	.9998	.9984	.9930	.9757
	15	1.0000	1.0000	1.0000	1.0000	1.0000	1.0000	.9997	.9981	.9920
	16	1.0000	1.0000	1.0000	1.0000	1.0000	1.0000	.9999	.9996	.9979
	17	1.0000	1.0000	1.0000	1.0000	1.0000	1.0000	1.0000	.9999	.9995
	18	1.0000	1.0000	1.0000	1.0000	1.0000	1.0000	1.0000	1.0000	.9999
	19	1.0000	1.0000	1.0000	1.0000	1.0000	1.0000	1.0000	1.0000	1.0000
	20	1.0000	1.0000	1.0000	1.0000	1.0000	1.0000	1.0000	1.0000	1.0000
	21	1.0000	1.0000	1.0000	1.0000	1.0000	1.0000	1.0000	1.0000	1.0000
	22	1.0000	1.0000	1.0000	1.0000	1.0000	1.0000	1.0000	1.0000	1.0000

TABLE A1
BINOMIAL DISTRIBUTION (CONTINUED)

n	x	p = .50	.55	.60	.65	.70	.75	.80	.85	.90	.95
21	0	.0000	.0000	.0000	.0000	.0000	.0000	.0000	.0000	.0000	.0000
	1	.0000	.0000	.0000	.0000	.0000	.0000	.0000	.0000	.0000	.0000
	2	.0001	.0000	.0000	.0000	.0000	.0000	.0000	.0000	.0000	.0000
	3	.0007	.0001	.0000	.0000	.0000	.0000	.0000	.0000	.0000	.0000
	4	.0036	.0008	.0002	.0000	.0000	.0000	.0000	.0000	.0000	.0000
	5	.0133	.0037	.0008	.0001	.0000	.0000	.0000	.0000	.0000	.0000
	6	.0392	.0132	.0036	.0007	.0001	.0000	.0000	.0000	.0000	.0000
	7	.0946	.0379	.0123	.0031	.0006	.0001	.0000	.0000	.0000	.0000
	8	.1917	.0908	.0352	.0108	.0024	.0004	.0000	.0000	.0000	.0000
	9	.3318	.1841	.0849	.0313	.0087	.0017	.0002	.0000	.0000	.0000
	10	.5000	.3210	.1744	.0772	.0264	.0064	.0010	.0001	.0000	.0000
	11	.6682	.4883	.3086	.1623	.0676	.0206	.0041	.0004	.0000	.0000
	12	.8083	.6587	.4763	.2941	.1477	.0561	.0144	.0020	.0001	.0000
	13	.9054	.8029	.6505	.4635	.2770	.1299	.0431	.0083	.0006	.0000
	14	.9608	.9036	.7998	.6433	.4495	.2564	.1085	.0287	.0033	.0000
	15	.9867	.9611	.9043	.7991	.6373	.4334	.2307	.0827	.0144	.0004
	16	.9964	.9874	.9630	.9076	.8016	.6326	.4140	.1975	.0522	.0032
	17	.9993	.9969	.9890	.9669	.9144	.8083	.6296	.3887	.1520	.0189
	18	.9999	.9994	.9976	.9914	.9729	.9255	.8213	.6295	.3516	.0849
	19	1.0000	.9999	.9997	.9986	.9944	.9810	.9424	.8450	.6353	.2830
	20	1.0000	1.0000	1.0000	.9999	.9994	.9976	.9908	.9671	.8906	.6594
	21	1.0000	1.0000	1.0000	1.0000	1.0000	1.0000	1.0000	1.0000	1.0000	1.0000
22	0	.0000	.0000	.0000	.0000	.0000	.0000	.0000	.0000	.0000	.0000
	1	.0000	.0000	.0000	.0000	.0000	.0000	.0000	.0000	.0000	.0000
	2	.0001	.0000	.0000	.0000	.0000	.0000	.0000	.0000	.0000	.0000
	3	.0004	.0001	.0000	.0000	.0000	.0000	.0000	.0000	.0000	.0000
	4	.0022	.0005	.0001	.0000	.0000	.0000	.0000	.0000	.0000	.0000
	5	.0085	.0021	.0004	.0001	.0000	.0000	.0000	.0000	.0000	.0000
	6	.0262	.0080	.0019	.0003	.0000	.0000	.0000	.0000	.0000	.0000
	7	.0669	.0243	.0070	.0016	.0002	.0000	.0000	.0000	.0000	.0000
	8	.1431	.0617	.0215	.0058	.0011	.0001	.0000	.0000	.0000	.0000
	9	.2617	.1328	.0551	.0180	.0043	.0007	.0001	.0000	.0000	.0000
	10	.4159	.2457	.1207	.0474	.0140	.0029	.0003	.0000	.0000	.0000
	11	.5841	.3963	.2280	.1070	.0387	.0100	.0016	.0001	.0000	.0000
	12	.7383	.5650	.3756	.2084	.0916	.0295	.0061	.0007	.0000	.0000
	13	.8569	.7236	.5460	.3534	.1865	.0746	.0201	.0030	.0001	.0000
	14	.9331	.8482	.7102	.5264	.3287	.1615	.0561	.0114	.0009	.0000
	15	.9738	.9295	.8416	.6978	.5058	.3006	.1330	.0368	.0044	.0001
	16	.9915	.9729	.9278	.8371	.6866	.4832	.2674	.0999	.0182	.0006
	17	.9978	.9917	.9734	.9284	.8355	.6765	.4571	.2262	.0621	.0040
	18	.9996	.9980	.9924	.9755	.9319	.8376	.6680	.4248	.1719	.0222
	19	.9999	.9997	.9984	.9939	.9793	.9394	.8455	.6618	.3800	.0948
	20	1.0000	1.0000	.9998	.9990	.9959	.9851	.9520	.8633	.6608	.3018
	21	1.0000	1.0000	1.0000	.9999	.9996	.9982	.9926	.9720	.9015	.6765
	22	1.0000	1.0000	1.0000	1.0000	1.0000	1.0000	1.0000	1.0000	1.0000	1.0000

n	x	p = .05	.10	.15	.20	.25	.30	.35	.40	.45
23	0	.3074	.0886	.0238	.0059	.0013	.0003	.0000	.0000	.0000
	1	.6794	.3151	.1204	.0398	.0116	.0030	.0007	.0001	.0000
	2	.8948	.5920	.3080	.1332	.0492	.0157	.0043	.0010	.0002
	3	.9742	.8073	.5396	.2965	.1370	.0538	.0181	.0052	.0012
	4	.9951	.9269	.7440	.5007	.2832	.1356	.0551	.0190	.0055
	5	.9992	.9774	.8811	.6947	.4685	.2688	.1309	.0540	.0186
	6	.9999	.9942	.9537	.8402	.6537	.4399	.2534	.1240	.0510
	7	1.0000	.9988	.9848	.9285	.8037	.6181	.4136	.2373	.1152
	8	1.0000	.9998	.9958	.9727	.9037	.7709	.5860	.3884	.2203
	9	1.0000	1.0000	.9990	.9911	.9592	.8799	.7408	.5562	.3636
	10	1.0000	1.0000	.9998	.9975	.9851	.9454	.8575	.7129	.5278
	11	1.0000	1.0000	1.0000	.9994	.9954	.9786	.9318	.8364	.6865
	12	1.0000	1.0000	1.0000	.9999	.9988	.9928	.9717	.9187	.8164
	13	1.0000	1.0000	1.0000	1.0000	.9997	.9979	.9900	.9651	.9063
	14	1.0000	1.0000	1.0000	1.0000	.9999	.9995	.9970	.9872	.9589
	15	1.0000	1.0000	1.0000	1.0000	1.0000	.9999	.9992	.9960	.9847
	16	1.0000	1.0000	1.0000	1.0000	1.0000	1.0000	.9998	.9990	.9952
	17	1.0000	1.0000	1.0000	1.0000	1.0000	1.0000	1.0000	.9998	.9988
	18	1.0000	1.0000	1.0000	1.0000	1.0000	1.0000	1.0000	1.0000	.9998
	19	1.0000	1.0000	1.0000	1.0000	1.0000	1.0000	1.0000	1.0000	1.0000
	20	1.0000	1.0000	1.0000	1.0000	1.0000	1.0000	1.0000	1.0000	1.0000
	21	1.0000	1.0000	1.0000	1.0000	1.0000	1.0000	1.0000	1.0000	1.0000
	22	1.0000	1.0000	1.0000	1.0000	1.0000	1.0000	1.0000	1.0000	1.0000
	23	1.0000	1.0000	1.0000	1.0000	1.0000	1.0000	1.0000	1.0000	1.0000
24	0	.2920	.0798	.0202	.0047	.0010	.0002	.0000	.0000	.0000
	1	.6608	.2925	.1059	.0331	.0090	.0022	.0005	.0001	.0000
	2	.8841	.5643	.2798	.1145	.0398	.0119	.0030	.0007	.0001
	3	.9702	.7857	.5049	.2639	.1150	.0424	.0133	.0035	.0008
	4	.9940	.9149	.7134	.4599	.2466	.1111	.0422	.0134	.0036
	5	.9990	.9723	.8606	.6559	.4222	.2288	.1044	.0400	.0127
	6	.9999	.9925	.9428	.8111	.6074	.3886	.2106	.0960	.0364
	7	1.0000	.9983	.9801	.9108	.7662	.5647	.3575	.1919	.0863
	8	1.0000	.9997	.9941	.9638	.8787	.7250	.5257	.3279	.1730
	9	1.0000	.9999	.9985	.9874	.9453	.8472	.6866	.4891	.2991
	10	1.0000	1.0000	.9997	.9962	.9787	.9258	.8167	.6502	.4539
	11	1.0000	1.0000	.9999	.9990	.9928	.9686	.9058	.7870	.6151
	12	1.0000	1.0000	1.0000	.9998	.9979	.9885	.9577	.8857	.7580
	13	1.0000	1.0000	1.0000	1.0000	.9995	.9964	.9836	.9465	.8659
	14	1.0000	1.0000	1.0000	1.0000	.9999	.9990	.9945	.9783	.9352
	15	1.0000	1.0000	1.0000	1.0000	1.0000	.9998	.9984	.9925	.9731
	16	1.0000	1.0000	1.0000	1.0000	1.0000	1.0000	.9996	.9978	.9905
	17	1.0000	1.0000	1.0000	1.0000	1.0000	1.0000	.9999	.9995	.9972
	18	1.0000	1.0000	1.0000	1.0000	1.0000	1.0000	1.0000	.9999	.9993
	19	1.0000	1.0000	1.0000	1.0000	1.0000	1.0000	1.0000	1.0000	.9999
	20	1.0000	1.0000	1.0000	1.0000	1.0000	1.0000	1.0000	1.0000	1.0000
	21	1.0000	1.0000	1.0000	1.0000	1.0000	1.0000	1.0000	1.0000	1.0000
	22	1.0000	1.0000	1.0000	1.0000	1.0000	1.0000	1.0000	1.0000	1.0000
	23	1.0000	1.0000	1.0000	1.0000	1.0000	1.0000	1.0000	1.0000	1.0000
	24	1.0000	1.0000	1.0000	1.0000	1.0000	1.0000	1.0000	1.0000	1.0000

TABLE A1
BINOMIAL DISTRIBUTION (CONTINUED)

n	x	p = .50	.55	.60	.65	.70	.75	.80	.85	.90	.95
23	0	.0000	.0000	.0000	.0000	.0000	.0000	.0000	.0000	.0000	.0000
	1	.0000	.0000	.0000	.0000	.0000	.0000	.0000	.0000	.0000	.0000
	2	.0000	.0000	.0000	.0000	.0000	.0000	.0000	.0000	.0000	.0000
	3	.0002	.0000	.0000	.0000	.0000	.0000	.0000	.0000	.0000	.0000
	4	.0013	.0002	.0000	.0000	.0000	.0000	.0000	.0000	.0000	.0000
	5	.0053	.0012	.0002	.0000	.0000	.0000	.0000	.0000	.0000	.0000
	6	.0173	.0048	.0010	.0002	.0000	.0000	.0000	.0000	.0000	.0000
	7	.0466	.0153	.0040	.0008	.0001	.0000	.0000	.0000	.0000	.0000
	8	.1050	.0411	.0128	.0030	.0005	.0001	.0000	.0000	.0000	.0000
	9	.2024	.0937	.0349	.0100	.0021	.0003	.0000	.0000	.0000	.0000
	10	.3388	.1836	.0813	.0283	.0072	.0012	.0001	.0000	.0000	.0000
	11	.5000	.3135	.1636	.0682	.0214	.0046	.0006	.0000	.0000	.0000
	12	.6612	.4722	.2871	.1425	.0546	.0149	.0025	.0002	.0000	.0000
	13	.7976	.6364	.4438	.2592	.1201	.0408	.0089	.0010	.0000	.0000
	14	.8950	.7797	.6116	.4140	.2291	.0963	.0273	.0042	.0002	.0000
	15	.9534	.8848	.7627	.5864	.3819	.1963	.0715	.0152	.0012	.0000
	16	.9827	.9490	.8760	.7466	.5601	.3463	.1598	.0463	.0058	.0001
	17	.9947	.9814	.9460	.8691	.7312	.5315	.3053	.1189	.0226	.0008
	18	.9987	.9945	.9810	.9449	.8644	.7168	.4993	.2560	.0731	.0049
	19	.9998	.9988	.9948	.9819	.9462	.8630	.7035	.4604	.1927	.0258
	20	1.0000	.9998	.9990	.9957	.9843	.9508	.8668	.6920	.4080	.1052
	21	1.0000	1.0000	.9999	.9993	.9970	.9884	.9602	.8796	.6849	.3206
	22	1.0000	1.0000	1.0000	1.0000	.9997	.9987	.9941	.9762	.9114	.6926
	23	1.0000	1.0000	1.0000	1.0000	1.0000	1.0000	1.0000	1.0000	1.0000	1.0000
24	0	.0000	.0000	.0000	.0000	.0000	.0000	.0000	.0000	.0000	.0000
	1	.0000	.0000	.0000	.0000	.0000	.0000	.0000	.0000	.0000	.0000
	2	.0000	.0000	.0000	.0000	.0000	.0000	.0000	.0000	.0000	.0000
	3	.0001	.0000	.0000	.0000	.0000	.0000	.0000	.0000	.0000	.0000
	4	.0008	.0001	.0000	.0000	.0000	.0000	.0000	.0000	.0000	.0000
	5	.0033	.0007	.0001	.0000	.0000	.0000	.0000	.0000	.0000	.0000
	6	.0113	.0028	.0005	.0001	.0000	.0000	.0000	.0000	.0000	.0000
	7	.0320	.0095	.0022	.0004	.0000	.0000	.0000	.0000	.0000	.0000
	8	.0758	.0269	.0075	.0016	.0002	.0000	.0000	.0000	.0000	.0000
	9	.1537	.0648	.0217	.0055	.0010	.0001	.0000	.0000	.0000	.0000
	10	.2706	.1341	.0535	.0164	.0036	.0005	.0000	.0000	.0000	.0000
	11	.4194	.2420	.1143	.0423	.0115	.0021	.0002	.0000	.0000	.0000
	12	.5806	.3849	.2130	.0942	.0314	.0072	.0010	.0001	.0000	.0000
	13	.7294	.5461	.3498	.1833	.0742	.0213	.0038	.0003	.0000	.0000
	14	.8463	.7009	.5109	.3134	.1528	.0547	.0126	.0015	.0001	.0000
	15	.9242	.8270	.6721	.4743	.2750	.1213	.0362	.0059	.0003	.0000
	16	.9680	.9137	.8081	.6425	.4353	.2338	.0892	.0199	.0017	.0000
	17	.9887	.9636	.9040	.7894	.6114	.3926	.1889	.0572	.0075	.0001
	18	.9967	.9873	.9600	.8956	.7712	.5778	.3441	.1394	.0277	.0010
	19	.9992	.9964	.9866	.9578	.8889	.7534	.5401	.2866	.0851	.0060
	20	.9999	.9992	.9965	.9867	.9576	.8850	.7361	.4951	.2143	.0298
	21	1.0000	.9999	.9993	.9970	.9881	.9602	.8855	.7202	.4357	.1159
	22	1.0000	1.0000	.9999	.9995	.9978	.9910	.9669	.8941	.7075	.3392
	23	1.0000	1.0000	1.0000	1.0000	.9998	.9990	.9953	.9798	.9202	.7080
	24	1.0000	1.0000	1.0000	1.0000	1.0000	1.0000	1.0000	1.0000	1.0000	1.0000

n	x	p = .05	.10	.15	.20	.25	.30	.35	.40	.45
25	0	.2774	.0718	.0172	.0038	.0008	.0001	.0000	.0000	.0000
	1	.6424	.2712	.0931	.0274	.0070	.0016	.0003	.0001	.0000
	2	.8729	.5371	.2537	.0982	.0321	.0090	.0021	.0004	.0001
	3	.9659	.7636	.4711	.2340	.0962	.0332	.0097	.0024	.0005
	4	.9928	.9020	.6821	.4207	.2137	.0905	.0320	.0095	.0023
	5	.9988	.9666	.8385	.6167	.3783	.1935	.0826	.0294	.0086
	6	.9998	.9905	.9305	.7800	.5611	.3407	.1734	.0736	.0258
	7	1.0000	.9977	.9745	.8909	.7265	.5118	.3061	.1536	.0639
	8	1.0000	.9995	.9920	.9532	.8506	.6769	.4668	.2735	.1340
	9	1.0000	.9999	.9979	.9827	.9287	.8106	.6303	.4246	.2424
	10	1.0000	1.0000	.9995	.9944	.9703	.9022	.7712	.5858	.3843
	11	1.0000	1.0000	.9999	.9985	.9893	.9558	.8746	.7323	.5426
	12	1.0000	1.0000	1.0000	.9996	.9966	.9825	.9396	.8462	.6937
	13	1.0000	1.0000	1.0000	.9999	.9991	.9940	.9745	.9222	.8173
	14	1.0000	1.0000	1.0000	1.0000	.9998	.9982	.9907	.9656	.9040
	15	1.0000	1.0000	1.0000	1.0000	1.0000	.9995	.9971	.9868	.9560
	16	1.0000	1.0000	1.0000	1.0000	1.0000	.9999	.9992	.9957	.9826
	17	1.0000	1.0000	1.0000	1.0000	1.0000	1.0000	.9998	.9988	.9942
	18	1.0000	1.0000	1.0000	1.0000	1.0000	1.0000	1.0000	.9997	.9984
	19	1.0000	1.0000	1.0000	1.0000	1.0000	1.0000	1.0000	.9999	.9996
	20	1.0000	1.0000	1.0000	1.0000	1.0000	1.0000	1.0000	1.0000	.9999
	21	1.0000	1.0000	1.0000	1.0000	1.0000	1.0000	1.0000	1.0000	1.0000
	22	1.0000	1.0000	1.0000	1.0000	1.0000	1.0000	1.0000	1.0000	1.0000
	23	1.0000	1.0000	1.0000	1.0000	1.0000	1.0000	1.0000	1.0000	1.0000
	24	1.0000	1.0000	1.0000	1.0000	1.0000	1.0000	1.0000	1.0000	1.0000
	25	1.0000	1.0000	1.0000	1.0000	1.0000	1.0000	1.0000	1.0000	1.0000
26	0	.2635	.0646	.0146	.0030	.0006	.0001	.0000	.0000	.0000
	1	.6241	.2513	.0817	.0227	.0055	.0011	.0002	.0000	.0000
	2	.8614	.5105	.2296	.0841	.0258	.0067	.0015	.0003	.0000
	3	.9613	.7409	.4385	.2068	.0802	.0260	.0070	.0016	.0003
	4	.9915	.8882	.6505	.3833	.1844	.0733	.0242	.0066	.0015
	5	.9985	.9601	.8150	.5775	.3371	.1626	.0649	.0214	.0058
	6	.9998	.9881	.9167	.7474	.5154	.2965	.1416	.0559	.0180
	7	1.0000	.9970	.9679	.8687	.6852	.4605	.2596	.1216	.0467
	8	1.0000	.9994	.9894	.9408	.8195	.6274	.4106	.2255	.1024
	9	1.0000	.9999	.9970	.9768	.9091	.7705	.5731	.3642	.1936
	10	1.0000	1.0000	.9993	.9921	.9599	.8747	.7219	.5213	.3204
	11	1.0000	1.0000	.9998	.9977	.9845	.9397	.8384	.6737	.4713
	12	1.0000	1.0000	1.0000	.9994	.9948	.9745	.9168	.8007	.6257
	13	1.0000	1.0000	1.0000	.9999	.9985	.9906	.9623	.8918	.7617
	14	1.0000	1.0000	1.0000	1.0000	.9996	.9970	.9850	.9482	.8650
	15	1.0000	1.0000	1.0000	1.0000	.9999	.9991	.9948	.9783	.9326
	16	1.0000	1.0000	1.0000	1.0000	1.0000	.9998	.9985	.9921	.9707
	17	1.0000	1.0000	1.0000	1.0000	1.0000	1.0000	.9996	.9975	.9890
	18	1.0000	1.0000	1.0000	1.0000	1.0000	1.0000	.9999	.9993	.9965
	19	1.0000	1.0000	1.0000	1.0000	1.0000	1.0000	1.0000	.9999	.9991
	20	1.0000	1.0000	1.0000	1.0000	1.0000	1.0000	1.0000	1.0000	.9998
	21	1.0000	1.0000	1.0000	1.0000	1.0000	1.0000	1.0000	1.0000	1.0000
	22	1.0000	1.0000	1.0000	1.0000	1.0000	1.0000	1.0000	1.0000	1.0000
	23	1.0000	1.0000	1.0000	1.0000	1.0000	1.0000	1.0000	1.0000	1.0000
	24	1.0000	1.0000	1.0000	1.0000	1.0000	1.0000	1.0000	1.0000	1.0000
	25	1.0000	1.0000	1.0000	1.0000	1.0000	1.0000	1.0000	1.0000	1.0000
	26	1.0000	1.0000	1.0000	1.0000	1.0000	1.0000	1.0000	1.0000	1.0000

n	x	p = .50	.55	.60	.65	.70	.75	.80	.85	.90	.95
25	0	.0000	.0000	.0000	.0000	.0000	.0000	.0000	.0000	.0000	.0000
	1	.0000	.0000	.0000	.0000	.0000	.0000	.0000	.0000	.0000	.0000
	2	.0000	.0000	.0000	.0000	.0000	.0000	.0000	.0000	.0000	.0000
	3	.0001	.0000	.0000	.0000	.0000	.0000	.0000	.0000	.0000	.0000
	4	.0005	.0001	.0000	.0000	.0000	.0000	.0000	.0000	.0000	.0000
	5	.0020	.0004	.0001	.0000	.0000	.0000	.0000	.0000	.0000	.0000
	6	.0073	.0016	.0003	.0000	.0000	.0000	.0000	.0000	.0000	.0000
	7	.0216	.0058	.0012	.0002	.0000	.0000	.0000	.0000	.0000	.0000
	8	.0539	.0174	.0043	.0008	.0001	.0000	.0000	.0000	.0000	.0000
	9	.1148	.0440	.0132	.0029	.0005	.0000	.0000	.0000	.0000	.0000
	10	.2122	.0960	.0344	.0093	.0018	.0002	.0000	.0000	.0000	.0000
	11	.3450	.1827	.0778	.0255	.0060	.0009	.0001	.0000	.0000	.0000
	12	.5000	.3063	.1538	.0604	.0175	.0034	.0004	.0000	.0000	.0000
	13	.6550	.4574	.2677	.1254	.0442	.0107	.0015	.0001	.0000	.0000
	14	.7878	.6157	.4142	.2288	.0978	.0297	.0056	.0005	.0000	.0000
	15	.8852	.7576	.5754	.3697	.1894	.0713	.0173	.0021	.0001	.0000
	16	.9461	.8660	.7265	.5332	.3231	.1494	.0468	.0080	.0005	.0000
	17	.9784	.9361	.8464	.6939	.4882	.2735	.1091	.0255	.0023	.0000
	18	.9927	.9742	.9264	.8266	.6593	.4389	.2200	.0695	.0095	.0002
	19	.9980	.9914	.9706	.9174	.8065	.6217	.3833	.1615	.0334	.0012
	20	.9995	.9977	.9905	.9680	.9095	.7863	.5793	.3179	.0980	.0072
	21	.9999	.9995	.9976	.9903	.9668	.9038	.7660	.5289	.2364	.0341
	22	1.0000	.9999	.9996	.9979	.9910	.9679	.9018	.7463	.4629	.1271
	23	1.0000	1.0000	.9999	.9997	.9984	.9930	.9726	.9069	.7288	.3576
	24	1.0000	1.0000	1.0000	1.0000	.9999	.9992	.9962	.9828	.9282	.7226
	25	1.0000	1.0000	1.0000	1.0000	1.0000	1.0000	1.0000	1.0000	1.0000	1.0000
26	0	.0000	.0000	.0000	.0000	.0000	.0000	.0000	.0000	.0000	.0000
	1	.0000	.0000	.0000	.0000	.0000	.0000	.0000	.0000	.0000	.0000
	2	.0000	.0000	.0000	.0000	.0000	.0000	.0000	.0000	.0000	.0000
	3	.0000	.0000	.0000	.0000	.0000	.0000	.0000	.0000	.0000	.0000
	4	.0003	.0000	.0000	.0000	.0000	.0000	.0000	.0000	.0000	.0000
	5	.0012	.0002	.0000	.0000	.0000	.0000	.0000	.0000	.0000	.0000
	6	.0047	.0009	.0001	.0000	.0000	.0000	.0000	.0000	.0000	.0000
	7	.0145	.0035	.0007	.0001	.0000	.0000	.0000	.0000	.0000	.0000
	8	.0378	.0110	.0025	.0004	.0000	.0000	.0000	.0000	.0000	.0000
	9	.0843	.0293	.0079	.0015	.0002	.0000	.0000	.0000	.0000	.0000
	10	.1635	.0674	.0217	.0052	.0009	.0001	.0000	.0000	.0000	.0000
	11	.2786	.1350	.0518	.0150	.0030	.0004	.0000	.0000	.0000	.0000
	12	.4225	.2383	.1082	.0377	.0094	.0015	.0001	.0000	.0000	.0000
	13	.5775	.3743	.1993	.0832	.0255	.0052	.0006	.0000	.0000	.0000
	14	.7214	.5287	.3263	.1616	.0603	.0155	.0023	.0002	.0000	.0000
	15	.8365	.6796	.4787	.2781	.1253	.0401	.0079	.0007	.0000	.0000
	16	.9157	.8064	.6358	.4269	.2295	.0909	.0232	.0030	.0001	.0000
	17	.9622	.8976	.7745	.5894	.3726	.1805	.0592	.0106	.0006	.0000
	18	.9855	.9533	.8784	.7404	.5395	.3148	.1313	.0321	.0030	.0000
	19	.9953	.9820	.9441	.8584	.7035	.4846	.2526	.0833	.0119	.0002
	20	.9988	.9942	.9786	.9351	.8374	.6629	.4225	.1850	.0399	.0015
	21	.9997	.9985	.9934	.9758	.9267	.8156	.6167	.3495	.1118	.0085
	22	1.0000	.9997	.9984	.9930	.9740	.9198	.7932	.5615	.2591	.0387
	23	1.0000	1.0000	.9997	.9985	.9933	.9742	.9159	.7704	.4895	.1386
	24	1.0000	1.0000	1.0000	.9998	.9989	.9945	.9973	.9183	.7487	.3759
	25	1.0000	1.0000	1.0000	1.0000	.9999	.9994	.9970	.9854	.9354	.7365
	26	1.0000	1.0000	1.0000	1.0000	1.0000	1.0000	1.0000	1.0000	1.0000	1.0000

n	x	p = .05	.10	.15	.20	.25	.30	.35	.40	.45
27	0	.2503	.0581	.0124	.0024	.0004	.0001	.0000	.0000	.0000
	1	.6061	.2326	.0716	.0187	.0042	.0008	.0001	.0000	.0000
	2	.8495	.4846	.2074	.0718	.0207	.0051	.0010	.0002	.0000
	3	.9563	.7179	.4072	.1823	.0666	.0202	.0051	.0011	.0002
	4	.9900	.8734	.6187	.3480	.1583	.0591	.0182	.0046	.0009
	5	.9981	.9529	.7903	.5387	.2989	.1358	.0507	.0155	.0038
	6	.9997	.9853	.9014	.7134	.4708	.2563	.1148	.0421	.0125
	7	1.0000	.9961	.9602	.8444	.6427	.4113	.2183	.0953	.0338
	8	1.0000	.9991	.9862	.9263	.7859	.5773	.3577	.1839	.0774
	9	1.0000	.9998	.9958	.9696	.8867	.7276	.5162	.3087	.1526
	10	1.0000	1.0000	.9989	.9890	.9472	.8434	.6698	.4585	.2633
	11	1.0000	1.0000	.9998	.9965	.9784	.9202	.7976	.6127	.4034
	12	1.0000	1.0000	1.0000	.9990	.9922	.9641	.8894	.7499	.5562
	13	1.0000	1.0000	1.0000	.9998	.9976	.9857	.9464	.8553	.7005
	14	1.0000	1.0000	1.0000	1.0000	.9993	.9950	.9771	.9257	.8185
	15	1.0000	1.0000	1.0000	1.0000	.9998	.9985	.9914	.9663	.9022
	16	1.0000	1.0000	1.0000	1.0000	1.0000	.9996	.9972	.9866	.9536
	17	1.0000	1.0000	1.0000	1.0000	1.0000	.9999	.9992	.9954	.9807
	18	1.0000	1.0000	1.0000	1.0000	1.0000	1.0000	.9998	.9986	.9931
	19	1.0000	1.0000	1.0000	1.0000	1.0000	1.0000	1.0000	.9997	.9979
	20	1.0000	1.0000	1.0000	1.0000	1.0000	1.0000	1.0000	.9999	.9995
	21	1.0000	1.0000	1.0000	1.0000	1.0000	1.0000	1.0000	1.0000	.9999
	22	1.0000	1.0000	1.0000	1.0000	1.0000	1.0000	1.0000	1.0000	1.0000
	23	1.0000	1.0000	1.0000	1.0000	1.0000	1.0000	1.0000	1.0000	1.0000
	24	1.0000	1.0000	1.0000	1.0000	1.0000	1.0000	1.0000	1.0000	1.0000
	25	1.0000	1.0000	1.0000	1.0000	1.0000	1.0000	1.0000	1.0000	1.0000
	26	1.0000	1.0000	1.0000	1.0000	1.0000	1.0000	1.0000	1.0000	1.0000
	27	1.0000	1.0000	1.0000	1.0000	1.0000	1.0000	1.0000	1.0000	1.0000
28	0	.2378	.0523	.0106	.0019	.0003	.0000	.0000	.0000	.0000
	1	.5883	.2152	.0627	.0155	.0033	.0006	.0001	.0000	.0000
	2	.8373	.4594	.1871	.0612	.0166	.0038	.0007	.0001	.0000
	3	.9509	.6946	.3772	.1602	.0551	.0157	.0037	.0007	.0001
	4	.9883	.8579	.5869	.3149	.1354	.0474	.0136	.0032	.0006
	5	.9977	.9450	.7646	.5005	.2638	.1128	.0393	.0111	.0025
	6	.9996	.9821	.8848	.6784	.4279	.2202	.0923	.0315	.0086
	7	1.0000	.9950	.9514	.8182	.5997	.3648	.1821	.0740	.0242
	8	1.0000	.9988	.9823	.9100	.7501	.5275	.3089	.1485	.0578
	9	1.0000	.9998	.9944	.9609	.8615	.6825	.4607	.2588	.1187
	10	1.0000	1.0000	.9985	.9851	.9321	.8087	.6160	.3986	.2135
	11	1.0000	1.0000	.9996	.9950	.9706	.8972	.7529	.5510	.3404
	12	1.0000	1.0000	.9999	.9985	.9888	.9509	.8572	.6950	.4875
	13	1.0000	1.0000	1.0000	.9996	.9962	.9792	.9264	.8132	.6356
	14	1.0000	1.0000	1.0000	.9999	.9989	.9923	.9663	.8975	.7654
	15	1.0000	1.0000	1.0000	1.0000	.9997	.9975	.9864	.9501	.8645
	16	1.0000	1.0000	1.0000	1.0000	.9999	.9993	.9952	.9785	.9304
	17	1.0000	1.0000	1.0000	1.0000	1.0000	.9998	.9985	.9919	.9685
	18	1.0000	1.0000	1.0000	1.0000	1.0000	1.0000	.9996	.9973	.9875
	19	1.0000	1.0000	1.0000	1.0000	1.0000	1.0000	.9999	.9992	.9957
	20	1.0000	1.0000	1.0000	1.0000	1.0000	1.0000	1.0000	.9998	.9988
	21	1.0000	1.0000	1.0000	1.0000	1.0000	1.0000	1.0000	1.0000	.9997
	22	1.0000	1.0000	1.0000	1.0000	1.0000	1.0000	1.0000	1.0000	.9999
	23	1.0000	1.0000	1.0000	1.0000	1.0000	1.0000	1.0000	1.0000	1.0000
	24	1.0000	1.0000	1.0000	1.0000	1.0000	1.0000	1.0000	1.0000	1.0000
	25	1.0000	1.0000	1.0000	1.0000	1.0000	1.0000	1.0000	1.0000	1.0000
	26	1.0000	1.0000	1.0000	1.0000	1.0000	1.0000	1.0000	1.0000	1.0000
	27	1.0000	1.0000	1.0000	1.0000	1.0000	1.0000	1.0000	1.0000	1.0000
	28	1.0000	1.0000	1.0000	1.0000	1.0000	1.0000	1.0000	1.0000	1.0000

TABLE A1
BINOMIAL DISTRIBUTION (CONTINUED)

n	x	p = .50	.55	.60	.65	.70	.75	.80	.85	.90	.95
27	0	.0000	.0000	.0000	.0000	.0000	.0000	.0000	.0000	.0000	.0000
	1	.0000	.0000	.0000	.0000	.0000	.0000	.0000	.0000	.0000	.0000
	2	.0000	.0000	.0000	.0000	.0000	.0000	.0000	.0000	.0000	.0000
	3	.0000	.0000	.0000	.0000	.0000	.0000	.0000	.0000	.0000	.0000
	4	.0002	.0000	.0000	.0000	.0000	.0000	.0000	.0000	.0000	.0000
	5	.0008	.0001	.0000	.0000	.0000	.0000	.0000	.0000	.0000	.0000
	6	.0030	.0005	.0001	.0000	.0000	.0000	.0000	.0000	.0000	.0000
	7	.0096	.0021	.0003	.0000	.0000	.0000	.0000	.0000	.0000	.0000
	8	.0261	.0069	.0014	.0002	.0000	.0000	.0000	.0000	.0000	.0000
	9	.0610	.0193	.0046	.0008	.0001	.0000	.0000	.0000	.0000	.0000
	10	.1239	.0464	.0134	.0028	.0004	.0000	.0000	.0000	.0000	.0000
	11	.2210	.0978	.0337	.0086	.0015	.0002	.0000	.0000	.0000	.0000
	12	.3506	.1815	.0743	.0229	.0050	.0007	.0000	.0000	.0000	.0000
	13	.5000	.2995	.1447	.0536	.0143	.0024	.0002	.0000	.0000	.0000
	14	.6494	.4438	.2501	.1106	.0359	.0078	.0010	.0000	.0000	.0000
	15	.7790	.5966	.3873	.2024	.0798	.0216	.0035	.0002	.0000	.0000
	16	.8761	.7367	.5415	.3302	.1566	.0528	.0110	.0011	.0000	.0000
	17	.9390	.8474	.6913	.4838	.2724	.1133	.0304	.0042	.0002	.0000
	18	.9739	.9226	.8161	.6423	.4227	.2141	.0737	.0138	.0009	.0000
	19	.9904	.9662	.9047	.7817	.5887	.3573	.1556	.0398	.0039	.0000
	20	.9970	.9875	.9579	.8852	.7437	.5292	.2866	.0986	.0147	.0003
	21	.9992	.9962	.9845	.9493	.8642	.7011	.4613	.2097	.0471	.0019
	22	.9998	.9991	.9954	.9818	.9409	.8417	.6520	.3813	.1266	.0100
	23	1.0000	.9998	.9989	.9949	.9798	.9334	.8177	.5928	.2821	.0437
	24	1.0000	1.0000	.9998	.9990	.9949	.9793	.9282	.7926	.5154	.1505
	25	1.0000	1.0000	1.0000	.9999	.9992	.9958	.9813	.9284	.7674	.3939
	26	1.0000	1.0000	1.0000	1.0000	.9999	.9996	.9976	.9876	.9419	.7497
	27	1.0000	1.0000	1.0000	1.0000	1.0000	1.0000	1.0000	1.0000	1.0000	1.0000
28	0	.0000	.0000	.0000	.0000	.0000	.0000	.0000	.0000	.0000	.0000
	1	.0000	.0000	.0000	.0000	.0000	.0000	.0000	.0000	.0000	.0000
	2	.0000	.0000	.0000	.0000	.0000	.0000	.0000	.0000	.0000	.0000
	3	.0000	.0000	.0000	.0000	.0000	.0000	.0000	.0000	.0000	.0000
	4	.0001	.0000	.0000	.0000	.0000	.0000	.0000	.0000	.0000	.0000
	5	.0005	.0001	.0000	.0000	.0000	.0000	.0000	.0000	.0000	.0000
	6	.0019	.0003	.0000	.0000	.0000	.0000	.0000	.0000	.0000	.0000
	7	.0063	.0012	.0002	.0000	.0000	.0000	.0000	.0000	.0000	.0000
	8	.0178	.0043	.0008	.0001	.0000	.0000	.0000	.0000	.0000	.0000
	9	.0436	.0125	.0027	.0004	.0000	.0000	.0000	.0000	.0000	.0000
	10	.0925	.0315	.0081	.0015	.0002	.0000	.0000	.0000	.0000	.0000
	11	.1725	.0696	.0215	.0048	.0007	.0001	.0000	.0000	.0000	.0000
	12	.2858	.1355	.0499	.0136	.0025	.0003	.0000	.0000	.0000	.0000
	13	.4253	.2346	.1025	.0337	.0077	.0011	.0001	.0000	.0000	.0000
	14	.5747	.3644	.1868	.0736	.0208	.0038	.0004	.0000	.0000	.0000
	15	.7142	.5125	.3050	.1428	.0491	.0112	.0015	.0001	.0000	.0000
	16	.8275	.6596	.4490	.2471	.1028	.0294	.0050	.0004	.0000	.0000
	17	.9075	.7865	.6014	.3840	.1913	.0679	.0149	.0015	.0000	.0000
	18	.9564	.8813	.7412	.5393	.3175	.1385	.0391	.0056	.0002	.0000
	19	.9822	.9422	.8515	.6911	.4725	.2499	.0900	.0177	.0012	.0000
	20	.9937	.9758	.9260	.8179	.6352	.4003	.1818	.0486	.0050	.0000
	21	.9981	.9914	.9685	.9077	.7798	.5721	.3216	.1152	.0179	.0004
	22	.9995	.9975	.9889	.9607	.8872	.7362	.4995	.2354	.0550	.0023
	23	.9999	.9994	.9968	.9864	.9526	.8646	.6851	.4131	.1421	.0117
	24	1.0000	.9999	.9993	.9963	.9843	.9449	.8398	.6228	.3054	.0491
	25	1.0000	1.0000	.9999	.9993	.9962	.9834	.9388	.8129	.5406	.1627
	26	1.0000	1.0000	1.0000	.9999	.9994	.9967	.9845	.9373	.7848	.4117
	27	1.0000	1.0000	1.0000	1.0000	1.0000	.9997	.9981	.9894	.9477	.7622
	28	1.0000	1.0000	1.0000	1.0000	1.0000	1.0000	1.0000	1.0000	1.0000	1.0000

TABLE A1
BINOMIAL DISTRIBUTION (CONTINUED)

n	x	p = .05	.10	.15	.20	.25	.30	.35	.40	.45
29	0	.2259	.0471	.0090	.0015	.0002	.0000	.0000	.0000	.0000
	1	.5708	.1989	.0549	.0128	.0025	.0004	.0001	.0000	.0000
	2	.8249	.4350	.1684	.0520	.0133	.0028	.0005	.0001	.0000
	3	.9452	.6710	.3487	.1404	.0455	.0121	.0026	.0005	.0001
	4	.9864	.8416	.5555	.2839	.1153	.0379	.0101	.0022	.0004
	5	.9973	.9363	.7379	.4634	.2317	.0932	.0303	.0080	.0017
	6	.9995	.9784	.8667	.6429	.3868	.1880	.0738	.0233	.0059
	7	.9999	.9938	.9414	.7903	.5568	.3214	.1507	.0570	.0172
	8	1.0000	.9984	.9777	.8916	.7125	.4787	.2645	.1187	.0427
	9	1.0000	.9997	.9926	.9507	.8337	.6360	.4076	.2147	.0913
	10	1.0000	.9999	.9978	.9803	.9145	.7708	.5617	.3427	.1708
	11	1.0000	1.0000	.9995	.9931	.9610	.8706	.7050	.4900	.2833
	12	1.0000	1.0000	.9999	.9978	.9842	.9348	.8207	.6374	.4213
	13	1.0000	1.0000	1.0000	.9994	.9944	.9707	.9022	.7659	.5689
	14	1.0000	1.0000	1.0000	.9999	.9982	.9883	.9524	.8638	.7070
	15	1.0000	1.0000	1.0000	1.0000	.9995	.9959	.9794	.9290	.8199
	16	1.0000	1.0000	1.0000	1.0000	.9999	.9987	.9921	.9671	.9008
	17	1.0000	1.0000	1.0000	1.0000	1.0000	.9997	.9973	.9865	.9514
	18	1.0000	1.0000	1.0000	1.0000	1.0000	.9999	.9992	.9951	.9790
	19	1.0000	1.0000	1.0000	1.0000	1.0000	1.0000	.9998	.9985	.9920
	20	1.0000	1.0000	1.0000	1.0000	1.0000	1.0000	1.0000	.9996	.9974
	21	1.0000	1.0000	1.0000	1.0000	1.0000	1.0000	1.0000	.9999	.9993
	22	1.0000	1.0000	1.0000	1.0000	1.0000	1.0000	1.0000	1.0000	.9998
	23	1.0000	1.0000	1.0000	1.0000	1.0000	1.0000	1.0000	1.0000	1.0000
	24	1.0000	1.0000	1.0000	1.0000	1.0000	1.0000	1.0000	1.0000	1.0000
	25	1.0000	1.0000	1.0000	1.0000	1.0000	1.0000	1.0000	1.0000	1.0000
	26	1.0000	1.0000	1.0000	1.0000	1.0000	1.0000	1.0000	1.0000	1.0000
	27	1.0000	1.0000	1.0000	1.0000	1.0000	1.0000	1.0000	1.0000	1.0000
	28	1.0000	1.0000	1.0000	1.0000	1.0000	1.0000	1.0000	1.0000	1.0000
	29	1.0000	1.0000	1.0000	1.0000	1.0000	1.0000	1.0000	1.0000	1.0000

TABLE A1
BINOMIAL DISTRIBUTION (CONTINUED)

n	x	p = .50	.55	.60	.65	.70	.75	.80	.85	.90	.95
29	0	.0000	.0000	.0000	.0000	.0000	.0000	.0000	.0000	.0000	.0000
	1	.0000	.0000	.0000	.0000	.0000	.0000	.0000	.0000	.0000	.0000
	2	.0000	.0000	.0000	.0000	.0000	.0000	.0000	.0000	.0000	.0000
	3	.0000	.0000	.0000	.0000	.0000	.0000	.0000	.0000	.0000	.0000
	4	.0001	.0000	.0000	.0000	.0000	.0000	.0000	.0000	.0000	.0000
	5	.0003	.0000	.0000	.0000	.0000	.0000	.0000	.0000	.0000	.0000
	6	.0012	.0002	.0000	.0000	.0000	.0000	.0000	.0000	.0000	.0000
	7	.0041	.0007	.0001	.0000	.0000	.0000	.0000	.0000	.0000	.0000
	8	.0121	.0026	.0004	.0000	.0000	.0000	.0000	.0000	.0000	.0000
	9	.0307	.0080	.0015	.0002	.0000	.0000	.0000	.0000	.0000	.0000
	10	.0680	.0210	.0049	.0008	.0001	.0000	.0000	.0000	.0000	.0000
	11	.1325	.0486	.0135	.0027	.0003	.0000	.0000	.0000	.0000	.0000
	12	.2291	.0992	.0329	.0079	.0013	.0001	.0000	.0000	.0000	.0000
	13	.3555	.1801	.0710	.0206	.0041	.0005	.0000	.0000	.0000	.0000
	14	.5000	.2930	.1362	.0476	.0117	.0018	.0001	.0000	.0000	.0000
	15	.6445	.4311	.2341	.0978	.0293	.0056	.0006	.0000	.0000	.0000
	16	.7709	.5787	.3626	.1793	.0652	.0158	.0022	.0001	.0000	.0000
	17	.8675	.7167	.5100	.2950	.1294	.0390	.0069	.0005	.0000	.0000
	18	.9320	.8292	.6573	.4383	.2292	.0855	.0197	.0022	.0001	.0000
	19	.9693	.9087	.7853	.5924	.3640	.1663	.0493	.0074	.0003	.0000
	20	.9879	.9573	.8813	.7355	.5213	.2875	.1084	.0223	.0016	.0000
	21	.9959	.9828	.9430	.8493	.6786	.4432	.2097	.0586	.0062	.0001
	22	.9988	.9941	.9767	.9262	.8120	.6132	.3571	.1333	.0216	.0005
	23	.9997	.9983	.9920	.9697	.9068	.7683	.5366	.2621	.0637	.0027
	24	.9999	.9996	.9978	.9899	.9621	.8847	.7161	.4445	.1584	.0136
	25	1.0000	.9999	.9995	.9974	.9879	.9545	.8596	.6513	.3290	.0548
	26	1.0000	1.0000	.9999	.9995	.9972	.9867	.9480	.8316	.5650	.1751
	27	1.0000	1.0000	1.0000	.9999	.9996	.9975	.9872	.9451	.8011	.4292
	28	1.0000	1.0000	1.0000	1.0000	1.0000	.9998	.9985	.9910	.9529	.7741
	29	1.0000	1.0000	1.0000	1.0000	1.0000	1.0000	1.0000	1.0000	1.0000	1.0000

TABLE A1
BINOMIAL DISTRIBUTION (CONTINUED)

n	x	p = .05	.10	.15	.20	.25	.30	.35	.40	.45
30	0	.2146	.0424	.0076	.0012	.0002	.0000	.0000	.0000	.0000
	1	.5535	.1837	.0480	.0105	.0020	.0003	.0000	.0000	.0000
	2	.8122	.4114	.1514	.0442	.0106	.0021	.0003	.0000	.0000
	3	.9392	.6474	.3217	.1227	.0374	.0093	.0019	.0003	.0000
	4	.9844	.8245	.5245	.2552	.0979	.0302	.0075	.0015	.0002
	5	.9967	.9268	.7106	.4275	.2026	.0766	.0233	.0057	.0011
	6	.9994	.9742	.8474	.6070	.3481	.1595	.0586	.0172	.0040
	7	.9999	.9922	.9302	.7608	.5143	.2814	.1238	.0435	.0121
	8	1.0000	.9980	.9722	.8713	.6736	.4315	.2247	.0940	.0312
	9	1.0000	.9995	.9903	.9389	.8034	.5888	.3575	.1763	.0694
	10	1.0000	.9999	.9971	.9744	.8943	.7304	.5078	.2915	.1350
	11	1.0000	1.0000	.9992	.9905	.9493	.8407	.6548	.4311	.2327
	12	1.0000	1.0000	.9998	.9969	.9784	.9155	.7802	.5785	.3592
	13	1.0000	1.0000	1.0000	.9991	.9918	.9599	.8737	.7145	.5025
	14	1.0000	1.0000	1.0000	.9998	.9973	.9831	.9348	.8246	.6448
	15	1.0000	1.0000	1.0000	.9999	.9992	.9936	.9699	.9029	.7691
	16	1.0000	1.0000	1.0000	1.0000	.9998	.9979	.9876	.9519	.8644
	17	1.0000	1.0000	1.0000	1.0000	.9999	.9994	.9955	.9788	.9286
	18	1.0000	1.0000	1.0000	1.0000	1.0000	.9998	.9986	.9917	.9666
	19	1.0000	1.0000	1.0000	1.0000	1.0000	1.0000	.9996	.9971	.9862
	20	1.0000	1.0000	1.0000	1.0000	1.0000	1.0000	.9999	.9991	.9950
	21	1.0000	1.0000	1.0000	1.0000	1.0000	1.0000	1.0000	.9998	.9984
	22	1.0000	1.0000	1.0000	1.0000	1.0000	1.0000	1.0000	1.0000	.9996
	23	1.0000	1.0000	1.0000	1.0000	1.0000	1.0000	1.0000	1.0000	.9999
	24	1.0000	1.0000	1.0000	1.0000	1.0000	1.0000	1.0000	1.0000	1.0000
	25	1.0000	1.0000	1.0000	1.0000	1.0000	1.0000	1.0000	1.0000	1.0000
	26	1.0000	1.0000	1.0000	1.0000	1.0000	1.0000	1.0000	1.0000	1.0000
	27	1.0000	1.0000	1.0000	1.0000	1.0000	1.0000	1.0000	1.0000	1.0000
	28	1.0000	1.0000	1.0000	1.0000	1.0000	1.0000	1.0000	1.0000	1.0000
	29	1.0000	1.0000	1.0000	1.0000	1.0000	1.0000	1.0000	1.0000	1.0000
	30	1.0000	1.0000	1.0000	1.0000	1.0000	1.0000	1.0000	1.0000	1.0000

TABLE A1
BINOMIAL DISTRIBUTION (CONTINUED)

n	x	p = .50	.55	.60	.65	.70	.75	.80	.85	.90	.95
30	0	.0000	.0000	.0000	.0000	.0000	.0000	.0000	.0000	.0000	.0000
	1	.0000	.0000	.0000	.0000	.0000	.0000	.0000	.0000	.0000	.0000
	2	.0000	.0000	.0000	.0000	.0000	.0000	.0000	.0000	.0000	.0000
	3	.0000	.0000	.0000	.0000	.0000	.0000	.0000	.0000	.0000	.0000
	4	.0000	.0000	.0000	.0000	.0000	.0000	.0000	.0000	.0000	.0000
	5	.0002	.0000	.0000	.0000	.0000	.0000	.0000	.0000	.0000	.0000
	6	.0007	.0001	.0000	.0000	.0000	.0000	.0000	.0000	.0000	.0000
	7	.0026	.0004	.0000	.0000	.0000	.0000	.0000	.0000	.0000	.0000
	8	.0081	.0016	.0002	.0000	.0000	.0000	.0000	.0000	.0000	.0000
	9	.0214	.0050	.0009	.0001	.0000	.0000	.0000	.0000	.0000	.0000
	10	.0494	.0138	.0029	.0004	.0000	.0000	.0000	.0000	.0000	.0000
	11	.1002	.0334	.0083	.0014	.0002	.0000	.0000	.0000	.0000	.0000
	12	.1808	.0714	.0212	.0045	.0006	.0001	.0000	.0000	.0000	.0000
	13	.2923	.1356	.0481	.0124	.0021	.0002	.0000	.0000	.0000	.0000
	14	.4278	.2309	.0971	.0301	.0064	.0008	.0001	.0000	.0000	.0000
	15	.5722	.3552	.1754	.0652	.0169	.0027	.0002	.0000	.0000	.0000
	16	.7077	.4975	.2855	.1263	.0401	.0082	.0009	.0000	.0000	.0000
	17	.8192	.6408	.4215	.2198	.0845	.0216	.0031	.0002	.0000	.0000
	18	.8998	.7673	.5689	.3452	.1593	.0507	.0095	.0008	.0000	.0000
	19	.9506	.8650	.7085	.4922	.2696	.1057	.0256	.0029	.0001	.0000
	20	.9786	.9306	.8237	.6425	.4112	.1966	.0611	.0097	.0005	.0000
	21	.9919	.9688	.9060	.7753	.5685	.3264	.1287	.0278	.0020	.0000
	22	.9974	.9879	.9565	.8762	.7186	.4857	.2392	.0698	.0078	.0001
	23	.9993	.9960	.9828	.9414	.8405	.6519	.3930	.1526	.0258	.0006
	24	.9998	.9989	.9943	.9767	.9234	.7974	.5725	.2894	.0732	.0033
	25	1.0000	.9998	.9985	.9925	.9698	.9021	.7448	.4755	.1755	.0156
	26	1.0000	1.0000	.9997	.9981	.9907	.9626	.8773	.6783	.3526	.0608
	27	1.0000	1.0000	1.0000	.9997	.9979	.9894	.9558	.8486	.5886	.1878
	28	1.0000	1.0000	1.0000	1.0000	.9997	.9980	.9895	.9520	.8163	.4465
	29	1.0000	1.0000	1.0000	1.0000	1.0000	.9998	.9988	.9924	.9576	.7854
	30	1.0000	1.0000	1.0000	1.0000	1.0000	1.0000	1.0000	1.0000	1.0000	1.0000

Note: For n larger than 30, the rth quantile x_r of a binomial random variable may be approximated using $x_r = np + z_r\sqrt{np(1 - p)}$, where z_r is the rth quantile of a standard normal random variable, obtained from Table A2.

TABLES A2 AND A2*
NORMAL DISTRIBUTION

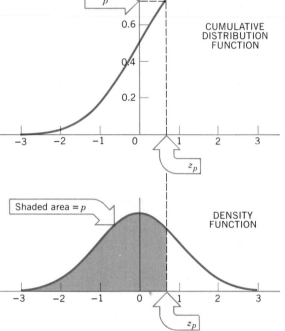

Note: Some useful selected quantiles are given as follows:

$$z_{.0001} = -3.7190 \quad z_{.0005} = -3.2905 \quad z_{.025} = -1.9600 \quad z_{.05} = -1.6449 \quad z_{.10} = -1.2816$$

$$z_{.9999} = 3.7190 \quad z_{.9995} = 3.2905 \quad z_{.975} = 1.9600 \quad z_{.95} = 1.6449 \quad z_{.90} = 1.2816$$

To find the pth quantile z_p use Table A2. First round p to three decimal places. The first two decimal places determine which row to use; the third decimal place determines which column to use to find z_p.

To find the cumulative probability p, use Table A2*. First round z to two decimal places. Use the first decimal place to determine which row to use, and the second decimal place to determine which column to use to find p.

Source: Table generated by R. L. Iman.

p	.000	.001	.002	.003	.004	.005	.006	.007	.008	.009
.00		-3.0902	-2.8782	-2.7478	-2.6521	-2.5758	-2.5121	-2.4573	-2.4089	-2.3656
.01	-2.3263	-2.2904	-2.2571	-2.2262	-2.1973	-2.1701	-2.1444	-2.1201	-2.0969	-2.0749
.02	-2.0537	-2.0335	-2.0141	-1.9954	-1.9774	-1.9600	-1.9431	-1.9268	-1.9110	-1.8957
.03	-1.8808	-1.8663	-1.8522	-1.8384	-1.8250	-1.8119	-1.7991	-1.7866	-1.7744	-1.7624
.04	-1.7507	-1.7392	-1.7279	-1.7169	-1.7060	-1.6954	-1.6849	-1.6747	-1.6646	-1.6546
.05	-1.6449	-1.6352	-1.6258	-1.6164	-1.6072	-1.5982	-1.5893	-1.5805	-1.5718	-1.5632
.06	-1.5548	-1.5464	-1.5382	-1.5301	-1.5220	-1.5141	-1.5063	-1.4985	-1.4909	-1.4833
.07	-1.4758	-1.4684	-1.4611	-1.4538	-1.4466	-1.4395	-1.4325	-1.4255	-1.4187	-1.4118
.08	-1.4051	-1.3984	-1.3917	-1.3852	-1.3787	-1.3722	-1.3658	-1.3595	-1.3532	-1.3469
.09	-1.3408	-1.3346	-1.3285	-1.3225	-1.3165	-1.3106	-1.3047	-1.2988	-1.2930	-1.2873
.10	-1.2816	-1.2759	-1.2702	-1.2646	-1.2591	-1.2536	-1.2481	-1.2426	-1.2372	-1.2319
.11	-1.2265	-1.2212	-1.2160	-1.2107	-1.2055	-1.2004	-1.1952	-1.1901	-1.1850	-1.1800
.12	-1.1750	-1.1700	-1.1650	-1.1601	-1.1552	-1.1503	-1.1455	-1.1407	-1.1359	-1.1311
.13	-1.1264	-1.1217	-1.1170	-1.1123	-1.1077	-1.1031	-1.0985	-1.0939	-1.0893	-1.0848
.14	-1.0803	-1.0758	-1.0714	-1.0669	-1.0625	-1.0581	-1.0537	-1.0494	-1.0450	-1.0407
.15	-1.0364	-1.0322	-1.0279	-1.0237	-1.0194	-1.0152	-1.0110	-1.0069	-1.0027	-.9986
.16	-.9945	-.9904	-.9863	-.9822	-.9782	-.9741	-.9701	-.9661	-.9621	-.9581
.17	-.9542	-.9502	-.9463	-.9424	-.9385	-.9346	-.9307	-.9269	-.9230	-.9192
.18	-.9154	-.9116	-.9078	-.9040	-.9002	-.8965	-.8927	-.8890	-.8853	-.8816
.19	-.8779	-.8742	-.8705	-.8669	-.8633	-.8596	-.8560	-.8524	-.8488	-.8452
.20	-.8416	-.8381	-.8345	-.8310	-.8274	-.8239	-.8204	-.8169	-.8134	-.8099
.21	-.8064	-.8030	-.7995	-.7961	-.7926	-.7892	-.7858	-.7824	-.7790	-.7756
.22	-.7722	-.7688	-.7655	-.7621	-.7588	-.7554	-.7521	-.7488	-.7454	-.7421
.23	-.7388	-.7356	-.7323	-.7290	-.7257	-.7225	-.7192	-.7160	-.7128	-.7095
.24	-.7063	-.7031	-.6999	-.6967	-.6935	-.6903	-.6871	-.6840	-.6808	-.6776
.25	-.6745	-.6713	-.6682	-.6651	-.6620	-.6588	-.6557	-.6526	-.6495	-.6464
.26	-.6433	-.6403	-.6372	-.6341	-.6311	-.6280	-.6250	-.6219	-.6189	-.6158
.27	-.6128	-.6098	-.6068	-.6038	-.6008	-.5978	-.5948	-.5918	-.5888	-.5858
.28	-.5828	-.5799	-.5769	-.5740	-.5710	-.5681	-.5651	-.5622	-.5592	-.5563
.29	-.5534	-.5505	-.5476	-.5446	-.5417	-.5388	-.5359	-.5330	-.5302	-.5273
.30	-.5244	-.5215	-.5187	-.5158	-.5129	-.5101	-.5072	-.5044	-.5015	-.4987
.31	-.4959	-.4930	-.4902	-.4874	-.4845	-.4817	-.4789	-.4761	-.4733	-.4705
.32	-.4677	-.4649	-.4621	-.4593	-.4565	-.4538	-.4510	-.4482	-.4454	-.4427
.33	-.4399	-.4372	-.4344	-.4316	-.4289	-.4261	-.4234	-.4207	-.4179	-.4152
.34	-.4125	-.4097	-.4070	-.4043	-.4016	-.3989	-.3961	-.3934	-.3907	-.3880
.35	-.3853	-.3826	-.3799	-.3772	-.3745	-.3719	-.3692	-.3665	-.3638	-.3611
.36	-.3585	-.3558	-.3531	-.3505	-.3478	-.3451	-.3425	-.3398	-.3372	-.3345
.37	-.3319	-.3292	-.3266	-.3239	-.3213	-.3186	-.3160	-.3134	-.3107	-.3081
.38	-.3055	-.3029	-.3002	-.2976	-.2950	-.2924	-.2898	-.2871	-.2845	-.2819
.39	-.2793	-.2767	-.2741	-.2715	-.2689	-.2663	-.2637	-.2611	-.2585	-.2559
.40	-.2533	-.2508	-.2482	-.2456	-.2430	-.2404	-.2378	-.2353	-.2327	-.2301
.41	-.2275	-.2250	-.2224	-.2198	-.2173	-.2147	-.2121	-.2096	-.2070	-.2045
.42	-.2019	-.1993	-.1968	-.1942	-.1917	-.1891	-.1866	-.1840	-.1815	-.1789
.43	-.1764	-.1738	-.1713	-.1687	-.1662	-.1637	-.1611	-.1586	-.1560	-.1535
.44	-.1510	-.1484	-.1459	-.1434	-.1408	-.1383	-.1358	-.1332	-.1307	-.1282
.45	-.1257	-.1231	-.1206	-.1181	-.1156	-.1130	-.1105	-.1080	-.1055	-.1030
.46	-.1004	-.0979	-.0954	-.0929	-.0904	-.0878	-.0853	-.0828	-.0803	-.0778
.47	-.0753	-.0728	-.0702	-.0677	-.0652	-.0627	-.0602	-.0577	-.0552	-.0527
.48	-.0502	-.0476	-.0451	-.0426	-.0401	-.0376	-.0351	-.0326	-.0301	-.0276
.49	-.0251	-.0226	-.0201	-.0175	-.0150	-.0125	-.0100	-.0075	-.0050	-.0025

p	.000	.001	.002	.003	.004	.005	.006	.007	.008	.009
.50	.0000	.0025	.0050	.0075	.0100	.0125	.0150	.0175	.0201	.0226
.51	.0251	.0276	.0301	.0326	.0351	.0376	.0401	.0426	.0451	.0476
.52	.0502	.0527	.0552	.0577	.0602	.0627	.0652	.0677	.0702	.0728
.53	.0753	.0778	.0803	.0828	.0853	.0878	.0904	.0929	.0954	.0979
.54	.1004	.1030	.1055	.1080	.1105	.1130	.1156	.1181	.1206	.1231
.55	.1257	.1282	.1307	.1332	.1358	.1383	.1408	.1434	.1459	.1484
.56	.1510	.1535	.1560	.1586	.1611	.1637	.1662	.1687	.1713	.1738
.57	.1764	.1789	.1815	.1840	.1866	.1891	.1917	.1942	.1968	.1993
.58	.2019	.2045	.2070	.2096	.2121	.2147	.2173	.2198	.2224	.2250
.59	.2275	.2301	.2327	.2353	.2378	.2404	.2430	.2456	.2482	.2508
.60	.2533	.2559	.2585	.2611	.2637	.2663	.2689	.2715	.2741	.2767
.61	.2793	.2819	.2845	.2871	.2898	.2924	.2950	.2976	.3002	.3029
.62	.3055	.3081	.3107	.3134	.3160	.3186	.3213	.3239	.3266	.3292
.63	.3319	.3345	.3372	.3398	.3425	.3451	.3478	.3505	.3531	.3558
.64	.3585	.3611	.3638	.3665	.3692	.3719	.3745	.3772	.3799	.3826
.65	.3853	.3880	.3907	.3934	.3961	.3989	.4016	.4043	.4070	.4097
.66	.4125	.4152	.4179	.4207	.4234	.4261	.4289	.4316	.4344	.4372
.67	.4399	.4427	.4454	.4482	.4510	.4538	.4565	.4593	.4621	.4649
.68	.4677	.4705	.4733	.4761	.4789	.4817	.4845	.4874	.4902	.4930
.69	.4959	.4987	.5015	.5044	.5072	.5101	.5129	.5158	.5187	.5215
.70	.5244	.5273	.5302	.5330	.5359	.5388	.5417	.5446	.5476	.5505
.71	.5534	.5563	.5592	.5622	.5651	.5681	.5710	.5740	.5769	.5799
.72	.5828	.5858	.5888	.5918	.5948	.5978	.6008	.6038	.6068	.6098
.73	.6128	.6158	.6189	.6219	.6250	.6280	.6311	.6341	.6372	.6403
.74	.6433	.6464	.6495	.6526	.6557	.6588	.6620	.6651	.6682	.6713
.75	.6745	.6776	.6808	.6840	.6871	.6903	.6935	.6967	.6999	.7031
.76	.7063	.7095	.7128	.7160	.7192	.7225	.7257	.7290	.7323	.7356
.77	.7388	.7421	.7454	.7488	.7521	.7554	.7588	.7621	.7655	.7688
.78	.7722	.7756	.7790	.7824	.7858	.7892	.7926	.7961	.7995	.8030
.79	.8064	.8099	.8134	.8169	.8204	.8239	.8274	.8310	.8345	.8381
.80	.8416	.8452	.8488	.8524	.8560	.8596	.8633	.8669	.8705	.8742
.81	.8779	.8816	.8853	.8890	.8927	.8965	.9002	.9040	.9078	.9116
.82	.9154	.9192	.9230	.9269	.9307	.9346	.9385	.9424	.9463	.9502
.83	.9542	.9581	.9621	.9661	.9701	.9741	.9782	.9822	.9863	.9904
.84	.9945	.9986	1.0027	1.0069	1.0110	1.0152	1.0194	1.0237	1.0279	1.0322
.85	1.0364	1.0407	1.0450	1.0494	1.0537	1.0581	1.0625	1.0669	1.0714	1.0758
.86	1.0803	1.0848	1.0893	1.0939	1.0985	1.1031	1.1077	1.1123	1.1170	1.1217
.87	1.1264	1.1311	1.1359	1.1407	1.1455	1.1503	1.1552	1.1601	1.1650	1.1700
.88	1.1750	1.1800	1.1850	1.1901	1.1952	1.2004	1.2055	1.2107	1.2160	1.2212
.89	1.2265	1.2319	1.2372	1.2426	1.2481	1.2536	1.2591	1.2646	1.2702	1.2759
.90	1.2816	1.2873	1.2930	1.2988	1.3047	1.3106	1.3165	1.3225	1.3285	1.3346
.91	1.3408	1.3469	1.3532	1.3595	1.3658	1.3722	1.3787	1.3852	1.3917	1.3984
.92	1.4051	1.4118	1.4187	1.4255	1.4325	1.4395	1.4466	1.4538	1.4611	1.4684
.93	1.4758	1.4833	1.4909	1.4985	1.5063	1.5141	1.5220	1.5301	1.5382	1.5464
.94	1.5548	1.5632	1.5718	1.5805	1.5893	1.5982	1.6072	1.6164	1.6258	1.6352
.95	1.6449	1.6546	1.6646	1.6747	1.6849	1.6954	1.7060	1.7169	1.7279	1.7392
.96	1.7507	1.7624	1.7744	1.7866	1.7991	1.8119	1.8250	1.8384	1.8522	1.8663
.97	1.8808	1.8957	1.9110	1.9268	1.9431	1.9600	1.9774	1.9954	2.0141	2.0335
.98	2.0537	2.0749	2.0969	2.1201	2.1444	2.1701	2.1973	2.2262	2.2571	2.2904
.99	2.3263	2.3656	2.4089	2.4573	2.5121	2.5758	2.6521	2.7478	2.8782	3.0902

TABLE A2*
NORMAL DISTRIBUTION (FOR FINDING PROBABILITIES)

z	.00	.01	.02	.03	.04	.05	.06	.07	.08	.09
−3.0	.0013	.0013	.0013	.0012	.0012	.0011	.0011	.0011	.0010	.0010
−2.9	.0019	.0018	.0018	.0017	.0016	.0016	.0015	.0015	.0014	.0014
−2.8	.0026	.0025	.0024	.0023	.0023	.0022	.0021	.0021	.0020	.0019
−2.7	.0035	.0034	.0033	.0032	.0031	.0030	.0029	.0028	.0027	.0026
−2.6	.0047	.0045	.0044	.0043	.0041	.0040	.0039	.0038	.0037	.0036
−2.5	.0062	.0060	.0059	.0057	.0055	.0054	.0052	.0051	.0049	.0048
−2.4	.0082	.0080	.0078	.0075	.0073	.0071	.0069	.0068	.0066	.0064
−2.3	.0107	.0104	.0102	.0099	.0096	.0094	.0091	.0089	.0087	.0084
−2.2	.0139	.0136	.0132	.0129	.0125	.0122	.0119	.0116	.0113	.0110
−2.1	.0179	.0174	.0170	.0166	.0162	.0158	.0154	.0150	.0146	.0143
−2.0	.0228	.0222	.0217	.0212	.0207	.0202	.0197	.0192	.0188	.0183
−1.9	.0287	.0281	.0274	.0268	.0262	.0256	.0250	.0244	.0239	.0233
−1.8	.0359	.0351	.0344	.0336	.0329	.0322	.0314	.0307	.0301	.0294
−1.7	.0446	.0436	.0427	.0418	.0409	.0401	.0392	.0384	.0375	.0367
−1.6	.0548	.0537	.0526	.0516	.0505	.0495	.0485	.0475	.0465	.0455
−1.5	.0668	.0655	.0643	.0630	.0618	.0606	.0594	.0582	.0571	.0559
−1.4	.0808	.0793	.0778	.0764	.0749	.0735	.0721	.0708	.0694	.0681
−1.3	.0968	.0951	.0934	.0918	.0901	.0885	.0869	.0853	.0838	.0823
−1.2	.1151	.1131	.1112	.1093	.1075	.1056	.1038	.1020	.1003	.0985
−1.1	.1357	.1335	.1314	.1292	.1271	.1251	.1230	.1210	.1190	.1170
−1.0	.1587	.1562	.1539	.1515	.1492	.1469	.1446	.1423	.1401	.1379
−0.9	.1841	.1814	.1788	.1762	.1736	.1711	.1685	.1660	.1635	.1611
−0.8	.2119	.2090	.2061	.2033	.2005	.1977	.1949	.1921	.1894	.1867
−0.7	.2420	.2389	.2358	.2327	.2296	.2266	.2236	.2206	.2177	.2148
−0.6	.2743	.2709	.2676	.2643	.2611	.2578	.2546	.2514	.2483	.2451
−0.5	.3085	.3050	.3015	.2981	.2946	.2912	.2877	.2843	.2810	.2776
−0.4	.3446	.3409	.3372	.3336	.3300	.3264	.3228	.3192	.3156	.3121
−0.3	.3821	.3783	.3745	.3707	.3669	.3632	.3594	.3557	.3520	.3483
−0.2	.4207	.4168	.4129	.4090	.4052	.4013	.3974	.3936	.3897	.3859
−0.1	.4602	.4562	.4522	.4483	.4443	.4404	.4364	.4325	.4286	.4247
−0.0	.5000	.4960	.4920	.4880	.4840	.4801	.4761	.4721	.4681	.4641

TABLE A2*
NORMAL DISTRIBUTION (FOR FINDING PROBABILITIES) (CONTINUED)

z	.00	.01	.02	.03	.04	.05	.06	.07	.08	.09
0.0	.5000	.5040	.5080	.5120	.5160	.5199	.5239	.5279	.5319	.5359
0.1	.5398	.5438	.5478	.5517	.5557	.5596	.5636	.5675	.5714	.5753
0.2	.5793	.5832	.5871	.5910	.5948	.5987	.6026	.6064	.6103	.6141
0.3	.6179	.6217	.6255	.6293	.6331	.6368	.6406	.6443	.6480	.6517
0.4	.6554	.6591	.6628	.6664	.6700	.6736	.6772	.6808	.6844	.6879
0.5	.6915	.6950	.6985	.7019	.7054	.7088	.7123	.7157	.7190	.7224
0.6	.7257	.7291	.7324	.7357	.7389	.7422	.7454	.7486	.7517	.7549
0.7	.7580	.7611	.7642	.7673	.7703	.7734	.7764	.7793	.7823	.7852
0.8	.7881	.7910	.7939	.7967	.7995	.8023	.8051	.8078	.8106	.8133
0.9	.8159	.8186	.8212	.8238	.8264	.8289	.8315	.8340	.8365	.8389
1.0	.8413	.8438	.8461	.8485	.8508	.8531	.8554	.8577	.8599	.8621
1.1	.8643	.8665	.8686	.8708	.8729	.8749	.8770	.8790	.8810	.8830
1.2	.8849	.8869	.8888	.8907	.8925	.8944	.8962	.8980	.8997	.9015
1.3	.9032	.9049	.9066	.9082	.9099	.9115	.9131	.9147	.9162	.9177
1.4	.9192	.9207	.9222	.9236	.9251	.9265	.9279	.9292	.9306	.9319
1.5	.9332	.9345	.9357	.9370	.9382	.9394	.9406	.9418	.9429	.9441
1.6	.9452	.9463	.9474	.9484	.9495	.9505	.9515	.9525	.9535	.9545
1.7	.9554	.9564	.9573	.9582	.9591	.9599	.9608	.9616	.9625	.9633
1.8	.9641	.9649	.9656	.9664	.9671	.9678	.9686	.9693	.9699	.9706
1.9	.9713	.9719	.9726	.9732	.9738	.9744	.9750	.9756	.9761	.9767
2.0	.9772	.9778	.9783	.9788	.9793	.9798	.9803	.9808	.9812	.9817
2.1	.9821	.9826	.9830	.9834	.9838	.9842	.9846	.9850	.9854	.9857
2.2	.9861	.9864	.9868	.9871	.9875	.9878	.9881	.9884	.9887	.9890
2.3	.9893	.9896	.9898	.9901	.9904	.9906	.9909	.9911	.9913	.9916
2.4	.9918	.9920	.9922	.9925	.9927	.9929	.9931	.9932	.9934	.9936
2.5	.9938	.9940	.9941	.9943	.9945	.9946	.9948	.9949	.9951	.9952
2.6	.9953	.9955	.9956	.9957	.9959	.9960	.9961	.9962	.9963	.9964
2.7	.9965	.9966	.9967	.9968	.9969	.9970	.9971	.9972	.9973	.9974
2.8	.9974	.9975	.9976	.9977	.9977	.9978	.9979	.9979	.9980	.9981
2.9	.9981	.9982	.9982	.9983	.9984	.9984	.9985	.9985	.9986	.9986
3.0	.9987	.9987	.9987	.9988	.9988	.9989	.9989	.9989	.9990	.9990

TABLE A3
STUDENT'S *t* DISTRIBUTION

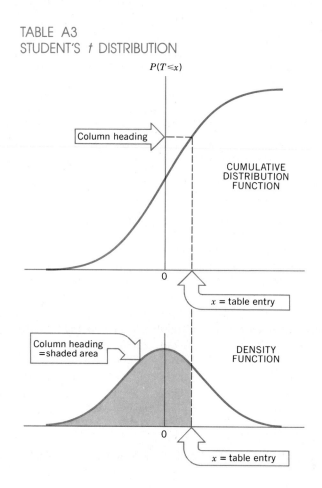

Source: Table generated by R. L. Iman.

TABLE A3
STUDENT'S t DISTRIBUTION

					Quantiles					
d.f.	.60	.75	.90	.95	.975	.99	.995	.999	.9995	.9999
1	.3249	1.0000	3.0777	6.3138	12.706	31.821	63.657	318.31	636.62	3183.1
2	.2887	.8165	1.8856	2.9200	4.3027	6.9646	9.9248	22.327	31.599	70.700
3	.2767	.7649	1.6377	2.3534	3.1824	4.5407	5.8409	10.215	12.924	22.204
4	.2707	.7407	1.5332	2.1318	2.7764	3.7469	4.6041	7.1732	8.6103	13.034
5	.2672	.7267	1.4759	2.0150	2.5706	3.3649	4.0321	5.8934	6.8688	9.6776
6	.2648	.7176	1.4398	1.9432	2.4469	3.1427	3.7074	5.2076	5.9588	8.0248
7	.2632	.7111	1.4149	1.8946	2.3646	2.9980	3.4995	4.7853	5.4079	7.0634
8	.2619	.7064	1.3968	1.8595	2.3060	2.8965	3.3554	4.5008	5.0413	6.4420
9	.2610	.7027	1.3830	1.8331	2.2622	2.8214	3.2498	4.2968	4.7809	6.0101
10	.2602	.6998	1.3722	1.8125	2.2281	2.7638	3.1693	4.1437	4.5869	5.6938
11	.2596	.6974	1.3634	1.7959	2.2010	2.7181	3.1058	4.0247	4.4370	5.4528
12	.2590	.6955	1.3562	1.7823	2.1788	2.6810	3.0545	3.9296	4.3178	5.2633
13	.2586	.6938	1.3502	1.7709	2.1604	2.6503	3.0123	3.8520	4.2208	5.1106
14	.2582	.6924	1.3450	1.7613	2.1448	2.6245	2.9768	3.7874	4.1405	4.9850
15	.2579	.6912	1.3406	1.7531	2.1314	2.6025	2.9467	3.7328	4.0728	4.8800
16	.2576	.6901	1.3368	1.7459	2.1199	2.5835	2.9208	3.6862	4.0150	4.7909
17	.2573	.6892	1.3334	1.7396	2.1098	2.5669	2.8982	3.6458	3.9651	4.7144
18	.2571	.6884	1.3304	1.7341	2.1009	2.5524	2.8784	3.6105	3.9216	4.6480
19	.2569	.6876	1.3277	1.7291	2.0930	2.5395	2.8609	3.5794	3.8834	4.5899
20	.2567	.6870	1.3253	1.7247	2.0860	2.5280	2.8453	3.5518	3.8495	4.5385
21	.2566	.6864	1.3232	1.7207	2.0796	2.5176	2.8314	3.5272	3.8193	4.4929
22	.2564	.6858	1.3212	1.7171	2.0739	2.5083	2.8188	3.5050	3.7921	4.4520
23	.2563	.6853	1.3195	1.7139	2.0687	2.4999	2.8073	3.4850	3.7676	4.4152
24	.2562	.6848	1.3178	1.7109	2.0639	2.4922	2.7969	3.4668	3.7454	4.3819
25	.2561	.6844	1.3163	1.7081	2.0595	2.4851	2.7874	3.4502	3.7251	4.3517
26	.2560	.6840	1.3150	1.7056	2.0555	2.4786	2.7787	3.4350	3.7066	4.3240
27	.2559	.6837	1.3137	1.7033	2.0518	2.4727	2.7707	3.4210	3.6896	4.2987
28	.2558	.6834	1.3125	1.7011	2.0484	2.4671	2.7633	3.4082	3.6739	4.2754
29	.2557	.6830	1.3114	1.6991	2.0452	2.4620	2.7564	3.3962	3.6594	4.2539
30	.2556	.6828	1.3104	1.6973	2.0423	2.4573	2.7500	3.3852	3.6460	4.2340
31	.2555	.6825	1.3095	1.6955	2.0395	2.4528	2.7440	3.3749	3.6335	4.2155
32	.2555	.6822	1.3086	1.6939	2.0369	2.4487	2.7385	3.3653	3.6218	4.1983
33	.2554	.6820	1.3077	1.6924	2.0345	2.4448	2.7333	3.3563	3.6109	4.1822
34	.2553	.6818	1.3070	1.6909	2.0322	2.4411	2.7284	3.3479	3.6007	4.1672
35	.2553	.6816	1.3062	1.6896	2.0301	2.4377	2.7238	3.3400	3.5911	4.1531
36	.2552	.6814	1.3055	1.6883	2.0281	2.4345	2.7195	3.3326	3.5821	4.1399
37	.2552	.6812	1.3049	1.6871	2.0262	2.4314	2.7154	3.3256	3.5737	4.1275
38	.2551	.6810	1.3042	1.6860	2.0244	2.4286	2.7116	3.3190	3.5657	4.1158
39	.2551	.6808	1.3036	1.6849	2.0227	2.4258	2.7079	3.3128	3.5581	4.1047
40	.2550	.6807	1.3031	1.6839	2.0211	2.4233	2.7045	3.3069	3.5510	4.0942
41	.2550	.6805	1.3025	1.6829	2.0195	2.4208	2.7012	3.3013	3.5442	4.0843
42	.2550	.6804	1.3020	1.6820	2.0181	2.4185	2.6981	3.2960	3.5377	4.0749
43	.2549	.6802	1.3016	1.6811	2.0167	2.4163	2.6951	3.2909	3.5316	4.0659
44	.2549	.6801	1.3011	1.6802	2.0154	2.4141	2.6923	3.2861	3.5258	4.0574
45	.2549	.6800	1.3006	1.6794	2.0141	2.4121	2.6896	3.2815	3.5203	4.0493
46	.2548	.6799	1.3002	1.6787	2.0129	2.4102	2.6870	3.2771	3.5150	4.0416
47	.2548	.6797	1.2998	1.6779	2.0117	2.4083	2.6846	3.2729	3.5099	4.0343
48	.2548	.6796	1.2994	1.6772	2.0106	2.4066	2.6822	3.2689	3.5051	4.0272
49	.2547	.6795	1.2991	1.6766	2.0096	2.4049	2.6800	3.2651	3.5004	4.0205
50	.2547	.6794	1.2987	1.6759	2.0086	2.4033	2.6778	3.2614	3.4960	4.0148

					Quantiles				
d.f.	.60	.75	.90	.95	.975	.99	.995	.999	.9995	.9999
51	.2547	.6793	1.2984	1.6753	2.0076	2.4017	2.6757	3.2579	3.4918	4.0079
52	.2546	.6792	1.2980	1.6747	2.0066	2.4002	2.6737	3.2545	3.4877	4.0020
53	.2546	.6791	1.2977	1.6741	2.0057	2.3988	2.6718	3.2513	3.4838	3.9963
54	.2546	.6791	1.2974	1.6736	2.0049	2.3974	2.6700	3.2481	3.4800	3.9908
55	.2546	.6790	1.2971	1.6730	2.0040	2.3961	2.6682	3.2451	3.4764	3.9856
56	.2546	.6789	1.2969	1.6725	2.0032	2.3948	2.6665	3.2423	3.4729	3.9805
57	.2545	.6788	1.2966	1.6720	2.0025	2.3936	2.6649	3.2395	3.4696	3.9757
58	.2545	.6787	1.2963	1.6716	2.0017	2.3924	2.6633	3.2368	3.4663	3.9710
59	.2545	.6787	1.2961	1.6711	2.0010	2.3912	2.6618	3.2342	3.4632	3.9664
60	.2545	.6786	1.2958	1.6706	2.0003	2.3901	2.6603	3.2317	3.4602	3.9621
61	.2545	.6785	1.2956	1.6702	1.9996	2.3890	2.6589	3.2293	3.4573	3.9579
62	.2544	.6785	1.2954	1.6698	1.9990	2.3880	2.6575	3.2270	3.4545	3.9538
63	.2544	.6784	1.2951	1.6694	1.9983	2.3870	2.6561	3.2247	3.4518	3.9499
64	.2544	.6783	1.2949	1.6690	1.9977	2.3860	2.6549	3.2225	3.4491	3.9461
65	.2544	.6783	1.2947	1.6686	1.9971	2.3851	2.6536	3.2204	3.4466	3.9424
66	.2544	.6782	1.2945	1.6683	1.9966	2.3842	2.6524	3.2184	3.4441	3.9389
67	.2544	.6782	1.2943	1.6679	1.9960	2.3833	2.6512	3.2164	3.4417	3.9354
68	.2543	.6781	1.2941	1.6676	1.9955	2.3824	2.6501	3.2145	3.4394	3.9321
69	.2543	.6781	1.2939	1.6672	1.9949	2.3816	2.6490	3.2126	3.4372	3.9288
70	.2543	.6780	1.2938	1.6669	1.9944	2.3808	2.6479	3.2108	3.4350	3.9257
71	.2543	.6780	1.2936	1.6666	1.9939	2.3800	2.6469	3.2090	3.4329	3.9226
72	.2543	.6779	1.2934	1.6663	1.9935	2.3793	2.6459	3.2073	3.4308	3.9197
73	.2543	.6779	1.2933	1.6660	1.9930	2.3785	2.6449	3.2057	3.4289	3.9168
74	.2543	.6778	1.2931	1.6657	1.9925	2.3778	2.6439	3.2041	3.4269	3.9140
75	.2542	.6778	1.2929	1.6654	1.9921	2.3771	2.6430	3.2025	3.4250	3.9113
76	.2542	.6777	1.2928	1.6652	1.9917	2.3764	2.6421	3.2010	3.4232	3.9086
77	.2542	.6777	1.2926	1.6649	1.9913	2.3758	2.6412	3.1995	3.4214	3.9061
78	.2542	.6776	1.2925	1.6646	1.9908	2.3751	2.6403	3.1980	3.4197	3.9036
79	.2542	.6776	1.2924	1.6644	1.9905	2.3745	2.6395	3.1966	3.4180	3.9011
80	.2542	.6776	1.2922	1.6641	1.9901	2.3739	2.6387	3.1953	3.4163	3.8988
81	.2542	.6775	1.2921	1.6639	1.9897	2.3733	2.6379	3.1939	3.4147	3.8964
82	.2542	.6775	1.2920	1.6636	1.9893	2.3727	2.6371	3.1926	3.4132	3.8942
83	.2542	.6775	1.2918	1.6634	1.9890	2.3721	2.6364	3.1913	3.4116	3.8920
84	.2542	.6774	1.2917	1.6632	1.9886	2.3716	2.6356	3.1901	3.4102	3.8899
85	.2541	.6774	1.2916	1.6630	1.9883	2.3710	2.6349	3.1889	3.4087	3.8878
86	.2541	.6774	1.2915	1.6628	1.9879	2.3705	2.6342	3.1877	3.4073	3.8857
87	.2541	.6773	1.2914	1.6626	1.9876	2.3700	2.6335	3.1866	3.4059	3.8837
88	.2541	.6773	1.2912	1.6624	1.9873	2.3695	2.6329	3.1854	3.4045	3.8818
89	.2541	.6773	1.2911	1.6622	1.9870	2.3690	2.6322	3.1843	3.4032	3.8799
90	.2541	.6772	1.2910	1.6620	1.9867	2.3685	2.6316	3.1833	3.4019	3.8780
91	.2541	.6772	1.2909	1.6618	1.9864	2.3680	2.6309	3.1822	3.4007	3.8762
92	.2541	.6772	1.2908	1.6616	1.9861	2.3676	2.6303	3.1812	3.3994	3.8745
93	.2541	.6771	1.2907	1.6614	1.9858	2.3671	2.6297	3.1802	3.3982	3.8727
94	.2541	.6771	1.2906	1.6612	1.9855	2.3667	2.6291	3.1792	3.3971	3.8710
95	.2541	.6771	1.2905	1.6611	1.9853	2.3662	2.6286	3.1782	3.3959	3.8694
96	.2541	.6771	1.2904	1.6609	1.9850	2.3658	2.6280	3.1773	3.3948	3.8678
97	.2540	.6770	1.2903	1.6607	1.9847	2.3654	2.6275	3.1764	3.3937	3.8662
98	.2540	.6770	1.2902	1.6606	1.9845	2.3650	2.6269	3.1755	3.3926	3.8646
99	.2540	.6770	1.2902	1.6604	1.9842	2.3646	2.6264	3.1746	3.3915	3.8631
100	.2540	.6770	1.2901	1.6602	1.9840	2.3642	2.6259	3.1737	3.3905	3.8616
∞	.2533	.6745	1.2816	1.6449	1.9600	2.3263	2.5758	3.0902	3.2905	3.7190

TABLE A4
CHI-SQUARE DISTRIBUTION

$P(T \leq x)$ for 10 d.f.

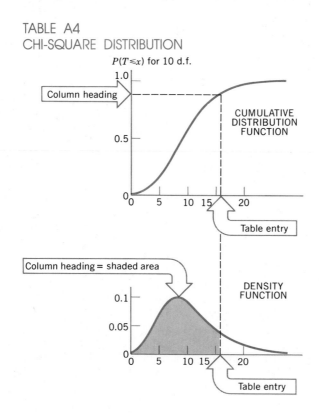

For d.f. > 100 use the approximation $x_p = (\frac{1}{2})(z_p + \sqrt{2k-1})^2$, or the more accurate $x_p = k[1 - 2/(9k) + z_p\sqrt{2/(9k)}]^3$, where k = d.f. and z_p is the value from the standardized normal distribution shown in the bottom of the table.

Source: Table generated by R. L. Iman.

TABLE A4
CHI-SQUARE DISTRIBUTION (CONTINUED)

					Quantiles					
d.f.	.0001	.0005	.001	.005	.01	.025	.05	.10	.25	.4
1	1.57E-8	3.93E-7	1.57E-6	3.93E-5	1.57E-4	9.83E-4	3.93E-3	1.58E-2	.102	.2
2	2.00E-4	1.00E-3	2.00E-3	1.00E-2	2.01E-2	5.06E-2	.103	.211	.575	1.0
3	5.21E-3	1.53E-2	2.43E-2	7.17E-2	.115	.216	.352	.584	1.213	1.8
4	2.84E-2	6.39E-2	9.08E-2	.207	.297	.484	.711	1.064	1.923	2.7
5	8.22E-2	.158	.210	.412	.554	.831	1.145	1.610	2.675	3.6
6	.172	.299	.381	.676	.872	1.237	1.635	2.204	3.455	4.5
7	.300	.485	.598	.989	1.239	1.690	2.167	2.833	4.255	5.4
8	.464	.710	.857	1.344	1.646	2.180	2.733	3.490	5.071	6.4
9	.661	.972	1.152	1.735	2.088	2.700	3.325	4.168	5.899	7.3
10	.889	1.265	1.479	2.156	2.558	3.247	3.940	4.865	6.737	8.2
11	1.145	1.587	1.834	2.603	3.053	3.816	4.575	5.578	7.584	9.2
12	1.427	1.934	2.214	3.074	3.571	4.404	5.226	6.304	8.438	10.
13	1.733	2.305	2.617	3.565	4.107	5.009	5.892	7.042	9.299	11.
14	2.061	2.697	3.041	4.075	4.660	5.629	6.571	7.790	10.17	12.
15	2.408	3.108	3.483	4.601	5.229	6.262	7.261	8.547	11.04	13.
16	2.774	3.536	3.942	5.142	5.812	6.908	7.962	9.312	11.91	13.
17	3.157	3.980	4.416	5.697	6.408	7.564	8.672	10.09	12.79	14
18	3.555	4.439	4.905	6.265	7.015	8.231	9.390	10.86	13.68	15.
19	3.968	4.912	5.407	6.844	7.633	8.907	10.12	11.65	14.56	16
20	4.395	5.398	5.921	7.434	8.260	9.591	10.85	12.44	15.45	17
21	4.835	5.896	6.447	8.034	8.897	10.28	11.59	13.24	16.34	18
22	5.286	6.404	6.983	8.643	9.542	10.98	12.34	14.04	17.24	19
23	5.749	6.924	7.529	9.260	10.20	11.69	13.09	14.85	18.14	20
24	6.223	7.453	8.085	9.886	10.86	12.40	13.85	15.66	19.04	21
25	6.707	7.991	8.649	10.52	11.52	13.12	14.61	16.47	19.94	22
26	7.200	8.538	9.222	11.16	12.20	13.84	15.38	17.29	20.84	23
27	7.702	9.093	9.803	11.81	12.88	14.57	16.15	18.11	21.75	24
28	8.213	9.656	10.39	12.46	13.56	15.31	16.93	18.94	22.66	25
29	8.731	10.23	10.99	13.12	14.26	16.05	17.71	19.77	23.57	26
30	9.258	10.80	11.59	13.79	14.95	16.79	18.49	20.60	24.48	27
31	9.792	11.39	12.20	14.46	15.66	17.54	19.28	21.43	25.39	28
32	10.33	11.98	12.81	15.13	16.36	18.29	20.07	22.27	26.30	29
33	10.88	12.58	13.43	15.82	17.07	19.05	20.87	23.11	27.22	30
34	11.44	13.18	14.06	16.50	17.79	19.81	21.66	23.95	28.14	31
35	12.00	13.79	14.69	17.19	18.51	20.57	22.47	24.80	29.05	32
36	12.56	14.40	15.32	17.89	19.23	21.34	23.27	25.64	29.97	33
37	13.13	15.02	15.97	18.59	19.96	22.11	24.07	26.49	30.89	34
38	13.71	15.64	16.61	19.29	20.69	22.88	24.88	27.34	31.81	35
39	14.29	16.27	17.26	20.00	21.43	23.65	25.70	28.20	32.74	36
40	14.88	16.91	17.92	20.71	22.16	24.43	26.51	29.05	33.66	37
41	15.48	17.54	18.58	21.42	22.91	25.21	27.33	29.91	34.58	38
42	16.07	18.19	19.24	22.14	23.65	26.00	28.14	30.77	35.51	39
43	16.68	18.83	19.91	22.86	24.40	26.79	28.96	31.63	36.44	40
44	17.28	19.48	20.58	23.58	25.15	27.57	29.79	32.49	37.36	41
45	17.89	20.14	21.25	24.31	25.90	28.37	30.61	33.35	38.29	42
46	18.51	20.79	21.93	25.04	26.66	29.16	31.44	34.22	39.22	42
47	19.13	21.46	22.61	25.77	27.42	29.96	32.27	35.08	40.15	43
48	19.75	22.12	23.29	26.51	28.18	30.75	33.10	35.95	41.08	44
49	20.38	22.79	23.98	27.25	28.94	31.55	33.93	36.82	42.01	45
50	21.01	23.46	24.67	27.99	29.71	32.36	34.76	37.69	42.94	46

ABLE A4
CHI-SQUARE DISTRIBUTION (CONTINUED)

d.f.	.60	.75	.90	.95	.975	.99	.995	.999	.9995	.9999
					Quantiles					
1	.708	1.323	2.706	3.841	5.024	6.635	7.879	10.83	12.12	15.14
2	1.833	2.773	4.605	5.991	7.378	9.210	10.60	13.82	15.20	18.42
3	2.946	4.108	6.251	7.815	9.348	11.34	12.84	16.27	17.73	21.11
4	4.045	5.385	7.779	9.488	11.14	13.28	14.86	18.47	20.00	23.51
5	5.132	6.626	9.236	11.07	12.83	15.09	16.75	20.52	22.11	25.74
6	6.211	7.841	10.64	12.59	14.45	16.81	18.55	22.46	24.10	27.86
7	7.283	9.037	12.02	14.07	16.01	18.48	20.28	24.32	26.02	29.88
8	8.351	10.22	13.36	15.51	17.53	20.09	21.95	26.12	27.87	31.83
9	9.414	11.39	14.68	16.92	19.02	21.67	23.59	27.88	29.67	33.72
10	10.47	12.55	15.99	18.31	20.48	23.21	25.19	29.59	31.42	35.56
11	11.53	13.70	17.28	19.68	21.92	24.72	26.76	31.26	33.14	37.37
12	12.58	14.85	18.55	21.03	23.34	26.22	28.30	32.91	34.82	39.13
13	13.64	15.98	19.81	22.36	24.74	27.69	29.82	34.53	36.48	40.87
14	14.69	17.12	21.06	23.68	26.12	29.14	31.32	36.12	38.11	42.58
15	15.73	18.25	22.31	25.00	27.49	30.58	32.80	37.70	39.72	44.26
16	16.78	19.37	23.54	26.30	28.85	32.00	34.27	39.25	41.31	45.92
17	17.82	20.49	24.77	27.59	30.19	33.41	35.72	40.79	42.88	47.57
18	18.87	21.60	25.99	28.87	31.53	34.81	37.16	42.31	44.43	49.19
19	19.91	22.72	27.20	30.14	32.85	36.19	38.58	43.82	45.97	50.80
20	20.95	23.83	28.41	31.41	34.17	37.57	40.00	45.31	47.50	52.39
21	21.99	24.93	29.62	32.67	35.48	38.93	41.40	46.80	49.01	53.96
22	23.03	26.04	30.81	33.92	36.78	40.29	42.80	48.27	50.51	55.52
23	24.07	27.14	32.01	35.17	38.08	41.64	44.18	49.73	52.00	57.07
24	25.11	28.24	33.20	36.42	39.36	42.98	45.56	51.18	53.48	58.61
25	26.14	29.34	34.38	37.65	40.65	44.31	46.93	52.62	54.95	60.14
26	27.18	30.43	35.56	38.89	41.92	45.64	48.29	54.05	56.41	61.66
27	28.21	31.53	36.74	40.11	43.19	46.96	49.64	55.48	57.86	63.16
28	29.25	32.62	37.92	41.34	44.46	48.28	50.99	56.89	59.30	64.66
29	30.28	33.71	39.09	42.56	45.72	49.59	52.34	58.30	60.73	66.15
30	31.32	34.80	40.26	43.77	46.98	50.89	53.67	59.70	62.16	67.63
31	32.35	35.89	41.42	44.99	48.23	52.19	55.00	61.10	63.58	69.11
32	33.38	36.97	42.58	46.19	49.48	53.49	56.33	62.49	65.00	70.57
33	34.41	38.06	43.75	47.40	50.73	54.78	57.65	63.87	66.40	72.03
34	35.44	39.14	44.90	48.60	51.97	56.06	58.96	65.25	67.80	73.48
35	36.47	40.22	46.06	49.80	53.20	57.34	60.27	66.62	69.20	74.93
36	37.50	41.30	47.21	51.00	54.44	58.62	61.58	67.99	70.59	76.36
37	38.53	42.38	48.36	52.19	55.67	59.89	62.88	69.35	71.97	77.80
38	39.56	43.46	49.51	53.38	56.90	61.16	64.18	70.70	73.35	79.22
39	40.59	44.54	50.66	54.57	58.12	62.43	65.48	72.05	74.73	80.65
40	41.62	45.62	51.81	55.76	59.34	63.69	66.77	73.40	76.09	82.06
41	42.65	46.69	52.95	56.94	60.56	64.95	68.05	74.74	77.46	83.47
42	43.68	47.77	54.09	58.12	61.78	66.21	69.34	76.08	78.82	84.88
43	44.71	48.84	55.23	59.30	62.99	67.46	70.62	77.42	80.18	86.28
44	45.73	49.91	56.37	60.48	64.20	68.71	71.89	78.75	81.53	87.68
45	46.76	50.98	57.51	61.66	65.41	69.96	73.17	80.08	82.88	89.07
46	47.79	52.06	58.64	62.83	66.62	71.20	74.44	81.40	84.22	90.46
47	48.81	53.13	59.77	64.00	67.82	72.44	75.70	82.72	85.56	91.84
48	49.84	54.20	60.91	65.17	69.02	73.68	76.97	84.04	86.90	93.22
49	50.87	55.27	62.04	66.34	70.22	74.92	78.23	85.35	88.23	94.60
50	51.89	56.33	63.17	67.50	71.42	76.15	79.49	86.66	89.56	95.97

TABLE A4
CHI-SQUARE DISTRIBUTION (CONTINUED)

	Quantiles									
d.f.	.0001	.0005	.001	.005	.01	.025	.05	.10	.25	.40
51	21.64	24.14	25.37	28.73	30.48	33.16	35.60	38.56	43.87	47.84
52	22.28	24.81	26.07	29.48	31.25	33.97	36.44	39.43	44.81	48.81
53	22.92	25.49	26.76	30.23	32.02	34.78	37.28	40.31	45.74	49.79
54	23.57	26.18	27.47	30.98	32.79	35.59	38.12	41.18	46.68	50.76
55	24.21	26.87	28.17	31.73	33.57	36.40	38.96	42.06	47.61	51.74
56	24.86	27.56	28.88	32.49	34.35	37.21	39.80	42.94	48.55	52.71
57	25.52	28.25	29.59	33.25	35.13	38.03	40.65	43.82	49.48	53.69
58	26.18	28.94	30.30	34.01	35.91	38.84	41.49	44.70	50.42	54.67
59	26.83	29.64	31.02	34.77	36.70	39.66	42.34	45.58	51.36	55.64
60	27.50	30.34	31.74	35.53	37.48	40.48	43.19	46.46	52.29	56.62
61	28.16	31.04	32.46	36.30	38.27	41.30	44.04	47.34	53.23	57.66
62	28.83	31.75	33.18	37.07	39.06	42.13	44.89	48.23	54.17	58.57
63	29.50	32.46	33.91	37.84	39.86	42.95	45.74	49.11	55.11	59.55
64	30.17	33.16	34.63	38.61	40.65	43.78	46.59	50.00	56.05	60.5
65	30.85	33.88	35.36	39.38	41.44	44.60	47.45	50.88	56.99	61.5
66	31.53	34.59	36.09	40.16	42.24	45.43	48.31	51.77	57.93	62.4
67	32.21	35.31	36.83	40.94	43.04	46.26	49.16	52.66	58.87	63.4
68	32.89	36.02	37.56	41.71	43.84	47.09	50.02	53.55	59.81	64.4
69	33.57	36.75	38.30	42.49	44.64	47.92	50.88	54.44	60.76	65.4
70	34.26	37.47	39.04	43.28	45.44	48.76	51.74	55.33	61.70	66.4
71	34.95	38.19	39.78	44.06	46.25	49.59	52.60	56.22	62.64	67.3
72	35.64	38.92	40.52	44.84	47.05	50.43	53.46	57.11	63.58	68.3
73	36.33	39.65	41.26	45.63	47.86	51.26	54.33	58.01	64.53	69.3
74	37.03	40.38	42.01	46.42	48.67	52.10	55.19	58.90	65.47	70.3
75	37.73	41.11	42.76	47.21	49.48	52.94	56.05	59.79	66.42	71.2
76	38.43	41.84	43.51	48.00	50.29	53.78	56.92	60.69	67.36	72.2
77	39.13	42.58	44.26	48.79	51.10	54.62	57.79	61.59	68.31	73.2
78	39.83	43.31	45.01	49.58	51.91	55.47	58.65	62.48	69.25	74.2
79	40.54	44.05	45.76	50.38	52.72	56.31	59.52	63.38	70.20	75.2
80	41.24	44.79	46.52	51.17	53.54	57.15	60.39	64.28	71.14	76.1
81	41.95	45.53	47.28	51.97	54.36	58.00	61.26	65.18	72.09	77.1
82	42.66	46.28	48.04	52.77	55.17	58.84	62.13	66.08	73.04	78.1
83	43.38	47.02	48.80	53.57	55.99	59.69	63.00	66.98	73.99	79.1
84	44.09	47.77	49.56	54.37	56.81	60.54	63.88	67.88	74.93	80.1
85	44.81	48.52	50.32	55.17	57.63	61.39	64.75	68.78	75.88	81.0
86	45.52	49.26	51.08	55.97	58.46	62.24	65.62	69.68	76.83	82.0
87	46.24	50.02	51.85	56.78	59.28	63.09	66.50	70.58	77.78	83.0
88	46.96	50.77	52.62	57.58	60.10	63.94	67.37	71.48	78.73	84.0
89	47.69	51.52	53.39	58.39	60.93	64.79	68.25	72.39	79.68	85.0
90	48.41	52.28	54.16	59.20	61.75	65.65	69.13	73.29	80.62	85.9
91	49.13	53.03	54.93	60.00	62.58	66.50	70.00	74.20	81.57	86.9
92	49.86	53.79	55.70	60.81	63.41	67.36	70.88	75.10	82.52	87.9
93	50.59	54.55	56.47	61.63	64.24	68.21	71.76	76.01	83.47	88.9
94	51.32	55.31	57.25	62.44	65.07	69.07	72.64	76.91	84.42	89.9
95	52.05	56.07	58.02	63.25	65.90	69.92	73.52	77.82	85.38	90.9
96	52.78	56.83	58.80	64.06	66.73	70.78	74.40	78.73	86.33	91.8
97	53.52	57.60	59.58	64.88	67.56	71.64	75.28	79.63	87.28	92.8
98	54.25	58.36	60.36	65.69	68.40	72.50	76.16	80.54	88.23	93.8
99	54.99	59.13	61.14	66.51	69.23	73.36	77.05	81.45	89.18	94.8
100	55.72	55.90	61.92	67.33	70.06	74.22	77.93	82.36	90.13	95.8
z_p	−3.7190	−3.2905	−3.0902	−2.5758	−2.3263	−1.9600	−1.6449	−1.2816	−.6745	−.25

TABLE A4
CHI-SQUARE DISTRIBUTION (CONTINUED)

					Quantiles					
d.f.	.60	.75	.90	.95	.975	.99	.995	.999	.9995	.9999
51	52.92	57.40	64.30	68.67	72.62	77.39	80.75	87.97	90.89	97.34
52	53.94	58.47	65.42	69.83	73.81	78.62	82.00	89.27	92.21	98.70
53	54.97	59.53	66.55	70.99	75.00	79.84	83.25	90.57	93.53	100.1
54	55.99	60.60	67.67	72.15	76.19	81.07	84.50	91.87	94.85	101.4
55	57.02	61.66	68.80	73.31	77.38	82.29	85.75	93.17	96.16	102.8
56	58.04	62.73	69.92	74.47	78.57	83.51	86.99	94.46	97.47	104.1
57	59.06	63.79	71.04	75.62	79.75	84.73	88.24	95.75	98.78	105.5
58	60.09	64.86	72.16	76.78	80.94	85.95	89.48	97.04	100.1	106.8
59	61.11	65.92	73.28	77.93	82.12	87.17	90.72	98.32	101.4	108.2
60	62.13	66.98	74.40	79.08	83.30	88.38	91.95	99.61	102.7	109.5
61	63.16	68.04	75.51	80.23	84.48	89.59	93.19	100.9	104.0	110.8
62	64.18	69.10	76.63	81.38	85.65	90.80	94.42	102.2	105.3	112.2
63	65.20	70.16	77.75	82.53	86.83	92.01	95.65	103.4	106.6	113.5
64	66.23	71.23	78.86	83.68	88.00	93.22	96.88	104.7	107.9	114.8
65	67.25	72.28	79.97	84.82	89.18	94.42	98.11	106.0	109.2	116.2
66	68.27	73.34	81.09	85.96	90.35	95.63	99.33	107.3	110.5	117.5
67	69.29	74.40	82.20	87.11	91.52	96.83	100.6	108.5	111.7	118.8
68	70.32	75.46	83.31	88.25	92.69	98.03	101.8	109.8	113.0	120.1
69	71.34	76.52	84.42	89.39	93.86	99.23	103.0	111.1	114.3	121.4
70	72.36	77.58	85.53	90.53	95.02	100.4	104.2	112.3	115.6	122.8
71	73.38	78.63	86.64	91.67	96.19	101.6	105.4	113.6	116.9	124.1
72	74.40	79.69	87.74	92.81	97.35	102.8	106.6	114.8	118.1	125.4
73	75.42	80.75	88.85	93.95	98.52	104.0	107.9	116.1	119.4	126.7
74	76.44	81.80	89.96	95.08	99.68	105.2	109.1	117.3	120.7	128.0
75	77.46	82.86	91.06	96.22	100.8	106.4	110.3	118.6	121.9	129.3
76	78.48	83.91	92.17	97.35	102.0	107.6	111.5	119.9	123.2	130.6
77	79.51	84.97	93.27	98.48	103.2	108.8	112.7	121.1	124.5	131.9
78	80.53	86.02	94.37	99.62	104.3	110.0	113.9	122.3	125.7	133.2
79	81.55	87.08	95.48	100.7	105.5	111.1	115.1	123.6	127.0	134.5
80	82.57	88.13	96.58	101.9	106.6	112.3	116.3	124.8	128.3	135.8
81	83.59	89.18	97.68	103.0	107.8	113.5	117.5	126.1	129.5	137.1
82	84.61	90.24	98.78	104.1	108.9	114.7	118.7	127.3	130.8	138.4
83	85.63	91.29	99.88	105.3	110.1	115.9	119.9	128.6	132.0	139.7
84	86.65	92.34	101.1	106.4	111.2	117.1	121.1	129.8	133.3	140.9
85	87.67	93.39	102.1	107.5	112.4	118.2	122.3	131.0	134.5	142.2
86	88.68	94.45	103.2	108.6	113.5	119.4	123.5	132.3	135.8	143.5
87	89.70	95.50	104.3	109.8	114.7	120.6	124.7	133.5	137.0	144.8
88	90.72	96.55	105.4	110.9	115.8	121.8	125.9	134.7	138.3	146.1
89	91.74	97.60	106.5	112.0	117.0	122.9	127.1	136.0	139.5	147.4
90	92.76	98.65	107.6	113.1	118.1	124.1	128.3	137.2	140.8	148.6
91	93.78	99.70	108.7	114.3	119.3	125.3	129.5	138.4	142.0	149.9
92	94.80	100.8	109.8	115.4	120.4	126.5	130.7	139.7	143.3	151.2
93	95.82	101.8	110.9	116.5	121.6	127.6	131.9	140.9	144.5	152.4
94	96.84	102.8	111.9	117.6	122.7	128.8	133.1	142.1	145.8	153.7
95	97.85	103.9	113.0	118.8	123.9	130.0	134.2	143.3	147.0	155.0
96	98.87	104.9	114.1	119.9	125.0	131.1	135.4	144.6	148.2	156.3
97	99.89	106.0	115.2	121.0	126.1	132.3	136.6	145.8	149.5	157.5
98	100.9	107.0	116.3	122.1	127.3	133.5	137.8	147.0	150.7	158.8
99	101.9	108.1	117.4	123.2	128.4	134.6	139.0	148.2	151.9	160.1
100	102.9	109.1	118.5	124.3	129.6	135.8	140.2	149.4	153.2	161.3
	.2533	.6745	1.2816	1.6449	1.9600	2.3263	2.5758	3.0902	3.2905	3.7190

TABLE A5
THE F DISTRIBUTION

$P(F \leq x)$ for (10,10) d.f.

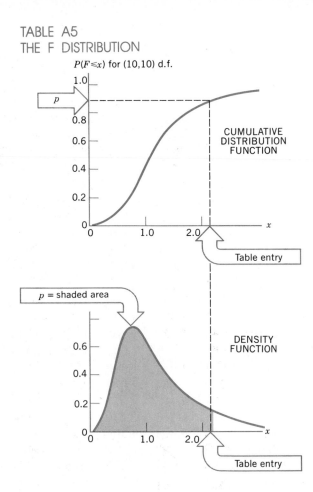

CUMULATIVE
DISTRIBUTION
FUNCTION

Table entry

p = shaded area

DENSITY
FUNCTION

Table entry

Source: Table generated by R. L. Iman.

TABLE A5
THE F DISTRIBUTION WITH k_1 AND k_2 DEGREES OF FREEDOM

k_2	p	k_1: 1	2	3	4	5	6	7	8	9
1	.75	5.828	7.500	8.200	8.581	8.820	8.983	9.102	9.192	9.263
	.90	39.86	49.50	53.59	55.83	57.24	58.20	58.91	59.44	59.86
	.95	161.4	199.5	215.7	224.6	230.2	234.0	236.8	238.9	240.5
	.975	647.8	799.5	864.2	899.6	921.8	937.1	948.2	956.7	963.3
	.99	4052.	5000.	5403.	5625.	5764.	5859.	5928.	5981.	6022.
2	.75	2.571	3.000	3.153	3.232	3.280	3.312	3.335	3.353	3.366
	.90	8.526	9.000	9.162	9.243	9.293	9.326	9.349	9.367	9.381
	.95	18.51	19.00	19.16	19.25	19.30	19.33	19.35	19.37	19.38
	.975	38.51	39.00	39.17	39.25	39.30	39.33	39.36	39.37	39.39
	.99	98.50	99.00	99.17	99.25	99.30	99.33	99.36	99.37	99.39
	.999	998.5	999.0	999.2	999.3	999.3	999.3	999.4	999.4	999.4
3	.75	2.024	2.280	2.356	2.390	2.409	2.422	2.430	2.436	2.441
	.90	5.538	5.462	5.391	5.343	5.309	5.285	5.266	5.252	5.240
	.95	10.13	9.552	9.277	9.117	9.013	8.941	8.887	8.845	8.812
	.975	17.44	16.04	15.44	15.10	14.88	14.73	14.62	14.54	14.47
	.99	34.12	30.82	29.46	28.71	28.24	27.91	27.67	27.49	27.35
	.999	167.0	148.5	141.1	137.1	134.6	132.8	131.6	130.6	129.9
4	.75	1.807	2.000	2.047	2.064	2.072	2.077	2.079	2.080	2.081
	.90	4.545	4.325	4.191	4.107	4.051	4.010	3.979	3.955	3.936
	.95	7.709	6.944	6.591	6.388	6.256	6.163	6.094	6.041	5.999
	.975	12.22	10.65	9.979	9.605	9.364	9.197	9.074	8.980	8.905
	.99	21.20	18.00	16.69	15.98	15.52	15.21	14.98	14.80	14.66
	.999	74.14	61.25	56.18	53.44	51.71	50.53	49.66	49.00	48.47
5	.75	1.692	1.853	1.884	1.893	1.895	1.894	1.894	1.892	1.891
	.90	4.060	3.780	3.619	3.520	3.453	3.405	3.368	3.339	3.316
	.95	6.608	5.786	5.409	5.192	5.050	4.950	4.876	4.818	4.772
	.975	10.01	8.434	7.764	7.388	7.146	6.978	6.853	6.757	6.681
	.99	16.26	13.27	12.06	11.39	10.97	10.67	10.46	10.29	10.16
	.999	47.18	37.12	33.20	31.09	29.75	28.83	28.16	27.65	27.24
6	.75	1.621	1.762	1.784	1.787	1.785	1.782	1.779	1.776	1.773
	.90	3.776	3.463	3.289	3.181	3.108	3.055	3.014	2.983	2.958
	.95	5.987	5.143	4.757	4.534	4.387	4.284	4.207	4.147	4.099
	.975	8.813	7.260	6.599	6.227	5.988	5.820	5.695	5.600	5.523
	.99	13.75	10.92	9.780	9.148	8.746	8.466	8.260	8.102	7.976
	.999	35.51	27.00	23.70	21.92	20.80	20.03	19.46	19.03	18.69
7	.75	1.573	1.701	1.717	1.716	1.711	1.706	1.701	1.697	1.693
	.90	3.589	3.257	3.074	2.961	2.883	2.827	2.785	2.752	2.725
	.95	5.591	4.737	4.347	4.120	3.972	3.866	3.787	3.726	3.677
	.975	8.073	6.542	5.890	5.523	5.285	5.119	4.995	4.899	4.823
	.99	12.25	9.547	8.451	7.847	7.460	7.191	6.993	6.840	6.719
	.999	29.25	21.69	18.77	17.20	16.21	15.52	15.02	14.63	14.33

TABLE A5
THE F DISTRIBUTION WITH k_1 AND k_2 DEGREES OF FREEDOM (CONTINUED)

k_2	p	k_1: 10	12	15	20	24	30	40	60	120
1	**.75**	9.320	9.406	9.493	9.581	9.625	9.670	9.714	9.759	9.804
	.90	60.19	60.71	61.22	61.74	62.00	62.26	62.53	62.79	63.06
	.95	241.9	243.9	245.9	248.0	249.1	250.1	251.1	252.2	253.3
	.975	968.6	976.7	984.9	993.1	997.2	1001.	1006.	1010.	1014.
	.99	6056.	6106.	6157.	6209.	6235.	6261.	6287.	6313.	6339.
2	**.75**	3.377	3.393	3.410	3.426	3.435	3.443	3.451	3.459	3.468
	.90	9.392	9.408	9.425	9.441	9.450	9.458	9.466	9.475	9.483
	.95	19.40	19.41	19.43	19.45	19.45	19.46	19.47	19.48	19.49
	.975	39.40	39.41	39.43	39.45	39.46	39.46	39.47	39.48	39.49
	.99	99.40	99.42	99.43	99.45	99.46	99.47	99.47	99.48	99.49
	.999	999.4	999.4	999.4	999.4	999.5	999.5	999.5	999.5	999.5
3	**.75**	2.445	2.450	2.455	2.460	2.463	2.465	2.467	2.470	2.472
	.90	5.230	5.216	5.200	5.184	5.176	5.168	5.160	5.151	5.143
	.95	8.786	8.745	8.703	8.660	8.639	8.617	8.594	8.572	8.549
	.975	14.42	14.34	14.25	14.17	14.12	14.08	14.04	13.99	13.95
	.99	27.23	27.05	26.87	26.69	26.60	26.50	26.41	26.32	26.22
	.999	129.2	128.3	127.4	126.4	125.9	125.4	125.0	124.5	124.0
4	**.75**	2.082	2.083	2.083	2.083	2.083	2.082	2.082	2.082	2.081
	.90	3.920	3.896	3.870	3.844	3.831	3.817	3.084	3.790	3.775
	.95	5.964	5.912	5.858	5.803	5.774	5.746	5.717	5.688	5.658
	.975	8.844	8.751	8.657	8.560	8.511	8.461	8.411	8.360	8.309
	.99	14.55	14.37	14.20	14.02	13.93	13.84	13.75	13.65	13.56
	.999	48.05	47.41	46.76	46.10	45.77	45.43	45.09	44.75	44.40
5	**.75**	1.890	1.888	1.885	1.882	1.880	1.878	1.876	1.874	1.872
	.90	3.297	3.268	3.238	3.207	3.191	3.174	3.157	3.140	3.123
	.95	4.735	4.678	4.619	4.558	4.527	4.496	4.464	4.431	4.398
	.975	6.619	6.525	6.428	6.329	6.278	6.227	6.175	6.123	6.069
	.99	10.05	9.888	9.722	9.553	9.466	9.379	9.291	9.202	9.112
	.999	26.92	26.42	25.91	25.39	25.13	24.87	24.60	24.33	24.06
6	**.75**	1.771	1.767	1.762	1.757	1.754	1.751	1.748	1.744	1.741
	.90	2.937	2.905	2.871	2.836	2.818	2.800	2.781	2.762	2.742
	.95	4.060	4.000	3.938	3.874	3.841	3.808	3.774	3.740	3.705
	.975	5.461	5.366	5.269	5.168	5.117	5.065	5.012	4.959	4.904
	.99	7.874	7.718	7.559	7.396	7.313	7.229	7.143	7.057	6.969
	.999	18.41	17.99	17.56	17.12	16.90	16.67	16.44	16.21	15.98
7	**.75**	1.690	1.684	1.678	1.671	1.667	1.663	1.659	1.655	1.650
	.90	2.703	2.668	2.632	2.595	2.575	2.555	2.535	2.514	2.493
	.95	3.637	3.575	3.511	3.445	3.410	3.376	3.340	3.304	3.267
	.975	4.761	4.666	4.568	4.467	4.415	4.362	4.309	4.254	4.199
	.99	6.620	6.469	6.314	6.155	6.074	5.992	5.908	5.824	5.737
	.999	14.08	13.71	13.32	12.93	12.73	12.53	12.33	12.12	11.91

TABLE A5
THE F DISTRIBUTION WITH k_1 AND k_2 DEGREES OF FREEDOM (CONTINUED)

k_2	p	k_1: 1	2	3	4	5	6	7	8	9
8	.75	1.538	1.657	1.668	1.664	1.658	1.651	1.645	1.640	1.635
	.90	3.458	3.113	2.924	2.806	2.726	2.668	2.624	2.589	2.561
	.95	5.318	4.459	4.066	3.838	3.687	3.581	3.500	3.438	3.388
	.975	7.571	6.059	5.416	5.053	4.817	4.652	4.529	4.433	4.357
	.99	11.26	8.649	7.591	7.006	6.632	6.371	6.178	6.029	5.911
	.999	25.41	18.49	15.83	14.39	13.48	12.86	12.40	12.05	11.77
9	.75	1.512	1.624	1.632	1.625	1.617	1.609	1.602	1.596	1.591
	.90	3.360	3.006	2.813	2.693	2.611	2.551	2.505	2.469	2.440
	.95	5.117	4.256	3.863	3.633	3.482	3.374	3.293	3.230	3.179
	.975	7.209	5.715	5.078	4.718	4.484	4.320	4.197	4.102	4.026
	.99	10.56	8.022	6.992	6.422	6.057	5.802	5.613	5.467	5.351
	.999	22.86	16.39	13.90	12.56	11.71	11.13	10.70	10.37	10.11
10	.75	1.491	1.598	1.603	1.595	1.585	1.576	1.569	1.562	1.556
	.90	3.285	2.924	2.728	2.605	2.522	2.461	2.414	2.377	2.347
	.95	4.965	4.103	3.708	3.478	3.326	3.217	3.135	3.072	3.020
	.975	6.937	5.456	4.826	4.468	4.236	4.072	3.950	3.855	3.779
	.99	10.04	7.559	6.552	5.994	5.636	5.386	5.200	5.057	4.942
	.999	21.04	14.91	12.55	11.28	10.48	9.926	9.517	9.204	8.956
11	.75	1.475	1.577	1.580	1.570	1.560	1.550	1.542	1.535	1.528
	.90	3.225	2.868	2.660	2.536	2.451	2.389	2.342	2.304	2.274
	.95	4.844	3.982	3.587	3.357	3.204	3.095	3.012	2.948	2.896
	.975	6.724	5.256	4.630	4.275	4.044	3.881	3.759	3.664	3.588
	.99	9.646	7.206	6.217	5.668	5.316	5.069	4.886	4.744	4.632
	.999	19.69	13.81	11.56	10.35	9.578	9.047	8.655	8.355	8.116
12	.75	1.461	1.560	1.561	1.550	1.539	1.529	1.520	1.512	1.505
	.90	3.177	2.807	2.606	2.480	2.394	2.331	2.283	2.245	2.214
	.95	4.747	3.885	3.490	3.259	3.106	2.996	2.913	2.849	2.796
	.975	6.554	5.096	4.474	4.121	3.891	3.728	3.607	3.512	3.436
	.99	9.330	6.927	5.953	5.412	5.064	4.821	4.640	4.499	4.388
	.999	18.64	12.97	10.80	9.633	8.892	8.379	8.001	7.710	7.480
13	.75	1.450	1.545	1.545	1.534	1.521	1.511	1.501	1.493	1.486
	.90	3.136	2.763	2.560	2.434	2.347	2.283	2.234	2.195	2.164
	.95	4.667	3.806	3.411	3.179	3.025	2.915	2.832	2.767	2.714
	.975	6.414	4.965	4.347	3.996	3.767	3.604	3.483	3.388	3.312
	.99	9.074	6.701	5.739	5.205	4.862	4.620	4.441	4.302	4.191
	.999	17.82	12.31	10.21	9.073	8.354	7.856	7.489	7.206	6.982
14	.75	1.440	1.533	1.532	1.519	1.507	1.495	1.485	1.477	1.470
	.90	3.102	2.726	2.522	2.395	2.307	2.243	2.193	2.154	2.122
	.95	4.600	3.739	3.344	3.112	2.958	2.848	2.764	2.699	2.646
	.975	6.298	4.857	4.242	3.892	3.663	3.501	3.380	3.285	3.209
	.99	8.862	6.515	5.564	5.035	4.695	4.456	4.278	4.140	4.030
	.999	17.14	11.78	9.729	8.622	7.922	7.436	7.077	6.802	6.583

TABLE A5
THE F DISTRIBUTION WITH k_1 AND k_2 DEGREES OF FREEDOM (CONTINUED)

k_2	p	k_1: 10	12	15	20	24	30	40	60	120
8	.75	1.631	1.624	1.617	1.609	1.604	1.600	1.595	1.589	1.584
	.90	2.538	2.502	2.464	2.425	2.404	2.383	2.361	2.339	2.316
	.95	3.347	3.284	3.218	3.150	3.115	3.079	3.043	3.005	2.967
	.975	4.295	4.200	4.101	3.999	3.947	3.894	3.840	3.784	3.728
	.99	5.814	5.667	5.515	5.359	5.279	5.198	5.116	5.032	4.946
	.999	11.54	11.19	10.84	10.48	10.30	10.11	9.919	9.727	9.532
9	.75	1.586	1.579	1.570	1.561	1.556	1.551	1.545	1.539	1.533
	.90	2.416	2.379	2.340	2.298	2.277	2.255	2.232	2.208	2.184
	.95	3.137	3.073	3.006	2.936	2.900	2.864	2.826	2.787	2.748
	.975	3.964	3.868	3.769	3.667	3.614	3.560	3.505	3.449	3.392
	.99	5.257	5.111	4.962	4.808	4.729	4.649	4.567	4.483	4.398
	.999	9.894	9.570	9.238	8.898	8.724	8.548	8.369	8.187	8.001
10	.75	1.551	1.543	1.534	1.523	1.518	1.512	1.506	1.499	1.492
	.90	2.323	2.284	2.244	2.201	2.178	2.155	2.132	2.107	2.082
	.95	2.978	2.913	2.845	2.774	2.737	2.700	2.661	2.621	2.580
	.975	3.717	3.621	3.522	3.419	3.365	3.311	3.255	3.198	3.140
	.99	4.849	4.706	4.558	4.405	4.327	4.247	4.165	4.082	3.996
	.999	8.754	8.445	8.129	7.804	7.638	7.469	7.297	7.122	6.944
11	.75	1.523	1.514	1.504	1.493	1.487	1.481	1.474	1.466	1.459
	.90	2.248	2.209	2.167	2.123	2.100	2.076	2.052	2.026	2.000
	.95	2.854	2.788	2.719	2.646	2.609	2.570	2.531	2.490	2.448
	.975	3.526	3.430	3.330	3.226	3.173	3.118	3.061	3.004	2.944
	.99	4.539	4.397	4.251	4.099	4.021	3.941	3.860	3.776	3.690
	.999	7.922	7.626	7.321	7.008	6.847	6.684	6.518	6.348	6.175
12	.75	1.500	1.490	1.480	1.468	1.461	1.454	1.447	1.439	1.431
	.90	2.188	2.147	2.105	2.060	2.036	2.011	1.986	1.960	1.932
	.95	2.753	2.687	2.617	2.544	2.505	2.466	2.426	2.384	2.341
	.975	3.374	3.277	3.177	3.073	3.019	2.963	2.906	2.848	2.787
	.99	4.296	4.155	4.010	3.858	3.780	3.701	3.619	3.535	3.449
	.999	7.292	7.005	6.709	6.405	6.249	6.090	5.928	5.762	5.593
13	.75	1.480	1.470	1.459	1.447	1.440	1.432	1.425	1.416	1.408
	.90	2.138	2.097	2.053	2.007	1.983	1.958	1.931	1.904	1.876
	.95	2.671	2.604	2.533	2.459	2.420	2.380	2.339	2.297	2.252
	.975	3.250	3.153	3.053	2.948	2.893	2.837	2.780	2.720	2.659
	.99	4.100	3.960	3.815	3.665	3.587	3.507	3.425	3.341	3.255
	.999	6.799	6.519	6.231	5.934	5.781	5.626	5.467	5.305	5.138
14	.75	1.463	1.453	1.441	1.428	1.421	1.414	1.405	1.397	1.387
	.90	2.095	2.054	2.010	1.962	1.938	1.912	1.885	1.857	1.828
	.95	2.602	2.534	2.463	2.388	2.349	2.308	2.266	2.223	2.178
	.975	3.147	3.050	2.949	2.844	2.789	2.732	2.674	2.614	2.552
	.99	3.939	3.800	3.656	3.505	3.427	3.348	3.266	3.181	3.094
	.999	6.404	6.130	5.848	5.557	5.407	5.254	5.098	4.938	4.773

TABLE A5
THE F DISTRIBUTION WITH k_1 AND k_2 DEGREES OF FREEDOM (CONTINUED)

k_2	p	k_1: 1	2	3	4	5	6	7	8	9
15	.75	1.432	1.523	1.520	1.507	1.494	1.482	1.472	1.463	1.456
	.90	3.073	2.695	2.490	2.361	2.273	2.208	2.158	2.119	2.086
	.95	4.543	3.682	3.287	3.056	2.901	2.790	2.707	2.641	2.588
	.975	6.200	4.765	4.153	3.804	3.576	3.415	3.293	3.199	3.123
	.99	8.683	6.359	5.417	4.893	4.556	4.318	4.142	4.004	3.895
	.999	16.59	11.34	9.335	8.253	7.567	7.092	6.741	6.471	6.256
16	.75	1.425	1.514	1.510	1.497	1.483	1.471	1.460	1.451	1.443
	.90	3.048	2.668	2.462	2.333	2.244	2.178	2.128	2.088	2.055
	.95	4.494	3.634	3.239	3.007	2.852	2.741	2.657	2.591	2.538
	.975	6.115	4.687	4.077	3.729	3.502	3.341	3.219	3.125	3.049
	.99	8.531	6.226	5.292	4.773	4.437	4.202	4.026	3.890	3.780
	.999	16.12	10.97	9.006	7.944	7.272	6.805	6.460	6.195	5.984
17	.75	1.419	1.506	1.502	1.487	1.473	1.460	1.450	1.441	1.433
	.90	3.026	2.645	2.437	2.308	2.218	2.152	2.102	2.061	2.028
	.95	4.451	3.592	3.197	2.965	2.810	2.699	2.614	2.548	2.494
	.975	6.042	4.619	4.011	3.665	3.438	3.277	3.156	3.061	2.985
	.99	8.400	6.112	5.185	4.669	4.336	4.102	3.927	3.791	3.682
	.999	15.72	10.66	8.727	7.683	7.022	6.562	6.223	5.962	5.754
18	.75	1.413	1.499	1.494	1.479	1.464	1.452	1.441	1.431	1.423
	.90	3.007	2.624	2.416	2.286	2.196	2.130	2.079	2.038	2.005
	.95	4.414	3.555	3.160	2.928	2.773	2.661	2.577	2.510	2.456
	.975	5.978	4.560	3.954	3.608	3.382	3.221	3.100	3.005	2.929
	.99	8.285	6.013	5.092	4.579	4.248	4.015	3.841	3.705	3.597
	.999	15.38	10.39	8.487	7.459	6.808	6.355	6.021	5.763	5.558
19	.75	1.408	1.493	1.487	1.472	1.457	1.444	1.432	1.423	1.414
	.90	2.990	2.606	2.397	2.266	2.176	2.109	2.058	2.017	1.984
	.95	4.381	3.522	3.127	2.895	2.740	2.628	2.544	2.477	2.423
	.975	5.922	4.508	3.903	3.559	3.333	3.172	3.051	2.956	2.880
	.99	8.185	5.926	5.010	4.500	4.171	3.939	3.765	3.631	3.523
	.999	15.08	10.16	8.280	7.265	6.622	6.175	5.845	5.590	5.388
20	.75	1.404	1.487	1.481	1.465	1.450	1.437	1.425	1.415	1.407
	.90	2.975	2.589	2.380	2.249	2.158	2.091	2.040	1.999	1.965
	.95	4.351	3.493	3.098	2.866	2.711	2.599	2.514	2.447	2.393
	.975	5.871	4.461	3.859	3.515	3.289	3.128	3.007	2.913	2.837
	.99	8.096	5.849	4.938	4.431	4.103	3.871	3.699	3.564	3.457
	.999	14.82	9.953	8.098	7.096	6.461	6.019	5.692	5.440	5.239
21	.75	1.400	1.482	1.475	1.459	1.444	1.430	1.419	1.409	1.400
	.90	2.961	2.575	2.365	2.233	2.142	2.075	2.023	1.982	1.948
	.95	4.325	3.467	3.072	2.840	2.685	2.573	2.488	2.420	2.366
	.975	5.827	4.420	3.819	3.475	3.250	3.090	2.969	2.874	2.798
	.99	8.017	5.780	4.874	4.369	4.042	3.812	3.640	3.506	3.398
	.999	14.59	9.772	7.938	6.947	6.318	5.881	5.557	5.308	5.109

TABLE A5
THE F DISTRIBUTION WITH k_1 AND k_2 DEGREES OF FREEDOM (CONTINUED)

k_2	p	k_1: 10	12	15	20	24	30	40	60	120
15	.75	1.449	1.438	1.426	1.413	1.405	1.397	1.389	1.380	1.370
	.90	2.059	2.017	1.972	1.924	1.899	1.873	1.845	1.817	1.787
	.95	2.544	2.475	2.403	2.328	2.288	2.247	2.204	2.160	2.114
	.975	3.060	2.963	2.862	2.756	2.701	2.644	2.585	2.524	2.461
	.99	3.805	3.666	3.522	3.372	3.294	3.214	3.132	3.047	2.959
	.999	6.081	5.812	5.535	5.248	5.101	4.950	4.796	4.638	4.475
16	.75	1.437	1.426	1.413	1.399	1.391	1.383	1.374	1.365	1.354
	.90	2.028	1.985	1.940	1.891	1.866	1.839	1.811	1.782	1.751
	.95	2.494	2.425	2.352	2.276	2.235	2.194	2.151	2.106	2.059
	.975	2.986	2.889	2.788	2.681	2.625	2.568	2.509	2.447	2.383
	.99	3.691	3.553	3.409	3.259	3.181	3.101	3.018	2.933	2.845
	.999	5.812	5.547	5.274	4.992	4.846	4.697	4.545	4.388	4.226
17	.75	1.426	1.414	1.401	1.387	1.379	1.370	1.361	1.351	1.341
	.90	2.001	1.958	1.912	1.862	1.836	1.809	1.781	1.751	1.719
	.95	2.450	2.381	2.308	2.230	2.190	2.148	2.104	2.058	2.011
	.975	2.922	2.825	2.723	2.616	2.560	2.502	2.442	2.380	2.315
	.99	3.593	3.455	3.312	3.162	3.084	3.003	2.920	2.835	2.746
	.999	5.584	5.324	5.054	4.775	4.631	4.484	4.332	4.177	4.016
18	.75	1.416	1.404	1.391	1.376	1.368	1.359	1.350	1.340	1.328
	.90	1.977	1.933	1.887	1.837	1.810	1.783	1.754	1.723	1.691
	.95	2.412	2.342	2.269	2.191	2.150	2.107	2.063	2.017	1.968
	.975	2.866	2.769	2.667	2.559	2.503	2.445	2.384	2.321	2.256
	.99	3.508	3.371	3.227	3.077	2.999	2.919	2.835	2.749	2.668
	.999	5.390	5.132	4.866	4.590	4.447	4.301	4.151	3.996	3.836
19	.75	1.407	1.395	1.382	1.367	1.358	1.349	1.339	1.329	1.317
	.90	1.956	1.912	1.865	1.814	1.787	1.759	1.730	1.699	1.666
	.95	2.738	2.308	2.234	2.155	2.114	2.071	2.026	1.980	1.930
	.975	2.817	2.720	2.617	2.509	2.452	2.394	2.333	2.270	2.203
	.99	3.434	3.257	3.153	3.003	2.925	2.844	2.761	2.674	2.584
	.999	5.222	4.967	4.704	4.430	4.288	4.143	3.994	3.840	3.680
20	.75	1.399	1.387	1.374	1.358	1.349	1.340	1.330	1.319	1.307
	.90	1.937	1.892	1.845	1.794	1.767	1.738	1.708	1.677	1.643
	.95	2.348	2.278	2.203	2.124	2.082	2.039	1.994	1.946	1.896
	.975	2.774	2.676	2.573	2.464	2.408	2.349	2.287	2.223	2.156
	.99	3.368	3.231	3.088	2.938	2.859	2.778	2.695	2.608	2.517
	.999	5.075	4.823	4.562	4.290	4.149	4.005	3.856	3.703	3.544
21	.75	1.392	1.380	1.366	1.350	1.341	1.332	1.322	1.311	1.298
	.90	1.920	1.875	1.827	1.776	1.748	1.719	1.689	1.657	1.623
	.95	2.321	2.250	2.176	2.096	2.054	2.010	1.965	1.916	1.866
	.975	2.735	2.637	2.534	2.425	2.368	2.308	2.246	2.182	2.144
	.99	3.310	3.173	3.030	2.880	2.801	2.720	2.636	2.548	2.457
	.999	4.946	4.696	4.437	4.167	4.027	3.884	3.736	3.583	3.424

TABLE A5
THE F DISTRIBUTION WITH k_1 AND k_2 DEGREES OF FREEDOM (CONTINUED)

k_2	p	k_1: 1	2	3	4	5	6	7	8	9
22	.75	1.396	1.477	1.470	1.454	1.438	1.424	1.413	1.402	1.394
	.90	2.949	2.561	2.351	2.219	2.128	2.060	2.008	1.967	1.933
	.95	4.301	3.443	3.049	2.817	2.661	2.549	2.464	2.397	2.342
	.975	5.786	4.383	3.783	3.440	3.215	3.055	2.934	2.839	2.763
	.99	7.945	5.719	4.817	4.313	3.988	3.758	3.587	3.453	3.346
	.999	14.38	9.612	7.796	6.814	6.191	5.758	5.438	5.190	4.993
23	.75	1.393	1.473	1.466	1.449	1.433	1.419	1.407	1.397	1.388
	.90	2.937	2.549	2.339	2.207	2.115	2.047	1.995	1.953	1.919
	.95	4.279	3.422	3.028	2.796	2.640	2.528	2.442	2.375	2.320
	.975	5.750	4.349	3.750	3.408	3.183	3.023	2.902	2.808	2.731
	.99	7.881	5.664	4.765	4.264	3.939	3.710	3.539	3.406	3.299
	.999	14.20	9.469	7.669	6.696	6.078	5.649	5.331	5.085	4.890
24	.75	1.390	1.470	1.462	1.445	1.428	1.414	1.402	1.392	1.383
	.90	2.927	2.538	2.327	2.195	2.103	2.035	1.983	1.941	1.906
	.95	4.260	3.403	3.009	2.776	2.621	2.508	2.423	2.355	2.300
	.975	5.717	4.319	3.721	3.379	3.155	2.995	2.874	2.779	2.703
	.99	7.823	5.614	4.718	4.218	3.895	3.667	3.496	3.363	3.256
	.999	14.03	9.339	7.554	6.589	5.977	5.550	5.235	4.991	4.797
25	.75	1.387	1.466	1.458	1.441	1.424	1.410	1.398	1.387	1.378
	.90	2.918	2.528	2.317	2.184	2.092	2.024	1.971	1.929	1.895
	.95	4.242	3.385	2.991	2.759	2.603	2.490	2.405	2.337	2.282
	.975	5.686	4.291	3.694	3.353	3.129	2.969	2.848	2.753	2.677
	.99	7.770	5.568	4.675	4.177	3.855	3.627	3.457	3.324	3.217
	.999	13.88	9.223	7.451	6.493	5.885	5.462	5.148	4.906	4.713
30	.75	1.376	1.452	1.443	1.424	1.407	1.392	1.380	1.369	1.359
	.90	2.881	2.489	2.276	2.142	2.049	1.980	1.927	1.884	1.849
	.95	4.171	3.316	2.922	2.690	2.534	2.421	2.334	2.266	2.211
	.975	5.568	4.182	3.589	3.250	3.026	2.867	2.746	2.651	2.575
	.99	7.562	5.390	4.510	4.018	3.699	3.473	3.304	3.173	3.067
	.999	13.29	8.773	7.054	6.125	5.534	5.122	4.817	4.581	4.393
40	.75	1.363	1.435	1.424	1.404	1.386	1.371	1.357	1.345	1.335
	.90	2.835	2.440	2.226	2.091	1.997	1.927	1.873	1.829	1.793
	.95	4.085	3.232	2.839	2.606	2.449	2.336	2.249	2.180	2.124
	.975	5.424	4.051	3.463	3.126	2.904	2.744	2.624	2.529	2.452
	.99	7.314	5.179	4.313	3.828	3.514	3.291	3.124	2.993	2.888
	.999	12.61	8.251	6.595	5.698	5.128	4.731	4.436	4.207	4.024
50	.75	1.355	1.425	1.413	1.393	1.374	1.358	1.344	1.332	1.321
	.90	2.809	2.412	2.197	2.061	1.966	1.895	1.840	1.796	1.760
	.95	4.034	3.183	2.790	2.557	2.400	2.286	2.199	2.130	2.073
	.975	5.340	3.975	3.390	3.054	2.833	2.674	2.553	2.458	2.381
	.99	7.171	5.057	4.199	3.720	3.408	3.186	3.020	2.890	2.785
	.999	12.22	7.956	6.336	5.459	4.901	4.512	4.222	3.998	3.818

TABLE A5
THE F DISTRIBUTION WITH k_1 AND k_2 DEGREES OF FREEDOM (CONTINUED)

k_2	p	k_1: 10	12	15	20	24	30	40	60	120
22	.75	1.386	1.374	1.359	1.343	1.334	1.324	1.314	1.303	1.290
	.90	1.904	1.859	1.811	1.759	1.731	1.702	1.671	1.639	1.604
	.95	2.297	2.226	2.151	2.071	2.028	1.984	1.938	1.889	1.838
	.975	2.700	2.602	2.498	2.389	2.331	2.272	2.210	2.145	2.076
	.99	3.258	3.121	2.978	2.827	2.749	2.667	2.583	2.495	2.403
	.999	4.832	4.583	4.326	4.058	3.919	3.776	3.629	3.476	3.317
23	.75	1.380	1.368	1.353	1.337	1.327	1.318	1.307	1.295	1.282
	.90	1.890	1.845	1.796	1.744	1.716	1.686	1.655	1.622	1.587
	.95	2.275	2.204	2.128	2.048	2.005	1.961	1.914	1.865	1.813
	.975	2.668	2.570	2.466	2.357	2.299	2.239	2.716	2.111	2.041
	.99	3.211	3.074	2.931	2.781	2.702	2.620	2.535	2.447	2.354
	.999	4.730	4.483	4.227	3.961	3.822	3.680	3.533	3.380	3.222
24	.75	1.375	1.362	1.347	1.331	1.321	1.311	1.300	1.289	1.275
	.90	1.877	1.832	1.783	1.730	1.702	1.672	1.641	1.607	1.571
	.95	2.255	2.183	2.108	2.027	1.984	1.939	1.892	1.842	1.790
	.975	2.640	2.541	2.437	2.327	2.269	2.209	2.146	2.080	2.010
	.99	3.168	3.032	2.889	2.738	2.659	2.577	2.492	2.403	2.310
	.999	4.638	4.393	4.139	3.873	3.735	3.593	3.447	3.295	3.136
25	.75	1.370	1.357	1.342	1.325	1.316	1.306	1.294	1.282	1.269
	.90	1.866	1.820	1.771	1.718	1.689	1.659	1.627	1.593	1.557
	.95	2.236	2.165	2.089	2.007	1.964	1.919	1.872	1.822	1.768
	.975	2.613	2.515	2.411	2.300	2.242	2.182	2.118	2.052	1.981
	.99	3.129	2.993	2.850	2.699	2.620	2.538	2.453	2.364	2.270
	.999	4.555	4.312	4.059	3.794	3.657	3.515	3.369	3.217	3.058
30	.75	1.351	1.337	1.321	1.303	1.293	1.282	1.270	1.257	1.242
	.90	1.819	1.773	1.722	1.667	1.638	1.606	1.573	1.538	1.499
	.95	2.165	2.092	2.015	1.932	1.887	1.841	1.792	1.740	1.683
	.975	2.511	2.412	2.307	2.195	2.136	2.074	2.009	1.940	1.866
	.99	2.979	2.843	2.700	2.549	2.469	2.386	2.299	2.208	2.111
	.999	4.239	4.001	3.753	3.493	3.357	3.217	3.072	2.920	2.760
40	.75	1.327	1.312	1.295	1.276	1.265	1.253	1.240	1.255	1.208
	.90	1.763	1.715	1.662	1.605	1.574	1.541	1.506	1.467	1.425
	.95	2.077	2.003	1.924	1.839	1.793	1.744	1.693	1.637	1.577
	.975	2.388	2.288	2.182	2.068	2.007	1.943	1.875	1.803	1.724
	.99	2.801	2.665	2.522	2.369	2.288	2.203	2.114	2.019	1.917
	.999	3.874	3.642	3.400	3.145	3.011	2.872	2.727	2.574	2.410
50	.75	1.312	1.297	1.280	1.259	1.248	1.235	1.221	1.205	1.186
	.90	1.729	1.680	1.627	1.568	1.536	1.502	1.465	1.424	1.379
	.95	2.026	1.952	1.871	1.784	1.737	1.687	1.634	1.576	1.511
	.975	2.317	2.216	2.109	1.993	1.931	1.866	1.796	1.721	1.639
	.99	2.698	2.562	2.419	2.265	2.183	2.098	2.007	1.909	1.803
	.999	3.671	3.443	3.204	2.951	2.817	2.679	2.533	2.378	2.211

TABLE A5
THE F DISTRIBUTION WITH k_1 AND k_2 DEGREES OF FREEDOM (CONTINUED)

k_2	p	k_1: 1	2	3	4	5	6	7	8	9
60	.75	1.349	1.419	1.405	1.385	1.366	1.349	1.335	1.323	1.312
	.90	2.791	2.393	2.177	2.041	1.946	1.875	1.819	1.775	1.738
	.95	4.001	3.150	2.758	2.525	2.368	2.254	2.167	2.097	2.040
	.975	5.286	3.925	3.343	3.008	2.786	2.627	2.507	2.412	2.334
	.99	7.077	4.977	4.126	3.649	3.339	3.119	2.953	2.823	2.718
	.999	11.97	7.768	6.171	5.307	4.757	4.372	4.086	3.865	3.687
70	.75	1.346	1.414	1.400	1.379	1.360	1.343	1.329	1.316	1.305
	.90	2.779	2.380	2.164	2.027	1.931	1.860	1.804	1.760	1.723
	.95	3.978	3.128	2.736	2.503	2.346	2.231	2.143	2.074	2.017
	.975	5.247	3.890	3.309	2.975	2.754	2.595	2.474	2.379	2.302
	.99	7.011	4.922	4.074	3.600	3.291	3.071	2.906	2.777	2.672
	.999	11.80	7.637	6.057	5.201	4.656	4.275	3.992	3.773	3.596
80	.75	1.343	1.411	1.396	1.375	1.355	1.338	1.324	1.311	1.300
	.90	2.769	2.370	2.154	2.016	1.921	1.849	1.793	1.748	1.711
	.95	3.960	3.111	2.719	2.486	2.329	2.214	2.126	2.056	1.999
	.975	5.218	3.864	3.284	2.950	2.730	2.571	2.450	2.355	2.277
	.99	6.963	4.881	4.036	3.563	3.255	3.036	2.871	2.742	2.637
	.999	11.67	7.540	5.972	5.123	4.582	4.204	3.923	3.705	3.530
90	.75	1.341	1.408	1.393	1.372	1.352	1.335	1.320	1.307	1.296
	.90	2.762	2.363	2.146	2.008	1.912	1.841	1.785	1.739	1.702
	.95	3.947	3.098	2.706	2.473	2.316	2.201	2.113	2.043	1.986
	.975	5.196	3.844	3.265	2.932	2.711	2.552	2.432	2.336	2.259
	.99	6.925	4.849	4.007	3.535	3.228	3.009	2.845	2.715	2.611
	.999	11.57	7.466	5.908	5.064	4.526	4.150	3.870	3.653	3.479
100	.75	1.339	1.406	1.391	1.369	1.349	1.332	1.317	1.304	1.293
	.90	2.756	2.356	2.139	2.002	1.906	1.834	1.778	1.732	1.695
	.95	3.936	3.087	2.696	2.463	2.305	2.191	2.103	2.032	1.975
	.975	5.179	3.828	3.250	2.917	2.696	2.537	2.417	2.321	2.244
	.99	6.895	4.824	3.984	3.513	3.206	2.988	2.823	2.694	2.590
	.999	11.50	7.408	5.857	5.017	4.482	4.107	3.829	3.612	3.439
120	.75	1.336	1.402	1.387	1.365	1.345	1.328	1.313	1.300	1.289
	.90	2.748	2.347	2.130	1.992	1.896	1.824	1.767	1.722	1.684
	.95	3.920	3.072	2.680	2.447	2.290	2.175	2.087	2.016	1.959
	.975	5.152	3.805	3.227	2.894	2.674	2.515	2.395	2.299	2.222
	.99	6.851	4.787	3.949	3.480	3.174	2.956	2.792	2.663	2.559
	.999	11.38	7.321	5.781	4.947	4.416	4.044	3.767	3.552	3.379
∞	.75	1.323	1.386	1.369	1.346	1.325	1.307	1.291	1.277	1.265
	.90	2.706	2.303	2.084	1.945	1.847	1.774	1.717	1.670	1.632
	.95	3.842	2.996	2.605	2.372	2.214	2.099	2.010	1.938	1.880
	.975	5.024	3.689	3.116	2.678	2.567	2.408	2.288	2.192	2.114
	.99	6.635	4.605	3.782	3.319	3.017	2.802	2.639	2.511	2.407
	.999	10.83	6.908	5.422	4.617	4.103	3.743	3.475	3.266	3.098

TABLE A5
THE F DISTRIBUTION WITH k_1 AND k_2 DEGREES OF FREEDOM (CONTINUED)

k_2	p	k_1: 10	12	15	20	24	30	40	60	120
60	.75	1.303	1.287	1.269	1.248	1.236	1.223	1.208	1.191	1.172
	.90	1.707	1.657	1.603	1.543	1.511	1.476	1.437	1.395	1.348
	.95	1.993	1.917	1.836	1.748	1.700	1.649	1.594	1.534	1.467
	.975	2.270	2.169	2.061	1.944	1.882	1.815	1.744	1.667	1.581
	.99	2.632	2.496	2.352	2.198	2.115	2.028	1.936	1.836	1.726
	.999	3.541	3.315	3.078	2.827	2.694	2.555	2.409	2.252	2.082
70	.75	1.296	1.280	1.262	1.240	1.228	1.214	1.199	1.181	1.161
	.90	1.691	1.641	1.587	1.526	1.493	1.457	1.418	1.374	1.325
	.95	1.969	1.893	1.812	1.722	1.674	1.622	1.566	1.505	1.435
	.975	2.237	2.136	2.028	1.910	1.847	1.779	1.707	1.628	1.539
	.99	2.585	2.450	2.306	2.150	2.067	1.980	1.886	1.785	1.672
	.999	3.452	3.227	2.991	2.741	2.608	2.469	2.322	2.164	1.991
80	.75	1.291	1.275	1.256	1.234	1.222	1.208	1.192	1.174	1.152
	.90	1.680	1.629	1.574	1.513	1.479	1.443	1.403	1.358	1.307
	.95	1.951	1.875	1.793	1.703	1.654	1.602	1.545	1.482	1.411
	.975	2.213	2.111	2.003	1.884	1.820	1.752	1.679	1.599	1.508
	.99	2.551	2.415	2.271	2.115	2.032	1.944	1.849	1.746	1.630
	.999	3.386	3.162	2.927	2.677	2.545	2.406	2.258	2.099	1.924
90	.75	1.287	1.270	1.252	1.229	1.217	1.202	1.186	1.168	1.145
	.90	1.670	1.620	1.564	1.503	1.468	1.432	1.391	1.346	1.293
	.95	1.938	1.861	1.779	1.688	1.639	1.586	1.528	1.465	1.391
	.975	2.194	2.092	1.983	1.864	1.800	1.731	1.657	1.576	1.483
	.99	2.524	2.389	2.244	2.088	2.004	1.916	1.820	1.716	1.598
	.999	3.336	3.113	2.879	2.629	2.497	2.357	2.209	2.049	1.871
100	.75	1.283	1.267	1.248	1.226	1.213	1.198	1.182	1.163	1.140
	.90	1.663	1.612	1.557	1.494	1.460	1.423	1.382	1.336	1.282
	.95	1.927	1.850	1.768	1.676	1.627	1.573	1.515	1.450	1.376
	.975	2.179	2.077	1.968	1.849	1.784	1.715	1.640	1.558	1.463
	.99	2.503	2.368	2.223	2.067	1.983	1.893	1.797	1.692	1.572
	.999	3.296	3.074	2.840	2.591	2.458	2.319	2.170	2.009	1.829
120	.75	1.279	1.262	1.243	1.220	1.207	1.192	1.175	1.156	1.131
	.90	1.652	1.601	1.545	1.482	1.447	1.409	1.368	1.320	1.265
	.95	1.910	1.834	1.750	1.659	1.608	1.554	1.495	1.429	1.352
	.975	2.157	2.055	1.945	1.825	1.760	1.690	1.614	1.530	1.433
	.99	2.472	2.336	2.192	2.035	1.950	1.860	1.763	1.656	1.533
	.999	3.237	3.016	2.783	2.534	2.402	2.262	2.113	1.950	1.767
∞	.75	1.255	1.237	1.216	1.191	1.177	1.160	1.140	1.116	1.084
	.90	1.599	1.546	1.487	1.421	1.383	1.342	1.295	1.240	1.169
	.95	1.831	1.752	1.666	1.571	1.517	1.459	1.394	1.318	1.221
	.975	2.048	1.945	1.833	1.709	1.640	1.566	1.484	1.388	1.268
	.99	2.321	2.185	2.039	1.878	1.791	1.696	1.592	1.473	1.325
	.999	2.959	2.743	2.513	2.266	2.132	1.990	1.835	1.660	1.447

Note: Table entries are the lower and upper limits for exact 90%, 95%, and 99% confidence intervals for the binomial parameter *p* when *x* successes are observed in *n* trials, *n* ≤ 30.

Source: Table generated by R. L. Iman.

TABLE A6
EXACT CONFIDENCE INTERVALS FOR THE BINOMIAL PARAMETER p

		90%		95%		99%	
n	x	Lower	Upper	Lower	Upper	Lower	Upper
1	0	.000	.950	.000	.975	.000	.995
	1	.050	1.000	.025	1.000	.005	1.000
2	0	.000	.776	.000	.842	.000	.929
	1	.025	.975	.013	.987	.003	.997
	2	.224	1.000	.158	1.000	.071	1.000
3	0	.000	.632	.000	.708	.000	.829
	1	.017	.865	.008	.906	.002	.959
	2	.135	.983	.094	.992	.041	.998
	3	.368	1.000	.292	1.000	.171	1.000
4	0	.000	.527	.000	.602	.000	.734
	1	.013	.751	.006	.806	.001	.889
	2	.098	.902	.068	.932	.029	.971
	3	.249	.987	.194	.994	.111	.999
	4	.473	1.000	.398	1.000	.266	1.000
5	0	.000	.451	.000	.522	.000	.653
	1	.010	.657	.005	.716	.001	.815
	2	.076	.811	.053	.853	.023	.917
	3	.189	.924	.147	.947	.083	.977
	4	.343	.990	.284	.995	.185	.999
	5	.549	1.000	.478	1.000	.347	1.000
6	0	.000	.393	.000	.459	.000	.586
	1	.009	.582	.004	.641	.001	.746
	2	.063	.729	.043	.777	.019	.856
	3	.153	.847	.118	.882	.066	.934
	4	.271	.937	.223	.957	.144	.981
	5	.418	.991	.359	.996	.254	.999
	6	.607	1.000	.541	1.000	.414	1.000
7	0	.000	.348	.000	.410	.000	.531
	1	.007	.521	.004	.579	.001	.685
	2	.053	.659	.037	.710	.016	.797
	3	.129	.775	.099	.816	.055	.882
	4	.225	.871	.184	.901	.118	.945
	5	.341	.947	.290	.963	.203	.984
	6	.479	.993	.421	.996	.315	.999
	7	.652	1.000	.590	1.000	.469	1.000

TABLE A6
EXACT CONFIDENCE INTERVALS FOR THE BINOMIAL
PARAMETER p (CONTINUED)

		90%		95%		99%	
n	x	Lower	Upper	Lower	Upper	Lower	Upper
8	**0**	.000	.312	.000	.369	.000	.484
	1	.006	.471	.003	.526	.001	.632
	2	.046	.600	.032	.651	.014	.742
	3	.111	.711	.085	.755	.047	.830
	4	.193	.807	.157	.843	.100	.900
	5	.289	.889	.245	.915	.170	.953
	6	.400	.954	.349	.968	.258	.986
	7	.529	.994	.474	.997	.368	.999
	8	.688	1.000	.631	1.000	.516	1.000
9	**0**	.000	.283	.000	.336	.000	.445
	1	.006	.429	.003	.482	.001	.585
	2	.041	.550	.028	.600	.012	.693
	3	.098	.655	.075	.701	.042	.781
	4	.169	.749	.137	.788	.087	.854
	5	.251	.831	.212	.863	.146	.913
	6	.345	.902	.299	.925	.219	.958
	7	.450	.959	.400	.972	.307	.988
	8	.571	.994	.518	.997	.415	.999
	9	.717	1.000	.664	1.000	.555	1.000
10	**0**	.000	.259	.000	.308	.000	.411
	1	.005	.394	.003	.445	.001	.544
	2	.037	.507	.025	.556	.011	.648
	3	.087	.607	.067	.652	.037	.735
	4	.150	.696	.122	.738	.077	.809
	5	.222	.778	.187	.813	.128	.872
	6	.304	.850	.262	.878	.191	.923
	7	.393	.913	.348	.933	.265	.963
	8	.493	.963	.444	.975	.352	.989
	9	.606	.995	.555	.997	.456	.999
	10	.741	1.000	.692	1.000	.589	1.000
11	**0**	.000	.238	.000	.285	.000	.382
	1	.005	.364	.002	.413	.000	.509
	2	.033	.470	.023	.518	.010	.608
	3	.079	.564	.060	.610	.033	.693
	4	.135	.650	.109	.692	.069	.767
	5	.200	.729	.167	.766	.115	.831
	6	.271	.800	.234	.833	.169	.885
	7	.350	.865	.308	.891	.233	.931
	8	.436	.921	.390	.940	.307	.967
	9	.530	.967	.482	.977	.392	.990
	10	.636	.995	.587	.998	.491	1.000
	11	.762	1.000	.715	1.000	.618	1.000

TABLE A6
EXACT CONFIDENCE INTERVALS FOR THE BINOMIAL
PARAMETER p (CONTINUED)

		90%		95%		99%	
n	x	Lower	Upper	Lower	Upper	Lower	Upper
12	**0**	.000	.221	.000	.265	.000	.357
	1	.004	.339	.002	.385	.000	.477
	2	.030	.438	.021	.484	.009	.573
	3	.072	.527	.055	.572	.030	.655
	4	.123	.609	.099	.651	.062	.728
	5	.181	.685	.152	.723	.103	.792
	6	.245	.755	.211	.789	.152	.848
	7	.315	.819	.277	.848	.208	.897
	8	.391	.877	.349	.901	.272	.938
	9	.473	.928	.428	.945	.345	.970
	10	.562	.970	.516	.979	.427	.991
	11	.661	.996	.615	.998	.523	1.000
	12	.779	1.000	.735	1.000	.643	1.000
13	**0**	.000	.206	.000	.247	.000	.335
	1	.004	.316	.002	.360	.000	.449
	2	.028	.410	.019	.454	.008	.541
	3	.066	.495	.050	.538	.028	.621
	4	.113	.573	.091	.614	.057	.691
	5	.166	.645	.139	.684	.094	.755
	6	.224	.713	.192	.749	.138	.811
	7	.287	.776	.251	.808	.189	.862
	8	.355	.834	.316	.861	.245	.906
	9	.427	.887	.386	.909	.309	.943
	10	.505	.934	.462	.950	.379	.972
	11	.590	.972	.546	.981	.459	.992
	12	.684	.996	.640	.998	.551	1.000
	13	.794	1.000	.753	1.000	.665	1.000
14	**0**	.000	.193	.000	.232	.000	.315
	1	.004	.297	.002	.339	.000	.424
	2	.026	.385	.018	.428	.008	.512
	3	.061	.466	.047	.508	.026	.589
	4	.104	.540	.084	.581	.053	.658
	5	.153	.610	.128	.649	.087	.720
	6	.206	.675	.177	.711	.127	.777
	7	.264	.736	.230	.770	.172	.828
	8	.325	.794	.289	.823	.223	.873
	9	.390	.847	.351	.872	.280	.913
	10	.460	.896	.419	.916	.342	.947
	11	.534	.939	.492	.953	.411	.974
	12	.615	.974	.572	.982	.488	.992
	13	.703	.996	.661	.998	.576	1.000
	14	.807	1.000	.768	1.000	.685	1.000

TABLE A6
EXACT CONFIDENCE INTERVALS FOR THE BINOMIAL
PARAMETER p (CONTINUED)

		90%		95%		99%	
n	x	Lower	Upper	Lower	Upper	Lower	Upper
15	**0**	.000	.181	.000	.218	.000	.298
	1	.003	.279	.002	.319	.000	.402
	2	.024	.363	.017	.405	.007	.486
	3	.057	.440	.043	.481	.024	.561
	4	.097	.511	.078	.551	.049	.627
	5	.142	.577	.118	.616	.080	.688
	6	.191	.640	.163	.677	.117	.744
	7	.244	.700	.213	.734	.159	.795
	8	.300	.756	.266	.787	.205	.841
	9	.360	.809	.323	.837	.256	.883
	10	.423	.858	.384	.882	.312	.920
	11	.489	.903	.449	.922	.373	.951
	12	.560	.943	.519	.957	.439	.976
	13	.637	.976	.595	.983	.514	.993
	14	.721	.997	.681	.998	.598	1.000
	15	.819	1.000	.782	1.000	.702	1.000
16	**0**	.000	.171	.000	.206	.000	.282
	1	.003	.264	.002	.302	.000	.381
	2	.023	.344	.016	.383	.007	.463
	3	.053	.417	.040	.456	.022	.534
	4	.090	.484	.073	.524	.045	.599
	5	.132	.548	.110	.587	.075	.658
	6	.178	.609	.152	.646	.109	.713
	7	.227	.667	.198	.701	.147	.764
	8	.279	.721	.247	.753	.190	.810
	9	.333	.773	.299	.802	.236	.853
	10	.391	.822	.354	.848	.287	.891
	11	.452	.868	.413	.890	.342	.925
	12	.516	.910	.476	.927	.401	.955
	13	.583	.947	.544	.960	.466	.978
	14	.656	.977	.617	.984	.537	.993
	15	.736	.997	.698	.998	.619	1.000
	16	.829	1.000	.794	1.000	.718	1.000

TABLE A6
EXACT CONFIDENCE INTERVALS FOR THE BINOMIAL
PARAMETER p (CONTINUED)

		90%		95%		99%	
n	x	Lower	Upper	Lower	Upper	Lower	Upper
17	0	.000	.162	.000	.195	.000	.268
	1	.003	.250	.001	.287	.000	.363
	2	.021	.326	.015	.364	.006	.441
	3	.050	.396	.038	.434	.021	.510
	4	.085	.461	.068	.499	.043	.573
	5	.124	.522	.103	.560	.070	.631
	6	.166	.580	.142	.617	.101	.685
	7	.212	.636	.184	.671	.137	.734
	8	.260	.689	.230	.722	.176	.781
	9	.311	.740	.278	.770	.219	.824
	10	.364	.788	.329	.816	.266	.863
	11	.420	.834	.383	.858	.315	.899
	12	.478	.876	.440	.897	.369	.930
	13	.539	.915	.501	.932	.427	.957
	14	.604	.950	.566	.962	.490	.979
	15	.674	.979	.636	.985	.559	.994
	16	.750	.997	.713	.999	.637	1.000
	17	.838	1.000	.805	1.000	.732	1.000
18	0	.000	.153	.000	.185	.000	.255
	1	.003	.238	.001	.273	.000	.346
	2	.020	.310	.014	.347	.006	.422
	3	.047	.377	.036	.414	.020	.488
	4	.080	.439	.064	.476	.040	.549
	5	.116	.498	.097	.535	.065	.605
	6	.156	.554	.133	.590	.095	.658
	7	.199	.608	.173	.643	.128	.707
	8	.244	.659	.215	.692	.165	.753
	9	.291	.709	.260	.740	.205	.795
	10	.341	.756	.308	.785	.247	.835
	11	.392	.801	.357	.827	.293	.872
	12	.446	.844	.410	.867	.342	.905
	13	.502	.884	.465	.903	.395	.935
	14	.561	.920	.524	.936	.451	.960
	15	.623	.953	.586	.964	.512	.980
	16	.690	.980	.653	.986	.578	.994
	17	.762	.997	.727	.999	.654	1.000
	18	.847	1.000	.815	1.000	.745	1.000

TABLE A6
EXACT CONFIDENCE INTERVALS FOR THE BINOMIAL
PARAMETER p (CONTINUED)

		90%		95%		99%	
n	x	Lower	Upper	Lower	Upper	Lower	Upper
19	**0**	.000	.146	.000	.176	.000	.243
	1	.003	.226	.001	.260	.000	.331
	2	.019	.296	.013	.331	.006	.404
	3	.044	.359	.034	.396	.019	.468
	4	.075	.419	.061	.456	.038	.527
	5	.110	.476	.091	.512	.062	.582
	6	.147	.530	.126	.565	.089	.633
	7	.188	.582	.163	.616	.121	.681
	8	.230	.632	.203	.665	.155	.726
	9	.274	.680	.244	.711	.192	.768
	10	.320	.726	.289	.756	.232	.808
	11	.368	.770	.335	.797	.274	.845
	12	.418	.813	.384	.837	.319	.879
	13	.470	.853	.435	.874	.367	.911
	14	.524	.890	.488	.909	.418	.938
	15	.581	.925	.544	.939	.473	.962
	16	.641	.956	.604	.966	.532	.981
	17	.704	.981	.669	.987	.596	.994
	18	.774	.997	.740	.999	.669	1.000
	19	.854	1.000	.824	1.000	.757	1.000
20	**0**	.000	.139	.000	.168	.000	.233
	1	.003	.216	.001	.249	.000	.317
	2	.018	.283	.012	.317	.005	.387
	3	.042	.344	.032	.379	.018	.449
	4	.071	.401	.057	.437	.036	.507
	5	.104	.456	.087	.491	.058	.560
	6	.140	.508	.119	.543	.085	.610
	7	.177	.558	.154	.592	.114	.657
	8	.217	.606	.191	.639	.146	.701
	9	.259	.653	.231	.685	.181	.743
	10	.302	.698	.272	.728	.218	.782
	11	.347	.741	.315	.769	.257	.819
	12	.394	.783	.361	.809	.299	.854
	13	.442	.823	.408	.846	.343	.886
	14	.492	.860	.457	.881	.390	.915
	15	.544	.896	.509	.913	.440	.942
	16	.599	.929	.563	.943	.493	.964
	17	.656	.958	.621	.968	.551	.982
	18	.717	.982	.683	.988	.613	.995
	19	.784	.997	.751	.999	.683	1.000
	20	.861	1.000	.832	1.000	.767	1.000

		90%		95%		99%	
n	x	Lower	Upper	Lower	Upper	Lower	Upper
21	0	.000	.133	.000	.161	.000	.223
	1	.002	.207	.001	.238	.000	.304
	2	.017	.271	.012	.304	.005	.372
	3	.040	.329	.030	.363	.017	.432
	4	.068	.384	.054	.419	.034	.488
	5	.099	.437	.082	.472	.055	.539
	6	.132	.487	.113	.522	.080	.588
	7	.168	.536	.146	.570	.108	.634
	8	.206	.583	.181	.616	.138	.677
	9	.245	.628	.218	.660	.171	.719
	10	.286	.672	.257	.702	.205	.758
	11	.328	.714	.298	.743	.242	.795
	12	.372	.755	.340	.782	.281	.829
	13	.417	.794	.384	.819	.323	.862
	14	.464	.832	.430	.854	.366	.892
	15	.513	.868	.478	.887	.412	.920
	16	.563	.901	.528	.918	.461	.945
	17	.616	.932	.581	.946	.512	.966
	18	.671	.960	.637	.970	.568	.983
	19	.729	.983	.696	.988	.628	.995
	20	.793	.998	.762	.999	.696	1.000
	21	.867	1.000	.839	1.000	.777	1.000
22	0	.000	.127	.000	.154	.000	.214
	1	.002	.198	.001	.228	.000	.292
	2	.016	.259	.011	.292	.005	.358
	3	.038	.316	.029	.349	.016	.416
	4	.065	.369	.052	.403	.032	.470
	5	.094	.420	.078	.454	.053	.520
	6	.126	.468	.107	.502	.076	.567
	7	.160	.515	.139	.549	.102	.612
	8	.196	.561	.172	.593	.131	.655
	9	.233	.605	.207	.636	.162	.695
	10	.271	.647	.244	.678	.195	.734
	11	.311	.689	.282	.718	.229	.771
	12	.353	.729	.322	.756	.266	.805
	13	.395	.767	.364	.793	.305	.838
	14	.439	.804	.407	.828	.345	.869
	15	.485	.840	.451	.861	.388	.898
	16	.532	.874	.498	.893	.433	.924
	17	.580	.906	.546	.922	.480	.947
	18	.631	.935	.597	.948	.530	.968
	19	.684	.962	.651	.971	.584	.984
	20	.741	.984	.708	.989	.642	.995
	21	.802	.998	.772	.999	.708	1.000
	22	.873	1.000	.846	1.000	.786	1.000

		90%		95%		99%	
n	x	Lower	Upper	Lower	Upper	Lower	Upper
23	**0**	.000	.122	.000	.148	.000	.206
	1	.002	.190	.001	.219	.000	.281
	2	.016	.249	.011	.280	.005	.345
	3	.037	.304	.028	.336	.015	.401
	4	.062	.355	.050	.388	.031	.453
	5	.090	.404	.075	.437	.050	.502
	6	.120	.451	.102	.484	.073	.548
	7	.152	.496	.132	.529	.097	.592
	8	.186	.540	.164	.573	.125	.634
	9	.222	.583	.197	.615	.154	.674
	10	.258	.625	.232	.655	.185	.712
	11	.296	.665	.268	.694	.218	.748
	12	.335	.704	.306	.732	.252	.782
	13	.375	.742	.345	.768	.288	.815
	14	.417	.778	.385	.803	.326	.846
	15	.460	.814	.427	.836	.366	.875
	16	.504	.848	.471	.868	.408	.903
	17	.549	.880	.516	.898	.452	.927
	18	.596	.910	.563	.925	.498	.950
	19	.645	.938	.612	.950	.547	.969
	20	.696	.963	.664	.972	.599	.985
	21	.751	.984	.720	.989	.655	.995
	22	.810	.998	.781	.999	.719	1.000
	23	.878	1.000	.852	1.000	.794	1.000
24	**0**	.000	.117	.000	.142	.000	.198
	1	.002	.183	.001	.211	.000	.271
	2	.015	.240	.010	.270	.004	.332
	3	.035	.292	.027	.324	.015	.387
	4	.059	.342	.047	.374	.029	.438
	5	.086	.389	.071	.422	.048	.485
	6	.115	.435	.098	.467	.069	.530
	7	.146	.479	.126	.511	.093	.573
	8	.178	.521	.156	.553	.119	.614
	9	.212	.563	.188	.594	.146	.653
	10	.246	.603	.221	.634	.176	.690
	11	.282	.642	.256	.672	.207	.726
	12	.319	.681	.291	.709	.240	.760
	13	.358	.718	.328	.744	.274	.793
	14	.397	.754	.366	.779	.310	.824
	15	.437	.788	.406	.812	.347	.854
	16	.479	.822	.447	.844	.386	.881
	17	.521	.854	.489	.874	.427	.907
	18	.565	.885	.533	.902	.470	.931
	19	.611	.914	.578	.929	.515	.952
	20	.658	.941	.626	.953	.562	.971
	21	.708	.965	.676	.973	.613	.985
	22	.760	.985	.730	.990	.668	.996
	23	.817	.998	.789	.999	.729	1.000
	24	.883	1.000	.858	1.000	.802	1.000

TABLE A6
EXACT CONFIDENCE INTERVALS FOR THE BINOMIAL
PARAMETER p (CONTINUED)

		90%		95%		99%	
n	x	Lower	Upper	Lower	Upper	Lower	Upper
25	0	.000	.113	.000	.137	.000	.191
	1	.002	.176	.001	.204	.000	.262
	2	.014	.231	.010	.260	.004	.321
	3	.034	.282	.025	.312	.014	.374
	4	.057	.330	.045	.361	.028	.424
	5	.082	.375	.068	.407	.046	.470
	6	.110	.420	.094	.451	.066	.514
	7	.139	.462	.121	.494	.089	.555
	8	.170	.504	.150	.535	.114	.595
	9	.202	.544	.180	.575	.140	.634
	10	.236	.583	.211	.613	.168	.670
	11	.270	.621	.244	.651	.197	.705
	12	.305	.659	.278	.687	.228	.739
	13	.341	.695	.313	.722	.261	.772
	14	.379	.730	.349	.756	.295	.803
	15	.417	.764	.387	.789	.330	.832
	16	.456	.798	.425	.820	.366	.860
	17	.496	.830	.465	.850	.405	.886
	18	.538	.861	.506	.879	.445	.911
	19	.580	.890	.549	.906	.486	.934
	20	.625	.918	.593	.932	.530	.954
	21	.670	.943	.639	.955	.576	.972
	22	.718	.966	.688	.975	.626	.986
	23	.769	.986	.740	.990	.679	.996
	24	.824	.998	.796	.999	.738	1.000
	25	.887	1.000	.863	1.000	.809	1.000

TABLE A6
EXACT CONFIDENCE INTERVALS FOR THE BINOMIAL
PARAMETER p (CONTINUED)

		90%		95%		99%	
n	x	Lower	Upper	Lower	Upper	Lower	Upper
26	0	.000	.109	.000	.132	.000	.184
	1	.002	.170	.001	.196	.000	.253
	2	.014	.223	.009	.251	.004	.310
	3	.032	.272	.024	.302	.013	.362
	4	.054	.318	.044	.349	.027	.410
	5	.079	.363	.066	.393	.044	.455
	6	.106	.405	.090	.436	.064	.498
	7	.134	.447	.116	.478	.085	.538
	8	.163	.487	.143	.518	.109	.578
	9	.194	.526	.172	.557	.134	.615
	10	.226	.564	.202	.594	.161	.651
	11	.258	.602	.234	.631	.189	.686
	12	.292	.638	.266	.666	.218	.719
	13	.327	.673	.299	.701	.249	.751
	14	.362	.708	.334	.734	.281	.782
	15	.398	.742	.369	.766	.314	.811
	16	.436	.774	.406	.798	.349	.839
	17	.474	.806	.443	.828	.385	.866
	18	.513	.837	.482	.857	.422	.891
	19	.553	.866	.522	.884	.462	.915
	20	.595	.894	.564	.910	.502	.936
	21	.637	.921	.607	.934	.545	.956
	22	.682	.946	.651	.956	.590	.973
	23	.728	.968	.698	.976	.638	.987
	24	.777	.986	.749	.991	.690	.996
	25	.830	.998	.804	.999	.747	1.000
	26	.891	1.000	.868	1.000	.816	1.000

TABLE A6
EXACT CONFIDENCE INTERVALS FOR THE BINOMIAL
PARAMETER p (CONTINUED)

		90%		95%		99%	
n	x	Lower	Upper	Lower	Upper	Lower	Upper
27	0	.000	.105	.000	.128	.000	.178
	1	.002	.164	.001	.190	.000	.245
	2	.013	.215	.009	.243	.004	.300
	3	.031	.263	.024	.292	.013	.351
	4	.052	.308	.042	.337	.026	.397
	5	.076	.351	.063	.381	.042	.441
	6	.101	.392	.086	.423	.061	.483
	7	.129	.432	.111	.463	.082	.523
	8	.157	.471	.138	.502	.104	.561
	9	.186	.509	.165	.540	.128	.597
	10	.217	.547	.194	.576	.154	.633
	11	.248	.583	.224	.612	.181	.667
	12	.280	.618	.255	.647	.209	.700
	13	.313	.653	.287	.681	.238	.731
	14	.347	.687	.319	.713	.269	.762
	15	.382	.720	.353	.745	.300	.791
	16	.417	.752	.388	.776	.333	.819
	17	.453	.783	.424	.806	.367	.846
	18	.491	.814	.460	.835	.403	.872
	19	.529	.843	.498	.862	.439	.896
	20	.568	.871	.537	.889	.477	.918
	21	.608	.899	.577	.914	.517	.939
	22	.649	.924	.619	.937	.559	.958
	23	.692	.948	.663	.958	.603	.974
	24	.737	.969	.708	.976	.649	.987
	25	.785	.987	.757	.991	.700	.996
	26	.836	.998	.810	.999	.755	1.000
	27	.895	1.000	.872	1.000	.822	1.000

TABLE A6
EXACT CONFIDENCE INTERVALS FOR THE BINOMIAL
PARAMETER p (CONTINUED)

		90%		95%		99%	
n	x	Lower	Upper	Lower	Upper	Lower	Upper
28	**0**	.000	.101	.000	.123	.000	.172
	1	.002	.159	.001	.183	.000	.237
	2	.013	.208	.009	.235	.004	.291
	3	.030	.254	.023	.282	.012	.340
	4	.050	.298	.040	.327	.025	.385
	5	.073	.339	.061	.369	.041	.428
	6	.098	.380	.083	.410	.059	.469
	7	.124	.419	.107	.449	.079	.508
	8	.151	.457	.132	.487	.100	.545
	9	.179	.494	.159	.524	.123	.581
	10	.208	.530	.186	.559	.148	.615
	11	.238	.565	.215	.594	.173	.649
	12	.269	.600	.245	.628	.200	.681
	13	.301	.634	.275	.661	.228	.713
	14	.333	.667	.306	.694	.257	.743
	15	.366	.699	.339	.725	.287	.772
	16	.400	.731	.372	.755	.319	.800
	17	.435	.762	.406	.785	.351	.827
	18	.470	.792	.441	.814	.385	.852
	19	.506	.821	.476	.841	.419	.877
	20	.543	.849	.513	.868	.455	.900
	21	.581	.876	.551	.893	.492	.921
	22	.620	.902	.590	.917	.531	.941
	23	.661	.927	.631	.939	.572	.959
	24	.702	.950	.673	.960	.615	.975
	25	.746	.970	.718	.977	.660	.988
	26	.792	.987	.765	.991	.709	.996
	27	.841	.998	.817	.999	.763	1.000
	28	.899	1.000	.877	1.000	.828	1.000

TABLE A6
EXACT CONFIDENCE INTERVALS FOR THE BINOMIAL
PARAMETER p (CONTINUED)

		90%		95%		99%	
n	x	Lower	Upper	Lower	Upper	Lower	Upper
29	0	.000	.098	.000	.119	.000	.167
	1	.002	.153	.001	.178	.000	.230
	2	.012	.202	.008	.228	.004	.282
	3	.029	.246	.022	.274	.012	.330
	4	.049	.288	.039	.317	.024	.374
	5	.070	.329	.058	.358	.039	.416
	6	.094	.368	.080	.397	.056	.455
	7	.119	.406	.103	.435	.076	.493
	8	.145	.443	.127	.472	.096	.530
	9	.172	.479	.153	.508	.119	.565
	10	.201	.514	.179	.543	.142	.599
	11	.229	.549	.207	.577	.167	.632
	12	.259	.583	.235	.611	.192	.664
	13	.289	.616	.264	.643	.219	.695
	14	.320	.648	.294	.675	.247	.724
	15	.352	.680	.325	.706	.276	.753
	16	.384	.711	.357	.736	.305	.781
	17	.417	.741	.389	.765	.336	.808
	18	.451	.771	.423	.793	.368	.833
	19	.486	.799	.457	.821	.401	.858
	20	.521	.828	.492	.847	.435	.881
	21	.557	.855	.528	.873	.470	.904
	22	.594	.881	.565	.897	.507	.924
	23	.632	.906	.603	.920	.545	.944
	24	.671	.930	.642	.942	.584	.961
	25	.712	.951	.683	.961	.626	.976
	26	.754	.971	.726	.978	.670	.988
	27	.798	.988	.772	.992	.718	.996
	28	.847	.998	.822	.999	.770	1.000
	29	.902	1.000	.881	1.000	.833	1.000

TABLE A6
EXACT CONFIDENCE INTERVALS FOR THE BINOMIAL
PARAMETER p (CONTINUED)

		90%		95%		99%	
n	x	Lower	Upper	Lower	Upper	Lower	Upper
30	**0**	.000	.095	.000	.116	.000	.162
	1	.002	.149	.001	.172	.000	.223
	2	.012	.195	.008	.221	.004	.274
	3	.028	.239	.021	.265	.012	.320
	4	.047	.280	.038	.307	.023	.363
	5	.068	.319	.056	.347	.038	.404
	6	.091	.357	.077	.386	.054	.443
	7	.115	.394	.099	.423	.073	.480
	8	.140	.430	.123	.459	.093	.516
	9	.166	.465	.147	.494	.114	.550
	10	.193	.499	.173	.528	.137	.583
	11	.221	.533	.199	.561	.160	.616
	12	.250	.566	.227	.594	.185	.647
	13	.279	.598	.255	.626	.211	.677
	14	.308	.630	.283	.657	.237	.707
	15	.339	.661	.313	.687	.265	.735
	16	.370	.692	.343	.717	.293	.763
	17	.402	.721	.374	.745	.323	.789
	18	.434	.750	.406	.773	.353	.815
	19	.467	.779	.439	.801	.384	.840
	20	.501	.807	.472	.827	.417	.863
	21	.535	.834	.506	.853	.450	.886
	22	.570	.860	.541	.877	.484	.907
	23	.606	.885	.577	.901	.520	.927
	24	.643	.909	.614	.923	.557	.946
	25	.681	.932	.653	.944	.596	.962
	26	.720	.953	.693	.962	.637	.977
	27	.761	.972	.735	.979	.680	.988
	28	.805	.988	.779	.992	.726	.996
	29	.851	.998	.828	.999	.777	1.000
	30	.905	1.000	.884	1.000	.838	1.000

TABLE A7
EXACT CONFIDENCE INTERVALS FOR THE POPULATION MEDIAN

For an ordered sample of size n, the table entries, S_1 and S_2, represent the location within the ordered sample of the values that form the lower and upper bounds, respectively, of 90%, 95%, and 99% target confidence intervals for the population median.

Source: Table generated by R. L. Iman.

TABLE A7
EXACT CONFIDENCE INTERVALS FOR THE POPULATION MEDIAN

				Target Confidence Level					
	90%			95%			99%		
n	S_1	S_2	Actual Percentage	S_1	S_2	Actual Percentage	S_1	S_2	Actual Percentage
4	1	4	87.50						
5	1	5	93.75						
6	2	5	78.13	1	6	96.87			
7	2	6	87.50	1	7	98.44			
8	3	6	71.09	2	7	92.97	1	8	99.22
9	3	7	82.03	2	8	96.09	1	9	99.61
10	3	8	89.06	2	9	97.85	1	10	99.80
11	3	9	93.46	2	10	98.83	1	11	99.90
12	4	9	85.40	3	10	96.14	2	11	99.37
13	4	10	90.77	3	11	97.75	2	12	99.66
14	5	10	82.04	4	11	94.26	3	12	98.71
15	5	11	88.15	4	12	96.48	3	13	99.26
16	5	12	92.32	4	13	97.87	3	14	99.58
17	6	12	85.65	5	13	95.10	4	14	98.73
18	6	13	90.37	5	14	96.91	4	15	99.25
19	6	14	93.64	5	15	98.08	4	16	99.56
20	7	14	88.47	6	15	95.86	5	16	98.82
21	7	15	92.16	6	16	97.34	5	17	99.28
22	8	15	86.62	7	16	94.75	5	18	99.57
23	8	16	90.69	7	17	96.53	6	18	98.94
24	8	17	93.61	7	18	97.73	6	19	99.34
25	9	17	89.22	8	18	95.67	7	19	98.54
26	9	18	92.45	8	19	97.10	7	20	99.06
27	10	18	87.79	9	19	94.78	7	21	99.41
28	10	19	91.28	9	20	96.43	8	21	98.75
29	11	19	86.40	10	20	93.86	8	22	99.19
30	11	20	90.13	10	21	95.72	8	23	99.48
31	11	21	92.92	10	22	97.06	9	23	98.93
32	12	21	88.98	11	22	94.99	9	24	99.30
33	12	22	91.99	11	23	96.49	10	24	98.65
34	13	22	87.86	12	23	94.24	10	25	99.10
35	13	23	91.05	12	24	95.90	10	26	99.40
36	14	23	86.75	13	24	93.48	11	26	98.87
37	14	24	90.11	13	25	95.30	11	27	99.24
38	14	25	92.70	13	26	96.64	12	27	98.61
39	15	25	89.19	14	26	94.67	12	28	99.05
40	15	26	91.93	14	27	96.15	12	29	99.36
41	16	26	88.27	15	27	94.04	13	29	98.85
42	16	27	91.16	15	28	95.64	13	30	99.21
43	17	27	87.37	16	28	93.40	14	30	98.63
44	17	28	90.39	16	29	95.12	14	31	99.04
45	17	29	92.75	16	30	96.43	14	32	99.34
46	18	29	89.62	17	30	94.59	15	32	98.86
47	18	30	92.11	17	31	96.00	15	33	99.21
48	19	30	88.86	18	31	94.05	16	33	98.67
49	19	31	91.46	18	32	95.56	16	34	99.06
50	20	31	88.11	19	32	93.51	16	35	99.34

TABLE A8
CRITICAL VALUES FOR THE MEDIAN TEST

The table entries, T_L and T_U, are the critical values used with the procedure outlined in Section 7.5 for testing hypotheses about the population median on the basis of a sample of size n.

Source: Table generated by R. L. Iman.

TABLE A8
CRITICAL VALUES FOR THE MEDIAN TEST

n	T_L	T_U	α-Level One-Tailed Test	α-Level Two-Tailed Test
4	1	3	.0625	.1250
5	1	4	.0312	.0625
6	1	5	.0156	.0313
	2	4	.1094	.2187
7	1	6	.0078	.0156
	2	5	.0625	.1250
8	1	7	.0039	.0078
	2	6	.0352	.0703
9	1	8	.0020	.0039
	2	7	.0195	.0391
	3	6	.0898	.1797
10	1	9	.0010	.0020
	2	8	.0107	.0215
	3	7	.0547	.1094
11	1	10	.0005	.0010
	2	9	.0059	.0117
	3	8	.0327	.0654
12	1	11	.0002	.0005
	2	10	.0032	.0063
	3	9	.0193	.0386
	4	8	.0730	.1460
13	1	12	.0001	.0002
	2	11	.0017	.0034
	3	10	.0112	.0225
	4	9	.0461	.0923
14	2	12	.0009	.0018
	3	11	.0065	.0129
	4	10	.0287	.0574
	5	9	.0898	.1796
15	2	13	.0005	.0010
	3	12	.0037	.0074
	4	11	.0176	.0352
	5	10	.0592	.1185
16	3	13	.0021	.0042
	4	12	.0106	.0213
	5	11	.0384	.0768
	6	10	.1051	.2101
17	3	14	.0012	.0023
	4	13	.0064	.0127
	5	12	.0245	.0490
	6	11	.0717	.1435

TABLE A8
CRITICAL VALUES FOR THE MEDIAN TEST (CONTINUED)

n	T_L	T_U	α-Level One-Tailed Test	α-Level Two-Tailed Test
18	4	14	.0038	.0075
	5	13	.0154	.0309
	6	12	.0481	.0963
	7	11	.1189	.2379
19	4	15	.0022	.0044
	5	14	.0096	.0192
	6	13	.0318	.0636
	7	12	.0835	.1671
20	4	16	.0013	.0026
	5	15	.0059	.0118
	6	14	.0207	.0414
	7	13	.0577	.1153
21	5	16	.0036	.0072
	6	15	.0133	.0266
	7	14	.0392	.0784
	8	13	.0946	.1892
22	5	17	.0022	.0043
	6	16	.0085	.0169
	7	15	.0262	.0525
	8	14	.0669	.1338
23	5	18	.0013	.0026
	6	17	.0053	.0106
	7	16	.0173	.0347
	8	15	.0466	.0931
	9	14	.1050	.2100
24	6	18	.0033	.0066
	7	17	.0113	.0227
	8	16	.0320	.0639
	9	15	.0758	.1516
25	6	19	.0020	.0041
	7	18	.0073	.0146
	8	17	.0216	.0433
	9	16	.0539	.1078
	10	15	.1148	.2295
26	7	19	.0047	.0094
	8	18	.0145	.0290
	9	17	.0378	.0755
	10	16	.0843	.1686
27	7	20	.0030	.0059
	8	19	.0096	.0192
	9	18	.0261	.0522
	10	17	.0610	.1221

TABLE A8
CRITICAL VALUES FOR THE MEDIAN TEST (CONTINUED)

n	T_L	T_U	α-Level One-Tailed Test	α-Level Two-Tailed Test
28	7	21	.0019	.0037
	8	20	.0063	.0125
	9	19	.0178	.0357
	10	18	.0436	.0872
	11	17	.0925	.1849
29	8	21	.0041	.0081
	9	20	.0121	.0241
	10	19	.0307	.0614
	11	18	.0680	.1360
30	8	22	.0026	.0052
	9	21	.0081	.0161
	10	20	.0214	.0428
	11	19	.0494	.0987
	12	18	.1002	.2005
31	8	23	.0017	.0033
	9	22	.0053	.0107
	10	21	.0147	.0294
	11	20	.0354	.0708
	12	19	.0748	.1496
32	9	23	.0035	.0070
	10	22	.0100	.0201
	11	21	.0251	.0501
	12	20	.0551	.1102
	13	19	.1077	.2153
33	9	24	.0023	.0046
	10	23	.0068	.0135
	11	22	.0175	.0351
	12	21	.0401	.0801
	13	20	.0814	.1628
34	10	24	.0045	.0090
	11	23	.0122	.0243
	12	22	.0288	.0576
	13	21	.0607	.1214
	14	20	.1147	.2295
35	10	25	.0030	.0060
	11	24	.0083	.0167
	12	23	.0205	.0410
	13	22	.0448	.0895
	14	21	.0877	.1755
36	10	26	.0020	.0039
	11	25	.0057	.0113
	12	24	.0144	.0288
	13	23	.0326	.0652
	14	22	.0662	.1325

TABLE A8
CRITICAL VALUES FOR THE MEDIAN TEST (CONTINUED)

n	T_L	T_U	α-Level One-Tailed Test	α-Level Two-Tailed Test
37	11	26	.0038	.0076
	12	25	.0100	.0201
	13	24	.0235	.0470
	14	23	.0494	.0989
	15	22	.0939	.1877
38	11	27	.0025	.0051
	12	26	.0069	.0139
	13	25	.0168	.0336
	14	24	.0365	.0730
	15	23	.0717	.1433
39	12	27	.0047	.0095
	13	26	.0119	.0237
	14	25	.0266	.0533
	15	24	.0541	.1081
	16	23	.0998	.1996
40	12	28	.0032	.0064
	13	27	.0083	.0166
	14	26	.0192	.0385
	15	25	.0403	.0807
	16	24	.0769	.1539
41	12	29	.0022	.0043
	13	28	.0058	.0115
	14	27	.0138	.0275
	15	26	.0298	.0596
	16	25	.0586	.1173
	17	24	.1055	.2110
42	13	29	.0040	.0079
	14	28	.0098	.0195
	15	27	.0218	.0436
	16	26	.0442	.0884
	17	25	.0821	.1641
43	13	30	.0027	.0054
	14	29	.0069	.0137
	15	28	.0158	.0315
	16	27	:0330	.0660
	17	26	.0631	.1263
44	14	30	.0048	.0096
	15	29	.0113	.0226
	16	28	.0244	.0488
	17	27	.0481	.0961
	18	26	.0871	.1742

TABLE A8
CRITICAL VALUES FOR THE MEDIAN TEST (CONTINUED)

n	T_L	T_U	α-Level One-Tailed Test	α-Level Two-Tailed Test
45	14	31	.0033	.0066
	15	30	.0080	.0161
	16	29	.0178	.0357
	17	28	.0362	.0725
	18	27	.0676	.1352
46	14	32	.0023	.0045
	15	31	.0057	.0114
	16	30	.0129	.0259
	17	29	.0270	.0541
	18	28	.0519	.1038
	19	27	.0920	.1839
47	15	32	.0040	.0079
	16	31	.0093	.0186
	17	30	.0200	.0400
	18	29	.0395	.0789
	19	28	.0719	.1439
48	15	33	.0028	.0055
	16	32	.0066	.0133
	17	31	.0147	.0293
	18	30	.0297	.0595
	19	29	.0557	.1114
	20	28	.0967	.1934
49	16	33	.0047	.0094
	17	32	.0106	.0213
	18	31	.0222	.0444
	19	30	.0427	.0854
	20	29	.0762	.1524
50	16	34	.0033	.0066
	17	33	.0077	.0153
	18	32	.0164	.0328
	19	31	.0325	.0649
	20	30	.0595	.1189
	21	29	.1013	.2026

ANSWERS TO SELECTED EXERCISES

1.3 **(a)** The expense of giving out free samples of the soap requires that only a portion of the potential customers receive free samples.

(b) The vaccine must be tested for effectiveness before it is released to the population at large. Also, this allows the possibility of side effects to show up, as was the case with the "swine flu" vaccine of 1976–1977.

(c) A sample will provide accurate estimates of nicotine content, and little, if any, additional information is gained by examining large quantities of the brand of cigarettes.

1.5 **(a)** Negative bias, because many low offers will not be accepted, while few high offers will be refused.

(b) Positive bias, because many of her neighbors will sell at amounts less than their asking prices, but none are likely to receive more than their asking price.

(c) Negative bias, because some unemployed people do not receive unemployment benefits for one reason or another.

1.9
Digit:	0	1	2	3	4	5	6	7	8	9
Frequency:	11	10	14	15	14	19	17	19	20	11

1.11 Sampling error

1.15 Age is quantitative, all others are qualitative.

1.17 **(a)** The experiment is the student taking the test and totaling the student's score on the test.

(b) {0, 5, 10, 15, . . . , 90, 95, 100}

(c) Answers will vary.

(d) {75, 80, 85, 90, 95, 100}

1.19 (a) There are four outcomes in the sample space: no unions, union A, union B, and both unions.

(b) Union A, union B.

(c) No unions $\rightarrow$ 0, union A $\rightarrow$ 1, union B $\rightarrow$ 1, both unions $\rightarrow$ 2.

1.21 Plural

1.24 The stratification forces all segments of the voting population to be represented.

1.26 (a) Systematic;   **(b)** convenience;   **(c)** random;   **(d)** stratified;   **(e)** cluster.

1.28 Stratified

1.30 One estimate would involve computing the average of the eight female weights, then the average of 24 male weights, and finally averaging those two averages. A stratified sample would be more appropriate, obtaining equal numbers of male and female students for the sample.

1.31 $\dfrac{6900}{2400} \times 100\% = 287.5$

1.33 $\dfrac{222 \times 7.6}{114 \times 7.8} \times 100\% = 189.7$

1.35 (a) $\dfrac{24.40}{23.60} \times 100\% = 103.4$

(b) $\dfrac{2(67.10) + 1(24.40) + 2(96.00)}{2(62.30) + 1(23.60) + 2(82.40)} \times 100\% = 112.0$

(c) $\dfrac{2(73.20) + 1(28.30) + 2(104.00)}{2(67.10) + 1(24.40) + 2(\ 96.00)} \times 100\% = 109.2$

Increase is 9.2%.

1.37 $\dfrac{11,600}{161.2} = 7196$ and $\dfrac{15,200}{217.4} = 6992$

Her purchasing power decreased by 2.8%

1.39 The increase is $125.3 - 121.3 = 4.0$ percentage points. The percentage increase is $(125.3/121.3) \times 100\% = 3.3\%$.

1.41 (a) Higher than the CPI.   **(b)** Higher than the CPI.

(c) Higher than the CPI.   **(d)** Less than the CPI.

(e) Less than the CPI.   **(f)** Less than the CPI.

1.47 (a) $I_t = \dfrac{7.24}{3.08} \times 100\% = 235.1$

(b) $I_t = \dfrac{4.81}{2.26} \times 100\% = 212.8$

(c) Construction wages increased by a larger percent over the 15-year period than did manufacturing wages.

1.51 This method has the advantage of being easy to use, but has the drawback of being likely to select only those customers who are extremely pleased or displeased.

1.53 This has the advantage of not allowing rigs to gear up for the safety inspection, and thus provides more reliable information about safety conditions. The disadvantage is that some rigs will not be inspected.

1.55 **1.** This could have been a chance occurrence that happens only rarely.

2. The fog may have been restricted to the waterfront, while the brochure may have meant that sun shines in some part of the city 360 days a year.

3. The 360 days of sunshine may not have meant sunshine all day long, but rather that the sun shines even for a brief period sometime during the day.

1.59 Return to the home and thus avoid introducing a bias into the survey.

1.61 Record the response in the respondent's exact words as nearly as possible.

2.1 (a) 5; (b) 10; (c) 22; (d) 20

2.3

```
12 | 5  8  1  3  6
13 | 3  4
14 | 8  3
15 | 3  1  9  8  3  0
16 | 2  5  6  2  6  8  0
17 | 8  4  8  5  0
18 | 6  7  6  8  4  0  5
19 |
20 | 1  0  2  8
21 | 8  5
```

2.5

```
1t | 39
1f | 59  56  54  47  41
1s | 75  75  67  66
1· | 95  86  81
2* | 12
2t | 28
2f | 58  58  44
2s | 66  61
```

(Multiply all numbers by 100 to get salary figures.) In this plot the last two significant digits were dropped after multiple splitting of the stems. Other options would be to have each leaf composed of either three digits after dropping the last significant digit or to have each leaf composed of the last four significant digits.

2.7 (a) From the suggested rule, five classes would be used for the 30 numbers at hand.

(b) $(60 - 21)/5 = 7.8$, which would be rounded up to 8.

(c) 20.5 to 28.5, 28.5 to 36.5, . . . , 52.5 to 60.5, or class boundaries could also be taken as 21 to 28, 29 to 36, . . . , 53 to 60

(d) For either set of boundaries given in part (c) the class marks would be 24.5, 32.5, . . . , 56.5.

2.9

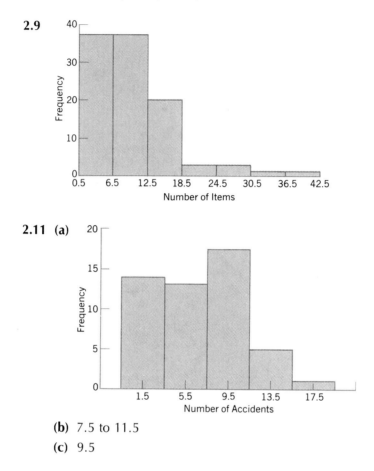

2.11 (a)

(b) 7.5 to 11.5

(c) 9.5

2.13 The areas occupied by the airplane are not in proportion to the relative frequencies.

2.15

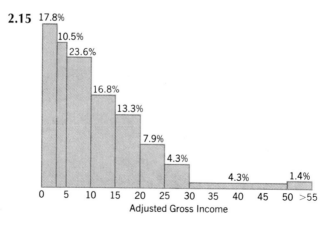

2.17

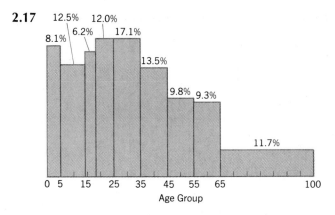

The 2000-year projection has a larger percentage in all classes representing 35 and older.

2.19

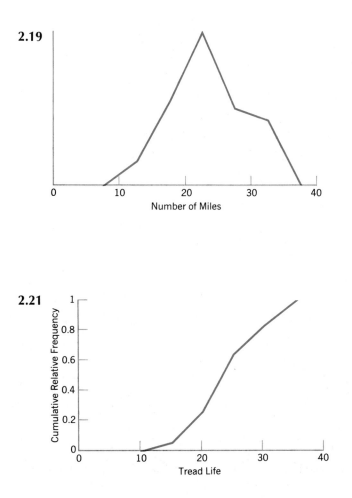

2.21

2.22

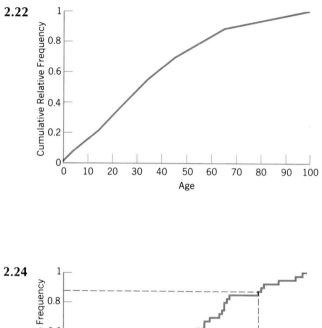

2.24

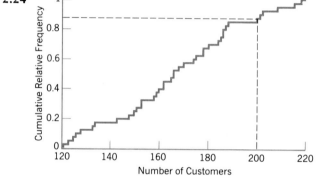

Twelve and a half percent of the time the number of customers was more than 200.

2.25

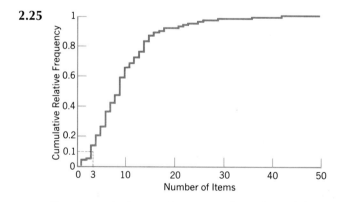

Two items is the maximum number allowed for using the fast lane.

2.27

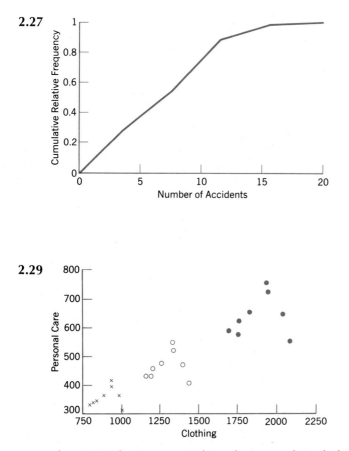

The scatterplot appears as three distinct and similarly shaped clusters of points. All points within each cluster are close to lying on a straight line except for those points corresponding to Cincinnati and Milwaukee.

2.31

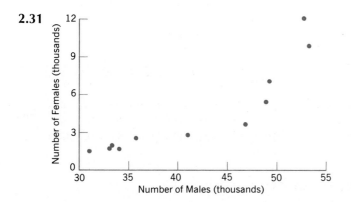

During the last few years covered by the study the number of female graduates grew at a faster rate than the number of male graduates.

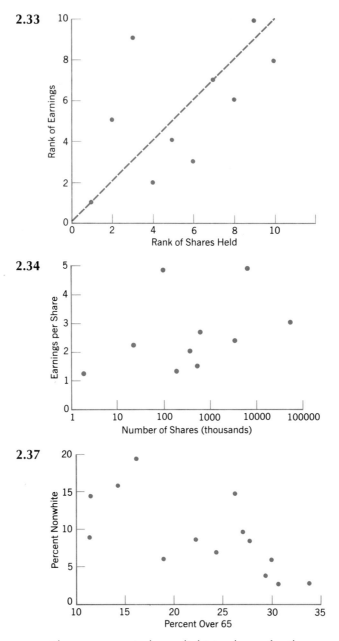

There appears to be a slight tendency for the counties with larger percentages of older citizens to have lower percentages of nonwhites. This may be a result of more retirement communities in those counties.

2.39 19; 15

2.41 Multivariate data with four or more variables.

2.43 From Figure 2.32 and variable 10, Shell has the largest gas production (2.60) and Sunoco has the smallest (0.30).

2.45 From Figure 2.36 the length of the eye represents variable 2 (the ratio of the number of leases won to the number bid on). Thus, a high ratio (near 1) represents a high success rate on getting leases, while a low ratio (near 0) represents a lesser success rate. This ratio is .21 in Figure 2.32 for Sunoco, and is the lowest of all 12 company groups. This fact is shown in Figure 2.37 by the small eyes for Sunoco.

2.48

4t	3
4f	4 5 5 5
4s	6 6 7 7 7 7 7
4˙	8 8 8 9 9 9 9 9 9
5*	0 0 0 0 0 0 0 1 1 1 1 1 1
5t	2 2 2 2 2 3 3 3 3
5f	4 4 5 5 5
5s	7
5˙	
6*	1

2.49

2.50

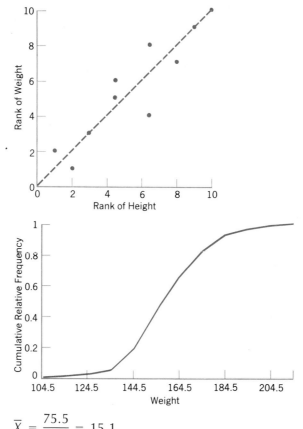

3.1 $\overline{X} = \dfrac{75.5}{5} = 15.1$

3.3 Data expressed in years: 4, 8/12, 2, 27, 14/12

$\overline{X} = \dfrac{34.83}{5} = 6.97$

3.5 $\overline{X} = [2(10) + 8(9) + 16(8) + 12(7) + 5(6)]/43 = 7.77$ from Eq. 3.2.
$\overline{X} = 10(2/43) + 9(8/43) + 8(16/43) + 7(12/43) + 6(5/43) = 7.77$ from Eq. 3.3.

3.7 $\overline{X} = \dfrac{1011}{102} = 9.91$

3.10 $\overline{X} = 9.8/7 = 1.4$ ft per roll, or 1400 ft per week. The percentage waste is $1.4/50 = 2.8\%$, constant for any period of time.

3.12 z scores: -1.591, -1.061, -1.061, 0, 0, 0, .530, .530, 1.061, 1.591
The mean of these z scores is 0 and the standard deviation is 1. These results are in agreement with those properties given for z scores.

3.15 $\overline{X} = \dfrac{6669}{40} = 166.73$

$s = \sqrt{\dfrac{1{,}138{,}039 - (6669)^2/40}{39}} = 25.9$

$\overline{X} \pm s$ is from 140.8 to 192.6 and contains 27 observations, or 68% of the total compared with the rule of thumb 67%, whereas $\overline{X} \pm 2s$ is from 114.9 to 218.5 and contains all 40 observations compared with the rule of thumb 95%.

3.17 $\Sigma f_i m_i = 2,340,000$ from Exercise 3.6.

$\Sigma f_i m_i^2 = 57,875,000,000$

$$s = \sqrt{\frac{57,875,000,000 - (2,340,000)^2/100}{99}} = 5613$$

3.19 $\Sigma f_i m_i = 2,500,000$ from Exercise 3.9.

$\Sigma f_i m_i^2 = 41,312,500,000$

$$s = \sqrt{\frac{41,312,500,000 - (2,500,000)^2/170}{169}} = 5187$$

3.20 81, 83, and 85 are all modes.

3.21 The modal class is 81–90.

3.22 **(a)** The median is $X_{.50} = (80 + 81)/2 = 80.5$.

    (b) $X_{.25} = 71$

    (c) $X_{.75} = 85$

    (d) $X_{.75} - X_{.25} = 85 - 71 = 14$

    (e) $X_{.90} = \dfrac{92 + 95}{2} = 93.5$

    (f) $X_{.80} = \dfrac{86 + 87}{2} = 86.5$

3.23 The value 37 is identified as an outlier in this boxplot.

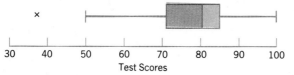

3.29 $\overline{X} = 50.12$; $s = 3.4796$

3.30 Data values less than, equal to, or greater than the sample mean of 50.12 will convert to negative, zero, or positive z scores, respectively. Thus, there are 28 negative z scores, no zeros, and 22 positive z scores.

3.31 $X_{.50} = 50$, $X_{.75} - X_{.25} = 52 - 48 = 4$

3.32 Mode $= 50$

3.33 The observation 61 is an outlier.

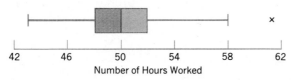
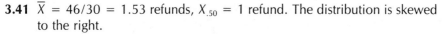

3.41 $\overline{X} = 46/30 = 1.53$ refunds, $X_{.50} = 1$ refund. The distribution is skewed to the right.

3.44 $\overline{X} = 1011/102 = 9.91$ from Exercise 3.7

$\Sigma f_i m_i^2 = 14{,}989$

$$s = \sqrt{\frac{14{,}989 - (1011)^2/102}{101}} = 7.01$$

4.1 **(a)**

	Married	Single	
Plan 1	.071	.179	.250
Plan 2	.500	.036	.536
Plan 3	.161	.054	.214
	.732	.268	1.000

(b) .250

(c) .732

(d) $P(\text{Married}) = .732$; $P(\text{Plan 1}) = .250$
$P(\text{Married and Plan 1}) = .071$
Since $.732 \times .250 = .183 \neq .071$, these events are not independent.

(e) $P(\text{Plan 1}|\text{Married}) = \dfrac{P(\text{Plan 1 and Married})}{P(\text{Married})} = \dfrac{.071}{.732} = .097$

(f) $P(\text{Plan 1}|\text{Single}) = \dfrac{P(\text{Plan 1 and Single})}{P(\text{Single})} = \dfrac{.179}{.268} = .668$

(g) $P(\text{Plan 2}|\text{Married}) = \dfrac{P(\text{Plan 2 and Married})}{P(\text{Married})} = \dfrac{.500}{.732} = .683$

(h) $P(\text{Plan 2}|\text{Single}) = \dfrac{P(\text{Plan 2 and Single})}{P(\text{Single})} = \dfrac{.036}{.268} = .134$

(i) No, plan selection depends to a great deal on marital status, as the probabilities in parts (e)–(h) demonstrate.

4.3 No, they are not independent events because

$$P(A \text{ on 1st}, A \text{ on 2nd}) \neq P(A \text{ on 1st}) \times P(A \text{ on 2nd})$$
$$.15 \neq .25 \times .25$$

4.5 **(a)**

	Married	Single	
Delinquent	.04	.08	.12
Nondelinquent	.62	.26	.88
	.66	.34	1.00

(b) .26

(c) $P(\text{Del.}|\text{Single}) = \dfrac{P(\text{Del. and Single})}{P(\text{Single})} = \dfrac{.08}{.34} = .24$

(d) No, since $P(\text{Married}) \times P(\text{Delinquent}) = .66 \times .12 = .07$, while $P(\text{Married and Delinquent}) = .04$.

4.9

	Microwave	None	
Color TV	.20	.60	.80
No color TV	.10	.10	.20
	.30	.70	1.00

Ten percent have neither.

4.12
                                            Branch Probabilities

                        Plan 1 (.097) .732 × .097 = .071

      Married (.732)   Plan 2 (.683) .732 × .683 = .500

                        Plan 3 (.220) .732 × .220 = .161

                        Plan 1 (.668) .268 × .668 = .179

      Single   (.268)   Plan 2 (.134) .268 × .134 = .036

                        Plan 3 (.201) .268 × .201 = .054

                                            Total 1.000

4.14
                                            Branch Probabilities

                Del   (.061) .66 × .061 = .040

      Married (.66)

                Not Del (.939) .66 × .939 = .620

                Del   (.235) .34 × .235 = .080

      Single   (.34)

                Not Del (.765) .34 × .765 = .260

                                            Total 1.000

4.15 x: 1   2   3   4   5   >5
        $P(X = x)$: .40   .23   .11   .09   .08   .09

(a) .40; (b) .74; (c) .23; (d) .74; (e) .26; (f) 0

4.16 .40 + .23 + .11 + .09 + .08 + .09 = 1.00

4.17 $P(X \leqslant x)$

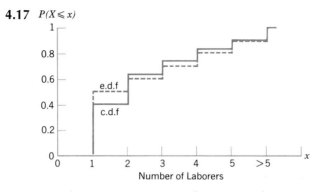

Number of Laborers

4.18 See plot in Exercise 4.17. The two graphs are quite similar.

4.22 **(a)** Neither (area = .5); **(b)** c.d.f.; **(c)** density; **(d)** c.d.f.

4.28 $f(x)$

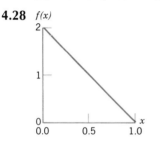

4.29 The area under the density function, between $x = 0$ and $x = .25$ is (width) $\times$ (average height) = .25(2 + 1.5)/2 = .4375.

4.30 $F(x)$

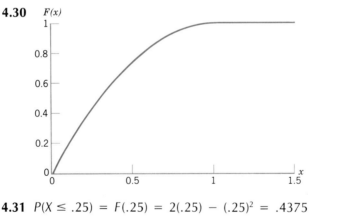

4.31 $P(X \leq .25) = F(.25) = 2(.25) - (.25)^2 = .4375$

4.36 $\mu = 50(.138) + 150(.158) + 250(.109) + 350(.079) + 450(.050) + 750(.168) + 1250(.069) + 1750(.040) + 2250(.030) + 3750(.059) + 7500(.040) + 30000(.050) + 75000(.010) = 3229$

4.37 $\sigma = [(50)^2(.138) + \cdots + (75,000)^2(.010) - (3229)^2]^{1/2}$
$= [104,871,045 - (3229)^2]^{1/2} = 9718$

4.42 $\mu = 1(.10) + 2(.30) + 3(.10) + 4(.20) + 5(.08) + 6(.11) + 7(.03) + 8(.08) = 3.71$

4.43   *F(x)*

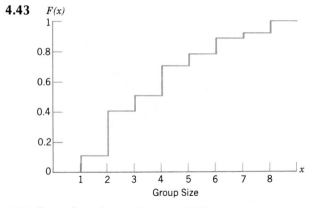

Group Size

4.44 From Exercise 4.42, $\mu = 3.71$.

$$\sigma = [1^2(.10) + 2^2(.30) + \cdots + 8^2(.08) - (3.71)^2]^{1/2} = 2.05$$

4.45 The population median is $(3 + 4)/2 = 3.5$.

4.46 The population interquartile range is $5 - 2 = 3$, which is slightly larger than the population standard deviation of 2.05.

4.55 $$P(\text{Accident}\,|\,\text{Uninsured}) = \frac{P(\text{Accident and Uninsured})}{P(\text{Accident})} = \frac{.3}{.5} = .6$$

4.56 **(a)** .41; **(b)** .54; **(c)** .18; **(d)** .18; **(e)** .05; **(f)** .14

4.57 *P(X = x)*

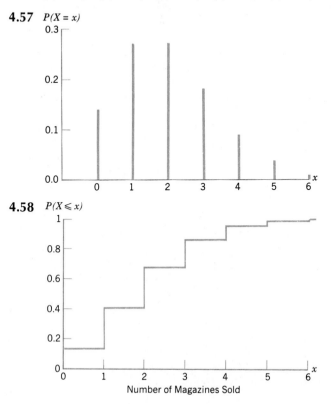

4.58 *P(X ⩽ x)*

Number of Magazines Sold

4.61 (a) $P(A|\text{No Defects}) = \dfrac{P(A \text{ and No Defects})}{P(\text{No Defects})} = \dfrac{.1250}{.1875} = .6667$

(b) $P(\text{No Defects}|A) = \dfrac{P(\text{No Defects and } A)}{P(A)} = \dfrac{.1250}{.5000} = .2500$

(c) $P(A|2 \text{ or More Defects}) = \dfrac{P(A \text{ and 2 or More Defects})}{P(2 \text{ or More Defects})}$

$= \dfrac{.3125}{.6875} = .4545$

4.63 The joint probability table is found first.

	Gasoline Only	Products Only	Both	
In State	.4(2/3)	.5(2/3)	.1(2/3)	2/3
Out of State	.25(1/3)	.30(1/3)	.45(1/3)	1/3
	.350	.433	.217	1.00

(a) $P(\text{Gasoline Only}) = .35$.

(b) $P(\text{Gasoline}) = .35 + .217 = .567$.

(c) No, because some customers purchase both types of products.

(d) No, because $P(\text{Gasoline and In-Store Products}) = .65/3 = .22$, and this is not equal to the product of $P(\text{Gasoline}) = .57$ from part (b) and $P(\text{In-Store Product}) = 1.3/3 + .65/3 = .65$. That is, $.22 \neq .57 \times .65$. Another way of looking at it is that if a customer does not purchase gasoline you know they will purchase an in-store product, otherwise they would not be a customer. Thus the two events are dependent.

(e)   Branch Probabilities

    Out of State (.333)   Gas Only (.25) .333 × .25 = .083
                           In Store (.30) .333 × .30 = .100
                           Both   (.45) .333 × .45 = .150
                           Gas Only (.40) .667 × .40 = .267
    In State   (.667)   In Store (.50) .667 × .50 = .333
                           Both   (.10) .667 × .10 = .067

                                                          Total 1.000

(f) $P(\text{Gasoline Only}) = .083 + .267 = .350$

$P(\text{Gasoline}) = .083 + .150 + .267 + .067 = .567$

Once the probability tree is constructed, it is very easy to find the probabilities from it; however, it does take some effort to construct the tree.

4.65 First the joint probability table is constructed.

	Smokers	Non-Smokers	
Male	16/36	8/36	24/36
Female	8/36	4/36	12/36
	24/36	12/36	1.00

(a) 24/36 = .667

(b) 24/36 = .667

(c) $P(\text{Smoker}|\text{Female}) = \dfrac{P(\text{Smoker and Female})}{P(\text{Female})} = \dfrac{8/36}{12/36} = .667$

(d) Yes the events are independent, because
$P(\text{Female}) \times P(\text{Smoker}) = (12/36)(24/36) = 2/9$, and also
$P(\text{Female and Smoker}) = 8/36 = 2/9$.

4.67 (a) (F,F,F,F),   (F,F,F,M),   (F,F,M,F),   (F,M,F,F),   (M,F,F,F),
(F,F,M,M),   (F,M,F,M),   (M,F,F,M),   (F,M,M,F),   (M,F,M,F),
(M,M,F,F),   (F,M,M,M),   (M,F,M,M),   (M,M,F,M),   (M,M,M,F),
(M,M,M,M)

(b) One sixteenth, because each sex has probability $\frac{1}{2}$ on each birth independently of each other, and $(\frac{1}{2})(\frac{1}{2})(\frac{1}{2})(\frac{1}{2}) = \frac{1}{16}$.

(c) X can equal 0, 1, 2, 3, and 4.

(d) $P(X = 0) = P(M,M,M,M) = \frac{1}{16}$; $P(X = 1) = P(\text{One Girl}) = \frac{4}{16}$, because four outcomes in the sample space have one girl and three boys.

(e)

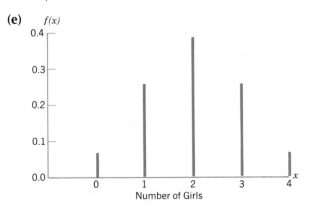

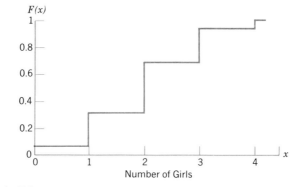

Number of Girls

(g) Discrete.

4.69 $\mu = [(-2.37) + (-3.19) + (-8.33) + (-4.18) + (-6.66)$
$+ (-14.35) + (-10.41) + 4.81 + 5.74 + 10.48 + 6.46$
$+ 8.07]/12 = -1.20$

$\lambda_{.5} = [(-3.19) + (-2.37)]/2 = -2.78$

5.1 (a) .7384; (b) .0060; (c) .0000; (d) .9502; (e) .0498; (f) .0085;
(g) .9663 $-$ .0393 $=$ .9270; (h) .9264 $-$ .0393 $=$ .8871;
(i) .9663 $-$ .0923 $=$ .8740; (j) .9264 $-$ .0923 $=$ .8341

5.3 $n = 80$; $p = .1$; $\mu = 80(.1) = 8$; $\sigma = [80(.1)(.9)]^{1/2} = 2.68$

5.5 $n = 30$; $p = .1$; expected number of defectives $= 30(.1) = 3$;
$P(X \geq 4) = 1 - P(X \leq 3) = 1 - .6474 = .3526$.

5.7 (a) $n = 20$; $p = .1$, from Table A1; $P(X \leq 3) = .8670$.
(b) $P(X \geq 3) = 1 - P(X \leq 2) = 1 - .6769 = .3231$.
(c) $P(X = 3) = P(X \leq 3) - P(X \leq 2) = .8670 - .6769 = .1901$.

5.9 $n = 15$; $p = .3$; $P(X \leq 8) = .9848$ from Table A1.

5.11 From Table A1 with $n = 8$ and $p = .3$
$$P(X \geq 2) = 1 - P(X \leq 1) = 1 - .2553 = .7447$$

5.13 From Table A1 with $n = 7$ and $p = .3$
$$P(X > 3) = 1 - P(X \leq 3) = 1 - .8740 = .1260$$

5.15 From Table A1 with $n = 6$ and $p = .4$
$$P(X = 3) = P(X \leq 3) - P(X \leq 2) = .8208 - .5443 = .2765$$
Or from the probability function in Eq. 5.1
$$P(X = 3) = \binom{6}{3}(.4)^3(.6)^3 = 20(.064)(.216) = .2765$$

5.17 $x_{.025} = 10 + 2z_{.025} = 10 + 2(-1.9600) = 6.0800$

$x_{.975} = 10 + 2z_{.975} = 10 + 2(1.9600) = 13.9200$

5.19 (a) $z = (6.08 - 10)/2 = -1.96$, from Table A2* the entry corresponding to -1.96 is .0250, therefore $P(X \leq 6.08) = .0250$.
(b) $z = (13.92 - 10)/2 = 1.96$, from Table A2* the entry corresponding to 1.96 is .9750, therefore $P(X \leq 13.92) = .9750$.

(c) $P(6.08 \leq X \leq 13.92) = P(X \leq 13.92) - P(X \leq 6.08)$
$= .9750 - .0250 = .9500.$

5.21 (a) $z = (.5 - .8)/.2 = -1.50$, from Table A2*, $P(X \leq .5) = .0668.$

(b) $P(X \geq 1.2) = 1 - P(X \leq 1.2) = 1 - .9772 = .0228$ as
$z = (1.2 - .8)/.2 = 2.00.$

(c) $P(.75 \leq X \leq 1.25) = P(X \leq 1.25) - P(X \leq .75)$
$= .9878 - .4013 = .5865$ as $z_1 = (.75 - .8)/.2 = -0.25$ and
$z_2 = (1.25 - .8)/.2 = 2.25.$

5.23 $x_{.95} = 125 + 10z_{.95} = 125 + 10(1.6449) = 141.45$; therefore, his
score of 140 is not quite high enough to be in the top 5%.

5.25 $P(X \leq -67) = P(Z \leq -67/45) = P(Z \leq -1.49) = .0681$ from Table
A2*, which gives basis to the manager's doubts.

5.27 From Table A1, $P(1 \leq X \leq 3) = P(X \leq 3) - P(X \leq 0) = .9961 -$
$.3164 = .6797$ as the exact answer. $P(1 \leq X \leq 3) \approx P(z_1 \leq Z \leq z_2)$,
where

$$z_1 = (1 - 4(.25) - .5)/\sqrt{4(.25)(.75)} = -0.58$$

and

$$z_2 = (3 - 4(.25) + .5)/\sqrt{4(.25)(.75)} = 2.89.$$

Therefore, $P(1 \leq X \leq 3) \approx .9981 - .2810 = .7171$. The error is
somewhat large, since np and nq are not both greater than 5.

5.29 (a) $50(.6) = 30$

(b) $P(26 \leq X \leq 35) \approx P(z_1 \leq Z \leq z_2)$
where

$$z_1 = (26 - 30 - .5)/\sqrt{50(.6)(.4)} = -1.30$$

and

$$z_2 = (35 - 30 + .5)/\sqrt{50(.6)(.4)} = 1.59.$$

Therefore, $P(26 \leq X \leq 35) \approx .9441 - .0968 = .8473.$

(c) $P(X \leq 25) \approx P(Z \leq -1.30) = .0968$

5.31 First note that these are discrete data that can be summarized in the
following frequency chart.

z score	f_i	z score	f_i	z score	f_i
-2.05	1	-0.61	3	0.83	4
-1.76	1	-0.32	6	1.12	2
-1.47	3	-0.03	7	1.40	3
-1.18	2	0.25	6	1.98	1
-0.90	5	0.54	5	3.13	1

Even though the data are discrete, the normal distribution appears to
provide a satisfactory approximate probability model, as shown by the
Lilliefors test. (See the Lilliefors graph on the next page.)

5.33 $P(X \geq 53)$ is found by using Eq. 5.13 as

$$P(53 \leq X \leq 87) \approx 1 - P(Z \leq z_1),$$

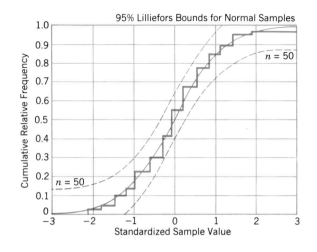

where

$$z_1 = (53 - 87(.5) - .5)/\sqrt{87(.5)(.5)} = 1 - P(Z \le 1.93)$$
$$= 1 - .9732 = .0268.$$

5.35 The number of calls X has a binomial distribution with $n = 1500$ and $p = .03$. The answer is x, where x is found by solving the equation $P(X \le x) = .95$. The normal approximation from Eq. 5.12 is used:

$$z_2 = \frac{x - 1500(.03) + .5}{\sqrt{1500(.03)(.97)}}$$

$P(Z \le z_2) = .95$, but from Table A2, $P(Z \le 1.6449) = .95$. Therefore, $z_2 = 1.6449$, or

$$\frac{x - 1500(.03) + .5}{\sqrt{1500(.03)(.97)}} = 1.6449$$

so $x = 55.37$, and the exchange needs to be able to handle 56 calls.

5.37 $\mu = 3$ and $\sigma = \sqrt{2}/\sqrt{4} = \sqrt{2}/2 = .707$.

5.39 $P(\bar{X} \le 160) = P\left(Z \le \dfrac{160 - 163}{3.5/\sqrt{10}}\right) = P(Z \le -2.71) = .0034$

5.44 $\bar{X} = 93/10 = 9.3$ minutes between arrivals, $\hat{\lambda} = 1/\bar{X} = 1/9.3 = .1075$ arrivals per minute.

5.45

```
0      10     20     30     40     50     60     70     80     90    100
| ---× |  ----  | ×--- | ×-- × | ---× |  ----  | ×--- |× - × ×|  ----  | -×-- |
    1       0       1      2       1       0       1       3       0       1
```
$3, 5, 1, 1, \hat{\lambda} = \bar{X} = [0(3) + 1(5) + 2(1) + 3(1)]/10 = 1$ arrival every 10 minutes, or .1 arrivals per minute.

5.47 $\mu = .75$ minutes, therefore $\lambda = 1/.75 = 1.33$
$$P(X \ge 3) = e^{-1.33(3)} = .0183$$

5.49 $\lambda = \frac{1}{2} = .5$; $P(X \ge 8) = e^{-.5(8)} = .0183$.

5.51 **(a)** .135;   **(b)** .271;   **(c)** .271   **(d)** .180;   **(e)** .143;   **(f)** 2;
(g) $\sqrt{2} = 1.414$.

5.55 The estimated number of vacancies is obtained as follows:

$$\left(\frac{59}{96}\right)(0) + \left(\frac{27}{96}\right)(1) + \left(\frac{9}{96}\right)(2) + \left(\frac{1}{96}\right)(3) = .5$$

If Y is a Poisson random variable with $\lambda = .5$ representing the number of appointments per year, then the probability of no appointments in a given year is found as

$$P(Y = 0) = \frac{e^{-.5}.5^0}{0!} = .6065$$

Assuming independence from year to year, the answer is given as $(.6065)^4 = .135$.

5.57 $P(X < 23{,}500) = P\left(Z \le \dfrac{23500 - 27500}{3000}\right)$

$= P(Z \le -1.33) = .0918$

$P(X > 30{,}000) = 1 - P\left(Z \le \dfrac{30000 - 27500}{3000}\right)$

$= 1 - P(Z \le 0.83) = 1 - .7967 = .2033$

5.59 **(a)** $P(\overline{X} \le 1230) = P\left(Z \le \dfrac{1230 - 1500}{900/\sqrt{25}}\right)$

$= P(Z \le -1.50) = .0668$

(b) $P(\overline{X} \le 1600) = P\left(Z \le \dfrac{1600 - 1500}{900/\sqrt{25}}\right)$

$= P(Z \le 0.56) = .7123$

(c) $P(1230 \le \overline{X} \le 1600) = P(\overline{X} \le 1600) - P(\overline{X} \le 1230)$

$= .7123 - .0668 = .6455$

5.61 $P(X < 16) = P\left(Z \le \dfrac{16 - 16.3}{.15}\right) = P(Z \le -2.00) = .0228$

5.63 $\lambda = 1, P(X \le 14/12) = 1 - e^{-1(14/12)} = .6886.$

5.65 $\overline{X} = 62.99, \hat{\lambda} = 1/62.99 = .0159.$

5.67 Let X = the number of passengers who need seats. Then X has a binomial distribution with unknown n and $p = .85$. Using the normal approximation, the value of n is needed such that $P(X \le 122) \ge .90$, or

$$P\left(Z \le \frac{122 - n(.85) + .5}{\sqrt{n(.85)(.15)}}\right) \ge .90$$

Since $P(Z \le 1.2816) = .90$ from Table A2, the following equation can be written and solved to find n:

$$\frac{122 - n(.85) + .5}{\sqrt{n(.85)(.15)}} \ge 1.2816$$

Either by trial and error or direct solution, $n = 137$ is the largest n that satisfies the preceding inequality; so the airlines can make 137 reservations. The actual probability associated with $n = 137$ is .9265, since $P(Z \leq 1.45) = .9265$.

5.69 Let X be the number of customers who buy a newspaper, of the first 1000 supermarket customers. Then X has a binomial distribution with $n = 1000$ and $p = .05$. Since $P(X \leq 2.3263) = .99$ from Table A2, it is necessary to find the value of x such that $P(X \leq x) = .99$. Based on the normal approximation, the following equation can be written and solved for x:

$$2.3263 = \frac{x - 1000(.05) + .5}{\sqrt{1000(.05)(.95)}}$$

which gives $x = 65.53$ or 66.

5.71 Solution using the Poisson distribution: Let $X =$ the number of customers arriving in 2 min, $\lambda = 4 =$ average number of arrivals in a *two*-minute interval.

$$P(X = 0) = \frac{e^{-\lambda}\lambda^0}{0!} = e^{-4} = .0183$$

Solution using the exponential distribution: Let $X =$ the time between arrivals. $\lambda = 2 =$ number of arrivals per minute

$$P(X > 2) = e^{-\lambda t} = e^{-2(2)} = e^{-4} = .0183$$

6.1 The estimator is $\bar{X}$ and the estimate is 16.3 lb.

6.3 The second plan involves more work, but also probably has the estimator with the smaller variance.

6.5 Medians: 61.55, 59.25, 58.10, 60.45, 58.65. The standard deviation of these medians is 1.40, which would indicate that the sample mean has a smaller standard deviation associated with it for these data.

6.7 From Table A6 with $n = 18$ and $x = 6$, Lower $= .133$, Upper $= .590$.

6.8 $6/18 \pm 1.9600\ [(6/18)(12/18)/18]^{1/2}$ or .116 to .551. This interval is slightly smaller than the one found using the exact method.

6.11 $6/50 \pm 1.9600\ [(6/50)(44/50)/50]^{1/2}$ or .030 to .210.

6.14 $n = \left(\dfrac{1.6449}{.04}\right)^2 = 1691$

6.16 $n = .36(.64)4\left(\dfrac{1.6449}{.04}\right)^2 = 1558$

6.18 **(a)** 1.7613.

(b) -1.7613.

(c) .75.

(d) By interpolation,

$$.90 + \frac{(.95 - .90)(1.52 - 1.3450)}{1.7613 - 1.3450} = .921$$

(e) $1 - .921 = .079$.

6.20 $\$2100 \pm 1.9600(400)/\sqrt{50}$ or $\$1989$ to $\$2211$. There is 95% confidence that the interval from \$1989 to \$2211 contains the true mean "average daily balance."

6.22 The required sample size is given as

$$n = [2(2.5758)(500)/400]^2 = 41.47$$

so approximately 41 households need to be in the survey.

6.25 $\bar{X} = 21{,}704$ and $s = 2733.37$. 95% confidence interval:

$$21{,}704 \pm \frac{2.7764(2733.37)}{\sqrt{5}} \qquad \text{or} \qquad \$18{,}310 \quad \text{to} \quad \$25{,}098$$

There is 95% confidence that the population mean is contained in this interval.

6.27 $980 \pm 1.8331(15.8)/\sqrt{10}$ or 970.8 to 989.2. The company's claim may not be justified, since 1000 is not covered by the interval.

6.30

Median	90% Confidence Interval
61.55	56.8 to 63.8
59.25	57.0 to 60.4
58.10	54.7 to 63.5
60.45	58.6 to 64.4
58.65	54.8 to 61.6

The confidence intervals for the mean are contained within the confidence intervals for the median in all cases except New York.

6.32 Median $= 24.43$, the 95% confidence interval is constructed from Table A7 where $S_1 = 6$ and $S_2 = 15$. Therefore, the confidence interval is from $X^{(6)}$ to $X^{(15)}$ or 24.14 to 24.62.

6.33 $S_1^* = (20 - 1.9600\sqrt{20})/2 = 5.62$ or $S_1 = 6$ and $S_2 = 20 - 6 + 1 = 15$. Therefore, the confidence interval is from $X^{(6)}$ to $X^{(15)}$ or 24.14 to 24.62. To find the approximate level of significance, compute $z = (2(6) - 20 - 1)/\sqrt{20} = -2.01$. From Table A2* $P(Z \leq -2.01) \approx .0222$, and the approximate level of significance is $100[1 - 2(.0222)]\% = 95.56\%$. From Table A7 the exact level of confidence is 95.86%.

6.36 From Table A7, $S_1 = 14$ and $S_2 = 27$. Therefore, the 95% confidence interval is from $X^{(14)}$ to $X^{(27)}$ or 75 to 84.

6.39 No, because X may be positive or negative, but a chi-square random variable is always positive.

6.41 $\chi^2_{.90,29} = 39.09$.

6.43 Approximately .286 by interpolation:

$$.25 + \frac{(.40 - .25)(16.93 - 16.34)}{18.77 - 16.34} = .286$$

6.45 50.8 is approximately the .902 quantile, by interpolation:

$$.90 + \frac{(.95 - .90)(50.8 - 50.66)}{54.57 - 50.66} = .902$$

The exceedence probability is $1 - .902 = .098$.

6.47 $(n - 1)s^2 = 7.51111 \times 10^{-5}$ $\chi^2_{.025,17} = 7.564$ $\chi^2_{.975,17} = 30.19$

$$\text{Lower limit} = \left(\frac{7.51111 \times 10^{-5}}{30.19}\right)^{1/2} = .00158$$

$$\text{Upper limit} = \left(\frac{7.51111 \times 10^{-5}}{7.564}\right)^{1/2} = .00315$$

6.49 $(n - 1)s^2 = 2190.37$ $\chi^2_{.025,29} = 16.05$ $\chi^2_{.975,29} = 45.72$

$$\text{Lower limit} = \left(\frac{2190.37}{45.72}\right)^{1/2} = 6.92$$

$$\text{Upper limit} = \left(\frac{2190.37}{16.05}\right)^{1/2} = 11.68$$

A standard deviation of at least 7%, and possibly as high as 11%, seems rather high under the circumstances. The point estimates $\overline{X} = 71.23\%$ and $s = 8.69\%$ suggest that as many as 5% of the capsules are outside the two standard deviation limit of $\overline{X} - 2s = 53.85\%$ to $\overline{X} + 2s = 88.61\%$, which is a large spread for percentage of active ingredient in a capsule.

6.50 Median $= 20$, the 95% confidence interval is constructed from Table A7, where $S_1 = 10$ and $S_2 = 21$. Therefore, the confidence interval is from $X^{(10)}$ to $X^{(21)}$ or 17 to 22.

6.51 $2870 \pm 1.9600(425)/\sqrt{40}$ or 2738 to 3002.

6.52 $84.2 \pm t_{.95,9}(12.2)/\sqrt{10}$ or $84.2 \pm 1.8331(12.2)/\sqrt{10}$ or 77.1 to 91.3.

6.53 $\overline{X} = 20.6$ $s = 5.27$

$20.6 \pm 2.0452(5.27)/\sqrt{30}$ or 18.6 to 22.6. The confidence interval for the mean is slightly smaller than the confidence interval for the median.

6.62 $1.43 \pm \dfrac{1.6449\,(.121)}{\sqrt{70}}$ or   1.406 to 1.454 oz

6.63 $(n - 1)s^2 = 1.010229$ $\chi^2_{.05,69} = 50.88$ $\chi^2_{.95,69} = 89.39$

$$\text{Lower limit} = \left(\frac{1.010229}{89.39}\right)^{1/2} = .1063$$

$$\text{Upper limit} = \left(\frac{1.010229}{50.88}\right)^{1/2} = .1409$$

7.2 Since she is interested in showing that the mean score μ for students who have completed her course is above 800, that becomes the alternative hypothesis. So H_1 is $\mu > 800$, and H_0 is $\mu = 800$.

7.4 **(a)** $H_0: \mu \geq 16$ and $H_1: \mu < 16$ **(b)** lower-tailed.

7.6 The p-value is $2P(T \geq 1.7613)$, since this is a two-tailed test. From Table A3, $P(T \geq 1.7613) = .05$, so the p-value is $2(.05) = .10$.

7.8 The null distribution is the binomial distribution with $n = 12$ and $p = .25$. From Table A1 with $p = .25$ and $n = 12$,

$$P(T \geq 7) = 1 - P(T \leq 6) = 1 - .9857 = .0143$$

so the p-value is .0143.

7.9   The upper 5% critical value T_U is the .95 quantile of the standard normal distribution, which is given in Table A2 as 1.6449. Therefore the decision rule is to reject H_0 if T exceeds 1.6449.

7.10 The upper 2.5% value T_U in a Student's t distribution with 17 degrees of freedom is the .975 quantile, which is given in Table A3 as 2.1098. Therefore the lower critical value is $T_L = -2.1098$, and the decision rule is to reject H_0 if $T > 2.1098$ or if $T < -2.1098$.

7.15 From Table A1 with $n = 24$ and $p = .8$,

$$P(T \leq 15) = .0362 \qquad \text{and} \qquad P(T > 22) = 1 - P(T \leq 22) = .0331$$

Therefore the decision rule is to reject H_0 if $T < 16$ or if $T > 22$. The actual size of the critical region is $.0362 + .0331 = .0693$. For $T = 11$ the decision is to reject H_0. The p-value is $2P(T \leq 11) = 2(.0002) = .0004$.

7.17 **(a)** H_1: The proportion of accounts qualifying for the "500 Club" is greater than 40%.

(b) Let T = the number of accounts qualifying for the "500 Club." Reject H_0 if $T > 11$, otherwise do not reject H_0. The exact level of significance for this test is $\alpha = .0565$.

(c) H_0 is not rejected.

(d) The p-value is $P(T \geq 11) = 1 - P(T \leq 10) = .1275$ from Table A1, with $n = 20$ and $p = .4$. The probability of getting a value of T as large as 11 when H_0 is true is .1275.

7.19 **(a)** $H_0: \sigma = 2.6$; $H_1: \sigma > 2.6$.

(b) The decision rule is to reject H_0 if $T > 37.92$, where $T_U = 37.92$ is the .90 quantile of the chi-square distribution with 28 degrees of freedom, found from Table A4.

(c) Since 41.1 is in the rejection region, the decision is to reject H_0.

(d) p-value $= P(T \geq 41.1) \approx .054$, using interpolation in Table A4 with 28 degrees of freedom.

7.20 Power $= P(T > 1.6449) =$
$$P\left(Z > \frac{1.6449 - 3}{1}\right) = 1 - .088 = .912$$

7.22 Power $= P(T < 4) + P(T > 11)$ for $n = 15$ and $p = .8$
$$= P(T \leq 3) + 1 - P(T \leq 11) \qquad \text{(See Table A1)}$$
$$= .0000 + 1 - .3518 = .6482$$

7.25 From Table A1 with $n = 8$, $p = .6$, $P(T \leq 2) = .0498$, so the decision rule at $\alpha = .0498$ is reject H_0 if $T < 3$. Since $T = 2$, H_0 is rejected. The p-value is .0498. The evidence is sufficient to show that the percentage of drivers wearing seat belts is less than 60%.

7.27 **(a)** $\alpha = P(T < 5) + P(T > 9)$ when $n = 10$ and $p = .7$. This is found in Table A1 as $.0473 + .0282 = .0755$.

(b) Power $= P(T < 5) + P(T > 9)$ when $n = 10$ and $p = .5$, which is found in Table A1 as $.3770 + .0010 = .3780$.

7.29 $H_0: p \geq .78$ versus $H_1: p < .78$. Reject H_0 if
$$T < 92(.78) - z_{.95}\sqrt{92(.78)(.22)} = 65.22.$$
Since $T = 64$, H_0 is rejected. The p-value is
$$P(T \leq 64) = P[Z < (64 - 92(.78) + .5)/\sqrt{92(.78)(.22)}]$$
$$= P(Z < -1.83) = .0336,$$
so the sample evidence refutes the claim.

7.31 $P(T \leq 1)$ for $n = 10$ and various values of p can be found from Table A1 to obtain the following graph.

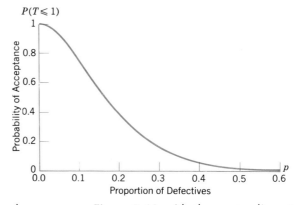

7.32 The graph appears as Figure 7.13 with the center line at .03 and the upper control line at
$$.03 + 2.3263\sqrt{\frac{(.03)(.97)}{50}} = .086.$$
For $\hat{p} = 5/50 = .10$ the process would be declared out of control.

7.36 **(a)** Let $T_1 = (\overline{X} - 16)/(.5/\sqrt{100})$. Reject H_0 if $T_1 < -z_{.95} = -1.6449$.

(b) For $\overline{X} = 15.5$, $T_1 = -10.00$; therefore, reject H_0 and readjust the equipment. For $\overline{X} = 15.95$, $T_1 = -1.00$; therefore, do not reject H_0 and do not readjust the equipment. For $\overline{X} = 16.1$, $T_1 = 2.00$; therefore, do not reject H_0 and do not readjust the equipment. For $\overline{X} = 15.90$, $T_1 = -2.00$; therefore reject H_0 and readjust the equipment.

7.37 **(a)** $n = 6$, $\Sigma X = 143.4$, and $\Sigma X^2 = 3431.18$, so $\overline{X} = 23.90$ and $s = .8854$
$$T_2 = \frac{23.9 - 25}{.8854/\sqrt{6}} = -3.0431$$
Reject H_0 if $T_2 < -t_{.95,5} = -2.0150$ or if $T_2 > t_{.95,5} = 2.0150$. H_0 is rejected.

(b) The p-value is between .02 and .05; therefore, the sample indicates the mean weight very likely has changed from 25.0 grams.

7.40 Since $\Sigma X = 138,000$ and $\Sigma X^2 = 3.196 \times 10^9$, $\overline{X} = 23,000$ and $s = 2097.6$
$$T_2 = \frac{23,000 - 25,000}{2097.6/\sqrt{6}} = -2.3355$$

Reject H_0 if $T_2 < -t_{.99,5} = -3.3649$. H_0 is not rejected. The p-value is between .025 and .05, so this sample does not indicate that the mean salary is less than \$25,000 at $\alpha = .01$, possibly because of the small sample size. However, the null hypothesis could be rejected if a value of $\alpha = .05$ were used.

7.41
$$T_1 = \frac{42.8 - 40}{6.89/\sqrt{200}} = 5.747$$

Reject H_0 if $T_1 > z_{.99} = 2.3263$. H_0 is rejected. The p-value is less than .001, so the evidence is quite strong that the average age is greater than 40.

7.44
$$T_1 = \frac{1.85 - 1.80}{.25/\sqrt{100}} = 2$$

Reject H_0 if $T_1 > z_{.95} = 1.6449$. H_0 is rejected. The p-value is approximately .0228. Note that if the population mean is really \$1.85 instead of \$1.80 the power is

$$P(T_1 > 1.6449) = P\left(Z > \frac{1.80 - 1.85}{.25/\sqrt{100}} + 1.6449\right)$$
$$= P(Z > -.3551) = .6406$$

7.46
$$T_3 = \frac{5(2097.6)^2}{1000^2} = 22$$

This is not between $\chi^2_{.025,5} = .831$ and $\chi^2_{.975,5} = 12.83$, so H_0 is rejected. The p-value is approximately $2(.0005) = .001$.

7.47
$$T_3 = \frac{199(6.89)^2}{8^2} = 147.61$$

This is less than $\chi^2_{.01,199} = 154.9$, which was calculated using the first equation given in Table A4, so H_0 is rejected. The p-value is .0031, which is obtained by solving the first equation in Table A4 for $z_p = -2.74$, and then using Table A2*.

7.50 $T_3 = 99(.25)^2/(.3)^2 = 68.75$, which is less than $\chi^2_{.01,99} = 69.23$, so H_0 is rejected. The p-value is between .005 and .01.

7.51 Reject H_0 if $T_1 < 7$ because $P(T_1 < 7) = .0577$ from Table A8 with $n = 20$. H_0 is rejected since $T_1 = 6$. The p-value is $P(T_1 \leq 6) = .0577$, so the sample evidence indicates the median shoe size is less than $7\frac{1}{2}$.

7.53 Reject H_0 if $T_2 > 13$ as $P(T_2 > 13) = .0577$ from Table A8 with $n = 20$. H_0 is rejected since $T_2 = 15$. The p-value is $P(T_2 \geq 15) = .0207$, so the sample evidence indicates that the median mpg is greater than 24.

7.55 $T_1 = T_2 = 2$. Reject H_0 if $T_1 < 3$ or if $T_2 > 7$, as $P(T_1 < 3) + P(T_2 > 7) = .1094$ from Table A8 with $n = 10$. H_0 is rejected. The p-value is also .1094, so the sample evidence indicates that the median is different from 7 min.

7.57 Reject H_0 if $T_2 > 10$, as $P(T_2 > 10) = .0112$ from Table A8 with $n = 13$. H_0 is rejected since $T_2 = 12$. The p-value is $P(T_2 \geq 12) = .0017$, so H_0 is strongly rejected.

7.58 Reject H_0 if $T_1 < 11$, as $P(T_1 < 11) = .0494$ from Table A8 with $n = 30$. H_0 is rejected since $T_1 = 10$. The p-value is $P(T_1 \leq 10) = .0494$, which indicates that the median age is less than 22.

7.59 $\overline{X} = 1497.5$, $s = 10.71$. Standardized sample values: -1.82, -1.26, -1.07, -0.89, -0.79, -0.42, -0.33, -0.05, 0.14, 0.42, 0.51, 0.61, 0.89, 1.07, 1.45, 1.54. Using the Lilliefors test, the assumption of normality is reasonable for these data.

7.60
$$T = \frac{1497.5 - 1492}{10.71/\sqrt{16}} = 2.0545$$

Reject H_0 if $T > t_{.95,15} = 1.7531$. H_0 is rejected and the p-value is between .025 and .05, so it is concluded that the mean melting point is greater than 1492.

7.61 $T_3 = 15(10.71)^2/(10)^2 = 17.2$, which is less than $\chi^2_{.95,15} = 25.00$, so H_0 is not rejected. The p value is between .25 and .40.

7.66 The upper and lower control lines are at $2.7 \pm 1.9600(.2)/\sqrt{15}$, which are at 2.60 and 2.80 cm.

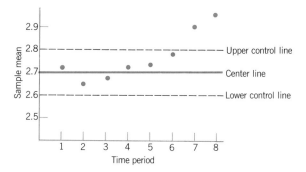

7.67 The process is judged to be out of control as indicated by the plot in Exercise 7.66.

8.1 $\Sigma D = 45$, $\Sigma D^2 = 473$, $\overline{D} = 4.5$, $s_D = 5.4823$, $z = -1.92$, -1.19, -0.64, -0.09, 0.09, 0.09, 0.64, 0.82, 1.00, 1.19. The data (i.e., the D_i) pass the Lilliefors test for normality.

H_0: $\mu_X = \mu_Y$ versus H_1: $\mu_X > \mu_Y$.

Reject H_0 if $T > t_{.975,9} = 2.2622$.

$T = 4.5\sqrt{10}/5.48 = 2.5957$, so H_0 is rejected.

The p-value is between .01 and .025. The 95% confidence interval is 4.5 ± 2.2622 $(5.48)/\sqrt{10}$ or .58 to 8.42.

8.3 $\Sigma D = 48$, $\Sigma D^2 = 772$, $\overline{D} = 2.4$, $s_D = 5.8795$.

H_0: $\mu_X = \mu_Y$ versus H_1: $\mu_X > \mu_Y$.

Reject H_0 if $T > t_{.95,19} = 1.7291$.

$T = 2.4\sqrt{20}/5.88 = 1.8255$, so H_0 is rejected.

The p-value is between .025 and .05.

8.4 $\Sigma D = -89$, $\Sigma D^2 = 1537$, $\overline{D} = -5.93$, $s_D = 8.4892$.

H_0: $\mu_X = \mu_Y$ versus H_1: $\mu_X < \mu_Y$.

Reject H_0 if $T < -t_{.99,14} = -2.6245$.

$T = -5.93 \sqrt{15}/8.49 = -2.7069$, so H_0 is rejected.

The p-value is between .005 and .01. The 90% confidence interval is $5.93 \pm 1.7613 \ (8.49)/\sqrt{15}$ or 2.07 to 9.79.

8.6 $\overline{D} = 6.20$, $s_D = 15.52$.

H_0: $\mu_X = \mu_Y$ versus H_1: $\mu_X \neq \mu_Y$.

Reject H_0 if $T < -t_{.975,19} = -2.0930$ or if $T > t_{.975,19} = 2.0930$.

$T = 6.20 \sqrt{20}/15.52 = 1.7864$, so H_0 is not rejected.

The p-value is between .05 and .10.

8.8 H_0: $\mu_X = \mu_Y$ versus H_1: $\mu_X \neq \mu_Y$.

Reject H_0 if $T < -t_{.975,59} = -2.0010$ or if $T > t_{.975,59} = 2.0010$.

$T = 58 \sqrt{60}/200 = 2.2463$, so H_0 is rejected.

The p-value is between .02 and .05.

8.10 $R = 4.5, 3, 7, -2, -6, 9, 1, 4.5, 10, 8$; $\Sigma R = 39$, $\Sigma R^2 = 384.5$.

H_0: $\mu_X = \mu_Y$ versus H_1: $\mu_X > \mu_Y$.

Reject H_0 if $T_R > t_{.975,9} = 2.2622$.

$T_R = 3.90 \sqrt{10}/5.08 = 2.4270$.

H_0 is rejected, which is in agreement with the paired t-test. The p-value is between .01 and .025, as with the paired t-test.

8.11 The new observation has no effect on the nonparametric analysis, since the ninth pair of scores already showed the largest absolute difference of all the pairs and therefore this difference receives a rank of 10 in either case.

8.13 $R = 14, -10.5, 3.5, 1.5, 10.5, 6, 18.5, -8, 18.5, 14, -3.5, 14, -18.5, 16, 10.5, -6, 6, 18.5, -10.5, 1.5$; $\Sigma R = 96$, $\Sigma R^2 = 2855$, $\overline{R} = 4.8$, $s_R = 11.2254$.

H_0: $\mu_X = \mu_Y$ versus H_1: $\mu_X > \mu_Y$.

Reject H_0 if $T_R > t_{.95,19} = 1.7291$.

$T_R = 4.8 \sqrt{20}/11.23 = 1.9123$.

H_0 is rejected, which is in agreement with the paired t-test. The p-value is between .025 and .05, as before.

8.15 D_i: $3, -2, 4, -1, 28, 17, 2, 18$

Three of the differences are quite large compared to the others and make the assumption of normality questionable; therefore, the nonparametric Wilcoxon signed-ranks test is used to analyze these data.

$R = 4, -2.5, 5, -1, 8, 6, 2.5, 7$; $\Sigma R = 29$, $\Sigma R^2 = 203.5$.

$\overline{R} = 3.625$, $s_R = 3.7488$.

H_0: $\mu_T = \mu_N$ versus H_1: $\mu_T > \mu_N$.

Reject H_0 if $T_R > t_{.95,7} = 1.8946$.

$T_R = 3.625 \sqrt{8}/3.7488 = 2.7350$.

H_0 is rejected. The p-value is between .01 and .025.

8.21 $\Sigma D = 107$, $\Sigma D^2 = 2259$, $\overline{D} = 7.133$, and $s_D = 10.336$. Standardized sample values: -1.66, -1.17, -0.88, -0.79, -0.50, -0.30, -0.30, -0.01, 0.08, 0.18, 0.37, 0.47, 0.76, 1.73, 2.02. The Lilliefors test indicates the assumption of normality to be reasonable for these data.

8.22 H_0: $\mu_N = \mu_P$ versus H_1: $\mu_N > \mu_P$.

Reject H_0 if $T > t_{.95,14} = 1.7613$.

$T = 7.13 \sqrt{15}/10.34 = 2.6729$.

H_0 is rejected. The p-value is between .005 and .01.

8.23 Since most of the points are below the diagonal, the decision seems to be justified.

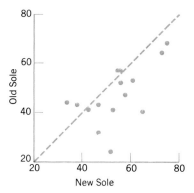

8.30 $\overline{D} = 1.3364$; $s_D = 2.0078$.

H_0: $\mu_P = \mu_A$ versus H_1: $\mu_P \neq \mu_A$.

Reject H_0 if $T < -t_{.975,13} = -2.1604$ or if $T > t_{.975,13} = 2.1604$.

$T = 1.34 \sqrt{14}/2.01 = 2.4906$.

H_0 is rejected. The p-value is between .02 and .05.

8.31 $1.34 \pm 2.1604(2.01)/\sqrt{14}$ or .18 to 2.50

8.32 $\overline{R} = 4.7857$; $s_R = 7.3084$.

H_0: $\mu_P = \mu_A$ versus H_1: $\mu_P \neq \mu_A$.

Reject H_0 if $T_R < -t_{.975,13} = -2.1604$ or if $T_R > t_{.975,13} = 2.1604$.

$T_R = 4.79 \sqrt{14}/7.31 = 2.4501$.

H_0 is rejected, which is in agreement with the paired t-test. The p-value is between .02 and .05, which is the same as with the paired t-test.

8.33 From Table A7 with $n = 14$, $S_1 = 4$, and $S_2 = 11$. The 94.26% confidence interval for the median difference is from $D^{(4)}$ to $D^{(11)}$ or -0.08 to 2.49.

9.1 $(28,000 - 22,000) \pm 1.9600 [(2100)^2/36 + (820)^2/32]^{1/2}$ or $5257 to $6743. The confidence is 95% that the true difference is contained in this interval.

9.3 H_0: $\mu_P = \mu_N$ versus H_1: $\mu_P \neq \mu_N$.

Reject H_0 if $T < -z_{.995} = -2.5758$ or if $T > z_{.995} = 2.5758$.

$$T = \frac{16.4 - 16.9}{[2.4025/40 + 2.7889/40]^{1/2}} = -1.3879$$

H_0 is not rejected and the p-value is about $2(.0823) = .1646$ from Table A2*.

9.5 H_0: $\mu_M = \mu_W$ versus H_1: $\mu_M > \mu_W$.

Reject H_0 if $T > z_{.90} = 1.2816$.

$$T = \frac{15.88 - 15.58}{[1.5625/45 + 1.4641/45]^{1/2}} = 1.1568$$

H_0 is not rejected and the p-value is about $.1230$ from Table A2*.

9.7 H_0: $\mu_M = \mu_W$ versus H_1: $\mu_M \neq \mu_W$.

Reject H_0 if $T < -z_{.975} = -1.9600$ or if $T > z_{.975} = 1.9600$.

$$T = \frac{28.3 - 26.7}{[10.89/35 + 24.01/40]^{1/2}} = 1.6760$$

H_0 is not rejected and the p-value is about $2(.0465) = .0930$ from Table A2*.

9.9 $\overline{X} = 70.78$ $s_x^2 = 310.63$ $\overline{Y} = 79.2$ $s_y^2 = 217.34$

The F-test shows the assumption of equal variances to be reasonable.

Reject H_0 if $T < -t_{.95,51} = -1.6753$.

$\overline{X} - \overline{Y} = -8.42$; $s_p^2 = 257.58$.

$$T = \frac{70.78 - 79.20}{[257.58(1/23 + 1/30)]^{1/2}} = -1.8924$$

H_0 is rejected and the p-value is between $.025$ and $.05$. A plot of the two e.d.f.s is as follows.

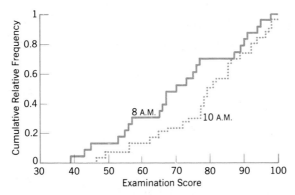

9.10 $-8.42 \pm 2.0076(16.0493)[1/23 + 1/30]^{1/2}$ or -17.35 to $.51$.

9.11 The test for equal variances shows $F = 2100^2/820^2 = 6.56$, which is compared with $F_{.975,30,30} = 2.074$, so the variances are declared to be significantly different. The degrees of freedom is

$$f = \frac{\left(\dfrac{2100^2}{36} + \dfrac{820^2}{32}\right)^2}{\dfrac{(2100^2/36)^2}{35} + \dfrac{(820^2/32)^2}{31}} = 46.5$$

Table A3 yields $t_{.975,46} = 2.0129$, which is only slightly different from the value of 1.9600 used in Exercise 9.1. The confidence interval is

$$(28,000 - 22,000) \pm 2.0129 \sqrt{\frac{2100^2}{36} + \frac{820^2}{32}} = \$5237 \text{ to } \$6763$$

only $40 wider than before. This confidence interval is more accurate if the populations are normal.

9.12 The F-test for unequal variances is not significant, so the two-sample t-test is used.

$H_0: \mu_G = \mu_E$ versus $H_1: \mu_G \neq \mu_E$.

Reject H_0 if $T < -t_{.975,8} = -2.3060$ or if $T > t_{.975,8} = 2.3060$.

$s_p^2 = [4(.09)^2 + 4(.16)^2]/8 = .01685$.

$$T = \frac{.90 - .70}{[.01685(1/5 + 1/5)]^{1/2}} = 2.4361$$

H_0 is rejected with a p-value between .02 and .05. 95% confidence interval:

$$(.90 - .70) \pm 2.3060[.01685(1/5 + 1/5)]^{1/2} = .0107 \text{ to } .3893 \text{ hr}$$

9.15 The F-test shows the assumption of equal variances is reasonable, so the two-sample t-test is used.

$H_0: \mu_X = \mu_Y$ versus $H_1: \mu_X \neq \mu_Y$.

Reject H_0 if $T < -t_{.975,38} = -2.0244$ or if $T > t_{.975,38} = 2.0244$. $s_p = 21.7830$.

$$T = \frac{355 - 375}{21.8[1/20 + 1/20]^{1/2}} = -2.9034$$

H_0 is rejected and the p-value is between .002 and .01.

9.17 (a) Location 1: $R_X = 3, 7, 17, 5, 2, 10, 8, 18, 1$; $n_X = 9$, $\Sigma R_X = 71$, $\Sigma R_X^2 = 865$, $\bar{R}_X = 7.89$, $s_{R_X} = 6.1734$.

Location 2: $R_Y = 6, 11, 13, 19, 4, 9, 20, 14.5, 12, 16, 14.5$; $n_Y = 11$, $\Sigma R_Y = 139$, $\Sigma R_Y^2 = 2004.5$, $\bar{R}_Y = 12.64$, $s_{R_Y} = 4.9804$.

$H_0: \mu_X = \mu_Y$ versus $H_1: \mu_X \neq \mu_Y$.

Reject H_0 if $T_R < -t_{.95,18} = -1.7341$ or if $T_R > t_{.95,18} = 1.7341$.

$$T_R = \frac{7.89 - 12.64}{5.54[1/9 + 1/11]^{1/2}} = -1.9057$$

H_0 is rejected and the p-value is between .05 and .10.

(b) Sample 1 $n_X = 9$, $\Sigma X = 5847$, $\Sigma X^2 = 3,873,151$, $\bar{X} = 649.7$, $s_x = 96.53$.

Standardized sample values $-1.09, -0.65, -0.60, -0.38, -0.26, -0.21, -0.13, 1.32, 2.00$. The assumption of normality is rejected based on the Lilliefors test.

Sample 2 $n_Y = 11$, $\Sigma Y = 7610$, $\Sigma Y^2 = 5,349,684$, $\bar{Y} = 691.8$, $s_y = 92.17$.

Standardized sample values $-0.91, -0.76, -0.62, -0.52, -0.46,$

-0.30, -0.26, -0.26, 0.24, 1.75, 2.10. The assumption of normality is rejected based on the Lilliefors test.

$$T = \frac{649.7 - 691.8}{94.13[1/9 + 1/11]^{1/2}} = -.9963$$

Use the same decision rule as in part (a). H_0 would not be rejected with the two-sample t-test; however, this test has no meaning, since the assumption of normality does not seem reasonable for these data.

(c) The graph of the e.d.f.s supports the WMW decision in part (a), as Sample 2 is shifted to the right of Sample 1.

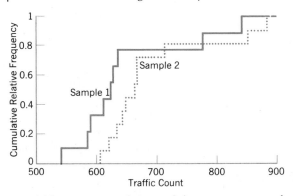

(d) This would have removed time of day as a source of variation and probably would have made a more accurate comparison, particularly since the sample sizes are so small.

9.19 H_0: $\mu_X = \mu_Y$ versus H_1: $\mu_X \neq \mu_Y$.

Reject H_0 if $T_R < -t_{.95,491} = -1.6449$ or if $T_R > t_{.95,491} = 1.6449$.

The test parallels the example in the text due to the large number of ties. The average ranks for the nine income classes are: 19, 69.5, 116.5, 158, 221.5, 294.5, 360, 412, and 464.

$$n_X = 233 \qquad \Sigma R_X = 56{,}293 \qquad \Sigma R_X^2 = 18{,}215{,}227.5$$
$$\overline{R}_X = 241.6 \qquad s_{R_X}^2 = 19{,}891.3$$

$$n_Y = 260 \qquad \Sigma R_Y = 65{,}478 \qquad \Sigma R_Y^2 = 21{,}700{,}053.5$$
$$\overline{R}_Y = 251.8 \qquad s_{R_Y}^2 = 20{,}116.5 \qquad s_p = 141.5$$

$$T_R = \frac{241.6 - 251.8}{141.30 \, [1/233 + 1/260]^{1/2}} = -.8022$$

H_0 is not rejected and the p-value is about $2(.2119) = .4238$ from Table A2*.

9.22 Reject H_0 if $T_R < -t_{.95,51} = -1.6753$.

$$\overline{R}_X = 22.54 \qquad s_{R_X}^2 = 250.45$$
$$\overline{R}_Y = 30.42 \qquad s_{R_Y}^2 = 209.38 \qquad s_p = 15.07$$

$$T_R = \frac{22.54 - 30.42}{15.07[1/23 + 1/30]^{1/2}} = -1.8851$$

H_0 is rejected and the p-value is between .025 and .05, which is in agreement with the results from the two-sample t-test.

9.25 $F = (.06)^2/(.03)^2 = 4$, which is less than $F_{.975,9,9} = 4.026$, so the null hypothesis is not rejected. The p-value is slightly larger than .05.

9.26 The F-test in Exercise 9.25 indicates the assumption of equal variances to be reasonable, so the two-sample t-test is used.

Reject H_0 if $T > t_{.99,18} = 2.5524$.

$$T = \frac{1.24 - 1.00}{.0474[1/10 + 1/10]^{1/2}} = 11.3137$$

H_0 is soundly rejected. The p-value is less than .0001.

9.27 95% confidence interval:

$$(1.24 - 1.00) \pm 2.1009(.0474)\sqrt{1/10 + 1/10} \quad \text{or} \quad .1955 \text{ to } .2845$$

9.32 Des Moines $R_X = 9, 13.5, 16, 10.5, 10.5, 17, 13.5, 15, 12$

$$n_X = 9 \qquad \Sigma R_X = 117 \qquad \Sigma R_X^2 = 1580$$

$$\overline{R}_X = 13 \qquad s_{R_X}^2 = 7.375$$

Spokane $R_Y = 2, 8, 6, 1, 3, 6, 4, 6$

$$n_Y = 8 \qquad \Sigma R_Y = 36 \qquad \Sigma R_Y^2 = 202$$

$$\overline{R}_Y = 4.5 \qquad s_{R_Y}^2 = 5.714 \qquad s_p = 2.57$$

Reject H_0 if $T_R > t_{.95,15} = 1.7531$.

$$T_R = \frac{13 - 4.5}{2.57[1/9 + 1/8]^{1/2}} = 6.8091$$

H_0 is rejected with a p-value less than .0001.

9.33 The separation of the graphs of the e.d.f.s would seem to justify the decision in Exercise 9.32.

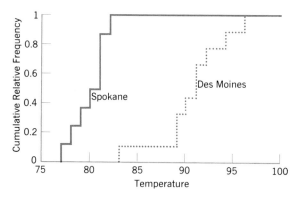

9.35 H_0: $\mu_A = \mu_N$ versus H_1: $\mu_A > \mu_N$.

The nonparametric procedure (Wilcoxon–Mann–Whitney rank-sum test) requires independent random samples taken from populations with identical distribution functions except for possibly different means.

Reject H_0 if $T_R > t_{.95,18} = 1.7341.$

$$\bar{R}_X = 14.6 \qquad s^2_{\bar{R}_X} = 16.88 \qquad \bar{R}_Y = 6.4 \qquad s^2_{\bar{R}_Y} = 19.49$$

$$s_p = 4.264$$

$$T_R = \frac{14.6 - 6.4}{4.264[1/10 + 1/10]^{1/2}} = 4.2999$$

H_0 is rejected with a p-value between .0001 and .0005. The graph of the e.d.f.s shows this decision to be justified.

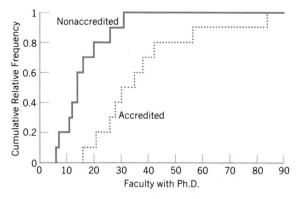

9.36 $\bar{X} = 37.5 \qquad s^2_X = 383.61 \qquad \bar{Y} = 15.7 \qquad s^2_Y = 63.34$

$F = 383.61/63.34 = 6.056$, which is greater than $F_{.975,9,9} = 4.026$, so the assumption of equal variances is not justified, and Satterthwaite's test is used. The degrees of freedom is

$$f = \frac{\left(\dfrac{383.61}{10} + \dfrac{63.34}{10}\right)^2}{\dfrac{(383.61/10)^2}{9} + \dfrac{(63.34/10)^2}{9}} = 11.9$$

The decision rule is to reject H_0 if $T > t_{.95,11} = 1.7959.$

$$T = \frac{37.5 - 15.7}{[383.61/10 + 63.34/10]^{1/2}} = 3.2608$$

Therefore H_0 is rejected, as before, and the p-value is between .001 and .005.

10.1

	Men	Women	
Union	48	22	70
Nonunion	15	15	30
	63	37	100

Reject H_0 if $\phi \sqrt{100} > z_{.95} = 1.6449.$

$$\phi \sqrt{100} = \frac{48 \cdot 15 - 15 \cdot 22}{\sqrt{63 \cdot 37 \cdot 30 \cdot 70}} \sqrt{100} = 1.763$$

H_0 is rejected and the p-value is about .0392 from Table A2*.

10.3 Reject H_0 if $T > \chi^2_{.95,1} = 3.841$.

$$T = \frac{200(5 \cdot 85 - 95 \cdot 15)^2}{100 \cdot 100 \cdot 20 \cdot 180} = 5.556$$

H_0 is rejected and the p-value is between .01 and .025.

10.7

	Direct Mail	Handbills	
Redeemed	43	28	71
Not Redeemed	207	172	379
	250	200	450

Reject H_0 if $T > \chi^2_{.95,1} = 3.841$.

$$T = \frac{450(43 \cdot 172 - 207 \cdot 28)^2}{250 \cdot 200 \cdot 379 \cdot 71} = .856$$

H_0 is not rejected and the p-value is between .25 and .40.

10.11 Reject H_0 if $T > \chi^2_{.95,4} = 9.488$.

$$R_i = 100, 48, 52 \qquad C_j = 124, 60, 16 \qquad n = 200$$

$$E_{ij} = 62, 30, 8; \; 29.8, 14.4, 3.8; \; 32.2, 15.6, 4.2$$

$$T = \Sigma \frac{O_{ij}^2}{E_{ij}} - n = 242.05 - 200 = 42.05$$

H_0 is soundly rejected and the p-value is less than .0001.

10.12 $\Phi = \dfrac{42.05}{200(2)} = .105$

10.15 Reject H_0 if $T > \chi^2_{.99,2} = 9.210$.

$$R_i = 167, 114, 25 \qquad C_j = 218, 88 \qquad n = 306$$

$$E_{ij} = 119, 48; \; 81.2, 32.8; \; 17.8, 7.2$$

$$T = \Sigma \frac{O_{ij}^2}{E_{ij}} - n = 308.28 - 306 = 2.28$$

H_0 is not rejected and the p-value is between .25 and .40.

10.16 $\Phi = \dfrac{2.28}{306} = .0075$

10.19 Reject H_0 if $T > \chi^2_{.99,8} = 20.09$.

$E_{ij} = 608.7, 688.9, 657.2, 687.9, 581.3; \; 694.2, 785.7, 749.6,$
784.6, 662.9; 540.1, 611.4, 583.2, 610.5, 515.8

$$T = \Sigma \frac{O_{ij}^2}{E_{ij}} - n = 9936.82 - 9762 = 174.82$$

H_0 is soundly rejected and the p-value is less than .0001.

10.21 Reject H_0 if $T > \chi^2_{.95,2} = 5.991$.

$$T = \frac{(37 - 46)^2}{46} + \frac{(47 - 70)^2}{70} + \frac{(116 - 84)^2}{84} = 21.51$$

H_0 is rejected with a p-value less than .0001. The accountant should conclude that more purchases are made on credit during the Christmas season than during the rest of the year.

10.23 Reject H_0 if $T > \chi^2_{.95,6} = 12.59$.

$np_i = 998.8, 549.34, 549.34, 799.04, 549.34, 549.34, 998.8$

$$T = \Sigma \frac{O_i^2}{np_i} - n = 5052.11 - 4994 = 58.11$$

H_0 is soundly rejected and the p-value is less than .0001.

10.25 Reject H_0 if $T > \chi^2_{.90,12} = 18.55$.

$np_i = 41.4, 47.4, 32.7, 23.7, 15.0, 50.4, 20.7, 12.0, 9.0,$
$\quad\quad 17.7, 12.0, 15.0, 3.0$

$$T = \Sigma \frac{O_i^2}{np_i} - n = 323.55 - 300 = 23.55$$

H_0 is rejected and the p-value is between .01 and .025.

10.27 $\hat{p} = \dfrac{\text{number of questions missed}}{\text{total number of questions}}$

$= \dfrac{2(0) + 8(1) + 11(2) + 12(3) + 11(4) + 0}{10(44)} = .25$

$P(X = x)$ is obtained from Table A1 with parameters 10 and .25.

Number Missed (x)	$P(X = x)$	O_i	$E_i = np_i$	O_i^2/E_i
0	.0563	2	2.48	1.61
1	.1877	8	8.26	7.75
2	.2816	11	12.39	9.77
3	.2503	12	11.01	13.08
4	.1460	11	6.42	18.84
5 or more	.0781	0	3.44	0.00
		44	44.00	51.04

Reject H_0 if $T > \chi^2_{.95,4} = 9.488$.

$$T = \Sigma \frac{O_i^2}{np_i} - n = 51.04 - 44 = 7.04$$

H_0 is not rejected and the p-value is between .10 and .25.

10.30 Reject H_0 if $T > \chi^2_{.99,7} = 18.48$.
Note that λ was given, not estimated, so there are 7 d.f. The respective Poisson probabilities are .1353, .2707, .2707, .1804, .0902, .0361, .0120, and .0046.

$$T = \Sigma \frac{O_i^2}{np_i} - n = 104.13 - 100 = 4.13$$

H_0 is not rejected and the p-value is greater than .4.

10.32 Reject H_0 if $T > \chi^2_{.95,6} = 12.59$.

$$E_{ij} = 12.3, 10.5, 16.5, 19.7; 18.4, 15.6, 24.6, 29.3;$$
$$16.3, 13.9, 21.8, 26.0$$

$$T = \Sigma \frac{O_{ij}^2}{E_{ij}} - n = 227.70 - 225 = 2.70$$

H_0 is not rejected and the p-value is greater than .40.

10.34

	Videotape	Reading	
Pass	35	43	78
Fail	13	9	22
	48	52	100

Reject the hypothesis of independence of training technique and performance on the exam if $T > \chi^2_{.95,1} = 3.841$.

$$T = \frac{100(35 \cdot 9 - 13 \cdot 43)^2}{48 \cdot 52 \cdot 22 \cdot 78} = 1.390$$

H_0 is not rejected and the p-value is between .10 and .25.

10.36

	A	B	C	D	
Too light	6	3	8	2	19
Not too light	34	47	32	43	156
	40	50	40	45	175

$E_{ij} = 4.3, 5.4, 4.3, 4.9; 35.7, 44.6, 35.7, 40.1$

Reject H_0 if $T > \chi^2_{.95,3} = 7.815$.

$$T = \Sigma \frac{O_{ij}^2}{E_{ij}} - n = 182.295 - 175 = 7.295$$

H_0 is not rejected and the p-value is between .05 and .10.

11.1 $r = .7583$.

11.2 Reject H_0 if $T > t_{.99,18} = 2.5524$ from Table A3. Since $T = 4.9351$, H_0 is rejected with a p-value of $< .0001$.

11.3 From Figure 11.6 for $r = .76$, $w = .9962$.

$$w_L = .9962 - \frac{1.9600}{\sqrt{17}} = .5208$$

$$w_U = .9962 + \frac{1.9600}{\sqrt{17}} = 1.4716$$

From Figure 11.6 w_L converts to $r_L = .48$ and w_U converts to $r_U = .90$. Therefore the 95% confidence interval for ρ is from .48 to .90.

11.7 $\Sigma X = 1200 \qquad \Sigma X^2 = 77,020 \qquad \Sigma Y = 149$

$\Sigma Y^2 = 1569 \qquad \Sigma XY = 8805$

$r = -.0889$ or $-.09$

so $w = -.0902$ from Figure 11.6 for $r = .09$, $w = .0902$

$$w_L = -.0902 - \frac{1.9600}{\sqrt{17}} = -.5656$$

$$w_U = -.0902 + \frac{1.9600}{\sqrt{17}} = .3852$$

Since w_L is negative, the value of r_L is found by changing the sign in Figure 11.6 for $w = .5656$ to give $r_L = -.51$. The value of r_U is read directly from Figure 11.6 as .37. The 95% confidence interval for ρ is an interval from $-.51$ to .37. There is 95% confidence that this interval contains the true value of the population correlation coefficient, and since this interval contains the value of zero it is doubtful that typing speed and the number of errors are correlated.

11.11 $\Sigma(R_X - R_Y)^2 = 68$, therefore from Eq. 11.6

$$r_R = 1 - \frac{6(68)}{8(63)} = .1905$$

Since this value of r_R would not be significant at any reasonable value of α given in Figure 11.9, there does not appear to be any advantage.

11.13 Reject H_0 if $|T_R| > t_{.975,8} = 2.3060$ from Table A3. Since $r_R = .7217$ and $T_R = 2.9491$, H_0 is rejected and the p-value is between .01 and .02.

11.17 Reject H_0 if $r_R > .619$ from Figure 11.9. $\Sigma(R_X - R_Y)^2 = 48$ and $r_R = .429$. H_0 is not rejected and the p-value is between .10 and .25.

11.19 Reject H_0 if $r_R > .771$ from Figure 11.9.

$$\Sigma(R_X - R_Y)^2 = 8 \qquad r_R = .771$$

H_0 is not rejected, with a p-value slightly greater than .05, due to the discrete nature of the test statistic.

11.20 For the AL East $r_R = .321$. From Figure 11.9 with $n = 7$ the .95 quantile is .679. Therefore there is no significant agreement. The p-value is slightly larger than .25, due to the discrete nature of r_R.

For the AL West there is one tie in the rankings requiring the use of Eq. 11.4 to calculate $r_R = -.3784$. $T_R = -.9141$ is compared with $t_{.95,5} = 2.0150$ from Table A3. Since T_R is not greater than 2.0150, there is no significant agreement and the p-value is between .75 and .90.

For the NL East $r_R = .600$. From Figure 11.9 with $n = 6$ the .95 quantile is .771. Therefore there is no significant agreement. The p-value is slightly larger than .10, due to the discrete nature of r_R.

For the NL West $r_R = -.143$. From Figure 11.9 with $n = 6$ the .95 quantile is .771. The p-value is much greater than .25. Therefore there is no significant agreement.

11.21 Reject H_0 if $T > t_{.95,8} = 1.8595$ from Table A3. Since $r = .9471$ and $T = 8.3466$, H_0 is rejected with a p-value $< .0001$.

11.22 $\Sigma R_X = 55$ $\qquad \Sigma R_X^2 = 384$ $\qquad \Sigma R_Y = 55$
$\Sigma R_Y^2 = 385$ $\qquad \Sigma R_X R_Y = 377.5$

Reject H_0 if $T_R > t_{.95,8} = 1.8595$ from Table A3. Since $r_R = .9147$ and $T_R = 6.3997$, H_0 is rejected with a p-value of approximately .0001.

11.25 Reject H_0 if $r_R > .552$ from Figure 11.9. $\Sigma(R_X - R_Y)^2 = 26$ and $r_R = .842$. H_0 is soundly rejected with a p-value between .005 and .001.

11.26 $\Sigma R_X = 210$ $\Sigma R_X^2 = 2869.5$ $\Sigma R_Y = 210$
$\Sigma R_Y^2 = 2863.5$ $\Sigma R_X R_Y = 2081$

Reject H_0 if $|T_R| > t_{.975,18} = 2.1009$ from Table A3. Since $r_R = -.1875$ and $T_R = -.8097$, H_0 is not rejected and the p-value is between .20 and .50.

11.31 $r = -.0354$. With a value of r this close to zero, there does not appear to be a relationship between the order in which the exam was handed in and the score.

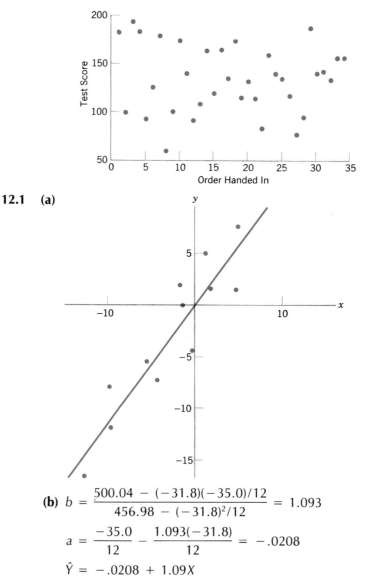

12.1 **(a)**

(b) $b = \dfrac{500.04 - (-31.8)(-35.0)/12}{456.98 - (-31.8)^2/12} = 1.093$

$a = \dfrac{-35.0}{12} - \dfrac{1.093(-31.8)}{12} = -.0208$

$\hat{Y} = -.0208 + 1.09X$

(c) See graph in part (a)

(d) This stock price tends to fluctuate more than the S&P index.

12.3 Total sum of squares = 64.94.

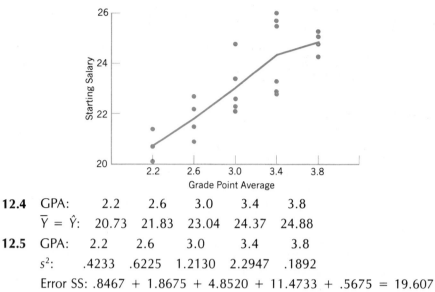

12.4 GPA: 2.2   2.6   3.0   3.4   3.8

$\overline{Y} = \hat{Y}$:   20.73   21.83   23.04   24.37   24.88

12.5 GPA: 2.2   2.6   3.0   3.4   3.8

s^2:   .4233   .6225   1.2130   2.2947   .1892

Error SS: .8467 + 1.8675 + 4.8520 + 11.4733 + .5675 = 19.607

12.6 Total SS = 64.94.

$r^2 = 1 - 19.607/64.94 = .698$, which implies that 69.8% of the variation in starting salaries is explained by the regression curve using GPA.

12.7 $\Sigma X = 67.6$; $\Sigma Y = 510.4$; $\Sigma XY = 1584.6$; $\Sigma X^2 = 213.68$; $\Sigma Y^2 = 11{,}906.22$; $r^2 = .684$; $\hat{\mu}_{Y|x} = 14.81 + 2.730X$

12.8 The values are not the same because two different methods of regression have been used.

12.9 $\hat{\mu}_{Y|x} = 14.81 + 2.730(3.45) = 24.2$ thousand dollars.

12.17 **(a)** The relationship does not appear to be linear.

(b) The variances do not appear to be equal.

12.19 **(a)** Residuals: -66.68, -72.33, 50.64, -38.87, 127.24.

(b) $\hat{\sigma}^2 = 29{,}943.43/3 = 9981.14$ from Eq. 12.8.

(c) $\hat{\sigma}^2 = (4/3)[226{,}334 - (30.9612)^2(228.3)] = 9982.07$, which is the same, except for round-off error.

12.21
$$\hat{\sigma} = \left(\frac{\Sigma e_i^2}{n-2}\right)^{1/2} = \left(\frac{80.7983}{10}\right)^{1/2} = 2.8425$$

Standardized residuals $(Y_i - \hat{Y}_i)/\hat{\sigma}$: -1.43, -1.39, -0.80, -0.77, -0.31, -0.12, 0.22, 0.54, 0.61, 1.06, 1.13, 1.26. These residuals pass the Lilliefors test for normality.

12.23 The variance associated with the residuals for the smallest six values of X is $s_1^2 = 6.492$, and the variance associated with the residuals for the largest six values of X is $s_2^2 = 9.437$. The F ratio is $9.437/6.492 = 1.454$. This value is compared with the .975 quantile of an F distribution with 5 and 5 degrees of freedom, which is $F_{.975,5,5} = 7.146$, so the assumption of homogeneity of variance seems reasonable for these regression data. The p-value is greater than .50.

12.29 Reject H_0 if $T > t_{.95,10} = 1.8125$. From the calculations in Exercise 12.1, $s_x = 5.8209$, $\hat{\sigma} = 2.8425$.

$$T = \frac{(1.093 - 1)(5.8209)\sqrt{11}}{2.8425} = .6301$$

H_0 is not rejected and the p-value is between .25 and .40.

12.30 From the calculations in Exercise 12.1, $s_x = 5.8209$, $\hat{\sigma} = 2.8425$.

$$1.093 \pm \frac{2.2281(2.8425)}{5.8209\sqrt{11}} \qquad \text{or} \qquad .765 \quad \text{to} \quad 1.421$$

12.31 From the calculations in Exercise 12.1

$$\hat{Y} = -.021 + 1.093(5) = 5.44$$

$$\mu_L = 5.44 - 2.2281(2.8425)\sqrt{\frac{1}{12} + \frac{[5 - (-2.650)]^2}{11(33.8827)}}$$

$$= 2.34$$

$$\mu_U = 5.44 + 2.2281(2.8425)\sqrt{\frac{1}{12} + \frac{[5 - (-2.650)]^2}{11(33.8827)}}$$

$$= 8.55$$

12.32 From the calculations in Exercise 12.1

$$\hat{Y} = -.021 + 1.093(2) = 2.165$$

$$Y_L = 2.165 - 2.2281(2.8425)\sqrt{1 + \frac{1}{12} + \frac{[2 - (-2.650)]^2}{11(33.8827)}}$$

$$= -4.60$$

$$Y_U = 2.165 + 2.2281(2.8425)\sqrt{1 + \frac{1}{12} + \frac{[2 - (-2.650)]^2}{11(33.8827)}}$$

$$= 8.93$$

12.37 H_0: $\beta = 25$ versus H_1: $\beta \neq 25$.

Reject H_0 if $|T| > t_{.975,3} = 3.1824$. From the calculations in Exercise 12.19, $s_x = 15.1096$, $\hat{\sigma} = 99.9057$.

$$T = \frac{(30.9612 - 25)(15.1096)\sqrt{4}}{99.9057} = 1.8031$$

H_0 is not rejected and the p-value is between .10 and .20.

12.38 From the calculations in Exercise 12.19, $s_x = 15.1096$, $\hat{\sigma} = 99.9057$.

$$30.9612 \pm \frac{3.1824(99.9057)}{15.1096\sqrt{4}} \qquad \text{or} \qquad 20.44 \quad \text{to} \quad 41.48$$

12.39 From the calculations in Exercise 12.19

$$\hat{Y} = 1047.35 + 30.9612(50) = \$2595$$

$$Y_L = 2595 - 3.1824(99.9057) \sqrt{1 + \frac{1}{5} + \frac{(50 - 55.4)^2}{4(228.30)}} = \$2243$$

$$Y_U = 2595 + 3.1824(99.9057) \sqrt{1 + \frac{1}{5} + \frac{(50 - 55.4)^2}{4(228.30)}} = \$2948$$

12.40 From Exercise 12.39 $\hat{Y} = \$2595$

$$\mu_L = 2595 - 3.1824(99.9057) \sqrt{\frac{1}{5} + \frac{(50 - 55.4)^2}{4(228.30)}} = \$2442$$

$$\mu_U = 2595 + 3.1824(99.9057) \sqrt{\frac{1}{5} + \frac{(50 - 55.4)^2}{4(228.30)}} = \$2749$$

12.41 The .95 quantile from Figure 11.9 with $n = 5$ is .800. The decision rule is reject H_0 if $r_R < -.800$ or if $r_R > .800$. The hypothesized residuals $(Y_i - 25X_i)$ are as follows:

$$1237, 1285, 1557, 1390, 1419$$

Rank of X:	2	3	5	4	1
Rank of residual:	1	2	5	3	4

$r_R = .400$, H_0 is not rejected, and the p-value is slightly larger than .50.

12.42 The .95 quantile from Figure 11.9 with $n = 12$ is .497. The decision rule is reject H_0 if $r_R > .497$. The hypothesized residuals $(Y_i - 1X_i)$ are as follows:

$$-2.7, -1.8, 0.1, -4.1, -3.5, -0.2, 3.3, 1.4, 2.2, 3.4, 2.1, -3.4$$

Rank of X:	5	3	4	8	11	10	9	7	12	6	2	1
Rank of residual:	4	5	7	1	2	6	11	8	10	12	9	3

$r_R = .098$, H_0 is not rejected, and the p-value is greater than .25. This result is in agreement with the result obtained in Exercise 12.29.

12.45 The assumption of linear regression does not appear to be true for these data. A nonlinear model would be preferred.

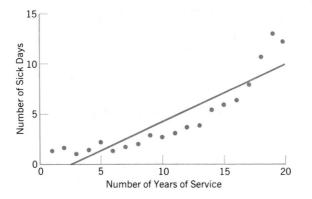

12.46 $\Sigma X = 210$; $\Sigma Y = 92$; $\Sigma XY = 1341.8$; $\Sigma X^2 = 2870$;
$\Sigma Y^2 = 685.56$; $\hat{\mu}_{Y|x} = -1.334 + .565X$.

12.47 The scatterplot for ranks appears much more linear in nature than the scatterplot for the original data.

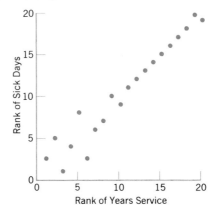

12.48 $\Sigma R_x = 210$; $\Sigma R_y = 210$; $\Sigma R_x R_y = 2848.5$; $\Sigma R_x^2 = 2870$;
$\Sigma R_y^2 = 2869.5$; $\hat{R}_y = .339 + .968R_x$.

12.49 From Eq. 12.24

$$R_{x_0} = 16.5 \qquad \text{and} \qquad \hat{R}_y = .339 + .968(16.5) = 16.31$$

From Eq. 12.25,

$$\hat{Y}_0 = 6.5 + \left(\frac{16.31 - 16}{17 - 16} \right) (8.0 - 6.5) = 6.96$$

12.50 For the original data, combining Eqs. 12.8 and 12.9 gives Error SS = $\Sigma(Y_i - \hat{Y}_i)^2 = (n - 1)(s_Y^2 - b^2 s_X^2) = 49.99$. For ranks Error SS = $\Sigma(Y_i - \hat{Y}_i)^2 = 3.42$ from using Eq. 12.25 to get

$$\hat{Y}_i = 1.16, 1.35, 1.45, 1.54, 1.72, 1.84, 2.12, 2.34, 2.81,$$
$$3.00, 3.20, 3.77, 3.98, 5.33, 5.93, 6.41, 7.68,$$
$$10.05, 11.79, 12.75$$

in order of increasing X values.

12.56 $\hat{\mu}_{Y|x} = -266.53 + 6.14X$

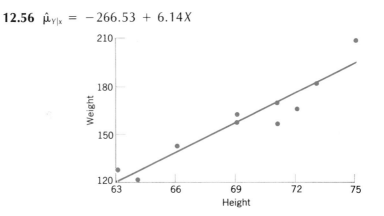

12.57 $\hat{\mu}_{Y|x} = -266.53 + 6.14(72) = 175.4$

$$\hat{\sigma} = \sqrt{\frac{9}{8}[644.40 - (6.14)^2(15.34)]} = 8.641$$

$$\mu_L = 175.4 - t_{.975,8}(8.641)\sqrt{\frac{1}{10} + \frac{(72 - 69.3)^2}{9(15.34)}} = 167.6$$

$$\mu_U = 175.4 + t_{.975,8}(8.641)\sqrt{\frac{1}{10} + \frac{(72 - 69.3)^2}{9(15.34)}} = 183.2$$

The 95% confidence interval is from 167.6 to 183.2.

13.1

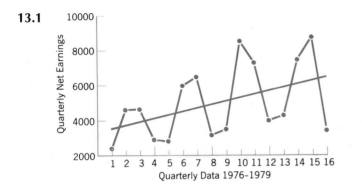

13.2 $\Sigma X = 136$; $\Sigma Y = 80,888$; $\Sigma XY = 753,812$; $\Sigma X^2 = 1496$; $\Sigma Y^2 = 475,245,210$; $\hat{Y} = 3398.9 + 194.89X$

13.7 The residuals are as follows:

Year	March 31	June 30	Sept. 30	Dec. 31
1976	−1129.8	879.3	708.4	−1223.5
1977	−1503.4	1411.7	1773.8	−1736.1
1978	−1600.9	3221.2	1788.3	−1740.6
1979	−1610.5	1369.6	2445.7	−3053.2

$r_1 = -.117$ $r_2 = -.907$ $r_3 = .059$ $r_4 = .887$

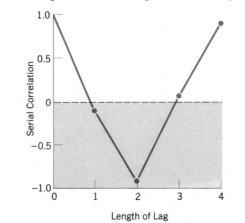

13.8   The ratio to trend is:

1976:	.687	1.233	1.179	.707
1977:	.656	1.308	1.371	.649
1978:	.688	1.600	1.320	.695
1979:	.727	1.237	1.383	.530

The four averages are .690, 1.342, 1.315, and .646, which lead to $S_1 = .691$, $S_2 = 1.344$, $S_3 = 1.317$, $S_4 = .647$.

13.9

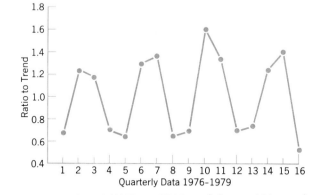

13.10 $\hat{Y} = 3398.9 + 194.89(17) = 6712$ and $S_1 = .691$, so the prediction is $6712(.691) = \$4638$, which is slightly below the actual value of $4790.

13.19 The $C_i \times I_i$ numbers are .992, .917, .894, 1.092, .949, .974, 1.042, 1.004, .997, 1.192, 1.004, 1.076, 1.054, .910, 1.053, .821.

13.20 The C_i numbers are .934, .968, .979, 1.005, .988, 1.006, 1.014, 1.064, 1.065, 1.091, 1.045, 1.013, 1.006, .928.

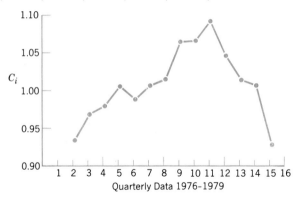

13.21 The I_i numbers are .981, .924, 1.116, .944, .985, 1.035, .990, .937, 1.120, .921, 1.030, 1.040, .905, 1.134.

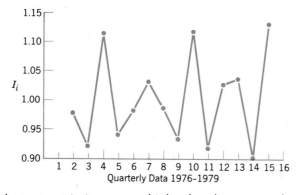

Quarterly Data 1976–1979

13.28 Sales bottom out in January and July of each year, reach a minor peak in April and May, and reach a maximum during the end of the year.

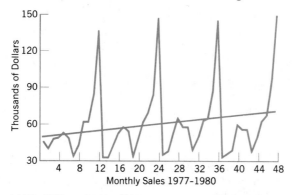

Monthly Sales 1977–1980

13.29 $\Sigma X = 1176$; $\Sigma Y = 2903$; $\Sigma XY = 75,313$; $\Sigma X^2 = 38,024$; $\Sigma Y^2 = 214,545$; $\hat{Y} = 49.34 + .455X$.

13.30 Jan: .649, Feb: .637

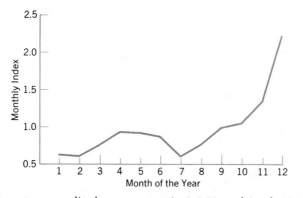

Month of the Year

13.31 The long-term cyclical component is 1.141 and is obtained by averaging the first nine values of $Y/(TS)$, which are 1.423, 1.249, 1.208, 1.001, 1.083, 1.035, 1.079, 1.055, and 1.138. The graph shows a decreasing cycle over the four-year period. The irregular component for May 1977 is

$$I = \frac{Y}{TSC} = \frac{53}{51.6(.949)(1.141)} = .949$$

The graph shows no apparent pattern to the irregular component.

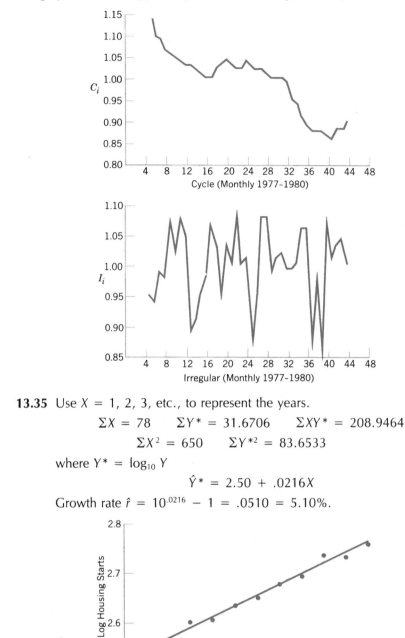

Cycle (Monthly 1977-1980)

Irregular (Monthly 1977-1980)

13.35 Use $X = 1, 2, 3$, etc., to represent the years.

$$\Sigma X = 78 \qquad \Sigma Y^* = 31.6706 \qquad \Sigma XY^* = 208.9464$$
$$\Sigma X^2 = 650 \qquad \Sigma Y^{*2} = 83.6533$$

where $Y^* = \log_{10} Y$

$$\hat{Y}^* = 2.50 + .0216X$$

Growth rate $\hat{r} = 10^{.0216} - 1 = .0510 = 5.10\%$.

Yearly Data 1970-1981

14.1 (a) $y = \beta_0 + \beta_1 x_1 + \beta_2 x_2 + \beta_3 x_3 + \beta_4 x_4 + \beta_5 x_5 + \beta_6 x_6$

(b) 7

(c) Least squares.

(d) .708

(e) $R^2 = 1236.38/1745.75 = .708$; 70.8% of the variation in Y is explained by the six independent variables.

14.2 14.4 min.

14.3 Models: (1) 10.0; (2) 22.0; (3) 26.7; (4) 29.6; (5) 26.9. The individual predictions vary greatly with Model 1 being the closest to the prediction in Exercise 14.2. This result is not surprising since Model 1 has the highest R^2 value in Figure 14.3. However, different inputs could produce different orderings.

14.4 **(a)** Decreases by 3.74 min.

(b) Increases by 1.09 min.

(c) Decreases by 0.50 min.

(d) Increases by 7.86 min.

(e) Decreases by 3.77 min.

14.12 $y = \beta_0 + \beta_1 x_1 + \beta_2 x_2 + \beta_3 D$, where $D = 0$ if parking is open and $D = 1$ if parking is covered.

14.13 $y = \beta_0 + \beta_1 x_2 + \beta_2 D_1 + \beta_3 D_2$

where

$$D_1 = 0, D_2 = 0 \text{ when the house has 1 bedroom}$$

$$D_1 = 1, D_2 = 0 \text{ when the house has 2 bedrooms}$$

$$D_1 = 0, D_2 = 1 \text{ when the house has 3 bedrooms}$$

14.14 $\hat{Y} = 3241.781 + 43.709X_2 + 1043.905D_1 - 1464.868D_2$

14.15 **One bedroom** $\hat{Y} = 3241.781 + 43.709X_2$.

Two bedrooms $\hat{Y} = 4285.686 + 43.709X_2$.

Three bedrooms $\hat{Y} = 1776.913 + 43.709X_2$.

Residuals 2294, 4514, 7825, −11624, −6056, −2803, 5360, −6329.

14.16 $y = \beta_0 + \beta_1 x_2 + \beta_2 D_1 + \beta_3 D_2 + \beta_4 D_1 x_2 + \beta_5 D_2 x_2$, with D_1 and D_2 defined as in Exercise 14.13.

14.17 $\hat{Y} = 35{,}133.05 + 2236.85t + 13{,}027.40D_1 + 16{,}357.05D_2 - 1785.55D_3$

Residuals: 3864, −2409, −3739, −1000.

14.19 $\hat{Y} = 3377.239 + 43.287X_2 + 1234.551D_1 - 1231.349D_2 + 332.545D_3$

$$D_1 = 0, D_2 = 0, D_3 = 0 \quad \hat{Y} = 3377.239 + 43.287X_2$$
$$D_1 = 1, D_2 = 0, D_3 = 0 \quad \hat{Y} = 4611.790 + 43.287X_2$$
$$D_1 = 0, D_2 = 1, D_3 = 0 \quad \hat{Y} = 2145.890 + 43.287X_2$$
$$D_1 = 0, D_2 = 0, D_3 = 1 \quad \hat{Y} = 3709.784 + 43.287X_2$$
$$D_1 = 1, D_2 = 0, D_3 = 1 \quad \hat{Y} = 4944.335 + 43.287X_2$$
$$D_1 = 0, D_2 = 1, D_3 = 1 \quad \hat{Y} = 2478.435 + 43.287X_2$$

14.20 $\hat{Y} = -10{,}723.481 + 58.925X_2 - 9341.631D_1 + 17{,}146.730D_2 + 11{,}760.776D_3 + 8.476D_1 X_2 - 18.657D_2 X_2 - 12.158D_3 X_2$

$$D_1 = 0, D_2 = 0, D_3 = 0 \quad \hat{Y} = -10{,}723.481 + 58.925X_2$$
$$D_1 = 1, D_2 = 0, D_3 = 0 \quad \hat{Y} = -20{,}065.112 + 67.401X_2$$
$$D_1 = 0, D_2 = 1, D_3 = 0 \quad \hat{Y} = 6423.249 + 40.268X_2$$

$$D_1 = 0, D_2 = 0, D_3 = 1 \qquad \hat{Y} = \qquad 1037.295 + 46.767X_2$$
$$D_1 = 1, D_2 = 0, D_3 = 1 \qquad \hat{Y} = \qquad -8304.336 + 55.243X_2$$
$$D_1 = 0, D_2 = 1, D_3 = 1 \qquad \hat{Y} = \qquad 18,184.025 + 28.110X_2$$

Whereas, each of the six equations in Exercise 14.19 had the same slope but different intercepts, these six equations all have both different slopes and intercepts

14.21 A second bedroom adds $1234.55 to the price, not counting the added square footage at $43.29/sq. ft, while covered parking adds $332.55. Both of these net effects need to be compared with the costs involved to see if they seem reasonable.

14.22 A second bedroom appears to lower the price by $9341.63, until the extra $8.48/sq. ft of the total area is added in. Covered parking by itself appears to add $11,760.78 to the sale price, until the $12.16/sq. ft is subtracted. An adjustment for both the addition of a second bedroom and covered parking makes more sense in Exercise 14.20 where they are both tied to the square footage of the condominium.

14.30 $\hat{Y} = -1719.288 + 49.392X_2 - 8254.317D_1$
$$+ 17,138.463D_2 + 7.659D_1X_2 - 18.085D_2X_2$$

When $D_1 = 0$ and $D_2 = 0$, the model reduces to $y = \beta_0 + \beta_1 x_2$, where β_1 is the amount by which the sales price will increase for each additional square foot added to the condominium. In the full model β_2 is the amount by which the sale price of a two bedroom will change from a one bedroom (ignoring square footage), β_3 is the amount by which the sales price of a three bedroom will change from a one bedroom (again ignoring square footage), β_4 represents the amount by which the cost per square foot β_1 will change for a two bedroom over a one bedroom, while β_5 has the same meaning only for a three bedroom.

14.31 $y = \beta_0 + \beta_1 x_1 + \beta_2 x_2 + \beta_3 D_1 + \beta_4 D_2 + \beta_5 D_3$

Six situations:
$$D_1 = 0, D_2 = 0, D_3 = 0 \qquad y = \beta_0 + \beta_1 x_1 + \beta_2 x_2$$
$$D_1 = 1, D_2 = 0, D_3 = 0 \qquad y = (\beta_0 + \beta_3) + \beta_1 x_1 + \beta_2 x_2$$
$$D_1 = 0, D_2 = 1, D_3 = 0 \qquad y = (\beta_0 + \beta_4) + \beta_1 x_1 + \beta_2 x_2$$
$$D_1 = 0, D_2 = 0, D_3 = 1 \qquad y = (\beta_0 + \beta_5) + \beta_1 x_1 + \beta_2 x_2$$
$$D_1 = 1, D_2 = 0, D_3 = 1 \qquad y = (\beta_0 + \beta_3 + \beta_5) + \beta_1 x_1 + \beta_2 x_2$$
$$D_1 = 0, D_2 = 1, D_3 = 1 \qquad y = (\beta_0 + \beta_4 + \beta_5) + \beta_1 x_1 + \beta_2 x_2$$

14.32 $y = \beta_0 + \beta_1 x_1 + \beta_2 x_2 + \beta_3 D_1 + \beta_4 D_2 + \beta_5 D_3 + \beta_6 D_1 x_1$
$$+ \beta_7 D_1 x_2 + \beta_8 D_2 x_1 + \beta_9 D_2 x_2 + \beta_{10} D_3 x_1 + \beta_{11} D_3 x_2$$

Individual models:
$$y = \beta_0 + \beta_1 x_1 + \beta_2 x_2$$
$$y = (\beta_0 + \beta_3) + (\beta_1 + \beta_6)x_1 + (\beta_2 + \beta_7)x_2$$
$$y = (\beta_0 + \beta_4) + (\beta_1 + \beta_8)x_1 + (\beta_2 + \beta_9)x_2$$
$$y = (\beta_0 + \beta_5) + (\beta_1 + \beta_{10})x_1 + (\beta_2 + \beta_{11})x_2$$
$$y = (\beta_0 + \beta_3 + \beta_5) + (\beta_1 + \beta_6 + \beta_{10})x_1 + (\beta_2 + \beta_7 + \beta_{11})x_2$$
$$y = (\beta_0 + \beta_4 + \beta_5) + (\beta_1 + \beta_8 + \beta_{10})x_1 + (\beta_2 + \beta_9 + \beta_{11})x_2$$

14.33 $y = \beta_0 + \beta_1 x_1 + \beta_2 x_2 + \beta_3 D_1 + \beta_4 D_2 + \beta_5 D_3 + \beta_6 D_1 x_1$
$+ \beta_7 D_1 x_2 + \beta_8 D_2 x_1 + \beta_9 D_2 x_2 + \beta_{10} D_3 x_1$
$+ \beta_{11} D_3 x_2 + \beta_{12} D_1 D_3 + \beta_{13} D_2 D_3 + \beta_{14} D_1 D_3 x_1$
$+ \beta_{15} D_1 D_3 x_2 + \beta_{16} D_2 D_3 x_1 + \beta_{17} D_2 D_3 x_2$

Individual models:

$y = \beta_0 + \beta_1 x_1 + \beta_2 x_2$
$y = (\beta_0 + \beta_3) + (\beta_1 + \beta_6)x_1 + (\beta_2 + \beta_7)x_2$
$y = (\beta_0 + \beta_4) + (\beta_1 + \beta_8)x_1 + (\beta_2 + \beta_9)x_2$
$y = (\beta_0 + \beta_5) + (\beta_1 + \beta_{10})x_1 + (\beta_2 + \beta_{11})x_2$
$y = (\beta_0 + \beta_3 + \beta_5 + \beta_{12}) + (\beta_1 + \beta_6 + \beta_{10} + \beta_{14})x_1$
$\quad + (\beta_2 + \beta_7 + \beta_{11} + \beta_{15})x_2$
$y = (\beta_0 + \beta_4 + \beta_5 + \beta_{13}) + (\beta_1 + \beta_8 + \beta_{10} + \beta_{16})x_1$
$\quad + (\beta_2 + \beta_9 + \beta_{11} + \beta_{17})x_2$

15.1 $\hat{Y} = 14.867 + 1.141X_2 + 0.0001774X_3 - 0.8088X_6 - 1.665X_8$

where X_2 is a dummy variable representing sex, 0 for females and 1 for males; hence the value 1.141 represents the mean amount by which job satisfaction is higher for males than it is for females; X_3 is salary, and the coefficient .0001774 represents the mean increase in job satisfaction score for every dollar increase in salary. Since X_6 represents education, the coefficient -0.8088 represents the mean amount by which job satisfaction score can be expected to decrease for every additional year of education. Since X_8 is a dummy variable for management position, 0 for nonmanagement and 1 for management, the value -1.665 represents the mean amount by which job satisfaction can be expected to decrease for management as opposed to nonmanagement.

15.3 $-0.8088 \pm 2.0141 (0.2465)$ or -1.3053 to -0.3123

This interval represents the amount by which mean job satisfaction score decreases for each additional year of education. This may run contrary to the opinion that job satisfaction increases with education.

15.5 **(a)** 3; **(b)** 41; **(c)** 44; **(d)** 1.181E04; **(e)** 9606.67; **(f)** 288.05; **(g)** 33.35; **(h)** .0001; **(i)** .709; **(j)** 5997.2; **(k)** 2.77; **(l)** ~.01; **(m)** ~.01; **(n)** ~.07; **(o)** 351.77.

15.7 SSE $= (1 - R^2)$Total SS, so SSE $= (1 - .966)1.104E06 = 3.754E04$
Error DF $= 25$

15.8
$$F = \frac{(3.754E04 - 8.006E03)/(25\text{-}21)}{8.006E03/21} = 19.4$$

which is compared with quantiles in Table A5 using 4 and 21 degrees of freedom, thus the p-value is smaller than .001. Hence, the additional terms X_1^2, X_2, X_2^2, and $X_1 X_2$ do make a significant contribution in the model.

15.11 $\hat{Y} = 14.464 + .000174(13500) - .7437(13) = 7.145$

15.12 There is 95% confidence that an individual with the characteristics described in Exercise 15.11 will have a job satisfaction score between 3.28 and 11.01, while there is 95% confidence that the mean job satisfaction for all employees with these characteristics will be between 6.36 and 7.93.

15.13 **(a)** Yes, since X_3 has a positive coefficient.

(b) No, since X_6 has a negative coefficient.

(c) Job satisfaction score of Person 1 = Job Satisfaction score of Person 2 or $14.46 + .000174(X_3 + c) - .7437(X_6 + 2) = 14.46 + .000174X_3 - .7437X_6$; thus $c = \$8548$.

15.16 X_2^2 with a p-value of .5052 would be eliminated first.

15.20 $\hat{Y} = 1196.165 - 1.732X_2 - 20.649X_4 + 3.195X_5$

$\hat{Y} = 1105.682 + 2.524X_1 - 20.554X_4$

$\qquad - 0.04887X_1X_2 + 0.2881X_4X_5$

15.21 The predictions are, respectively, 912.212 and 911.599. The disagreement is caused by the use of different variables in each of the prediction equations. Note, however, that these two predictions are in good agreement with each other.

15.22 The interpretation is that there is 95% confidence that an individual city with those characteristics will have a mortality rate between 850.85 and 973.59, but the mean mortality rate for all cities with those characteristics (a hypothetical population) is between 889.05 and 935.40 with 95% confidence.

15.31 $\hat{Y} = 14.464 + 0.000174(58,000) - .7437(19) = 10.4$

15.32 There is 95% confidence that an individual with the characteristics described in Exercise 15.31 will have a job satisfaction score between 6.41 and 14.44, while there is 95% confidence that the mean job satisfaction for all individuals with these characteristics will be between 9.09 and 11.76.

15.35 $1.298E - 09$

15.36 $\hat{Y} = 14.116 + 1.108X_2 + 4.361E - 04X_3 - 0.993X_6$

$\qquad - 3.869E - 09X_3^2$

(a) 9.03; **(b)** 6.39; **(c)** 8.64.

15.37 $\hat{Y} = 1196.165 - 1.732(38.6) - 20.649(11.5) + 3.195(9.9)$

$\qquad = 923.477$

$\hat{Y} = 1105.682 + 2.524(20.4) - 20.554(11.5) - 0.04887(20.4)(38.6)$

$\qquad + 0.2881(11.5)(9.9) = 915.119.$

These two prediction equations use different variables, and this in turn causes the predictions to differ.

15.38 There is 95% confidence that a city with the characteristics described in Exercise 15.37 will have a mortality rate between 865.18 and 981.79, while there is a 95% confidence that the mean mortality rate for all cities with these characteristics will be between 910.43 and 936.54.

15.42 $-.7437 \pm 2.0117(0.253)$ or -1.2527 to $-.2347$

There is 95% confidence that the mean job satisfaction score will decrease by 1.25 points to 0.23 points for each additional year of education, if the salary remains the same.

16.1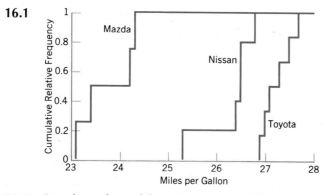

16.2 Based on the e.d.f.s in Exercise 16.1, it would seem reasonable to reject H_0 with $\mu_M < \mu_N < \mu_T$.

16.5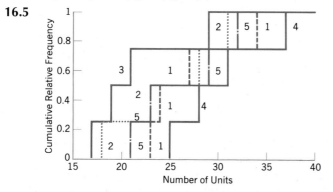

16.6 There is so much overlap present in the e.d.f.s in Exercise 16.5 that the null hypothesis seems reasonable.

16.9 The data yield the following calculations:

$$n_T = 6 \qquad n_N = 5 \qquad n_M = 4 \qquad N = 15$$

$$\text{Total SS} = \Sigma X^2 - \frac{(\Sigma X)^2}{N} = 10{,}172.94 - \frac{(390)^2}{15} = 32.94$$

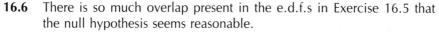

$$\text{SST} = \sum_j \frac{\left(\sum_i X_{ij}\right)^2}{n_j} - \frac{(\Sigma X)^2}{N}$$

$$= \frac{(163.5)^2}{6} + \frac{(131.5)^2}{5} + \frac{(95.0)^2}{4} - \frac{(390)^2}{15} = 30.075$$

$$\text{Error SS} = \text{Total SS} - \text{SST} = 32.94 - 30.075 = 2.865$$

$$F = \frac{30.075/2}{2.865/12} = 62.98$$

The null hypothesis is rejected with a p-value of .0001.

16.10 The conclusion in Exercise 16.9 to reject H_0 is in agreement with the first tentative guess made in Exercise 16.2.

16.13 Based on the computer printout, the null hypothesis is not rejected. The *p*-value is .2610.

16.14 The conclusion of Exercise 16.13 is not to reject H_0, which is in agreement with the first tentative guess made in Exercise 16.6.

16.17 $\Sigma X = 3753.59$ $\Sigma X^2 = 282{,}038.3497$ $\Sigma X_{SL} = 740.93$

$\Sigma X_{SF} = 764.82$ $\Sigma X_O = 725.44$ $\Sigma X_W = 772.64$

$\Sigma X_D = 749.76$

From Eq. 16.1,

$$\text{Total SS} = 282{,}038.3497 - \frac{(3753.59)^2}{50} = 249.5919$$

From Eq. 16.2,

$$\text{SST} = \frac{(740.93)^2}{10} + \frac{(764.82)^2}{10} + \frac{(725.44)^2}{10}$$
$$+ \frac{(772.64)^2}{10} + \frac{(749.76)^2}{10} - \frac{(3753.59)^2}{50} = 141.5141$$

From Eq. 16.3,

$$\text{SSE} = 249.5919 - 141.5141 = 108.0778$$

16.19 $X_i = \beta_0 + \beta_1 D_1 + \beta_2 D_2 + \epsilon_i$ $i = 1, \ldots, n_j$

$D_1 = 1,$ if X_i is from the Toyota population

$\quad = 0,$ otherwise.

$D_2 = 1,$ if X_i is from the Nissan population

$\quad = 0,$ otherwise.

The mean for the Toyota population is $\beta_0 + \beta_1$; the mean for the Nissan population is $\beta_0 + \beta_2$; and the mean for the Mazda population is β_0. Thus, β_1 is the amount by which the Toyota mean differs from the Mazda mean, and β_2 is the amount by which the Nissan mean differs from the Mazda mean.

16.20 Refer to the computations in Exercise 16.9.

$$\overline{X}_T = 27.25 \qquad \overline{X}_N = 26.30 \qquad \overline{X}_M = 23.75$$

For Toyota and Nissan:

$$\text{LSD}_{.05} = t_{.975,12} \sqrt{.23875} \; [\tfrac{1}{6} + \tfrac{1}{5}]^{1/2} = .645$$

Since $\overline{X}_T - \overline{X}_N = .95$ exceeds .645, these means are declared to be significantly different.

For Toyota and Mazda:

$$\text{LSD}_{.05} = t_{.975,12} \sqrt{.23875} \; [\tfrac{1}{6} + \tfrac{1}{4}]^{1/2} = .687$$

Since $\overline{X}_T - \overline{X}_M = 3.5$ exceeds .687, these means are declared to be significantly different.

For Nissan and Mazda:

$$\text{LSD}_{.05} = t_{.975,12} \sqrt{.23875} \; [\tfrac{1}{5} + \tfrac{1}{4}]^{1/2} = .714$$

Since $\overline{X}_N - \overline{X}_M = 2.55$ exceeds .714, these means are declared to be significantly different.

In summary, $\mu_M < \mu_N < \mu_T$.

16.21 The conclusion in Exercise 16.20 is the same as the ordering made on the second tentative guess in Exercise 16.2.

16.24 Since the null hypothesis was not rejected in Exercise 16.13, multiple comparisons should not be made.

16.25 No multiple comparisons were made in Exercise 16.24 and this is in agreement with the conclusion in Exercise 16.6 that the null hypothesis is not rejected.

16.28 $\quad n_T = 6 \qquad n_N = 5 \qquad n_M = 4 \qquad N = 15 \qquad \Sigma R_x = 120$

$\qquad \Sigma R_x^2 = 1239.5 \qquad \Sigma R_T = 75 \qquad \Sigma R_N = 35 \qquad \Sigma R_M = 10$

Based on the following computer printout, the null hypothesis is rejected with a p-value of .0001. This is in agreement with Exercise 16.9.

SOURCE	DF	SUM OF SQUARES	MEAN SQUARE	F VALUE	PR > F
TREATMENT	2	247.5	123.75	46.41	.0001
ERROR	12	32.0	2.67		
TOTAL	14	279.5			

16.29 Refer to the computations in Exercise 16.28.

$$\bar{R}_T = 12.5 \qquad \bar{R}_N = 7 \qquad \bar{R}_M = 2.5$$

The LSD values are given by the formula:

$$LSD_{.05} = t_{.975,12} \sqrt{2.67} \left[\frac{1}{n_i} + \frac{1}{n_j} \right]^{1/2}$$

These values are as follows.

	n_i	n_j	$LSD_{.05}$	$|\bar{R}_i - \bar{R}_j|$
T, N	6	5	2.154	5.5
T, M	6	4	2.297	10.0
N, M	5	4	2.387	4.5

Based on these LSD values, all pairs of means are declared to be significantly different and the summary is $\mu_M < \mu_N < \mu_T$. This is the same as the conclusion in Exercise 16.20.

16.32 Based on the computer printout, the null hypothesis is not rejected. The p-value is .3096. This is in agreement with Exercise 16.13.

16.33 Since the null hypothesis was not rejected in Exercise 16.32, multiple comparisons should not be made. This is in agreement with the decision in Exercise 16.24.

16.36 $\quad n_1 = 9 \qquad n_2 = 8 \qquad n_3 = 10 \qquad n_4 = 9 \qquad N = 36$

$\qquad \Sigma R_x = 666 \qquad \Sigma R_x^2 = 16,156.5 \qquad \Sigma R_{i1} = 162.5$

$\qquad \Sigma R_{i2} = 170.5 \qquad \Sigma R_{i3} = 79 \qquad \Sigma R_{i4} = 254$

Based on the following computer printout, the null hypothesis is rejected with a p-value of .0001. This is in agreement with the analysis on the original data.

SOURCE	DF	SUM OF SQUARES	MEAN SQUARE	F VALUE	PR > F
TREATMENT	3	2039.353	679.784	12.11	.0001
ERROR	32	1796.147	56.130		
TOTAL	35	3835.500			

16.37 Refer to the computations in Exercise 16.36.

$$\overline{R}_1 = 18.056 \quad \overline{R}_2 = 21.313 \quad \overline{R}_3 = 7.9 \quad \overline{R}_4 = 28.222$$

The LSD values are given by the formula:

$$\text{LSD}_{.05} = t_{.975,32} \sqrt{56.130} \left[\frac{1}{n_i} + \frac{1}{n_j} \right]^{1/2}$$

These values are as follows.

i, j	n_i	n_j	$\text{LSD}_{.05}$	$|\overline{R}_i - \overline{R}_j|$
1, 2	9	8	7.415	3.257
1, 3	9	10	7.012	10.156
1, 4	9	9	7.194	10.166
2, 3	8	10	7.239	13.413
2, 4	8	9	7.415	6.909
3, 4	10	9	7.012	20.322

Based on these LSD values, the summary of ordered means is $\mu_3 < \mu_1$, $\mu_3 < \mu_2$, and $\mu_3 < \mu_4$; also, $\mu_1 < \mu_4$. So far this is in agreement with the analysis on the original data. However, the analysis on ranks fails to show μ_2 and μ_4 to be significantly different, while the analysis on the original data showed a significant difference for these means.

16.40

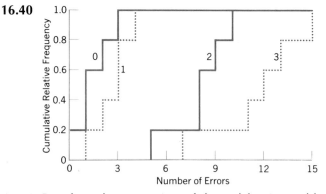

16.41 Based on the separation of the e.d.f.s, it would seem reasonable to reject H_0 with $\mu_0 = \mu_1 < \mu_2 < \mu_3$.

16.42 The test assumptions are reasonable for these data. The data yield the following calculations. All $n_j \doteq 5$, $N = 20$

$$\Sigma X = 118 \quad \Sigma X^2 = 1096 \quad \Sigma X_{i0} = 7$$
$$\Sigma X_{i1} = 13 \quad \Sigma X_{i2} = 40 \quad \Sigma X_{i3} = 58$$

Based on the following computer printout, the null hypothesis is rejected with a p-value of .0001.

SOURCE	DF	SUM OF SQUARES	MEAN SQUARE	F VALUE	PR > F
TREATMENT	3	340.2	113.400	30.44	.0001
ERROR	16	59.6	3.725		
TOTAL	19	399.8			

16.43 The decision in Exercise 16.42 is in agreement with the first tentative guess in Exercise 16.41.

16.44 Refer to the computations in Exercise 16.42.

$$\overline{X}_0 = 1.4 \qquad \overline{X}_1 = 2.6 \qquad \overline{X}_2 = 8 \qquad \overline{X}_3 = 11.6$$

The LSD value for all pairs of comparisons is

$$\text{LSD}_{.05} = t_{.975,16} \sqrt{3.725} \, [\tfrac{1}{5} + \tfrac{1}{5}]^{1/2} = 2.588.$$

The pairs of comparisons for pairs (i, j) are as follows.

i, j	$\lvert \overline{X}_i - \overline{X}_j \rvert$	i, j	$\lvert \overline{X}_i - \overline{X}_j \rvert$
0, 1	1.2	1, 2	5.4
0, 2	6.6	1, 3	9.0
0, 3	10.2	2, 3	3.6

The summary is $\mu_0 = \mu_1 < \mu_2 < \mu_3$.

16.45 The decision in Exercise 16.44 is in agreement with the second tentative guess in Exercise 16.41.

17.7 $T = \Sigma X = 207 \qquad \Sigma X^2 = 2961 \qquad C_1 = 61 \qquad C_2 = 49$
$C_3 = 44 \qquad C_4 = 53 \qquad R_1 = 68 \qquad R_2 = 28 \qquad R_3 = 60$
$R_4 = 51 \qquad b = 4 \qquad k = 4$

SOURCE	DF	SUM OF SQUARES	MEAN SQUARE	F VALUE	PR > F
TREATMENT	3	38.69	12.90	5.79	.0174
BLOCK	3	224.19	74.73	33.52	.0001
ERROR	9	20.06	2.23		
TOTAL	15	282.94			

$F = 5.79$ with a p-value $= .0174$, so H_0 is rejected.

17.8 $\text{LSD}_{.05} = 2.2622 \sqrt{2.23(2/4)} = 2.39$

$$\overline{X}_1 = 15.25 \qquad \overline{X}_2 = 12.25 \qquad \overline{X}_3 = 11.00 \qquad \overline{X}_4 = 13.25$$

Conclude $\underline{\mu_3 \quad \mu_2 \quad \mu_4 \quad \mu_1}$

17.13 The z scores are:

Level 1 $-1.37, \; -0.76, \; -1.49, \; -1.14, \; -0.25, \; -0.25$

Level 2 $-0.46, \; -0.46, \; -0.09, \; -0.44, \; -1.24, \; -0.74$

Level 3 $0.46, \; -0.15, \; 0.79, \; 0.09, \; 0.00, \; -0.50$

Level 4 $1.67, \; 1.06, \; 1.31, \; 0.96, \; 1.73, \; 1.24$

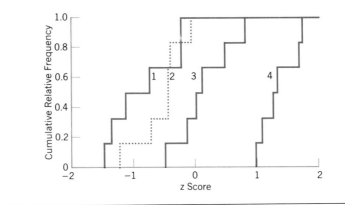

17.14

SOURCE	DF	SUM OF SQUARES	MEAN SQUARE	F VALUE	PR > F
TREATMENT	3	324.00	108.00	49.85	.0001
BLOCK	2	1.33	0.67	0.31	.7408
INTERACTION	6	68.00	11.33	5.23	.0073
ERROR	12	26.00	2.17		
TOTAL	23	419.33			

$F = 49.85$ with a p-value $= .0001$, so H_0 is rejected.

17.15 $LSD_{.05} = 2.1788 \sqrt{2.17(2/6)} = 1.85$

Conclude $\mu_1 = \mu_2 < \mu_3 < \mu_4$

17.16 $F = 5.23$ with a p-value $= .0073$, so H_0 is rejected.

17.21 The z scores are:

Dryer A $1.03, -0.09, 2.15, 1.59, 0.47, -0.52, -1.45, 0.72, -0.83$
-0.52

Dryer B $-0.66, 0.19, -0.94, -1.22, -0.66, -0.21, 1.03, -1.14,$
$-0.52, -0.21$

Dryer C $-0.37, -1.22, -0.09, -0.66, 0.47, 0.72, -0.52, 1.66,$
$-0.21, 1.97$

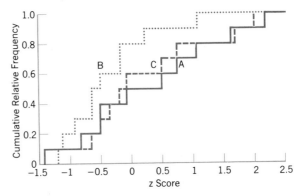

17.22 Nonadditivity of block effects is assumed along with the ϵ_{ij} being independent and identically distributed normal random variables with mean 0 and variance σ^2. $F = 2.39$ with a p-value of .1134, so H_0 is not rejected.

17.23 Since H_0 was not rejected, multiple comparisons should not be made.

17.24 Nonadditivity of block effects is assumed along with the ϵ_{ij} being independent and identically distributed normal random variables with mean 0 and variance σ^2. $F = 7.61$ with a p-value of .0028, so H_0 is rejected. While there is no significant difference among clothes dryers, some are better with gas and others are better with electricity.

17.25 $X_{ijm} = \mu_j + \beta_i + \gamma_{ij} + \epsilon_{ijm} \qquad j = 1,2,3 \quad i = 1,2 \quad m = 1, \ldots, 5$

where μ_1, μ_2, and μ_3 are population means for the dryers, β_1 and β_2 represent the block effects, and the γ_{ij} represent the interaction between block i and treatment j.

17.26 $X_{ijm} = \beta_0 + \beta_1 D_1 + \beta_2 D_2 + \beta_3 D_3 + \beta_4 D_1 D_3 + \beta_5 D_2 D_3 + \epsilon_{ijm}$

The dummy variables D_1 and D_2 represent Dryers B and C and D_3 represents Electric; β_0 corresponds to the population mean of A; $\beta_0 + \beta_1$ represents the population mean of B and $\beta_0 + \beta_2$ represents the population mean of C; β_3 is the block effect for Electric; $\beta_4 D_1 D_3$ is the interaction of B with dryer type; and $\beta_5 D_2 D_3$ is the interaction of C with dryer type.

17.27 $A = 209.50 \qquad B = 186.36$

SOURCE	DF	SUM OF SQUARES	MEAN SQUARE	F VALUE	PR > F
TREATMENT	3	11.36	3.79	2.94	.0608
BLOCK	6	0.00	0.00	0.00	1.0000
ERROR	18	23.14	1.29		
TOTAL	27	34.50			

$F_R = 2.94$ with a p-value of .0608, so H_0 is not rejected.

17.28 Since H_0 was not rejected, multiple comparisons are not made.

17.31 $F = 3.19$ with a p-value of .0362, so H_0 is rejected.

17.32 $\text{LSD}_{.05} = 2.0345 \sqrt{1.33(2/12)} = 0.96$

$\bar{R}_1 = 3.17 \qquad \bar{R}_2 = 1.96 \qquad \bar{R}_3 = 2.04 \qquad \bar{R}_4 = 2.83$

Conclude that there is no preference among grass types 2, 3, and 4; that there is no preference between grass types 4 and 1; but that grass types 2 and 3 are both preferred over grass type 1.

17.33

SOURCE	DF	SUM OF SQUARES	MEAN SQUARE	F VALUE	PR > F
TREATMENT	3	584.68	194.89	4.74	.0132
BLOCK	6	29082.21	4847.04	117.89	.0001
ERROR	18	740.07	41.12		
TOTAL	27	30406.96			

$F = 4.74$ with a p-value of .0132, so H_0 is rejected. This compares with a p-value of .0608 in Exercise 17.27.

17.34 $LSD_{.05} = 2.1009 \sqrt{41.12(2/7)} = 7.20$

$$\overline{X}_1 = 48.43 \qquad \overline{X}_2 = 57.14 \qquad \overline{X}_3 = 47.57 \qquad \overline{X}_4 = 45$$

Conclude $\mu_4 = \mu_3 = \mu_1 < \mu_2$. Multiple comparisons were not made in Exercise 17.28 because the p-value was greater than .05.

17.37 $A = 83.50 \qquad B = 75.58$

SOURCE	DF	SUM OF SQUARES	MEAN SQUARE	F VALUE	PR > F
TREATMENT	2	3.58	1.79	2.26	.1546
BLOCK	5	0.00	0.00	0.00	1.0000
ERROR	10	7.92	0.79		
TOTAL	17	11.50			

$F = 2.26$ with a p-value of .1546, so H_0 is not rejected. This compares with a p-value of .045 in Section 17.1.

17.38 Since the p-value was greater than .05, no multiple comparisons are made.

17.41 $A = 120 \qquad B = 111.5$

SOURCE	DF	SUM OF SQUARES	MEAN SQUARE	F VALUE	PR > F
TREATMENT	3	11.50	3.83	4.06	.0444
BLOCK	3	0.00	0.00	0.00	1.0000
ERROR	9	8.50	0.94		
TOTAL	15	20.00			

$F = 4.06$ with a p-value of .0444. This compares with a p-value of .0174 in Exercise 17.7.

17.42 $LSD_{.05} = 2.2622 \sqrt{.687} = 1.55$

$$\overline{R}_1 = 3.75 \qquad \overline{R}_2 = 2.00 \qquad \overline{R}_3 = 1.50 \qquad \overline{R}_4 = 2.75$$

Conclude that there is no difference in the output of workers 3, 2, and 4; likewise for workers 4 and 1; but that workers 3 and 2 produce less than worker 1. This is the same conclusion as in Exercise 17.8.

17.45

SOURCE	DF	SUM OF SQUARES	MEAN SQUARE	F VALUE	PR > F
TREATMENT	2	80.89	40.44	4.20	.0318
BLOCK	2	32.89	16.44	1.71	.2094
INTERACTION	4	184.89	46.22	4.80	.0082
ERROR	18	173.33	9.63		
TOTAL	26	472.00			

$F = 4.20$ with a p-value of .0318, so H_0 is rejected.

17.46 $\text{LSD}_{.05} = 2.1009 \sqrt{9.63(2/9)} = 3.07$

$\bar{X}_1 = 37.33 \qquad \bar{X}_2 = 41.56 \qquad \bar{X}_3 = 39.11$

Conclude $\underline{\mu_1 \quad \mu_2} \quad \mu_3$

17.47 The z scores are as follows:

Operator 1 1.69, -0.21, 0.42, 0.26, -0.68, -1.15, -0.86, -0.86, -1.63

Operator 2 1.05, -0.84, -0.21, 1.20, 1.67, 0.26, 0.30, 1.07, -0.09

Operator 3 -0.84, -1.48, 0.42, -0.68, -1.15, 0.26, -0.09, 0.69, 1.46

These graphs make the previous conclusions seem reasonable.

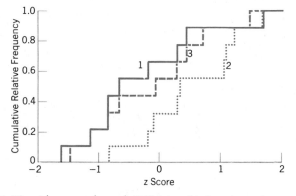

17.48 $F = 4.80$ with a p-value of .0082, so H_0 is rejected.

17.49 The cell means are:

	Operator		
Machine	1	2	3
A	40.7	38.7	36.7
B	38.7	45.3	38.7
C	32.7	40.7	42.0

Based on this graph, the results in Exercise 17.48 seem quite reasonable.

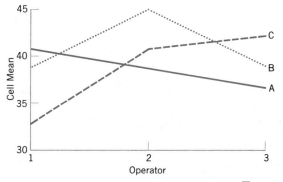

18.1 The numbers 22, 24, 31, 30, 17, and 17 have $\bar{X} = 23.5$ and $s = 6.091$, leading to z scores -0.25, 0.08, 1.23, 1.07, -1.07, -1.07 as in the first row of cells in Figure 18.2.

18.4 All hypotheses are rejected based on the small p-values in the computer printout.

18.5 Glue: $LSD_{.05} = 2.0281\sqrt{120.81(2/16)} = 7.88$

$\overline{X}_1 = 377.125 \qquad \overline{X}_2 = 390 \qquad \overline{X}_3 = 428.563$

Conclude $\mu_1 < \mu_2 < \mu_3$. Since there are only two wood types and two widths, multiple comparisons are not necessary because in both cases F-tests have shown those two means to be significantly different.

18.8 $F = [(55.48 - 29.14)/(18 - 14)]/2.08 = 3.16$, which is compared with quantiles in Table A5 using 4 and 14 degrees of freedom. The p-value is slightly less than .05, so H_0 is rejected.

18.9 H_0 is strongly rejected with a p-value of .0001. After an adjustment for the covariate, the mean productivity rates are barely declared significantly different at the .05 level.

18.10 The first graph indicates a strong separation of productivity rate. The second graph shows that after an adjustment for the covariate, a great deal of overlap is present.

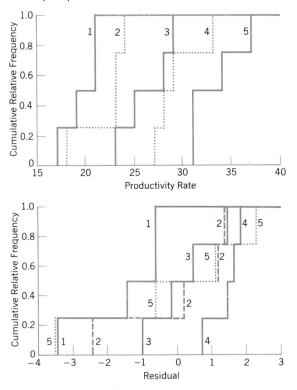

18.14 All hypotheses are rejected based on the small p-values in the computer printout and all results are in agreement with those for Exercise 18.4.

18.15 Glue: $LSD_{.05} = 2.0281\sqrt{10.95(2/16)} = 2.37$

$\overline{R}_1 = 16.69 \qquad \overline{R}_2 = 21.75 \qquad \overline{R}_3 = 35.06$

Conclude $\mu_1 < \mu_2 < \mu_3$. Since there are only two types of wood and two widths, significant F-tests for those factors imply that the two means are different in both cases. This is in agreement with the results of Exercise 18.5.

18.18 $F = [(16.93 - 10.86)/(18 - 14)]/0.78 = 1.96$, which is compared with quantiles in Table A5 with 4 and 14 degrees of freedom. The p-value is between .10 and .25 compared to a p-value slightly less than .05 in Exercise 18.8. A plot of Y versus X using the data in Exercise 18.8 shows that the five treatments do not appear to have the same regression slope, due to a nonlinear pattern. On the other hand, a plot of the rank of Y versus the rank of X shows a linear relationship with equal slopes across treatments. For this reason the analysis on the ranks is more justifiable than an analysis on the original data.

18.21 All hypotheses are rejected except for the fourth and last.

18.22 Piece: $LSD_{.05} = 2.0049 \sqrt{1.34(2/24)} = 0.67$

$$\overline{X}_1 = 9.21 \qquad \overline{X}_2 = 10.17 \qquad \overline{X}_3 = 10.71$$

Conclude $\mu_1 < \mu_2 = \mu_3$
Shift: $LSD_{.05} = 2.0049 \sqrt{1.34(2/24)} = 0.67$

$$\overline{X}_1 = 9.29 \qquad \overline{X}_2 = 10.38 \qquad \overline{X}_3 = 10.42$$

Conclude $\mu_1 < \mu_2 = \mu_3$. No multiple comparisons are necessary for Years because there are only two means.

18.23 The conclusions are the same as in Exercise 18.21 except that the fifth hypothesis is no longer rejected.

18.24 Piece: $LSD_{.05} = 2.0049 \sqrt{163.45(2/24)} = 7.40$

$$\overline{R}_1 = 27.46 \qquad \overline{R}_2 = 38.19 \qquad \overline{R}_3 = 43.85$$

Conclude $\mu_1 < \mu_2 = \mu_3$
Shift: $LSD_{.05} = 2.0049 \sqrt{163.45(2/24)} = 7.40$

$$\overline{R}_1 = 29.48 \qquad \overline{R}_2 = 41.83 \qquad \overline{R}_3 = 38.19$$

Conclude $\mu_1 < \mu_2 = \mu_3$. No multiple comparisons are necessary for Years because there are only two means. These results are in agreement with those in Exercise 18.22.

18.25 The twelve observations with the same years of experience and the same shift need to be grouped together. For example, the group for less-than-one-year experience and the swing shift contains the observations 9, 8, 9, 8, 9, 8, 9, 8, 11, 9, 9, and 9.

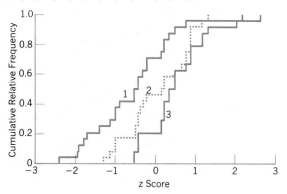

18.30 $F = [(1227.57 - 1172.70)/(19 - 17)]/68.98 = 0.40$, which has an associated p-value greater than .25, so H_0 is not rejected.

18.31 $F = [(626.99 - 602.60)/(19 - 17)]/35.45 = 0.34$, which has an associated p-value greater than .25, so H_0 is not rejected, which is in agreement with Exercise 18.30.

18.32 $F = 0.02$ with a p-value of .9785 indicating H_0 is not rejected, which is in agreement with Exercise 18.30.

19.1 a_1

19.2 a_4

19.3 a_2

19.4

	State of Nature		
Action	s_1	s_2	s_3
a_1	0	10	8
a_3	3	0	2
a_4	12	8	0

19.5

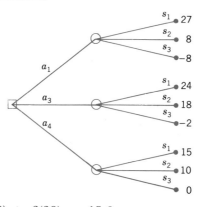

19.11 $.6(15) + .2(10) + .2(20) = 15.0$
$.6(27) + .2(8) + .2(-8) = 16.2$
$.6(18) + .2(24) + .2(-2) = 15.2$

$.2(15) + .6(10) + .2(20) = 13.0$
$.2(27) + .6(8) + .2(-8) = 8.6$
$.2(18) + .6(24) + .2(-2) = 17.6$

$.2(15) + .2(10) + .6(20) = 17.0$
$.2(27) + .2(8) + .6(-8) = 2.2$
$.2(18) + .2(24) + .6(-2) = 7.2$

19.12 $E[r(a_1, s)] = 9000(.01) + 9000(.6) + 9000(.39) = 9000$
$E[R(a_2, s)] = 14{,}700(.01) + 10{,}500(.6) + 6300(.39) = 8904$
$E[r(a_3, s)] = 20{,}400(.01) + 12{,}000(.6) + 3600(.39) = 8808$
The maximum expected payoff is \$9000 for action a_1.

19.13 $E[OL(a_1)] = 11,400(.01) + 3000(.6) + 0(.39) = 1914$

$E[OL(a_2)] = 5700(.01) + 1500(.6) + 2700(.39) = 2010$

$E[OL(a_3)] = 0(.01) + 0(.6) + 5400(.39) = 2106$

The minimum expected opportunity loss is \$1914, associated with action a_1, which is the same as the decision in Exercise 19.12.

19.14 $E[r(a_1, s)] = .1(27) + .1(8) + .8(-8) = -2.9$

$E[r(a_3, s)] = .1(24) + .1(18) + .8(-2) = 2.6$

$E[r(a_4, s)] = .1(15) + .1(10) + .8(0) = 2.5$

The maximum expected payoff decision is a_3.

19.15 $E[OL(a_1)] = .7(0) + .2(10) + .1(8) = 2.8$

$E[OL(a_3)] = .7(3) + .2(0) + .1(2) = 2.3$

$E[OL(a_4)] = .7(12) + .2(8) + .1(0) = 10.0$

The minimum expected opportunity loss decision is a_3.

19.23 This person is risk averse.

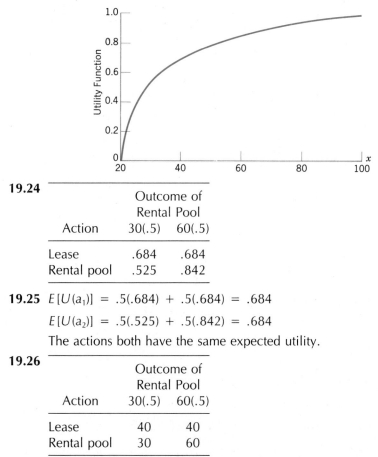

19.24

Action	Outcome of Rental Pool	
	30(.5)	60(.5)
Lease	.684	.684
Rental pool	.525	.842

19.25 $E[U(a_1)] = .5(.684) + .5(.684) = .684$

$E[U(a_2)] = .5(.525) + .5(.842) = .684$

The actions both have the same expected utility.

19.26

Action	Outcome of Rental Pool	
	30(.5)	60(.5)
Lease	40	40
Rental pool	30	60

$E[r(a_1, s)] = .5(40) + .5(40) = 40$

$E[r(a_2, s)] = .5(30) + .5(60) = 45$

The action corresponding to the maximum expected payoff is to put the condo into the rental pool.

19.31

	s_1	s_2	s_3
a_1	0	0	1000
a_2	1000	500	0

19.32 a_2

19.33 a_1

19.34 $E[r(a_1, s)] = .3(-1000) + .4(2000) + .3(3500) = 1550$

$E[r(a_2, s)] = .3(-2000) + .4(1500) + .3(4500) = 1350$

The maximum expected payoff decision is a_1, which agrees with the maximin criterion.

19.35 $E[OL(a_1)] = .3(0) + .4(0) + .3(1000) = 300$

$E[OL(a_2)] = .3(1000) + .4(500) + .3(0) = 500$

The minimum expected opportunity loss decision is a_1.

19.40 a_2 is inadmissible.

19.41 a_1

19.42 a_5

19.43 $E[r(a_1, s)] = .3(15) + .4(60) + .2(-20) + .1(16) = 26.1$

$E[r(a_3, s)] = .3(30) + .4(5) + .2(40) + .1(-5) \quad = 18.5$

$E[r(a_4, s)] = .3(20) + .4(-15) + .2(9) + .1(38) \quad = 5.6$

$E[r(a_5, s)] = .3(50) + .4(20) + .2(25) + .1(12) \quad = 29.2$

The maximum expected payoff decision is a_5.

19.44

a_1	35	0	60	22
a_3	20	55	0	43
a_4	30	75	31	0
a_5	0	40	15	26

$E[OL(a_1)] = .3(35) + .4(0) + .2(60) + .1(22) = 24.7$

$E[OL(a_3)] = .3(20) + .4(55) + .2(0) + .1(43) = 32.3$

$E[OL(a_4)] = .3(30) + .4(75) + .2(31) + .1(0) = 45.2$

$E[OL(a_5)] = .3(0) + .4(40) + .2(15) + .1(26) = 21.6$

The minimum expected opportunity loss decision is a_5, which agrees with that in Exercise 19.43.

20.1 $P(\text{oil}|\text{no dome}) = \dfrac{P(\text{oil and no dome})}{P(\text{no dome})} = \dfrac{.12}{.68} = .18$

$P(\text{dry hole}|\text{no dome}) = \dfrac{P(\text{dry hole and no dome})}{P(\text{no dome})} = \dfrac{.56}{.68} = .82$

These posterior probabilities are quite different from those of .56 and .44 that were worked out in the text.

20.3

	Oil	Dry Hole	
Dome	.6(.5)	.2(.5)	.4
No dome	.4(.5)	.8(.5)	.6
	.5	.5	1.0

$$P(\text{oil}|\text{dome}) = \frac{.6(.5)}{.4} = .75$$

$$P(\text{dry hole}|\text{dome}) = \frac{.2(.5)}{.4} = .25$$

20.4

	Oil	Dry Hole	
Dome	.75(.3)	.10(.7)	.295
No dome	.25(.3)	.90(.7)	.705
	.3	.7	1.000

$$P(\text{oil}|\text{dome}) = \frac{.75(.3)}{.295} = .7627$$

$$P(\text{dry hole}|\text{dome}) = \frac{.10(.7)}{.295} = .2373$$

20.5

	Oil	Dry Hole	
Dome	.5(.3)	.5(.7)	.5
No dome	.5(.3)	.5(.7)	.5
	.3	.7	1.0

$$P(\text{oil}|\text{dome}) = \frac{.5(.3)}{.5} = .3$$

$$P(\text{dry hole}|\text{dome}) = \frac{.5(.7)}{.5} = .7$$

But these are also the prior probabilities. This will always be the case when the conditional probability, $P(B|s_i)$, and the unconditional probability, $P(B)$, are equal in Eq. 20.3. That is, in this case Eq. 20.3 becomes $P(s_i|B) = P(s_i)$.

20.6

Forecast increase	.15(.5)	.30(.3)	.75(.2)	.315
Forecast no increase	.85(.5)	.70(.3)	.25(.2)	.685
	.5	.3	.2	1.000

$$P(s_1|\text{firm had forecast an increase in PIR}) = \frac{.15(.5)}{.315} = .2381$$

$$P(s_2|\text{firm had forecast an increase in PIR}) = \frac{.30(.3)}{.315} = .2857$$

$$P(s_3|\text{firm had forecast an increase in PIR}) = \frac{.75(.2)}{.315} = .4762$$

20.7 $E[r(a_1, s)] = 9000(.2381) + 9000(.2857) + 9000(.4762) = \$9000$
$E[r(a_2, s)] = 14,700(.2381) + 10,500(.2857) + 6300(.4762) = \$9500$
$E[r(a_3, s)] = 20,400(.2381) + 12,000(.2857) + 3600(.4762) = \$10,000$
The decision associated with the maximum expected payoff is a_3.

20.12

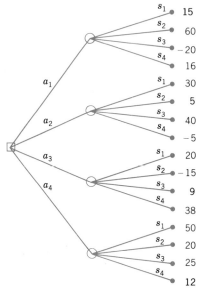

20.13 $50(.3) + 60(.4) + 40(.2) + 38(.1) = 50.8$

20.14 The expected value of perfect information represents the maximum amount one would be willing to pay for perfect information. It is found as the difference between the expected payoff under perfect information and the maximum expected payoff, or $50.8 - 29.2 = 21.6$.

20.15 The minimum expected opportunity loss from Exercise 19.44 is 21.6, which is the same as the expected value of perfect information found in Exercise 20.14.

20.23 $E[r(a_1, s)] = -20(.2) + 10(.5) + 80(.3) = 25$

$E[r(a_2, s)] = -100(.2) - 20(.5) + 150(.3) = 15$

The maximum expected payoff is associated with action a_1.

20.24

Sample Outcome	Percent of Market			Totals
	5	10	15	
$X = 0$	.8145(.2)	.6561(.5)	.5220(.3)	.6476
$X = 1$	.1715(.2)	.2916(.5)	.3685(.3)	.2906
$X = 2$	.0135(.2)	.0486(.5)	.0975(.3)	.0563
$X = 3$	.0005(.2)	.0036(.5)	.0115(.3)	.0053
$X = 4$	.0000(.2)	.0001(.5)	.0005(.3)	.0002
	.2	.5	.3	1.0000

20.25

Sample	Percent of Market		
Outcome	5	10	15
$X = 0$	.2516	.5066	.2418
$X = 1$	.1180	.5017	.3803
$X = 2$	.0481	.4316	.5200
$X = 3$	.0178	.3372	.6450
$X = 4$	.0062	.2462	.7477

20.26

Sample Outcome	Maximum Expected Payoff	Best Action
$X = 0$	19.4	a_1
$X = 1$	35.2	a_2
$X = 2$	64.6	a_2
$X = 3$	88.2	a_2
$X = 4$	106.6	a_2

20.27 $.6476(19.4) + .2906(35.2) + .0563(64.6) + .0053(88.2)$
$+ .0002(106.6) = 26.9$

20.28 From Exercises 20.23 and 20.27 the expected value of sample information is $26.9 - 25 = 1.9$. This value represents the maximum amount that one would be willing to pay for sample information.

INDEX